ETALE COHOMOLOGY THEORY

NANKAI TRACTS IN MATHEMATICS

Published

Vol. 1 Scissors Congruences, Group Homology and Characteristic Classes
by J. L. Dupont

Vol. 2 The Index Theorem and the Heat Equation Method
by Y. L. Yu

Vol. 3 Least Action Principle of Crystal Formation of Dense Packing Type and Kepler's Conjecture
by W. Y. Hsiang

Vol. 4 Lectures on Chern–Weil Theory and Witten Deformations
by W. P. Zhang

Vol. 5 Contemporary Trends in Algebraic Geometry and Algebraic Topology
edited by Shiing-Shen Chern, Lei Fu & Richard Hain

Vol. 6 Riemann–Finsler Geometry
by Shiing-Shen Chern & Zhongmin Shen

Vol. 7 Iterated Integrals and Cycles on Algebraic Manifolds
by Bruno Harris

Vol. 8 Minimal Submanifolds and Related Topics
by Yuanlong Xin

Vol. 9 Introduction to Several Complex Variables: Three Methods of Several Complex Variables
by Yum-Tong Siu et al.

Vol. 10 Differential Geometry and Physics
edited by Mo-Lin Ge & Weiping Zhang

Vol. 11 Inspired by S S Chern
edited by Phillip A. Griffiths

Vol. 12 Topology and Physics
edited by Kevin Lin, Zhenghan Wang & Weiping Zhang

Vol. 13 Etale Cohomology Theory
by Lei Fu

Nankai Tracts in Mathematics – Vol. 13

ETALE COHOMOLOGY THEORY

Lei Fu

Chern Institute of Mathematics, Nankai University, China

World Scientific

NEW JERSEY • LONDON • SINGAPORE • BEIJING • SHANGHAI • HONG KONG • TAIPEI • CHENNAI

Published by

World Scientific Publishing Co. Pte. Ltd.
5 Toh Tuck Link, Singapore 596224
USA office: 27 Warren Street, Suite 401-402, Hackensack, NJ 07601
UK office: 57 Shelton Street, Covent Garden, London WC2H 9HE

British Library Cataloguing-in-Publication Data
A catalogue record for this book is available from the British Library.

ETALE COHOMOLOGY THEORY
Nankai Tracts in Mathematics — Vol. 13

ISBN-13 978-981-4307-72-7
ISBN-10 981-4307-72-6

Printed in Singapore.

Preface

The manuscript of this book was written in 1998–1999. At that time, I was invited to give a series of talks at the Morningside Center of Mathematics on Deligne's proof of the Weil conjecture. To prepare the talks, I wrote the book [Fu (2006)] on algebraic geometry covering the main materials in [EGA] I–III, and the current book covering the main materials in [SGA] 1, 4, $4\frac{1}{2}$, and 5 related to etale cohomology theory. I hope this book provides adequate preparation for reading more advanced papers such as [Beilinson, Bernstein and Deligne (1982)], [Deligne (1974)], [Deligne (1980)] and [Laumon (1987)].

The prerequisites for reading this book are [Fu (2006)] and the book [Matsumura (1970)] on commutative algebra. As [Fu (2006)] may not be widely available, whenever a result from it is quoted, a corresponding result in [EGA] or [Hartshorne (1977)] is also indicated. A result used in this book but not covered in these books is Artin's approximation theorem [Artin (1969)]. A nice account can be found in [Bosch, Lütkebohmert and Raynaud (1990)].

At the beginning of each section, I give a list of references related to the content of this section. I strongly encourage the reader to go through these references, especially [SGA], for more general and thorough treatment. When I was a graduate student, the books [Freitag and Kiehl (1988)] and [Milne (1980)] on etale cohomology theory gave me great help for reading [SGA]. It is inevitable that some treatments in this book are influenced by them.

During the preparation of this book, I am supported by the Qiu Shi Science & Technologies Foundation and the NSFC.

Lei Fu
Chern Institute of Mathematics

Contents

Chapter 1

Descent Theory

Unless otherwise stated, rings in this book are commutative with the identity element 1, and homomorphisms of rings map 1 to 1. For any ring A and any A-module M, we assume $1 \cdot x = x$ for all $x \in M$. For any scheme $(S, \mathscr{O}_S)$ and any $s \in S$, denote the maximal ideal pf $\mathscr{O}_{S,s}$ by $\mathfrak{m}_s$, and denote the residue field of $\mathscr{O}_{S,s}$ by $k(s)$. For any nonnegative integer n, denote by $\mathbb{A}^n_S$ the affine space $\mathbf{Spec}\, \mathscr{O}_S[t_1, \ldots, t_n]$ over S, and by $\mathbb{P}^n_S$ the projective space $\mathbf{Proj}\, \mathscr{O}_S[t_0, t_1, \ldots, t_n]$ over S. We identify $\mathbb{A}^n_S$ with the open subscheme $\mathbf{Spec}\, \mathscr{O}_S[\frac{t_1}{t_0}, \ldots, \frac{t_n}{t_0}]$ of $\mathbb{P}^n_S$.

1.1 Flat Modules

([SGA 1] IV 1.)

Let A be a ring. An A-module M is called *flat* if the functor $N \mapsto M \otimes_A N$ on the category of A-modules is exact. We also say that M is flat over A, or A-flat. Let $A \to B$ be a homomorphism of rings. If M is a flat A-module, then $B \otimes_A M$ is a flat B-module. If N is a flat B-module, and B is flat over A, then N is flat over A.

Proposition 1.1.1. *Let A be a ring and M an A-module. The following conditions are equivalent:*

(i) *M is flat.*

(ii) *For any A-modules N, we have $\mathrm{Tor}_i^A(M, N) = 0$ for all $i \geq 1$.*

(iii) *For any finitely generated A-module N, we have $\mathrm{Tor}_i^A(M, N) = 0$ for all $i \geq 1$.*

(iv) *For any A-module N, we have $\mathrm{Tor}_1^A(M, N) = 0$.*

(v) *For any finitely generated A-module N, we have $\mathrm{Tor}_1^A(M, N) = 0$.*

(vi) *For any ideal I of A, we have $\mathrm{Tor}_1^A(M, A/I) = 0$.*

(vii) *For any finitely generated ideal I of A, we have* $\mathrm{Tor}_1^A(M, A/I) = 0$.

(viii) *For any ideal I of A, the canonical homomorphism*

$$I \otimes_A M \to M, \quad a \otimes x \mapsto ax$$

is injective, that is, it induces an isomorphism $I \otimes_A M \cong IM$.

(ix) *For any finitely generated ideal I of A, the canonical homomorphism*

$$I \otimes_A M \to M, \quad a \otimes x \mapsto ax$$

is injective.

Let M be a flat A-module, N an A-module, N' and N'' submodules of N. Then $M \otimes_A N'$ and $M \otimes_A N''$ can be regarded as submodules of $M \otimes_A N$. We have

$$M \otimes_A (N' \cap N'') \cong (M \otimes_A N') \cap (M \otimes_A N''),$$
$$M \otimes_A (N' + N'') \cong (M \otimes_A N') + (M \otimes_A N''),$$

where on the right-hand side, we take the intersection and the summation inside $M \otimes_A N$.

Proposition 1.1.2.

(i) *Let A be a ring and let S be a multiplicative subset in A. Then $S^{-1}A$ is flat over A. If M is a flat A-module, then $S^{-1}M$ is a flat $S^{-1}A$-module.*

(ii) *Let $A \to B$ be a homomorphism of rings, let S (resp. T) be a multiplicative subset in A (resp. B) such that the image of S in B is contained in T, and let N be a B-module. If N is flat over A, then $T^{-1}N$ is flat over A and over $S^{-1}A$.*

(iii) *Let $A \to B$ be a homomorphism of rings and let N be a B-module. Suppose for every maximal ideal $\mathfrak{n}$ of B, $N_{\mathfrak{n}}$ is flat over A. Then N is flat over A.*

Proof. Let us prove (ii). For any A-module M, we have

$$T^{-1}N \otimes_A M \cong T^{-1}(N \otimes_A M).$$

If N is flat over A, the functor $T^{-1}(N \otimes_A -)$ on the category of A-modules is exact. It follows that $T^{-1}N$ is flat over A. By (i), $S^{-1}T^{-1}N$ is flat over $S^{-1}A$. We have $S^{-1}T^{-1}N \cong T^{-1}N$. □

Proposition 1.1.3.

(i) *Let A be a ring and let M be a flat A-module. If $a \in A$ is not a zero divisor, then the canonical homomorphism*

$$M \to M, \quad x \mapsto ax$$

is injective. In particular, if A is an integral domain, then M has no torsion.

(ii) *Let A be an integral domain such that $A_{\mathfrak{m}}$ is a discrete valuation ring for every maximal ideal $\mathfrak{m}$ of A. Then an A-module M is flat if and only if it has no torsion.*

Proof. Let us prove the "if" part of (ii). Suppose M has no torsion. To prove M is A-flat, it suffices to show $M_{\mathfrak{m}}$ is $A_{\mathfrak{m}}$-flat for any maximal ideal $\mathfrak{m}$ of A. Let I be an ideal of $A_{\mathfrak{m}}$. By our assumption, I is principal, say generated by some element $r \in A$. The canonical map

$$A_{\mathfrak{m}} \to I, \quad a \mapsto ra$$

is an isomorphism. So we have an isomorphism

$$M_{\mathfrak{m}} \stackrel{\cong}{\to} I \otimes_{A_{\mathfrak{m}}} M_{\mathfrak{m}}, \quad x \mapsto r \otimes x.$$

The composite of this isomorphism with the canonical homomorphism

$$I \otimes_{A_{\mathfrak{m}}} M_{\mathfrak{m}} \to M_{\mathfrak{m}}, \quad a \otimes x \to ax$$

is

$$M_{\mathfrak{m}} \to M_{\mathfrak{m}}, \quad x \mapsto rx,$$

which is injective since M has no torsion. We then apply 1.1.1 (viii). □

1.2 Faithfully Flat Modules

([SGA 1] IV 2–4.)

Let $\mathscr{C}$ and $\mathscr{D}$ be two categories, and let $F : \mathscr{C} \to \mathscr{D}$ be a functor. We say that F is *faithful* if for all objects X and Y in $\mathscr{C}$, the map

$$\mathrm{Hom}_{\mathscr{C}}(X, Y) \to \mathrm{Hom}_{\mathscr{D}}(F(X), F(Y)), \quad f \mapsto F(f)$$

is injective. If $\mathscr{C}$ and $\mathscr{D}$ are additive categories and F is an additive functor, then the above condition is equivalent to saying that the condition $F(u) = 0$ implies the condition $u = 0$ for any $u \in \mathrm{Hom}_{\mathscr{C}}(X, Y)$. In this case, the condition $F(X) = 0$ implies the condition $X = 0$ for any object X in $\mathscr{C}$. Indeed, we have $F(\mathrm{id}_X) = \mathrm{id}_{F(X)} = 0$, and hence $\mathrm{id}_X = 0$.

A functor $F : \mathscr{C} \to \mathscr{D}$ is called *fully faithful* if the map

$$\mathrm{Hom}_{\mathscr{C}}(X, Y) \to \mathrm{Hom}_{\mathscr{D}}(F(X), F(Y)), \quad f \mapsto F(f)$$

is bijective for all objects $X, Y \in \mathrm{ob}\,\mathscr{C}$. F is called *essentially surjective* if for any object Z in $\mathscr{D}$, there exists an object X in $\mathscr{C}$ such that $F(X) \cong Z$.

We say that F is an *equivalence of categories* if F is fully faithful and essentially surjective.

Proposition 1.2.1. *Let $\mathscr{C}$ and $\mathscr{D}$ be abelian categories and let $F : \mathscr{C} \to \mathscr{D}$ be an additive functor. The following conditions are equivalent:*

(i) *F is exact and faithful.*

(ii) *A sequence*

$$M' \to M \to M''$$

in $\mathscr{C}$ is exact if and only if

$$F(M') \to F(M) \to F(M'')$$

is exact.

(iii) *F is exact and the condition $F(X) = 0$ implies the condition $X = 0$.*

Suppose furthermore that there exists a family of nonzero objects $\{Z_i\}$ in $\mathscr{C}$ such that for any nonzero object X in $\mathscr{C}$, there exist some Z_i and some object Y in $\mathscr{C}$ admitting a monomorphism $Y \to X$ and an epimorphism $Y \to Z_i$. Then the above conditions are equivalent to the following:

(iv) *F is exact and $F(Z_i) \neq 0$ for all Z_i.*

Proof.
(i)⇒(ii) Given a sequence

$$M' \stackrel{u}{\to} M \stackrel{v}{\to} M''$$

in $\mathscr{C}$, suppose

$$F(M') \stackrel{F(u)}{\to} F(M) \stackrel{F(v)}{\to} F(M'')$$

is exact. We have $F(vu) = F(v)F(u) = 0$. Since F is faithful, we have $vu = 0$. Hence $\mathrm{im}\, u \subset \ker v$. Since F is exact, we have

$$F(\ker v/\mathrm{im}\, u) \cong F(\ker v)/F(\mathrm{im}\, u) \cong \ker F(v)/\mathrm{im}\, F(u) = 0.$$

Hence $\ker v/\mathrm{im}\, u = 0$, that is, $\ker v = \mathrm{im}\, u$.

(ii)⇒(iii) If $F(X) = 0$, then

$$F(0) \to F(X) \to F(0)$$

is exact. Our condition implies that

$$0 \to X \to 0$$

is exact. So $X = 0$.

(iii)⇒(i) Let $u : X \to Y$ be a morphism in $\mathscr{C}$. If $F(u) = 0$, then $\mathrm{im}\, F(u) = 0$. Since F is exact, we have $F(\mathrm{im}\, u) \cong \mathrm{im}\, F(u) = 0$. By our condition, we have $\mathrm{im}\, u = 0$, that is, $u = 0$.

(iii)$\Rightarrow$ (iv) is obvious.

(iv)$\Rightarrow$(iii) For any nonzero object X in $\mathscr{C}$, choose an object Y admitting a monomorphism $Y \to X$ and an epimorphism $Y \to Z_i$. Then $F(Y) \to F(Z_i)$ is an epimorphism. As $F(Z_i) \neq 0$, we have $F(Y) \neq 0$. The morphism $F(Y) \to F(X)$ is a monomorphism. It follows that $F(X) \neq 0$. □

Corollary 1.2.2. *Let A be a ring and M an A-module. The following conditions are equivalent:*

(i) *The functor $N \mapsto M \otimes_A N$ on the category of A-modules is exact and faithful.*

(ii) *A sequence of A-modules*

$$N' \to N \to N''$$

is exact if and only if

$$M \otimes_A N' \to M \otimes_A N \to M \otimes_A N''$$

is exact.

(iii) *M is flat and the condition $M \otimes_A N = 0$ implies the condition $N = 0$.*

(iv) *M is flat and $M \otimes_A A/\mathfrak{m} \neq 0$ for any maximal ideal $\mathfrak{m}$ of A.*

When M satisfies the above equivalent conditions, we say that M is *faithfully flat.*

Corollary 1.2.3. *Let $(A, \mathfrak{m}) \to (B, \mathfrak{n})$ be a local homomorphism of local rings and let M be a finitely generated B-module. Then M is faithfully flat over A if and only if it is flat over A and nonzero.*

Indeed, by Nakayama's lemma, the condition $M \otimes_A A/\mathfrak{m} \neq 0$ is equivalent to the condition $M \neq 0$.

Proposition 1.2.4. *Let $A \to B$ be a homomorphism of rings. If there exists a B-module M faithfully flat over A, then the map* $\mathrm{Spec}\, B \to \mathrm{Spec}\, A$ *is onto.*

Proof. It suffices to show that for any $\mathfrak{p} \in \mathrm{Spec}\, A$, the fiber $\mathrm{Spec}\,(B \otimes_A A_\mathfrak{p}/\mathfrak{p}A_\mathfrak{p})$ of the map $\mathrm{Spec}\, B \to \mathrm{Spec}\, A$ over $\mathfrak{p}$ is not empty, or equivalently, $B \otimes_A A_\mathfrak{p}/\mathfrak{p}A_\mathfrak{p}$ is nonzero. Indeed, since M is faithfully flat over A, $M \otimes_A A_\mathfrak{p}/\mathfrak{p}A_\mathfrak{p}$ is faithfully flat over $A_\mathfrak{p}/\mathfrak{p}A_\mathfrak{p}$. This implies that $M \otimes_A A_\mathfrak{p}/\mathfrak{p}A_\mathfrak{p} \neq 0$. But $M \otimes_A A_\mathfrak{p}/\mathfrak{p}A_\mathfrak{p}$ is a $(B \otimes_A A_\mathfrak{p}/\mathfrak{p}A_\mathfrak{p})$-module. So $B \otimes_A A_\mathfrak{p}/\mathfrak{p}A_\mathfrak{p} \neq 0$. □

Corollary 1.2.5. *Let $\phi : A \to B$ be a homomorphism of rings and let M be a finitely generated B-module flat over A. Suppose* $\operatorname{Supp} M = \operatorname{Spec} B$. *Then for any $\mathfrak{p} \in \operatorname{Spec} A$, and any prime ideal $\mathfrak{q} \in \operatorname{Spec} B$ which is minimal among those prime ideals of B containing $\mathfrak{p}B$, we have $\phi^{-1}(\mathfrak{q}) = \mathfrak{p}$. In particular, for any minimal prime ideal $\mathfrak{q}$ of B, $\phi^{-1}(\mathfrak{q})$ is a minimal prime ideal of A.*

Proof. By 1.2.3, $M_{\mathfrak{q}}$ is faithfully flat over $A_{\phi^{-1}(\mathfrak{q})}$. By 1.2.4, the map $\operatorname{Spec} B_{\mathfrak{q}} \to \operatorname{Spec} A_{\phi^{-1}(\mathfrak{q})}$ is onto. We have $\mathfrak{p}A_{\phi^{-1}(\mathfrak{q})} \in \operatorname{Spec} A_{\phi^{-1}(\mathfrak{q})}$. By the minimality of $\mathfrak{q}$, the preimage of $\mathfrak{p}A_{\phi^{-1}(\mathfrak{q})}$ in $\operatorname{Spec} B_{\mathfrak{q}}$ must be $\mathfrak{q}B_{\mathfrak{q}}$. It follows that $\phi^{-1}(\mathfrak{q}) = \mathfrak{p}$. □

Proposition 1.2.6. *Let $\phi : A \to B$ be a homomorphism of rings. The following conditions are equivalent:*

(i) *B is faithfully flat over A.*

(ii) *B is flat over A and* $\operatorname{Spec} B \to \operatorname{Spec} A$ *is onto.*

(iii) *B is flat over A, and for every maximal ideal $\mathfrak{m}$ of A, there exists a maximal ideal $\mathfrak{n}$ of B such that $\phi^{-1}(\mathfrak{n}) = \mathfrak{m}$.*

(iv) *B is flat over A, and for any A-module M, the canonical homomorphism*

$$M \to M \otimes_A B, \quad x \mapsto x \otimes 1$$

is injective.

(v) *For every ideal I of A, the canonical homomorphism*

$$I \otimes_A B \to B, \quad x \otimes b \to bx$$

is injective and $\phi^{-1}(IB) = I$.

(vi) *ϕ is injective and* $\operatorname{coker} \phi$ *is flat over A.*

Proof.

(i)⇒(ii) follows from 1.2.4.

(ii)⇒(iii) Let $\mathfrak{m}$ be a maximal ideal of A. Suppose $\operatorname{Spec} B \to \operatorname{Spec} A$ is onto. Then there exists a prime ideal $\mathfrak{q}$ of B such that $\phi^{-1}(\mathfrak{q}) = \mathfrak{m}$. Let $\mathfrak{n}$ be a maximal ideal of B containing $\mathfrak{q}$. Then $\phi^{-1}(\mathfrak{n}) = \mathfrak{m}$.

(iii)⇒(i) For any maximal ideal $\mathfrak{m}$ of A, let $\mathfrak{n}$ be a maximal ideal of B such that $\phi^{-1}(\mathfrak{n}) = \mathfrak{m}$. We have $B \otimes_A A/\mathfrak{m} \cong B/\mathfrak{m}B$, and $B/\mathfrak{m}B$ has a quotient $B/\mathfrak{n}$ which is nonzero. It follows that $B \otimes_A A/\mathfrak{m} \neq 0$. We then apply 1.2.2.

(i)⇒(iv) Suppose B is faithfully flat over A. To show $M \to M \otimes_A B$ is injective, it suffices to show that the homomorphism

$$M \otimes_A B \to M \otimes_A B \otimes_A B, \quad x \otimes b \to x \otimes 1 \otimes b$$

is injective. Indeed, this homomorphism has a left inverse

$$M \otimes_A B \otimes_A B \to M \otimes_A B, \quad x \otimes b_1 \otimes b_2 \to x \otimes b_1 b_2.$$

(iv)⇒(v) Taking $M = A/I$, we see that the canonical homomorphism

$$A/I \to A/I \otimes_A B, \quad \bar{a} \mapsto \bar{a} \otimes 1$$

is injective. This is equivalent to saying that the canonical homomorphism $A/I \to B/IB$ is injective. Hence $\phi^{-1}(IB) = I$. By 1.1.1 (viii), the canonical homomorphism $I \otimes_A B \to B$ is injective.

(v)⇒ (iii) By 1.1.1 (viii), B is flat over A. For any maximal ideal $\mathfrak{m}$ of A, we have $\phi^{-1}(\mathfrak{m}B) = \mathfrak{m}$. In particular, $\mathfrak{m}B$ is a proper ideal of B. Let $\mathfrak{n}$ be a maximal ideal of B containing $\mathfrak{m}B$. Then we have $\phi^{-1}(\mathfrak{n}) = \mathfrak{m}$.

(iv)⇒(vi) In (iv), if we take $M = A$, we see ϕ is injective. We thus have a short exact sequence

$$0 \to A \to B \to \operatorname{coker} \phi \to 0.$$

For any A-module M, we have an exact sequence

$$0 \to \operatorname{Tor}_1^A(M, B) \to \operatorname{Tor}_1^A(M, \operatorname{coker} \phi) \to M \to M \otimes_A B.$$

Since B is flat over A, we have $\operatorname{Tor}_1^A(M, B) = 0$. As $M \to M \otimes_A B$ is injective, the above exact sequence shows that $\operatorname{Tor}_1^A(M, \operatorname{coker} \phi) = 0$. By 1.1.1 (iv), $\operatorname{coker} \phi$ is flat over A.

(vi)⇒(iv) We have a short exact sequence

$$0 \to A \to B \to \operatorname{coker} \phi \to 0.$$

For any A-module M, we have an exact sequence

$$0 \to \operatorname{Tor}_1^A(M, B) \to \operatorname{Tor}_1^A(M, \operatorname{coker} \phi) \to M \to M \otimes_A B.$$

Since $\operatorname{coker} \phi$ is flat over A, we have $\operatorname{Tor}_1^A(M, \operatorname{coker} \phi) = 0$. The above exact sequence then shows that $M \to M \otimes_A B$ is injective and $\operatorname{Tor}_1^A(M, B) = 0$. □

Proposition 1.2.7. *Let A be a noetherian ring, I an ideal of A, and $\hat{A}$ the I-adic completion of A. Then $\hat{A}$ is flat over A. It is faithfully flat over A if and only if I is contained in the radical of A.*

See [Fu (2006)] 2.1.23, or [Matsumura (1970)] (23.L) Corollary 1, and (24.A) Theorem 56.

Proposition 1.2.8. *Let A be a ring, I an ideal of A, and M an A-module. Suppose one of the following conditions holds:*

(a) *I is nilpotent.*

(b) *A is noetherian, I is contained in the radical of A, and M is finitely generated.*

Then the following conditions are equivalent:

(i) *M is free over A.*

(ii) *$M \otimes_A A/I$ is free over A/I and $\mathrm{Tor}_1^A(M, A/I) = 0$.*

(iii) *$M \otimes_A A/I$ is free over A/I and the canonical homomorphism*

$$M/IM \otimes_{A/I} \left(\bigoplus_{n=0}^{\infty} I^n/I^{n+1}\right) \to \bigoplus_{n=0}^{\infty} I^n M/I^{n+1}M$$

is an isomorphism.

Proof. (i)⇒(ii) and (i)⇒(iii) are obvious.

(ii)⇒(i) Let x_λ ($\lambda \in \Lambda$) be a family of elements in M such that their images in $M \otimes_A A/I$ form a basis. By Nakayama's lemma, $\{x_\lambda\}$ generates M. Let L be a free A-module with rank $|\Lambda|$, let $L \to M$ be an epimorphism mapping a basis of L to $\{x_\lambda\}$, and let R be its kernel. We have a short exact sequence

$$0 \to R \to L \to M \to 0.$$

Since $\mathrm{Tor}_1^A(M, A/I) = 0$, the sequence

$$0 \to R \otimes_A A/I \to L \otimes_A A/I \to M \otimes_A A/I \to 0$$

is exact. The homomorphism $L \otimes_A A/I \to M \otimes_A A/I$ is an isomorphism by the definition of the homomorphism $L \to M$. So $R \otimes_A A/I = 0$, that is, $R/IR = 0$. This implies that $R = 0$ by Nakayama's lemma. Hence $L \cong M$ and M is free.

(iii)⇒(i) Keep the notation above. We have a commutative diagram

$$\begin{array}{ccc} L/IL \otimes_{A/I} \left(\bigoplus_{n=0}^{\infty} I^n/I^{n+1}\right) & \to & \bigoplus_{n=0}^{\infty} I^n L/I^{n+1}L \\ \downarrow & & \downarrow \\ M/IM \otimes_{A/I} \left(\bigoplus_{n=0}^{\infty} I^n/I^{n+1}\right) & \to & \bigoplus_{n=0}^{\infty} I^n M/I^{n+1}M. \end{array}$$

By the definition of the homomorphism $L \to M$, we have $L/IL \cong M/IM$, and hence the first vertical arrow is an isomorphism. The two horizontal arrows are isomorphisms by our assumption and the fact that L is free. It follows that

$$\bigoplus_{n=0}^{\infty} I^n L/I^{n+1}L \to \bigoplus_{n=0}^{\infty} I^n M/I^{n+1}M$$

is an isomorphism. This implies that $R \subset \bigcap_{n=0}^{\infty} I^n L$. Since L is free and $\bigcap_{n=0}^{\infty} I^n = 0$ under the condition (a) or (b), we have $R = 0$. Hence $L \cong M$ and M is free. □

Corollary 1.2.9. *Let $\mathfrak{m}$ be a maximal ideal of a ring A and let M be an A-module. Suppose one of the following conditions holds:*

(a) $\mathfrak{m}$ *is nilpotent.*

(b) A *is noetherian and local, and M is finitely generated.*

Then the following conditions are equivalent:

(i) M *is free.*

(ii) M *is projective.*

(iii) M *is flat.*

(iv) $\mathrm{Tor}_1^A(M, A/\mathfrak{m}) = 0$.

(v) *The canonical homomorphism*

$$M/\mathfrak{m}M \otimes_{A/\mathfrak{m}} \left(\bigoplus_{n=0}^{\infty} \mathfrak{m}^n/\mathfrak{m}^{n+1} \right) \to \bigoplus_{n=0}^{\infty} \mathfrak{m}^n M/\mathfrak{m}^{n+1} M$$

is an isomorphism.

If $(A, \mathfrak{m})$ is a local noetherian integral domain with residue field k and fraction field K, and M is finitely generated, then the above conditions are equivalent to

(vi) $\dim_K(M \otimes_A K) = \dim_k(M \otimes_A k)$.

Proof. The equivalence of (i)–(v) follows directly from 1.2.8. (i)⇒(vi) is obvious. Suppose (vi) holds. Choose $x_i \in M$ $(i = 1, \ldots, n)$ such that their images in $M/\mathfrak{m}M \cong M \otimes_A k$ form a basis. By Nakayama's lemma, $\{x_i\}$ generates M. Let L be a free A-module of rank n, let $L \to M$ be an epimorphism mapping a basis of L to $\{x_i\}$, and let R be the kernel of $L \to M$. Then we have an exact sequence

$$0 \to R \otimes_A K \to L \otimes_A K \to M \otimes_A K \to 0.$$

Since $M \otimes_A K$ has the same dimension as $L \otimes_A K$, we have $R \otimes_A K = 0$. But R is contained in the free A-module L, and hence has no torsion. It follows that $R = 0$. So $L \cong M$ and M is free. □

1.3 Local Criteria for Flatness

([SGA 1] IV 5.)

Proposition 1.3.1. *Let A be a ring, I be an ideal of A, and M an A-module. If $\mathrm{Tor}_1^A(M, A/I^n) = 0$ for any $n \geq 0$, then the canonical homomorphism*

$$M/IM \otimes_{A/I} \Big(\bigoplus_{n=0}^{\infty} I^n/I^{n+1}\Big) \to \bigoplus_{n=0}^{\infty} I^n M/I^{n+1}M$$

is an isomorphism. The converse is true if I is nilpotent.

Proof. Suppose $\mathrm{Tor}_1^A(M, A/I^n) = 0$ for any $n \geq 0$. Then for each n, we have an exact sequence

$$0 \to M \otimes_A I^n/I^{n+1} \to M \otimes_A A/I^{n+1} \to M \otimes_A A/I^n \to 0.$$

The homomorphism

$$M \otimes_A A/I^{n+1} \to M \otimes_A A/I^n$$

can be identified with the canonical homomorphism

$$M/I^{n+1}M \to M/I^n M,$$

and the kernel of the latter is $I^n M/I^{n+1}M$. It follows that the canonical homomorphism

$$M \otimes_A I^n/I^{n+1} \to I^n M/I^{n+1}M$$

is an isomorphism. But

$$M \otimes_A I^n/I^{n+1} \cong M/IM \otimes_{A/I} I^n/I^{n+1}.$$

Hence

$$M/IM \otimes_{A/I} I^n/I^{n+1} \cong I^n M/I^{n+1}M.$$

In general, we have an exact sequence

$$\mathrm{Tor}_1^A(M, A/I^{n+1}) \to \mathrm{Tor}_1^A(M, A/I^n) \to$$
$$M \otimes_A I^n/I^{n+1} \to M \otimes_A A/I^{n+1} \to M \otimes_A A/I^n \to 0.$$

If the homomorphism

$$M/IM \otimes_{A/I} I^n/I^{n+1} \to I^n M/I^{n+1}M$$

is an isomorphism, then the homomorphism

$$M \otimes_A I^n/I^{n+1} \to M \otimes_A A/I^{n+1}$$

is injective, and hence the homomorphism

$$\mathrm{Tor}_1^A(M, A/I^{n+1}) \to \mathrm{Tor}_1^A(M, A/I^n)$$

is onto. If I is nilpotent, then $\mathrm{Tor}_1^A(M, A/I^{n+1}) = 0$ for large n. This implies that $\mathrm{Tor}_1^A(M, A/I^n) = 0$ for all $n \geq 0$. $\square$

Proposition 1.3.2. *Let $A \to B$ be a homomorphism of rings and let M be an A-module. The following conditions are equivalent:*

(i) *For every B-module N, we have $\mathrm{Tor}_1^A(M,N)=0$.*

(ii) *$M \otimes_A B$ is flat over B, and $\mathrm{Tor}_1^A(M,B)=0$.*

Proof.

(i)⇒(ii) Taking $N=B$, we see $\mathrm{Tor}_1^A(M,B)=0$. Let

$$0 \to N' \to N \to N'' \to 0$$

be an exact sequence of B-modules. Since $\mathrm{Tor}_1^A(M,N'')=0$, the sequence

$$0 \to M \otimes_A N' \to M \otimes_A N \to M \otimes_A N'' \to 0$$

is exact, that is, the sequence

$$0 \to (M \otimes_A B) \otimes_B N' \to (M \otimes_A B) \otimes_B N \to (M \otimes_A B) \otimes_B N'' \to 0$$

is exact. Hence $M \otimes_A B$ is flat over B.

(ii)⇒(i) Let $L \to N$ be an epimorphism such that L is a free B-module, and let R be the kernel of this epimorphism. We have an exact sequence

$$\mathrm{Tor}_1^A(M,L) \to \mathrm{Tor}_1^A(M,N) \to M \otimes_A R \to M \otimes_A L \to M \otimes_A N \to 0.$$

Since $\mathrm{Tor}_1^A(M,B)=0$ and L is free over B, we have $\mathrm{Tor}_1^A(M,L)=0$. Since $M \otimes_A B$ is flat over B, the homomorphism

$$M \otimes_A R \to M \otimes_A L$$

is injective. So we have $\mathrm{Tor}_1^A(M,N)=0$. □

Corollary 1.3.3. *Let A be a ring, I an ideal of A, and M an A-module. The following conditions are equivalent:*

(i) $\mathrm{Tor}_1^A(M,N)=0$ *for any A/I-module N.*

(ii) *$M \otimes_A A/I$ is flat over A/I, and $\mathrm{Tor}_1^A(M,A/I)=0$.*

(iii) $\mathrm{Tor}_1^A(M,N)=0$ *for any A-module N annihilated by some power of I.*

If these conditions are satisfied, then the canonical homomorphism

$$M/IM \otimes_{A/I} \left(\bigoplus_{n=0}^{\infty} I^n/I^{n+1}\right) \to \bigoplus_{n=0}^{\infty} I^n M/I^{n+1}M$$

is an isomorphism.

Proof.

(i)⇔(ii) follows from 1.3.2.

(iii)⇒(i) is obvious.

(i)⇒(iii) We have $\mathrm{Tor}_1^A(M, I^nN/I^{n+1}N) = 0$ for any $n \geq 0$. By induction on n and using the long exact sequence for Tor, we see $\mathrm{Tor}_1^A(M, N/I^{n+1}N) = 0$ for all $n \geq 0$. Taking n sufficiently large, we get $\mathrm{Tor}_1^A(M, N) = 0$.

The last assertion follows from 1.3.1. □

Proposition 1.3.4. *Let A be a ring, I and ideal of A, and M an A-module. Consider the following conditions:*

(i) *M is flat over A.*

(ii) *$M \otimes_A A/I$ is flat over A/I and $\mathrm{Tor}_1^A(M, A/I) = 0$.*

(iii) *$M \otimes_A A/I$ is flat over A/I and the canonical homomorphism*

$$M/IM \otimes_{A/I} \left(\bigoplus_{n=0}^{\infty} I^n/I^{n+1}\right) \to \bigoplus_{n=0}^{\infty} I^nM/I^{n+1}M$$

is an isomorphism.

(iv) *For any $n \geq 0$, $M \otimes_A A/I^n$ is flat over A/I^n.*

We have (i)⇒(ii)⇒(iii)⇒(iv)*. Suppose that either I is nilpotent, or A is noetherian and $M \otimes_A N$ is separated with respect to the I-adic topology for any finitely generated A-module N. Then* (iv)⇒(i).

Proof.

(i)⇒(ii) is trivial.

(ii)⇒(iii) follows from 1.3.3.

(iii)⇒(iv) By 1.3.1, we have $\mathrm{Tor}_1^{A/I^n}(M/I^nM, A/I) = 0$. The condition 1.3.3 (ii) holds if we take A to be A/I^n and M to be M/I^nM. It follows from 1.3.3 (iii) that we have $\mathrm{Tor}_1^{A/I^n}(M/I^nM, N) = 0$ for any A/I^n-module N. Hence M/I^nM is flat over A/I^n.

If I is nilpotent, then taking n sufficiently large, we see (iv)⇒(i). Suppose A is noetherian and $M \otimes_A N$ is separated with respect to the I-adic topology for any finitely generated A-module N. Let us prove M is flat over A under the condition (iv). We need to prove that the canonical homomorphism

$$\mathrm{id}_M \otimes i : M \otimes_A N' \to M \otimes_A N$$

is injective for any monomorphism $i : N' \hookrightarrow N$ between finitely generated A-modules. It suffices to show that $\ker(\mathrm{id}_M \otimes i)$ is contained in

$$I^n(M \otimes_A N') = \mathrm{im}\,(M \otimes_A I^nN' \to M \otimes_A N')$$

for all $n \geq 0$, or equivalently, that $\ker(\mathrm{id}_M \otimes i)$ is contained in

$$\mathrm{im}\,(M \otimes_A V \to M \otimes_A N') = \ker\,(M \otimes_A N' \to M \otimes_A N'/V),$$

where V goes over a family of submodules of N' which form a base of neighborhoods of N' at 0 with respect to the I-adic topology. By the Artin–Rees theorem ([Matsumura (1970)] (11.C) Theorem 15), we can take $V = N' \cap I^n N$ $(n \geq 0)$. We have a commutative diagram

$$\begin{array}{ccc} M \otimes_A N' & \to & M \otimes_A (N'/N' \cap I^n N) \\ \downarrow & & \downarrow \\ M \otimes_A N & \to & M \otimes_A (N/I^n N). \end{array}$$

Since the canonical homomorphism $N'/N' \cap I^n N \to N/I^n N$ is injective and $M \otimes_A A/I^n$ is flat over A/I^n, the right vertical arrow is injective. This implies that $\ker(\mathrm{id}_M \otimes i)$ is contained in $\ker\,(M \otimes_A N' \to M \otimes_A (N'/N' \cap I^n N))$ for all n. Our assertion follows. □

Theorem 1.3.5. *Let $A \to B$ be a homomorphism of noetherian rings, I an ideal of A such that IB is contained in the radical of B, and M a finitely generated B-module. The following conditions are equivalent:*

(i) *M is flat over A.*

(ii) *$M \otimes_A A/I$ is flat over A/I and $\mathrm{Tor}_1^A(M, A/I) = 0$.*

(iii) *$M \otimes_A A/I$ is flat over A/I and the canonical homomorphism*

$$M/IM \otimes_{A/I} \Big(\bigoplus_{n=0}^{\infty} I^n/I^{n+1}\Big) \to \bigoplus_{n=0}^{\infty} I^n M/I^{n+1} M$$

is an isomorphism.

(iv) *$M \otimes_A A/I^n$ is flat over A/I^n for all $n \geq 0$.*

Proof. We use 1.3.4. Let us check that $M \otimes_A N$ is separated with respect to the I-adic topology for any finitely generated A-module N. Note that $M \otimes_A N$ is a finitely generated B-module. Since IB is contained in the radical of B, $M \otimes_A N$ is separated with respect to the IB-adic topology by [Fu (2006)] 1.5.8, or [Matsumura (1970)] (11.D) Corollary 1. Our assertion follows. □

Proposition 1.3.6. *Let $(A, \mathfrak{m}) \to (B, \mathfrak{n})$ be a local homomorphism between noetherian local rings, and let $u : M' \to M$ be a homomorphism between finitely generated B-modules. Suppose M is flat over A. The following conditions are equivalent:*

(i) *u is injective, and $\mathrm{coker}\, u$ is flat over A.*

(ii) *$u \otimes \mathrm{id} : M' \otimes_A A/\mathfrak{m} \to M \otimes_A A/\mathfrak{m}$ is injective.*

Proof.

(i)⇒(ii) Since u is injective, we have an exact sequence

$$0 \to M' \to M \to \operatorname{coker} u \to 0.$$

Since $\operatorname{coker} u$ is flat over A, the sequence

$$0 \to M' \otimes_A A/\mathfrak{m} \to M \otimes_A A/\mathfrak{m} \to \operatorname{coker} u \otimes_A A/\mathfrak{m} \to 0$$

is exact. So $u \otimes \mathrm{id}$ is injective.

(ii)⇒(i) Consider the commutative diagram

$$\begin{array}{ccc} M'/\mathfrak{m}M' \otimes_{A/\mathfrak{m}} \left(\bigoplus_{n=0}^{\infty} \mathfrak{m}^n/\mathfrak{m}^{n+1} \right) & \to & M/\mathfrak{m}M \otimes_{A/\mathfrak{m}} \left(\bigoplus_{n=0}^{\infty} \mathfrak{m}^n/\mathfrak{m}^{n+1} \right) \\ \downarrow & & \downarrow \\ \bigoplus_{n=0}^{\infty} \mathfrak{m}^n M'/\mathfrak{m}^{n+1} M' & \to & \bigoplus_{n=0}^{\infty} \mathfrak{m}^n M/\mathfrak{m}^{n+1} M. \end{array}$$

Suppose $u \otimes \mathrm{id}$ is injective. Then the top horizontal arrow is injective. Since M is flat over A, the right vertical arrow is an isomorphism by 1.3.1. One can check the left vertical arrow is surjective. These facts imply that the bottom horizontal arrow is injective, which then implies that u is injective. We thus have a short exact sequence

$$0 \to M' \to M \to \operatorname{coker} u \to 0.$$

It gives rise to an exact sequence

$$\mathrm{Tor}_1^A(M, A/\mathfrak{m}) \to \mathrm{Tor}_1^A(\operatorname{coker} u, A/\mathfrak{m}) \to M' \otimes_A A/\mathfrak{m} \stackrel{u\otimes\mathrm{id}}{\to} M \otimes_A A/\mathfrak{m}.$$

Since M is flat over A, we have $\mathrm{Tor}_1^A(M, A/\mathfrak{m}) = 0$. Since $u \otimes \mathrm{id}$ is injective, the above exact sequence shows that $\mathrm{Tor}_1^A(\operatorname{coker} u, A/\mathfrak{m}) = 0$. By 1.3.5, $\operatorname{coker} u$ is flat over A. □

Proposition 1.3.7. *Let*

$$(C, \mathfrak{l}) \to (A, \mathfrak{m}) \to (B, \mathfrak{n})$$

be local homomorphisms of noetherian local rings, and let M be a finitely generated B-module. Suppose A is flat over C. The following conditions are equivalent:

(i) *M is flat over A.*

(ii) *M is flat over C and $M \otimes_C C/\mathfrak{l}$ is flat over $A \otimes_C C/\mathfrak{l}$.*

Proof.

(i)⇒(ii) is obvious.

(ii)⇒(i) Suppose M is flat over C, and $M \otimes_C C/\mathfrak{l}$ is flat over $A \otimes_C C/\mathfrak{l}$, that is, $M/\mathfrak{l}M$ is flat over $A/\mathfrak{l}A$. By 1.3.5, to prove M is flat over A, it suffices to show the canonical homomorphism

$$M/\mathfrak{l}M \otimes_{A/\mathfrak{l}A} \Big(\bigoplus_{n=0}^{\infty} \mathfrak{l}^n A/\mathfrak{l}^{n+1} A\Big) \to \bigoplus_{n=0}^{\infty} \mathfrak{l}^n M/\mathfrak{l}^{n+1} M$$

is an isomorphism. Since A and M are flat over C, the canonical homomorphisms

$$A/\mathfrak{l}A \otimes_{C/\mathfrak{l}} \Big(\bigoplus_{n=0}^{\infty} \mathfrak{l}^n/\mathfrak{l}^{n+1}\Big) \to \bigoplus_{n=0}^{\infty} \mathfrak{l}^n A/\mathfrak{l}^{n+1} A,$$

$$M/\mathfrak{l}M \otimes_{C/\mathfrak{l}} \Big(\bigoplus_{n=0}^{\infty} \mathfrak{l}^n/\mathfrak{l}^{n+1}\Big) \to \bigoplus_{n=0}^{\infty} \mathfrak{l}^n M/\mathfrak{l}^{n+1} M$$

are isomorphisms by 1.3.1. We thus have the following isomorphisms

$$\begin{aligned} & M/\mathfrak{l}M \otimes_{A/\mathfrak{l}A} \Big(\bigoplus_{n=0}^{\infty} \mathfrak{l}^n A/\mathfrak{l}^{n+1} A\Big) \\ \cong\ & M/\mathfrak{l}M \otimes_{A/\mathfrak{l}A} \Big(A/\mathfrak{l}A \otimes_{C/\mathfrak{l}} \Big(\bigoplus_{n=0}^{\infty} \mathfrak{l}^n/\mathfrak{l}^{n+1}\Big)\Big) \\ \cong\ & M/\mathfrak{l}M \otimes_{C/\mathfrak{l}} \Big(\bigoplus_{n=0}^{\infty} \mathfrak{l}^n/\mathfrak{l}^{n+1}\Big) \\ \cong\ & \bigoplus_{n=0}^{\infty} \mathfrak{l}^n M/\mathfrak{l}^{n+1} M. \end{aligned}$$

Our assertion follows. □

1.4 Constructible Sets

([EGA] 0 9.1–9.2, IV 1.8–1.10.)

Let X be a topological space. A subset of X is called *locally closed* if it is the intersection of an open subset and a closed subset. Suppose X is a noetherian topological space. We say that a subset of X is *constructible* if it is a union of finitely many locally closed subsets. If X_1 and X_2 are constructible subsets, then $X_1 \cup X_2$, $X_1 \cap X_2$ and $X_1 - X_2$ are constructible.

Proposition 1.4.1. *Let X be a noetherian topological space. A subset Y of X is constructible if and only if for any irreducible closed subset F of X such that $Y \cap F$ is dense in F, the set $Y \cap F$ contains a nonempty open subset of F.*

Proof.

($\Rightarrow$) Suppose Y is constructible, F is irreducible and closed such that $Y \cap F$ is dense in F. Note that $Y \cap F$ is constructible. Write

$$Y \cap F = (U_1 \cap F_1) \cup \cdots \cup (U_n \cap F_n),$$

where each U_i (resp. F_i) is open (resp. closed). Replacing F_i by $F_i \cap F$, we may assume $F_i \subset F$. We have

$$F = \overline{Y \cap F} = \overline{(U_1 \cap F_1)} \cup \cdots \cup \overline{(U_n \cap F_n)}.$$

Since F is irreducible, we have $F = \overline{U_i \cap F_i}$ for some i. But $\overline{U_i \cap F_i} \subset F_i$. These facts imply that $F_i = F$, and $Y \cap F$ contains the nonempty open subset $U_i \cap F_i$ of F.

($\Leftarrow$) We use noetherian induction. Suppose that for any irreducible closed subset F of X such that $Y \cap F$ is dense in F, $Y \cap F$ contains a nonempty open subset of F. If Y is not constructible, then the set

$$\mathscr{S} = \{Z | \emptyset \neq Z \subset X, Z \text{ is closed, and } Y \cap Z \text{ is not constructible}\}$$

is nonempty. Since X is a noetherian topological space, $\mathscr{S}$ contains a minimal element, say Z_0. If Z_0 is not irreducible, we can write $Z_0 = Z_1 \cup Z_2$, where Z_1 and Z_2 are proper closed subsets of Z_0. By the minimality of Z_0, we have $Z_1, Z_2 \notin \mathscr{S}$. Hence $Y \cap Z_1$ and $Y \cap Z_2$ are constructible. But then $Y \cap Z_0 = (Y \cap Z_1) \cup (Y \cap Z_2)$ is constructible. This contradicts $Z_0 \in \mathscr{S}$. So Z_0 must be irreducible. If $Y \cap Z_0$ is not dense in Z_0, then by the minimality of Z_0, $Y \cap \overline{(Y \cap Z_0)}$ is constructible. But $Y \cap \overline{(Y \cap Z_0)} = Y \cap Z_0$. This contradicts $Z_0 \in \mathscr{S}$. So $Y \cap Z_0$ is dense in Z_0. By our assumption, there exists a nonempty open subset U of Z_0 contained in $Y \cap Z_0$. By the minimality of Z_0, $Y \cap (Z_0 - U)$ is constructible. We have $Y \cap Z_0 = U \cup (Y \cap (Z_0 - U))$. This shows that $Y \cap Z_0$ is constructible. Again this contradicts $Z_0 \in \mathscr{S}$. So Y must be constructible. □

Theorem 1.4.2 (Chevalley). *Let $f : X \to Y$ be a morphism of finite type between noetherian schemes. Then $f(X)$ is constructible.*

Proof. Cover Y by finitely many affine open subsets V_i and cover each $f^{-1}(V_i)$ by finitely many affine open subsets U_{ij}. We have $f(X) = \cup_{i,j} f(U_{ij})$. It suffices to show that each $f(U_{ij})$ is a constructible subset of V_i. Replacing X by U_{ij} and Y by V_i, we are reduced to the case where $X = \operatorname{Spec} B$ and $Y = \operatorname{Spec} A$. Any irreducible closed subset of $\operatorname{Spec} A$ is of the form $V(\mathfrak{p}) = \{\mathfrak{p}' \in \operatorname{Spec} A | \mathfrak{p} \subset \mathfrak{p}'\}$ for some prime idea $\mathfrak{p}$ of A. The set $V(\mathfrak{p})$ can be identified with $\operatorname{Spec} A/\mathfrak{p}$, and $f(\operatorname{Spec} B) \cap V(\mathfrak{p})$ can be identified with the image of the morphism $\operatorname{Spec} B/\mathfrak{p}B \to \operatorname{Spec} A/\mathfrak{p}$ obtained from f by base change. The closure of the image of this morphism consists of those prime ideals of $A/\mathfrak{p}$ containing the kernel of the homomorphism $A/\mathfrak{p} \to B/\mathfrak{p}B$. If $f(\operatorname{Spec} B) \cap V(\mathfrak{p})$ is dense in $V(\mathfrak{p})$, then the homomorphism $A/\mathfrak{p} \to B/\mathfrak{p}B$ is injective. By 1.4.1, to prove the theorem, it suffices to prove the following statement: Let A be a noetherian integral domain, and let $A \to B$ be an injective homomorphism such that B is a finitely generated A-algebra. Then there exists a nonzero element $a \in A$ such that every prime ideal of A lying in the set $D(a) = \{\mathfrak{p} \in \operatorname{Spec} A | a \notin \mathfrak{p}\}$ is the inverse image of a prime ideal of B.

Let us prove this statement. Write $B = A[x_1, \ldots, x_n]$, where $x_1, \ldots, x_r$ are algebraically independent over A and x_j $(r+1 \leq j \leq n)$ are algebraic over $A[x_1, \ldots, x_r]$. For each $r+1 \leq j \leq n$, find a relation

$$p_{j0}x_j^{d_j} + p_{j1}x_j^{d_j-1} + \cdots + p_{jd_j} = 0,$$

where $p_{ji} \in A[x_1, \ldots, x_r]$ $(0 \leq i \leq d_j)$, $d_j \geq 1$, and $p_{j0} \neq 0$. Then $\prod_{j=r+1}^{n} p_{j0}$ is a nonzero polynomial in $x_1, \ldots, x_r$ with coefficients in A. Let a be one of the nonzero coefficients of this polynomial. For any prime ideal $\mathfrak{p}$ of A with $a \notin \mathfrak{p}$, let $\mathfrak{p}' = \mathfrak{p}[x_1, \ldots, x_r]$, which is a prime ideal of $A[x_1, \ldots, x_r]$. Then $\prod_{j=r+1}^{n} p_{j0} \notin \mathfrak{p}'$. So $B_{\mathfrak{p}'}$ is integral over $A[x_1, \ldots, x_r]_{\mathfrak{p}'}$. Hence there exists a prime ideal $\mathfrak{q}$ of B whose inverse image in $A[x_1, \ldots, x_r]$ is $\mathfrak{p}'$. The inverse image of $\mathfrak{q}$ in A is then $\mathfrak{p}$. □

Corollary 1.4.3. *Let $f : X \to Y$ be a morphism of finite type between noetherian schemes. If Z is a constructible subset of X, then $f(Z)$ is constructible.*

Proof. Z is a union of finitely many locally closed subset. It suffices to consider the case where Z is locally closed. Then there exists a subscheme structure on Z, and we can apply 1.4.2 to the morphism $Z \to Y$. □

Let X be a topological space and let $x, y \in X$. We say that y is a *generalization* of x, and that x is a *specialization* of y if $x \in \overline{\{y\}}$.

Proposition 1.4.4. *Suppose X is a noetherian topological space such that every irreducible closed subset has a generic point. Let U be a constructible subset of X and let x be a point in U. Then U contains an open neighborhood of x if and only if every generalization y of x lies in U.*

Proof. The "only if" part is clear. Let us prove the "if" part. We use noetherian induction. Suppose U contains every generalization of x. If U

does not contain any open neighborhood of x, then the set

$$\mathscr{S} = \{Z | x \in Z \subset X, Z \text{ is closed}, U \cap Z \text{ is not a neighborhood of } x \text{ in } Z\}$$

is not empty. Since X is noetherian, $\mathscr{S}$ contains a minimal element, say Z_0. If Z_0 is not irreducible, we can write $Z_0 = Z_1 \cup Z_2$, where Z_1 and Z_2 are proper closed subsets of Z_0. If $x \in Z_1 \cap Z_2$, then by the minimality of Z_0, we can find open neighborhoods V_i $(i = 1, 2)$ of x such that $V_i \cap Z_i \subset U \cap Z_i$. Then $U \cap Z_0$ contains the neighborhood $(V_1 \cap V_2) \cap Z_0$ of x in Z_0. This contradicts $Z_0 \in \mathscr{S}$. If $x \in Z_1 - Z_2$, then we can find an open neighborhood V_1 of x such that $V_1 \cap Z_1 \subset U \cap Z_1$. But then $U \cap Z_0$ contains the neighborhood $(V_1 - Z_2) \cap Z_0$ of x in Z_0. This contradicts $Z_0 \in \mathscr{S}$. Similarly, if $x \in Z_2 - Z_1$, we are again led to contradiction. So Z_0 must be irreducible. Let y be the generic point of Z_0. Then y is a generalization of x. By our assumption, we have $y \in U$. But then $U \cap Z_0$ is dense in Z_0. By 1.4.1, $U \cap Z_0$ contains a nonempty open subset V of Z_0. Since $Z_0 \notin \mathscr{S}$, we have $x \notin V$. By the minimality of Z_0, we have $Z_0 - V \in \mathscr{S}$. So there exists an open neighborhood W of x in X such that $W \cap (Z_0 - V) \subset U \cap (Z_0 - V)$. But then $W \cap Z_0$ is neighborhood of x in Z_0 contained in $U \cap Z_0$. This contradicts $Z_0 \in \mathscr{S}$. So U contains an open neighborhood of x. $\square$

Corollary 1.4.5. *Let X be a noetherian topological space such that every irreducible closed subset has a generic point. A subset U of X is open if and only if the following two conditions hold:*

(a) *U contains generalizations of points in U.*

(b) *For every $x \in U$, $U \cap \overline{\{x\}}$ contains a nonempty open subset of $\overline{\{x\}}$.*

Proof. The "only if" part is clear. Let us prove the "if" part. Let F be an irreducible closed subset of X. Then $F = \overline{\{x\}}$ for some $x \in X$. If $U \cap F$ is dense in F, then $x \in U$ by condition (a). But then $U \cap F$ contains a nonempty open subset of F by condition (b). So U is constructible by 1.4.1. By condition (a) and 1.4.4, U is open. $\square$

One can also study constructible sets on non-noetherian schemes. Confer [EGA] 0 9.1–9.2 and IV 1.8–1.10.

1.5 Flat Morphisms

Proposition 1.5.1. *Let $f : X \to Y$ be a morphism of finite type between noetherian schemes, $x \in X$, and $y = f(x)$. Then the following conditions are equivalent:*

(i) *f maps every neighborhood of x to a neighborhood of y.*

(ii) *For any generalization y' of y, there exists a generalization x' of x such that $f(x') = y'$.*

(iii) *The morphism* $\mathrm{Spec}\, \mathscr{O}_{X,x} \to \mathrm{Spec}\, \mathscr{O}_{Y,y}$ *induced by f is surjective on the underlying topological spaces.*

Proof.

(i)⇒(ii) Suppose $y \in \overline{\{y'\}}$. Let F be the union of those irreducible components of $f^{-1}(\overline{\{y'\}})$ not containing x. Then $X - F$ is a neighborhood of x. So $f(X - F)$ is a neighborhood of y, and hence $y' \in f(X - F)$. Let $x_1 \in X - F$ such that $f(x_1) = y'$. Then x_1 lies in an irreducible component C of $f^{-1}(\overline{\{y'\}})$ containing x. Let x' be the generic point of C. Then x' is a generalization of x. We have $x_1 \in \overline{\{x'\}}$. Hence $f(x_1) \in \overline{\{f(x')\}}$, that is, $y' \in \overline{\{f(x')\}}$. On the other hand, we have $x' \in C \subset f^{-1}(\overline{\{y'\}})$, and hence $f(x') \in \overline{\{y'\}}$. Therefore $f(x') = y'$

(ii)⇒(i) Let U be an open neighborhood of x in X. Then U is constructible. By 1.4.3, $f(U)$ is a constructible subset in Y. The condition (ii) implies that $f(U)$ contains every generalization of y. By 1.4.4, $f(U)$ is a neighborhood of y.

(ii)⇔(iii) is clear. □

Theorem 1.5.2. *Let $f : X \to Y$ be a morphism of finite type between noetherian schemes and let $\mathscr{F}$ be a coherent $\mathscr{O}_X$-module with* $\mathrm{Supp}\, \mathscr{F} = X$. *Suppose $\mathscr{F}_x$ is flat over $\mathscr{O}_{Y,f(x)}$ for any $x \in X$. Then f is an open mapping.*

Proof. For any $x \in X$, since $\mathscr{F}_x$ is flat over $\mathscr{O}_{Y,f(x)}$ and $\mathscr{F}_x \neq 0$, it is faithfully flat over $\mathscr{O}_{Y,f(x)}$ by 1.2.3. By 1.2.4, the morphism $\mathrm{Spec}\, \mathscr{O}_{X,x} \to \mathrm{Spec}\, \mathscr{O}_{Y,y}$ induced by f is surjective. So by 1.5.1, f is an open mapping. □

Lemma 1.5.3. *Let A be a noetherian integral domain, B a finitely generated A-algebra, and M a finitely generated B-module. Then there exists a nonzero element $f \in A$ such that M_f is free over A_f.*

For a proof, see [Matsumura (1970)] (22.A) Lemma 1.

Lemma 1.5.4. *Let A be a noetherian ring, B a finitely generated A-algebra, M a finitely generated B-module, $\mathfrak{q}$ a prime ideal of B, and $\mathfrak{p}$ the inverse image of $\mathfrak{q}$ in A. Suppose $M_{\mathfrak{q}}$ is flat over $A_{\mathfrak{p}}$. Then there exists $g \in B - \mathfrak{q}$ such that $(M/\mathfrak{p}M)_g$ is flat over $A/\mathfrak{p}$ and* $\mathrm{Tor}_1^A(M, A/\mathfrak{p})_g = 0$.

Proof. By 1.5.3, there exists $f \in A - \mathfrak{p}$ such that $(M/\mathfrak{p}M)_f$ is flat over

$A/\mathfrak{p}$. Since $M_\mathfrak{q}$ is flat over A, we have

$$\mathrm{Tor}_1^A(M, A/\mathfrak{p})_\mathfrak{q} \cong \mathrm{Tor}_1^A(M_\mathfrak{q}, A/\mathfrak{p}) = 0.$$

From the exact sequence

$$0 \to \mathrm{Tor}_1^A(M, A/\mathfrak{p}) \to M \otimes_A \mathfrak{p} \to M \to M \otimes_A A/\mathfrak{p} \to 0,$$

we deduce that $\mathrm{Tor}_1^A(M, A/\mathfrak{p})$ is a finitely generated B-module. So there exists $g' \in B - \mathfrak{q}$ such that $\mathrm{Tor}_1^A(M, A/\mathfrak{p})_{g'} = 0$. We then take $g = fg'$. □

Lemma 1.5.5. *Under the condition of 1.5.4, for every prime ideal $\mathfrak{q}'$ of B containing $\mathfrak{q}$ but not containing g, $M_{\mathfrak{q}'}$ is flat over A.*

Proof. We apply 1.3.5 to the homomorphism $A \to B_{\mathfrak{q}'}$, the ideal $I = \mathfrak{p}$ of A, and the $B_{\mathfrak{q}'}$-module $M_{\mathfrak{q}'}$. □

Theorem 1.5.6. *Let $f : X \to Y$ be a morphism of finite type between noetherian schemes and let $\mathscr{F}$ be a coherent $\mathscr{O}_X$-module.*

(i) *If Y is integral, then there exists a nonempty open subset V in Y such that for any $x \in f^{-1}(V)$, $\mathscr{F}_x$ is flat over $\mathscr{O}_{Y,f(x)}$.*

(ii) *In general, the set U of points $x \in X$ such that $\mathscr{F}_x$ are flat over $\mathscr{O}_{Y,f(x)}$ is open.*

Proof.

(i) follows from 1.5.3.

(ii) We may assume that $X = \operatorname{Spec} B$ and $Y = \operatorname{Spec} A$ are affine, and $\mathscr{F} = M^\sim$ for some finitely generated B-module M. By 1.1.2 (ii), the set U contains generalizations of points in U. Let $x \in U$. By 1.5.5, $U \cap \overline{\{x\}}$ contains a nonempty open subset of $\overline{\{x\}}$. By 1.4.5, U is open. □

A morphism $f : X \to Y$ of schemes is called *flat* at $x \in X$ if $\mathscr{O}_{X,x}$ is flat over $\mathscr{O}_{Y,f(x)}$. It is called a *flat morphism* if it is flat at every point of X. It is called a *faithfully flat morphism* if it is flat and surjective on the underlying topological spaces. If f is a morphism of finite type and X and Y are noetherian schemes, then the set of points in X, where f is flat, is open by 1.5.6. If f is a flat morphism of finite type between noetherian schemes, then f is an open mapping by 1.5.2.

1.6 Descent of Quasi-coherent Sheaves

([SGA 1] VIII 1.)

A morphism of schemes $f : X \to Y$ is called *quasi-compact* if the inverse image in X of any quasi-compact open subset of Y is quasi-compact. This is equivalent to saying that there exists a covering $\{U_i\}$ of Y by affine open subsets such that each $f^{-1}(U_i)$ is a union of finitely many affine open subsets in X. Let $g_1, g_2 : B \to C$ be maps of sets. We define

$$\ker(B \underset{g_2}{\overset{g_1}{\rightrightarrows}} C) = \{b \in B | g_1(b) = g_2(b)\}.$$

Let $f : A \to B$ be another map of set. We say the sequence

$$A \xrightarrow{f} B \underset{g_2}{\overset{g_1}{\rightrightarrows}} C$$

is exact if f is injective and $\operatorname{im} f = \ker(B \underset{g_2}{\overset{g_1}{\rightrightarrows}} C)$.

Theorem 1.6.1. *Let $g : S' \to S$ be a quasi-compact faithfully flat morphism, and let $S'' = S' \times_S S'$. For any $1 \leq i \leq 2$, let*

$$p_i : S'' = S' \times_S S' \to S'$$

be the projection to the i-th factor, and for any $1 \leq i < j \leq 3$, let

$$p_{ij} : S' \times_S S' \times_S S' \to S'' = S' \times_S S'$$

be the projection to the (i,j)-factor.

(i) *Let $\mathscr{F}$ and $\mathscr{G}$ be quasi-coherent $\mathscr{O}_S$-modules, and let $\mathscr{F}'$ and $\mathscr{G}'$ (resp. $\mathscr{F}''$ and $\mathscr{G}''$) be their inverse images on S' (resp. S''), respectively. Then the sequence*

$$\mathrm{Hom}_{\mathscr{O}_S}(\mathscr{F}, \mathscr{G}) \xrightarrow{g^*} \mathrm{Hom}_{\mathscr{O}_{S'}}(\mathscr{F}', \mathscr{G}') \underset{p_2^*}{\overset{p_1^*}{\rightrightarrows}} \mathrm{Hom}_{\mathscr{O}_{S''}}(\mathscr{F}'', \mathscr{G}'')$$

is exact, that is, if $u' \in \mathrm{Hom}_{\mathscr{O}_{S'}}(\mathscr{F}', \mathscr{G}')$ satisfies $p_1^*(u) = p_2^*(u)$, then there exists a unique $u \in \mathrm{Hom}_{\mathscr{O}_S}(\mathscr{F}, \mathscr{G})$ such that $g^*u = u'$.

(ii) *For any $1 \leq i \leq 3$, let*

$$\pi_i : S' \times_S S' \times_S S' \to S'$$

be the projection to the i-th factor. Let $\mathscr{F}'$ be a quasi-coherent sheaf on S' such that there exists an isomorphism $\sigma : p_1^*\mathscr{F}' \xrightarrow{\cong} p_2^*\mathscr{F}'$ satisfying

$$p_{13}^*(\sigma) = p_{23}^*(\sigma) \circ p_{12}^*(\sigma).$$

Here we identify

$$p_{13}^*(\sigma)\ (\textit{resp. } p_{23}^*(\sigma),\ \textit{resp. } p_{12}^*(\sigma))$$

with a morphism

$$\pi_1^*\mathscr{F}' \to \pi_3^*\mathscr{F}'\ (\textit{resp. } \pi_2^*\mathscr{F}' \to \pi_3^*\mathscr{F}',\ \textit{resp. } \pi_1^*\mathscr{F}' \to \pi_2^*\mathscr{F}')$$

through the canonical natural transformations

$$\begin{aligned} p_{13}^* \circ p_1^* &\cong \pi_1^*, & p_{13}^* \circ p_2^* &\cong \pi_3^*, \\ p_{23}^* \circ p_1^* &\cong \pi_2^*, & p_{23}^* \circ p_2^* &\cong \pi_3^*, \\ p_{12}^* \circ p_1^* &\cong \pi_1^*, & p_{12}^* \circ p_2^* &\cong \pi_2^*. \end{aligned}$$

Then there exist a quasi-coherent $\mathscr{O}_S$-module $\mathscr{F}$ and an isomorphism $\tau : g^*\mathscr{F} \stackrel{\cong}{\to} \mathscr{F}'$ such that the diagram

$$\begin{array}{ccc} p_1^* g^* \mathscr{F} & \stackrel{p_1^*\tau}{\to} & p_1^*\mathscr{F}' \\ \cong\downarrow & & \cong\downarrow \sigma \\ p_2^* g^* \mathscr{F} & \stackrel{p_2^*\tau}{\to} & p_2^*\mathscr{F}' \end{array}$$

commutes, where the left vertical arrow is the canonical isomorphism. Moreover $\mathscr{F}$ is unique up to unique isomorphism.

Any isomorphism $\sigma : p_1^*\mathscr{F}' \to p_2^*\mathscr{F}'$ satisfying $p_{13}^*(\sigma) = p_{23}^*(\sigma) \circ p_{12}^*(\sigma)$ is called a *descent datum* for $\mathscr{F}'$.

Proof. We leave it for the reader to reduce the proof of the theorem to the case where S and S' are affine. So we assume $S = \operatorname{Spec} A$, $S' = \operatorname{Spec} A'$, and A' is a faithfully flat A-algebra.

(i) Let M and N be A-modules such that $\mathscr{F} = M^\sim$ and $\mathscr{G} = N^\sim$, respectively. Set

$$M' = M \otimes_A A', \quad M'' = M \otimes_A A' \otimes_A A',$$

$$N' = N \otimes_A A', \quad N'' = N \otimes_A A' \otimes_A A'.$$

We need to show that the sequence

$$\operatorname{Hom}_A(M, N) \to \operatorname{Hom}_{A'}(M', N') \rightrightarrows \operatorname{Hom}_{A''}(M'', N'')$$

is exact. By 1.2.6 (iv), $N \to N'$ is injective. It follows that

$$\operatorname{Hom}_A(M, N) \to \operatorname{Hom}_{A'}(M', N')$$

is injective. Let

$$u' \in \ker(\operatorname{Hom}_{A'}(M', N') \rightrightarrows \operatorname{Hom}_{A''}(M'', N'')).$$

To show that

$$u' \in \operatorname{im}(\operatorname{Hom}_A(M,N) \to \operatorname{Hom}_{A'}(M',N')),$$

it suffices to show that u' maps the A-submodule M of M' to the A-submodule N of N'. For any $x \in M$, $u'(x)$ lies in $\ker(N' \rightrightarrows N'')$. So it suffices to prove

$$N \to N' \rightrightarrows N''$$

is exact, that is,

$$0 \to N \stackrel{\phi}{\to} N \otimes_A A' \stackrel{\psi}{\to} N \otimes_A A' \otimes_A A''$$

is exact, where

$$\phi(x) = x \otimes 1, \quad \psi(x \otimes a') = x \otimes a' \otimes 1 - x \otimes 1 \otimes a'$$

for any $x \in N$ and $a' \in A'$. Since A' is faithfully flat over A, it suffices to prove the sequence

$$0 \to N \otimes_A A' \stackrel{\phi\otimes\mathrm{id}}{\to} N \otimes_A A' \otimes_A A' \stackrel{\psi\otimes\mathrm{id}}{\to} N \otimes_A A' \otimes_A A' \otimes_A A'$$

is exact. Consider the diagram

$$\begin{array}{ccccccc} 0 \to N \otimes_A A' & \stackrel{\phi\otimes\mathrm{id}}{\to} & N \otimes_A A' \otimes_A A' & \stackrel{\psi\otimes\mathrm{id}}{\to} & N \otimes_A A' \otimes_A A' \otimes_A A', \\ \swarrow \quad \downarrow \mathrm{id} & D_1 \swarrow & \downarrow \mathrm{id} & D_2 \swarrow & \\ 0 \to N \otimes_A A' & \stackrel{\phi\otimes\mathrm{id}}{\to} & N \otimes_A A' \otimes_A A' & \stackrel{\psi\otimes\mathrm{id}}{\to} & N \otimes_A A' \otimes_A A' \otimes_A A', \end{array}$$

where

$$D_1(x \otimes a_1 \otimes a_2) = x \otimes a_1a_2, \quad D_2(x \otimes a_1 \otimes a_2 \otimes a_3) = x \otimes a_1 \otimes a_2a_3$$

for any $x \in N$ and $a_i \in A'(i = 1, 2, 3)$. Then

$$D_1 \circ (\phi \otimes \mathrm{id}) = \mathrm{id}, \quad (\phi \otimes \mathrm{id}) \circ D_1 + D_2 \circ (\psi \otimes \mathrm{id}) = \mathrm{id},$$

that is, D_i $(i = 1, 2)$ define a homotopy between the identity and the zero chain map. Using these identities, one can verify that the sequence is exact.

(ii) The uniqueness of $\mathscr{F}$ follows from (i). Assume $\mathscr{F}' = N'^{\sim}$ for some A'-module N'. The isomorphism $\sigma : p_1^*\mathscr{F}' \stackrel{\cong}{\to} p_2^*\mathscr{F}'$ is induced by an isomorphism

$$\alpha : N' \otimes_A A' \stackrel{\cong}{\to} A' \otimes_A N'$$

of $A' \otimes_A A'$-modules. Set

$$N = \{x \in N' | \alpha(x \otimes 1) = 1 \otimes x\}.$$

Then N is an A-module and we have an exact sequence

$$N \to N' \underset{\delta}{\overset{\gamma}{\rightrightarrows}} A' \otimes_A N',$$

where

$$\gamma(x') = \alpha(x' \otimes 1), \quad \delta(x') = 1 \otimes x'$$

for any $x' \in N'$. Since A' is faithfully flat over A, the sequence

$$N \otimes_A A' \to N' \otimes_A A' \underset{\delta\otimes\mathrm{id}}{\overset{\gamma\otimes\mathrm{id}}{\rightrightarrows}} A' \otimes_A N' \otimes_A A'$$

is exact. The homomorphism

$$A' \to A' \otimes_A A', \quad a' \mapsto 1 \otimes a'$$

makes $A' \otimes_A A'$ a faithfully flat A'-algebra. By the proof of (i), we have an exact sequence

$$N' \to N' \otimes_{A'} (A' \otimes_A A') \rightrightarrows N' \otimes_{A'} ((A' \otimes_A A') \otimes_{A'} (A' \otimes_A A')),$$

that is,

$$N' \to A' \otimes_A N' \underset{\eta}{\overset{\xi}{\rightrightarrows}} A' \otimes_A A' \otimes_A N'$$

is exact, where

$$\xi(a' \otimes x') = a' \otimes 1 \otimes x', \quad \eta(a' \otimes x') = 1 \otimes a' \otimes x'$$

for any $a' \in A'$ and $x' \in N'$. Consider the diagram

$$\begin{array}{ccccc} N \otimes_A A' & \to & N' \otimes_A A' & \underset{\delta\otimes\mathrm{id}}{\overset{\gamma\otimes\mathrm{id}}{\rightrightarrows}} & A' \otimes_A N' \otimes_A A' \\ \beta \downarrow & & \alpha \downarrow & & p_{23}^*(\alpha) \downarrow \\ N' & \to & A' \otimes_A N' & \underset{\eta}{\overset{\xi}{\rightrightarrows}} & A' \otimes_A A' \otimes_A N'. \end{array}$$

Using the assumption $p_{13}^*(\sigma) = p_{23}^*(\sigma) \circ p_{12}^*(\sigma)$, one checks that the squares on the right-hand side commute. So there exists a homomorphism of A'-modules

$$\beta : N \otimes_A A' \to N'$$

making the square on the left-hand side commute. It induces an isomorphism $\tau : g^*\mathscr{F} \overset{\cong}{\to} \mathscr{F}'$, where $\mathscr{F} = N^{\sim}$. By the commutativity of the diagram above, we have

$$1 \otimes \beta(x \otimes 1) = \alpha(x \otimes 1)$$

for any $x \in N$. But $\alpha(x \otimes 1) = 1 \otimes x$ for any $x \in N$. So

$$1 \otimes \beta(x \otimes 1) = 1 \otimes x.$$

The images of $1 \otimes \beta(x \otimes 1)$ and $1 \otimes x$ under the map

$$A' \otimes_A N' \to N', \quad a' \otimes x' \to a'x'$$

are $\beta(x \otimes 1)$ and x, respectively. So $\beta(x \otimes 1) = x$ for any $x \in N$. Using this fact, one checks that $\sigma \circ p_1^*\tau$ is identified with $p_2^*\tau$. □

Corollary 1.6.2. *Let $g : S' \to S$ be a quasi-compact faithfully flat morphism, let $h : S' \times_S S' \to S''$ be the canonical morphism, and let $\mathscr{G}$ be a quasi-coherent $\mathscr{O}_S$-module. Then the sequence*

$$\mathscr{G} \to g_* g^* \mathscr{G} \rightrightarrows h_* h^* \mathscr{G}$$

is exact, where the two morphisms $g_* g^* \mathscr{G} \rightrightarrows h_* h^* \mathscr{G}$ are the canonical morphisms

$$g_* g^* \mathscr{G} \to g_* p_{i*} p_i^* g^* \mathscr{G} \cong h_* h^* \mathscr{G} \quad (i = 1, 2)$$

induced by the two projections $p_1, p_2 : S' \times_S S' \to S'$.

Proof. For any open subset U of S, the sequence

$$\begin{aligned} &\mathrm{Hom}_{\mathscr{O}_U}(\mathscr{O}_U, \mathscr{G}|_U) \to \mathrm{Hom}_{\mathscr{O}_{g^{-1}(U)}}(\mathscr{O}_{g^{-1}(U)}, g^*\mathscr{G}|_{g^{-1}(U)}) \\ &\quad \rightrightarrows \mathrm{Hom}_{\mathscr{O}_{h^{-1}(U)}}(\mathscr{O}_{h^{-1}(U)}, h^*\mathscr{G}|_{h^{-1}(U)}) \end{aligned}$$

is exact by 1.6.1, that is, the sequence

$$\mathscr{G}(U) \to (g_* g^* \mathscr{G})(U) \rightrightarrows (h_* h^* \mathscr{G})(U)$$

is exact. □

Corollary 1.6.3. *Let $g : S' \to S$ be a quasi-compact faithfully flat morphism, let $\mathscr{F}$ be a quasi-coherent $\mathscr{O}_S$-module, and let $\mathscr{F}'$ (resp. $\mathscr{F}''$) be the inverse image of $\mathscr{F}$ on S' (resp. $S' \times_S S'$). Denote by* Quot($\mathscr{F}$) *(resp.* Quot($\mathscr{F}'$)*, resp.* Quot($\mathscr{F}''$)*) the set of isomorphic classes of quasi-coherent quotient sheaves of $\mathscr{F}$ (resp. $\mathscr{F}'$, resp. $\mathscr{F}''$). Then the sequence*

$$\mathrm{Quot}(\mathscr{F}) \xrightarrow{g^*} \mathrm{Quot}(\mathscr{F}') \underset{p_2^*}{\overset{p_1^*}{\rightrightarrows}} \mathrm{Quot}(\mathscr{F}'')$$

is exact.

Proof. Let $\mathscr{G}' \in \ker(\mathrm{Quot}(\mathscr{F}') \rightrightarrows \mathrm{Quot}(\mathscr{F}''))$. Then we have an isomorphism $\sigma : p_1^* \mathscr{G}' \xrightarrow{\cong} p_2^* \mathscr{G}'$ such that the diagram

$$\begin{array}{ccc} p_1^*\mathscr{F}' & \to & p_1^*\mathscr{G}' \\ \cong\downarrow & & \\ \mathscr{F}'' & & \downarrow\sigma \\ \cong\downarrow & & \\ p_2^*\mathscr{F}' & \to & p_2^*\mathscr{G}' \end{array}$$

commutes. So the following diagrams commute, where $\mathscr{F}'''$ is the inverse image of $\mathscr{F}$ on $S' \times_S S' \times_S S'$:

$$\begin{array}{ccc} p_{12}^*p_1^*\mathscr{F}' \to p_{12}^*p_1^*\mathscr{G}' & p_{23}^*p_1^*\mathscr{F}' \to p_{23}^*p_1^*\mathscr{G}' & p_{13}^*p_1^*\mathscr{F}' \to p_{13}^*p_1^*\mathscr{G}' \\ \cong\downarrow \qquad\qquad & \cong\downarrow \qquad\qquad & \cong\downarrow \qquad\qquad \\ \mathscr{F}''' \qquad \downarrow p_{12}^*(\sigma) & \mathscr{F}''' \qquad \downarrow p_{23}^*(\sigma) & \mathscr{F}''' \qquad \downarrow p_{13}^*(\sigma) \\ \cong\downarrow \qquad\qquad & \cong\downarrow \qquad\qquad & \cong\downarrow \qquad\qquad \\ p_{12}^*p_2^*\mathscr{F}' \to p_{12}^*p_2^*\mathscr{G}', & p_{23}^*p_2^*\mathscr{F}' \to p_{23}^*p_2^*\mathscr{G}', & p_{13}^*p_2^*\mathscr{F}' \to p_{13}^*p_2^*\mathscr{G}'. \end{array}$$

Since the horizontal arrows are surjective, this implies that

$$p_{13}^*(\sigma) = p_{23}^*(\sigma) \circ p_{12}^*(\sigma).$$

By 1.6.1 (ii), there exists a quasi-coherent $\mathscr{O}_S$-module $\mathscr{G}$ whose inverse image on S' is isomorphic to $\mathscr{G}'$. By 1.6.1 (i), there exists a morphism $\mathscr{F} \to \mathscr{G}$ inducing the epimorphism $\mathscr{F}' \to \mathscr{G}'$. Since $g : S' \to S$ is faithfully flat, $\mathscr{F} \to \mathscr{G}$ is surjective. So $\mathscr{G}$ is a quotient of $\mathscr{F}$. The uniqueness of $\mathscr{G}$ follows from 1.6.1 (i). □

Corollary 1.6.4. *Let $g : S' \to S$ be a quasi-compact faithfully flat morphism, let $\mathscr{F}$ be a quasi-coherent $\mathscr{O}_S$-module, and let $\mathscr{F}'$ (resp. $\mathscr{F}''$) be the inverse image of $\mathscr{F}$ on S' (resp. $S' \times_S S'$). Denote by* Sub($\mathscr{F}$) *(resp.* Sub($\mathscr{F}'$), *resp.* Sub($\mathscr{F}''$)) *the set of isomorphic classes of quasi-coherent subsheaves of $\mathscr{F}$ (resp. $\mathscr{F}'$, resp. $\mathscr{F}''$). Then the sequence*

$$\mathrm{Sub}(\mathscr{F}) \xrightarrow{g^*} \mathrm{Sub}(\mathscr{F}') \underset{p_2^*}{\overset{p_1^*}{\rightrightarrows}} \mathrm{Sub}(\mathscr{F}'')$$

is exact.

We leave the proof to the reader.

Corollary 1.6.5. *Let $g : S' \to S$ be a quasi-compact faithfully flat morphism, let $S'' = S' \times_S S'$, and let* Sub(S) *(resp.* Sub(S'), *resp.* Sub(S'')) *be the set of isomorphic classes of closed subschemes of S (resp. S', resp. S''). Then the sequence*

$$\mathrm{Sub}(S) \xrightarrow{g^*} \mathrm{Sub}(S') \underset{p_2^*}{\overset{p_1^*}{\rightrightarrows}} \mathrm{Sub}(S'')$$

is exact, where g^*, p_i^* $(i = 1, 2)$ denote base changes of closed subschemes.

Proof. This follows from 1.6.3 or 1.6.4 by taking $\mathscr{F} = \mathscr{O}_S$. □

Proposition 1.6.6. *Let $g : S' \to S$ be a quasi-compact faithfully flat morphism, and let $\mathscr{F}$ be a quasi-coherent $\mathscr{O}_S$-module. Then $g^*\mathscr{F}$ is locally of finite type (resp. locally of finite presentation, resp. locally free of finite rank) if and only if $\mathscr{F}$ is so.*

Proof. The "if" part is clear. Let us prove the "only if" part. We leave it to the reader to reduce the proof to the case where $S = \operatorname{Spec} A$ and $S' = \operatorname{Spec} A'$ are affine and A' is a faithfully flat A-algebra. Let M be an A-module. Write $M = \varinjlim_\lambda M_\lambda$, where M_λ goes over the set of finitely generated submodules of M. We have $M \otimes_A A' = \varinjlim_\lambda M_\lambda \otimes_A A'$. Suppose $M \otimes_A A'$ is a finitely generated A'-module. Then there exists some M_λ such that $M_\lambda \otimes_A A' \to M \otimes_A A'$ is surjective. Since A' is faithfully flat over A, we must have $M_\lambda = M$. So M is a finitely generated A-module. If $M \otimes_A A'$ has finite presentation, then M must be a finitely generated A-module. So we can find an exact sequence

$$0 \to R \to L \to M \to 0$$

such that L is a free A-module of finite rank. As A' is faithfully flat over A, the sequence

$$0 \to R \otimes_A A' \to L \otimes_A A' \to M \otimes_A A' \to 0$$

is exact. Since $M \otimes_A A'$ has finite presentation, $R \otimes_A A'$ is a finitely generated A'-module. This implies that R is a finitely generated A-module. So M has finite presentation. The statement about local freeness follows from 1.6.7 below and the fact that M is flat over A if and only if $M \otimes_A A'$ is flat over A'. □

Lemma 1.6.7. *Let M be an A-module. The following conditions are equivalent:*

(i) *$M^\sim$ is a locally free $\mathscr{O}_{\operatorname{Spec} A}$-module of finite rank.*

(ii) *M is a finitely generated projective A-module.*

(iii) *M is a flat A-module of finite presentation.*

Proof.

(i)⇒(ii) Since $M^\sim$ is locally free of finite rank, we can find a finite affine open covering $\{U_i\}$ of $\operatorname{Spec} A$ such that $M^\sim|_{U_i}$ are free of finite rank. The canonical morphism $g : \coprod_i U_i \to S$ is quasi-compact and faithfully flat, and $g^*M^\sim$ has finite presentation. So M has finite presentation by 1.6.6. For any A-module N, we then have

$$\mathscr{H}om_{\mathscr{O}_X}(M^\sim, N^\sim) \cong (\operatorname{Hom}_A(M, N))^\sim.$$

Since $M^\sim$ is locally free, the functor $\mathscr{H}om_{\mathscr{O}_X}(M^\sim, -)$ is exact on the category of $\mathscr{O}_{\mathrm{Spec}\, A}$-modules. So the functor $\mathrm{Hom}_A(M, -)$ is exact on the category of A-modules. Hence M is a projective A-module.

(ii)⇒(iii) There exists an A-module R such that $A^n \cong M \oplus R$ for some finite n. It follows that M is flat and has finite presentation.

(iii)⇒(i) For any $\mathfrak{p} \in \mathrm{Spec}\, A$, let us prove $M^\sim$ is free in a neighborhood of $\mathfrak{p}$. $M_\mathfrak{p}$ is a flat $A_\mathfrak{p}$-module with finite presentation. This implies that $M_\mathfrak{p}$ is a free $A_\mathfrak{p}$-module of finite rank. (Confer the proof of (ii)⇒(i) of 1.2.8.) Choose $x_1, \ldots, x_k \in M$ so that their images in $M_\mathfrak{p}$ form a basis. Consider the homomorphism $\phi : A^k \to M$ which maps a basis of A^k to $\{x_1, \ldots, x_k\}$. We have

$$(\ker \phi)_\mathfrak{p} = (\mathrm{coker}\, \phi)_\mathfrak{p} = 0.$$

Since M is finitely generated, so it is with $\mathrm{coker}\, \phi$. It follows that there exists an affine open neighborhood of $\mathfrak{p}$ in $\mathrm{Spec}\, A$ on which $(\mathrm{coker}\, \phi)^\sim$ vanishes. Replacing $\mathrm{Spec}\, A$ by this neighborhood, we may assume $\mathrm{coker}\, \phi = 0$. Then ϕ is an epimorphism. Since M has finite presentation, $\ker \phi$ is finitely generated. There exists an affine open neighborhood of $\mathfrak{p}$ in $\mathrm{Spec}\, A$ on which $(\ker \phi)^\sim$ vanishes. Replacing $\mathrm{Spec}\, A$ by this neighborhood, we may assume $\ker \phi = 0$. Then ϕ is an isomorphism, and $M^\sim$ is free. □

1.7 Descent of Properties of Morphisms

([SGA 1] VIII 3, 4)

Let $f : X \to S$ be a morphism of schemes. We say that f is surjective (resp. injective) if f is surjective (resp. injective) on the underlying topological spaces. f is called *radiciel* if it is universally injective, that is, for any morphism $S' \to S$, the base change $f' : X \times_S S' \to S'$ of f is injective.

Proposition 1.7.1. *Let $f : X \to S$ be a morphism. The following conditions are equivalent:*

(i) *f is radiciel.*

(ii) *For any algebraically closed field K, the map $X(K) \to S(K)$ induced by f is injective, where $X(K) = \mathrm{Hom}(\mathrm{Spec}\, K, X)$ and $S(K) = \mathrm{Hom}(\mathrm{Spec}\, K, S)$ are the sets K-points in X and S, respectively.*

(iii) *f is injective, and for any $x \in X$, the residue field $k(x)$ of X at x is a purely inseparable algebraic extension of the residue field $k(f(x))$ of Y at $f(x)$.*

(iv) *The diagonal morphism $\Delta : X \to X \times_S X$ is surjective.*

In particular, any radiciel morphism is separated.

Proof.

(i)⇒(ii) Suppose f is universally injective and $t_1, t_2 : \operatorname{Spec} K \to X$ are two morphisms such that $ft_1 = ft_2$. Let us prove $t_1 = t_2$. Regard $\operatorname{Spec} K$ as an S-scheme through the morphism $ft_1 = ft_2$. Let

$$\Gamma_{t_i} : \operatorname{Spec} K \to X \times_S \operatorname{Spec} K \quad (i = 1, 2)$$

be the graphs of t_i, and let $f_K : X \times_S \operatorname{Spec} K \to \operatorname{Spec} K$ be the base change of f. It suffices to show $\Gamma_{t_1} = \Gamma_{t_2}$. We have

$$f_K \Gamma_{t_1} = f_K \Gamma_{t_2} = \mathrm{id}_{\operatorname{Spec} K}.$$

Since f_K is injective, Γ_{t_1} and Γ_{t_2} have the same image in $X \times_Y \operatorname{Spec} K$. Let x' be their common image. Then the composites

$$K \stackrel{f_K^\natural}{\to} k(x') \stackrel{\Gamma_{t_i}^\natural}{\to} K \quad (i = 1, 2)$$

are the identity, where $f_K^\natural$ and $\Gamma_{t_i}^\natural$ are the homomorphisms on residue fields induced by the morphisms f_K and Γ_{t_i}, respectively. It follows that $\Gamma_{t_i}^\natural$ $(i = 1, 2)$ are isomorphisms, and are inverse to $f_K^\natural$, and hence they coincide. So we have $\Gamma_{t_1} = \Gamma_{t_2}$.

(ii)⇒(i) Let $g : S' \to S$ be a morphism, and let x_1' and x_2' be two points in $X \times_S S'$ having the same image s' in S'. The homomorphism $k(s') \to k(x_1')$ is injective. Since $k(x_2')$ is flat over $k(s')$, the canonical homomorphism

$$k(x_2') \to k(x_1') \otimes_{k(s')} k(x_2')$$

is injective. It follows that $0 \neq 1$ in $k(x_1') \otimes_{k(s')} k(x_2')$. So $k(x_1') \otimes_{k(s')} k(x_2')$ has a maximal ideal $\mathfrak{m}$. Let K be an algebraic closure of $(k(x_1') \otimes_{k(s')} k(x_2'))/\mathfrak{m}$. Then we have a commutative diagram

$$\begin{array}{ccc} k(s') & \to & k(x_1') \\ \downarrow & & \downarrow \\ k(x_2') & \to & K. \end{array}$$

For each $i \in \{1, 2\}$, the homomorphism $k(x_i') \to K$ defines a morphism $s_i : \operatorname{Spec} K \to X \times_S S'$ having the image x_i' in $X \times_S S'$. Let $g' : X \times_S S' \to X$ and $f' : X \times_S S' \to S'$ be the projections. We have $f's_1 = f's_2$. It follows that

$$fg's_1 = gf's_1 = gf's_2 = fg's_2.$$

By our assumption, we then have $g's_1 = g's_2$. Together with the identity $f's_1 = f's_2$, this implies that $s_1 = s_2$. In particular, we have $x_1' = x_2'$. So f is universally injective.

(i)⇔(iii) Suppose f is injective. Let x be a point in X and let s be its image in S. Then $f^{-1}(s)$ consists of only one point x. Consider a Cartesian diagram

$$\begin{array}{ccc} X \times_S S' & \xrightarrow{g'} & X \\ {\scriptstyle f'}\downarrow & & \downarrow{\scriptstyle f} \\ S' & \xrightarrow{g} & S. \end{array}$$

For any point s' in S' with image s in S, we have a one-to-one correspondence between the underlying set of $f'^{-1}(s')$ and the underlying set of the scheme $\mathrm{Spec}\,(k(x) \otimes_{k(s)} k(s'))$. It follows that f is universally injective if and only if for any field k' containing $k(s)$, the scheme $\mathrm{Spec}\,(k(x) \otimes_{k(s)} k')$ contains only one point. Let us prove that this last condition is equivalent to saying that $k(x)$ is a purely inseparable algebraic extension of $k(s)$. For convenience, we denote $k(s)$ by k, and $k(x)$ by K.

If K is not algebraic over k, then we can find an element t in K which is transcendental over k. We have $k(t) \otimes_k k[T] \cong k(t)[T]$. Let $\mathfrak{m} = (T-t)$ be the prime ideal of $k(t)[T]$ generated by $T-t$. By the transcendence of t, we have $\mathfrak{m} \cap k[T] = 0$. On the other hand, the zero prime ideal $\mathfrak{p}$ of $k(t)[T]$ also has the property $\mathfrak{p} \cap k[T] = 0$. It follows that there are at least two points in the fiber of

$$\mathrm{Spec}\, k(t)[T] \to \mathrm{Spec}\, k[T]$$

over the zero prime ideal of $k[T]$. Hence $\mathrm{Spec}\,(k(t) \otimes_k k(T))$ contains at least two points. Since

$$\mathrm{Spec}\,(K \otimes_k k(T)) \to \mathrm{Spec}\,(k(t) \otimes_k k(T))$$

is surjective, $\mathrm{Spec}\,(K \otimes_k k(T))$ contains at least two points.

If K is algebraic over k, but not pure inseparable over k, then we can find a subfield F of K strictly containing k and separable over k. Let $\bar{k}$ be an algebraic closure of k. Then $\mathrm{Spec}\,(F \otimes_k \bar{k})$ contains more than one point. As

$$\mathrm{Spec}\,(K \otimes_k \bar{k}) \to \mathrm{Spec}\,(F \otimes_k \bar{k})$$

is surjective, $\mathrm{Spec}\,(K \otimes_k \bar{k})$ contains more than one point.

If K is a purely inseparable algebraic extension of k, then for any $t \in K$, we have $t^{p^n} \in k$ for a large integer n, where p is the characteristic of k.

For any field k' containing k, and any $t' \in K \otimes_k k'$, we have $t'^{p^n} \in k'$ for a large integer n. Let $\mathfrak{p}$ be a prime ideal of $K \otimes_k k'$. We have $\mathfrak{p} \cap k' = 0$. For any $t' \in \mathfrak{p}$, suppose $t'^{p^n} \in k'$. Then $t'^{p^n} \in \mathfrak{p} \cap k' = 0$. So any prime ideal of $K \otimes_k k'$ is nilpotent. This implies that $K \otimes_k k'$ has only one prime ideal.

(ii)⇒(iv) Given a point z in $X \times_S X$, let K be an algebraic closure of $k(z)$. We then have a morphism $t : \operatorname{Spec} K \to X \times_S X$ with image z. Let $p_1, p_2 : X \times_S X \to X$ be the projections. We have $fp_1t = fp_2t$. So $p_1t = p_2t$ by our assumption. Let $t' = p_1t = p_2t$. Then we have

$$p_i(\Delta t') = (p_i\Delta)t' = t' = p_it \quad (i = 1, 2).$$

So we have $\Delta t' = t$. In particular, z lies in the image of Δ. So Δ is surjective.

(iv)⇒(ii) Let $t_1, t_2 : \operatorname{Spec} K \to X$ be two morphisms such that $ft_1 = ft_2$. Then there exists a morphism $t : \operatorname{Spec} K \to X \times_S X$ such that $p_it = t_i$ for $i = 1, 2$. Let z be the image of t. Since Δ is surjective, there exists a point $x \in X$ such that $\Delta(x) = z$. Since Δ is an immersion, it induces an isomorphism $\Delta^\natural : k(z) \xrightarrow{\cong} k(x)$. So there exists a morphism $t' : \operatorname{Spec} K \to X$ such that $\Delta t' = t$. But then

$$t_i = p_it = p_i\Delta t' = t' \quad (i = 1, 2).$$

So we have $t_1 = t_2$. □

Proposition 1.7.2. *Consider a Cartesian diagram*

$$\begin{array}{ccc} X \times_S S' & \xrightarrow{g'} & X \\ {\scriptstyle f'} \downarrow & & \downarrow {\scriptstyle f} \\ S' & \xrightarrow{g} & S. \end{array}$$

Assume g is surjective.

(i) *f is surjective if and only if f' is so.*

(ii) *If f' is injective, then so is f.*

(iii) *f is radiciel if and only if f' is so.*

Proof. For any $s' \in S'$, we have

$$f'^{-1}(s') \cong f^{-1}(g(s')) \otimes_{k(g(s'))} k(s').$$

The projection $f'^{-1}(s') \to f^{-1}(g(s'))$ is surjective by 1.2.4 and the fact that $k(s')$ is faithfully flat over $k(g(s'))$. So $f'^{-1}(s') \neq \emptyset$ if and only if $f^{-1}(g(s')) \neq \emptyset$. Since g is surjective, this implies (i). If f' is injective, then $f'^{-1}(s')$ contains at most one point. So $f^{-1}(g(s'))$ contains at most one point, and hence f is injective. This shows (ii). (iii) follows from (ii). □

Proposition 1.7.3. *Consider a Cartesian diagram*

$$\begin{array}{ccc} X \times_S S' & \xrightarrow{g'} & X \\ {\scriptstyle f'}\downarrow & & \downarrow{\scriptstyle f} \\ S' & \xrightarrow{g} & S. \end{array}$$

Assume g is quasi-compact and surjective. Then f is quasi-compact if and only if f' is so.

Proof. Suppose f is quasi-compact. To prove that f' is quasi-compact, it suffices to show that for any affine open subset V' of S' such that $g(V')$ is contained in an affine open subset V of S, $f'^{-1}(V')$ is a union of finitely many affine open subsets of $X \times_S S'$. We can write $f^{-1}(V) = \bigcup_{i=1}^n U_i$ for finitely many affine open subsets U_i in X. Then we have

$$f'^{-1}(V') = \bigcup_{i=1}^{n} (V' \times_V U_i),$$

where each $V' \times_V U_i$ is regarded as an affine open subscheme of $X \times_S S'$.

Conversely suppose f' is quasi-compact. Let V be a quasi-compact open subset of S. Since g is surjective, we have $V = g(g^{-1}(V))$. So

$$f^{-1}(V) = f^{-1}(g(g^{-1}(V)) = g'(f'^{-1}g^{-1}(V)).$$

Since g and f' are quasi-compact, $f'^{-1}g^{-1}(V)$ is quasi-compact, and hence $f^{-1}(V) = g'(f'^{-1}g^{-1}(V))$ is quasi-compact. So f is quasi-compact. □

Proposition 1.7.4. *Consider a Cartesian diagram*

$$\begin{array}{ccc} X \times_S S' & \xrightarrow{g'} & X \\ {\scriptstyle f'}\downarrow & & \downarrow{\scriptstyle f} \\ S' & \xrightarrow{g} & S. \end{array}$$

Assume g is quasi-compact and faithfully flat. Then f is of finite type if and only if so is f'.

The proof is similar to that of 1.6.6.

Proposition 1.7.5. *Let $g : Y' \to Y$ be a flat morphism and let Z be a subset of Y. Assume there exists a quasi-compact morphism $f : X \to Y$ such that $Z = f(X)$. Then $g^{-1}(\overline{Z}) = \overline{g^{-1}(Z)}$.*

Proof. The inclusion $\overline{g^{-1}(Z)} \subset g^{-1}(\overline{Z})$ is clear. Let $y' \in g^{-1}(\overline{Z})$. We need to show $y' \in \overline{g^{-1}(Z)}$. Choose an affine open neighborhood $U' = \operatorname{Spec} A'$ of y' and an affine open neighborhood $U = \operatorname{Spec} A$ of $g(y')$ such that $g(U') \subset U$. We have $y' \in U' \cap g^{-1}(\overline{Z \cap U} \cap U)$. To prove $y' \in \overline{g^{-1}(Z)}$, it suffices to show $y' \in \overline{g^{-1}(Z \cap U)} \cap U'$. Moreover, $Z \cap U$ is the image of the quasi-compact morphism $f^{-1}(U) \to U$ induced from f. We are thus reduced to the case where $Y' = \operatorname{Spec} A'$ and $Y = \operatorname{Spec} A$ are affine and A' is a flat A-algebra. Since $f : X \to Y$ is quasi-compact and Y is now affine, we can cover X by finitely many open affine subsets V_i. Replacing X by $\coprod_i V_i$, we may assume $X = \operatorname{Spec} B$ for some A-algebra B. Let I be the kernel of $A \to B$. We have

$$g^{-1}(\overline{Z}) = g^{-1}(\overline{f(X)}) = g^{-1}(V(I)) = V(IA').$$

Let

$$f' : X' = \operatorname{Spec}(A' \otimes_A B) \to Y' = \operatorname{Spec} A'$$

be the base change of f. Since A' is flat over A, the kernel of $A' \to A' \otimes_A B$ is IA'. It follows that $\overline{f'(X')} = V(IA')$. So we have

$$\overline{g^{-1}(Z)} = \overline{g^{-1}(f(X))} = \overline{f'(X')} = V(IA').$$

We thus have $g^{-1}(\overline{Z}) = \overline{g^{-1}(Z)}$. This proves our assertion. □

Corollary 1.7.6. *Let $g : Y' \to Y$ be a quasi-compact flat morphism, and let Z' be a closed subset of Y' satisfying $Z' = g^{-1}(g(Z'))$. Then $Z' = g^{-1}(\overline{g(Z')})$. Moreover, the subspace topology on $g(Y')$ induced from Y coincides with the quotient topology induced from Y'.*

Proof. The first assertion follows from 1.7.5. To prove the second assertion, note that since the morphism $g : Y' \to Y$ is continuous, every subset of $g(Y')$ that is closed with respect to the subspace topology induced from Y is closed with respect to the quotient topology induced from Y'. Conversely, let Z be a subset of $g(Y')$ closed with respect to the quotient topology induced from Y'. Then $g^{-1}(Z)$ is closed. Applying the first assertion to $Z' = g^{-1}(Z)$, we get $g^{-1}(Z) = g^{-1}(\overline{g(g^{-1}(Z))})$, that is, $g^{-1}(Z) = g^{-1}(\overline{Z})$. Hence $Z = \overline{Z} \cap g(Y')$. So Z is closed with respect to the topology induced from Y. □

Corollary 1.7.7. *Assume $g : Y' \to Y$ is a quasi-compact faithfully flat morphism. Then the topology on Y coincides with the quotient topology induced from Y'.*

Corollary 1.7.8. *Let $g : S' \to S$ be a quasi-compact faithfully flat morphism, $S'' = S' \times_S S'$, $\mathrm{O}(S)$ (resp. $\mathrm{O}(S')$, resp. $\mathrm{O}(S'')$) the set of open subsets of S (resp. S', resp. S''), and $\mathrm{F}(S)$ (resp. $\mathrm{F}(S')$, resp. $\mathrm{F}(S'')$) the* *set of closed subsets of S (resp. S', resp. S''). Then the following sequences are exact:*

$$\mathrm{O}(S) \to \mathrm{O}(S') \overset{p_1^{-1}}{\underset{p_2^{-1}}{\rightrightarrows}} \mathrm{O}(S''), \quad \mathrm{F}(S) \to \mathrm{F}(S') \overset{p_1^{-1}}{\underset{p_2^{-1}}{\rightrightarrows}} \mathrm{F}(S'').$$

Corollary 1.7.9. *Assume $g : Y' \to Y$ is a quasi-compact faithfully flat morphism. Let Z be a subset of Y which is the image of a quasi-compact morphism. (For example, Y is noetherian and Z is constructible.) Then Z is locally closed if and only if $g^{-1}(Z)$ is so.*

Proof. Consider the Cartesian diagram

$$\begin{array}{ccc} g^{-1}(\overline{Z}) & \to & \overline{Z} \\ \downarrow & & \downarrow \\ Y' & \xrightarrow{g} & Y, \end{array}$$

wherein we put a closed subscheme structure on $\overline{Z}$. The morphism $g^{-1}(\overline{Z}) \to \overline{Z}$ is quasi-compact and faithfully flat. By 1.7.5, we have $g^{-1}(\overline{Z}) = \overline{g^{-1}(Z)}$. If $g^{-1}(Z)$ is locally closed, then it is open in $g^{-1}(\overline{Z})$. By 1.7.7, Z is open in $\overline{Z}$. So Z is locally closed. □

Corollary 1.7.10. *Consider a Cartesian diagram*

$$\begin{array}{ccc} X \times_S S' & \xrightarrow{g'} & X \\ {\scriptstyle f'}\downarrow & & \downarrow{\scriptstyle f} \\ S' & \xrightarrow{g} & S. \end{array}$$

Assume g is quasi-compact and faithfully flat. If f' is an open mapping (resp. a closed mapping, resp. a quasi-compact embedding, resp. a homeomorphism), then f is so.

Proof. For any subset Z of X, we have $g^{-1}(f(Z)) = f'(g'^{-1}(Z))$. If Z is open (resp. closed) and f' is an open mapping (resp. a closed mapping), then $f'g'^{-1}(Z)$ is open (resp. closed). So $g^{-1}(f(Z))$ is open (resp. closed). By 1.7.7, $f(Z)$ is open (resp. closed). So f is an open (resp. closed) mapping. Suppose f' is a quasi-compact embedding. Then f is injective by 1.7.2, and quasi-compact by 1.7.3. Let Z be a closed subset of X. By 1.7.5, we have

$$g^{-1}(\overline{f(Z)}) = \overline{g^{-1}(f(Z))} = \overline{f'(g'^{-1}(Z))}.$$

Since f' is an embedding, we have

$$f'^{-1}(\overline{f'(g'^{-1}(Z))}) = g'^{-1}(Z).$$

It follows that

$$g'^{-1}(f^{-1}(\overline{f(Z)})) = f'^{-1}(g^{-1}(\overline{f(Z)})) = f'^{-1}(\overline{f'(g'^{-1}(Z))}) = g'^{-1}(Z).$$

Since g' is surjective, this implies that $f^{-1}(\overline{f(Z)}) = Z$. Hence f is an embedding. If f' is a homeomorphism, then f is an embedding by what we have proved, and f is surjective by 1.7.2. So f is a homeomorphism. □

Corollary 1.7.11. *Consider a Cartesian diagram*

$$\begin{array}{ccc} X \times_S S' & \xrightarrow{g'} & X \\ f' \downarrow & & \downarrow f \\ S' & \xrightarrow{g} & S. \end{array}$$

Assume g is quasi-compact and faithfully flat. Then f is universally open, universally closed, or universally a homeomorphism, if and only if f' is so.

Corollary 1.7.12. *Consider a Cartesian diagram*

$$\begin{array}{ccc} X \times_S S' & \xrightarrow{g'} & X \\ f' \downarrow & & \downarrow f \\ S' & \xrightarrow{g} & S. \end{array}$$

Assume g is quasi-compact and faithfully flat. Then f is separated or proper if and only if so is f'.

Proof. f is separated if and only if the diagonal morphism $X \to X \times_S X$ is a closed mapping. f is proper if it is of finite type, separated, and universally closed. Our assertions follow from 1.7.10, 1.7.4, and 1.7.11. □

1.8 Descent of Schemes

([SGA 1] VIII 2, 5.)

Proposition 1.8.1. *Let $g : S' \to S$ be a morphism of schemes.*

(i) *Suppose g is surjective and $\mathscr{O}_S \to g_*\mathscr{O}_{S'}$ is injective. Then g is an epimorphism in the category of schemes.*

(ii) *Suppose g is surjective and the topology on S is the quotient topology induced from S'. Let $S'' = S' \times_S S'$, let $p_1, p_2 : S'' = S' \times_S S' \to S'$ be the projections, and let $h : S'' \to S$ be the canonical morphism. Assume*

$$\mathscr{O}_S \to g_*\mathscr{O}_{S'} \underset{p_2^\natural}{\overset{p_1^\natural}{\rightrightarrows}} h_*\mathscr{O}_{S''}$$

is exact. Then for any scheme Z, the sequence

$$\mathrm{Hom}(S,Z)\xrightarrow{g^*}\mathrm{Hom}(S',Z)\underset{p_2^*}{\overset{p_1^*}{\rightrightarrows}}\mathrm{Hom}(S'',Z)$$

is exact, where g^* maps a morphism $f:S\to Z$ to fg, and p_i^* $(i=1,2)$ map a morphism $f:S'\to Z$ to $fp_i:S''\to Z$.

Proof. We leave the proof of (i) to the reader. Let us prove (ii). By (i), the map $g^*:\mathrm{Hom}(S,Z)\to\mathrm{Hom}(S',Z)$ is injective. Let $f':S'\to Z$ be a morphism of schemes such that $f'p_1=f'p_2$. For any $s_1',s_2'\in S'$ such that $g(s_1')=g(s_2')$, there exists $s''\in S''$ such that $p_i(s'')=s_i'$ $(i=1,2)$. We then have

$$f'(s_1')=f'p_1(s'')=f'p_2(s'')=f'(s_2').$$

As g is surjective, there exists a map $f:S\to Z$ on the underlying set such that $fg=f'$. Since the topology on S is the quotient topology induced from S', f is continuous. We have a commutative diagram

$$\begin{array}{ccccc} & & f'_*\mathcal{O}_{S'} & \to & f'_*p_{1*}\mathcal{O}_{S''} \\ & \nearrow & \wr\| & & \wr\| \\ \mathcal{O}_Z & \to & f_*g_*\mathcal{O}_{S'} & \underset{p_2^\natural}{\overset{p_1^\natural}{\rightrightarrows}} & f_*h_*\mathcal{O}_{S''} \\ & \searrow & \wr\| & & \wr\| \\ & & f'_*\mathcal{O}_{S'} & \to & f'_*p_{2*}\mathcal{O}_{S''}. \end{array}$$

Since $f'p_1=f'p_2$, the image of the morphism $\mathcal{O}_Z\to f_*g_*\mathcal{O}_{S'}$ lies in the kernel of $f_*g_*\mathcal{O}_{S'}\rightrightarrows f_*h_*\mathcal{O}_{S''}$. By our assumption, we have an exact sequence

$$f_*\mathcal{O}_S\to f_*g_*\mathcal{O}_{S'}\rightrightarrows f_*h_*\mathcal{O}_{S''}.$$

So the morphism $\mathcal{O}_Z\to f_*g_*\mathcal{O}_{S'}$ induces a morphism $\mathcal{O}_Z\to f_*\mathcal{O}_S$. The pair formed by the continuous map $f:S\to Z$ and the morphism $\mathcal{O}_Z\to f_*\mathcal{O}_S$ of sheaves of rings defines a morphism $f:S\to Z$ of ringed spaces such that $f'=fg$. For any $s'\in S'$, the composite of the homomorphisms

$$\mathcal{O}_{Z,fg(s')}\xrightarrow{f^\natural_{g(s')}}\mathcal{O}_{S,g(s')}\xrightarrow{g^\natural_{s'}}\mathcal{O}_{S',s'}$$

coincides with the homomorphism $f'^\natural_{s'}:\mathcal{O}_{Z,fg(s')}\to\mathcal{O}_{S',s'}$. Since f' and g are morphisms of schemes, $f'^\natural_{s'}$ and $g^\natural_{s'}$ are local homomorphisms of local rings. It follows that $f^\natural_{g(s')}$ is also a local homomorphism. Hence f is a morphism of schemes. □

Corollary 1.8.2. *Let $g : S' \to S$ be a quasi-compact faithfully flat morphism, and let $S'' = S' \times_S S'$. For any scheme Z, the sequence*

$$\mathrm{Hom}(S, Z) \to \mathrm{Hom}(S', Z) \rightrightarrows \mathrm{Hom}(S'', Z)$$

is exact.

Proof. Use 1.6.2, 1.7.7, and 1.8.1. □

Corollary 1.8.3. *Let $g : S' \to S$ be a quasi-compact faithfully flat morphism, $S'' = S' \times_S S'$, X and Y two S-schemes, $X' = X \times_S S'$, $Y' = Y \times_S S'$, $X'' = X \times_S S''$, and $Y'' = Y \times_S S''$. Then the sequence*

$$\mathrm{Hom}_S(X, Y) \to \mathrm{Hom}_{S'}(X', Y') \rightrightarrows \mathrm{Hom}_{S''}(X'', Y'')$$

is exact.

Proof. By 1.8.2, the sequences

$$\mathrm{Hom}(X, Y) \to \mathrm{Hom}(X', Y) \rightrightarrows \mathrm{Hom}(X'', Y)$$
$$\mathrm{Hom}(X, S) \to \mathrm{Hom}(X', S) \rightrightarrows \mathrm{Hom}(X'', S)$$

are exact. It follows that the sequence

$$\mathrm{Hom}_S(X, Y) \to \mathrm{Hom}_S(X', Y) \rightrightarrows \mathrm{Hom}_S(X'', Y)$$

is exact. The above sequence can be identified with the sequence in the corollary. □

Corollary 1.8.4. *In the notation of 1.8.3, let $f : X \to Y$ be an S-morphism, and let $f' : X' \to Y'$ be the base change of f. Then f is an isomorphism if and only if so is f'.*

Proof. Suppose f' is an isomorphism. Let g' be its inverse. Then g' lies in the kernel of $\mathrm{Hom}_{S'}(Y', X') \rightrightarrows \mathrm{Hom}_{S''}(Y'', X'')$ since its images under these two maps are both the inverse of the base change $f'' : X'' \to Y''$ of f. Thus there exists an S-morphism $g : Y \to X$ so that g' is its base change. Since $g'f' = \mathrm{id}_{X'}$ and the map $\mathrm{Hom}_S(X, X) \to \mathrm{Hom}_{S'}(X', X')$ is injective, we have $gf = \mathrm{id}_X$. Similarly, we have $fg = \mathrm{id}_Y$. So f is an isomorphism. □

Corollary 1.8.5. *Under the assumption of 1.8.4, f is a closed immersion, an open immersion, or a quasi-compact immersion, if and only if f' is so.*

Proof. Suppose f' is a closed immersion. By 1.6.5, there exists a closed immersion $f_0 : X_0 \to Y$ such that we have an isomorphism $\phi' : X' \stackrel{\cong}{\to} X_0'$ with the property that $f_0'\phi' = f'$, where $f_0' : X_0' = X_0 \times_S S' \to Y'$ is the base change of f_0. Let $p_i^*(\phi')$ $(i = 1, 2)$ be the images of the ϕ' under the maps $\mathrm{Hom}_{Y'}(X', X_0') \rightrightarrows \mathrm{Hom}_{Y''}(X'', X_0'')$. We have

$$f_0'' p_1^*(\phi') = f'' = f_0'' p_2^*(\phi'),$$

where f'' and f_0'' are the base changes of f and f_0, respectively. Since f_0'' is a closed immersion, this implies that $p_1^*(\phi') = p_2^*(\phi')$. By 1.8.3, there exists a morphism $\phi : X \to X_0$ such that ϕ' is its base change and $f_0\phi = f$. By 1.8.4, ϕ is an isomorphism. As f_0 is a closed immersion, so is f.

Using the same argument and 1.7.8, one can show if f' is an open immersion, then so is f.

Suppose f' is a quasi-compact immersion. Then f is quasi-compact by 1.7.3, and $f(X)$ is locally closed in Y by 1.7.9. Let U be an open subset of Y so that $f(X)$ is closed in U, and let U' be the inverse image of U in Y'. Then f' induces a closed immersion $X' \to U'$. By the above discussion, f induces a closed immersion $X \to U$. So f is an immersion. □

Proposition 1.8.6. *Let $g : S' \to S$ be a quasi-compact faithfully flat morphism, $S'' = S' \times_S S'$, $S''' = S' \times_S S' \times S'$, $p_1, p_2 : S'' \to S'$ and $p_{12}, p_{13}, p_{23} : S''' \to S''$ the projections. Suppose X' is an S'-scheme provided with an S''-isomorphism $\sigma : p_1^*X' \stackrel{\cong}{\to} p_2^*X'$ satisfying $p_{13}^*(\sigma) = p_{23}^*(\sigma) \circ p_{12}^*(\sigma)$, where p_i^*X' $(i = 1, 2)$ are defined by the Cartesian diagrams*

$$\begin{array}{ccc} p_i^*X' & \to & X' \\ \downarrow & & \downarrow \\ S'' & \stackrel{p_i}{\to} & S'. \end{array}$$

If X' is affine over S', then there exist an S-scheme X and an S'-isomorphism $\tau : g^*X \stackrel{\cong}{\to} X'$ such that the diagram

$$\begin{array}{ccc} p_1^*g^*X & \stackrel{p_1^*\tau}{\to} & p_1^*X' \\ \wr\| & & \downarrow \sigma \\ p_2^*g^*X & \stackrel{p_2^*\tau}{\to} & p_2^*X' \end{array}$$

commutes, where $g^*X = X \times_S S'$. The S-scheme X is unique up to a unique isomorphism, and X is affine over S.

Any S''-isomorphism $\sigma : p_1^*X' \stackrel{\cong}{\to} p_2^*X'$ satisfying $p_{13}^*(\sigma) = p_{23}^*(\sigma) \circ p_{12}^*(\sigma)$ is called a *descent datum* for the S'-scheme X'.

Proof. Let $f' : X' \to S'$ be the structure morphism and let $\mathscr{A}' = f'_*\mathscr{O}_{X'}$. Then $\mathscr{A}'$ is a quasi-coherent $\mathscr{O}_{S'}$-algebra. The descent datum on X' defines a descent datum on $\mathscr{A}'$. By 1.6.1 (ii), there exist a quasi-coherent $\mathscr{O}_S$-module $\mathscr{A}$ and an isomorphism $\tau : g^*\mathscr{A} \stackrel{\cong}{\to} \mathscr{A}'$ such that the diagram

$$\begin{array}{ccc} p_1^* g^* \mathscr{A} & \stackrel{p_1^*\tau}{\to} & p_1^*\mathscr{A}' \\ \wr\| & & \downarrow\sigma \\ p_2^* g^* \mathscr{A} & \stackrel{p_2^*\tau}{\to} & p_2^*\mathscr{A}' \end{array}$$

commutes. By 1.6.1 (i), the morphism $\mathscr{A}' \otimes_{\mathscr{O}'_S} \mathscr{A}' \to \mathscr{A}'$ defining the multiplication on $\mathscr{A}'$ can be descended down to a morphism $\mathscr{A} \otimes_{\mathscr{O}_S} \mathscr{A} \to \mathscr{A}$, and this defines an $\mathscr{O}_S$-algebra structure on $\mathscr{A}$. Indeed, the associative law, the commutative law, and the distributive law for $\mathscr{A}$ are equivalent to the commutativity of some diagrams involving tensor products of $\mathscr{A}$, and these diagrams commute by 1.6.1 (i) and the commutativity of the corresponding diagrams for $\mathscr{A}'$. Set $X = \mathbf{Spec}\,\mathscr{A}$. Then X has the required property. □

Corollary 1.8.7. *Under the assumption of 1.8.4, f is affine if and only if f' is affine.*

Proof. Suppose f' is affine. Then f' is quasi-compact and separated. By 1.7.3 and 1.7.12, f is quasi-compact and separated. So $f_*\mathscr{O}_X$ is a quasi-coherent $\mathscr{O}_Y$-module ([Fu (2006)] 1.4.9 (iii), [EGA] I 9.2.1, [Hartshorne (1977)] II 5.8 (c)). Since the base change is flat, the inverse image of $f_*\mathscr{O}_X$ on Y' is isomorphic to $f'_*\mathscr{O}_{X'}$ ([Fu (2006)] 2.4.10, [EGA] III 1.4.15, [Hartshorne (1977)] III 9.3). The base change of the canonical Y-morphism $X \to \mathbf{Spec}\, f_*\mathscr{O}_X$ can be identified with the canonical Y'-morphism $X' \to \mathbf{Spec}\, f'_*\mathscr{O}_{X'}$. Since f' is affine, the morphism $X' \to \mathbf{Spec}\, f'_*\mathscr{O}_{X'}$ is an isomorphism. By 1.8.4, the morphism $X \to \mathbf{Spec}\, f_*\mathscr{O}_X$ is an isomorphism. So f is affine. □

Corollary 1.8.8. *Under the assumption of 1.8.4, f is integral, finite, or finite and locally free, if and only if f' is so.*

Proof. We prove the statement about integral morphisms, and leave the rest to the reader. Suppose f' is integral. By 1.8.7, f is affine. To prove f is integral, we may reduce to the case where $X = \mathrm{Spec}\, B$, $Y = \mathrm{Spec} A$, and $Y' = \mathrm{Spec}\, A'$ are affine. We have $B = \varinjlim_i B_i$, where B_i are finitely generated A-subalgebras of B. Since A' is flat over A, $B_i \otimes_A A'$ are subalgebras of $B \otimes_A A'$, and they are finitely generated over A'. Moreover, we have $B \otimes_A A' \cong \varinjlim_i B_i \otimes_A A'$. Since $B \otimes_A A'$ is integral over A', $B_i \otimes_A A'$

are finitely generated as A'-modules. By 1.6.6, B_i are finitely generated as A-modules. So B is integral over A. Hence f is integral. □

A scheme is called *quasi-affine* if it is isomorphic to a quasi-compact open subscheme of an affine scheme. A morphism $f : X \to Y$ is called *quasi-affine* if there exists an affine open covering $\{V_i\}$ of Y such that $f^{-1}(V_i)$ are quasi-affine. Any quasi-affine morphism is quasi-compact and separated.

Proposition 1.8.9. *A morphism $f : X \to Y$ is quasi-affine if and only if it is quasi-compact, separated, and the canonical Y-morphism $X \to$ $\mathbf{Spec}\, f_*\mathscr{O}_X$ is an open immersion.*

Proof. Note that if f is quasi-compact and separated, then $f_*\mathscr{O}_X$ is a quasi-coherent $\mathscr{O}_Y$-algebra, and we can talk about the Y-scheme $\mathbf{Spec}\, f_*\mathscr{O}_X$. Suppose f is quasi-affine, and let us prove that the Y-morphism $X \to \mathbf{Spec}\, f_*\mathscr{O}_X$ is an open immersion. We may reduce to the case where Y is affine, and prove that if X is a quasi-affine scheme, then the canonical morphism $X \to \mathrm{Spec}\,\Gamma(X, \mathscr{O}_X)$ is an open immersion. Suppose X is a quasi-compact open subscheme of an affine scheme $\mathrm{Spec}\, A$. We can find $a_1, \ldots, a_n \in A$ such that $X = D(a_1) \cup \cdots \cup D(a_n)$. We have a commutative diagram

$$\begin{array}{ccc} X & & \hookrightarrow \mathrm{Spec}\, A, \\ \downarrow & \nearrow & \\ \mathrm{Spec}\,\Gamma(X, \mathscr{O}_X) & & \end{array}$$

where the morphism $\mathrm{Spec}\,\Gamma(X, \mathscr{O}_X) \to \mathrm{Spec}\, A$ is the morphism induced by the homomorphism $A \to \Gamma(X, \mathscr{O}_X)$. For each i, let $a_i' \in \Gamma(X, \mathscr{O}_X)$ be the image of a_i, and let

$$X_{a_i'} = \{x \in X | a_i' \text{ is a unit in } \mathscr{O}_{X,x}\}.$$

Restricted to $D(a_i)$, the above diagram induces a commutative diagram

$$\begin{array}{ccc} X_{a_i'} & & = D(a_i). \\ \downarrow & \nearrow & \\ \mathrm{Spec}\,\Gamma(X, \mathscr{O}_X)_{a_i'} & & \end{array}$$

So $X_{a_i'} = D(a_i)$ is affine, and hence we have

$$X_{a_i'} \cong \mathrm{Spec}\,\Gamma(X_{a_i'}, \mathscr{O}_X).$$

On the other hand, we have

$$\Gamma(X_{a_i'}, \mathscr{O}_X) \cong \Gamma(X, \mathscr{O}_X)_{a_i'}.$$

([Fu (2006)] 1.3.9 (iii), [EGA] I 9.3.3, [Hartshorne (1977)] II 5.14.) It follows that $X_{a_i'} \to \mathrm{Spec}\,\Gamma(X, \mathscr{O}_X)_{a_i'}$ is an isomorphism. Hence $X \to$ $\mathrm{Spec}\,\Gamma(X, \mathscr{O}_X)$ is an open immersion. □

Corollary 1.8.10. *Under the assumption of 1.8.4, f is quasi-affine if and only if f' is so.*

Proof. Suppose f' is quasi-affine. Then f' is quasi-compact and separated. This implies that f is quasi-compact and separated. The base change of the canonical Y-morphism $X \to \mathbf{Spec}\, f_*\mathscr{O}_X$ can be identified with the canonical Y'-morphism $X' \to \mathbf{Spec}\, f'_*\mathscr{O}_X$, which is an open immersion. By 1.8.5, $X \to \mathbf{Spec}\, f_*\mathscr{O}_X$ is an open immersion. So f is quasi-affine. □

Proposition 1.8.11. *The same as 1.8.6, except that we assume X' is quasi-affine over S'.*

Proof. Using 1.6.1, one can show there exists a quasi-coherent $\mathscr{O}_S$-algebra $\mathscr{A}$ such that $g^*\mathscr{A} \cong f'_*\mathscr{O}_{X'}$. By 1.7.8, there exists an open immersion $X \hookrightarrow \mathbf{Spec}\, \mathscr{A}$ whose base change is the open immersion $X' \hookrightarrow \mathbf{Spec}\, f'_*\mathscr{O}_{X'}$. Then X has the required property. □

1.9 Quasi-finite Morphisms

([SGA 1] I 2, [EGA] II 6.2.)

Let $(A, \mathfrak{m}) \to (B, \mathfrak{n})$ be a local homomorphism of local rings. We say that B is *quasi-finite* over A if $B/\mathfrak{m}B$ is finite dimensional over $A/\mathfrak{m}$.

Lemma 1.9.1. *Let $f : X \to Y$ be a morphism of finite type and let $x \in X$. The following conditions are equivalent:*
(i) *x is isolated in $f^{-1}(f(x))$.*
(ii) *$\mathscr{O}_{X,x}$ is quasi-finite over $\mathscr{O}_{Y,f(x)}$.*

If these conditions hold, we say f is *quasi-finite at x*.

Proof. Replacing the morphism $X \to Y$ by $f^{-1}(f(x)) \to \mathrm{Spec}\, k(f(x))$ does not change $\mathscr{O}_{X,x}/\mathfrak{m}_{f(x)}\mathscr{O}_{Y,f(x)}$ and $\mathscr{O}_{Y,f(x)}/\mathfrak{m}_{f(x)}\mathscr{O}_{Y,f(x)}$. So we may assume $Y = \mathrm{Spec}\, k$ for some field k.

(i)⇒(ii) Replacing X by a neighborhood of x, we may assume that X consists of one point. Since f is of finite type, we have $X = \mathrm{Spec}\, B$ for some finitely generated k-algebra B with a single prime ideal $\mathfrak{n}$. It follows that $B/\mathfrak{n}$ is finite dimensional over k, and $\mathfrak{n}$ is nilpotent. Suppose $\mathfrak{n}^n = 0$. Since B is a noetherian ring, for each $1 \leq i \leq n$, $\mathfrak{n}^{i-1}/\mathfrak{n}^i$ is finite dimensional over $B/\mathfrak{n}$, and hence finite dimensional over k. It follows that B is finite dimensional over k.

(ii)⇒(i) Replacing X by an affine open neighborhood of x, we may assume $X = \operatorname{Spec} B$ for some finitely generated k-algebra B. Let $\mathfrak{q}$ be the prime ideal of B corresponding to the point x. Then $B_\mathfrak{q}$ is finite dimensional over k. This implies that $B_\mathfrak{q}$ is an artinian ring, and hence $\mathfrak{q}$ is a minimal prime ideal of B. Thus x is the generic point of an irreducible component of X. Moreover, $B_\mathfrak{q}/\mathfrak{q}B_\mathfrak{q}$ is finite dimensional over k. Hence $B/\mathfrak{q}$ is finite dimensional over k. So $B/\mathfrak{q}$ is a field, and $\mathfrak{q}$ is a maximal ideal of B. Thus x is a closed point of X, and the irreducible component with generic point x consists of only one point. Therefore x is isolated in X. □

Lemma 1.9.2. *Let k be a field and let A be a finitely generated k-algebra. The following conditions are equivalent:*

(i) $\operatorname{Spec} A$ *is discrete as a topological space.*

(ii) $\operatorname{Spec} A$ *is finite as a set.*

(iii) *A has only finitely many maximal ideals.*

(iv) *A is a finite dimensional k-algebra.*

Proof.

(i)⇒(ii) $\operatorname{Spec} A$ is quasi-compact. If it is discrete, it must be finite.

(ii)⇒(iii) Trivial.

(iii)⇒(i) Let $\mathfrak{m}_1, \ldots, \mathfrak{m}_n$ be all the maximal ideals of A. Since A is a Jacobson ring, any prime ideal of A is the intersection of some maximal ideals. Suppose $\mathfrak{p} = \cap_j \mathfrak{m}_{i_j}$ is a prime ideal. Then we have $\mathfrak{p} \supset \prod_j \mathfrak{m}_{i_j}$. Hence $\mathfrak{p} \supset \mathfrak{m}_{i_j}$ for some j. By the maximality of $\mathfrak{m}_{i_j}$, we have $\mathfrak{p} = \mathfrak{m}_{i_j}$. So any prime ideal of A is a maximal ideal. Hence any point in $\operatorname{Spec} A$ is a closed point. As $\operatorname{Spec} A$ is finite, it must be discrete.

(i)⇒(iv) We have shown that $\operatorname{Spec} A$ is a finite discrete topological space. Let $\mathfrak{m}_1, \ldots, \mathfrak{m}_n$ be all the maximal ideals of A. We then have

$$\operatorname{Spec} A \cong \operatorname{Spec} A_{\mathfrak{m}_1} \coprod \cdots \coprod \operatorname{Spec} A_{\mathfrak{m}_n}.$$

Hence

$$A \cong A_{\mathfrak{m}_1} \times \cdots \times A_{\mathfrak{m}_n}.$$

By the proof of 1.9.1, each $A_{\mathfrak{m}_i}$ is finite dimensional over k. So A is finite dimensional over k.

(iv)⇒(ii) If A is finite dimensional over k, then it is an artinian ring. Hence $\operatorname{Spec} A$ is finite. □

Proposition 1.9.3. *Let $f : X \to Y$ be a morphism of finite type. The following conditions are equivalent:*

(i) *For any* $y \in Y$, $f^{-1}(y)$ *is discrete as a topological space.*
(ii) *For any* $x \in X$, $\mathscr{O}_{X,x}$ *is quasi-finite over* $\mathscr{O}_{Y,f(x)}$.
(iii) *For any* $y \in Y$, $f^{-1}(y)$ *is a finite set.*

If these conditions hold, we say that f is *quasi-finite.*

Proof.
(i)⇔(ii) follows from 1.9.1.
(i)⇔(iii) follows from 1.9.2. □

Proposition 1.9.4. *Consider a Cartesian diagram*

$$\begin{array}{ccc} X \times_Y Y' & \stackrel{g'}{\to} & X \\ {\scriptstyle f'}\downarrow & & \downarrow{\scriptstyle f} \\ Y' & \stackrel{g}{\to} & Y. \end{array}$$

Suppose g *is quasi-compact and faithfully flat. Then* f *is quasi-finite if and only if* f' *is so.*

Proof. Apply 1.7.4 and 1.9.2. □

Lemma 1.9.5. *Let* $(A, \mathfrak{m})$ *be a local noetherian ring, and let* M *be an* A*-module such that* $M/\mathfrak{m}M$ *is finite dimensional over* $A/\mathfrak{m}$*. Then* $\widehat{M} = \varprojlim_k M/\mathfrak{m}^k M$ *is finitely generated over* $\widehat{A} = \varprojlim_k A/\mathfrak{m}^k$.

Proof. Let $\{x_1^{(0)}, \ldots, x_k^{(0)}\}$ be a family of generators for the $A/\mathfrak{m}$-module $M/\mathfrak{m}M$. Choose $x_1^{(i)}, \ldots, x_k^{(i)} \in M/\mathfrak{m}^{i+1}M$ by induction on i so that their images in $M/\mathfrak{m}^i M$ are $x_1^{(i-1)}, \ldots, x_k^{(i-1)}$, respectively. For each i, consider the homomorphism

$$(A/\mathfrak{m}^{i+1})^k \to M/\mathfrak{m}^{i+1}M, \quad (a_1, \ldots, a_k) \mapsto a_1 x_1^{(i)} + \cdots + a_k x_k^{(i)}.$$

It is surjective by Nakayama's lemma. Let R_i be its kernel. We have an exact sequence

$$0 \to R_i \to (A/\mathfrak{m}^{i+1})^k \to M/\mathfrak{m}^{i+1}M \to 0.$$

Each R_i is an $(A/\mathfrak{m}^{i+1})$-module of finite length since $A/\mathfrak{m}^{i+1}$ is artinian and R_i is finitely generated. By [Fu (2006)] 1.5.1 or [EGA] 0 13.2.2, the sequence

$$0 \to \varprojlim_i R_i \to \varprojlim_i (A/\mathfrak{m}^{i+1})^k \to \varprojlim_i M/\mathfrak{m}^{i+1}M \to 0$$

is exact. In particular, we have an epimorphism $\widehat{A}^k \to \widehat{M}$. So $\widehat{M}$ is a finitely generated $\widehat{A}$-module. □

Lemma 1.9.6. *Let $(A,\mathfrak{m}) \to (B,\mathfrak{n})$ be a local homomorphism of local noetherian rings, let $\widehat{A} = \varprojlim_k A/\mathfrak{m}^k$, and let $\widehat{B} = \varprojlim_k B/\mathfrak{n}^k$.*

(i) *B is quasi-finite over A if and only if $B/\mathfrak{n}$ is finite dimensional over $A/\mathfrak{m}$, and there exists a positive integer n such that $\mathfrak{n}^n \subset \mathfrak{m}B \subset \mathfrak{n}$.*

(ii) *If B is quasi-finite over A, then $\widehat{B}$ is finite over $\widehat{A}$.*

Proof.

(i) Suppose $B/\mathfrak{m}B$ is finite dimensional over $A/\mathfrak{m}$. Then $B/\mathfrak{n}$ is finite dimensional over $A/\mathfrak{m}$. Moreover $B/\mathfrak{m}B$ is an artinian ring. So its maximal ideal $\mathfrak{n}/\mathfrak{m}B$ is nilpotent. Hence there exists a natural number n such that $\mathfrak{n}^n \subset \mathfrak{m}B \subset \mathfrak{n}$.

Conversely, suppose $B/\mathfrak{n}$ is finite dimensional over $A/\mathfrak{m}$, and there exists a natural number n such that $\mathfrak{n}^n \subset \mathfrak{m}B \subset \mathfrak{n}$. To prove that $B/\mathfrak{m}B$ is finite dimensional over $A/\mathfrak{m}$, it suffices to prove that $B/\mathfrak{n}^n$ is an A-module of finite length. We are then reduced to proving that $\mathfrak{n}^{i-1}/\mathfrak{n}^i$ is an A-module of finite length for each $1 \leq i \leq n$. Since B is noetherian, each $\mathfrak{n}^{i-1}/\mathfrak{n}^i$ is finite dimensional over $B/\mathfrak{n}$, and hence finite dimensional over $A/\mathfrak{m}$. Our assertion follows.

(ii) By (i), we have

$$\widehat{B} = \varprojlim_k B/\mathfrak{n}^k \cong \varprojlim_k B/\mathfrak{m}^k B.$$

We then apply 1.9.5. □

Proposition 1.9.7. *Let A be a complete noetherian local ring, $Y = \operatorname{Spec} A$, $f : X \to Y$ a separated quasi-finite morphism, and $x \in X$ a point over the closed point of Y. Then $\mathscr{O}_{X,x}$ is finite over A, and the canonical morphism $\operatorname{Spec} \mathscr{O}_{X,x} \to X$ is an open and closed immersion.*

Proof. By 1.9.6, $\widehat{\mathscr{O}}_{X,x}$ is finite over A. So $\mathscr{O}_{X,x}$ is finite over A. Let g be the canonical morphism $\operatorname{Spec} \mathscr{O}_{X,x} \to X$. Then fg is finite. Since f is separated, g is finite. So $g(\operatorname{Spec} \mathscr{O}_{X,x})$ is closed in X. On the other hand, g induces an isomorphism of the local ring of X at x with the local ring of $\operatorname{Spec} \mathscr{O}_{X,x}$ at its closed point. By [Fu (2006)] 1.3.13 (iii) or [EGA] I 6.5.4 (ii), g is an open immersion when restricted to a sufficiently small open neighborhood of the closed point of $\operatorname{Spec} \mathscr{O}_{X,x}$. But $\operatorname{Spec} \mathscr{O}_{X,x}$ is the only open neighborhood of the closed point. Hence g is an open and closed immersion. □

Corollary 1.9.8. *Let A be a complete noetherian local ring, $Y = \operatorname{Spec} A$, and $f : X \to Y$ a separated quasi-finite morphism. Then X is the disjoint union of two open and closed subsets X' and X'' such that X' is finite over Y, and that the fiber of $X'' \to Y$ over the closed point of Y is empty.*

1.10 Passage to Limit

([EGA] IV 8, [SGA 1] VIII 6.)

Throughout this section, we fix a scheme S_0, a direct set I, and a direct system $(\mathscr{A}_\lambda, \phi_{\lambda\mu})$, where $\mathscr{A}_\lambda$ $(\lambda \in I)$ are quasi-coherent $\mathscr{O}_{S_0}$-algebras, and $\phi_{\lambda\mu} : \mathscr{A}_\lambda \to \mathscr{A}_\mu$ $(\lambda \leq \mu)$ are morphisms of $\mathscr{O}_{S_0}$-algebras. Let $\mathscr{A} = \varinjlim_\lambda \mathscr{A}_\lambda$ and let $\phi_\lambda : \mathscr{A}_\lambda \to \mathscr{A}$ be the canonical morphisms. Set $S_\lambda = \mathbf{Spec}\, \mathscr{A}_\lambda$ and $S = \mathbf{Spec}\, \mathscr{A}$. They are S_0-schemes. Let $u_{\lambda\mu} : S_\mu \to S_\lambda$ be the S_0-morphisms induced by $\phi_{\lambda\mu}$ and $u_\lambda : S \to S_\lambda$ the S_0-morphisms induced by ϕ_λ. We denote an object over S_0 by a symbol with subscript 0, and denote the corresponding object over S_λ (resp. S) induced by base change by the same symbol with subscript λ (resp. without subscript). For example, for any $\mathscr{O}_{S_0}$-module $\mathscr{F}_0$, let $\mathscr{F}_\lambda$ (resp. $\mathscr{F}$) be the inverse images of $\mathscr{F}_0$ on S_λ (resp. S). For any morphism $f_0 : \mathscr{F}_0 \to \mathscr{G}_0$ of $\mathscr{O}_{S_0}$-modules, let $f_\lambda : \mathscr{F}_\lambda \to \mathscr{G}_\lambda$ and $f : \mathscr{F} \to \mathscr{G}$ be the morphisms induced by f_0. For any S_0-scheme X_0, let $X_\lambda = X_0 \times_{S_0} S_\lambda$ and let $X = X_0 \times_{S_0} S$. For any S_0-morphism $f_0 : X_0 \to Y_0$ of S_0-schemes, let $f_\lambda : X_\lambda \to Y_\lambda$ and $f : X \to Y$ be the morphisms induced by f_0.

In this section, we prove that given an object over S_0, if its base change to S has some property, then its base changes to S_λ have the same property for sufficiently large λ. We often apply results of this section to the following two situations:

(a) Let A be a ring and let $\{A_\lambda\}$ be the direct system of subalgebras of A finitely generated by $\mathbb{Z}$. We have $A = \varinjlim_\lambda A_\lambda$. Using results of this section, we can often reduce problems over a general base scheme $S = \operatorname{Spec} A$ to problems over a noetherian base schemes $S_\lambda = \operatorname{Spec} A_\lambda$.

(b) Let S_0 be an affine scheme, let x be a point in S_0, let $\{S_\lambda\}$ be the inverse system of affine open neighborhoods of x in S_0, and let $S = \operatorname{Spec} \mathscr{O}_{S_0,x}$. We have $\mathscr{O}_{S_0,x} = \varinjlim_\lambda \Gamma(S_\lambda, \mathscr{O}_{S_\lambda})$. Using results in this section, we can often prove that if the base change to $\operatorname{Spec} \mathscr{O}_{S_0,x}$ of an object over S_0 has some property, then its base change to a neighborhood of x has the same property.

Proposition 1.10.1.

(i) *S is the inverse limit of the inverse system $(S_\lambda, u_{\lambda\mu})$ in the category of schemes.*

(ii) *For any quasi-compact open subset U of S, there exists a quasi-compact open subset U_λ of S_λ for some λ such that $u_\lambda^{-1}(U_\lambda) = U$.*

(iii) *The underlying topological space of S is the inverse limit of the* *inverse system* $(S_\lambda, u_{\lambda\mu})$ *in the category of topological spaces.*

(iv) *If* S_0 *is quasi-compact and* $S = \emptyset$, *then* $S_\lambda = \emptyset$ *for sufficiently large* λ.

(v) *Suppose* S_0, S_λ, S *are noetherian schemes. Let* E_0 *be a constructible subset of* S_0 *and let* E_λ *(resp.* E*) be the inverse images of* E_0 *in* S_λ *(resp.* S*). If* $E = \emptyset$ *(resp.* $E = S$*), then* $E_\lambda = \emptyset$ *(resp.* $E_\lambda = S_\lambda$*) for sufficiently large* λ.

Proof.

(i) First we show S is the inverse limit of $(S_\lambda, u_{\lambda\mu})$ in the category of S_0-schemes. Let $f : T \to S_0$ be an S_0-scheme. We need to show the canonical map

$$\mathrm{Hom}_{S_0}(T, S) \to \varprojlim_{\lambda} \mathrm{Hom}_{S_0}(T, S_\lambda)$$

is bijective. We have

$$\begin{aligned}\mathrm{Hom}_{S_0}(T, S_\lambda) &\cong \mathrm{Hom}_{\mathscr{O}_{S_0}}(\mathscr{A}_\lambda, f_*\mathscr{O}_T),\\ \mathrm{Hom}_{S_0}(T, S) &\cong \mathrm{Hom}_{\mathscr{O}_{S_0}}(\mathscr{A}, f_*\mathscr{O}_T).\end{aligned}$$

Since $\mathscr{A} = \varinjlim_\lambda \mathscr{A}_\lambda$, the canonical map

$$\mathrm{Hom}_{\mathscr{O}_{S_0}}(\mathscr{A}, f_*\mathscr{O}_T) \to \varprojlim_{\lambda} \mathrm{Hom}_{\mathscr{O}_{S_0}}(\mathscr{A}_\lambda, f_*\mathscr{O}_T)$$

is bijective. Our assertion follows.

Next we show that S is the inverse limit of $(S_\lambda, u_{\lambda\mu})$ in the category of schemes. Let T be a scheme. We need to show that the canonical map

$$\mathrm{Hom}(T, S) \to \varprojlim_{\lambda} \mathrm{Hom}(T, S_\lambda)$$

is bijective. Any $f \in \mathrm{Hom}(T, S_0)$ defines an S_0-scheme structure on T. Let $\mathrm{Hom}_f(T, S_\lambda)$ (resp. $\mathrm{Hom}_f(T, S)$) be the set of S_0-morphisms with respect to the S_0-scheme structure on T defined by f. We have

$$\begin{aligned}\mathrm{Hom}(T, S_\lambda) &= \bigcup_{f\in \mathrm{Hom}(T,S_0)} \mathrm{Hom}_f(T, S_\lambda),\\ \mathrm{Hom}(T, S) &= \bigcup_{f\in \mathrm{Hom}(T,S_0)} \mathrm{Hom}_f(T, S).\end{aligned}$$

We have shown that the canonical map

$$\mathrm{Hom}_f(T, S) \to \varprojlim_{\lambda} \mathrm{Hom}_f(T, S_\lambda)$$

is bijective. Our assertion follows.

(ii) It suffices to show that there exists a base of topology for S consisting of open subsets of the form $u_\lambda^{-1}(U_\lambda)$, where U_λ are quasi-compact open subsets of S_λ. We may reduce to the case where S_0 is affine. Let $A_\lambda = \Gamma(S_0, \mathscr{A}_\lambda)$ and let $A = \Gamma(S_0, \mathscr{A})$. Then $S_\lambda \cong \operatorname{Spec} A_\lambda$ and $S \cong \operatorname{Spec} A$. Open subsets of $\operatorname{Spec} A$ of the form $D(f)$ $(f \in A)$ form a base of topology for $\operatorname{Spec} A$. Any $f \in A$ is the image of some $f_\lambda \in A_\lambda$ for some λ. Then $D(f)$ is the inverse image of the open subset $D(f_\lambda)$ of $\operatorname{Spec} A_\lambda$. Our assertion follows.

(iii) The morphisms $u_\lambda : S \to S_\lambda$ induce a map $S \to \varprojlim_\lambda S_\lambda$ on the underlying topological spaces. To prove that it is a homeomorphism, we may reduce to the case where S_0 is affine. Keep the notation in the proof of (ii). The canonical map

$$\operatorname{Spec} A \to \varprojlim_\lambda \operatorname{Spec} A_\lambda$$

is bijective on the underlying sets. Indeed, given any system of prime ideals $\mathfrak{p}_\lambda$ of A_λ with $\phi_{\lambda\mu}^{-1}(\mathfrak{p}_\mu) = \mathfrak{p}_\lambda$ $(\lambda \leq \mu)$, let $\mathfrak{p} = \varinjlim_\lambda \mathfrak{p}_\lambda$. Then $\mathfrak{p}$ is the unique prime ideal of A with the property $\phi_\lambda^{-1}(\mathfrak{p}) = \mathfrak{p}_\lambda$. We conclude from (ii) that $\operatorname{Spec} A \to \varprojlim_\lambda \operatorname{Spec} A_\lambda$ is a homeomorphism.

(iv) We may reduce to the case where S_0 is affine. Keep the above notation. Since $S = \emptyset$, we have $0 = 1$ in A. Then we have $0 = 1$ in A_λ for sufficiently large λ. So $S_\lambda = \emptyset$ for sufficiently large λ.

(v) Write $E_0 = \cup_{i=1}^n (U_i \cap F_i)$, where U_i (resp. F_i) are open (resp. closed) subsets of S_0. Put the reduced subscheme structures on $U_i \cap F_i$. Let $X_0 = \coprod_{i=1}^n (U_i \cap F_i)$ and let $f_0 : X_0 \to S_0$ be the canonical morphism. We have $E_0 = f(X_0)$ and hence $E_\lambda = f_\lambda(X_\lambda)$ and $E = f(X)$. If $E = \emptyset$, then $X = \emptyset$. With (iv) applied to the inverse system (X_λ), we get $X_\lambda = \emptyset$ for sufficiently large λ. Then $E_\lambda = \emptyset$ for sufficiently large λ. The second part of (v) follows from the first part applied to the constructible subset $S_0 - E_0$. □

Proposition 1.10.2. *Assume S_0 is quasi-compact and quasi-separated.*

(i) *Let $\mathscr{F}_0$ and $\mathscr{G}_0$ be quasi-coherent $\mathscr{O}_{S_0}$-modules. Suppose $\mathscr{F}_0$ locally has finite presentation. Then the canonical map*

$$\varinjlim_\lambda \operatorname{Hom}_{\mathscr{O}_{S_\lambda}}(\mathscr{F}_\lambda, \mathscr{G}_\lambda) \to \operatorname{Hom}_{\mathscr{O}_S}(\mathscr{F}, \mathscr{G})$$

is bijective.

(ii) *Suppose $\mathscr{F}_0$ and $\mathscr{G}_0$ are quasi-coherent $\mathscr{O}_{S_0}$-modules locally of finite presentation. If there exists an isomorphism $f : \mathscr{F} \stackrel{\cong}{\to} \mathscr{G}$, then for a sufficiently large λ, there exists an isomorphism $f_\lambda : \mathscr{F}_\lambda \stackrel{\cong}{\to} \mathscr{G}_\lambda$ inducing f.*

(iii) *For any quasi-coherent $\mathscr{O}_S$-module $\mathscr{F}$ locally with finite presentation, there exists a quasi-coherent $\mathscr{O}_{S_\lambda}$-module $\mathscr{F}_\lambda$ locally with finite presentation for a sufficiently large λ such that $\mathscr{F} \cong u_\lambda^* \mathscr{F}_\lambda$.*

We leave the proof to the reader. (Confer the proof of 1.10.9.) Applying 1.10.2 (i) to the case where $\mathscr{F} = \mathscr{O}_{S_0}$, we get the following:

Corollary 1.10.3. *Assume S_0 is quasi-compact and quasi-separated. For any quasi-coherent $\mathscr{O}_{S_0}$-module $\mathscr{G}_0$, the canonical map*

$$\varinjlim_\lambda \Gamma(S_\lambda, \mathscr{G}_\lambda) \to \Gamma(S, \mathscr{G})$$

is bijective.

Let A be a ring and let B be an A-algebra. We say that B is an A-algebra of *finite presentation* if it is isomorphic to $A[t_1, \ldots, t_n]/I$ for some nonnegative integer n and some finitely generated ideal I of $A[t_1, \ldots, t_n]$. Note that for any $s \in A$, $A_s \cong A[t]/(st - 1)$ is an A-algebra of finite presentation.

Lemma 1.10.4. *Let A_0 be a ring, $(A_\lambda, \phi_{\lambda\mu})$ a direct system of A_0-algebras, and $A = \varinjlim_\lambda A_\lambda$.*

(i) *Let B_0 be an A_0-algebra, $B_\lambda = B_0 \otimes_{A_0} A_\lambda$, $B = B_0 \otimes_{A_0} A$, (C_λ) a direct system of A_0-algebras such that we have a morphism of direct systems from (A_λ) to (C_λ), and $C = \varinjlim_\lambda C_\lambda$. If B_0 is an A_0-algebra of finite type (resp. finite presentation), then the canonical map*

$$\varinjlim_\lambda \mathrm{Hom}_{A_\lambda}(B_\lambda, C_\lambda) \to \mathrm{Hom}_A(B, C)$$

is injective (resp. bijective).

(ii) *For any A-algebra B with finite presentation, there exists an A_λ-algebra B_λ with finite presentation for a sufficiently large λ such that $B \cong B_\lambda \otimes_{A_\lambda} A$.*

Proof.

(i) We have one-to-one correspondences

$$\mathrm{Hom}_{A_\lambda}(B_\lambda, C_\lambda) \cong \mathrm{Hom}_{A_0}(B_0, C_\lambda), \quad \mathrm{Hom}_A(B, C) \cong \mathrm{Hom}_{A_0}(B_0, C).$$

It suffices to show that

$$\varinjlim_\lambda \mathrm{Hom}_{A_0}(B_0, C_\lambda) \to \mathrm{Hom}_{A_0}(B_0, C)$$

is injective (resp. bijective).

Suppose B_0 is of finite type over A_0, and let $x_1, \ldots, x_n$ be a finite family of generators of B_0 over A_0. If $\alpha_1, \alpha_2 : B_0 \to C_\lambda$ are A_0-homomorphisms inducing the same homomorphism $B_0 \to C$, then $\alpha_1(x_i)$ and $\alpha_2(x_i)$ have the same image in C for each i. Since $C = \varinjlim_{\mu \geq \lambda} C_\mu$, there exists $\mu \geq \lambda$ such that $\alpha_1(x_i)$ and $\alpha_2(x_i)$ have the same image in C_μ for each i. But then α_1 and α_2 induce the same homomorphism $B_0 \to C_\mu$. This proves that our map is injective.

Suppose B_0 is of finite presentation over A_0. Then there exists an epimorphism

$$\pi : A_0[t_1, \ldots, t_n] \to B_0$$

of A_0-algebras such that $\ker \pi$ is a finitely generated ideal of $A_0[t_1, \ldots, t_n]$. Let $f_j(t_1, \ldots, t_n)$ $(j = 1, \ldots, m)$ be a finite family of generators of $\ker \pi$. Given an A_0-homomorphism $\alpha : B_0 \to C$, we have

$$f_j(\alpha\pi(t_1), \ldots, \alpha\pi(t_n)) = \alpha\pi(f_j(t_1, \ldots, t_n)) = 0.$$

Since $C = \varinjlim_\lambda C_\lambda$, we can find $y_{\lambda 1}, \ldots, y_{\lambda n} \in C_\lambda$ for some λ such that $\alpha\pi(t_i)$ is the image of $y_{\lambda i}$ in C for each i. Choosing λ sufficiently large, we may assume $f_j(y_{\lambda 1}, \ldots, y_{\lambda n}) = 0$ $(j = 1, \ldots, m)$. Define an A_0-homomorphism $A_0[t_1, \ldots, t_n] \to C_\lambda$ by mapping t_i to $y_{\lambda i}$ for each i. Then $\ker \pi$ is contained in the kernel of this homomorphism. So it induces an A_0-homomorphism $\alpha_\lambda : B \to C_\lambda$. Its composite with $C_\lambda \to C$ is α. This proves that our map is surjective.

(ii) We have $B \cong A[t_1, \ldots, t_n]/I$ for some n and some finitely generated ideal I of $A[t_1, \ldots, t_n]$. Let $f_j(t_1, \ldots, t_n)$ $(j = 1, \ldots, m)$ be a finite family of generators of I. As $A = \varinjlim_\lambda A_\lambda$, we can find a sufficiently large λ and $f_{\lambda j}(t_1, \ldots, t_n) \in A_\lambda[t_1, \ldots, t_n]$ whose images in $A[t_1, \ldots, t_n]$ are $f_j(t_1, \ldots, t_n)$, respectively. Let $B_\lambda = A_\lambda[t_1, \ldots, t_n]/(f_{\lambda 1}, \ldots, f_{\lambda m})$. Then $B \cong B_\lambda \otimes_{A_\lambda} A$. □

A morphism $f : X \to Y$ of schemes is called *locally of finite presentation* if for any $x \in X$, there exists an affine open neighborhood $V = \operatorname{Spec} B$ of $f(x)$ in Y, and an affine open neighborhood $U = \operatorname{Spec} C$ of x in $f^{-1}(V)$ such that C is a B-algebra with finite presentation. We leave the proof of the following proposition to the reader.

Proposition 1.10.5.

(i) *Local isomorphisms are locally of finite presentation.*

(ii) *If two morphisms $f : X \to Y$ and $g : Y \to Z$ are locally of finite presentation, then so is gf.*

(iii) *If $f : X \to Y$ is a morphism locally of finite presentation, then for any morphism $g : Y' \to Y$, the base change $f' : X \times_Y Y' \to Y'$ of f is locally of finite presentation.*

(iv) *Given two morphisms $f : X \to Y$ and $g : Y \to Z$, suppose g is locally of finite type and gf is locally of finite presentation. Then f is locally of finite presentation.*

Proof. (iv) follows from (ii), (iii), and the fact that if g is locally of finite type, then the diagonal morphism $\Delta : Y \to Y \times_Z Y$ is locally of finite presentation. Indeed, if B is a finitely generated A-algebra, and $\{b_1, \ldots, b_n\} \subset B$ is a finite family of generators, then the kernel of the epimorphism

$$B \otimes_A B \to B, \quad b \otimes b' \mapsto bb'$$

is generated by $1 \otimes b_i - b_i \otimes 1$ $(i = 1, \ldots, n)$ as an ideal of $B \otimes_A B$. □

Lemma 1.10.6. *Let C be a ring and I an ideal of C. Suppose C/I is a C-algebra of finite presentation. Then I is a finitely generated ideal of C.*

Proof. Let $u : C[t_1, \ldots, t_n] \to C/I$ be an epimorphism of C-algebras so that $\ker u$ is a finitely generated ideal of $C[t_1, \ldots, t_n]$. Choose $c_i \in C$ $(i = 1, \ldots, n)$ such that $u(t_i)$ are the images of c_i in C/I respectively. Let $v : C[t_1, \ldots, t_n] \to C$ be the C-homomorphism defined by $v(t_i) = c_i$ and let $p : C \to C/I$ be the projection. We have $u = pv$, and hence $\ker u = v^{-1}(I)$. It is obvious that v is surjective. So we have

$$I = v(v^{-1}(I)) = v(\ker u).$$

Since $\ker u$ is a finitely generated ideal of $C[t_1, \ldots, t_n]$, I is a finitely generated ideal of C. □

Corollary 1.10.7. *Let A be a ring and B an A-algebra. If* $\operatorname{Spec} B \to \operatorname{Spec} A$ *is locally of finite presentation, then B is an A-algebra of finite presentation. Moreover, for any epimorphism $u : A[t_1, \ldots, t_n] \to B$,* $\ker u$ *is a finitely generated ideal of $A[t_1, \ldots, t_n]$.*

Proof. Note that $\operatorname{Spec} B \to \operatorname{Spec} A$ is locally of finite type. This implies that B is an A-algebra of finite type. So we have an epimorphism $A[t_1, \ldots, t_n] \to B$. Let I be the kernel of this epimorphism. By 1.10.5, the closed immersion

$$\operatorname{Spec} A[t_1, \ldots, t_n]/I \to \operatorname{Spec} A[t_1, \ldots, t_n]$$

is locally of finite presentation. This implies that there exist $s_i \in A[t_1, \ldots, t_n]$ ($i = 1, \ldots, k$) such that $D(s_i)$ form an open covering of $\operatorname{Spec} A[t_1, \ldots, t_n]$ and each $A[t_1, \ldots, t_n]_{s_i}/I_{s_i}$ is an $A[t_1, \ldots, t_n]_{s_i}$-algebra of finite presentation. By Lemma 1.10.6, I_{s_i} is a finitely generated ideal of $A[t_1, \ldots, t_n]_{s_i}$. This implies that I is a finitely generated ideal of $A[t_1, \ldots, t_n]$. So B is an A-algebra of finite presentation. □

Proposition 1.10.8. *Let A be a ring and B a finite A-algebra. Then B is an A-algebra of finite presentation if and only if it is an A-module of finite presentation.*

Proof. Let x_i ($i = 1, \ldots, n$) be a finite family of generators of B as an A-module. For each x_i, let $f_i(t) \in A[t]$ be a monic polynomial such that $f_i(x_i) = 0$. Then we have an epimorphism of A-algebras

$$u : A[t_1, \ldots, t_n]/(f_1(t_1), \ldots, f_n(t_n)) \to B, \quad t_i \mapsto x_i.$$

Note that

$$A[t_1, \ldots, t_n]/(f_1(t_1), \ldots, f_n(t_n)) \cong A[t_1]/(f_1(t_1)) \otimes_A \cdots \otimes_A A[t_n]/(f_n(t_n)).$$

Hence $A[t_1, \ldots, t_n]/(f_1(t_1), \ldots, f_n(t_n))$ is a free A-module of finite rank. If B is an A-module of finite presentation, then $\ker u$ is a finitely generated A-module, and hence a finitely generated ideal. It follows that B is an A-algebra of finite presentation. Conversely, if B is an A-algebra of finite presentation, then by 1.10.7, $\ker u$ is a finitely generated ideal. But $A[t_1, \ldots, t_n]/(f_1(t_1), \ldots, f_n(t_n))$ is a finitely generated A-module. So $\ker u$ is a finitely generated A-module. Hence B is an A-module of finite presentation. □

A morphism $f : X \to Y$ is said to be *of finite presentation* if it is quasi-compact, quasi-separated, and locally of finite presentation.

Proposition 1.10.9. *Assume S_0 is quasi-compact and quasi-separated.*

(i) *Let X_0 and Y_0 be S_0-schemes such that $X_0 \to S_0$ is quasi-compact and quasi-separated, and that $Y_0 \to S_0$ is locally of finite presentation. Then the canonical map*

$$\varinjlim_{\lambda} \operatorname{Hom}_{S_\lambda}(X_\lambda, Y_\lambda) \to \operatorname{Hom}_S(X, Y)$$

is bijective.

(ii) *Suppose X_0 and Y_0 are S_0-schemes of finite presentation. If there exists an S-isomorphism $f : X \xrightarrow{\cong} Y$, then for a sufficiently large λ, there exists an S_λ-isomorphism $f_\lambda : X_\lambda \xrightarrow{\cong} Y_\lambda$ inducing f.*

(iii) *For any S-scheme X of finite presentation, there exists an S_λ-scheme X_λ of finite presentation for a sufficiently large λ such that $X \cong X_\lambda \times_{S_\lambda} S$.*

Proof.

(i) Identifying a morphism with its graph, we get one-to-one correspondences

$$\mathrm{Hom}_{S_\lambda}(X_\lambda, Y_\lambda) \cong \mathrm{Hom}_{X_\lambda}(X_\lambda, X_\lambda \times_{S_\lambda} Y_\lambda),$$
$$\mathrm{Hom}_S(X, Y) \cong \mathrm{Hom}_X(X, X \times_S Y).$$

It suffices to show that

$$\varinjlim_{\lambda} \mathrm{Hom}_{X_\lambda}(X_\lambda, X_\lambda \times_{S_\lambda} Y_\lambda) \to \mathrm{Hom}_X(X, X \times_S Y)$$

is bijective. We are thus reduced to the case where $X_0 = S_0$, to prove that

$$\varinjlim_{\lambda} \mathrm{Hom}_{S_\lambda}(S_\lambda, Y_\lambda) \to \mathrm{Hom}_S(S, Y)$$

is bijective.

To prove that the above map is injective, we may assume S_0 is affine since the problem is local. Let $f_\lambda, f'_\lambda : S_\lambda \to Y_\lambda$ be two S_λ-morphisms inducing the same S-morphism $S \to Y$ by base change. Cover $f_\lambda(S_\lambda) \cup f'_\lambda(S_\lambda)$ by affine open subsets $V_\lambda^{(i)}$ $(i \in I)$ in Y_λ and cover each $f_\lambda^{-1}(V_\lambda^{(i)}) \cap f'^{-1}_\lambda(V_\lambda^{(i)})$ by affine open subsets $U_\lambda^{(ij)}$ $(j \in J_i)$ in S_λ. Then $U^{(ij)}$ $(i \in I,\ j \in J_i)$ form an open covering of S. By 1.10.1 (iv), $U_\mu^{(ij)}$ $(i \in I,\ j \in J_i)$ form an open covering of S_μ for some $\mu \geq \lambda$. Since S_μ is quasi-compact, this open covering has a finite subcovering. Changing notation, we may assume that there exist finitely many affine open subsets $V_\mu^{(i)}$ $(i = 1, \ldots, n)$ in Y_μ, and affine open subsets $U_\mu^{(i)}$ $(i = 1, \ldots, n)$ in S_μ, such that $U_\mu^{(i)}$ cover S_μ and

$$U_\mu^{(i)} \subset f_\mu^{-1}(V_\mu^{(i)}) \cap f'^{-1}_\mu(V_\mu^{(i)})$$

for all i. Note that the morphisms $f_\mu|_{U_\mu^{(i)}}, f'_\mu|_{U_\mu^{(i)}} : U_\mu^{(i)} \to V_\mu^{(i)}$ induce the same morphism $U^{(i)} \to V^{(i)}$ by base change. Since S_λ and $V_\lambda^{(i)}$ are affine, for any $\nu \geq \mu$, we have

$$\mathrm{Hom}_{S_\nu}(U_\nu^{(i)}, V_\nu^{(i)}) \cong \mathrm{Hom}_{\Gamma(S_\nu, \mathscr{O}_{S_\nu})}\Big(\Gamma(V_\nu^{(i)}, \mathscr{O}_{V_\nu^{(i)}}), \Gamma(U_\nu^{(i)}, \mathscr{O}_{U_\nu^{(i)}})\Big),$$
$$\mathrm{Hom}_S(U^{(i)}, V^{(i)}) \cong \mathrm{Hom}_{\Gamma(S, \mathscr{O}_S)}\Big(\Gamma(V^{(i)}, \mathscr{O}_{V^{(i)}}), \Gamma(U^{(i)}, \mathscr{O}_{U^{(i)}})\Big).$$

Moreover, by 1.10.3, we have

$$\Gamma(U^{(i)}, \mathscr{O}_{U^{(i)}}) \cong \varinjlim_{\nu \geq \mu} \Gamma(U_\nu^{(i)}, \mathscr{O}_{U_\nu^{(i)}}).$$

So by 1.10.4 (i), the map

$$\varinjlim_{\nu\geq\mu} \mathrm{Hom}_{S_\nu}(U_\nu^{(i)}, V_\nu^{(i)}) \to \mathrm{Hom}_S(U^{(i)}, V^{(i)})$$

is injective. Hence there exists some $\nu \geq \mu$ such that $f_\nu|_{U_\nu^{(i)}} = f'_\nu|_{U_\nu^{(i)}}$ for all i. Since $U_\nu^{(i)}$ form an open covering of S_ν, we have $f_\nu = f'_\nu$. This proves that our map is injective.

Next we prove that our map is surjective. Let $f : S \to Y$ be an S-morphism. For each $s \in S$, choose an affine open neighborhood $V_{0,s}$ of the image of $f(s)$ in Y_0, and an affine open neighborhood $W_{0,s}$ of the image of s in S_0 such that the image of $V_{0,s}$ in S_0 is contained in $W_{0,s}$. Let V_s and W_s be the inverse images of $V_{0,s}$ and $W_{0,s}$ in Y and S, respectively. Cover each $f^{-1}(V_s)$ by quasi-compact open subsets $U_s^{(i)}$ ($i \in I_s$) in S. Then $U_s^{(i)}$ ($s \in S$, $i \in I_s$) form an open covering of S. Since S is quasi-compact, this open covering has a finite subcovering. Changing notation, we may assume that there exist finitely many affine open subsets $V_0^{(i)}$ ($i = 1, \ldots, n$) of Y_0, affine open subsets $W_0^{(i)}$ ($i = 1, \ldots, n$) of S_0, and quasi-compact open subsets $U^{(i)}$ of S such that $U^{(i)}$ cover S, $f(U^{(i)}) \subset V^{(i)}$, and the images $V_0^{(i)}$ in S_0 are respectively contained in $W_0^{(i)}$. By 1.10.1, we can find quasi-compact open subsets $U_\lambda^{(i)}$ of $W_\lambda^{(i)}$ for some λ such that $U^{(i)} = u_\lambda^{-1}(U_\lambda^{(i)})$ and that $U_\lambda^{(i)}$ form an open covering of S_λ. By 1.10.3, we have

$$\Gamma(U^{(i)}, \mathscr{O}_{U^{(i)}}) \cong \varinjlim_{\mu\geq\lambda} \Gamma(U_\mu^{(i)}, \mathscr{O}_{U_\mu^{(i)}}).$$

By 1.10.4 (i), there exist $W_\mu^{(i)}$-morphisms $f_\mu^{(i)} : U_\mu^{(i)} \to V_\mu^{(i)}$ for some $\mu \geq \lambda$ inducing the morphisms $f|_{U^{(i)}} : U^{(i)} \to V^{(i)}$. By the injectivity of our map that we have already shown, we have $f_\mu^{(i)}|_{U_\mu^{(i)}\cap U_\mu^{(j)}} = f_\mu^{(j)}|_{U_\mu^{(i)}\cap U_\mu^{(j)}}$ for all pairs (i, j) if μ is sufficiently large. So we can glue $f_\mu^{(i)}$ together to get a morphism $f_\mu : S_\mu \to Y_\mu$ inducing f. This proves that our map is surjective.

(ii) Let $g : Y \to X$ be the inverse of f. By (i), for a sufficiently large λ, we can find S_λ-morphisms $f_\lambda : X_\lambda \to Y_\lambda$ and $g_\lambda : Y_\lambda \to X_\lambda$ inducing f and g, respectively. We have $fg = \mathrm{id}_Y$ and $gf = \mathrm{id}_X$. Choosing λ sufficiently large, we may assume $f_\lambda g_\lambda = \mathrm{id}_{Y_\lambda}$ and $g_\lambda f_\lambda = \mathrm{id}_{X_\lambda}$. Then f_λ is an isomorphism.

(iii) Cover S_0 by finitely many affine open subsets. They induce a finite affine open covering of S. Cover X by finitely many affine open subsets $U^{(i)}$ ($i = 1, \ldots, n$) so that their images in S are contained in members of the above affine open covering of S. By 1.10.4 (ii), there exist S_λ-schemes $U_\lambda^{(i)}$ for some λ such that $U_\lambda^{(i)} \times_{S_\lambda} S$ are S-isomorphic to $U^{(i)}$, respectively.

By 1.10.1 (ii), if λ is sufficiently large, we may find quasi-compact open subsets $U_\lambda^{(ij)}$ of $U_\lambda^{(i)}$ such that $U_\lambda^{(ij)}\times_{S_\lambda} S$ are S-isomorphic to $U^{(i)}\cap U^{(j)}$, respectively. By (ii), if λ is sufficiently large, we have isomorphisms $\phi_\lambda^{(ij)}$: $U_\lambda^{(ij)}\stackrel{\cong}{\to} U_\lambda^{(ji)}$ such that $\phi_\lambda^{(ij)}=(\phi_\lambda^{(ji)})^{-1}$, $\phi_\lambda^{(ij)}(U_\lambda^{(ij)}\cap U_\lambda^{(ik)})=U_\lambda^{(ji)}\cap U_\lambda^{(jk)}$, and $\phi_\lambda^{(jk)}\phi_\lambda^{(ij)}=\phi_\lambda^{(ik)}$ on $U_\lambda^{(ij)}\cap U_\lambda^{(ik)}$. Gluing $U_\lambda^{(i)}$ together using these isomorphisms, we get an S_λ-scheme X_λ such that $X_\lambda\times_{S_\lambda} S$ is S-isomorphic to X. □

Proposition 1.10.10. *Assume S_0 is quasi-compact and quasi-separated. Let $f_0 : X_0\to S_0$ be a morphism of finite presentation. If f is a morphism of one of the types listed below, then for sufficiently large λ, f_λ is of the same type.*

(i) *Open immersion.*
(ii) *Closed immersion.*
(iii) *Separated.*
(iv) *Finite.*
(v) *Affine*
(vi) *Surjective.*
(vii) *Radiciel.*
(viii) *Immersion.*
(ix) *Quasi-affine.*
(x) *Quasi-finite.*
(xi) *Proper.*

Proof.

(i) follows from 1.10.1 (ii).

(ii) By 1.10.8, $f_*\mathscr{O}_X$ is locally of finite presentation as an $\mathscr{O}_S$-module, and $f^\natural : \mathscr{O}_S\to f_*\mathscr{O}_X$ is surjective. By 1.10.2 (iii), there exists a quasi-coherent $\mathscr{O}_{S_\lambda}$-module $\mathscr{B}_\lambda$ locally of finite presentation for some λ so that $f_*\mathscr{O}_X\cong u_\lambda^*\mathscr{B}_\lambda$. By 1.10.2 (i), the morphism

$$f_*\mathscr{O}_X\otimes_{\mathscr{O}_S} f_*\mathscr{O}_X\to f_*\mathscr{O}_X$$

defining the multiplication on $f_*\mathscr{O}_X$ is induced by a morphism

$$\mathscr{B}_\lambda\otimes_{\mathscr{O}_{S_\lambda}}\mathscr{B}_\lambda\to\mathscr{B}_\lambda$$

if λ is sufficiently large. We may assume that this morphism defines an $\mathscr{O}_{S_\lambda}$-algebra structure on $\mathscr{B}_\lambda$. Indeed, the diagrams expressing the associative law, the commutative law and the distributive law commute by 1.10.2 (i) if λ is sufficiently large. Moreover, we can assume that there exists a morphism $\mathscr{O}_{S_\lambda}\to\mathscr{B}_\lambda$ of sheaves of rings inducing $\mathscr{O}_S\to f_*\mathscr{O}_X$. Let $\mathscr{C}_\lambda$ be the cokernel of $\mathscr{O}_{S_\lambda} \to \mathscr{B}_\lambda$. Then $u_\lambda^* \mathscr{C}_\lambda$ is the cokernel of $\mathscr{O}_S \to f_* \mathscr{O}_X$, and hence $u_\lambda^* \mathscr{C}_\lambda = 0$. By 1.10.2 (ii), we may assume $\mathscr{C}_\lambda = 0$. Then $\mathbf{Spec}\, \mathscr{B}_\lambda \to S_\lambda$ is a closed immersion and it induces $f : X \to S$. By 1.10.9 (ii), we may identify $\mathbf{Spec}\, \mathscr{B}_\lambda \to S_\lambda$ with $f_\lambda : X_\lambda \to S_\lambda$ if λ is sufficiently large. So f_λ is a closed immersion.

(iii) Apply (ii) to the diagonal morphism.

(iv) Use the same argument as the proof of (ii).

(v) The problem is local. We may assume S_0 is affine. Then there exists a closed immersion

$$i : X \to \mathbb{A}_S^n = \mathbf{Spec}\, \mathscr{O}_S[t_1, \ldots, t_n]$$

such that $\pi i = f$, where $\pi : \mathbb{A}_S^n \to S$ is the projection. By 1.10.9 (i), there exists an S_λ-morphism $i_\lambda : X_\lambda \to \mathbb{A}_{S_\lambda}^n$ inducing i for some λ. By (ii), we may assume that i_λ is a closed immersion. Then f_λ is affine.

We prove (vi)–(xi) under the extra condition that S_0, S_λ and S are noetherian. The general case is treated in [EGA] IV 8.10.5 and 9.3.3.

(vi) Let $E_\lambda = f_\lambda(X_\lambda)$ and let $E = f(X)$. They are constructible by 1.4.2. We have $u_\lambda^{-1}(E_\lambda) = E$. If f is surjective, then $E = S$. By 1.10.1(v), we have $E_\lambda = S_\lambda$ for a sufficiently large λ. Then f_λ is surjective.

(vii) Apply (vi) to the diagonal morphisms and use 1.7.1.

(viii) f can be factorized as the composite of a closed immersion and an open immersion. Under our extra noetherian condition, we can assume that these two immersions in the factorization are of finite presentation. We then use 1.10.9, (i) and (ii).

(ix) The problem is local. We may assume S_0 is affine. Let

$$A_\lambda = \Gamma(S_\lambda, \mathscr{O}_{S_\lambda}), \quad A = \Gamma(S, \mathscr{O}_S), \quad B = \Gamma(X, \mathscr{O}_X).$$

By 1.8.9, the canonical morphism $X \to \operatorname{Spec} B$ is an open immersion. Choose $b_1, \ldots, b_n \in B$ so that X can be identified with the open subset $D(b_1) \cup \cdots \cup D(b_n)$ of $\operatorname{Spec} B$. Since X is of finite type over S, each $\Gamma(D(b_i), \mathscr{O}_X) \cong B_{b_i}$ is a finitely generated A-algebra. Let $\{\frac{b_{i1}}{b_i^{k_{i1}}}, \ldots, \frac{b_{in_i}}{b_i^{k_{in_i}}}\}$ be a finite family of generators for B_{b_i} and let C be the A-subalgebra of B generated by b_i, b_{ij} $(i = 1, \ldots, n, j = 1, \ldots, n_i)$. Then $C_{b_i} \cong B_{b_i}$. Consider the canonical morphism $g : X \to \operatorname{Spec} C$. The image of g is contained in the open subset $D(b_1) \cup \cdots \cup D(b_n)$ of $\operatorname{Spec} C$. Moreover, for each i, the homomorphism $\Gamma(D(b_i), \mathscr{O}_{\operatorname{Spec} C}) \to \Gamma(D(b_i), \mathscr{O}_X)$ is an isomorphism. So $g : X \to \operatorname{Spec} C$ is an open immersion. Since C is a finitely generated A-algebra, there exists a closed immersion from $\operatorname{Spec} C$ to some $\mathbb{A}_S^n$. Composed with g, we get an immersion $h : X \to \mathbb{A}_S^n$ which is an S-morphism.

Using 1.10.9 and (viii), one can show that for a sufficiently large λ, there exists an S_λ-morphism $h_\lambda : X_\lambda \to \mathbb{A}^n_{S_\lambda}$ inducing h, and h_λ is an immersion. Then f_λ is quasi-affine.

(x) The problem is local. We may assume X_0 and S_0 are affine. Then by the Zariski Main Theorem ([Fu (2006)] 2.5.14, [EGA] III 4.4.3), there exist a finite morphism $\bar{f} : \overline{X} \to S$ and an open immersion $j : X \hookrightarrow \overline{X}$ such that $f = \bar{f}j$. By 1.10.9, we can find an S_λ-scheme $\overline{X}_\lambda$ and an S_λ-morphism $j_\lambda : X_\lambda \to \overline{X}_\lambda$ inducing j for some λ. By (i) and (iv), we may assume that $\overline{X}_\lambda$ is finite over S_λ and j_λ is an open immersion. Then f_λ is quasi-finite.

(xi) By Chow's lemma ([Fu (2006)] 1.4.18, [EGA] II 5.6.1, [Hartshorne (1977)] Exer. II 4.10), there exists a surjective projective morphism $g : X' \to X$ such that fg is projective. By 1.10.9, for some λ, we can find an S_λ-scheme X'_λ of finite presentation, and an S_λ-morphism $g_\lambda : X'_\lambda \to X_\lambda$ such that X' and g are induced from X'_λ and g_λ by base change. Let

$$X' \to \mathbb{P}^n_S = \operatorname{Proj} \mathscr{O}_S[t_0, \ldots, t_n]$$

be a closed immersion so that its composite with the projection $\mathbb{P}^n_S \to S$ coincides with fg. By 1.10.9 and (ii), if we choose λ sufficiently large, then we can find a closed immersion

$$X'_\lambda \to \mathbb{P}^n_{S_\lambda} = \operatorname{Proj} \mathscr{O}_{S_\lambda}[t_0, \ldots, t_n]$$

so that its composite with the projection $\mathbb{P}^n_S \to S$ coincides with $f_\lambda g_\lambda$. This implies that $f_\lambda g_\lambda$ is proper. Similarly, we can show that g_λ is a projective morphism and hence proper if λ is sufficiently large. On the other hand, we may assume that g_λ is surjective by (vi). It follows that f_λ is proper if λ is sufficiently large. □

Lemma 1.10.11. *Let S be a noetherian scheme, and let $f : X \to S$ be a quasi-affine morphism of finite type. Then there exist a projective morphism $\bar{f} : \overline{X} \to S$ and an open immersion $j : X \hookrightarrow \overline{X}$ such that $f = \bar{f}j$.*

Proof. By 1.8.9, the canonical S-morphism $X \to \mathbf{Spec}\, f_*\mathscr{O}_X$ is an open immersion. By [Fu (2006)] 2.5.2 or [EGA] I 9.4.9, or [Hartshorne (1977)] Exer. II 5.15, we have $f_*\mathscr{O}_X = \varinjlim_i \mathscr{E}_i$, where $\mathscr{E}_i$ are coherent $\mathscr{O}_S$-submodule of $f_*\mathscr{O}_X$. Let $\mathscr{A}(\mathscr{E}_i)$ be the $\mathscr{O}_S$-subalgebra of $f_*\mathscr{O}_X$ generated by $\mathscr{E}_i$. If $S(\mathscr{E}_i) = \bigoplus_{k=1}^\infty \operatorname{Sym}^k(\mathscr{E}_i)$ is the symmetric product of $\mathscr{E}_i$, then $\mathscr{A}_i$ is the image of the canonical morphism $S(\mathscr{E}_i) \to f_*\mathscr{O}_X$. In particular, $\mathscr{A}(\mathscr{E}_i)$ is quasi-coherent. Using the same argument as the proof of 1.10.10 (ix), one can show that when $\mathscr{E}_i$ is sufficiently large, the canonical morphism $X \to \mathbf{Spec}\, \mathscr{A}(\mathscr{E}_i)$ is an open immersion. We have a closed immersion $\mathbf{Spec}\,\mathscr{A}(\mathscr{E}_i) \to \mathbf{Spec}\,S(\mathscr{E}_i)$. So there exists an immersion $X \to \mathbf{Spec}\,S(\mathscr{E}_i)$. Composed with the canonical open immersion $\mathbf{Spec}\,S(\mathscr{E}_i) \to \mathbf{Proj}\,S(\mathscr{E}_i \oplus \mathscr{O}_S)$, we get an immersion $X \to \mathbf{Proj}\,S(\mathscr{E}_i \oplus \mathscr{O}_S)$, which is an S-morphism. Let $\overline{X}$ be the scheme theoretic image of this morphism. Then $\overline{X}$ is projective over S, and X is an open subscheme of $\overline{X}$. □

Lemma 1.10.12. *Let S be a noetherian scheme, and let $f : X \to S$ be a separated quasi-finite morphism. Then f is quasi-affine.*

Proof. We use induction on $\dim S$. To show that f is quasi-affine, it suffices to show that for any $s \in S$, there exists a neighborhood V of s such that $f^{-1}(V) \to V$ is quasi-affine. Let $\{V_i\}$ be the family of affine open neighborhoods of s in S. We have $\mathscr{O}_{S,s} \cong \varinjlim_i \Gamma(V_i, \mathscr{O}_S)$. If we can show that

$$X \times_S \mathrm{Spec}\,\mathscr{O}_{S,s} \to \mathrm{Spec}\,\mathscr{O}_{S,s}$$

is quasi-affine, then by 1.10.10 (ix), there exists some V_i such that $f^{-1}(V_i) \to V_i$ is quasi-affine. So it suffices to treat the case where $S = \mathrm{Spec}\,A$ for some local noetherian ring A. The morphism $\mathrm{Spec}\,\widehat{A} \to \mathrm{Spec}\,A$ is quasi-compact and faithfully flat, where $\widehat{A}$ is the completion of A. By 1.8.10, it suffices to show that

$$X \times_S \mathrm{Spec}\,\widehat{A} \to \mathrm{Spec}\,\widehat{A}$$

is quasi-affine. We are thus reduced to the case where A is a complete local noetherian ring. By 1.9.8, we have $X = X' \coprod X''$ such that $X' \to S$ is finite, and the fiber of $X'' \to S$ above the closed point of S is empty. The image of $X'' \to S$ is contained in an open subset of S of dimension $\leq \dim S - 1$. By the induction hypothesis, $X'' \to S$ is quasi-affine. So $X \to S$ is quasi-affine. □

Theorem 1.10.13 (Zariski Main Theorem). *Let S be a noetherian scheme, and let $f : X \to S$ be a separated quasi-finite morphism. Then there exist a finite morphism $\bar{f} : \overline{X} \to S$ and an open immersion $j : X \hookrightarrow \overline{X}$ such that $f = \bar{f}j$.*

Proof. Use 1.10.11, 1.10.12 and the Zariski Main Theorem in [Fu (2006)] 2.5.14 or [EGA] III 4.4.3. □

The Zariski Main Theorem 1.10.13 still holds if we only assume S is quasi-compact and quasi-separated. Confer [EGA] IV 18.12.13.

Remark 1.10.14. Let $(S_\lambda, u_{\lambda\mu})_{\lambda\in I}$ be an inverse system in the category of schemes such that $u_{\lambda\mu} : S_\mu \to S_\lambda$ $(\lambda \leq \mu)$ are affine. Then the inverse limit S of this system exists, and all the results in this section hold for such a system. Indeed, fix $\lambda_0 \in I$ and take $S_0 = S_{\lambda_0}$. We can apply the results of this section to the inverse system $(S_\lambda, u_{\lambda\mu})_{\lambda\geq\lambda_0}$ and we have $\varprojlim_{\lambda\in I} S_\lambda = \varprojlim_{\lambda\in I} S_\lambda$.

Chapter 2

Etale Morphisms and Smooth Morphisms

2.1 The Sheaf of Relative Differentials

Let $A \to B$ be a homomorphism of rings, and let M be a B-module. An A-derivation is a map $D : B \to M$ satisfying

$$D(b_1 + b_2) = D(b_1) + D(b_2), \quad D(b_1 b_2) = b_1 D(b_2) + b_2 D(b_1), \quad D(a) = 0$$

for any $a \in A$ and $b_1, b_2 \in B$. The *module of relative differentials* of B over A is a B-module $\Omega_{B/A}$ together with an A-derivation $d : B \to \Omega_{B/A}$ such that for any B-module M and any A-derivation $D : B \to M$, there exists a unique B-module homomorphism $\phi : \Omega_{B/A} \to M$ such that $D = \phi \circ d$. Let $\mathrm{Der}_A(B, M)$ be the set of A-derivations from B to M. The functor $M \mapsto \mathrm{Der}_A(B, M)$ is represented by $\Omega_{B/A}$. Recall that a covariant (resp. contravariant) functor $F : \mathscr{C} \to$ (**Sets**) from a category $\mathscr{C}$ to the category of sets is called *representable* if there exists an object X in $\mathscr{C}$ such that F is isomorphic to the functor $\mathrm{Hom}_{\mathscr{C}}(X, -)$ (resp. $\mathrm{Hom}_{\mathscr{C}}(-, X)$). In this case, we say F is represented by X.

Let I be the kernel of the epimorphism

$$B \otimes_A B \to B, \quad b_1 \otimes b_2 \to b_1 b_2.$$

We have three ways to put a B-module structure on I/I^2.

(a) I/I^2 is a $(B \otimes_A B)/I$-module. We have a canonical isomorphism $(B \otimes_A B)/I \cong B$. This gives a B-module structure on I/I^2.

(b) By multiplication on the left, $B \otimes_A B$ is a B-module, and I and I^2 are submodules. This induces a B-module structure on I/I^2.

(c) By multiplication on the right, $B \otimes_A B$ is a B-module, and I and I^2 are submodules. This induces a B-module structure on I/I^2.

One verifies that these three B-module structures are the same. Define $d : B \to I/I^2$ by

$$d(b) = 1 \otimes b - b \otimes 1 \mod I^2.$$

Then d is an A-derivation.

Proposition 2.1.1. *Notation as above. The pair $(I/I^2, d)$ is a module of relative differentials of B over A.*

Proposition 2.1.2. *Let A be a ring, let A' and B be A-algebras, and let $B' = B \otimes_A A'$. We have*

$$\Omega_{B'/A'} \cong \Omega_{B/A} \otimes_B B'.$$

For any multiplicative subset S of B, we have

$$\Omega_{S^{-1}B/A} \cong S^{-1}\Omega_{B/A}.$$

Proof. Let us prove the second statement. Let $\phi : B \to S^{-1}B$ be the canonical homomorphism. For any $S^{-1}B$-module N, the canonical map

$$\mathrm{Der}_A(S^{-1}B, N) \to \mathrm{Der}_A(B, N), \quad D \mapsto D \circ \phi$$

is bijective. In fact, for any derivation $D \in \mathrm{Der}_A(B, N)$, set

$$D'\left(\frac{b}{s}\right) = \frac{sD(b) - bD(s)}{s^2}.$$

Then $D' \in \mathrm{Der}_A(S^{-1}B, N)$ and $D \mapsto D'$ defines an inverse of the above map. We have

$$\mathrm{Der}_A(S^{-1}B, N) \cong \mathrm{Hom}_{S^{-1}B}(\Omega_{S^{-1}B/A}, N),$$
$$\mathrm{Der}_A(B, N) \cong \mathrm{Hom}_B(\Omega_{B/A}, N) \cong \mathrm{Hom}_{S^{-1}B}(S^{-1}\Omega_{B/A}, N).$$

It follows that we have a one-to-one correspondence

$$\mathrm{Hom}_{S^{-1}B}(\Omega_{S^{-1}B/A}, N) \cong \mathrm{Hom}_{S^{-1}B}(S^{-1}\Omega_{B/A}, N)$$

which is functorial in N. Hence $\Omega_{S^{-1}B/A} \cong S^{-1}\Omega_{B/A}$. □

Proposition 2.1.3. *Let $A \to B \to C$ be homomorphisms of rings. Then we have a natural exact sequence of C-modules*

$$\Omega_{B/A} \otimes_B C \to \Omega_{C/A} \to \Omega_{C/B} \to 0.$$

Proof. Denote the homomorphism $B \to C$ by ϕ. For any C-module N, the kernel of the homomorphism

$$\mathrm{Der}_A(C, N) \to \mathrm{Der}_A(B, N), \quad D \mapsto D \circ \phi$$

is exactly $\mathrm{Der}_B(C, N)$. So we have an exact sequence

$$0 \to \mathrm{Der}_B(C, N) \to \mathrm{Der}_A(C, N) \to \mathrm{Der}_A(B, N).$$

Hence we have an exact sequence

$$0 \to \mathrm{Hom}_C(\Omega_{C/B}, N) \to \mathrm{Hom}_C(\Omega_{C/A}, N) \to \mathrm{Hom}_C(\Omega_{B/A} \otimes_B C, N).$$

This sequence is functorial in N. Our assertion follows. □

Proposition 2.1.4. *Let $A \to B$ be a homomorphism of rings, let I be an ideal of B, and let $C = B/I$. Then we have a natural exact sequence of C-modules*

$$I/I^2 \xrightarrow{\delta} \Omega_{B/A} \otimes_B C \to \Omega_{C/A} \to 0,$$

where δ is induced by the map

$$I \to \Omega_{B/A} \otimes_B C, \quad b \mapsto d(b) \otimes 1.$$

Proof. Let N be a B/I-module. For any derivation $D : B \to N$, we have $D|_{I^2} = 0$ and D induces a B/I-module homomorphism $I/I^2 \to N$. We thus have an exact sequence

$$0 \to \mathrm{Der}_A(B/I, N) \to \mathrm{Der}_A(B, N) \to \mathrm{Hom}_{B/I}(I/I^2, N).$$

So we have an exact sequence

$$0 \to \mathrm{Hom}_C(\Omega_{C/A}, N) \to \mathrm{Hom}_C(\Omega_{B/A} \otimes_A C, N) \to \mathrm{Hom}_C(I/I^2, N).$$

This sequence is functorial in N. Our assertion follows. □

Proposition 2.1.5. *Let k be a field, and let $(B, \mathfrak{n})$ be a local k-algebra such that the homomorphism $k \to B$ induces an isomorphism $k \cong B/\mathfrak{n}$. Then the canonical map*

$$\delta : \mathfrak{n}/\mathfrak{n}^2 \to \Omega_{B/k} \otimes_B B/\mathfrak{n}$$

is an isomorphism.

Proof. Since $k \cong B/\mathfrak{n}$, the map δ is surjective by 2.1.4. To show it is injective, it suffices to show that

$$\delta^* : \mathrm{Hom}_k(\Omega_{B/k} \otimes_B B/\mathfrak{n}, k) \to \mathrm{Hom}_k(\mathfrak{n}/\mathfrak{n}^2, k)$$

is surjective, that is,

$$\mathrm{Der}_k(B, k) \to \mathrm{Hom}_k(\mathfrak{n}/\mathfrak{n}^2, k)$$

is surjective. Let $\phi : \mathfrak{n}/\mathfrak{n}^2 \to k$ be a k-homomorphism. For any $x \in B$, we can find $a \in k$ such that $x \equiv a \mod \mathfrak{n}$. Define $D(x) = \phi(x - a)$. Then $D : B \to k$ is a derivation and $\delta^*(D) = \phi$. □

Proposition 2.1.6. *If $B = A[t_1, \ldots, t_n]$ is a polynomial ring over A, then $\Omega_{B/A}$ is a free B-module of rank n with basis $\{dt_1, \ldots, dt_n\}$. Suppose A is an algebra over a ring K, then the sequence*

$$0 \to \Omega_{A/K} \otimes_A B \to \Omega_{B/K} \to \Omega_{B/A} \to 0$$

is split exact.

Proof. The map

$$d : B \to B^n, \quad d(f) = \left(\frac{\partial f}{\partial t_1}, \ldots, \frac{\partial f}{\partial t_n}\right)$$

is a B-derivation. Let us show that the pair (B^n, d) satisfies the universal property for the module of relative differentials. Let M be a B-module, let $D : B \to M$ be an A-derivation, and let $x_i = D(t_i)$. Then for any $f \in B$, we have

$$D(f) = \sum_{i=1}^{n} \frac{\partial f}{\partial t_i} x_i.$$

The homomorphism

$$\phi : B^n \to M, \quad (b_1, \ldots, b_n) \mapsto b_1 x_1 + \cdots + b_n x_n$$

is the unique homomorphism such that $D = \phi \circ d$.

To get the required split short exact sequence, it suffices to show that any K-derivation $D_0 : A \to M$ can be extended to a K-derivation $D : B \to M$. Indeed, fix $x_i \in M$ $(i = 1, \ldots, n)$. We can define

$$\begin{aligned} &D\left(\sum_{i_1,\ldots,i_n} a_{i_1\ldots,i_n} t_1^{i_1} \cdots t_n^{i_n}\right) \\ &= \sum_{i_1,\ldots,i_n} D_0(a_{i_1\ldots,i_n}) t_1^{i_1} \cdots t_n^{i_n} + \sum_{i=1}^{n} x_i \frac{\partial}{\partial t_i}\left(\sum_{i_1,\ldots,i_n} a_{i_1\ldots,i_n} t_1^{i_1} \cdots t_n^{i_n}\right) \end{aligned}$$

for any $\sum_{i_1,\ldots,i_n} a_{i_1\ldots,i_n} t_1^{i_1} \cdots t_n^{i_n} \in B$. □

Corollary 2.1.7. *Let C be an A-algebra of finite type (resp. of finite presentation), then $\Omega_{C/A}$ is a C-module of finite type (resp. of finite presentation).*

Proof. We may assume $C = A[t_1, \ldots, t_n]/I$ for some n and some ideal I of $A[t_1, \ldots, t_n]$. Let $B = A[t_1, \ldots, t_n]$. By 2.1.4, we have an exact sequence

$$I/I^2 \xrightarrow{\delta} \Omega_{B/A} \otimes_B C \to \Omega_{C/A} \to 0.$$

By 2.1.6, $\Omega_{B/A} \otimes_B C$ is a free C-module of finite rank. When C has finite presentation over A, I/I^2 is a finitely generated C-module. Our assertion follows. □

Proposition 2.1.8. *Let L/K be a finite separable extension of fields. Then $\Omega_{L/K} = 0$.*

Proof. Let M be an L-module, and let $D : L \to M$ be a K-derivation. For any $x \in L$, let $f(t)$ be the minimal polynomial of x over K. Then we have $f(x) = 0$ and $f'(x) \neq 0$. On the other hand, we have

$$0 = D(f(x)) = f'(x)D(x).$$

It follows that $D(x) = 0$. So $\mathrm{Der}_K(L, M) = 0$. This is true for any L-module M. So $\Omega_{L/K} = 0$. $\square$

Let $f : X \to Y$ be a morphism of schemes, and let $\Delta : X \to X \times_Y X$ be the diagonal morphism. There exists an open subset U of $X \times_Y X$ containing the image of Δ such that Δ induces a closed immersion $\Delta : X \to U$. Let $\mathscr{I}$ be the ideal sheaf of this closed immersion. We define the *sheaf of relative differentials* of X over Y to be the sheaf $\Omega_{X/Y} = \Delta^{-1}(\mathscr{I}/\mathscr{I}^2)$. For all affine open subsets $W = \mathrm{Spec}\, B$ of X and $V = \mathrm{Spec}\, A$ of Y such that $f(W) \subset V$, $\Omega_{X/Y}|_W$ can be identified with $\Omega_{B/A}^{\sim}$. $\Omega_{X/Y}$ is a quasi-coherent $\mathscr{O}_X$-module. If f is locally of finite type (resp. locally of finite presentation), then $\Omega_{X/Y}$ is an $\mathscr{O}_X$-module locally of finite type (resp. locally of finite presentation).

Proposition 2.1.9. *Consider a Cartesian diagram*

$$\begin{array}{ccc} X' = X \times_Y Y' & \stackrel{p_1}{\to} & X \\ \downarrow & & \downarrow \\ Y' & \to & Y. \end{array}$$

We have $p_1^*\Omega_{X/Y} \cong \Omega_{X'/Y'}$.

Proposition 2.1.10. *Let* $f : X \to Y$ *and* $Y \to Z$ *be morphism of schemes. We have an exact sequence of* $\mathscr{O}_X$*-modules*

$$f^*\Omega_{Y/Z} \to \Omega_{X/Z} \to \Omega_{X/Y} \to 0.$$

Proposition 2.1.11. *Let* $X \to Y$ *be a morphism and let* $i : Z \to X$ *be a closed immersion with ideal sheaf* $\mathscr{I}$. *Then we have an exact sequence of* $\mathscr{O}_Z$*-modules*

$$i^*(\mathscr{I}/\mathscr{I}^2) \to i^*\Omega_{X/Y} \to \Omega_{Z/Y} \to 0.$$

Proposition 2.1.12. *Let* S *be a scheme, and let* $\mathbb{A}_S^n = \mathbf{Spec}\, \mathscr{O}_S[t_1, \ldots, t_n]$. *Then we have* $\Omega_{\mathbb{A}_S^n/S} \cong \mathscr{O}_{\mathbb{A}_S^n}^n$.

2.2 Unramified Morphisms

([SGA 1] I 3, [EGA] IV 17.4.)

Proposition 2.2.1. *Let $f : X \to Y$ be a morphism locally of finite type, x a point in X, and $y = f(x)$. The following conditions are equivalent:*

(i) $\mathfrak{m}_y\mathscr{O}_{X,x} = \mathfrak{m}_x$ *and the residue field $k(x)$ is a finite separable extension of the residue field $k(y)$.*

(ii) $(\Omega_{X/Y})_x = 0$.

(iii) *The diagonal morphism $\Delta : X \to X \times_Y X$ is an open immersion in a neighborhood of x.*

Proof.

(i)⇒(ii) We have

$$\mathscr{O}_{X,x} \otimes_{\mathscr{O}_{Y,y}} k(y) \cong \mathscr{O}_{X,x}/\mathfrak{m}_y\mathscr{O}_{X,x} = \mathscr{O}_{X,x}/\mathfrak{m}_x = k(x).$$

By 2.1.2 and 2.1.8, we have

$$(\Omega_{X/Y})_x \otimes_{\mathscr{O}_{X,x}} \mathscr{O}_{X,x}/\mathfrak{m}_y\mathscr{O}_{X,x} \cong \Omega_{k(x)/k(y)} = 0.$$

Since f is locally of finite type, $(\Omega_{X/Y})_x$ is a finitely generated $\mathscr{O}_{X,x}$-module by 2.1.7. We then apply Nakayama's lemma.

(ii)⇒(iii) Let V be an open subset containing $\mathrm{im}(\Delta)$ such that Δ induces a closed immersion $X \to V$, and let $\mathscr{I}$ be the ideal sheaf of this closed immersion. We have

$$(\mathscr{I}/\mathscr{I}^2)_{\Delta(x)} \cong (\Omega_{X/Y})_x = 0.$$

Using the assumption that f is locally of finite type, one can verify that $\mathscr{I}_{\Delta(x)}$ is a finitely generated $\mathscr{O}_{X\times_Y X,\Delta(x)}$-module. (Confer the proof of 1.10.5 (iv).) By Nakayama's lemma, we have $\mathscr{I}_{\Delta(x)} = 0$. But then $\mathscr{I}$ vanishes in a neighborhood of $\Delta(x)$. So Δ is an isomorphism in a neighborhood of x.

(iii)⇒(i) By the base change $\mathrm{Spec}\, k(y) \to Y$, we are reduced to the case where $Y = \mathrm{Spec}\, k$ for some field k. Replacing X by a neighborhood of x, we may assume Δ is an open immersion.

First consider the case where k is algebraically closed. For any closed point t of X, we have a k-morphism $t : \mathrm{Spec}\, k \to X$ with image t. Let Γ_t be its graph. We have a Cartesian diagram

$$\begin{array}{ccc} \mathrm{Spec}\, k & \stackrel{\Gamma_t}{\to} & \mathrm{Spec}\, k \times_{\mathrm{Spec}\, k} X \\ \downarrow & & \downarrow \\ X & \stackrel{\Delta}{\to} & X \times_{\mathrm{Spec}\, k} X. \end{array}$$

Since Δ is an open immersion, Γ_t is also an open immersion. We have an isomorphism $\operatorname{Spec} k \times_{\operatorname{Spec} k} X \cong X$, and through this isomorphism, Γ_t is identified with $t : \operatorname{Spec} k \to X$. So t is an open immersion. This shows that every closed point of X is isolated and $\Gamma(t, \mathscr{O}_X) = k$. This implies that X is a disjoint union of copies of $\operatorname{Spec} k$.

In general, let $\bar{k}$ be an algebraic closure of k. The above argument shows that after shrinking X, $X \otimes_k \bar{k}$ is isomorphic to a disjoint union of finitely many copies of $\operatorname{Spec} \bar{k}$. So X is finite as a set. Replacing X by a neighborhood of x, we may assume that X consists of a single point. By 1.9.2, $X = \operatorname{Spec} A$ for some finite dimensional k-algebra A. We then have $A \otimes_k \bar{k} \cong \bar{k} \times \cdots \times \bar{k}$. So $A \otimes_k \bar{k}$ is reduced and hence A is reduced. As X consists of a single point, A is necessarily a field that is finite separable over k. □

Let $f : X \to Y$ be a morphism locally of finite presentation and let $x \in X$. If the conditions in 2.2.1 hold, then we say f is *unramified at* x. The set of points where f is unramified is open. If f is unramified at every point, we say f is unramified.

Proposition 2.2.2. *In the following, morphisms are assumed to be locally of finite presentation.*

(i) *Unramified morphisms are locally quasi-finite.*
(ii) *Immersions are unramified.*
(iii) *Composites of unramified morphisms are unramified.*
(iv) *Base changes of unramified morphisms are unramified.*
(v) *If the composite $X \to Y \to Z$ is unramified, then so is $X \to Y$.*

Proposition 2.2.3. *Let S_0 be a quasi-compact quasi-separated scheme, $(\mathscr{A}_\lambda, \phi_{\lambda\mu})$ a direct system of quasi-coherent $\mathscr{O}_{S_0}$-algebras, $\mathscr{A} = \varinjlim_\lambda \mathscr{A}_\lambda$, $S_\lambda = \mathbf{Spec}\, \mathscr{A}_\lambda$, $S = \mathbf{Spec}\, \mathscr{A}$, X_0 an S_0-scheme, $X_\lambda = X_0 \times_{S_0} S_\lambda$, and $X = X \times_{S_0} S$. Assume that X_0 is of finite presentation over S_0. If X is unramified over S, then X_λ is unramified over S_λ for some λ.*

Proof. Let $u_\lambda : X \to X_\lambda$ be the canonical morphism. We have $u_\lambda^*(\Omega^1_{X_\lambda/S_\lambda}) \cong \Omega_{X/S}$. Since X/S is unramified, we have $\Omega_{X/S} = 0$. By 1.10.2 (ii), we have $\Omega_{X_\lambda/S_\lambda} = 0$ for some λ. Then X_λ is unramified over S_λ. □

Proposition 2.2.4. *Let $X \xrightarrow{f} Y \to Z$ be two morphisms. If $Y \to Z$ if unramified, then the graph $\Gamma_f : X \to X \times_Z Y$ is an open immersion.*

Proof. We have a Cartesian diagram

$$\begin{array}{ccc} X & \xrightarrow{\Gamma_f} & X\times_Z Y \\ \downarrow & & \downarrow \\ Y & \xrightarrow{\Delta} & Y\times_Z Y. \end{array}$$

We then apply 2.2.1 (iii). □

Proposition 2.2.5. *Consider a Cartesian diagram*

$$\begin{array}{ccc} X\times_Y Y' & \xrightarrow{g'} & X \\ f'\downarrow & & \downarrow f \\ Y' & \xrightarrow{g} & Y. \end{array}$$

Suppose $f : X \to Y$ is locally of finite presentation and $g : Y' \to Y$ is faithfully flat. If f' is unramified, then so is f.

Proof. We have

$$g'^*\Omega_{X/Y} \cong \Omega_{X\times_Y Y'/Y'} = 0.$$

Since g' is faithfully flat, this implies that $\Omega_{X/Y} = 0$. □

2.3 Etale Morphisms

([SGA 1] I 4, [EGA] IV 17.6–9, 18.1–4.)

Let $f : X \to Y$ be a morphism locally of finite presentation and let x be a point in X. We say f is *etale at* x if f is flat at x and unramified at x. If f is etale at every point, we say f is etale.

Proposition 2.3.1. *In the following, morphisms are assumed to be locally of finite presentation.*

(i) *Open immersions are etale.*

(ii) *Composites of etale morphisms are etale.*

(iii) *Base changes of etale morphisms are etale.*

(iv) *Let $f : X \to Y$ and $g : Y \to Z$ be two morphisms such that gf is etale and g is unramified. Then f is etale.*

Proof. Let us prove (iv). By 2.2.4, the graph $\Gamma_f : X \to X\times_Z Y$ of f is an open immersion. By (iii), the projection $p_2 : X\times_Z Y \to Y$ is etale. By (ii), $f = p_2\Gamma_f$ is etale. □

Remark 2.3.2. The same argument as the proof of 2.3.1 (iv) shows that the following is true: Let $f : X \to Y$ and $g : Y \to Z$ be two morphisms such that gf is flat at $x \in X$ and g is unramified. Then f is flat at x.

Proposition 2.3.3. *Let A be a ring, $F(t) \in A[t]$ a monic polynomial, and $B = A[t]/(F(t))$. The canonical morphism $\mathrm{Spec}\, B \to \mathrm{Spec}\, A$ is etale at a prime ideal $\mathfrak{q} \in \mathrm{Spec}\, B$ if and only if $\mathfrak{q}$ does not contain the image of $F'(t)$ in B. Hence $\mathrm{Spec}\, B \to \mathrm{Spec}\, A$ is etale if and only if the image of $F'(t)$ in B is a unit, or equivalently, the ideal generated by $F(t)$ and $F'(t)$ is $A[t]$.*

Proof. Since $F(t)$ is monic, $A[t]/(F(t))$ is flat over A. By 2.1.4 and 2.1.6, we have

$$\Omega_{B/A} \cong A[t]/(F(t), F'(t)).$$

It follows that $(\Omega_{B/A})_{\mathfrak{q}} = 0$ if and only if $\mathfrak{q}$ does not contain the image of $F'(t)$ in B. □

Lemma 2.3.4. *Let $(A, \mathfrak{m})$ be a local ring, B' a finite A-algebra, $\mathfrak{n}'$ a maximal ideal of B', u an element of B' not lying in $\mathfrak{n}'$ but lying in all other maximal ideals of B' distinct from $\mathfrak{n}'$, $B = A[u]$, and $\mathfrak{n} = \mathfrak{n}' \cap B$. Assume that $B'_{\mathfrak{n}'}/\mathfrak{m}B'_{\mathfrak{n}'}$ is generated by the image of u in $B'_{\mathfrak{n}'}/\mathfrak{m}B'_{\mathfrak{n}'}$ as an $A/\mathfrak{m}$-algebra. Then the canonical homomorphism $B_{\mathfrak{n}} \to B'_{\mathfrak{n}'}$ is an isomorphism.*

Proof. Let $S = B - \mathfrak{n}$ and $S' = B' - \mathfrak{n}'$. We need to show that $S^{-1}B \to S'^{-1}B'$ is an isomorphism. First let us show $S^{-1}B' \to S'^{-1}B'$ is an isomorphism. It suffices to show that every element in S' is invertible in $S^{-1}B'$, or equivalently, the images in $S^{-1}B'$ of elements in S' do not lie in any maximal ideal of $S^{-1}B'$. Since $S^{-1}B'$ is finite over $S^{-1}B$, any maximal ideal of $S^{-1}B'$ lies above the maximal ideal of $S^{-1}B = B_{\mathfrak{n}}$, and hence above the maximal ideal $\mathfrak{n}$ of B. But B' is finite over B. So any maximal ideal of $S^{-1}B'$ lies above a maximal ideal of B', and hence is of the form $S^{-1}\mathfrak{n}''$ for some maximal ideal $\mathfrak{n}''$ of B' satisfying $\mathfrak{n}'' \cap B = \mathfrak{n}$. If $u \in \mathfrak{n}''$, then $u \in \mathfrak{n}'' \cap B = \mathfrak{n} \subset \mathfrak{n}'$. This contradicts to our assumption that $u \notin \mathfrak{n}'$. So $u \notin \mathfrak{n}''$. As u lies in maximal ideals of B' distinct from $\mathfrak{n}'$, we must have $\mathfrak{n}' = \mathfrak{n}''$. Hence $S^{-1}\mathfrak{n}'$ is the only maximal ideal of $S^{-1}B'$. It is clear that the images in $S^{-1}B'$ of elements in S' do not lie in $S^{-1}\mathfrak{n}'$. So $S^{-1}B' \to S'^{-1}B'$ is an isomorphism. Since B is a subring of B', $S^{-1}B \to S^{-1}B'$ is injective. Hence $S^{-1}B \to S'^{-1}B'$ is injective. To prove it is surjective, it suffices to show that $B_{\mathfrak{n}}/\mathfrak{m}B_{\mathfrak{n}} \to B'_{\mathfrak{n}'}/\mathfrak{m}B'_{\mathfrak{n}'}$ is surjective by Nakayama's lemma. This follows from the assumption that the image of u in $B'_{\mathfrak{n}'}/\mathfrak{m}B'_{\mathfrak{n}'}$ is a generator over $A/\mathfrak{m}$. □

Theorem 2.3.5 (Chevalley). *Let $f : X \to Y$ be a morphism locally of finite presentation, x a point in X, and $A = \mathscr{O}_{Y,f(x)}$. Then f is etale at x if and only if we can find a monic polynomial $F(t) \in A[t]$, a maximal ideal $\mathfrak{n}$ of $B = A[t]/(F(t))$ not containing the image of $F'(t)$ in B such that $\mathscr{O}_{X,x}$ is A-isomorphic to $B_{\mathfrak{n}}$.*

Proof. The "if" part follows from 2.3.3. To prove the "only if" part, we make the base change $\mathrm{Spec}\, A \to Y$ to reduce to the case where $Y = \mathrm{Spec}\, A$. Let $\mathfrak{m}$ be the maximal ideal of A. First suppose f is unramified at x. Let U be an affine open neighborhood of x in X such that $f|_U : U \to \mathrm{Spec}\, A$ is unramified. By the Zariski Main Theorem 1.10.13, $f|_U$ can be factorized as a composite

$$U \stackrel{j}{\hookrightarrow} \mathrm{Spec} B' \stackrel{\bar{f}}{\to} \mathrm{Spec} A$$

such that j is an open immersion and $\bar{f}$ is a finite morphism. (We only prove the Zariski Main Theorem 1.10.13 under the noetherian assumption. Write $A = \varinjlim_\lambda A_\lambda$, where A_λ are subalgebras of A finitely generated over $\mathbb{Z}$. By 1.10.4 (ii) and 2.2.3, we can find an unramified affine morphism $f_\lambda : U_\lambda \to \mathrm{Spec}\, A_\lambda$ of finite type such that $f|_U$ is obtained from f_λ by base change. Applying 1.10.13 to the morphism f_λ and taking base change, we get the above factorization of $f|_U$.) Let $\mathfrak{n}'$ be the prime ideal of B' corresponding to the image of x in $\mathrm{Spec}\, B'$. Since it is above the maximal ideal $\mathfrak{m}$ of A, $\mathfrak{n}'$ is a maximal ideal of B'. Note that $\mathscr{O}_{X,x}$ is A-isomorphic to $B'_{\mathfrak{n}'}$. Let $\mathfrak{n}'_1, \ldots, \mathfrak{n}'_r$ be all the maximal ideals of B' distinct from $\mathfrak{n}'$. Since $B'/\mathfrak{n}'$ is a finite separable extension of $A/\mathfrak{m}$, it is generated by a single element. By the Chinese remainder theorem, we can find $u \in \mathfrak{n}'_1 \cap \cdots \cap \mathfrak{n}'_r$ such that the image of u in $B'/\mathfrak{n}'$ is nonzero and generates $B'/\mathfrak{n}'$ over $A/\mathfrak{m}$. Let $B = A[u]$ and let $\mathfrak{n} = \mathfrak{n}' \cap B$. By 2.3.4, $\mathscr{O}_{X,x}$ is A-isomorphic to $B_{\mathfrak{n}}$. Let $\bar{u}$ be the image of u in $B/\mathfrak{m}B$, let $f(t) \in (A/\mathfrak{m})[t]$ be the minimal polynomial of $\bar{u}$ over $A/\mathfrak{m}$, and let $d = \deg f$. We have

$$B/\mathfrak{m}B \cong (A/\mathfrak{m})[t]/(f(t)).$$

By Nakayama's lemma, B is generated by $1, u, \ldots, u^{d-1}$ as an A-module. So there exists a monic polynomial $F(t) \in A[t]$ of degree d such that $F(u) = 0$. The image of $F(t)$ in $(A/\mathfrak{m})[t]$ is necessarily $f(t)$. Since $A/\mathfrak{m}$ is a field, $\mathrm{Spec}\, B/\mathfrak{m}B \to \mathrm{Spec}\, A/\mathfrak{m}$ is etale at $\mathfrak{n}/\mathfrak{m}B$. By 2.3.3, we have $f'(\bar{u}) \notin \mathfrak{n}/\mathfrak{m}B$. So we have $F'(u) \notin \mathfrak{n}$.

Suppose furthermore that f is flat at x. Then $B_{\mathfrak{n}}$ is flat over A. Let $\mathfrak{n}_0$ be the inverse image of $\mathfrak{n}$ under the epimorphism

$$A[t]/(F(t)) \to B, \quad t \mapsto u$$

of A-algebras. Then $\mathfrak{n}_0$ does not contain the image of $F'(t)$ in $A[t]/(F(t))$. By 2.3.3, $\operatorname{Spec} A[t]/(F(t)) \to \operatorname{Spec} A$ is etale at $\mathfrak{n}_0$, and in particular unramified in a neighborhood of $\mathfrak{n}_0$. Applying 2.3.2 to the composite

$$\operatorname{Spec} B \to \operatorname{Spec} A[t]/(F(t)) \to \operatorname{Spec} A,$$

we see that the morphism

$$\operatorname{Spec} B \to \operatorname{Spec} A[t]/(F(t))$$

is flat at $\mathfrak{n}$. So $(A[t]/(F(t)))_{\mathfrak{n}_0} \to B_{\mathfrak{n}}$ is faithfully flat and hence injective. But $A[t]/(F(t)) \to B$ is surjective, so we must have $(A[t]/(F(t)))_{\mathfrak{n}_0} \cong B_{\mathfrak{n}}$. Hence $(A[t]/(F(t)))_{\mathfrak{n}_0}$ is A-isomorphic to $\mathscr{O}_{X,x}$. □

Corollary 2.3.6. *Let $f : X \to Y$ be a morphism locally of finite presentation. The set of points in X where f is etale is open.*

Proof. Suppose f is etale at $x \in X$. By 2.3.5, we can find a monic polynomial $F(t) \in \mathscr{O}_{Y,f(x)}[t]$ and a maximal ideal of $\mathfrak{n}$ of $B = \mathscr{O}_{Y,f(x)}[t]/(F(t))$ not containing the image of $F'(t)$ in B such that $\mathscr{O}_{X,x}$ is $\mathscr{O}_{Y,f(x)}$-isomorphic to $B_{\mathfrak{n}}$. Let $V = \operatorname{Spec} A$ be an affine open neighborhood of $f(x)$ in Y such that $F(t)$ is the image of a monic polynomial in $A[t]$ under the canonical homomorphism $A[t] \to \mathscr{O}_{Y,f(x)}[t]$, and denote such a monic polynomial in $A[t]$ also by $F(t)$. Let $\mathfrak{p}$ be the inverse image of $\mathfrak{n}$ under the canonical homomorphism $A[t]/(F(t)) \to \mathscr{O}_{Y,f(x)}[t]/(F(t))$. Then $\operatorname{Spec} A[t]/(F(t)) \to V$ is unramified at $\mathfrak{p}$, and hence unramified in an open neighborhood W of $\mathfrak{p}$ in $\operatorname{Spec} A[t]/(F(t))$. As $A[t]/(F(t))$ is a flat A-algebra, $\operatorname{Spec} A[t]/(F(t)) \to V$ is etale in W. Note that $\mathscr{O}_{X,x}$ is $\mathscr{O}_{Y,f(x)}$-isomorphic to $(A[t]/(F(t))_{\mathfrak{p}}$. By [Fu (2006)] 1.3.13 (iii) or [EGA] I 6.5.4 (ii), there exists an open neighborhood U of x in X and an open neighborhood W' of $\mathfrak{p}$ in W such that U and W' are Y-isomorphic. Then f is etale in U. □

Proposition 2.3.7. *Let S_0 be a quasi-compact quasi-separated scheme, $(\mathscr{A}_\lambda, \phi_{\lambda\mu})$ a direct system of quasi-coherent $\mathscr{O}_{S_0}$-algebras, $\mathscr{A} = \varinjlim_\lambda \mathscr{A}_\lambda$, $S_\lambda = \mathbf{Spec}\, \mathscr{A}_\lambda$, $S = \mathbf{Spec}\, \mathscr{A}$, X_0 an S_0-scheme, $X_\lambda = X_0 \times_{S_0} S_\lambda$, and $X = X \times_{S_0} S$. Assume that X_0 is of finite presentation over S_0. If X is etale over S, then X_λ is etale over S_λ for some λ.*

Proof. Let x be a point in X and let s be its image in S. By 2.3.5, we can find a monic polynomial $F(t) \in \mathscr{O}_{S,s}[t]$ and a maximal ideal $\mathfrak{n}$ of $B = \mathscr{O}_{S,s}[t]/(F(t))$ not containing the image of $F'(t)$ in B such that $\mathscr{O}_{X,x}$ is $\mathscr{O}_{S,s}$-isomorphic to $B_{\mathfrak{n}}$. Let $V = \operatorname{Spec} A$ be an affine open neighborhood of s in S such that $F(t)$ is the image of a monic polynomial in $A[t]$ under the canonical homomorphism $A[t] \to \mathscr{O}_{S,s}[t]$, and denote such a monic polynomial in $A[t]$ also by $F(t)$. Let $\mathfrak{p}$ be the inverse image of $\mathfrak{n}$ under the canonical homomorphism $A[t]/(F(t)) \to \mathscr{O}_{S,s}[t]/(F(t))$. Then $\mathscr{O}_{X,x}$ is $\mathscr{O}_{S,s}$-isomorphic to $(A[t]/(F(t)))_{\mathfrak{p}}$. We can find a quasi-compact open neighborhood U of x in X whose image in S is contained in V, and a quasi-compact open neighborhood W of $\mathfrak{p}$ in $\operatorname{Spec} A[t]/(F(t))$ such that U and W are V-isomorphic by [Fu (2006)] 1.3.13 (iii) or [EGA] I 6.5.4 (ii). By 1.10.1 (ii), there exists an open subset V_λ of S_λ for some λ whose inverse image in S is V. Shrinking V_λ and V, we may assume $V_\lambda = \operatorname{Spec} A_\lambda$ is affine. Taking λ sufficiently large, we may assume $F(t)$ is the image in $A[t]$ of a monic polynomial $F_\lambda(t) \in A_\lambda[t]$ such that the inverse image $\mathfrak{p}_\lambda$ of $\mathfrak{p}$ in $A_\lambda[t]/(F_\lambda(t))$ does not contain the image of $F'_\lambda(t)$ in $A_\lambda[t]/(F_\lambda(t))$. Again by 1.10.1 (ii), we may assume that U (resp. W) is the inverse image of an open subset U_λ (resp. W_λ) of X_λ (resp. $\operatorname{Spec} A_\lambda[t]/(F_\lambda(t))$) in X (resp. in $\operatorname{Spec} A[t]/(F(t))$), and by 1.10.9 (ii), we may assume that U_λ is V_λ-isomorphic to W_λ. But $W_\lambda \to V_\lambda$ is etale at $\mathfrak{p}_\lambda$ by 2.3.3. So X_λ is etale over S_λ at x_λ, where x_λ is the image of x in X_λ. For each λ, let O_λ be the largest open subset of X_λ on which X_λ is etale over S_λ, and let $u_\lambda : X \to X_\lambda$ be the projection. The above discussion shows that $X = \cup_\lambda u_\lambda^{-1}(O_\lambda)$. Since X is quasi-compact and $u_\lambda^{-1}(O_\lambda) \subset u_\mu^{-1}(O_\mu)$ for any pair $\lambda \leq \mu$, we have $X = u_\lambda^{-1}(O_\lambda)$ for some λ, that is, $u_\lambda^{-1}(X_\lambda - O_\lambda) = \emptyset$. By 10.1.1 (iv), there exists $\mu \geq \lambda$ such that $X_\mu - O_\mu = \emptyset$. Then X_μ is etale over S_μ. □

Proposition 2.3.8. *Etale morphisms are open mappings.*

Proof. The problem is local. It suffices to prove that any affine etale morphism $f : \operatorname{Spec} B \to \operatorname{Spec} A$ is an open mapping. By 1.10.4 (ii) and 2.3.7, we can find a subalgebra A_0 of A finitely generated over $\mathbb{Z}$, and an algebra B_0 etale and finitely generated over A_0 such that we have an A-isomorphism $B \cong B_0 \otimes_{A_0} A$. Let us prove that the morphism $\operatorname{Spec}(B_0 \otimes_{A_0} A) \to \operatorname{Spec} A$ is an open mapping. Let $\{A_\lambda\}$ be the family of subalgebras of A finitely generated over A_0. Fix the notation by the following diagram:

$$\begin{array}{ccc} \operatorname{Spec}(B_0 \otimes_{A_0} A) & \stackrel{v_\lambda}{\to} & \operatorname{Spec}(B_0 \otimes_{A_0} A_\lambda) \\ f \downarrow & & \downarrow f_\lambda \\ \operatorname{Spec} A & \stackrel{u_\lambda}{\to} & \operatorname{Spec} A_\lambda. \end{array}$$

Each f_λ is etale and hence flat. By 1.5.2, f_λ is an open mapping. By 1.10.1 (ii), any quasi-compact open subset U of $\operatorname{Spec}(B_0 \otimes_{A_0} A)$ is of the form $U = v_\lambda^{-1}(U_\lambda)$ for a large λ and an open subset U_λ of $\operatorname{Spec}(B_0 \otimes_{A_0} A_\lambda)$. Since $f_\lambda(U_\lambda)$ is open and

$$f(U) = f v_\lambda^{-1}(U_\lambda) = u_\lambda^{-1} f_\lambda(U_\lambda),$$

$f(U)$ is also open. □

Proposition 2.3.9. *A morphism $f : X \to Y$ is an open immersion if and only if it is etale and radiciel.*

Proof. The "only if" part is clear. Let us prove the "if" part. Since f is injective, it suffices to show that f is locally an open immersion. So we may assume that $f : X \to Y$ is induced by a base change $Y \to \operatorname{Spec} A_0$ from an etale morphism $f_0 : \operatorname{Spec} B_0 \to \operatorname{Spec} A_0$, where A_0 is a finitely generated $\mathbb{Z}$-algebra and B_0 is a finitely generated A_0-algebra. By the Zariski Main Theorem 1.10.13, f_0 can be factorized as the composite of an open immersion and a finite morphism. So we have a factorization $f = \bar{f}j$ such that $j : X \hookrightarrow \overline{X}$ is an open immersion and $\bar{f} : \overline{X} \to Y$ is finite. By 2.3.8, $f(X)$ is open. Replacing Y by $f(X)$ and $\overline{X}$ by $\bar{f}^{-1}(f(X))$, we may assume that f is surjective. Since f is radiciel and etale, it is universally injective and universally open, and hence universally a homeomorphism. It follows that f is proper. Then j is proper. Hence j is an open and closed immersion, and in particular a finite morphism. So f is a finite morphism. Let us prove this implies that f is an isomorphism. We may reduce to the case where $Y = \operatorname{Spec} A$ is affine. Then $X = \operatorname{Spec} B$ for some finite A-algebra B. Since f is radiciel, for any prime ideal $\mathfrak{p}$ of A, $B_{\mathfrak{p}}$ is local. Since f is also etale, we have $A_{\mathfrak{p}}/\mathfrak{p}A_{\mathfrak{p}} \cong B_{\mathfrak{p}}/\mathfrak{p}B_{\mathfrak{p}}$. By Nakayama's lemma, $A_{\mathfrak{p}} \to B_{\mathfrak{p}}$ is surjective. It is injective since it is faithfully flat. So $A_{\mathfrak{p}} \cong B_{\mathfrak{p}}$ for any prime ideal $\mathfrak{p}$ of A. This implies that $A \cong B$. □

Corollary 2.3.10. *Let $f : X \to Y$ be a morphism.*

(i) *If f is unramified and separated and Y is connected, then there is a one-to-one correspondence between the set of sections of f and the set of connected components of X on which f induces an isomorphism. In particular, a section of f is completely determined by its value at one point of Y.*

(ii) *If f is etale, then there is a one-to-one correspondence between the set of sections of f and the set of open subsets of X on which f is radiciel and surjective.*

Proof. Suppose that $f : X \to Y$ is unramified and $s : Y \to X$ is a section of f. We have $fs = \mathrm{id}$. It follows that s must be an immersion and etale. By 2.3.9, s is an open immersion, and hence it induces an isomorphism between Y and the open subset $s(Y)$ of X. If f is separated and Y is connected, then s is an open and closed immersion and it induces an isomorphism between Y and a connected component of X. □

Lemma 2.3.11. *Let S be a scheme, S_0 a subscheme of S, and $X_0 \to S_0$ an etale morphism. For any $x \in X_0$, there exists an open neighborhood U_0 of x in X_0 such that U_0 is S_0-isomorphic to $U \times_S S_0$ for some etale S-scheme U.*

Proof. Let s be the image of x in S_0. By 2.3.5, there exists a monic polynomial $F_0(t) \in \mathscr{O}_{S_0,s}[t]$ and a maximal ideal $\mathfrak{n}_0$ of $\mathscr{O}_{S_0,s}[t]/(F_0(t))$ such that $\mathscr{O}_{X_0,x}$ is $\mathscr{O}_{S_0,s}$-isomorphic to $(\mathscr{O}_{S_0,s}[t]/(F_0(t)))_{\mathfrak{n}_0}$ and the image of $F_0'(t)$ in $\mathscr{O}_{S_0,y}[t]/(F_0(t))$ does not lie in $\mathfrak{n}_0$. Let $F(t) \in \mathscr{O}_{S,s}[t]$ be a monic polynomial whose image under the homomorphism $\mathscr{O}_{S,s}[t] \to \mathscr{O}_{S_0,s}[t]$ is $F_0(t)$, and let $\mathfrak{n}$ be the inverse image of $\mathfrak{n}_0$ under this homomorphism. Then

$$\operatorname{Spec} \mathscr{O}_{S,s}[t]/(F(t)) \to \operatorname{Spec} \mathscr{O}_{S,s}$$

is etale at $\mathfrak{n}$. We have

$$(\mathscr{O}_{S,s}[t]/(F(t)))_{\mathfrak{n}} \otimes_{\mathscr{O}_{S,s}} \mathscr{O}_{S_0,s} \cong (\mathscr{O}_{S_0,s}[t]/(F_0(t)))_{\mathfrak{n}_0}.$$

There exists an S-scheme U locally of finite presentation and a point $z \in U$ above s such that $\mathscr{O}_{U,z}$ is $\mathscr{O}_{S,s}$-isomorphic to $(\mathscr{O}_{S,s}[t]/(F(t)))_{\mathfrak{n}}$. We may assume that $U \to S$ is etale. We have an $\mathscr{O}_{S_0,s}$-isomorphism $\mathscr{O}_{U\times_S S_0,z} \cong \mathscr{O}_{X_0,x}$. It extends to an isomorphism between a neighborhood U_0 of x in X_0 and a neighborhood W_0 of z in $U \times_S S_0$. Shrinking U, we may assume $W_0 = U \times_S S_0$. $\square$

Proposition 2.3.12. *Let S be a scheme, and let S_0 be a closed subscheme of S with the same underlying topological space as S. The functor $X \mapsto X \times_S S_0$ from the category of etale S-schemes to the category of etale S_0-schemes is an equivalence of categories.*

Proof. Let X and Y be S-schemes. Suppose that Y is etale over S. Let us prove that the canonical map

$$\operatorname{Hom}_S(X, Y) \to \operatorname{Hom}_{S_0}(X \times_S S_0, Y \times_S S_0)$$

is bijective. An S-morphism $f : X \to Y$ is completely determined by its graph $\Gamma_f : X \to X \times_S Y$, and Γ_f is a section of the projection $p : X \times_S Y \to X$. Since S_0 is a closed subscheme of S with the same underlying topological space as S, by 2.3.10 (ii), there is a one-to-one correspondence between the set of sections of p and the set of sections of the base change p_0 of p by $S_0 \to S$. Our assertion follows.

Let $X_0 \to S_0$ be an etale morphism. By 2.3.11, there exists an open covering $\{U_{\alpha 0}\}$ of X_0 such that each $U_{\alpha 0}$ is S_0-isomorphic to $U_\alpha \times_S S_0$ for some etale S-scheme U_α. Note that each U_α has the same underlying topological space as $U_{\alpha 0}$. By the discussion above, for each pair α, β, the S-scheme structure on $U_{\alpha 0} \cap U_{\beta 0}$ induced from U_α is isomorphic to that induced from U_β. We can glue $\{U_\alpha\}$ together to get an S-scheme X so that $X \times_S S_0$ is S_0-isomorphic to X_0. □

2.4 Smooth Morphisms

([SGA 1] II 1–3, [EGA] IV 17.5.)

A morphism $f : X \to Y$ is called *smooth* at a point x in X if there exists an open neighborhood U of x such that $f|_U$ can be factorized as a composite of an etale morphism $U \to \mathbb{A}^n_Y$ and the projection $\mathbb{A}^n_Y \to Y$ for some integer n. Note that n is unique. It is equal to $\inf_V \dim (f^{-1}f(x) \cap V)$, where V goes over the family of open neighborhoods of x in X. We call $\inf_V \dim (f^{-1}f(x) \cap V)$ the *relative dimension* of f at x. If the relative dimension of f is the constant n on X, we say f is pure of relative dimension n. If f is smooth at every point, we say that f is smooth. Smooth morphisms are flat.

Proposition 2.4.1.
(i) *A morphism is etale if and only if it is smooth and locally quasi-finite.*
(ii) *Composites of smooth morphisms are smooth.*
(iii) *Base changes of smooth morphisms are smooth.*
(iv) *Let $f : X \to Y$ and $g : Y \to Z$ be morphisms such that g is unramified and gf is smooth. Then f is smooth.*

Proposition 2.4.2. *Let S be a noetherian scheme, X and Y two S-schemes locally of finite type, $f : X \to Y$ an S-morphism, x a point in X, s the image of x in S, $X_s = X \otimes_{\mathcal{O}_S} k(s)$, $Y_s = Y \otimes_{\mathcal{O}_S} k(s)$, and $f_s : X_s \to Y_s$ the morphism induced by f.*
(i) *f is quasi-finite (resp. unramified) at x if and only if f_s is quasi-finite (resp. unramified) at x.*
(ii) *Suppose that X and Y are flat over S. Then f is flat (resp. etale, resp. smooth) at x if and only if f_s is flat (resp. etale, resp. smooth) at x.*

Proof. (i) is clear since the property of a morphism being quasi-finite or unramified depends only on fibers of the morphism. The statement for the flat morphism follows from 1.3.7. The statement for the etale morphism follows from that for the flat morphism and for the unramified morphism. The statement for the smooth morphism follows from 2.4.3 below. □

Lemma 2.4.3. *Let $f: X \to Y$ be a morphism locally of finite type between noetherian schemes, x a point in X and $y = f(x)$. Then f is smooth at x if and only if f is flat at x and $f_y : X_y = X \otimes_{\mathscr{O}_Y} k(y) \to \mathrm{Spec}\, k(y)$ is smooth at x.*

Proof. The "only if" part is clear. Let us prove the "if" part. Shrinking X, we may assume there exists an etale $k(y)$-morphism

$$X_y \to \mathbb{A}^n_{k(y)} = \mathrm{Spec}\, k(y)[t_1, \ldots, t_n].$$

Let s_i $(i = 1, \ldots, n)$ be the images of t_i in $\Gamma(X_y, \mathscr{O}_{X_y})$. Shrinking X again, we may assume that s_i lie in the image of the canonical homomorphism $\Gamma(X, \mathscr{O}_X) \to \Gamma(X_y, \mathscr{O}_{X_y})$. Choose a preimage $s_i' \in \Gamma(X, \mathscr{O}_X)$ of s_i for each i. Consider the Y-morphism

$$X \to \mathbb{A}^n_Y = \mathbf{Spec}\, \mathscr{O}_Y[t_1, \ldots, t_n]$$

corresponding to the $\mathscr{O}_Y$-algebra morphism $\mathscr{O}_Y[t_1, \ldots, t_n] \to f_*\mathscr{O}_X$ mapping t_i to s_i'. By the statement for the etale morphism in 2.4.2 (ii), this morphism is etale at x. So $X \to Y$ is smooth at x. □

Proposition 2.4.4. *Let $f : X \to Y$ be a morphism locally of finite type between noetherian schemes, and let x be a point in X. Suppose f is smooth at x. Then $\mathscr{O}_{X,x}$ is reduced (resp. regular, resp. normal) if and only if $\mathscr{O}_{Y,f(x)}$ is so.*

Proof. The problem is local. We may assume there exists an etale Y-morphism $X \to \mathbb{A}^n_Y$. Let y' be the image of x in $\mathbb{A}^n_Y$. Note that $\mathscr{O}_{\mathbb{A}^n_Y, y'}$ is reduced (resp. regular, resp. normal) if and only if $\mathscr{O}_{Y,f(x)}$ is so. (Confer [Matsumura (1970)] (17.B), (17.I) Theorem 40, (21.D) Theorem 51.) We are thus reduced to the case where f is etale. The statement about regularity follows from [Matsumura (1970)] (21.D) Theorem 51. Recall that a noetherian ring is reduced (resp. normal) if it satisfies (R_0) and (S_1) (resp. (R_1) and (S_2)). (Confer [Matsumura (1970)] (17.I) Theorem 39.) The statements about being reduced and being normal follow from [Matsumura (1970)] (21.C) Corollary 2. □

Proposition 2.4.5. *Let S_0 be a quasi-compact quasi-separated scheme, $(\mathscr{A}_\lambda, \phi_{\lambda\mu})$ a direct system of quasi-coherent $\mathscr{O}_{S_0}$-algebras, $\mathscr{A} = \varinjlim_\lambda \mathscr{A}_\lambda$, $S_\lambda = \mathbf{Spec}\, \mathscr{A}_\lambda$, $S = \mathbf{Spec}\, \mathscr{A}$, X_0 an S_0-scheme of finite presentation, $X_\lambda = X_0 \times_{S_0} S_\lambda$, and $X = X_0 \times_{S_0} S$. If X is smooth over S, then X_λ is smooth over S_λ for some λ.*

Proof. The problem is local. We may assume there exists an etale S-morphism $X \to \mathbb{A}^n_S$. By 1.10.9 (i), this S-morphism is induced by an S_λ-morphism $X_\lambda \to \mathbb{A}^n_S$ for some λ. If λ is sufficiently large, $X_\lambda \to \mathbb{A}^n_S$ is etale by 2.3.7. Then X_λ is smooth over S_λ. □

2.5 Jacobian Criterion

([SGA 1] II 4, [EGA] IV 17.11–12.)

Proposition 2.5.1. *Let S be a scheme, X and Y two S-schemes, and $f : X \to Y$ an S-morphism locally of finite presentation. Then f is unramified if and only if the canonical morphism $f^*\Omega_{Y/S} \to \Omega_{X/S}$ is surjective.*

Proof. Use the exact sequence

$$f^*\Omega_{Y/S} \to \Omega_{X/S} \to \Omega_{X/Y} \to 0.$$

□

Lemma 2.5.2. *Let $A \to B$ be a local homomorphism of local noetherian rings, let M be a finitely generated B-module, and let t be an element in the maximal ideal of A which is not a zero divisor in A. Then M is flat over A if and only if the homomorphism*

$$t : M \to M, \quad x \mapsto tx$$

is injective and M/tM is flat over A/tA.

Proof. If M is flat over A, then M/tM is flat over A/tA. Since t is not a zero divisor, the multiplication by t is injective on A. So $t : M \to M$ is injective.

Conversely, from the short exact sequence

$$0 \to A \xrightarrow{t} A \to A/tA \to 0,$$

we get a long exact sequence

$$0 \to \mathrm{Tor}_1^A(M, A/tA) \to M \xrightarrow{t} M \to M/tM \to 0.$$

If $t : M \to M$ is injective, then we have $\mathrm{Tor}_1^A(M, A/tA) = 0$. Furthermore, if M/tM is flat over A/tA, then M is flat over A by 1.3.5. □

Proposition 2.5.3. *Let S be a scheme, X and Y two S-schemes, and $f : X \to Y$ an S-morphism locally of finite presentation. If f is etale, then the canonical morphism $f^*\Omega_{Y/S} \to \Omega_{X/S}$ is an isomorphism. The converse holds if X and Y are smooth over S.*

Proof. Suppose f is etale and let us prove $f^*\Omega_{Y/S} \cong \Omega_{X/S}$. The problem is local. We may assume that X and Y are separated over S. The diagonal morphisms $\Delta_{X/S}$, $\Delta_{Y/S}$ and $\Delta_{X/Y}$ are closed immersions. We have a commutative diagram

$$\begin{array}{ccccc} X & \stackrel{\Delta_{X/Y}}{\to} & X \times_Y X & \stackrel{q}{\to} & X \times_S X \\ & \searrow_f & \downarrow p & & \downarrow f\times f \\ & & Y & \stackrel{\Delta_{Y/S}}{\to} & Y \times_S Y, \end{array}$$

where the square is Cartesian and $q\Delta_{X/Y} = \Delta_{X/S}$. Let $\mathscr{I}$ be the $\mathscr{O}_{Y\times_S Y}$-ideal defining the closed immersion $\Delta_{Y/S}$. Since $f \times f$ is flat, the $\mathscr{O}_{X\times_S X}$-ideal defining the closed immersion q is $(f \times f)^*\mathscr{I}$. Since f is unramified, $\Delta_{X/Y}$ is an open immersion. It follows that

$$\begin{aligned} \Omega_{X/S} &\cong \Delta^*_{X/Y}q^*\left((f \times f)^*\mathscr{I}/(f \times f)^*\mathscr{I}^2\right) \cong \Delta^*_{X/Y}q^*(f \times f)^*(\mathscr{I}/\mathscr{I}^2) \\ &\cong f^*\Delta^*_{Y/S}(\mathscr{I}/\mathscr{I}^2) \cong f^*\Omega_{Y/S}. \end{aligned}$$

Suppose X and Y are smooth over S and $f^*\Omega_{Y/S} \cong \Omega_{X/S}$. Let us prove f is etale. The problem is local. We may reduce to the case where X, Y and S are affine. Using 1.10.2 (ii), 1.10.4 (ii) and 2.4.5, we can then reduce to the case where S is noetherian. By 2.5.1, f is unramified. We need to show that f is flat. By 2.4.2 (ii), it suffices to show $f_s : X_s \to Y_s$ is flat for any $s \in S$. So we may assume $S = \operatorname{Spec} k$ for some field k. By base change to an algebraic closure of k, we may assume that k is algebraically closed. Since X and Y are smooth over k, they are regular. Let x be a closed point of X, and let $y = f(x)$. We need to show $\mathscr{O}_{X,x}$ is flat over $\mathscr{O}_{Y,y}$. By 2.1.5, we have canonical isomorphisms

$$\begin{aligned} \mathfrak{m}_x/\mathfrak{m}_x^2 &\cong \Omega_{\mathscr{O}_{X,x}/k} \otimes_{\mathscr{O}_{X,x}} \mathscr{O}_{X,x}/\mathfrak{m}_x, \\ \mathfrak{m}_y/\mathfrak{m}_y^2 &\cong \Omega_{\mathscr{O}_{Y,y}/k} \otimes_{\mathscr{O}_{Y,y}} \mathscr{O}_{Y,y}/\mathfrak{m}_y. \end{aligned}$$

As $f^*\Omega_{Y/S} \cong \Omega_{X/S}$, we have $\mathfrak{m}_y/\mathfrak{m}_y^2 \cong \mathfrak{m}_x/\mathfrak{m}_x^2$. So if $t_1, \ldots, t_n \in \mathfrak{m}_y$ form a regular system of parameters of $\mathscr{O}_{Y,y}$, then their images in $\mathscr{O}_{X,x}$ form a regular system of parameters of $\mathscr{O}_{X,x}$. Since $\mathscr{O}_{X,x}/(t_1, \ldots, t_n)\mathscr{O}_{X,x}$ is flat over $\mathscr{O}_{Y,y}/(t_1, \ldots, t_n) \cong k$, by 2.5.2 and induction on n, $\mathscr{O}_{X,x}$ is flat over $\mathscr{O}_{Y,y}$. □

Proposition 2.5.4. *Let S be a scheme, X and Y two S-schemes, and $f : X \to Y$ a smooth S-morphism.*

(i) *$\Omega_{X/Y}$ is a locally free $\mathscr{O}_X$-module, and its rank is equal to the relative dimension of f.*

(ii) *The sequence*

$$0 \to f^*\Omega_{Y/S} \to \Omega_{X/S} \to \Omega_{X/Y} \to 0$$

is exact.

Proof. The problem is local. We may assume that $f : X \to Y$ is a composite

$$X \xrightarrow{g} \mathbb{A}_Y^n \xrightarrow{\pi} Y,$$

where g is etale, and π is the projection. By 2.5.3,we have

$$\Omega_{X/Y} \cong g^*\Omega_{\mathbb{A}_Y^n/Y}, \quad \Omega_{X/S} \cong g^*\Omega_{\mathbb{A}_Y^n/S}.$$

The sequence in (ii) can be identified with

$$0 \to g^*\pi^*\Omega_{Y/S} \to g^*\Omega_{\mathbb{A}_Y^n/S} \to g^*\Omega_{\mathbb{A}_Y^n/Y} \to 0.$$

This reduces the proof to the case where f is the morphism $\pi : \mathbb{A}_Y^n \to Y$. We can then apply 2.1.6. □

Corollary 2.5.5. *Let S be a scheme, X and Y two S-schemes locally of finite presentation, and $f : X \to Y$ an etale S-morphism. If X is etale over S, then $f(X)$ is etale over S.*

Proof. By 2.5.4, we have an exact sequence

$$0 \to f^*\Omega_{Y/S} \to \Omega_{X/S} \to \Omega_{X/Y} \to 0.$$

Since X is etale over S, we have $\Omega_{X/S} = 0$. So $f^*\Omega_{Y/S} = 0$. But f is flat. It follows that $\Omega_{Y/S}|_{f(X)} = 0$. Hence $f(X)$ is unramified over S. For any $x \in X$, let y and s be its images in Y and S, respectively. Then $\mathscr{O}_{X,x}$ is faithfully flat both over $\mathscr{O}_{Y,y}$ and over $\mathscr{O}_{S,s}$. This implies that $\mathscr{O}_{Y,y}$ is flat over $\mathscr{O}_{S,s}$. So $f(X)$ is etale over S. □

Lemma 2.5.6. *Let X be a scheme, $\mathscr{F}$ and $\mathscr{G}$ quasi-coherent $\mathscr{O}_X$-modules, x a point in X, and $u : \mathscr{F} \to \mathscr{G}$ a morphism. Suppose that $\mathscr{G}$ is free of finite rank in a neighborhood of x. The following conditions are equivalent:*

(i) *$\mathscr{F}$ has finite presentation in a neighborhood of x, $u_x : \mathscr{F}_x \to \mathscr{G}_x$ is injective and* $\operatorname{coker} u_x$ *is a free $\mathscr{O}_{X,x}$-module.*

(ii) *There exists a neighborhood U of x in X such that $u|_U$ induces an isomorphism between $\mathscr{F}|_U$ and a direct factor of $\mathscr{G}|_U$.*

(iii) *$\mathscr{F}$ has finite presentation in a neighborhood of x and u_x is universally injective, that is, for any $\mathscr{O}_{X,x}$-module M, $u_x \otimes \mathrm{id} : \mathscr{F}_x \otimes_{\mathscr{O}_{X,x}} M \to \mathscr{G}_x \otimes_{\mathscr{O}_{X,x}} M$ is injective.*

(iv) *$\mathscr{F}$ has finite presentation in a neighborhood of x and $u_x \otimes \mathrm{id} : \mathscr{F}_x \otimes_{\mathscr{O}_{X,x}} k(x) \to \mathscr{G}_x \otimes_{\mathscr{O}_{X,x}} k(x)$ is injective.*

(v) *$\mathscr{F}$ is free in a neighborhood of x, and the transpose $u_x^\vee : \mathscr{G}_x^\vee \to \mathscr{F}_x^\vee$ of u_x is surjective, where for any free $\mathscr{O}_{X,x}$-module M of finite rank, we define $M^\vee = \mathrm{Hom}_{\mathscr{O}_{X,x}}(M, \mathscr{O}_{X,x})$.*

Proof.

(i)⇒(ii) The short exact sequence

$$0 \to \mathscr{F}_x \stackrel{u_x}{\to} \mathscr{G}_x \to \mathrm{coker}\, u_x \to 0$$

is split. So there exists a homomorphism $v_x : \mathscr{G}_x \to \mathscr{F}_x$ such that $v_x u_x = \mathrm{id}$. Since $\mathscr{F}$ and $\mathscr{G}$ have finite presentation in a neighborhood of x, we can find an open neighborhood U of x in X and a morphism $v : \mathscr{G}|_U \to \mathscr{F}|_U$ extending v_x such that $v \circ (u|_U) = \mathrm{id}$. Then $u|_U$ induces an isomorphism between $\mathscr{F}|_U$ and a direct factor of $\mathscr{G}|_U$.

(ii)⇒(iii) and (iii)⇒(iv) are obvious.

(iv)⇒(v) Choose $s_1, \ldots, s_m \in \mathscr{F}_x$ so that their images in $\mathscr{F}_x \otimes_{\mathscr{O}_{X,x}} k(x)$ form a basis of $\mathscr{F}_x \otimes_{\mathscr{O}_{X,x}} k(x)$. By Nakayama's lemma, $s_1, \ldots, s_m$ generate $\mathscr{F}_x$ as an $\mathscr{O}_{X,x}$-module. Let $s'_1, \ldots, s'_m \in \mathscr{G}_x$ be the images of $s_1, \ldots, s_m$, respectively. The images of $s'_1, \ldots, s'_m$ in $\mathscr{G}_x \otimes_{\mathscr{O}_{X,x}} k(x)$ form a part of a basis of $\mathscr{G}_x \otimes_{\mathscr{O}_{X,x}} k(x)$ since $\mathscr{F}_x \otimes_{\mathscr{O}_{X,x}} k(x) \to \mathscr{G}_x \otimes_{\mathscr{O}_{X,x}} k(x)$ is injective. Choose $s'_{m+1}, \ldots, s'_n \in \mathscr{G}_x$ so that the images of $s'_1, \ldots, s'_n$ in $\mathscr{G}_x \otimes_{\mathscr{O}_{X,x}} k(x)$ form a basis of $\mathscr{G}_x \otimes_{\mathscr{O}_{X,x}} k(x)$. By Nakayama's lemma and the assumption that $\mathscr{G}_x$ is free, $s'_1, \ldots, s'_n$ form a basis of $\mathscr{G}_x$. This implies that $s_1, \ldots, s_m$ are linearly independent. Since $s_1, \ldots, s_m$ generate $\mathscr{F}_x$, they form a basis of $\mathscr{F}_x$. Hence $\mathscr{F}_x$ is a free $\mathscr{O}_{X,x}$-module. Since $\mathscr{F}$ is of finite presentation in a neighborhood of x, $\mathscr{F}$ is free in a neighborhood of x.

The homomorphism

$$(u_x \otimes \mathrm{id})^\vee : \mathrm{Hom}(\mathscr{G}_x \otimes_{\mathscr{O}_{X,x}} k(x), k(x)) \to \mathrm{Hom}(\mathscr{F}_x \otimes_{\mathscr{O}_{X,x}} k(x), k(x))$$

is surjective. So the homomorphism

$$u_x^\vee \otimes \mathrm{id} : \mathscr{G}_x^\vee \otimes_{\mathscr{O}_{X,x}} k(x) \to \mathscr{F}_x^\vee \otimes_{\mathscr{O}_{X,x}} k(x)$$

is surjective. By Nakayama's lemma, $u_x^\vee$ is surjective.

(v)⇒(i) Since $u_x^\vee$ is surjective, and $\mathscr{F}_x$ is a free $\mathscr{O}_{X,x}$-module, the map $\mathrm{Hom}(\mathscr{G}_x, \mathscr{F}_x) \to \mathrm{Hom}(\mathscr{F}_x, \mathscr{F}_x)$ induced by u_x is surjective. So there exists $v_x : \mathscr{G}_x \to \mathscr{F}_x$ such that $v_x u_x = \mathrm{id}$. Thus the sequence

$$0 \to \mathscr{F}_x \stackrel{u_x}{\to} \mathscr{G}_x \to \mathrm{coker}\, u_x \to 0$$

is exact and split. It follows that u_x is injective and $\mathrm{coker}\, u_x$ is a direct factor of $\mathscr{G}_x$. As $\mathscr{G}_x$ is a free $\mathscr{O}_{X,x}$-module, so is $\mathrm{coker}\, u_x$. □

Proposition 2.5.7. *Let S be a scheme, X and Y two S-schemes, $f : X \to Y$ an S-morphism, x a point in X, and y and s the images of x in Y and S, respectively. Suppose that Y is smooth over S at y. The following conditions are equivalent:*

(i) *f is smooth at x.*

(ii) *X is smooth over S at x and the morphism $f^*\Omega_{Y/S} \to \Omega_{X/S}$ satisfies the equivalent conditions in 2.5.6 at x.*

Proof.

(i)$\Rightarrow$(ii) follows from 2.4.1 (ii) and 2.5.4.

(ii)$\Rightarrow$(i) The cokernel of $f^*\Omega_{Y/S} \to \Omega_{X/S}$ is isomorphic to $\Omega_{X/Y}$. So $\Omega_{X/Y}$ is free in a neighborhood of x. Shrinking X and Y, we may assume X and Y are smooth over S and $\Omega_{X/Y}$ is a free $\mathscr{O}_X$-module with basis $dg_1, \ldots, dg_n$ for some $g_1, \ldots, g_n \in \Gamma(X, \mathscr{O}_X)$. Let $\pi : \mathbb{A}^n_Y \to Y$ be the projection, and let

$$g : X \to \mathbb{A}^n_Y = \mathbf{Spec}\, \mathscr{O}_Y[t_1, \ldots, t_n]$$

be the Y-morphism defined by the $\mathscr{O}_Y$-algebra morphism

$$\mathscr{O}_Y[t_1, \ldots, t_n] \to f_*\mathscr{O}_X, \quad t_i \mapsto g_i \quad (i = 1, \ldots, n).$$

By 2.5.4, the sequence

$$0 \to \pi^*\Omega_{Y/S} \to \Omega_{\mathbb{A}^n_Y/S} \to \Omega_{\mathbb{A}^n_Y/Y} \to 0$$

is exact. As $\Omega_{\mathbb{A}^n_Y/Y}$ is free, the above sequence is split locally, and hence the sequence

$$0 \to g^*\pi^*\Omega_{Y/S} \to g^*\Omega_{\mathbb{A}^n_Y/S} \to g^*\Omega_{\mathbb{A}^n_Y/Y} \to 0$$

is also exact. We have a commutative diagram

$$\begin{array}{ccccccc} 0 \to & g^*\pi^*\Omega_{Y/S} & \to & g^*\Omega_{\mathbb{A}^n_Y/S} & \to & g^*\Omega_{\mathbb{A}^n_Y/Y} & \to 0 \\ & \cong\downarrow & & \downarrow & & \downarrow & \\ 0 \to & f^*\Omega_{Y/S} & \to & \Omega_{X/S} & \to & \Omega_{X/Y} & \to 0. \end{array}$$

After shrinking X, the sequence on the second line is exact by our assumption. By the choice of g, the last vertical arrow $g^*\Omega_{\mathbb{A}^n_Y/Y} \to \Omega_{X/Y}$ is an isomorphism. By the five lemma, we have $g^*\Omega_{\mathbb{A}^n_Y/S} \cong \Omega_{X/S}$. By 2.5.3, g is etale. So f is smooth. $\square$

Theorem 2.5.8. *Let S be a scheme, $f : X \to S$ a smooth morphism, $i : Y \to X$ a closed immersion defined by a quasi-coherent $\mathscr{O}_X$-ideal $\mathscr{I}$ locally of finite type, x a point of Y, n and m the relative dimensions of X and Y respectively at x, and*

$$k : \mathbb{A}^m_S = \mathbf{Spec}\, \mathscr{O}_S[t_1, \ldots, t_m] \to \mathbb{A}^n_S = \mathbf{Spec}\, \mathscr{O}_S[t_1, \ldots, t_n]$$

the closed immersion defined by the $\mathscr{O}_S$-algebra epimorphism

$$\mathscr{O}_S[t_1,\ldots,t_n]\to\mathscr{O}_S[t_1,\ldots,t_m],\quad t_j\mapsto\begin{cases}t_j & \text{if } 1\leq j\leq m,\\ 0 & \text{if } m+1\leq j\leq n.\end{cases}$$

The following conditions are equivalent:

(i) *$fi:Y\to S$ is smooth at x.*

(ii) *There exists an open neighborhood U of x in X admitting an etale S-morphism $g:U\to\mathbb{A}^n_S=$ **Spec** $\mathscr{O}_S[t_1,\ldots,t_n]$ such that the closed immersion $i_U:Y\cap U\to U$ induced by i can be obtained from the closed immersion $k:\mathbb{A}^m_S\to\mathbb{A}^n_S$ by the base change g, that is, we have a Cartesian diagram*

$$\begin{array}{ccc}Y\cap U & \stackrel{i_U}{\to} & U\\ g'\downarrow & & \downarrow g\\ \mathbb{A}^m_S & \stackrel{k}{\to} & \mathbb{A}^n_S.\end{array}$$

(iii) *There exist generators $g_{m+1},\ldots,g_n$ of the ideal $\mathscr{I}_x$ of $\mathscr{O}_{X,x}$ such that $dg_{m+1},\ldots,dg_n$ form a part of a basis of $(\Omega_{X/S})_x$, or equivalently, the images of $dg_{m+1},\ldots,dg_n$ in $(\Omega_{X/S})_x\otimes_{\mathscr{O}_{X,x}}k(x)$ are linearly independent.*

(iv) *The canonical morphism*

$$\delta:i^*(\mathscr{I}/\mathscr{I}^2)\to i^*\Omega_{X/S}$$

satisfies the equivalent conditions in 2.5.6 at x.

Proof.

(i)⇒(ii) We have an exact sequence

$$i^*(\mathscr{I}/\mathscr{I}^2)\stackrel{\delta}{\to}i^*\Omega_{X/S}\to\Omega_{Y/S}\to 0.$$

Since X and Y are smooth over S at x, $i^*\Omega_{X/S}$ and $\Omega_{Y/S}$ are free in a neighborhood x. Choose $g_1,\ldots,g_m\in\mathscr{O}_{X,x}$ so that the images of $dg_1,\ldots,dg_m$ in $(\Omega_{Y/S})_x\otimes_{\mathscr{O}_{Y,x}}k(x)$ form a basis. Choose $g_{m+1},\ldots,g_n\in\mathscr{I}_x$ so that the images of $dg_1,\ldots,dg_n$ in $(\Omega_{X/S})_x\otimes_{\mathscr{O}_{X,x}}k(x)$ form a basis. Shrinking X, we may assume $g_{m+1},\ldots,g_n\in\Gamma(X,\mathscr{I})$ and $g_1,\ldots,g_n\in\Gamma(X,\mathscr{O}_X)$. The $\mathscr{O}_S$-algebra morphisms

$$\begin{aligned}\mathscr{O}_S[t_1,\ldots,t_n]&\to f_*\mathscr{O}_X, &\quad t_j\mapsto g_j\quad(1\leq j\leq n),\\ \mathscr{O}_S[t_1,\ldots,t_m]&\to(fi)_*\mathscr{O}_Y, &\quad t_j\mapsto g_j|_Y\quad(1\leq j\leq m),\end{aligned}$$

define S-morphisms $g:X\to\mathbb{A}^n_S$ and $g':Y\to\mathbb{A}^m_S$ such that the diagram

$$\begin{array}{ccc}Y & \stackrel{i}{\to} & X\\ g'\downarrow & & \downarrow g\\ \mathbb{A}^m_S & \stackrel{k}{\to} & \mathbb{A}^n_S\end{array}$$

commutes. Since $\Omega_{X/S}$ is free in a neighborhood of x and the images of $dg_1, \ldots, dg_n$ in $(\Omega_{X/S})_x \otimes_{\mathscr{O}_{X,x}} k(x)$ form a basis, the morphism $g : X \to \mathbb{A}^n_S$ is etale at x by 2.5.3. Similarly the morphism $g' : Y \to \mathbb{A}^m_S$ is also etale at x. Shrinking X, we may assume that these two morphisms are etale. The morphisms $i : Y \to X$ and $g' : Y \to \mathbb{A}^m_S$ induce a morphism $i' : Y \to X \times_{\mathbb{A}^n_S} \mathbb{A}^m_S$ such that $\pi_1 i' = i$ and $\pi_2 i' = g'$, where $\pi_1 : X \times_{\mathbb{A}^n_S} \mathbb{A}^m_S \to X$ and $\pi_2 : X \times_{\mathbb{A}^n_S} \mathbb{A}^m_S \to \mathbb{A}^m_S$ are the projections. Since π_1 and i are closed immersions, i' is a closed immersion. Since π_2 and g' are etale, i' is etale by 2.3.1 (iv). By 2.3.9, $i' : Y \to X \times_{\mathbb{A}^n_S} \mathbb{A}^m_S$ is an open and closed immersion. Shrinking X again, we may assume that it is an isomorphism.

(ii)$\Rightarrow$(i) is obvious.

(ii)$\Rightarrow$(iv) Let $\mathscr{J}$ be the ideal of $\mathscr{O}_S[t_1, \ldots, t_n]$ generated by $t_{m+1}, \ldots, t_n$. Since g is flat, we have $\mathscr{I}|_U \cong g^* \mathscr{J}$. Since g is etale, we have

$$g^*\Omega_{\mathbb{A}^n_S/S} \cong (\Omega_{X/S})|_U, \quad g'^*\Omega_{\mathbb{A}^m_S/S} \cong (\Omega_{Y/S})|_{Y\cap U}.$$

One can check that the following sequence is exact:

$$0 \to k^*(\mathscr{J}/\mathscr{J}^2) \to k^*\Omega_{\mathbb{A}^n_S/S} \to \Omega_{\mathbb{A}^m_S/S} \to 0.$$

Applying g'^* to this exact sequence, we get an exact sequence

$$0 \to (i^*(\mathscr{I}/\mathscr{I}^2))|_{Y\cap U} \stackrel{\delta}{\to} (i^*\Omega_{X/S})|_{Y\cap U} \to (\Omega_{Y/S})|_{Y\cap U} \to 0.$$

The conditions of 2.5.6 thus hold for $\delta : i^*(\mathscr{I}/\mathscr{I}^2) \to i^*\Omega_{X/S}$ at x.

(iv)$\Rightarrow$(iii) Let $g_{m+1}, \ldots, g_n \in \mathscr{I}_x$ so that their images in $\mathscr{I}_x/\mathscr{I}_x^2 \otimes_{\mathscr{O}_{X,x}} k(x)$ form a basis. By Nakayama's lemma, $g_{m+1}, \ldots, g_n$ generate $\mathscr{I}_x$. Since

$$\mathscr{I}_x/\mathscr{I}_x^2 \otimes_{\mathscr{O}_{X,x}} k(x) \to (\Omega_{X/S})_x \otimes_{\mathscr{O}_{X,x}} k(x)$$

is injective, the images of $dg_{m+1}, \ldots, dg_n$ in $(\Omega_{X/S})_x \otimes_{\mathscr{O}_{X,x}} k(x)$ form a basis.

(iii)$\Rightarrow$(ii) Since $\Omega_{X/S}$ is free in a neighborhood of x, we can find $g_1, \ldots, g_m \in \mathscr{O}_{X,x}$ such that $dg_1, \ldots, dg_n$ form a basis for $(\Omega_{X/S})_x$. Shrinking X, we may assume that $g_1, \ldots, g_n \in \Gamma(X, \mathscr{O}_X)$, $g_{m+1}, \ldots, g_n \in \Gamma(X, \mathscr{I})$, $\Omega_{X/S}$ is free with basis $dg_1, \ldots, dg_n$, and $g_{m+1}, \ldots, g_n$ generate $\mathscr{I}$. The family $g_1, \ldots, g_n$ defines an S-morphism $g : X \to \mathbb{A}^n_S$ which is etale by 2.5.3, and $i : Y \to X$ is induced from $k : \mathbb{A}^m_S \to \mathbb{A}^n_S$ by the base change $g : X \to \mathbb{A}^n_S$. □

Corollary 2.5.9 (Jacobian Criterion). *Let A be a ring, I a finitely generated ideal of $A[t_1, \ldots, t_n]$, and $\mathfrak{p}$ a prime ideal of $A[t_1, \ldots, t_n]$ containing I. If there exist polynomials $g_{m+1}, \ldots, g_n \in A[t_1, \ldots, t_n]$ $(0 \leq m \leq n)$*

such that their images in the ideal $I_{\mathfrak{p}}$ *generate* $I_{\mathfrak{p}}$ *and there exists a subset* $\{i_{m+1},\ldots,i_n\}$ *of* $\{1,\ldots,n\}$ *such that*

$$\det\begin{pmatrix} \frac{\partial g_{m+1}}{\partial t_{i_{m+1}}} & \cdots & \frac{\partial g_{m+1}}{\partial t_{i_n}} \\ \vdots & & \vdots \\ \frac{\partial g_n}{\partial t_{i_{m+1}}} & \cdots & \frac{\partial g_n}{\partial t_{i_n}} \end{pmatrix} \notin \mathfrak{p},$$

then $\mathrm{Spec}\,(A[t_1,\ldots,t_n]/I) \to \mathrm{Spec}\,A$ *is smooth at* $\mathfrak{p}/I$*. Furthermore, if* $m=0$*, then* $\mathrm{Spec}\,(A[t_1,\ldots,t_n]/I) \to \mathrm{Spec}\,A$ *is etale at* $\mathfrak{p}/I$*.*

Proof. Use the equivalence of (i) and (iii) in 2.5.8. □

Proposition 2.5.10. *Consider a Cartesian diagram*

$$\begin{array}{ccc} X' = X\times_Y Y' & \stackrel{g'}{\to} & X \\ f'\downarrow & & \downarrow f \\ Y' & \stackrel{g}{\to} & Y. \end{array}$$

Suppose that $f: X\to Y$ *is locally of finite presentation and* $g: Y'\to Y$ *is faithfully flat. If* f' *is unramified (resp. etale, resp. smooth), then so is* f*.*

Proof. The statement for unramified morphisms is 2.2.5. Suppose f' is smooth. Let us prove f is smooth. The problem is local. We may assume that $X=\mathrm{Spec}\,B$ and $Y=\mathrm{Spec}\,A$ are affine, and $B\cong A[t_1,\ldots,t_n]/I$ for some finitely generated ideal I of $A[t_1,\ldots,t_n]$. We have a closed immersion $i: X\to \mathbb{A}^n_Y$ with ideal sheaf $I^\sim$. Since g is flat, we have a closed immersion $i': X'\to \mathbb{A}^n_{Y'}$ with ideal sheaf $g''^*I^\sim$, where $g'': \mathbb{A}^n_{Y'}\to\mathbb{A}^n_Y$ is obtained from $g: Y'\to Y$ by base change. Since f' is smooth, by 2.5.8 (iv), we have an exact sequence

$$0\to i'^*g''^*(I/I^2)^\sim \to i'^*\Omega_{\mathbb{A}^n_{Y'}/Y'}\to \Omega_{X'/Y'}\to 0$$

and $\Omega_{X'/Y'}$ is locally free. Since g is faithfully flat, the sequence

$$0\to i^*(I/I^2)^\sim \to i^*\Omega_{\mathbb{A}^n_Y/Y}\to\Omega_{X/Y}\to 0$$

is exact. Since f is locally of finite presentation, $\Omega_{X/Y}$ is an $\mathscr{O}_X$-module locally of finite presentation. For any $x\in X$, let $x'\in X'$ be a preimage of x. Applying 1.6.6 to the quasi-compact faithfully flat morphism

$$\mathrm{Spec}\,\mathscr{O}_{X',x'}\to \mathrm{Spec}\,\mathscr{O}_{X,x}$$

and using the fact that $(\Omega_{X'/Y'})_{x'}$ is a free $\mathscr{O}_{X',x'}$-module of finite rank, we see that $(\Omega_{X/Y})_x$ is a free $\mathscr{O}_{X,x}$-module of finite rank. So $\Omega_{X/Y}$ is locally free. By 2.5.8, f is smooth.

The statement about etale morphisms follows from the statements about unramified morphisms and smooth morphisms since a morphism is etale if and only if it is unramified and smooth. □

2.6 Infinitesimal Liftings of Morphisms

([SGA 1] III, [EGA] IV 17.14.)

Lemma 2.6.1. *Let B be a ring, A and C two B-algebras, I an ideal of A with the property $I^2 = 0$, $p : A \to A/I$ the projection, and $\phi : C \to A$ a B-homomorphism. Then there exists a one-to-one correspondence between the set*

$$\{\phi' : C \to A \,|\, \phi' \text{ is a } B\text{-homomorphism satisfying } p\phi' = p\phi\}$$

and the set $\mathrm{Hom}_{A/I}(\Omega_{C/B} \otimes_C A/I, I)$, where A/I is regarded as a C-algebra through the homomorphism $p\phi : C \to A/I$.

$$\begin{array}{ccc} A/I & \stackrel{p\phi}{\leftarrow} & C \\ {\scriptstyle p}\uparrow & \swarrow {\scriptstyle \phi'} & \uparrow \\ A & \leftarrow & B \end{array}$$

Proof. Since $I^2 = 0$, I can be regarded as an A/I-module and hence a C-module. Let $\phi' : C \to A$ be a B-homomorphism with the property $p\phi' = p\phi$. Then the image of $\phi' - \phi$ is contained in I. We claim that $\phi' - \phi : C \to I$ is a B-derivation. Indeed, since ϕ and ϕ' are B-homomorphisms, the restriction of $\phi' - \phi$ to the image of B in C is 0. For any $x_1, x_2 \in C$, we have

$$\begin{aligned}(\phi' - \phi)(x_1x_2) &= \phi'(x_1)\phi'(x_2) - \phi(x_1)\phi(x_2) \\ &= \phi(x_1)(\phi'(x_2) - \phi(x_2)) + \phi'(x_2)(\phi'(x_1) - \phi(x_1)) \\ &= x_1 \cdot (\phi'(x_2) - \phi(x_2)) + x_2 \cdot (\phi'(x_1) - \phi(x_1)).\end{aligned}$$

This proves our claim. Conversely, if $D : C \to I$ is a B-derivation, then one can verify $\phi + D : C \to A$ is a B-homomorphism with the property $p(\phi + C) = p\phi$. Our assertion then follows from the fact that the set of B-derivations $\mathrm{Der}_B(C, I)$ is in one-to-one correspondence with the set $\mathrm{Hom}_C(\Omega_{C/B}, I) \cong \mathrm{Hom}_{A/I}(\Omega_{C/B} \otimes_C A/I, I)$. □

Theorem 2.6.2. *Let $f : X \to S$ be a morphism locally of finite presentation. The following conditions are equivalent:*

(i) *f is unramified (resp. etale, resp. smooth).*

(ii) *For any S-scheme* Spec A *and any ideal I of A with the property $I^2 = 0$, the canonical map*

$$\mathrm{Hom}_S(\mathrm{Spec}\, A, X) \to \mathrm{Hom}_S(\mathrm{Spec}\, A/I, X)$$

is injective (resp. bijective, resp. surjective).

Proof.

(i)⇒(ii) We have one-to-one correspondences

$$\mathrm{Hom}_S(\mathrm{Spec}\,A, X) \cong \mathrm{Hom}_{\mathrm{Spec}\,A}(\mathrm{Spec}\,A, \mathrm{Spec}\,A \times_S X),$$
$$\mathrm{Hom}_S(\mathrm{Spec}\,A/I, X) \cong \mathrm{Hom}_{\mathrm{Spec}\,A}(\mathrm{Spec}\,A/I, \mathrm{Spec}\,A \times_S X).$$

It suffices to show that

$$\mathrm{Hom}_{\mathrm{Spec}\,A}(\mathrm{Spec}\,A, \mathrm{Spec}\,A \times_S X) \to \mathrm{Hom}_{\mathrm{Spec}\,A}(\mathrm{Spec}\,A/I, \mathrm{Spec}\,A \times_S X)$$

is injective (resp. bijective, resp. surjective). Making the base change $\mathrm{Spec}\,A \to S$, we are reduced to prove that if $f : X \to \mathrm{Spec}\,A$ is unramified (resp. etale, resp. smooth), then the canonical map

$$\mathrm{Hom}_{\mathrm{Spec}\,A}(\mathrm{Spec}\,A, X) \to \mathrm{Hom}_{\mathrm{Spec}\,A}(\mathrm{Spec}\,A/I, X)$$

is injective (resp. bijective, resp. surjective).

Suppose $f : X \to \mathrm{Spec}\,A$ is unramified. An element g in $\mathrm{Hom}_{\mathrm{Spec}\,A}(\mathrm{Spec}\,A, X)$ is a section of f and hence is an open immersion. (Confer the proof of 2.3.10.) We have $g = (f|_{\mathrm{im}\,g})^{-1}$. So g is completely determined by $\mathrm{im}\,g$. Let g' be the composite

$$\mathrm{Spec}\,A/I \to \mathrm{Spec}\,A \xrightarrow{g} X.$$

Since $I^2 = 0$, $\mathrm{Spec}\,A/I$ and $\mathrm{Spec}\,A$ have the same underlying topological space. So g and g' have the same image in X. It follows that g is completely determined by g', and the map

$$\mathrm{Hom}_{\mathrm{Spec}\,A}(\mathrm{Spec}\,A, X) \to \mathrm{Hom}_{\mathrm{Spec}\,A}(\mathrm{Spec}\,A/I, X)$$

is injective.

The statement for etale morphisms follows from 2.3.12.

Suppose $f : X \to \mathrm{Spec}\,A$ is smooth. Let $\{U_\alpha\}$ be a covering of X by affine open subsets such that we have etale A-morphisms $g_\alpha : U_\alpha \to \mathbb{A}_A^{n_\alpha}$. Let $i : \mathrm{Spec}\,A/I \to \mathrm{Spec}\,A$ be the closed immersion. Given an A-morphism $g' : \mathrm{Spec}\,A/I \to X$,

$$\begin{array}{ccc} \mathrm{Spec}\,A/I & \xrightarrow{g'} & X \\ {\scriptstyle i}\downarrow & & \downarrow \\ \mathrm{Spec}\,A & = & \mathrm{Spec}\,A, \end{array}$$

regard each $g'^{-1}(U_\alpha)$ as an open subset of $\mathrm{Spec}\,A$, and cover it by affine open subschemes $V_{\alpha\beta}$ of $\mathrm{Spec}\,A$. Let $g'_{\alpha\beta} : V_{\alpha\beta} \otimes_A A/I \to U_\alpha$ be the morphisms induced by g', and let $i : V_{\alpha\beta} \otimes_A A/I \to V_{\alpha\beta}$ be the closed immersions. We can construct A-morphisms $h_{\alpha\beta} : V_{\alpha\beta} \to \mathbb{A}_A^n$ such that $h_{\alpha\beta} i = g_\alpha g'_{\alpha\beta}$.

Since g_α is etale, by the etale case that we have just treated, there exist morphisms $g_{\alpha\beta} : V_{\alpha\beta} \to U_\alpha$ such that the following diagram commutes:

$$\begin{array}{ccc} V_{\alpha\beta} \otimes_A A/I & \stackrel{g'_{\alpha\beta}}{\to} & U_\alpha \\ i \downarrow & \stackrel{g_{\alpha\beta}}{\nearrow} & \downarrow g_\alpha \\ V_{\alpha\beta} & \stackrel{h_{\alpha\beta}}{\to} & \mathbb{A}_A^{n_\alpha}. \end{array}$$

For convenience, we change our notation on indices: We have shown that there exist a covering $\{U_\lambda\}$ (resp. $\{V_\lambda\}$) of X (resp. Spec A) by affine open subsets and A-morphisms $g_\lambda : V_\lambda \to U_\lambda$ such that $g'(V_\lambda \otimes_A A/I) \subset U_\lambda$, and $g_\lambda i = g'_\lambda$, where $i : V_\lambda \otimes_A A/I \to V_\lambda$ are the closed immersions, and $g'_\lambda : V_\lambda \otimes_A A/I \to U_\lambda$ the morphisms induced by g'. For each pair (λ, μ) of indices, consider the commutative diagram

$$\begin{array}{ccc} (V_\lambda \cap V_\mu) \otimes_A A/I & \stackrel{g'_{\lambda\mu}}{\to} & U_\lambda \cap U_\mu \\ i \downarrow & \stackrel{g_{\lambda\mu}}{\nearrow} & \downarrow f \\ V_\lambda \cap V_\mu & \to & \mathrm{Spec}\, A, \end{array}$$

where

$$g'_{\lambda\mu} = g'_\lambda|_{(V_\lambda \cap V_\mu)\otimes_A A/I} = g'_\mu|_{(V_\lambda \cap V_\mu)\otimes_A A/I}, \quad g_{\lambda\mu} = g_\lambda|_{V_\lambda \cap V_\mu}.$$

If we replace $g_{\lambda\mu}$ in this diagram by $g_{\mu\lambda} = g_\mu|_{V_\lambda \cap V_\mu}$, the diagram still commutes. By 2.6.1, $g_{\mu\lambda}$ corresponds to an element

$$s_{\lambda\mu} \in \Gamma((V_\lambda \cap V_\mu) \otimes_A A/I, \mathscr{H}om_{\mathscr{O}_{\mathrm{Spec}\, A/I}}(g'^*\Omega_{X/\mathrm{Spec}\, A}, I^\sim)).$$

One can check that on $(V_\lambda \cap V_\mu \cap V_\nu) \otimes_A A/I$, we have

$$s_{\lambda\mu} + s_{\mu\nu} + s_{\nu\lambda} = 0.$$

Since Spec A/I is affine, we have

$$\check{H}^1(\mathrm{Spec}\, A/I, \mathscr{H}om_{\mathscr{O}_{\mathrm{Spec}\, A/I}}(g'^*\Omega_{X/\mathrm{Spec}\, A}, I^\sim)) = 0.$$

Replacing $\{V_\lambda\}$ by a refinement, we may assume there exist

$$s_\lambda \in \Gamma(V_\lambda \otimes_A A/I, \mathscr{H}om_{\mathscr{O}_{\mathrm{Spec}\, A/I}}(g'^*\Omega_{X/\mathrm{Spec}\, A}, I^\sim))$$

such that $s_{\lambda\mu} = s_\lambda - s_\mu$. Consider the commutative diagram

$$\begin{array}{ccc} V_\lambda \otimes_A A/I & \stackrel{g'_\lambda}{\to} & U_\lambda \\ i \downarrow & \stackrel{g_\lambda}{\nearrow} & \downarrow f \\ V_\lambda & \to & \mathrm{Spec}\, A. \end{array}$$

By 2.6.1, each s_λ corresponds to an A-morphism $\tilde{g}_\lambda : V_\lambda \to U_\lambda$ such that $\tilde{g}_\lambda i = g'_\lambda$. One can check that $\tilde{g}_\lambda$ and $\tilde{g}_\mu$ coincide on $V_\lambda \cap V_\mu$. So we can glue $\tilde{g}_\lambda$ together to get an A-morphism $g : \operatorname{Spec} A \to X$ such that $gi = g'$. This proves that the map

$$\mathrm{Hom}_{\operatorname{Spec} A}(\operatorname{Spec} A, X) \to \mathrm{Hom}_{\operatorname{Spec} A}(\operatorname{Spec} A/I, X)$$

is surjective.

(ii)⇒(i) The problem is local. We may assume that $X = \operatorname{Spec} C$ and $S = \operatorname{Spec} B$ are affine. Suppose that for any B-algebra A and any ideal I of A satisfying $I^2 = 0$, the canonical map

$$\mathrm{Hom}_{\operatorname{Spec} B}(\operatorname{Spec} A, \operatorname{Spec} C) \to \mathrm{Hom}_{\operatorname{Spec} B}(\operatorname{Spec} A/I, \operatorname{Spec} C)$$

is injective. Then by 2.6.1, we have $\mathrm{Hom}_{A/I}(\Omega_{C/B} \otimes_C A/I, I) = 0$ whenever there exists a B-homomorphism from C to A. Take $A = (C \otimes_B C)/J^2$ and $I = J/J^2$, where J is the kernel of the epimorphism

$$C \otimes_B C \to C, \quad x \otimes y \mapsto xy.$$

We have a B-algebra homomorphism

$$C \to A, \quad x \mapsto x \otimes 1 + J^2.$$

So $\mathrm{Hom}_{A/I}(\Omega_{C/B} \otimes_C A/I, I) = 0$, that is, $\mathrm{Hom}_C(\Omega_{C/B}, \Omega_{C/B}) = 0$. Hence $\Omega_{C/B} = 0$ and f is unramified.

Suppose for any B-algebra A and any ideal I of A satisfying $I^2 = 0$, the canonical map

$$\mathrm{Hom}_{\operatorname{Spec} B}(\operatorname{Spec} A, \operatorname{Spec} C) \to \mathrm{Hom}_{\operatorname{Spec} B}(\operatorname{Spec} A/I, \operatorname{Spec} C)$$

is surjective. We may assume $C = B[t_1, \ldots, t_n]/J$ for some finitely generated ideal J of $B[t_1, \ldots, t_n]$. Taking $A = B[t_1, \ldots, t_n]/J^2$ and $I = J/J^2$, we see that the map

$$\begin{aligned} &\mathrm{Hom}_B(B[t_1, \ldots, t_n]/J, B[t_1, \ldots, t_n]/J^2) \\ &\to \mathrm{Hom}_B(B[t_1, \ldots, t_n]/J, B[t_1, \ldots, t_n]/J) \end{aligned}$$

is surjective. So there exists a B-homomorphism

$$\phi : B[t_1, \ldots, t_n]/J \to B[t_1, \ldots, t_n]/J^2$$

which is a section of the projection

$$p : B[t_1, \ldots, t_n]/J^2 \to B[t_1, \ldots, t_n]/J.$$

Consider $\mathrm{id} - \phi p$. As

$$p(\mathrm{id} - \phi p) = p - (p\phi)p = p - p = 0,$$

the image of $\mathrm{id} - \phi p$ lies in J/J^2. We claim that

$$\mathrm{id} - \phi p : B[t_1, \ldots, t_n]/J^2 \to J/J^2$$

is a B-derivation. Indeed, we have $(\mathrm{id} - \phi p)(b) = 0$ for any $b \in B$ since ϕ is a B-homomorphism. For any $x_1, x_2 \in B[t_1, \ldots, t_n]/J^2$, we have

$$\begin{aligned}
&(\mathrm{id} - \phi p)(x_1 x_2) \\
&\quad = x_1 x_2 - \phi p(x_1)\phi p(x_2) \\
&\quad = x_1(x_2 - \phi p(x_2)) + (x_1 - \phi p(x_1))\phi p(x_2) \\
&\quad = x_1(x_2 - \phi p(x_2)) + x_2(x_1 - \phi p(x_1)) - (x_1 - \phi p(x_1))(x_2 - \phi p(x_2)) \\
&\quad = x_1(x_2 - \phi p(x_2)) + x_2(x_1 - \phi p(x_1)).
\end{aligned}$$

This proves our claim. Let

$$q : B[t_1, \ldots, t_n] \to B[t_1, \ldots, t_n]/J^2$$

be the projection. Then

$$(\mathrm{id} - \phi p)q : B[t_1, \ldots, t_n] \to J/J^2$$

is a B-derivation. It corresponds to a $(B[t_1, \ldots, t_n]/J)$-module homomorphism

$$\psi : \Omega_{B[t_1,\ldots,t_n]/B} \otimes_{B[t_1,\ldots,t_n]} (B[t_1, \ldots, t_n]/J) \to J/J^2.$$

We claim that ψ is a left inverse of the canonical homomorphism

$$\delta : J/J^2 \to \Omega_{B[t_1,\ldots,t_n]/B} \otimes_{B[t_1,\ldots,t_n]} (B[t_1, \ldots, t_n]/J).$$

Indeed, for any $x \in J$, we have

$$\psi\delta(x + J^2) = \psi(dx \otimes 1) = (\mathrm{id} - \phi p)q(x) = x$$

since $pq|_J = 0$. This proves our claim. So the sequence

$$\begin{aligned}
0 \to J/J^2 \to \Omega_{B[t_1,\ldots,t_n]/B} &\otimes_{B[t_1,\ldots,t_n]} (B[t_1, \ldots, t_n]/J) \\
&\to \Omega_{(B[t_1,\ldots,t_n]/J)/B} \to 0
\end{aligned}$$

is split exact. By 2.5.8 (iv), $\operatorname{Spec} B[t_1, \ldots, t_n]/J \to \operatorname{Spec} B$ is smooth.

The statement for the etale morphism follows from that for the unramified and smooth morphism. □

2.7 Direct Limits and Inverse Limits

Let I and $\mathscr{C}$ be two categories, and let $F : I \to \mathscr{C}$ be a covariant (resp. contravariant) functor. We often write F as $(F(i))_{i \in \mathrm{ob}\, I}$, or simply $(F(i))$. For every object X in $\mathscr{C}$, let $C_X : I \to \mathscr{C}$ be the constant functor that maps each object in I to X, and each morphism in I to id_X. Consider the covariant (resp. contravariant) functor

$$\mathscr{C} \to (\mathbf{Sets}), \ \ X \mapsto \mathrm{Hom}(F, C_X) \ \ (\text{resp. } \mathscr{C} \to (\mathbf{Sets}), \ \ X \mapsto \mathrm{Hom}(C_X, F))$$

from $\mathscr{C}$ to the category of sets. If this functor is representable, we call the object in $\mathscr{C}$ representing this functor the *direct limit* (resp. *inverse limit*) of F, and denote it by $\varinjlim_{i\in \mathrm{ob}\, I} F(i)$ (resp. $\varprojlim_{i\in \mathrm{ob}\, I} F(i)$). We can also define $\varinjlim_{i\in \mathrm{ob}\, I} F(i)$ (resp. $\varprojlim_{i\in \mathrm{ob}\, I} F(i)$) as an object in $\mathscr{C}$ together with morphisms

$$\phi_i : F(i) \to \varinjlim_{i\in \mathrm{ob}\, I} F(i) \quad (\text{resp. } \phi_i : \varprojlim_{i\in \mathrm{ob}\, I} F(i) \to F(i))$$

with the property

$$\phi_j \circ F(f_{ij}) = \phi_i \quad (\text{resp. } F(f_{ij}) \circ \phi_j = \phi_i)$$

whenever there exists a morphism $f_{ij} : i \to j$ in I. Moreover, these data satisfy the following universal property: For any object X in $\mathscr{C}$ together with morphisms $\psi_i : F(i) \to X$ (resp. $\psi_i : X \to F(i)$) with the property $\psi_j \circ F(f_{ij}) = \psi_i$ (resp. $F(f_{ij}) \circ \psi_j = \psi_i$) whenever there exists a morphism $f_{ij} : i \to j$ in I, there exists a unique morphism $\psi : \varinjlim_{i\in \mathrm{ob}\, I} F(i) \to X$ (resp. $\psi : X \to \varprojlim_{i\in \mathrm{ob}\, I} F(i)$) in $\mathscr{C}$ such that $\psi\phi_i = \psi_i$ (resp. $\phi_i\psi = \psi_i$).

In the following, $\mathscr{C}$ is the category of sets, or the category of groups.

Suppose objects in I form a set. Then for any contravariant functor $F : I \to \mathscr{C}$, $\varprojlim_i F(i)$ exists. In fact, $\varprojlim_i F(i)$ is the subspace of $\prod_{i\in \mathrm{ob}\, I} F(i)$ consisting of those elements $(x_i) \in \prod_{i\in \mathrm{ob}\, I} F(i)$ such that $F(f_{ij})(x_j) = x_i$ whenever there exists a morphism $f_{ij} : i \to j$. The canonical morphisms $\phi_i : \varprojlim_i F(i) \to F(i)$ are induced by the projections $\prod_i F(i) \to F(i)$.

Consider the following conditions of the category I:

(I1) Given two morphisms $i \to j$ and $i \to j'$ in I, there exist two morphisms $j \to k$ and $j' \to k$ such that the following diagram commutes:

$$\begin{array}{ccc} i & \to & j \\ \downarrow & & \downarrow \\ j' & \to & k. \end{array}$$

(I2) Given two morphisms $i \rightrightarrows j$ in I, there exists a morphism $j \to k$ so that its composite with these two morphisms are the same.

(I3) Given two objects i and j in I, there exists an object k in I admitting morphisms $i \to k$ and $j \to k$.

Note that (I2) + (I3) $\Rightarrow$ (I1).

Suppose objects in I form a set and I satisfies (I2) and (I3). For any covariant functor $F: I \to \mathscr{C}$, $\varinjlim_i F(i)$ exists. It can be defined as the set (or group)

$$\Big(\coprod_{i\in I} F(i)\Big)/R,$$

where R is the equivalence relation on $\coprod_{i\in I} F(i)$ defined as follows: For any $x \in F(i)$ and $y \in F(j)$, we say $x \sim y$ if there exist morphisms $i \to k$ and $j \to k$ in I such that x and y have the same image in $F(k)$. In the case where $\mathscr{C}$ is the category of groups, the group structure on $\Big(\coprod_{i\in I} F(i)\Big)/R$ is defined as follows: For any $x \in F(i)$ and $y \in F(j)$, let $i \to k$ and $j \to k$ be morphisms in I. Then xy is defined to be the product of the images of x and y in $F(k)$, and passed to the quotient. The canonical morphisms $F(i) \to \varinjlim_i F(i)$ are induced by the inclusions $F(i) \to \coprod_{i\in I} F(i)$. Note that $\varinjlim_I$ is an exact functor, that is, if $F \to G \to H$ is a sequence of functors such that

$$F(i) \to G(i) \to H(i)$$

are exact for all i, then

$$\varinjlim_i F(i) \to \varinjlim_i G(i) \to \varinjlim_i H(i)$$

is exact.

Suppose objects in I form a set and I satisfies (I1) and (I2). Define an equivalence relation on I as follows: We say that $i \sim j$ if there exists a sequence of objects

$$i = i_0, i_1, \ldots, i_n = j$$

in I such that for each $0 \leq m \leq n-1$, there exists a morphism $i_m \to i_{m+1}$ or a morphism $i_{m+1} \to i_m$. Let I_λ $(\lambda \in \Lambda)$ be the equivalence classes of I with respect to this relation. Then each I_λ satisfies (I2) and (I3). For any covariant functor $F: I \to \mathscr{C}$, $\varinjlim_{i\in \mathrm{ob}\, I} F(i)$ exists, and

$$\varinjlim_{i\in \mathrm{ob}\, I} F(i) \cong \begin{cases} \coprod_{\lambda\in\Lambda} \big(\varinjlim_{i\in \mathrm{ob}\, I_\lambda} F(i)\big) & \text{if } \mathscr{C} = (\textbf{Sets}), \\ \bigoplus_{\lambda\in\Lambda} \big(\varinjlim_{i\in \mathrm{ob}\, I_\lambda} F(i)\big) & \text{if } \mathscr{C} = (\textbf{Groups}). \end{cases}$$

$\varinjlim_I$ is an exact functor.

Assume that objects in I form a set. For any covariant functor $F: I \to \mathscr{C}$, $\varinjlim_{i\in \mathrm{ob}\, I} F(i)$ exists, and

$$\varinjlim_{i\in \mathrm{ob}\, I} F(i) \cong \begin{cases} \Big(\coprod_{i\in \mathrm{ob}\, I} F(i)\Big)/R & \text{if } \mathscr{C} = (\textbf{Sets}), \\ \Big(\bigoplus_{i\in \mathrm{ob}\, I} F(i)\Big)/R' & \text{if } \mathscr{C} = (\textbf{Groups}), \end{cases}$$

where R (resp. R') is the equivalence relation generated by $x \sim y$ (resp. the subgroup generated by $x - y$), where $x \in F(i)$, $Y \in F(j)$, and there exists a morphism $i \to j$ such that the image of x in $F(j)$ is y.

Let J be a full subcategory of I. We say that J is *cofinal* in I if for any object i in I, there exists a morphism $i \to j$ in I such that j is an object in J. If I satisfies (I1) (resp. (I2), resp. (I3)) and J is cofinal in I, then J satisfies the same condition. Suppose I satisfies (I1) and J is cofinal in I, then for any covariant (resp. contravariant) functor $F : I \to \mathscr{C}$ and any object X in $\mathscr{C}$, the canonical map

$$\mathrm{Hom}(F, C_X) \to \mathrm{Hom}(F|_J, C_X|_J) \quad (\text{resp. } \mathrm{Hom}(C_X, F) \to \mathrm{Hom}(C_X|_J, F|_J))$$

is bijective. So we have

$$\varinjlim_{i\in \mathrm{ob}\, I} F(i) \cong \varinjlim_{j\in \mathrm{ob}\, J} F(j) \quad (\text{resp. } \varprojlim_{i\in \mathrm{ob}\, I} F(i) \cong \varprojlim_{j\in \mathrm{ob}\, J} F(j)).$$

If I satisfies (I1) and has a cofinal full subcategory J whose objects form a set, we can apply our previous descriptions of $\varprojlim_{j\in \mathrm{ob}\, J} F(j)$ and $\varinjlim_{j\in \mathrm{ob}\, J} F(j)$ to get descriptions of $\varprojlim_{i\in \mathrm{ob}\, I} F(i)$ and $\varinjlim_{i\in \mathrm{ob}\, I} F(i)$. An object j in I is called a *final object* if for any object i in I, there exists one and only one morphism $i \to j$. In this case, the category J consisting of only one object j and only one morphism id_j is a cofinal full subcategory of I, and we have

$$F(j) \cong \varinjlim_{i\in \mathrm{ob}\, I} F(i) \quad (\text{resp. } \varprojlim_{i\in \mathrm{ob}\, I} F(i) \cong F(j).)$$

2.8 Henselization

([EGA IV] 18.5–8.)

Lemma 2.8.1.

(i) *Let A be a ring. The map*

$$a \mapsto D(a)$$

defines a one-to-one correspondence between the set of idempotent elements in A and the set of open and closed subsets of Spec A.

(ii) *Let A be a ring. The map*

$$a \mapsto (\mathrm{Ann}(a), \mathrm{Ann}(1-a))$$

defines a one-to-one correspondence between the set of idempotent elements in A and the set of pairs $(\mathfrak{a}_1, \mathfrak{a}_2)$ of ideals of A with the properties

$$\mathfrak{a}_1 \cap \mathfrak{a}_2 = 0, \quad \mathfrak{a}_1 + \mathfrak{a}_2 = A.$$

(iii) *Let $(R, \mathfrak{m})$ be a local ring, let $f(t) \in R[t]$ be a monic polynomial, and let $A = R[t]/(f(t))$. Then the map*

$$(g(t), h(t)) \mapsto (Ag(t), Ah(t))$$

defines a one-to-one correspondence between the set of pairs of monic polynomials $(g(t), h(t))$ in $R[t]$ with the properties

$$f(t) = g(t)h(t), \quad g(t)R[t] + h(t)R[t] = R[t],$$

and the set of pairs $(\mathfrak{a}_1, \mathfrak{a}_2)$ of ideals of A with the properties

$$\mathfrak{a}_1 \cap \mathfrak{a}_2 = 0, \quad \mathfrak{a}_1 + \mathfrak{a}_2 = A.$$

Proof.

(i) If $a \in A$ is idempotent, then we have

$$a(1-a) = 0, \quad a + (1-a) = 1.$$

It follows that

$$D(a) \cap D(1-a) = \emptyset, \quad D(a) \cup D(1-a) = \operatorname{Spec} A.$$

So $D(a)$ is an open and closed subset of $\operatorname{Spec} A$. Suppose that U is an open and closed subset of A. Let $a \in A$ be the element corresponding to the section in $\Gamma(\operatorname{Spec} A, \mathscr{O}_{\operatorname{Spec} A})$ such that $a|_U = 1$ and $a|_{\operatorname{Spec} A - U} = 0$. Then a is idempotent and $U = D(a)$. So the map $a \mapsto D(a)$ is surjective. To prove it is injective, let a_1 and a_2 be idempotent elements in A such that $D(a_1) = D(a_2)$, and let us prove that $a_1 = a_2$. It suffices to show that for any prime ideal $\mathfrak{p}$ of A, a_1 and a_2 have the same image in $A_{\mathfrak{p}}$. So we may assume that A is a local ring. Then $\operatorname{Spec} A$ is connected. It follows that either

$$D(a_1) = D(a_2) = \emptyset$$

or

$$D(a_1) = D(a_2) = \operatorname{Spec} A.$$

In the first case, a_1 and a_2 are nilpotent. Since they are idempotent, we have $a_1 = a_2 = 0$. In the second case, we have

$$D(1-a_1) = D(1-a_2) = \emptyset,$$

and the same argument shows that $1 - a_1 = 1 - a_2 = 0$. So we always have $a_1 = a_2$.

(ii) For any $a \in A$, if $x \in \mathrm{Ann}(a) \cap \mathrm{Ann}(1-a)$, then

$$x = x \cdot (a + (1-a)) = 0.$$

If a is idempotent, then $a(1-a) = 0$. So $1-a \in \mathrm{Ann}(a)$ and $a \in \mathrm{Ann}(1-a)$, and hence

$$1 = (1-a) + a \in \mathrm{Ann}(a) + \mathrm{Ann}(1-a).$$

It follows that

$$\mathrm{Ann}(a) \cap \mathrm{Ann}(1-a) = 0, \quad \mathrm{Ann}(a) + \mathrm{Ann}(1-a) = A.$$

Suppose that a_1 and a_2 are idempotent elements in A such that $\mathrm{Ann}(a_1) = \mathrm{Ann}(a_2)$. We have $1 - a_1 \in \mathrm{Ann}(a_1)$. It follows that $(1-a_1)a_2 = 0$, that is, $a_2 = a_1 a_2$. Similarly, we have $a_1 = a_1 a_2$. So $a_1 = a_2$. Thus the map $a \mapsto (\mathrm{Ann}(a), \mathrm{Ann}(1-a))$ is injective.

Suppose that $(\mathfrak{a}_1, \mathfrak{a}_2)$ is a pair of ideals of A with the properties $\mathfrak{a}_1 \cap \mathfrak{a}_2 = 0$ and $\mathfrak{a}_1 + \mathfrak{a}_2 = A$. By the Chinese remainder theorem, we have

$$A \cong A/\mathfrak{a}_1 \times A/\mathfrak{a}_2.$$

Note that

$$\mathrm{Ann}(1,0) = \{0\} \times A/\mathfrak{a}_2, \quad \mathrm{Ann}(0,1) = A/\mathfrak{a}_1 \times \{0\}.$$

Let a be the element in A corresponding to the element $(1,0)$ in $A/\mathfrak{a}_1 \times A/\mathfrak{a}_2$. Then we have

$$(\mathrm{Ann}(a), \mathrm{Ann}(1-a)) = (\mathfrak{a}_1, \mathfrak{a}_2).$$

So the map $a \mapsto (\mathrm{Ann}(a), \mathrm{Ann}(1-a))$ is surjective.

(iii) Suppose that $(g(t), h(t))$ is a pair of monic polynomials in $R[t]$ with the properties $f(t) = g(t)h(t)$ and $g(t)R[t] + h(t)R[t] = R[t]$. Then

$$Ag(t) + Ah(t) = A.$$

Choose polynomials $a(t), b(t) \in R[t]$ such that

$$a(t)g(t) + b(t)h(t) = 1.$$

If $r(t) \in g(t)R[t] \cap h(t)R[t]$, then we have

$$r(t) = g(t)p(t) = h(t)q(t)$$

for some polynomials $p(t)$ and $q(t)$. We then have

$$\begin{aligned} r(t) &= a(t)g(t)r(t) + b(t)h(t)r(t) \\ &= a(t)g(t)h(t)q(t) + b(t)h(t)g(t)p(t) \\ &= f(t)(a(t)q(t) + b(t)p(t)). \end{aligned}$$

So $r(t) \in f(t)R[t]$. It follows that $Ag(t) \cap Ah(t) = 0$.

Suppose $(g_1(t), h_1(t))$ is another pair of monic polynomials in $R[t]$ with the properties $f(t) = g_1(t)h_1(t)$ and $g_1(t)R[t] + h_1(t)R[t] = R[t]$. If $Ag(t) = Ag_1(t)$, then $g(t)R[t] = g_1(t)R[t]$. Since $g(t)$ and $g_1(t)$ are monic, this implies that $g(t) = g_1(t)$. Similarly $h(t) = h_1(t)$.

Given a pair $(\mathfrak{a}_1, \mathfrak{a}_2)$ of ideals of A with the properties $\mathfrak{a}_1 \cap \mathfrak{a}_2 = 0$ and $\mathfrak{a}_1 + \mathfrak{a}_2 = A$, let $A_1 = A/\mathfrak{a}_1$ and $A_2 = A/\mathfrak{a}_2$. By the Chinese remainder theorem, we have

$$R[t]/(f(t)) \cong A_1 \times A_2.$$

Let u be the image of t in A_1. Then $A_1/\mathfrak{m}A_1$ is generated by the image $\bar{u}$ of u in $A_1/\mathfrak{m}A_1$ as a k-algebra, where $k = R/\mathfrak{m}$. Let $\bar{g}(t) \in k[t]$ be the minimal polynomial of $\bar{u}$, and let $d = \deg \bar{g}(t)$. Then $\{1, \bar{u}, \ldots, \bar{u}^{d-1}\}$ is a basis of $A_1/\mathfrak{m}A_1$ over k. By Nakayama's lemma, A_1 is generated by $1, u, \ldots, u^{d-1}$ as an R-algebra. So there exists a monic polynomial $g(t) \in R[t]$ of degree d such that $g(u) = 0$. $g(t)$ is necessarily a lifting of $\bar{g}(t)$. Since A_1 is a direct factor of the free R-module $R[t]/(f(t))$ and R is a local ring, A_1 is a free R-module. As $\{1, \bar{u}, \ldots, \bar{u}^{d-1}\}$ is a basis of $A_1/\mathfrak{m}A_1$ over k, $\{1, u, \ldots, u^{d-1}\}$ is a basis of A_1 over R. So the R-algebra homomorphism

$$R[t]/(g(t)) \to A_1, \quad t \mapsto u$$

is an isomorphism. This is equivalent to saying that $\mathfrak{a}_1 = Ag(t)$. Similarly, there exists a monic polynomial $h(t) \in R[t]$ such that $\mathfrak{a}_2 = Ah(t)$. We have

$$R[t]/(f(t)) \cong R[t]/(g(t)) \times R[t]/(h(t)).$$

The image of $g(t)h(t)$ in $R[t]/(g(t)) \times R[t]/(h(t))$ is 0. So its image in $R[t]/(f(t))$ is 0, that is, $f(t)|g(t)h(t)$. On the other hand, we have

$$\begin{aligned}\deg f(t) &= \operatorname{rank}\big(R[t]/(f(t))\big) = \operatorname{rank}\big(R[t]/(g(t)) \times R[t]/(h(t))\big)\\ &= \deg\,(g(t)h(t)).\end{aligned}$$

As $f(t)$ and $g(t)h(t)$ are monic polynomials, we have $f(t) = g(t)h(t)$. So we have

$$R[t]/(g(t)h(t)) \cong R[t]/(g(t)) \times R[t]/(h(t)).$$

It is clear that the kernels of the projections

$$\begin{aligned}&R[t]/(g(t)) \times R[t]/(h(t)) \to R[t]/(g(t)),\\ &R[t]/(g(t)) \times R[t]/(h(t)) \to R[t]/(h(t))\end{aligned}$$

generate the ideal $R[t]/(g(t)) \times R[t]/(h(t))$. So the kernels of the projections

$$\begin{aligned}&R[t]/(g(t)h(t)) \to R[t]/(g(t)),\\ &R[t]/(g(t)h(t)) \to R[t]/(h(t))\end{aligned}$$

generate the ideal $R[t]/(g(t)h(t))$, that is, $(g(t))/(g(t)h(t))$ and $(h(t))/(g(t)h(t))$ generate the ideal $R[t]/(g(t)h(t))$. This implies that $g(t)$ and $h(t)$ generate the ideal $R[t]$. □

Lemma 2.8.2. *Let $(R, \mathfrak{m})$ be a local ring and let A be an R-algebra which is free of finite rank as an R-module. For any R-algebra A', let $\mathrm{Idem}(A' \otimes_R A)$ be the set of idempotent elements in $A' \otimes_R A$. Then there exists an etale R-algebra B of finite presentation such that the functor $A' \mapsto \mathrm{Idem}(A' \otimes_R A)$ is represented by B, that is, we have one-to-one correspondences*

$$\mathrm{Idem}(A' \otimes_R A) \cong \mathrm{Hom}_R(B, A')$$

functorial in A'.

Proof. Let $\{e_1, \ldots, e_n\}$ be a basis of A over R. We have

$$e_i e_j = \sum_k a_{ijk} e_k$$

for some $a_{ijk} \in R$. Let

$$P_k(t_1, \ldots, t_n) = \sum_{i,j} a_{ijk} t_i t_j - t_k \quad (k = 1, \ldots, n).$$

An element $\sum_i a_i e_i$ $(a_i \in R)$ in A is idempotent if and only if $P_k(a_1, \ldots, a_n) = 0$ for all k. For any R-algebra A', $\{1 \otimes e_1, \ldots, 1 \otimes e_n\}$ is a basis of $A' \otimes_R A$ over A'. An element $\sum_i a'_i \otimes e_i$ $(a'_i \in A')$ in $A' \otimes_R A$ is idempotent if and only if $P_k(a'_1, \ldots, a'_n) = 0$ for all k. So we have a one-to-one correspondence

$$\mathrm{Idem}(A' \otimes_R A) \cong \mathrm{Hom}_R(R[t_1, \ldots, t_n]/(P_1, \ldots, P_n), A').$$

Let us prove $R[t_1, \ldots, t_n]/(P_1, \ldots, P_n)$ is etale over R. By 2.6.2, it suffices to show for any R-algebra A' and any ideal I of A' satisfying $I^2 = 0$, that the canonical map

$$\begin{aligned} &\mathrm{Hom}_R(R[t_1, \ldots, t_n]/(P_1, \ldots, P_n), A') \\ &\to \mathrm{Hom}_R(R[t_1, \ldots, t_n]/(P_1, \ldots, P_n), A'/I) \end{aligned}$$

is bijective, that is, the map

$$\mathrm{Idem}(A' \otimes_R A) \to \mathrm{Idem}\big((A'/I) \otimes_R A\big)$$

is bijective. This is equivalent to saying that we have a one-to-one correspondence between the set of open and closed subsets of $\mathrm{Spec}\,(A' \otimes_R A)$ and the set of open and closed subsets of $\mathrm{Spec}\,\big((A'/I) \otimes_R A\big)$. This follows from the fact that $\mathrm{Spec}\,(A' \otimes_R A)$ has the same underlying topological space as $\mathrm{Spec}\,\big((A'/I) \otimes_R A\big)$ □

Proposition 2.8.3. *Let $(R, \mathfrak{m})$ be a local ring, $k = R/\mathfrak{m}$ its residue field, $S = \operatorname{Spec} R$, and s the closed point of S. The following conditions are equivalent.*

(i) *Every finite R-algebra A is a direct product of local rings.*

(ii) *The condition* (i) *holds for $A = R[t]/(f(t))$ for any monic polynomial $f(t) \in R[t]$.*

(iii) *For any finite R-algebra A, the canonical homomorphism $A \to A/\mathfrak{m}A$ induces a one-to-one correspondence between the set of idempotent elements in A and the set of idempotent elements in $A/\mathfrak{m}A$.*

(iv) *The condition* (iii) *holds for $A = R[t]/(f(t))$ for any monic polynomial $f(t) \in R[t]$.*

(v) *For any monic polynomial $f(t) \in R[t]$ and any factorization $\bar{f}(t) = \bar{g}(t)\bar{h}(t)$, where $\bar{f}(t)$ is the image of $f(t)$ in $k[t]$, and $\bar{g}(t)$ and $\bar{h}(t)$ are relatively prime monic polynomials in $k[t]$, there exist uniquely determined polynomials $g(t)$ and $h(t)$ in $R[t]$ such that $f(t) = g(t)h(t)$, $\bar{g}(t)$ and $\bar{h}(t)$ are images of $g(t)$ and $h(t)$ in $k[t]$, respectively, and the ideal generated by $g(t)$ and $h(t)$ is $R[t]$.*

(vi) *The condition* (v) *holds for any factorization $\bar{f}(t) = (t - \bar{a})\bar{h}(t)$ such that $t - \bar{a}$ and $\bar{h}(t)$ are relatively prime monic polynomials in $k[t]$.*

(vii) *For any etale morphism $g : X \to S$, any section of $g_s : X \otimes_R k \to \operatorname{Spec} k$ is induced by a section g.*

When $(R, \mathfrak{m})$ satisfies the above equivalent conditions, we say R is *henselian.* If R is henselian and $k = R/\mathfrak{m}$ is separably closed, we say R is *strictly local* or *strictly henselian.*

Proof.

(i)⇒(iii) Let A be a finite R-algebra. Any maximal ideal of A lies over the maximal ideal $\mathfrak{m}$ of R. Since $A/\mathfrak{m}A$ is finite over $R/\mathfrak{m}$, it is artinian and has finitely many maximal ideals. It follows that A has finitely many maximal ideals. For any maximal ideal $\mathfrak{n}$ of A, the canonical morphism $\operatorname{Spec} A_{\mathfrak{n}} \to \operatorname{Spec} A$ is an embedding of topological spaces and $\operatorname{Spec} A_{\mathfrak{n}}$ is connected. When $\mathfrak{n}$ goes over the finite set of maximal ideals of A, the images of $\operatorname{Spec} A_{\mathfrak{n}} \to \operatorname{Spec} A$ cover $\operatorname{Spec} A$. So $\operatorname{Spec} A$ has finitely many connected components. Similarly, $\operatorname{Spec} A/\mathfrak{m}A$ has finitely many connected components. Note that if a topological space has finitely many connected components, then open and closed subsets are exactly unions of connected components. By 2.8.1 (i), to prove (iii), it suffices to show that taking inverse image under the morphism $\operatorname{Spec} A/\mathfrak{m}A \to \operatorname{Spec} A$ defines a one-to-one correspondence between the set of open and closed subsets of $\operatorname{Spec} A$

and the set of open and closed subsets of $\mathrm{Spec}\, A/\mathfrak{m}A$. By assumption, we have

$$A \cong \prod_{\mathfrak{n}} A_{\mathfrak{n}}.$$

So we have

$$\mathrm{Spec}\, A \cong \coprod_{\mathfrak{n}} \mathrm{Spec}\, A_{\mathfrak{n}}.$$

Each $\mathrm{Spec}\, A_{\mathfrak{n}}$ can be regarded as a connected open and closed subset of $\mathrm{Spec}\, A$. So any open closed subset of $\mathrm{Spec}\, A$ is a union of some $\mathrm{Spec}\, A_{\mathfrak{n}}$. On the other hand, we have

$$\mathrm{Spec}\, A/\mathfrak{m}A \cong \mathrm{Spec}\, (\prod_{\mathfrak{n}} A_{\mathfrak{n}}/\mathfrak{m}A_{\mathfrak{n}}) \cong \coprod_{\mathfrak{n}} \mathrm{Spec}\, A_{\mathfrak{n}}/\mathfrak{m}A_{\mathfrak{n}},$$

and each $\mathrm{Spec}\, A_{\mathfrak{n}}/\mathfrak{m}A_{\mathfrak{n}}$ consists of only one point. Any open and closed subset of $\mathrm{Spec}\, A/\mathfrak{m}A$ is a union of some $\mathrm{Spec}\, A_{\mathfrak{n}}/\mathfrak{m}A_{\mathfrak{n}}$. So taking inverse image under the morphism $\mathrm{Spec}\, A/\mathfrak{m}A \to \mathrm{Spec}\, A$ defines a one-to-one correspondence between the set of open and closed subsets of $\mathrm{Spec}\, A$ and the set of open and closed subsets of $\mathrm{Spec}\, A/\mathfrak{m}A$.

(iii)⇒(i) If a topological space has finitely many connected components, then connected components are exactly minimal open and closed subsets. So taking inverse images under the morphism $\mathrm{Spec}\, A/\mathfrak{m}A \to \mathrm{Spec}\, A$ defines a one-to-one correspondence between the set of connected components of $\mathrm{Spec}\, A$ and the set of connected components of $\mathrm{Spec}\, A/\mathfrak{m}A$. Note that $\mathrm{Spec}\, A/\mathfrak{m}A$ is discrete and its connected components are of the form $\{\mathfrak{n}/\mathfrak{m}A\}$ for maximal ideals $\mathfrak{n}$ of A. So different maximal ideals of A lie in different connected components of $\mathrm{Spec}\, A$. Let $\mathfrak{n}_1$ and $\mathfrak{n}_2$ be two different maximal ideals of A. We claim that $\mathrm{Spec}\, A_{\mathfrak{n}_1}$ and $\mathrm{Spec}\, A_{\mathfrak{n}_2}$ are disjoint considered as subsets of $\mathrm{Spec}\, A$. Indeed, if they have a nonempty intersection, then $\mathrm{Spec}\, A_{\mathfrak{n}_1} \cup \mathrm{Spec}\, A_{\mathfrak{n}_2}$ is connected. This contradicts that the fact that $\mathfrak{n}_1$ and $\mathfrak{n}_2$ lie in different connected components. So as a set, $\mathrm{Spec}\, A$ is a disjoint union of $\mathrm{Spec}\, A_{\mathfrak{n}}$. One can show that each $\mathrm{Spec}\, A_{\mathfrak{n}}$ is a maximal connected subset of $\mathrm{Spec}\, A$. So each $\mathrm{Spec}\, A_{\mathfrak{n}}$ is a connected component of $\mathrm{Spec}\, A$. It follows that $\mathrm{Spec}\, A = \coprod_{\mathfrak{n}} \mathrm{Spec}\, A_{\mathfrak{n}}$ as schemes. So $A = \prod_{\mathfrak{n}} A_{\mathfrak{n}}$.

(i)⇒(ii) Trivial.

(ii)⇒(iv) Use the same argument as (i)⇒(iii).

(iv)⇒(iii) Let $e_1, e_2 \in A$ be idempotent elements such that $e_1 \equiv e_2 \mod \mathfrak{m}A$. We have

$$(e_1 - e_2)^3 = e_1^3 - 3e_1^2 e_2 + 3e_1 e_2^2 - e_2^3 = e_1 - 3e_1 e_2 + 3e_1 e_2 - e_2 = e_1 - e_2.$$

So we have

$$(e_1 - e_2)((e_1 - e_2)^2 - 1) = 0.$$

Since $e_1 - e_2 \in \mathfrak{m}A$, $(e_1 - e_2)^2 - 1$ is a unit in A. It follows that $e_1 - e_2 = 0$. So the map $A \to A/\mathfrak{m}A$ is always injective when restricted to the set of idempotent elements in A.

Let $\bar{e} \in A/\mathfrak{m}A$ be an idempotent element. We need to show it can be lifted to an idempotent element in A. Let $a \in A$ be an arbitrary lift of $\bar{e}$, and let $A' = R[a]$. We claim that $\bar{e}$ is the image of an idempotent element $\bar{e}'$ in $A'/\mathfrak{m}A'$. Let $\bar{a}$ be the image of a in $A'/\mathfrak{m}A'$ and let

$$f : \operatorname{Spec} A/\mathfrak{m}A \to \operatorname{Spec} A'/\mathfrak{m}A'$$

be the canonical morphism. Then

$$f^{-1}(D(\bar{a})) = D(\bar{e}).$$

Since $\operatorname{Spec} A'/\mathfrak{m}A'$ is discrete, $D(\bar{a})$ is open and closed. So there exists an idempotent element $\bar{e}'$ in $A'/\mathfrak{m}A'$ such that $D(\bar{e}') = D(\bar{a})$. We have

$$f^{-1}(D(\bar{e}')) = D(\bar{e}).$$

This implies that the image of $\bar{e}'$ in $A/\mathfrak{m}A$ is $\bar{e}$. To prove (ii), it suffices to show that $\bar{e}'$ can be lifted to an idempotent element in A'. Replacing A by A', we are reduced to the case where $A = R[a]$. Since a is integral over R, there exists a monic polynomial $f(t) \in R[t]$ such that $f(a) = 0$. We then have an epimorphism of R-algebras

$$R[t]/(f(t)) \to A, \quad t \mapsto a.$$

We claim that $\bar{e}$ is the image of an idempotent element in $k[t]/(\bar{f}(t))$, where $\bar{f}(t)$ is the image of $f(t)$ in $k[t]$. Indeed, the homomorphism

$$k[t]/(\bar{f}(t)) \to A/\mathfrak{m}A$$

is surjective. The corresponding morphism

$$\operatorname{Spec} A/\mathfrak{m}A \to \operatorname{Spec} k[t]/(\bar{f}(t))$$

is a closed immersion. Since $\operatorname{Spec} k[t]/(\bar{f}(t))$ is discrete, the open and closed subset $D(\bar{e})$ of $\operatorname{Spec} A/\mathfrak{m}A$ is also open and closed in $\operatorname{Spec} k[t]/(\bar{f}(t))$. So there exists an idempotent element $\bar{e}''$ in $k[t]/(\bar{f}(t))$ such that $D(\bar{e}) = D(\bar{e}'')$. The image $\bar{e}''$ in $A/\mathfrak{m}A$ is $\bar{e}$. If the condition (iv) holds, there exists an idempotent element e'' in $R[t]/(f(t))$ lifting $\bar{e}''$. Then the image e of e'' in A is an idempotent element lifting $\bar{e}$.

(iv)$\Rightarrow$(v) Use 2.8.1 (ii), (iii).

(v)⇒(vi) Trivial.

(vi)⇒(vii) Given a section of g_s, let x be its image. By 2.3.5, there exists a monic polynomial $f(t) \in R[t]$ and a maximal ideal $\mathfrak{n}$ of $B = R[t]/(f(t))$ not containing the image of $f'(t)$ in B such that $\mathscr{O}_{X,x}$ is R-isomorphic to $B_\mathfrak{n}$. Let $\bar{f}(t)$ be the image of $f(t)$ in $k[t]$. The given section of g_s defines a k-morphism

$$\mathrm{Spec}\, k \to \mathrm{Spec}\, \mathscr{O}_{X,x}/\mathfrak{m}\mathscr{O}_{X,x}.$$

Composed with the canonical morphisms

$$\mathrm{Spec}\, \mathscr{O}_{X,x}/\mathfrak{m}\mathscr{O}_{X,x} \cong \mathrm{Spec}\, B_\mathfrak{n}/\mathfrak{m}B_\mathfrak{n} \to \mathrm{Spec}\, B/\mathfrak{m}B \cong \mathrm{Spec}\, k[t]/(\bar{f}(t)),$$

we get a k-morphism

$$\mathrm{Spec}\, k \to \mathrm{Spec}\, k[t]/(\bar{f}(t)).$$

It corresponds to a k-algebra homomorphism

$$k[t]/(\bar{f}(t)) \to k.$$

Let $\bar{a}$ be the image of t under this homomorphism. Then $\bar{f}(\bar{a}) = 0$ and $\bar{f}'(\bar{a}) \neq 0$. So $\bar{a}$ is a simple root of $\bar{f}(t)$, and we have

$$\bar{f}(t) = (t - \bar{a})\bar{h}(t)$$

for some monic polynomial $\bar{h}(t) \in k[t]$ relatively prime to $t - \bar{a}$. If the condition (vi) holds, this factorization can be lifted to a factorization

$$f(t) = (t - a)h(t)$$

for some $a \in R$ lifting $\bar{a}$ and some monic polynomial $h(t) \in R[t]$ lifting $\bar{h}(t)$ such that the ideal generated by $t - a$ and $h(t)$ is $R[t]$. The R-algebra homomorphism

$$R[t]/(f(t)) \to R, \quad t \mapsto a$$

induces an R-algebra homomorphism

$$B_\mathfrak{n} = (R[t]/(f(t)))_\mathfrak{n} \to R$$

and hence an S-morphism

$$S = \mathrm{Spec}\, R \to \mathrm{Spec}\, B_\mathfrak{n} \cong \mathrm{Spec}\, \mathscr{O}_{X,x}.$$

Composed with the canonical morphism $\mathrm{Spec}\, \mathscr{O}_{X,x} \to X$, we get a section of $g : X \to S$ inducing the given section of g_s.

(vii)⇒(iv) Let B be the etale R-algebra in 2.8.2 for the free R-algebra of finite rank $A = R[t]/(f(t))$. In the proof of (iv)⇒(iii), we have shown that the map

$$\mathrm{Idem}(A) \to \mathrm{Idem}(A/\mathfrak{m}A)$$

is always injective. We need to show it is surjective. It suffices to show that the canonical map

$$\mathrm{Hom}_R(B, R) \to \mathrm{Hom}_R(B, R/\mathfrak{m})$$

is surjective. $\mathrm{Hom}_R(B, R)$ (resp. $\mathrm{Hom}_R(B, R/\mathfrak{m})$) can be identified with the set of sections of the morphism $g : \mathrm{Spec}\, B \to \mathrm{Spec}\, R$ (resp. $g_s : \mathrm{Spec}\,(B \otimes_R k) \to \mathrm{Spec}\, k$). Since g is etale, our assertion follows if condition (vii) holds. □

Proposition 2.8.4. *Suppose $(R, \mathfrak{m})$ is a complete local noetherian ring. Then R is henselian.*

Proof.

First Method: Let us verify that the condition 2.8.3 (i) holds. Let A be a finite R-algebra. Then A is complete with respect to the $\mathfrak{m}$-adic topology. Any maximal ideal of A lies over the maximal ideal $\mathfrak{m}$ of R. Since $A/\mathfrak{m}A$ is finite over $R/\mathfrak{m}$, it is artinian. It follows that A has finitely many maximal ideals. Let $\mathfrak{m}_1, \ldots, \mathfrak{m}_n$ be all the maximal ideals of A. Then the nilpotent radical of $A/\mathfrak{m}A$ is $(\mathfrak{m}_1 \cap \cdots \cap \mathfrak{m}_n)/\mathfrak{m}A$. So there exists a positive integer k such that

$$(\mathfrak{m}_1 \cap \cdots \cap \mathfrak{m}_n)^k \subset \mathfrak{m}A \subset \mathfrak{m}_1 \cap \cdots \cap \mathfrak{m}_n.$$

Combined with the Chinese remainder theorem, we get

$$A \cong \varprojlim_i A/\mathfrak{m}^i A \cong \varprojlim_i A/\mathfrak{m}_1^i \cdots \mathfrak{m}_n^i \cong \varprojlim_i (A/\mathfrak{m}_1^i \times \cdots \times A/\mathfrak{m}_n^i) \cong \widehat{A}_{\mathfrak{m}_1} \times \cdots \times \widehat{A}_{\mathfrak{m}_n}.$$

So A is a direct product of local rings.

Second Method: Let us verify that the condition 2.8.3 (vi) holds. Let $f(t) \in R[t]$ be a monic polynomial and let

$$\bar{f}(t) = (t - \bar{a})\bar{h}(t)$$

be a factorization of the image $\bar{f}(t)$ of $f(t)$ in $k[t]$, where $t - \bar{a}$ and $\bar{h}(t)$ are relatively prime monic polynomials in $k[t]$. We lift $\bar{a}$ to a root of $f(t)$ in R using a method similar to Newton's iteration method of finding a root of a smooth function in calculus. Note that $\bar{a}$ is a simple root of $\bar{f}(t)$. So we have

$$\bar{f}(\bar{a}) = 0, \quad \bar{f}'(\bar{a}) \neq 0.$$

Let $a_0 \in R$ be an arbitrary lift of $\bar{a}$. Then $f(a_0) \in \mathfrak{m}$ and $f'(a_0)$ is a unit in R. Suppose that we have found a lift $a_i \in R$ of $\bar{a}$ such that $f(a_i) \in \mathfrak{m}^{i+1}$. Then $f'(a_i)$ is a unit in R. Define

$$a_{i+1} = a_i - \frac{f(a_i)}{f'(a_i)}.$$

We have

$$a_{i+1} - a_i = -\frac{f(a_i)}{f'(a_i)} \in \mathfrak{m}^{i+1}.$$

So a_{i+1} is also a lift of $\bar{a}$. Note that for any $a, \delta \in R$, we can find $b \in R$ such that

$$f(a+\delta) = f(a) + f'(a)\delta + b\delta^2.$$

In particular, we can find $b_i \in R$ such that

$$\begin{aligned} f(a_{i+1}) &= f(a_i) + f'(a_i) \cdot \Big(-\frac{f(a_i)}{f'(a_i)} \Big) + b_i \Big(-\frac{f(a_i)}{f'(a_i)} \Big)^2 \\ &= b_i \Big(-\frac{f(a_i)}{f'(a_i)} \Big)^2 \in \mathfrak{m}^{2i+2} \subset \mathfrak{m}^{i+2}. \end{aligned}$$

Since R is complete, the limit

$$a = \lim_{i\to\infty} a_i$$

exists in R. It is a lift of $\bar{a}$, and $f(a) = 0$. So we have

$$f(t) = (t-a)h(t)$$

for some monic polynomial $h(t)$ lifting $\bar{h}(t)$. Since the ideal generated by $t-\bar{a}$ and $\bar{h}(t)$ is $k[t]$, by Nakayama's lemma applied to the finitely generated R-module $R[t]/(f(t))$, we see the ideal generated by $t-a$ and $h(t)$ is $R[t]$.

Finally let us prove that the lift of $\bar{a}$ to a root of $f(t)$ in R is unique. Let $a, a' \in R$ be two roots of $f(t)$ lifting $\bar{a}$. We can find $b \in R$ such that

$$f(a') = f(a) + f'(a)(a'-a) + b(a'-a)^2,$$

that is,

$$a' - a = -\frac{b}{f'(a)}(a'-a)^2.$$

We have $a' - a \in \mathfrak{m}$. Together with the above equality, this implies that

$$a' - a \in \bigcap_{i=1}^{\infty} \mathfrak{m}^i.$$

So we have $a' - a = 0$. □

Let $A \to A'$ be a local homomorphism of local rings. We say A' is *essentially etale* over A if there exists an etale A-algebra B and a prime ideal $\mathfrak{p}$ of B lying above the maximal ideal of A such that A' is A-isomorphic to $B_{\mathfrak{p}}$. Using the fact that $\operatorname{Spec} B \to \operatorname{Spec} A$ is quasi-finite, one can show the maximal ideal of A' is the only prime ideal of A' lying above the maximal ideal of A. So if A' and A'' are two local essentially etale A-algebras, then any A-homomorphism $A' \to A''$ is local. A local homomorphism $A \to A'$ of locally rings is called *strictly essentially etale* if it is essentially etale and the homomorphism $A \to A'$ induces an isomorphism on the residue field. Using 1.10.9 and 2.3.7, one can show if $A \to A'$ and $A' \to A''$ are essentially etale local homomorphisms, then their composite $A \to A''$ is also essentially etale.

Lemma 2.8.5. *Let A and A' be local rings, $A \to A'$ a strictly essentially etale local homomorphism, and $(R, \mathfrak{m})$ a henselian local ring. Then the canonical map*

$$\text{loc.Hom}(A', R) \to \text{loc.Hom}(A, R)$$

is bijective, where loc.Hom *denotes the set of local homomorphisms.*

Proof. Given a local homomorphism $A \to R$, we need to show it can be extended uniquely to a local homomorphism $A' \to R$. Consider $\operatorname{Spec} R$ as a scheme over $\operatorname{Spec} A$. We need to show that there exists a unique A-morphism $\operatorname{Spec} R \to \operatorname{Spec} A'$ that maps the closed point of $\operatorname{Spec} R$ to the closed point of $\operatorname{Spec} A'$, or equivalently, that there exists a unique section for the projection

$$\operatorname{Spec}(A' \otimes_A R) \to \operatorname{Spec} R$$

that maps the closed point of $\operatorname{Spec} R$ to the point in $\operatorname{Spec}(A' \otimes_A R)$ that is above the closed points in $\operatorname{Spec} A'$ and $\operatorname{Spec} R$. Note that since $A \to A'$ induces an isomorphism of residue fields, there is one and only one point in $\operatorname{Spec}(A' \otimes_A R)$ that is above the closed points in $\operatorname{Spec} A'$ and $\operatorname{Spec} R$. Let B be an etale A-algebra such that A' is A-isomorphic to $B_{\mathfrak{p}}$ for some prime ideal $\mathfrak{p}$ of B. It suffices to show there exists a unique section for the projection

$$\operatorname{Spec}(B \otimes_A R) \to \operatorname{Spec} R$$

that maps the closed point of $\operatorname{Spec} R$ to the unique point in $\operatorname{Spec}(B \otimes_A R)$ that is above the closed point in $\operatorname{Spec} R$ and the point $\mathfrak{p}$ in $\operatorname{Spec} B$. This point has the same residue field as R and hence is the image of a unique section for the projection

$$\operatorname{Spec}(B \otimes_A R/\mathfrak{m}) \to \operatorname{Spec} R/\mathfrak{m}.$$

By 2.8.3 (vii), this section is induced by a section of the projection $\operatorname{Spec}(B \otimes_A R) \to \operatorname{Spec} R$, which has the required property. The uniqueness follows from 2.3.10 (i). □

Lemma 2.8.6. *Let A be a local ring, and let A_1 and A_2 be strictly essentially etale local A-algebras.*

(i) *There exists a strictly essentially etale local A-algebra A_3 dominating A_1 and A_2, that is, there exist A-homomorphisms (necessarily local) from A_i $(i = 1, 2)$ to A_3.*

(ii) *There exists at most one A-homomorphism from A_1 to A_2.*

Proof.

(i) We can find etale A-algebras B_i $(i = 1, 2)$ and prime ideals $\mathfrak{p}_i$ of B_i such that A_i are A-isomorphic to $(B_i)_{\mathfrak{p}_i}$, respectively. There exists a unique prime ideal $\mathfrak{p}$ of $B_1 \otimes_A B_2$ lying above $\mathfrak{p}_1$ and $\mathfrak{p}_2$. We can take $A_3 = (B_1 \otimes_A B_2)_{\mathfrak{p}}$.

(ii) Keep the above notation. It suffices to show that there exists at most one A-morphism from $\mathrm{Spec}\, A_2$ to $\mathrm{Spec}\, B_1$ that maps the closed point of $\mathrm{Spec}\, A_2$ to $\mathfrak{p}_1$, or equivalently, that there exists at most one section for the projection

$$\mathrm{Spec}\,(A_2 \otimes_A B_1) \to \mathrm{Spec}\, A_2$$

that maps the closed point of $\mathrm{Spec}\, A_2$ to the (unique) point of $\mathrm{Spec}\,(A_2 \otimes_A B_1)$ that is above $\mathfrak{p}_1 \in \mathrm{Spec}\, B_1$ and the closed point of $\mathrm{Spec}\, A_2$. This follows from 2.3.10 (i). □

Let A be a local ring. A local homomorphism $A \to A'$ is called essentially of finite type if there exists a finitely generated A-algebra B such that A' is A-isomorphic to $B_{\mathfrak{p}}$ for some prime ideal $\mathfrak{p}$ of B lying above the maximal ideal of A. There exists a set of local A-algebras essentially of finite type such that every local A-algebra essentially of finite type is isomorphic to a member of this set. For example, we can take this set to be the family of A-algebras of the form $(A[t_1, \ldots, t_n]/I)_{\mathfrak{p}}$, where n goes over the set of nonnegative integers, I goes over the set of ideals of $A[t_1, \ldots, t_n]$, and $\mathfrak{p}$ goes over the set of prime ideals of $A[t_1, \ldots, t_n]/I$ lying above the maximal ideal of A. Therefore there exists a set of strictly essentially etale local A-algebras such that any strictly essentially etale local A-algebra is isomorphic to a member of this set. Let $\mathscr{S}$ be the category whose objects are strictly essentially etale local A-algebras and whose morphisms are A-homomorphisms. By 2.8.6, $\mathscr{S}$ satisfies the condition (I2) and (I3) in 2.7. We define the *henselization* A^h of A to be

$$A^h = \varinjlim_{A' \in \mathrm{ob}\, \mathscr{S}} A'.$$

Lemma 2.8.7. *Let A be a ring, I a finitely generated ideal of A, $\widehat{A} = \varprojlim_n A/I^n$, and $J_n = \ker(\widehat{A} \to A/I^n)$. Then*

(i) *The canonical homomorphism $A \to \widehat{A}$ induces isomorphisms $J_n = I^n\widehat{A}$, $A/I^n \cong \widehat{A}/I^n\widehat{A}$, and $I^n/I^{n+1} \cong I^n\widehat{A}/I^{n+1}\widehat{A}$.*

(ii) *If A/I is noetherian, then $\widehat{A}$ is noetherian.*

Proof.

(i) It is clear that the projection $\widehat{A} \to A/I^n$ is surjective. Its kernel is J_n. It follows that $\widehat{A}/J_n \cong A/I^n$. Consider the commutative diagram

$$\begin{array}{ccccccccc}
0 \to & I^n/I^{m+n} & \to & A/I^{m+n} & \to & A/I^n & \to 0 \\
& \downarrow & & \downarrow & & \downarrow & \\
0 \to & J_n/J_{m+n} & \to & \widehat{A}/J_{m+n} & \to & \widehat{A}/J_n & \to 0 \\
& \downarrow & & \cong\downarrow & & \cong\downarrow & \\
0 \to & I^n/I^{m+n} & \to & A/I^{m+n} & \to & A/I^n & \to 0,
\end{array}$$

where the first three vertical arrows are induced by the canonical homomorphism $i : A \to \widehat{A}$, and the last three vertical arrows are induced by the projection $\widehat{A} \to A/I^{m+n}$. By the five lemma, the vertical arrow$J_n/J_{m+n} \to I^n/I^{m+n}$ is an isomorphism. It follows that the vertical arrow $I^n/I^{m+n} \to J_n/J_{m+n}$ is an isomorphism. So we have $J_n = i(I^n) + J_{m+n}$. Let $x_1, \ldots, x_k \in I$ be a finite family of generators of I. For any $z \in J_n$, since $J_n = i(I^n) + J_{n+1}$, we have

$$z = i\Big(\sum_{i_1+\cdots+i_k=n} a^{(0)}_{i_1\ldots i_k} x_1^{i_1}\cdots x_k^{i_k}\Big) + z_{n+1}$$

for some $a^{(0)}_{i_1\ldots i_k} \in A$ and $z_{n+1} \in J_{n+1}$. Since $J_{n+1} = i(I^{n+1}) + J_{n+2}$, we have

$$z_{n+1} = i\Big(\sum_{i_1+\cdots+i_k=n} a^{(1)}_{i_1\ldots i_k} x_1^{i_1}\cdots x_k^{i_k}\Big) + z_{n+2}$$

for some $a^{(1)}_{i_1\ldots i_k} \in I$ and $z_{n+2} \in J_{n+2}$. Repeating this process, we get a family $a^{(m)}_{i_1\ldots i_k} \in I^m$ and $z_{n+m+1} \in J_{n+m+1}$ ($i_1 + \cdots + i_k = n$, $m = 0, 1, \ldots$) such that

$$z = i\Big(\sum_{i_1+\cdots+i_k=n} (a^{(0)}_{i_1\ldots i_k} + \cdots + a^{(m)}_{i_1\ldots i_k}) x_1^{i_1}\cdots x_k^{i_k}\Big) + z_{n+m+1}$$

for any $m \geq 0$. Taking limits on both sides, we get

$$z = \sum_{i_1+\cdots+i_k=n} (i(a^{(0)}_{i_1\ldots i_k}) + i(a^{(1)}_{i_1\ldots i_k}) + \cdots) i(x_1^{i_1}\cdots x_k^{i_k}).$$

So we have $z \in I^n\widehat{A}$. Hence $J_n \subset I^n\widehat{A}$. It is clear that $I^n\widehat{A} \subset J_n$. So we have $J_n = I^n\widehat{A}$. Since $I^n/I^{m+n} \cong J_n/J_{m+n}$, we have $I^n/I^{m+n} \cong I^n\widehat{A}/I^{m+n}\widehat{A}$. In particular, we have $A/I^n \cong \widehat{A}/I^n\widehat{A}$ and $I^n/I^{n+1} \cong I^n\widehat{A}/I^{n+1}\widehat{A}$.

(ii) By [Atiyah and Macdonald (1969)] 10.25, it suffices to show that $\bigoplus_{n=0}^{\infty}(I^n\widehat{A}/I^{n+1}\widehat{A})$ is a noetherian ring, that is, $\bigoplus_{n=0}^{\infty} I^n/I^{n+1}$ is a noetherian ring. This follows from [Atiyah and Macdonald (1969)] 10.7. □

Proposition 2.8.8. *Let $(A_\lambda, \phi_{\lambda\mu})$ be a direct system of local rings with $\phi_{\lambda\mu}$ being local homomorphisms. For each λ, let $\mathfrak{m}_\lambda$ be the maximal ideal of A_λ and let $k_\lambda = A_\lambda/\mathfrak{m}_\lambda$ be the residue field.*

(i) *$A = \varinjlim_\lambda A_\lambda$ is a local ring with the maximal ideal $\mathfrak{m} = \varinjlim_\lambda \mathfrak{m}_\lambda$, and its residue field $A/\mathfrak{m}$ is isomorphic to $\varinjlim_\lambda k_\lambda$.*

(ii) *If $\mathfrak{m}_\mu = \mathfrak{m}_\lambda A_\mu$ for any pair $\mu \geq \lambda$, then $\mathfrak{m} = \mathfrak{m}_\lambda A$ for any λ.*

(iii) *If A_μ is flat over A_λ for any pair $\mu \geq \lambda$, then A is flat over A_λ for any λ.*

(iv) *If $\mathfrak{m}_\mu = \mathfrak{m}_\lambda A_\mu$ and A_μ is flat over A_λ for any pair $\mu \geq \lambda$, and if A_λ is noetherian for any λ, then A is noetherian.*

Proof. Note that $\mathfrak{m} = \varinjlim_\lambda \mathfrak{m}_\lambda$ is an ideal of A, and we have

$$A/\mathfrak{m} \cong \varinjlim_{\lambda} A_\lambda/\mathfrak{m}_\lambda = \varinjlim_{\lambda} k_\lambda,$$

which is a field. It follows that $\mathfrak{m}$ is a maximal ideal of A. We have

$$A - \mathfrak{m} = \varinjlim_{\lambda}(A_\lambda - \mathfrak{m}_\lambda).$$

As $A_\lambda - \mathfrak{m}_\lambda$ are the groups of units of A_λ, $A - \mathfrak{m}$ is the group of units of A. So A is a local ring with the maximal ideal $\mathfrak{m}$. If $\mathfrak{m}_\mu = \mathfrak{m}_\lambda A_\mu$ for any pair $\mu \geq \lambda$, then we have

$$\mathfrak{m} = \varinjlim_{\mu\geq\lambda} \mathfrak{m}_\lambda A_\mu = \mathfrak{m}_\lambda A$$

for any λ. If A_μ is flat over A_λ for any pair $\mu \geq \lambda$, then $A \cong \varinjlim_{\mu\geq\lambda} A_\mu$ is flat over A_λ for any λ. Suppose that all the above conditions hold and suppose furthermore that each A_λ is noetherian. Then each $\mathfrak{m}_\lambda$ is finitely generated. So $\mathfrak{m} = \mathfrak{m}_\lambda A$ is finitely generated. By 2.8.7, $\widehat{A} = \varprojlim_n A/\mathfrak{m}^n$ is noetherian, and $\widehat{A}/\mathfrak{m}^n\widehat{A} \cong A/\mathfrak{m}^n$. Since

$$A/\mathfrak{m}^n = A/\mathfrak{m}_\lambda^n A \cong A_\lambda/\mathfrak{m}_\lambda^n \otimes_{A_\lambda} A$$

and A is flat over A_λ, $A/\mathfrak{m}^n$ is flat over $A_\lambda/\mathfrak{m}_\lambda^n$. So $\widehat{A}/\mathfrak{m}^n\widehat{A}$ is flat over $A_\lambda/\mathfrak{m}_\lambda^n$. By 1.3.5 (iv), $\widehat{A}$ is flat over A_λ. Let I be a finitely generated ideal of A. Then $I = I_\lambda A$ for some ideal I_λ of A_λ. Since A is flat over A_λ, we have $I_\lambda \otimes_{A_\lambda} A \cong I$. Since $\widehat{A}$ is flat over A_λ, the canonical homomorphism $I_\lambda \otimes_{A_\lambda} \widehat{A} \to \widehat{A}$ is injective, that is,

$$I_\lambda \otimes_{A_\lambda} A \otimes_A \widehat{A} \to \widehat{A}$$

is injective. So the canonical homomorphism $I \otimes_A \widehat{A} \to \widehat{A}$ is injective. By 1.1.1 (ix), $\widehat{A}$ is flat over A. Note that $\widehat{A}$ is local and the canonical homomorphism $A \to \widehat{A}$ is local. Indeed, $\widehat{A}/\mathfrak{m}\widehat{A}$ is a field since it is isomorphic to $A/\mathfrak{m}A$. For any $x \in \widehat{A} - \mathfrak{m}\widehat{A}$, the image of x under each projection $\varprojlim_n A/\mathfrak{m}^n \to A/\mathfrak{m}^n$ is a unit, and hence x is a unit. So $\widehat{A}$ is faithfully flat over A. If

$$I_1 \subset I_2 \subset \cdots$$

is an ascending chain of ideals of A, then the chain

$$I_1\widehat{A} \subset I_2\widehat{A} \subset \cdots$$

is stationary since $\widehat{A}$ is noetherian. But $I_i\widehat{A} \cap A = I_i$ for all i by 1.2.6 (v). So we have

$$I_{i_0} = I_{i_0+1} = \cdots$$

for a large i_0. Hence A is noetherian. □

Proposition 2.8.9. *Let $(A, \mathfrak{m})$ be a local ring.*

(i) *A^h is a henselian local ring, the canonical homomorphism $A \to A^h$ is local and faithfully flat, $\mathfrak{m}A^h$ is the maximal ideal of A^h, and the homomorphism $A/\mathfrak{m} \to A^h/\mathfrak{m}A^h$ is an isomorphism.*

(ii) *For any henselian local ring R, the canonical map*

$$\text{loc.Hom}(A^h, R) \to \text{loc.Hom}(A, R)$$

is bijective.

(iii) *If A is henselian, then $A \cong A^h$.*

(iv) *The canonical homomorphism $\widehat{A} \to \widehat{A^h}$ is an isomorphism.*

(v) *If A is noetherian, then so is A^h.*

Proof.

(i) For every strictly essentially etale local A-algebra A', the homomorphism $A \to A'$ is local and faithfully flat, $\mathfrak{m}A'$ is the maximal ideal of A', and $A/\mathfrak{m} \to A'/\mathfrak{m}A'$ is an isomorphism. By 2.8.8, A^h is a local ring, the canonical homomorphism $A \to A^h$ is local and faithfully flat, $\mathfrak{m}A^h$ is the maximal ideal of A^h, and the homomorphism $A/\mathfrak{m} \to A^h/\mathfrak{m}A^h$ is an isomorphism.

Let us verify that the condition 2.8.3 (vii) holds for A^h. Given an etale morphism $g : X \to \operatorname{Spec} A^h$ and a section $s : \operatorname{Spec} A^h/\mathfrak{m}A^h \to X \otimes_{A^h} A^h/\mathfrak{m}A^h$ of the fiber of g over the closed point of $\operatorname{Spec} A^h$, let us prove that this section can be lifted to a section of g. Shrinking X, we may assume that g has finite presentation. By 1.10.9 and 2.3.7, we can find a strictly essentially etale local A-algebra A' and an etale morphism $g' : X' \to \operatorname{Spec} A'$ inducing g after base change. Let $x' \in X'$ be the image of the composite

$$\operatorname{Spec} A^h/\mathfrak{m}A^h \xrightarrow{s} X \otimes_{A^h} A^h/\mathfrak{m}A^h \to X \to X'.$$

Then $\mathscr{O}_{X',x'}$ is a strictly essentially etale local A-algebra. So we have an A'-homomorphism $\mathscr{O}_{X',x'} \to A^h$. It induces an A'-morphism $\operatorname{Spec} A^h \to X'$. The graph of this morphism is a section of g extending s.

(ii) follows from 2.8.5.

(iii) By 2.8.3 (vii), if A is henselian, then any strictly essentially etale local A-algebra is isomorphic to A. So $A \cong A^h$.

(iv) Since A^h is flat over A, we have

$$\begin{aligned} \mathfrak{m}^n A^h/\mathfrak{m}^{n+1}A^h &\cong \mathfrak{m}^n/\mathfrak{m}^{n+1} \otimes_A A^h \\ &\cong \mathfrak{m}^n/\mathfrak{m}^{n+1} \otimes_{A/\mathfrak{m}} A^h/\mathfrak{m}A^h \\ &\cong \mathfrak{m}^n/\mathfrak{m}^{n+1} \otimes_{A/\mathfrak{m}} A/\mathfrak{m} \\ &\cong \mathfrak{m}^n/\mathfrak{m}^{n+1} \end{aligned}$$

for all n. By induction on n, this implies that

$$A/\mathfrak{m}^n \cong A^h/\mathfrak{m}^n A^h$$

for all n. So $\widehat{A} \cong \widehat{A^h}$.

(v) follows from 2.8.8. □

Proposition 2.8.10. *Let A be a local noetherian ring. Then A is reduced (resp. regular, resp. normal) if and only if A^h is so.*

Proof. Use the same argument as the proof of 2.4.4. □

Proposition 2.8.11. *Let $(R_\lambda, \phi_{\lambda\mu})$ be a direct system of local rings such that $\phi_{\lambda\mu}$ are local homomorphisms.*

(i) *If each R_λ is henselian, so is $\varinjlim_\lambda R_\lambda$.*

(ii) *In general, we have*

$$(\varinjlim_\lambda R_\lambda)^h \cong \varinjlim_\lambda R_\lambda^h.$$

Proof.

(i) For each λ, let $\mathfrak{m}_\lambda$ be the maximal ideal of R_λ, let $R = \varinjlim_\lambda R_\lambda$ and let $\mathfrak{m} = \varinjlim_\lambda \mathfrak{m}_\lambda$. Then R is a local ring with the maximal ideal $\mathfrak{m}$. Let us check that the condition 2.8.3 (iv) holds. Let $A = R[t]/(f(t))$, where $f(t) \in R[t]$ is a monic polynomial. We need to show that the canonical homomorphism $A \to A/\mathfrak{m}A$ induces a one-to-one correspondence on the sets of idempotent elements in A and in $A/\mathfrak{m}A$. By the proof of 2.8.3 (iv)$\Rightarrow$(iii), $A \to A/\mathfrak{m}A$ is injective when restricted to the set of idempotent elements in A. Let $\bar{e}$ be an idempotent element in $A/\mathfrak{m}A$. We can find λ such that there exists a monic polynomial $f_\lambda(t) \in R_\lambda[t]$ whose image in $R[t]$ is $f(t)$ and an idempotent element $\bar{e}_\lambda \in A_\lambda/\mathfrak{m}_\lambda A_\lambda$ whose image in $A/\mathfrak{m}A$ is $\bar{e}$, where $A_\lambda = R_\lambda[t]/(f_\lambda(t))$. Since R_λ is henselian, there exists an idempotent element e_λ in A_λ lifting $\bar{e}_\lambda$. Let e be the image of e_λ in A. Then e is idempotent element lifting $\bar{e}$.

(ii) By 2.8.9 (ii), the canonical homomorphisms $R_\lambda \to \varinjlim_\lambda R_\lambda$ induce homomorphisms $R_\lambda^h \to (\varinjlim_\lambda R_\lambda)^h$ and hence a homomorphism

$$\Phi : \varinjlim_{\lambda} R_\lambda^h \to (\varinjlim_{\lambda} R_\lambda)^h.$$

By (i), $\varinjlim_\lambda R_\lambda^h$ is henselian. By 2.8.9 (ii), the canonical homomorphism $\varinjlim_\lambda R_\lambda \to \varinjlim_\lambda R_\lambda^h$ induce a homomorphism

$$\Psi : (\varinjlim_{\lambda} R_\lambda)^h \to \varinjlim_{\lambda} R_\lambda^h.$$

One can verify that Φ and Ψ are inverse to each other. $\square$

Proposition 2.8.12. *Let* $(A, \mathfrak{m})$ *be a local ring,* B *a finite* A*-algebra, and* $\mathfrak{n}_1, \ldots, \mathfrak{n}_k$ *all the maximal ideals of* B*. Then we have*

$$B \otimes_A A^h \cong (B_{\mathfrak{n}_1})^h \times \cdots \times (B_{\mathfrak{n}_k})^h.$$

Proof. Since $A \to A^h$ induces an isomorphism on residue fields, the fibers of $\operatorname{Spec} B \to \operatorname{Spec} A$ and $\operatorname{Spec}(B \otimes_A A^h) \to \operatorname{Spec} A^h$ above the closed points of $\operatorname{Spec} A$ and $\operatorname{Spec} A^h$ are isomorphic. Since B (resp. $B \otimes_A A^h$) is finite over A (resp. A^h), the fiber of $\operatorname{Spec} B \to \operatorname{Spec} A$ (resp. $\operatorname{Spec}(B \otimes_A A^h) \to \operatorname{Spec} A^h$) above the closed point consists of maximal ideals of B (resp. $B \otimes_A A^h$). So $\operatorname{Spec}(B \otimes_A A^h) \to \operatorname{Spec} B$ induces a one-to-one correspondence between the set of maximal ideals of $B \otimes_A A^h$ and the set of maximal ideals of B. Let $\mathfrak{n}_1', \ldots, \mathfrak{n}_k'$ be the maximal ideals of $B \otimes_A A^h$ lying above $\mathfrak{n}_1, \ldots, \mathfrak{n}_k$, respectively. By 2.8.3 (i), we have

$$B \otimes_A A^h \cong (B \otimes_A A^h)_{\mathfrak{n}_1'} \times \cdots \times (B \otimes_A A^h)_{\mathfrak{n}_k'}.$$

To prove our assertion, it suffices to show $(B \otimes_A A^h)_{\mathfrak{n}'_i} \cong (B_{\mathfrak{n}_i})^h$ for all i. For each i, let e_i be the element in $B \otimes_A A^h$ so that its projection to $(B \otimes_A A^h)_{\mathfrak{n}'_j}$ is 0 if $j \neq i$ and 1 if $j = i$. We have

$$e_i^2 = e_i, \quad e_i e_j = 0 \quad (i \neq j), \quad e_1 + \cdots + e_k = 1.$$

Since $B \otimes_A A^h = \varinjlim_{A' \in \mathrm{ob}\,\mathscr{S}} B \otimes_A A'$, there exists a strictly essentially etale local A-algebra A' such that $e_1, \ldots, e_k$ are images of some elements $e'_1, \ldots, e'_k \in B \otimes_A A'$ with the property

$$e_i'^2 = e'_i, \quad e'_i e'_j = 0 \quad (i \neq j), \quad e'_1 + \cdots + e'_k = 1.$$

We then have

$$B \otimes_A A' \cong (B \otimes_A A')e'_1 \times \cdots \times (B \otimes_A A')e'_k.$$

On the other hand, we have

$$(B \otimes_A A')e'_i \otimes_{A'} A^h \cong (B \otimes_A A^h)e_i \cong (B \otimes_A A^h)_{\mathfrak{n}'_i}.$$

As $\mathrm{Spec}\,(B \otimes_A A')e'_i \to \mathrm{Spec}\,A'$ and $\mathrm{Spec}\,(B \otimes_A A')e'_i \otimes_{A'} A^h \to \mathrm{Spec}\,A^h$ have the same fiber over the closed point of $\mathrm{Spec}\,A'$ and $\mathrm{Spec}\,A^h$, it follows that $(B \otimes_A A')e'_i$ is local and its maximal ideal lies above $\mathfrak{n}_i$. Moreover $(B \otimes_A A')e'_i$ is strictly essentially etale over $B_{\mathfrak{n}_i}$, so we have

$$(B_{\mathfrak{n}_i})^h \cong ((B \otimes_A A')e'_i)^h.$$

We also have $A^h \cong A'^h$. Replacing A by A' and B by $B \otimes_A A'$, we are reduced to the case where B is a product of local rings. It suffices to treat the case where B is local. In this case, $B \otimes_A A^h$ is local. One can check that it satisfies the condition 2.8.3 (i). So it is henselian. Moreover, we have $B \otimes_A A^h \cong \varinjlim_{A' \in \mathrm{ob}\,\mathscr{S}} (B \otimes_A A')$ and each $B \otimes_A A'$ is a strictly essentially etale local B-algebra. So we have $B \otimes_A A^h \cong B^h$. □

Proposition 2.8.13. *Let R be a henselian local ring, $S = \mathrm{Spec}\,R$, and s the closed point of S. For any smooth morphism $g : X \to S$, any section of $g_s : X_s = X \otimes_R k(s) \to \mathrm{Spec}\,k(s)$ can be lifted to a section of g.*

Proof. Let $h_s : \mathrm{Spec}\,k(s) \to X_s$ be a section of g_s, and let x be the image of h_s. Since g is smooth, there exists an open neighborhood U of x admitting an etale S-morphism

$$j : U \to \mathbb{A}^n_S = \mathrm{Spec}\,R[t_1, \ldots, t_n].$$

Let $y = j(x)$. We have $k(x) \cong k(y) \cong k(s)$. So $\mathscr{O}_{X,x}$ and $\mathscr{O}_{\mathbb{A}^n_S,y}$ have the same henselization. Denote the common henselization by A. Consider the commutative diagram

$$\begin{array}{ccc}
\operatorname{Spec} A & & \\
\downarrow & \searrow & \\
U & \xrightarrow{j} & \mathbb{A}^n_S. \\
{\scriptstyle g|_U}\downarrow & \swarrow & \\
S & &
\end{array}$$

It suffices to construct a section for $\operatorname{Spec} A \to S$ because such a section composed with the canonical morphism $\operatorname{Spec} A \to X$ gives rise to a section of g inducing h_s. Since $k(y) \cong k(s)$, there exist $\bar{a}_1, \ldots, \bar{a}_n \in k(s)$ such that the image of y in $\mathbb{A}^n_s = \operatorname{Spec} k(s)[t_1, \ldots, t_n]$ corresponds to the maximal ideal $(t_1 - \bar{a}_1, \ldots, t_n - \bar{a}_n)$ of $k(s)[t_1, \ldots, t_n]$. Let $a_i \in R$ $(i = 1, \ldots n)$ be liftings of $\bar{a}_i$, respectively. Then y corresponds to the maximal ideal of $R[t_1, \ldots, t_n]$ generated by the maximal ideal of R and $t_1 - a_1, \ldots, t_n - a_n$. The R-homomorphism

$$R[t_1, \ldots, t_n] \to R, \quad t_i \mapsto a_i$$

induces an R-homomorphism $A \to R$ by passing to localization and then to henselization. The corresponding morphism $\operatorname{Spec} R \to \operatorname{Spec} A$ is a section for $\operatorname{Spec} A \to S$. □

Proposition 2.8.14. *Let $(R, \mathfrak{m})$ be a local ring, $k = R/\mathfrak{m}$ the residue field, $S = \operatorname{Spec} R$, and s the closed point of S. The following conditions are equivalent:*

(i) *R is strictly henselian.*

(ii) *R is henselian, and any finite etale S-scheme X is isomorphic to a disjoint union of copies of S.*

(iii) *For any etale morphism $g : X \to S$ and any point $x \in X$ lying above s, there exists a section $h : S \to X$ of g such that $h(s) = x$.*

Proof.

(i)⇒(iii) The residue field of X at x is isomorphic to k since it is finite separable over k and k is separably closed. So there exists a section of $g_s : X \otimes_R k \to \operatorname{Spec} k$ with image x. We then apply 2.8.3 (vii).

(iii)⇒(ii) By 2.8.3, R is henselian. Suppose that $X = \operatorname{Spec} B$ for some finite etale R-algebra B. Then B is a direct product of finitely many local rings. Without loss of generality, assume B local. The closed point of X is above the closed point of S. If the condition (iii) holds, then $X \to S$ admits a section. Since X and S are connected, we must have $X \cong S$ by 2.3.10 (i).

(ii)⇒(i) We need to show that k is separably closed. If this is not true, then there exists a monic irreducible polynomial $\bar{f}(t) \in k[t]$ of degree > 1 such that $\bar{f}(t)$ and $\bar{f}'(t)$ generate the ideal $k[t]$. Let $f(t) \in R[t]$ be a monic polynomial lifting $\bar{f}(t)$. Applying Nakayama's lemma to $R[t]/(f(t))$, we see $f(t)$ and $f'(t)$ generate the ideal $R[t]$. By 2.3.3, $R[t]/(f(t))$ is etale and finite over R. If the condition (ii) holds, then $R[t]/(f(t))$ is isomorphic to a direct product of copies of R as an R-algebra. This implies that $f(t)$ is a product of linear polynomials. Then $\bar{f}(t)$ has the same property. This contradicts the fact that $\bar{f}(t)$ is irreducible and has degree > 1. □

We leave the proof of 2.8.15–19 below to the reader.

Lemma 2.8.15. *Let A be a local ring, A' an essentially etale local A-algebra, R a henselian local ring, and $k(A)$, $k(A')$ and $k(R)$ the residue fields of A, A' and R, respectively. Given a local homomorphism $\phi : A \to R$ and a $k(A)$-homomorphism $\alpha : k(A') \to k(R)$, there exists a unique local A-homomorphism $\phi' : A' \to R$ inducing the homomorphism α on residue fields.*

Lemma 2.8.16. *Let A be a local ring, let A_1 and A_2 be essentially etale local A-algebras, let $k(A)$, $k(A_1)$ and $k(A_2)$ be the residue fields of A, A_1 and A_2 respectively, and let k be a field containing $k(A)$.*

(i) *For $k(A)$-homomorphisms $\beta_i : k(A_i) \to k$ $(i = 1, 2)$, there exist an essentially etale local A-algebra A_3 and a $k(A)$-homomorphism $\beta_3 : k(A_3) \to k$ such that (A_3, β_3) dominates (A_i, β_i) $(i = 1, 2)$, that is, there exist A-homomorphisms (necessarily local) $\phi_i : A_i \to A_3$ such that $\beta_i = \beta_3\bar{\phi}_i$, where $\bar{\phi}_i$ are the homomorphisms induced by ϕ_i on residue fields.*

(ii) *For any $k(A)$-homomorphism $\gamma : k(A_1) \to k(A_2)$, there exists at most one A-homomorphism from A_1 to A_2 inducing the homomorphism γ on residue fields.*

Let $(A, \mathfrak{m})$ be a local ring, $k = A/\mathfrak{m}$ its residue fields, Ω a separable closure of k, and $i : k \to \Omega$ the inclusion. Let $\mathscr{T}$ be the category defined as follows: Objects in $\mathscr{T}$ are pairs $(A', \beta_{A'})$, where A' are essentially etale local A-algebras and $\beta_{A'} : k(A') \to \Omega$ are k-homomorphisms from the residue fields $k(A')$ of A' to Ω. A morphism $(A_1, \beta_{A_1}) \to (A_2, \beta_{A_2})$ in $\mathscr{T}$ is an A-algebra homomorphism $\phi : A_1 \to A_2$ such that $\beta_{A_1} = \beta_{A_2}\bar{\phi}$, where $\bar{\phi}$ is the homomorphism induced by ϕ on residue fields. By 2.8.16, $\mathscr{T}$ satisfies the condition (I2) and (I3) in 2.7. There exists a full subcategory of $\mathscr{T}$ whose objects form a set such that any object in $\mathscr{T}$ is isomorphic to an object in this subcategory. We define the *strict henselization* A_i^{hs} of A relative to i to be

$$A_i^{hs} = \varinjlim_{(A',\beta_{A'})\in \mathrm{ob}\,\mathscr{T}} A'.$$

We also call A_i^{hs} the *strict localization* of A relative to i. We often denote A_i^{hs} by A^{hs} or $\widetilde{A}$.

Proposition 2.8.17. *Let $(A, \mathfrak{m})$ be a local ring, $k = A/\mathfrak{m}$ the residue field, Ω a separable closure of k, and $i : k \to \Omega$ the inclusion.*

(i) *A_i^{hs} is a strictly henselian local ring, the canonical homomorphism $A \to A_i^{hs}$ is local and faithfully flat, $\mathfrak{m}A_i^{hs}$ is the maximal ideal of A_i^{hs}, and the residue field of A_i^{hs} is k-isomorphic to Ω.*

(ii) *Let R be a henselian local ring with residue field $k(R)$, $\phi : A \to R$ a local homomorphism, and $\alpha : \Omega \to k(R)$ a homomorphism such that $\alpha i : k \to k(R)$ coincides with the homomorphism induced by ϕ. Then there exists a unique local A-homomorphism $\phi' : A_i^{hs} \to R$ inducing the homomorphism α on residue fields.*

(iii) *If A is strictly henselian, then $A \cong A_i^{hs}$.*

(iv) *If A is noetherian, then so is A_i^{hs}.*

(v) *Suppose $i' : k \to \Omega'$ is the inclusion of k into another separable closure of k. Then for any k-isomorphism $\sigma : \Omega \to \Omega'$, the A-homomorphism $A_i^{hs} \to A_{i'}^{hs}$ defined in* (ii) *is an isomorphism. We have*

$$\mathrm{Aut}(A_i^{hs}/A) \cong \mathrm{Gal}(\Omega/k).$$

Proposition 2.8.18. *Let A be a local noetherian ring. Then A is reduced (resp. regular, resp. normal) if and only if A^{hs} is so.*

Proposition 2.8.19. *Let $(R_\lambda, \phi_{\lambda\mu})$ be a direct system of local rings such that $\phi_{\lambda\mu}$ are local homomorphisms.*

(i) *If each R_λ is strictly henselian, so is $\varinjlim_\lambda R_\lambda$.*

(ii) *In general, let i be the inclusion of the residue field of $\varinjlim_\lambda R_\lambda$ in one of its separable closure Ω, and for each λ, let i_λ be the inclusion of the residue field of R_λ in its separable closure in Ω. Then we have*

$$(\varinjlim_\lambda R_\lambda)_i^{hs} \cong \varinjlim_\lambda (R_\lambda)_{i_\lambda}^{hs}.$$

Proposition 2.8.20. *Let $(A, \mathfrak{m})$ be a local ring, B a finite A-algebra, $\tilde{\mathfrak{n}}$ a maximal ideal of $B \otimes_A A^{hs}$, and $\mathfrak{n}$ the inverse image of $\tilde{\mathfrak{n}}$ in B. Then $\mathfrak{n}$ is a maximal ideal of B, the residue field $k(\tilde{\mathfrak{n}}) = (B \otimes_A A^{hs})/\tilde{\mathfrak{n}}$ is a separable* *closure of the residue field* $k(\mathfrak{n}) = B/\mathfrak{n}$, *and* $(B \otimes_A A^{hs})_{\tilde{\mathfrak{n}}}$ *is isomorphic to the strict henselization of* $B_{\mathfrak{n}}$ *with respect to the separable closure* $k(\tilde{\mathfrak{n}})$ *of* $k(\mathfrak{n})$.

Proof. Since $B \otimes_A A^{hs}$ is finite over A^{hs}, $\tilde{\mathfrak{n}}$ lies above the maximal ideal of A^{hs}, and hence above the maximal ideal of A. So $\mathfrak{n}$ is above the maximal ideal of A. As B is finite over A, $\mathfrak{n}$ is a maximal ideal of B. Note that $k(\tilde{\mathfrak{n}})$ is isomorphic to the residue field of $k(\mathfrak{n}) \otimes_{k(A)} k(A^{hs})$ at a prime ideal, where $k(A)$ and $k(A^{hs})$ are the residue fields of A and A^{hs}, respectively. Since $k(\mathfrak{n})$ is finite over $k(A)$ and $k(A^{hs})$ is a separable closure of $k(A)$, $k(\tilde{\mathfrak{n}})$ is a separable closure of $k(\mathfrak{n})$. Since A^{hs} is henselian and $B \otimes_A A^{hs}$ is finite over A^{hs}, we have

$$B \otimes_A A^{hs} \cong (B \otimes_A A^{hs})_{\tilde{\mathfrak{n}}_1} \times \cdots \times (B \otimes_A A^{hs})_{\tilde{\mathfrak{n}}_k},$$

where $\tilde{\mathfrak{n}}_i$ $(i = 1, \ldots, k)$ are all the maximal ideals of $B \otimes_A A^{hs}$. For each i, let e_i be the element in $B \otimes_A A^{hs}$ so that its projection to $(B \otimes_A A^{hs})_{\tilde{\mathfrak{n}}_j}$ is 0 if $j \neq i$ and 1 if $j = i$. We have

$$e_i^2 = e_i, \quad e_i e_j = 0 \quad (i \neq j), \quad e_1 + \cdots + e_k = 1.$$

Since $B \otimes_A A^{hs} = \varinjlim_{A' \in \mathrm{ob}\, \mathscr{T}} B \otimes_A A'$, there exists an essentially etale local A-algebra A' in $\mathscr{T}$ such that $e_1, \ldots, e_k$ are images of some elements $e'_1, \ldots, e'_k \in B \otimes_A A'$ with the property

$$e_i'^2 = e'_i, \quad e'_i e'_j = 0 \quad (i \neq j), \quad e'_1 + \cdots + e'_k = 1.$$

We then have

$$B \otimes_A A' \cong (B \otimes_A A')e'_1 \times \cdots \times (B \otimes_A A')e'_k.$$

On the other hand, we have

$$(B \otimes_A A')e'_i \otimes_{A'} A^{hs} \cong (B \otimes_A A^{hs})e_i \cong (B \otimes_A A^{hs})_{\tilde{\mathfrak{n}}_i}.$$

In particular, $(B \otimes_A A')e'_i \otimes_{A'} A^{hs}$ is local. Note that the canonical morphism

$$\mathrm{Spec}\,((B \otimes_A A')e'_i \otimes_{A'} A^{hs}) \otimes_{A^{hs}} A^{hs}/\mathfrak{m}A^{hs} \to \mathrm{Spec}\,(B \otimes_A A')e'_i \otimes_{A'} A'/\mathfrak{m}A'$$

is surjective. It follows that $(B \otimes_A A')e'_i$ is local and its maximal ideal lies above the maximal ideal $\mathfrak{n}_i = \tilde{\mathfrak{n}}_i \cap B$ of B. Moreover $(B \otimes_A A')e'_i$ is essentially etale over $B_{\mathfrak{n}_i}$, so we have

$$(B_{\mathfrak{n}_i})^{hs} \cong ((B \otimes_A A')e'_i)^{hs}.$$

We also have $A^{hs} \cong A'^{hs}$. Replacing A by A' and B by $(B \otimes_A A')e'_i$, we are reduced to the case where B and $B \otimes_A A^{hs}$ are local. In this case, $B \otimes_A A^{hs}$ is a strictly henselian local ring. We have $B \otimes_A A^{hs} \cong \varinjlim_{A' \in \mathrm{ob}\, \mathscr{T}} (B \otimes_A A')$ and each $B \otimes_A A'$ is an essentially etale local B-algebra. So we have $B \otimes_A A^{hs} \cong B^{hs}$. □

2.9 Etale Morphisms between Normal Schemes

([SGA 1] I 10, [EGA IV] 18.10.)

Proposition 2.9.1. *Let $f : X \to Y$ be a morphism locally of finite type between noetherian schemes. Assume that for any $y \in f(X)$, $\mathscr{O}_{Y,y}$ is normal. The following conditions are equivalent:*

(i) f is etale.

(ii) f is unramified and the canonical homomorphism $\mathscr{O}_{Y,f(x)} \to \mathscr{O}_{X,x}$ is injective for any $x \in X$.

(iii) f is unramified and $\dim \mathscr{O}_{Y,f(x)} \leq \dim \mathscr{O}_{X,x}$ for any $x \in X$.

Proof.

(i)⇒(iii) is clear.

(iii)⇒(ii) Let I be the kernel of the homomorphism $\mathscr{O}_{Y,f(x)} \to \mathscr{O}_{X,x}$. Shrinking Y, we may assume $I = \mathscr{I}_{f(x)}$ for some coherent ideal $\mathscr{I}$ of $\mathscr{O}_Y$. Let Y_1 be the closed subscheme of Y with the ideal sheaf $\mathscr{I}$. Shrinking X and Y, we may assume that f can be factorized as a composite

$$X \to Y_1 \to Y.$$

We have

$$\dim \mathscr{O}_{X,x} \leq \dim \mathscr{O}_{Y_1,f(x)} + \dim \mathscr{O}_{X,x}/\mathfrak{m}_{Y_1,f(x)}\mathscr{O}_{X,x}$$

by [Matsumura (1970)] (13.B) Theorem 19 (1). Note that $X \to Y_1$ is unramified and hence $\dim \mathscr{O}_{X,x}/\mathfrak{m}_{Y_1,f(x)}\mathscr{O}_{X,x} = 0$. So we have $\dim \mathscr{O}_{X,x} \leq \dim \mathscr{O}_{Y_1,f(x)}$, and hence $\dim \mathscr{O}_{Y,f(x)} \leq \dim \mathscr{O}_{Y_1,f(x)}$. Since $\mathscr{O}_{Y,f(x)}$ is normal and hence an integral domain, this implies that $I = 0$. So $\mathscr{O}_{Y,f(x)} \to \mathscr{O}_{X,x}$ is injective.

(ii)⇒(i) For any $x \in X$, let $\widetilde{\mathscr{O}}_{Y,f(x)}$ be the strict henselization of $\mathscr{O}_{Y,f(x)}$ with respect to a separable closure of $k(f(x))$, let $\widetilde{X} = X \times_Y \operatorname{Spec} \widetilde{\mathscr{O}}_{Y,f(x)}$, and let $x' \in \widetilde{X}$ be a point lying above $x \in X$ and the closed point of $\operatorname{Spec} \widetilde{\mathscr{O}}_{Y,f(x)}$. We claim that the homomorphism $\widetilde{\mathscr{O}}_{Y,f(x)} \to \mathscr{O}_{\widetilde{X},x'}$ is injective. Indeed, $\mathscr{O}_{Y,f(x)}$ is normal and hence an integral domain. Let $y_1 \in Y$ be the point corresponding to the generic point of $\operatorname{Spec} \mathscr{O}_{Y,f(x)}$. Since $\mathscr{O}_{Y,f(x)} \to \mathscr{O}_{X,x}$ is injective, the ring $\mathscr{O}_{X,x} \otimes_{\mathscr{O}_{Y,f(x)}} \mathscr{O}_{Y,y_1}$ is nonzero. Hence there exists $x_1 \in X$ lying above y_1 such that $x \in \overline{\{x_1\}}$. Since $\mathscr{O}_{\widetilde{X},x'}$ is faithfully flat over $\mathscr{O}_{X,x}$, there exists $x_1' \in \widetilde{X}$ lying above x_1 such that $x' \in \overline{\{x_1'\}}$. By 2.8.18, $\widetilde{\mathscr{O}}_{Y,f(x)}$ is normal and hence an integral domain. Since the generic point of $\operatorname{Spec} \widetilde{\mathscr{O}}_{Y,f(x)}$ is the only point lying above the generic point in Spec $\mathscr{O}_{Y,f(x)}$, and x'_1 lies above the generic point of Spec $\mathscr{O}_{Y,f(x)}$, it follows that x'_1 lies above the generic point of Spec $\widetilde{\mathscr{O}}_{Y,f(x)}$. This implies that the homomorphism $\widetilde{\mathscr{O}}_{Y,f(x)} \to \mathscr{O}_{\widetilde{X},x'}$ is injective.

Note that $\widetilde{X} \to \operatorname{Spec} \widetilde{\mathscr{O}}_{Y,f(x)}$ is unramified and hence quasi-finite. Let U be an affine open neighborhood of x' in $\widetilde{X}$. By the Zariski Main Theorem 1.10.13, $U \to \operatorname{Spec} \mathscr{O}_{Y,f(x)}$ can be factorized as

$$U \stackrel{j}{\hookrightarrow} \overline{U} \to \operatorname{Spec} \widetilde{\mathscr{O}}_{Y,f(x)}$$

such that j is an open immersion and $\overline{U} \to \operatorname{Spec} \widetilde{\mathscr{O}}_{Y,f(x)}$ is finite. By 2.8.3 (i), we have

$$\overline{U} = \coprod_{\bar{x}} \operatorname{Spec} \mathscr{O}_{\overline{U},\bar{x}},$$

where $\bar{x}$ goes over the set of points of $\overline{U}$ above the closed point of Spec $\widetilde{\mathscr{O}}_{Y,f(x)}$. Note that $j(x')$ is such a point. Since j is an open immersion, we have

$$\mathscr{O}_{\widetilde{X},x'} \cong \mathscr{O}_{U,x'} \cong \mathscr{O}_{\overline{U},j(x')}.$$

Since $\overline{U} \to \operatorname{Spec} \widetilde{\mathscr{O}}_{Y,f(x)}$ is finite, $\mathscr{O}_{\overline{U},j(x')}$ is finite over $\widetilde{\mathscr{O}}_{Y,f(x)}$. So $\mathscr{O}_{\widetilde{X},x'}$ is finite over $\widetilde{\mathscr{O}}_{Y,f(x)}$. Let $\mathfrak{m}$ be the maximal ideal of $\widetilde{\mathscr{O}}_{Y,f(x)}$. Since $\widetilde{X} \to \operatorname{Spec} \widetilde{\mathscr{O}}_{Y,f(x)}$ is unramified and the residue field of $\widetilde{\mathscr{O}}_{Y,f(x)}$ is separably closed, we have

$$\widetilde{\mathscr{O}}_{Y,f(x)}/\mathfrak{m} \cong \mathscr{O}_{\widetilde{X},x'}/\mathfrak{m}\mathscr{O}_{\widetilde{X},x'}.$$

By Nakayama's lemma, $\widetilde{\mathscr{O}}_{Y,f(x)} \to \mathscr{O}_{\widetilde{X},x'}$ is surjective. We have shown it is injective. So we have

$$\widetilde{\mathscr{O}}_{Y,f(x)} \cong \mathscr{O}_{\widetilde{X},x'}.$$

Consider the commutative diagram

$$\begin{array}{ccc} \mathscr{O}_{Y,f(x)} & \to & \mathscr{O}_{X,x} \\ \downarrow & & \downarrow \\ \widetilde{\mathscr{O}}_{Y,f(x)} & \stackrel{\cong}{\to} & \mathscr{O}_{\widetilde{X},x'}. \end{array}$$

Since $\widetilde{\mathscr{O}}_{Y,f(x)}$ is faithfully flat over $\mathscr{O}_{Y,f(x)}$, $\mathscr{O}_{\widetilde{X},x'}$ is faithfully flat over $\mathscr{O}_{Y,f(x)}$. On the other hand, $\mathscr{O}_{\widetilde{X},x'}$ is faithfully flat over $\mathscr{O}_{X,x}$. It follows that $\mathscr{O}_{X,x}$ is faithfully flat over $\mathscr{O}_{Y,f(x)}$. This is true for all $x \in X$. So $f : X \to Y$ is flat. Since it is unramified, it is etale. □

Proposition 2.9.2. *Let Y be a normal connected noetherian scheme and let X be an etale separated Y-scheme of finite type. Then each connected component X_i of X is normal and is an open subscheme of the normalization of Y in the function field of X_i. X_i is finite over Y if and only if it is equal to the this normalization.*

Proof. Let K_i and K be the function fields of X_i and Y, respectively, and let Y_i' be the normalization of Y in K_i. Note that Y_i' is finite over Y since K_i is finite separable over K. By 2.4.4, X is normal. So we have a birational Y-morphism $X_i \to Y_i'$. Since X_i is quasi-finite and separated over Y, $X_i \to Y_i'$ is also quasi-finite and separated. By the Zariski Main Theorem 1.10.13, it can be factorized as a composite

$$X_i \hookrightarrow X_i' \to Y_i'$$

such that $X_i \hookrightarrow X_i'$ is an open immersion, and $X_i' \to Y_i'$ is finite. Replacing X_i' by the closure of X_i with the reduced closed subscheme structure, we may assume that X_i' is integral. Note that $X_i' \to Y_i'$ is a finite birational morphism. Since Y_i' is normal, we have $X_i' \cong Y_i'$. So $X_i \to Y_i'$ is an open immersion. If X_i is finite over Y, then X_i is finite over Y_i' and hence $X_i \cong Y_i'$. □

Let Y be a normal connected noetherian scheme, K its function field, and L a finite separable extension of K. We say L is unramified over Y if the normalization of Y in L is unramified over Y, or equivalently, etale over Y. By 2.9.2, the category of connected finite etale Y-schemes is equivalent to the category of finite separable extensions of $K(Y)$ unramified over Y. Suppose $L = L_1 \times \cdots \times L_n$ for some finite separable extensions L_i of K. We say that L is unramified over Y if each L_i is unramified over Y.

Proposition 2.9.3. *Let Y be a normal connected noetherian scheme and let K be its function field.*

(i) *K is unramified over Y.*

(ii) *Let L be an extension of K unramified over Y, M an extension of L unramified over the normalization of Y in L. Then M is unramified over Y.*

(iii) *Let Y' be a normal connected noetherian Y-scheme dominating Y and let K' be its function field. If L is an extension of K unramified over Y, then $L \otimes_K K'$ is unramified over Y'. If $Y = \operatorname{Spec} A$ and $Y' = \operatorname{Spec} A'$ are affine and B is the integral closure of A in L, then $B \otimes_A A'$ is the integral closure of A' in $L \otimes_K K'$.*

(iv) *If L_1 and L_2 are extensions of K unramified over Y, then $L_1 \otimes_K L_2$ is unramified over Y.*

Proof. (i) is obvious. (ii) follows from 2.2.2 (iii).

(iii) It suffices to consider the case where $Y = \operatorname{Spec} A$ and $Y' = \operatorname{Spec} A'$ are affine. Note that $\operatorname{Spec}(B \otimes_A A')$ is finite and etale over $\operatorname{Spec} A'$. So $\operatorname{Spec}(B \otimes_A A')$ is normal. It function ring is isomorphic to

$$B \otimes_A A' \otimes_{A'} K' \cong B \otimes_A K \otimes_K K' \cong L \otimes_K K'.$$

So $B \otimes_A A'$ is the integral closure of A' in $L \otimes_K K'$, and $L \otimes_K K'$ is unramified over $\operatorname{Spec} A'$.

(iv) follows from (ii) and (iii). $\square$

Chapter 3

Etale Fundamental Groups

3.1 Finite Group Actions on Schemes

([SGA 1] V 1.)

Let X be a scheme on which a finite group G acts on the right. If $X = \operatorname{Spec} A$, this is equivalent to saying G acts on A on the left. For any scheme Z, G acts on the left on the set $\operatorname{Hom}(X, Z)$. Let $\operatorname{Hom}(X, Z)^G$ be the subset of $\operatorname{Hom}(X, Z)$ consisting of morphisms invariant under G. A morphism $X \to Y$ invariant under the action of G is called the *quotient* of X by G if the canonical map

$$\operatorname{Hom}(Y, Z) \to \operatorname{Hom}(X, Z)^G$$

is bijective for any scheme Z, that is, Y represents the functor $Z \mapsto \operatorname{Hom}(X, Z)^G$. We often denote Y by X/G.

Proposition 3.1.1. *Let A be a ring on which a finite group G acts on the left, $B = A^G$, $X = \operatorname{Spec} A$, $Y = \operatorname{Spec} B$, and $\pi : X \to Y$ the morphism corresponding to the homomorphism $A^G \hookrightarrow A$.*

(i) *X is integral over Y.*

(ii) *π is surjective. Its fibers are orbits of G, and the topology on Y is the quotient topology induced from X.*

(iii) *Given $x \in X$, let $y = \pi(x)$ and let*

$$G_x = \{g \in G | gx = x\}$$

be the stabilizer of x. Then the residue field $k(x)$ is a normal algebraic extension of the residue field $k(y)$, and the canonical homomorphism $G_x \to \operatorname{Gal}(k(x)/k(y))$ is surjective.

(iv) *The canonical morphism $\mathscr{O}_Y \to (\pi_* \mathscr{O}_X)^G$ is an isomorphism, and Y is the quotient of X by G.*

Proof.

(i) For any $a \in A$, the polynomial $\prod_{g\in G}(t - ga)$ in $A[t]$ is invariant under the action of G. So it lies in $B[t]$. a is a root of this polynomial. So A is integral over B.

(ii) By [Atiyah and Macdonald (1969)] 5.10, π is surjective, and $\pi(V(I)) = V(I \cap B)$ for any ideal I of A. So π is a closed map, and hence the topology on Y is the quotient topology induced from X. Let $\mathfrak{q}$ be a prime ideal of B, and let $\mathfrak{p}$ and $\mathfrak{p}'$ be prime ideals of A such that

$$\mathfrak{q} = \mathfrak{p} \cap B = \mathfrak{p}' \cap B.$$

Then for any $a \in \mathfrak{p}'$, we have

$$\prod_{g\in G} ga \in \mathfrak{p}' \cap B = \mathfrak{q} \subset \mathfrak{p}.$$

So $ga \in \mathfrak{p}$ for some $g \in G$. It follows that $\mathfrak{p}' \subset \cup_{g\in G} g\mathfrak{p}$. This implies that $\mathfrak{p}' \subset g\mathfrak{p}$ for some $g \in G$ by [Matsumura (1970)] (1.B). We have

$$g\mathfrak{p} \cap B = g(\mathfrak{p} \cap B) = \mathfrak{p} \cap B = \mathfrak{p}' \cap B.$$

By [Atiyah and Macdonald (1969)] 5.9, we have $\mathfrak{p}' = g\mathfrak{p}$. So the fibers of π are orbits of G.

(iii) Let $\mathfrak{p}$ be the prime ideal of A corresponding to x and let $\mathfrak{q} = \mathfrak{p} \cap B$ be the prime ideal corresponding to y. Replacing A and B by $A \otimes_B B_\mathfrak{q}$ and $B_\mathfrak{q}$, respectively, we may assume B is a local ring with the maximal ideal $\mathfrak{q}$. Then $\mathfrak{p}$ is a maximal ideal of A. Any $a \in A$ is a root of the polynomial of $\prod_{g\in G}(t - ga) \in B[t]$. So every element in $k(x)$ is a root of a polynomial in $k(y)[t]$ splitting in $k(x)$ with degree $|G|$. Hence $k(x)/k(y)$ is a normal algebraic extension. Since any finite separable extension is generated by one element, the separable closure $k(y)_s$ of $k(y)$ in $k(x)$ has degree $\leq |G|$. Let $a \in A$ so that its image $\bar{a}$ in $k(x)$ lies in $k(y)_s$ and generates $k(y)_s$ over $k(y)$. For any $g \in G - G_x$, $g\mathfrak{p}$ is a maximal ideal of A distinct from $\mathfrak{p}$. By the Chinese remainder theorem, there exists $a' \in A$ such that $a' - a \in \mathfrak{p}$ and $a' \in g\mathfrak{p}$ for any $g \notin G - G_x$. Let $\overline{ga'}$ be the image of ga' in $k(x)$, and let

$$f(t) = \prod_{g\in G} (t - \overline{ga'}).$$

We have $f(t) \in k(y)[t]$ and $f(\overline{a'}) = 0$. Given $\tau \in \mathrm{Gal}(k(x)/k(y))$, $\tau(\overline{a'})$ is a root of $f(t)$. So $\tau(\overline{a'}) = \overline{ga'}$ for some $g \in G$. If $g \notin G_x$, we have $ga' \in \mathfrak{p}$ and hence $\overline{ga'} = 0$. So $\tau(\overline{a'}) = 0$ and hence $\overline{a'} = 0$. But $\overline{a'} = \bar{a}$ generates $k(y)_s$ over $k(x)$. So we have $k(y)_s = k(y)$ and $\mathrm{Gal}(k(x)/k(y)) = \{e\}$. The homomorphism $G_x \to \mathrm{Gal}(k(x)/k(y))$ is trivially surjective in this case. If $g \in G_x$, then τ is the image of g under this homomorphism. So this homomorphism is always surjective.

(iv) For any $f \in B$, we have $B_f \cong (A_f)^G$, that is,

$$\mathscr{O}_Y(D(f)) \cong (\pi_*\mathscr{O}_X)^G(D(f)).$$

So $\mathscr{O}_Y \to (\pi_*\mathscr{O}_X)^G$ is an isomorphism. Using this fact and (ii), one checks that Y is the quotient of X by G. □

Let X be a scheme on which a finite group G acts on the right. We say the action is *admissible* if there exists an affine morphism $\pi : X \to Y$ invariant under G such that $\mathscr{O}_Y = (\pi_*\mathscr{O}_X)^G$. Then 3.1.1 still holds and every open subset U of Y is the quotient of $\pi^{-1}(U)$ by G.

Proposition 3.1.2. *Let X be a scheme on which a finite group G acts on the right. The following conditions are equivalent:*

(i) *The action is admissible.*

(ii) *X is a union of affine open subsets stable under the action of G.*

(iii) *Each orbit of G is contained in an affine open subset.*

Proof.

(i)⇒(iii) Let $\pi : X \to X/G$ be the canonical morphism. It is affine. Cover X/G by affine open subsets U_i. Then $\{\pi^{-1}(U_i)\}$ is an affine open covering of X. Each orbit of G is contained in one of the affine $\pi^{-1}(U_i)$.

(iii)⇒(ii) Let S be an orbit of G and let U be an affine open subset of X containing S. Then $S \subset \cap_{g\in G} gU \subset U$. Since U is affine and S is finite, there exists an affine open subset V such that $S \subset V \subset \cap_{g\in G} gU$. Indeed, suppose $U = \mathrm{Spec}\, A$, $S = \{\mathfrak{p}_1, \ldots, \mathfrak{p}_n\}$, and $\cap_{g\in G} gU = \mathrm{Spec}\, A - V(\mathfrak{a})$, where $\mathfrak{p}_i$ are prime ideals, and $\mathfrak{a}$ is an ideal of A. We have $\mathfrak{a} \not\subset p_i$. By [Matsumura (1970)] (1.B), we have $\mathfrak{a} \not\subset \cup_i \mathfrak{p}_i$. Take $f \in \mathfrak{a} - \cup_i \mathfrak{p}_i$. Then $V = D(f)$ has the required property. Note that $gV \subset U$ for all $g \in G$ and gV are affine. It follows that $\cap_{g\in G} gV$ is affine. It is also open, stable under G, and contains S. Since S is an arbitrary orbit, X is a union of affine open subsets stable under G.

(ii)⇒(i) Let Y be the quotient topological space of X by G, and let $\pi : X \to Y$ be the canonical continuous map. Then the ringed space $(Y, (\pi_*\mathscr{O}_X)^G)$ is a scheme and is the quotient of X by G. Indeed, if $X = \cup_i U_i$, where each $U_i = \mathrm{Spec}\, A_i$ is affine open and stable under the action of G, then Y can be covered by the affine schemes $\mathrm{Spec}\, A_i^G$. □

Corollary 3.1.3. *Let X be a scheme on which a finite group G acts on the right.*

(i) *If G acts admissibly on X, then any subgroup of G acts admissibly on X.*

(ii) *If there exists an affine morphism $f : X \to Z$ invariant under G, then G acts admissibly on X and X/G is Z-isomorphic to $\mathbf{Spec}\,(f_*\mathscr{O}_X)^G$.*

Proposition 3.1.4. *Let X be a scheme on which a finite group G acts on the right and let $X \to Y$ a morphism invariant under G. For any morphism $Y' \to Y$, let G act on $X \times_Y Y'$ by base change.*

(i) *If G acts admissibly on X, $Y \cong X/G$, and $Y' \to Y$ is flat, then G acts admissibly on $X \times_Y Y'$ and $Y' \cong (X \times_Y Y')/G$.*

(ii) *If $Y' \to Y$ is quasi-compact and faithfully flat, G acts admissibly on $X \times_Y Y'$, and $Y' \cong (X \times_Y Y')/G$, then G acts admissibly on X and $Y \cong X/G$.*

Proof.

(i) We may assume $X = \operatorname{Spec} A$, where A is a ring on which G acts on the left. Let $B = A^G$. Then $Y = \operatorname{Spec} B$. We may assume that $Y' = \operatorname{Spec} B'$ is affine, where B' is a flat B-algebra. The sequence

$$0 \to B \to A \to \prod_{g\in G} A$$

is exact, where the homomorphism $A \to \prod_{g\in G} A$ is defined by

$$a \mapsto (a - ga).$$

Since B' is flat over B, the sequence

$$0 \to B' \to A \otimes_B B' \to \prod_{g\in G} (A \otimes_B B')$$

is exact. It follows that $B' \cong (A \otimes_B B')^G$. So G acts admissibly on $X \times_Y Y'$, and $Y' \cong (X \times_Y Y')/G$.

(ii) We may reduce to the case where $Y = \operatorname{Spec} B$, $Y' = \operatorname{Spec} B'$ and B' is a faithfully flat B-algebra. By 1.8.7, $X \to Y$ is affine. Let $X = \operatorname{Spec} A$. Then the sequence

$$0 \to B' \to A \otimes_B B' \to \prod_{g\in G} (A \otimes_B B')$$

is exact. Since B' is faithfully flat over B, the sequence

$$0 \to B \to A \to \prod_{g\in G} A$$

is exact. So $B \cong A^G$. Hence G acts admissibly on X and $Y \cong X/G$. $\square$

3.2 Etale Covering Spaces and Fundamental Groups

([SGA 1] V 2–5.)

Let X be a scheme. A *geometric point* of X is a morphism $\gamma : s \to X$ such that s is the spectrum of a separably closed field. We often write $k(s)$ for the separably closed field such that $s = \operatorname{Spec} k(s)$. Giving a geometric point $\gamma : s \to X$ with image $x \in X$ is equivalent to giving a separably closed extension $k(s)$ of the residue field $k(x)$. Let $f : X \to X'$ be a morphism. Then $f\gamma : s \to X'$ is a geometric point of X'. A *pointed scheme* is a pair (X, γ) such that X is a scheme, and $\gamma : s \to X$ is a geometric point of X. A morphism $f : (X, \gamma) \to (X', \gamma')$ between pointed schemes is a morphism $f : X \to X'$ such that $f\gamma = \gamma'$.

Let X be a scheme on which a finite group G acts on the right and let $x \in X$. The stabilizer

$$G_x = \{g \in G | gx = x\}$$

of x is called the *decomposition subgroup* at x and is denoted by $G_d(x)$. This subgroup acts on the residue field $k(x)$. The subgroup of elements in $G_d(x)$ acting trivially on $k(x)$ is called the *inertia subgroup* at x and is denoted by $G_i(x)$. Let $\gamma : s \to X$ be a geometric point with image x. G acts on the set $\operatorname{Hom}(s, X)$. $G_i(x)$ is exactly the stabilizer of $\gamma \in \operatorname{Hom}(s, X)$:

$$G_i(x) = \{g \in G | g\gamma = \gamma\}.$$

Using this interpretation of the inertia subgroup, one can prove the following:

Proposition 3.2.1. *Consider a Cartesian diagram*

$$\begin{array}{ccc} X \times_S S' & \to & X \\ \downarrow & & \downarrow \\ S' & \to & S. \end{array}$$

Suppose a finite group G acts on X on the right, and suppose $X \to S$ is invariant under G. Let G act on $X \times_S S'$ by base change. For any $x' \in X \times_S S'$ with image x in X, we have $G_i(x) = G_i(x')$.

An *etale covering space* of a scheme S is a finite etale morphism $X \to S$. The group $\operatorname{Aut}(X/S)$ of S-automorphisms on X acts on X on the left. Let $G = \operatorname{Aut}(X/S)^\circ$ be the opposite group of $\operatorname{Aut}(X/S)$. Its elements are S-automorphisms of X, and for any S-automorphisms g_1 and g_2, the product $g_1 g_2$ in G is defined to be the composite $g_2 \circ g_1$. Then G acts on X on the right. If $S = X/G$, we say $X \to S$ is a *galois etale covering space* with galois group G.

Lemma 3.2.2. *Let k be a field, and let A be a finite dimensional k-algebra. Then $A \cong \prod_i L_i$ for finitely many fields L_i finite separable over k if and only if the bilinear form*

$$A \times A \to k, \quad (x, y) \mapsto \mathrm{Tr}_{A/k}(xy)$$

is nondegenerate.

Proof. Suppose the bilinear form is nondegenerate. If $x \in A$ is nilpotent, then for any $y \in A$, xy is also nilpotent, and hence $\mathrm{Tr}(xy) = 0$. This implies $x = 0$. So A is reduced. Let K be any field containing k. Then the bilinear form

$$(A \otimes_k K) \times (A \otimes_k K) \to K, \quad (x, y) \mapsto \mathrm{Tr}(xy)$$

is nondegenerate. So $A \otimes_k K$ is reduced. As A is a finitely dimensional k-algebra, it is artinian and hence $A = \prod_i L_i$ for finitely many local artinian ring L_i. As A is reduced, each L_i must be a field. As $A \otimes_k K$ is reduced for any field K containing k, each L_i is a separable extension of k.

To prove the converse, it suffices to show for any field K finite separable over k, that the bilinear form

$$K \times K \to k, \quad (x, y) \mapsto \mathrm{Tr}(xy)$$

is nondegenerate. Choose $x \in K$ so that K is generated by x over k. Let $f(t) = t^n + a_1 t^{n-1} + \cdots + a_n$ be the minimal polynomial of x, and let $x_1, x_2, \ldots, x_n$ be all the roots of $f(t)$ in an algebraic closure of k. They are distinct. Since $\{1, x, \ldots, x^{n-1}\}$ is a basis of K over k, to prove that the above bilinear form is nondegenerate, it suffices to show

$$\det \begin{pmatrix} \mathrm{Tr}(1 \cdot 1) & \mathrm{Tr}(1 \cdot x) & \ldots & \mathrm{Tr}(1 \cdot x^{n-1}) \\ \mathrm{Tr}(x \cdot 1) & \mathrm{Tr}(x \cdot x) & \ldots & \mathrm{Tr}(x \cdot x^{n-1}) \\ \vdots & \vdots & & \vdots \\ \mathrm{Tr}(x^{n-1} \cdot 1) & \mathrm{Tr}(x^{n-1} \cdot x) & \ldots & \mathrm{Tr}(x^{n-1} \cdot x^{n-1}) \end{pmatrix} \neq 0.$$

This determinant is equal to

$$\det \begin{pmatrix} 1 & 1 & \ldots & 1 \\ x_1 & x_2 & \ldots & x_n \\ \vdots & \vdots & & \vdots \\ x_1^{n-1} & x_2^{n-1} & \ldots & x_n^{n-1} \end{pmatrix} \det \begin{pmatrix} 1 & x_1 & \ldots & x_1^{n-1} \\ 1 & x_2 & \ldots & x_2^{n-1} \\ \vdots & \vdots & \ldots & \vdots \\ 1 & x_n & \ldots & x_n^{n-1} \end{pmatrix} = \prod_{i<j} (x_i - x_j)^2.$$

Since $x_1, \ldots, x_n$ are distinct, we have $\prod_{i<j}(x_i - x_j)^2 \neq 0$. $\square$

Proposition 3.2.3. *Let $f : X \to S$ be a finite flat morphism between noetherian schemes. Then $\mathscr{A} = f_*\mathscr{O}_X$ is a locally free $\mathscr{O}_S$-module of finite rank. f is etale if and only if the bilinear form*

$$\mathscr{A} \times \mathscr{A} \to \mathscr{O}_S, \quad (x, y) \mapsto \mathrm{Tr}(xy)$$

is nondegenerate.

Proof. We may assume that $X = \operatorname{Spec} A$ and $S = \operatorname{Spec} B$ are affine. Then as a B-module, A is finitely generated and flat. By 1.6.7, $\mathscr{A} = A^\sim$ is a locally free $\mathscr{O}_S$-module of finite rank. As f is known to be flat, it is etale if and only if for any prime ideal $\mathfrak{q}$ of B, $A_\mathfrak{q}/\mathfrak{q}A_\mathfrak{q}$ can be written as $\prod_i L_i$, where L_i are finitely many fields finite separable over $B_\mathfrak{q}/\mathfrak{q}B_\mathfrak{q}$. By 3.2.2, the later condition is equivalent to saying that the bilinear form

$$A_\mathfrak{q}/\mathfrak{q}A_\mathfrak{q} \times A_\mathfrak{q}/\mathfrak{q}A_\mathfrak{q} \to B_\mathfrak{q}/\mathfrak{q}B_\mathfrak{q}, \quad (x, y) \mapsto \mathrm{Tr}(xy)$$

is nondegenerate. One then checks this last condition is equivalent to saying the bilinear form in the proposition is nondegenerate. □

Proposition 3.2.4. *Let (S, γ) be a pointed scheme, and let $(X_i, \alpha_i) \to (S, \gamma)$ $(i = 1, 2)$ be two morphisms of pointed schemes.*

(i) *If X_1 is connected and X_2 is unramified and separated over S, then there exists at most one S-morphism from (X_1, α_1) to (X_2, α_2).*

(ii) *Suppose X_1 and X_2 are etale covering spaces of S. There exists a connected pointed etale covering space (X_3, α_3) of (S, γ) dominating (X_i, α_i) $(i = 1, 2)$, that is, there exist S-morphisms from (X_3, α_3) to (X_i, α_i).*

Proof.

(i) An S-morphism f from X_1 to X_2 is completely determined by its graph $\Gamma_f : X_1 \to X_1 \times_S X_2$, which is a section of the projection $\pi_1 : X_1 \times_S X_2 \to X_1$. If $f(\alpha_1) = \alpha_2$, then $\Gamma_f(\alpha_1) = (\alpha_1, \alpha_2)$. Such Γ_f is unique if it exists by 2.3.10 (i).

(ii) α_1 and α_2 define a point $\alpha_3 = (\alpha_1, \alpha_2)$ of $X_1 \times_S X_2$. Let X_3 be the connected component of $X_1 \times_S X_2$ containing the image of α_3. Then (X_3, α_3) dominates (X_i, α_i) $(i = 1, 2)$. □

Proposition 3.2.5. *Let X be a scheme on which a finite group G acts admissibly on the right and let $Y = X/G$ be the quotient of X by G.*

(i) *Suppose X is of finite presentation over Y. If the inertia subgroup $G_i(x)$ is trivial for any $x \in X$, then X is etale over Y.*

(ii) *Suppose that X is connected, S is a noetherian scheme, $X \to S$ is an etale covering space, and $G \subset \mathrm{Aut}(X/S)^\circ$. Then $G_i(x)$ is trivial for any $x \in X$, $X \to Y$ and $Y \to S$ are etale covering spaces, and $G = \mathrm{Aut}(X/Y)^\circ$.*

Proof.

(i) For any $y \in Y$, let $\widetilde{\mathscr{O}}_{Y,\bar{y}}$ be the strict henselization of $\mathscr{O}_{Y,y}$ with respect to a separable closure of $k(y)$. By 2.5.10, it suffices to show $X \times_Y \operatorname{Spec} \widetilde{\mathscr{O}}_{Y,\bar{y}}$ is etale over $\operatorname{Spec} \widetilde{\mathscr{O}}_{Y,\bar{y}}$. By 3.1.4, $\operatorname{Spec} \widetilde{\mathscr{O}}_{Y,\bar{y}}$ is the quotient of $X \times_Y \operatorname{Spec} \widetilde{\mathscr{O}}_{Y,\bar{y}}$ by G. Making the base $\operatorname{Spec} \widetilde{\mathscr{O}}_{Y,\bar{y}} \to Y$, we are reduced to the case where $Y = \operatorname{Spec} A$ for some strictly henselian local ring A. Since X is of finite presentation and integral over Y, we have $X = \operatorname{Spec} B$ for some finite A-algebra B. Let $\mathfrak{n}_1, \ldots, \mathfrak{n}_k$ be all the maximal ideals of B. By 2.8.3, we have $B \cong B_{\mathfrak{n}_1} \times \cdots \times B_{\mathfrak{n}_k}$. Let $G_d(\mathfrak{n}_j)$ and $G_i(\mathfrak{n}_j)$ be the decomposition subgroups and inertia subgroups at $\mathfrak{n}_j$, respectively. Since $A = B^G$, and G acts transitively on $\{\mathfrak{n}_1, \ldots, \mathfrak{n}_k\}$ by 3.1.1 (ii), we have $A \cong (B_{\mathfrak{n}_j})^{G_d(\mathfrak{n}_j)}$. Since the residue field of A is separably closed, we have $G_d(\mathfrak{n}_j) = G_i(\mathfrak{n}_j) = \{e\}$. So we have $A \cong B_{\mathfrak{n}_j}$. Thus $\operatorname{Spec} B \to \operatorname{Spec} A$ is etale.

(ii) Let α be a geometric point of X with image $x \in X$. Any $g \in G_i(x)$ induces an S-automorphism of (X, α). Since X is connected and $G \subset \operatorname{Aut}(X/S)^\circ$, we have $g = e$ by 3.2.4 (i). So $G_i(x)$ is trivial. Note that $X \to Y$ and $Y \to S$ are finite. To prove this, we may assume that $S = \operatorname{Spec} C$ is affine. Then $X = \operatorname{Spec} B$ for some finite C-algebra B and $Y \cong \operatorname{Spec} B^G$. Since C is noetherian, B^G is finite over C. By (i), $X \to Y$ is etale. By 2.5.5, $Y \to S$ is etale. Let $\sigma : X \to X$ be a Y-automorphism of X. Since G acts transitively on the fibers of $X \to Y$, there exists $g_1 \in G$ such that $g_1(x) = \sigma(x)$. Let y be the image of x in Y. Then σ and g_1 induce two $k(y)$-isomorphisms from $k(\sigma(x))$ to $k(x)$, which we denote by $\sigma^\natural$ and $g_1^\natural$, respectively. We have $\sigma^\natural (g_1^\natural)^{-1} \in \operatorname{Gal}(k(x)/k(y))$. By 3.1.1 (iii), there exists $g_2 \in G_d(x)$ such that $\sigma^\natural (g_1^\natural)^{-1}$ is the image of g_2 in $\operatorname{Gal}(k(x)/k(y))$. By 3.2.4 (i), we must have $\sigma = g_2 g_1 \in G$. Hence $G = \operatorname{Aut}(X/Y)^\circ$. □

Corollary 3.2.6. *Let S be a noetherian scheme, let $X \to S$ be an etale covering space, and let G be a finite group acting on the right of X such that $X \to S$ is invariant under G. Then the action is admissible and the quotient X/G is an etale covering space of S.*

Proof. By 3.1.3 (ii), the action is admissible. G acts on connected components of X. Let $X_1, \ldots, X_n$ be connected components of X so that any connected component of X is of the form gX_j for some $g \in G$ and a unique $j \in \{1, \ldots, n\}$. Let $G_j = \{g \in G | gX_j = X_j\}$ be the stabilizer of X_j. Then we have $X/G \cong \coprod_j X_j/G_j$. Working with each X_j and G_j, we are reduced to the case where X is connected. Replacing G by its image in $\operatorname{Aut}(X/S)^\circ$, we may assume $G \subset \operatorname{Aut}(X/S)^\circ$. We then apply 3.2.5 (ii). □

Proposition 3.2.7. *Let (S, γ) be a pointed connected noetherian scheme, X_1 and X_2 two etale covering spaces of S, $u : X_1 \to X_2$ an S-morphism, and $X_i(\gamma)$ $(i = 1, 2)$ the sets of geometric points of X_i lying above γ. If the map $X_1(\gamma) \to X_2(\gamma)$ induced by u is bijective, then u is an isomorphism.*

Proof. The image of any connected component of X_2 in S is both open and closed. Since S is connected, the image is S. Replacing X_2 by its connected components and X_1 by the inverse images of these components, we are reduced to the case where X_2 is connected. Note that u is finite and etale. So $u_*\mathcal{O}_{X_1}$ is a locally free $\mathcal{O}_{X_2}$-module of constant finite rank. Let $s \in S$ be the image of γ and let $x_2 \in X_2$ be a point above s. Then

$$X_1 \times_{X_2} \operatorname{Spec} \mathcal{O}_{X_2,x_2} \cong \operatorname{Spec} A$$

for some $\mathcal{O}_{X_2,x_2}$-algebra A which is free of finite rank as an $\mathcal{O}_{X_2,x_2}$-module. Since $X_1(\gamma) \to X_2(\gamma)$ is bijective, there is one and only one point x_1 in X_1 lying above x_2. So A is a local ring. Since u is etale, $\mathfrak{m}_{x_2}A$ is the maximal ideal of A and $A/\mathfrak{m}_{x_2}A$ is finite separable over $\mathcal{O}_{X_2,x_2}/\mathfrak{m}_{x_2}$. Again because $X_1(\gamma) \to X_2(\gamma)$ is bijective, we must have $\mathcal{O}_{X_2,x_2}/\mathfrak{m}_{x_2} \cong A/\mathfrak{m}_2A$. It follows that the rank of $u_*\mathcal{O}_{X_1}$ is 1. Let x_2' be an arbitrary point of X_2 and let A' be an $\mathcal{O}_{X_2,x_2'}$-algebra such that

$$X_1 \times_{X_2} \operatorname{Spec} \mathcal{O}_{X_2,x_2'} \cong \operatorname{Spec} A'.$$

Since $\operatorname{rank}(u_*\mathcal{O}_{X_1}) = 1$, A' is a free $\mathcal{O}_{X_2,x_2'}$-module of rank 1. The homomorphism $\mathcal{O}_{X_2,x_2'}/\mathfrak{m}_{x_2'} \to A'/\mathfrak{m}_{x_2'}A$ is a nonzero homomorphism of one dimensional vectors spaces. It is necessarily surjective. By Nakayama's lemma, the homomorphism $\mathcal{O}_{X_2,x_2'} \to A'$ is also surjective. It is injective since it is faithfully flat. So we have $\mathcal{O}_{X_2,x_2'} \cong A'$. Hence $\mathcal{O}_{X_2} \cong u_*\mathcal{O}_{X_1}$, and u is an isomorphism. □

Proposition 3.2.8. *Let (S, γ) be a pointed connected noetherian scheme, X a connected etale covering space of S, $X(\gamma)$ the set of geometric points in X lying above γ, and $G = \operatorname{Aut}(X/S)^\circ$. The following conditions are equivalent:*

(i) *$X/G \cong S$, that is, X is a galois covering of S.*
(ii) *G acts transitively on $X(\gamma)$.*
(iii) *G and $X(\gamma)$ have the same number of elements.*

Proof. Let α be a geometric point of X above γ. For any $g_1, g_2 \in G$, if $g_1\alpha = g_2\alpha$, then $g_1 = g_2$ by 3.2.4 (i). It follows that $\#G$ is equal to the number of orbits of G on α. So (ii) and (iii) are equivalent.

Let $\beta : s \to X/G$ be a geometric point in X/G lying above γ, let $\alpha_1, \alpha_2 : s \to X$ be two geometric points in X lying above β, and let y, x_1, x_2 be the images of $\beta, \alpha_1, \alpha_2$, respectively. By 3.1.1 (ii), there exists $g_1 \in G$ such that $g_1 x_1 = x_2$. Consider the commutative diagram

$$\begin{array}{ccc} k(y) & \to & k(x_1) \\ \downarrow & \searrow_{\beta^\natural} & \downarrow \alpha_1^\natural \\ k(x_2) & \stackrel{\alpha_2^\natural}{\to} & k(s), \end{array}$$

where $\beta^\natural, \alpha_1^\natural, \alpha_2^\natural$ are the homomorphisms on residue fields induced by $\beta, \alpha_1, \alpha_2$, respectively. Let $g_1^\natural : k(x_2) \to k(x_1)$ be the $k(y)$-homomorphism induced by g_1. Since $k(x_2)$ is a normal algebraic extension of $k(y)$ by 3.1.1 (iii), $\alpha_1^\natural g_1^\natural$ and $\alpha_2^\natural$ have the same image in $k(s)$. So there exists $\tau \in \mathrm{Gal}(k(x_2)/k(y))$ such that $\alpha_1^\natural g_1^\natural \tau = \alpha_2^\natural$. By 3.1.1 (iii), there exists $g_2 \in G_{x_2}$ such that $\tau = g_2^\natural$. We then have $\alpha_1^\natural (g_1 g_2)^\natural = \alpha_2^\natural$. Hence $g_1 g_2$ maps α_1 to α_2. This shows that G acts transitively on the set $X(\beta)$ of geometric points in X lying above β. So (i) implies (ii).

Suppose (ii) holds. Then $(X/G)(\gamma)$ has only one element since two distinct elements in $(X/G)(\gamma)$ can be lifted to two elements in $X(\gamma)$ that are not in the same orbit. By 3.2.5 (ii), X/G is an etale covering space of S. By 3.2.7, we have $X/G \cong S$. □

Proposition 3.2.9. *let S be a connected noetherian scheme and let X be a connected etale covering space of S. Then any S-morphism $u : X \to X$ is an isomorphism.*

Proof. Note that u is necessarily etale and finite. So $u(X)$ is both open and closed. As X is connected, we have $u(X) = X$. Let γ be a geometric point of S, let α be a geometric point of X lying above γ, and let x be the image of α. Then there exists $x' \in X$ such that $u(x') = x$. Moreover the residue field $k(x')$ is a finite separable extension of the residue field $k(x)$. So there exists a geometric point α' of X with image x' such that $u\alpha' = \alpha$. Thus u induces a surjective map $X(\gamma) \to X(\gamma)$. As $X(\gamma)$ is finite. This map is bijective. By 3.2.7, u is an isomorphism. □

Proposition 3.2.10. *Let (S, γ) be a pointed connected noetherian scheme, and let (Y, β) be a pointed etale covering space of (S, γ). There exists a pointed connected galois etale covering space (X, α) of (S, γ) dominating (Y, β), that is, there exists an S-morphism from (X, α) to (Y, β).*

Proof. Let $\beta_1, \ldots, \beta_k$ be all the distinct geometric points of Y lying above γ. They define a geometric point α of $Y^k = Y \times_S \cdots \times_S Y$ such that

$p_i\alpha = \beta_i$ $(i = 1, \ldots, k)$, where $p_i : Y^k \to Y$ are the projections. Let X be the connected component of Y^k containing the image of α. It suffices to show that X is a galois covering of S. By 3.2.8 and 3.2.9, it suffices to show that for any geometric point α' of X lying above γ, there exists an S-morphism $u : X \to X$ such that $u\alpha = \alpha'$. Let $j : X \hookrightarrow Y^k$ be the open immersion. Since $p_i j\alpha = \beta_i$ $(i = 1, \ldots, k)$ are distinct, $p_i j\alpha'$ are distinct by 3.2.4 (i). Let σ be a permutation of $\{1, \ldots, k\}$ such that $p_i j\alpha' = p_{\sigma(i)} j\alpha$, and let $v : Y^k \to Y^k$ be the S-morphism with the property $p_i v = p_{\sigma(i)}$. We have

$$p_i v j\alpha = p_{\sigma(i)} j\alpha = p_i j\alpha'.$$

It follows that $vj\alpha = j\alpha'$. Since X is the connected component of Y^k containing the images of α and α', v induces a morphism $u : X \to X$ with the property $u(\alpha) = \alpha'$. This proves our assertion. □

Let (S, γ) be a pointed connected noetherian scheme, and let $\mathbf{Et}(S)$ the category whose objects are etale covering spaces of S, and whose morphisms are S-morphisms between these covering spaces. We have a functor F from $\mathbf{Et}(S)$ to the category of finite sets that maps each etale covering space X of S to the set $X(\gamma)$ of geometric points in X lying above γ. We call F the *fiber functor* for (S, γ). Let I be the opposite category of the category of pointed connected galois etale covering spaces of (S, γ). For any $i \in \mathrm{ob}\, I$, denote by (X_i, α_i) the corresponding pointed connected galois etale covering space of (S, γ). Given two objects i and j in I, there exists a morphism $j \to i$ in I if there exists a morphism $f_{ij} : (X_i, \alpha_i) \to (X_j, \alpha_j)$. Note that such a morphism is unique if it exists by 3.2.4 (i). By 3.2.4 and 3.2.10, I satisfies the conditions (I2) and (I3) in 2.7, and for any $X \in \mathrm{ob}\, \mathbf{Et}(S)$, the map

$$\varinjlim_{i \in \mathrm{ob}\, I} \mathrm{Hom}_S(X_i, X) \to F(X), \quad f \mapsto f(\alpha_i) \text{ for any } f \in \mathrm{Hom}_S(X_i, X)$$

is bijective. So F is pro-represented by $\varprojlim_{i \in \mathrm{ob}\, I} X_i$.

Let $f_{ij} : (X_i, \alpha_i) \to (X_j, \alpha_j)$ be a morphism in I. We have bijections

$$\mathrm{Aut}(X_v/S) \to F(X_v), \quad \sigma \mapsto \sigma(\alpha_v) \quad (v = i, j).$$

Through these bijections, the map

$$F(X_i) \to F(X_j), \quad \alpha \mapsto f_{ij}(\alpha)$$

is identified with a map

$$\phi_{ij} : \mathrm{Aut}(X_i/S) \to \mathrm{Aut}(X_j/S).$$

We have

$$\phi_{ij}(\sigma)(\alpha_j) = f_{ij}(\sigma(\alpha_i))$$

for any $\sigma \in \mathrm{Aut}(X_i/S)$. By 3.2.4 (i), we have

$$\phi_{ij}(\sigma)f_{ij} = f_{ij}\sigma,$$

that is, the following diagram commutes:

$$\begin{array}{ccc} X_i & \stackrel{\sigma}{\to} & X_i \\ f_{ij} \downarrow & & \downarrow f_{ij} \\ X_j & \stackrel{\phi_{ij}(\sigma)}{\to} & X_j. \end{array}$$

One can check that ϕ_{ij} is an epimorphism of groups. We thus get an inverse system of groups $(\mathrm{Aut}(X_i/S),\ \phi_{ij})_{i\in I}$. We define the *fundamental group* of (S,γ) to be

$$\pi_1(S,\gamma) = \varprojlim_i \mathrm{Aut}(X_i/S)^\circ.$$

Note that $\mathrm{Aut}(X_i/S)^\circ$ acts on the left on the set $\mathrm{Hom}(X_i, X)$ for any object X in $\mathbf{Et}(S)$. So $\pi_1(S,\gamma)$ acts on the left on the sets

$$\varinjlim_i \mathrm{Hom}_S(X_i, X) \cong F(X).$$

Since $F(X)$ is finite, the map

$$\mathrm{Hom}_S(X_i, X) \to F(X)$$

is surjective for some i. Then the action of $\pi_1(S,\gamma)$ on $F(X)$ factors through the finite quotient $\mathrm{Aut}(X_i/S)^\circ$. Put the discrete topology on $\mathrm{Aut}(X_i/S)^\circ$, and put the product topology on $\pi_1(S,\gamma) = \varprojlim_i \mathrm{Aut}(X_i/S)^\circ$. Then $\pi_1(S,\gamma)$ acts continuously on the discrete finite set $F(X)$.

Proposition 3.2.11. *An etale covering space X of a pointed connected noetherian scheme (S,γ) is connected if and only if $\pi_1(S,\gamma)$ acts transitively on $F(X)$.*

Proof. Let $X_1,\ldots,X_k$ be all the connected components of X. Then we have $F(X) = \coprod_{v=1}^{k} F(X_v)$. Each $F(X_v)$ is stable under the action of $\pi_1(S,\gamma)$. So if X is not connected, then $\pi_1(S,\gamma)$ does not act transitively on $F(X)$. Let α and α' be two geometric points in X lying above γ, and let (X_i,α_i) be a pointed galois etale covering space of (S,γ) dominating (X,α). If X is connected, then the morphism $(X_i,\alpha_i) \to (X,\alpha)$ is surjective. So there exists a geometric point α_i' of X_i lying above α'. By 3.2.8, $\mathrm{Aut}(X_i/S)$ acts transitively on $F(X_i)$. Let $\sigma \in \mathrm{Aut}(X_i/S)$ such that $\sigma(\alpha_i) = \alpha_i'$. Then $\sigma(\alpha) = \alpha'$. So $\mathrm{Aut}(X_i/S)$ acts transitively on $F(X)$. $\square$

Theorem 3.2.12. *Let (S,γ) be a pointed connected noetherian scheme. Then the fiber functor $F : X \mapsto X(\gamma)$ defines an equivalence from the category* **Et**(S) *of etale covering spaces of S to the category of finite sets on which $\pi_1(S,\gamma)$ acts continuously on the left.*

Proof. Given etale covering spaces X and Y of S, let us prove that

$$\mathrm{Hom}_S(X,Y) \to \mathrm{Hom}_{\pi_1(S,\gamma)}(F(X),F(Y))$$

is bijective. We may assume X and Y are connected. By 3.2.4 (i), this map is injective. Fix $\alpha \in F(X)$ and $\beta \in F(Y)$. Let (X_i,α_i) be a pointed connected galois etale covering space of (S,γ) dominating both (X,α) and (Y,β), and let $q : X_i \to Y$ be the S-morphism with the property $q(\alpha_i) = \beta$. Since Y is connected, q is onto. So for any $\phi \in \mathrm{Hom}_{\pi_1(S,\gamma)}(F(X_i),F(Y))$, there exists $\alpha_i' \in F(X_i)$ such that $q(\alpha_i') = \phi(\alpha_i)$. Since X_i is galois over S, there exists $\sigma \in \mathrm{Aut}(X_i/S)$ such that $\sigma(\alpha_i) = \alpha_i'$. Then $q\sigma(\alpha_i) = \phi(\alpha_i)$. Since $\pi_1(S,\gamma)$ acts transitively on $F(X_i)$, $q\sigma$ is mapped to ϕ under the map

$$\mathrm{Hom}_S(X_i,Y) \to \mathrm{Hom}_{\pi_1(S,\gamma)}(F(X_i),F(Y)).$$

Hence this map is surjective. Let $X_i(\alpha)$ be the set of geometric points in X_i lying above the geometric point α in X. Given two elements α_1 and α_2 in $X_i(\alpha)$, since X_i is galois over S, there exists $\sigma \in \mathrm{Aut}(X_i/S)$ such that $\sigma(\alpha_1) = \alpha_2$. By 3.2.4 (i), σ is an X-morphism. So $H = \mathrm{Aut}(X_i/X)^\circ$ acts transitively on $X_i(\alpha)$. By 3.2.8, we have $X = X_i/H$. For any $\phi \in \mathrm{Hom}_{\pi_1(S,\gamma)}(F(X),F(Y))$, by the above discussion, there exists an S-morphism $f' : X_i \to Y$ such that $f'(\alpha_i) = \phi(\alpha)$. For any $h \in \mathrm{Aut}(X_i/X)$, we have

$$(f'h)(\alpha_i) = h(f'(\alpha_i)) = h(\phi(\alpha)) = \phi(h(\alpha)) = \phi(\alpha) = f'(\alpha_i).$$

By 3.2.4 (i), we must have $f'h = f'$. So there exists a morphism $f : X \to Y$ such that $fp = f'$, where $p : X_i \to X$ is the S-morphism such that $p(\alpha_i) = \alpha$. We have

$$f(\alpha) = fp(\alpha_i) = f'(\alpha_i) = \phi(\alpha).$$

Since $\pi_1(S,\gamma)$ acts transitively on $F(X)$, f is mapped to ϕ under the map

$$\mathrm{Hom}_S(X,Y) \to \mathrm{Hom}_{\pi_1(S,\gamma)}(F(X),F(Y)).$$

So this map is surjective. This proves that the functor F is fully faithful.

Let A be a finite set on which $\pi_1(S,\gamma)$ acts continuously on the left. Let us prove there exists an etale covering space X of S such that $F(X) \cong A$ as sets with $\pi_1(S,\gamma)$-actions. Writing A as a disjoint union of orbits, and working with each orbit, we are reduced to the case where $\pi_1(S,\gamma)$ acts on A transitively. Let (X_i,α_i) be a connected pointed galois covering space of S such that the action of $\pi_1(S,\gamma)$ on A factors through $\mathrm{Aut}(X_i/S)^\circ$. Then there exists a subgroup H of $\mathrm{Aut}(X_i/S)^\circ$ such that A is isomorphic to $\mathrm{Aut}(X_i/S)^\circ/H$. Let $X = X_i/H$. Then we have $F(X)\cong A$. □

Proposition 3.2.13. *Let S be a noetherian scheme and let γ and γ' be two geometric points in S. If S is connected, then we have an isomorphism*

$$\pi_1(S,\gamma') \stackrel{\cong}{\to} \pi_1(S,\gamma),$$

and any two such isomorphisms differ by an inner automorphism of $\pi_1(S,\gamma)$.

Proof. Keep our previous notation. For each $i\in \mathrm{ob}\, I$, the set of geometric points $X_i(\gamma')$ in X_i lying above γ' is finite, nonempty, and each f_{ij} induces a surjective map from $X_i(\gamma')$ to $X_j(\gamma')$. We may find $\alpha_i'\in X_i(\gamma')$ for each i such that $f_{ij}(\alpha_i')=\alpha_j'$ for each f_{ij}. Let (X,α') be a pointed etale covering space of (S,γ') and let X' be the connected component of X containing the image of α'. Since S is connected, the morphism $X'\to S$ is surjective. So we can find a geometric point α of X lying above γ so that the image of α also lies in X'. We can find some $i\in\mathrm{ob}\, I$ such that (X_i,α_i) dominates (X,α). Let $p:(X_i,\alpha_i)\to(X,\alpha)$ be the morphism with the property $p(\alpha_i)=\alpha$. Then X_i is mapped onto X' and we can find a geometric point α_i'' of X_i over α'. Since X_i is galois over S, there exists $g\in\mathrm{Aut}(X_i/S)$ such that $g(\alpha_i')=\alpha_i''$. Then pg is a morphism from (X_i,α_i') to (X,α'). Thus any pointed etale covering space of (S,γ') is dominated by some (X_i,α_i'). This implies that the family $\{(X_i,\alpha_i')\}$ is cofinal in the category of connected pointed galois etale covering spaces of (S,γ'). So we have

$$\pi_1(S,\gamma')\cong \varprojlim_i \mathrm{Aut}(X_i/S)^\circ \cong \pi_1(S,\gamma).$$

The isomorphism $\pi_1(S,\gamma')\stackrel{\cong}{\to}\pi_1(S,\gamma)$ depends on the choice of the family of geometric points (α_i'). Let (α_i'') be another family so that α_i'' are geometric points of X_i lying above γ' and $f_{ij}(\alpha_i'')=\alpha_j''$. For each i, there exists a unique S-automorphism g_i of X_i such that $g_i(\alpha_i')=\alpha_i''$. One can show $\phi_{ij}(g_i)=g_j$. So (g_i) is an element in $\pi_1(S,\gamma)$. Given an element $\sigma\in\pi_1(S,\gamma')$, suppose its image under the isomorphism $\pi_1(S,\gamma')\stackrel{\cong}{\to}\pi_1(S,\gamma)$ defined by the family (α_i') is (σ_i), where $\sigma_i\in\mathrm{Aut}(X_i/S)^\circ$. Then the image of σ under the isomorphism $\phi'':\pi_1(S,\gamma')\stackrel{\cong}{\to}\pi_1(S,\gamma)$ defined by the family

(α_i'') is $(g_i^{-1}\sigma_i g_i)$. It follows that the two isomorphisms differ by an inner automorphism of $\pi_1(S,\gamma)$. □

Proposition 3.2.14. *Let K be a field, Ω a separably closed field containing K, $\gamma : \operatorname{Spec}\Omega \to \operatorname{Spec} K$ the corresponding geometric point, and K_s the separable closure of K contained in Ω. Then we have a canonical isomorphism*

$$\pi_1(\operatorname{Spec} K,\gamma) \cong \operatorname{Gal}(K_s/K).$$

Proof. Let $\{K_i\}$ be the family of finite galois extensions of K contained in Ω and let $\alpha_i : \operatorname{Spec}\Omega \to \operatorname{Spec} K_i$ be the corresponding geometric points. Then $\{(\operatorname{Spec} K_i,\alpha_i)\}$ is cofinal in the category of pointed connected galois etale covering spaces of $(\operatorname{Spec} K,\gamma)$. We have

$$\operatorname{Aut}(\operatorname{Spec} K_i/\operatorname{Spec} K)^\circ = \operatorname{Gal}(K_i/K).$$

So we have

$$\pi_1(\operatorname{Spec} K,\gamma) \cong \varprojlim_i \operatorname{Gal}(K_i/K) \cong \operatorname{Gal}(K_s/K).$$

□

3.3 Functorial Properties of Fundamental Groups

([SGA 1] V 6.)

Let $(S',\gamma') \to (S,\gamma)$ be a morphism of pointed connected noetherian schemes, let (X_i,α_i) be the family of pointed connected galois etale covering spaces of (S,γ), and let (α_i,γ') be the geometric point of $X_i\times_S S'$ lying above α_i and γ'. Then $(X_i\times_S S',(\alpha_i,\gamma'))$ is a pointed etale covering space of (S',γ'). Let F and F' be the fiber functors for (S,γ) and (S',γ'), respectively. We have a canonical one-to-one correspondence

$$F'(X_i\times_S S') \cong F(X_i),$$

through which $\pi_1(S',\gamma')$ acts on $F(X_i)$. We can define a map

$$\phi_i : \pi_1(S',\gamma') \to \operatorname{Aut}(X_i/S)^\circ$$

such that

$$\phi_i(g')(\alpha_i) = g'\alpha_i$$

for any $g' \in \pi_1(S',\gamma')$. Suppose that X_i' is the connected component of $X_i\times_S S'$ containing (α_i,γ'). One can check that $\operatorname{Aut}(X_i'/S)$ acts transitively on $F'(X_i')$. So X_i' is galois over S'. Let g_i' be the image of $g' \in \pi_1(S',\gamma')$

under the homomorphism $\pi_1(S',\gamma') \to \mathrm{Aut}(X_i'/S')^\circ$, and let $p : X_i \times_S S' \to X_i$ be the projection. We have

$$\phi_i(g')(\alpha_i) = (p|_{X_i'})(g_i'(\alpha_i,\gamma')).$$

On the other hand, we have

$$\phi_i(g')(\alpha_i) = \phi_i(g')(p|_{X_i'})((\alpha_i,\gamma')).$$

So we have

$$(p|_{X_i'})(g_i'(\alpha_i,\gamma')) = \phi_i(g')(p|_{X_i'})((\alpha_i,\gamma')).$$

This implies that

$$(p|_{X_i'})g_i' = \phi_i(g')(p|_{X_i'}).$$

Using this equality, one can check that ϕ_i are homomorphisms of groups, and are compatible with the canonical homomorphisms $\phi_{ij} : \mathrm{Aut}(X_i/S)^\circ \to \mathrm{Aut}(X_j/S)^\circ$ whenever (X_i,α_i) dominates (X_j,α_j). So we have a continuous homomorphism

$$\phi = \varprojlim_i \phi_i : \pi_1(S',\gamma') \to \pi_1(S,\gamma).$$

It follows from the definition that the following holds:

Proposition 3.3.1. *Let $f : (S',\gamma') \to (S,\gamma)$ be a morphism of pointed connected noetherian schemes, and let $\phi : \pi_1(S',\gamma') \to \pi_1(S,\gamma)$ be the induced homomorphism on fundamental groups. Denote by $(\pi_1(S',\gamma')$-**sets**$)$ (resp. $(\pi_1(S,\gamma)$-**sets**$)$) the category of finite sets on which $\pi_1(S',\gamma')$ (resp. $\pi_1(S,\gamma)$) acts continuously on the left, and by F' (resp. F) the fiber functor for (S',γ') (resp. (S,γ)). Consider the functor*

$$H_f : \mathbf{Et}(S) \to \mathbf{Et}(S'), \quad X \mapsto X \times_S S'$$

and the functor

$$H_\phi : (\pi_1(S,\gamma)\text{-}\mathbf{sets}) \to (\pi_1(S',\gamma')\text{-}\mathbf{sets})$$

induced by the homomorphism ϕ. The following diagram commutes:

$$\begin{array}{ccc} \mathbf{Et}(S) & \stackrel{H_f}{\to} & \mathbf{Et}(S') \\ {\scriptstyle F}\downarrow & & \downarrow{\scriptstyle F'} \\ (\pi_1(S,\gamma)\text{-}\mathbf{sets}) & \stackrel{H_\phi}{\to} & (\pi_1(S',\gamma')\text{-}\mathbf{sets}). \end{array}$$

Suppose that $(S', \gamma') \to (S, \gamma)$ is an etale covering space. Let $\{(X_i, \alpha_i)\}$ be the family of pointed connected galois etale covering spaces of (S, γ) dominating (S', γ'). By 3.2.4 and 3.2.10, $\{(X_i, \alpha_i)\}$ is cofinal in the category of pointed connected galois etale covering spaces of (S, γ). So we have

$$\pi_1(S, \gamma) \cong \varprojlim_i \mathrm{Aut}(X_i/S)^\circ.$$

One can show that $\mathrm{Aut}(X_i/S')^\circ$ acts transitively on the set $X_i(\gamma')$ of geometric points in X_i lying above γ'. So X_i are galois over S'. Again by 3.2.4 and 3.2.10, $\{(X_i, \alpha_i)\}$ is cofinal in the category of pointed connected galois etale covering spaces of (S', γ'). So we have

$$\pi_1(S', \gamma') \cong \varprojlim_i \mathrm{Aut}(X_i/S')^\circ.$$

The homomorphism $\pi_1(S', \gamma') \to \pi_1(S, \gamma)$ can be identified with the canonical inclusion

$$\varprojlim_i \mathrm{Aut}(X_i/S')^\circ \to \varprojlim_i \mathrm{Aut}(X_i/S)^\circ.$$

In particular, $\pi_1(S', \gamma') \to \pi_1(S, \gamma)$ is injective.

A topological group π is called a *profinite group* if it is isomorphic to a topological group of the form $\varprojlim_{i \in I} G_i$, where I is a direct set, and $\{G_i\}$ is an inverse system of finite groups with discrete topology, and $\varprojlim_{i \in I} G_i$ is provided with the product topology. Confer 4.2 for more properties of profinite groups.

Lemma 3.3.2. *Let $\phi : \pi' \to \pi$ be a continuous homomorphism of profinite groups,* (π'-**sets**) *(resp.* (π-**sets**)*) the category of finite sets on which π (resp. π') acts continuously on the left, and H_ϕ :* (π-**sets**) $\to$ (π'-**sets**) *the functor induced by ϕ.*

(i) *ϕ is surjective if and only if for any $A \in \mathrm{ob}$* (π-**sets**) *on which π acts transitively, π' acts transitively on $H_\phi(A)$.*

(ii) *ϕ is trivial if and only if for any $A \in \mathrm{ob}$* (π-**sets**)*, π acts trivially on $H_\phi(A)$.*

(iii) *Let H be an open subgroup of π. Then $\mathrm{im}\,\phi \subset H$ if and only if the element eH in $H_\phi(\pi/H)$ is fixed by π'.*

(iv) *Let H' be an open subgroup of π'. Then $\ker \phi \subset H'$ if and only if there exists an object A in* (π-**sets**) *on which π acts transitively such that there exists a π'-morphism from the π'-orbit of an element in $H_\phi(A)$ to π'/H'. If $\ker\phi \subset H'$ and ϕ is surjective, then there exists an object A in* (π-**sets**) *on which π acts transitively and $\pi'/H' \cong H_\phi(A)$.*

(v) ϕ *is injective if and only if for any* $A' \in \mathrm{ob}\,(\pi'\text{-}\mathbf{sets})$, *there exists* $A \in \mathrm{ob}\,(\pi\text{-}\mathbf{sets})$ *on which* π *acts transitively such that there exists a* π'-*morphism from the* π'-*orbit of an element in* $H_\phi(A)$ *to* A'.

Proof.

(i) Let us prove the "if" part. Write $\pi = \varprojlim_i \pi_i$, where π_i are finite groups such that $\pi \to \pi_i$ are surjective. Then π acts transitively on π_i. If π' acts transitively on $H_\phi(\pi_i)$, then the composites

$$\pi' \xrightarrow{\phi} \pi \to \pi_i$$

are surjective. Hence $\mathrm{im}\,\phi$ is dense in π. But ϕ is a continuous map from a compact space to a Hausdorff space. So ϕ is a closed mapping. It follows that $\mathrm{im}\,\phi = \pi$.

(ii) and (iii) are obvious.

(iv) Suppose $\ker\phi \subset H'$. We have $\{e\} = \bigcap\limits_H H$, where H goes over the set of open subgroups of π. It follows that $\bigcap\limits_H \phi^{-1}(H) = \ker\phi \subset H'$, that is, $H'^c \subset \bigcup\limits_H \phi^{-1}(H)^c$. Since H'^c is compact and $\phi^{-1}(H)^c$ are open, there exist finitely many open subgroups $H_1, \ldots, H_k$ of π such that

$$H'^c \subset \phi^{-1}(H_1)^c \cup \cdots \cup \phi^{-1}(H_k)^c.$$

Let $H = H_1 \cap \cdots \cap H_k$. Then we have $\phi^{-1}(H) \subset H'$. Let $A = \pi/H$ and let A' be the π'-orbit of eH in $H_\phi(A)$. Then

$$A' \to \pi'/H', \quad \phi(g')H \mapsto g'H' \text{ for any } g' \in \pi'$$

is a well-defined π'-morphism.

Suppose that $\ker\phi \subset H'$ and ϕ is surjective. Then we have $\pi'/H' \cong \pi/\phi(H')$. In particular, $\phi(H')$ has finite index in π. Since $\phi(H')$ is compact, and hence closed in π, it is open. Take $A = \pi/\phi(H')$. Then we have $\pi'/H' \cong H_\phi(A)$.

Suppose that A is an object in $\mathrm{ob}\,(\pi\text{-}\mathbf{sets})$ on which π acts transitively such that there exists a π'-morphism from the π'-orbit of an element a in A to π'/H'. We may choose a so that it is mapped to eH' in π'/H'. Let $H \subset \pi$ be the stabilizer of a. Then H is an open subgroup of π, and the π'-orbit of a is isomorphic to $\pi'/\phi^{-1}(H)$ so that a is identified with $e\phi^{-1}(H)$. So we have a π'-morphism $\pi'/\phi^{-1}(H) \to \pi'/H'$ mapping $e\phi^{-1}(H)$ to eH'. This is possible only if $\phi^{-1}(H) \subset H'$. In particular, we have $\ker\phi \subset H'$.

(v) follows from (iv). □

Lemma 3.3.3. *Let $\phi : \pi' \to \pi$ and $\psi : \pi \to \pi''$ be two continuous homomorphisms of profinite groups,* (π-**sets**) *(resp.* (π'-**sets**), *resp.* (π''-**sets**)*) the category of finite sets on which π (resp. π', resp. π'') acts continuously on the left, and H_ϕ :* (π-**sets**) $\to$ (π'-**sets**) *(resp. H_ψ :* (π''-**sets**) $\to$ (π-**sets**)*) the functor induced by ϕ (resp. ψ).*

(i) *$\psi\phi$ is trivial if and only if for any object A'' in* (π''-**sets**), *π' acts trivially on $H_\phi H_\psi(A'')$.*

(ii) $\ker\psi \subset \mathrm{im}\phi$ *if and only if for any object A in* (π-**sets**) *such that π acts transitively and that there exists an element in A fixed by* $\mathrm{im}\,\phi$, *there exists an object A'' in* (π''-**sets**) *on which π'' acts transitively such that there exists a π-morphism from the π-orbit of an element in $H_\psi(A'')$ to A.*

Proof. (i) follows from 3.3.2 (ii). (ii) follows from 3.3.2 (iii) and (iv) and the fact that $\ker\psi \subset \mathrm{im}\,\phi$ if and only if any open subgroup H of π containing $\mathrm{im}\,\phi$ contains $\ker\psi$. Indeed, for any closed subgroup C of a profinite group G, we have $C = \bigcap_{H\supset C} H$, where H goes over the family of open subgroups of G containing C. This can be proved as follows. If we have $x \notin C$, then by 4.2.6 in the next chapter, there exists a normal open subgroup N of G such that $xN \cap C = \emptyset$. Then $H = NC$ is an open subgroup of G containing C and $x \notin H$. □

¿From 3.3.1–3.3.3, we get the following two propositions:

Proposition 3.3.4. *Let $(S', \gamma') \to (S, \gamma)$ be a morphism of pointed connected noetherian schemes.*

(i) *$\pi_1(S', \gamma') \to \pi_1(S, \gamma)$ is surjective if and only if for any connected etale covering space X of S, $X \times_S S'$ is connected.*

(ii) *$\pi_1(S', \gamma') \to \pi_1(S, \gamma)$ is injective if and only if for any connected etale covering space X' of S', there exists a connected etale covering space X of S such that there exists an S'-morphism from a connected component of $X \times_S S'$ to X'.*

(iii) *$\pi_1(S', \gamma') \to \pi_1(S, \gamma)$ is trivial if and only if for any connected etale covering space X of S, $X \times_S S'$ is S'-isomorphic to a disjoint union of copies of S'.*

Proposition 3.3.5. *Let $(S', \gamma') \to (S, \gamma) \to (S'', \gamma'')$ be morphisms of pointed connected noetherian schemes, and let*

$$\pi_1(S', \gamma') \stackrel{\phi}{\to} \pi_1(S, \gamma) \stackrel{\psi}{\to} \pi_1(S'', \gamma'')$$

be the corresponding homomorphisms.

(i) *$\psi\phi$ is trivial if and only if for any connected etale covering space X'' of S'', $X'' \times_{S''} S'$ is S'-isomorphic to a disjoint union of copies of S'.*

(ii) $\ker\psi \subset \mathrm{im}\phi$ *if and only if for any connected etale covering space X of S such that $X \times_S S' \to S'$ admits a section, there exists a connected etale covering space X'' of S'' and an S-morphism from a connected component of $X'' \times_{S''} S$ to X.*

Proposition 3.3.6. *Let S be a normal connected noetherian scheme, K its function field, Ω a separably closed field containing K, and K_s the separable closure of K in Ω. Denote by γ both the geometric point* $\mathrm{Spec}\,\Omega \to \mathrm{Spec}\,K$ *and the geometric point defined by the composite* $\mathrm{Spec}\,\Omega \to \mathrm{Spec}\,K \to S$. *Then the canonical homomorphism*

$$\pi_1(\mathrm{Spec}\,K, \gamma) \to \pi_1(S, \gamma)$$

is an epimorphism. Through the isomorphism $\pi_1(\mathrm{Spec}\,K, \gamma) \cong \mathrm{Gal}(K_s/K)$, the kernel of the above epimorphism is identified with the subgroup $\mathrm{Gal}(K_s/K_{ur})$, where K_{ur} is the subfield of K_s generated by all finite separable extensions of K contained in K_s that are unramified over S. In particular, we have

$$\pi_1(S, \gamma) \cong \mathrm{Gal}(K_{ur}/K).$$

Proof. Let X be a connected etale covering space of S. Then X is normal and integral. So $X \otimes_S K$ is integral and hence connected. By 3.3.4 (i), the homomorphism $\pi_1(\mathrm{Spec}\,K, \gamma) \to \pi_1(S, \gamma)$ is surjective. Let $\{(X_i, \alpha_i)\}$ be the family of pointed connected galois etale covering spaces of (S, γ), and let K_i be the function field of X_i. Then K_i are finite galois extensions of K unramified over S, and $\mathrm{Aut}(X_i/S)^\circ \cong \mathrm{Gal}(K_i/K)$ by 3.1.4 (i). The geometric points α_i define embeddings of K_i into Ω. By 2.9.2, K_{ur} is generated by the images of K_i in Ω, and we have

$$\pi_1(S, \gamma) \cong \varprojlim_i \mathrm{Gal}(K_i/K) \cong \mathrm{Gal}(K_{ur}/K).$$

□

Proposition 3.3.7. *Let k be a field, k_s a separable closure of k, S a connected k-scheme of finite type, α a geometric point of $S \otimes_k k_s$, and β (resp. γ) the geometric points in S (resp.* $\mathrm{Spec}\,k$*) defined by the composite of α with $S \otimes_k k_s \to S$ (resp. $S \otimes_k k_s \to \mathrm{Spec}\,k$). Suppose $S \otimes_k k_s$ is connected. Then the following sequence is exact:*

$$1 \to \pi_1(S \otimes_k k_s, \alpha) \to \pi_1(S, \beta) \to \pi_1(\mathrm{Spec}\,k, \gamma) \to 1.$$

Proof. Any connected etale covering space of $\mathrm{Spec}\,k$ is isomorphic to $\mathrm{Spec}\,K \to \mathrm{Spec}\,k$ for some finite separable extension K of k contained in

k_s. Since $S \otimes_k k_s$ is connected and the projection $S \otimes_k k_s \to S \otimes_k K$ is surjective, $S \otimes_k K$ is also connected. By 3.3.4 (i), $\pi_1(S, \beta) \to \pi_1(\operatorname{Spec} k, \gamma)$ is surjective. We have

$$(S \otimes_k k_s) \otimes_k K \cong S \otimes_k (k_s \otimes_k K).$$

Since $k_s \otimes_k K$ is a direct product of finitely many copies of k_s, $(S \otimes_k k_s) \otimes_k K$ is a disjoint union of copies of $S \otimes_k k_s$. By 3.3.5 (i), the composite

$$\pi_1(S \otimes_k k_s, \alpha) \to \pi_1(S, \beta) \to \pi_1(\operatorname{Spec} k, \gamma)$$

is trivial.

Let $X \to S$ be a connected etale covering space such that $X \otimes_k k_s \to S \otimes_k k_s$ admits a section. Then by 1.10.9, there exists a finite separable extension K of k contained in k_s such that $X \otimes_k K \to S \otimes_k K$ admits a section. Taking the composite of this section with the projection $X \otimes_k K \to X$, we get an S-morphism $S \otimes_k K \to X$. By 3.3.5, the sequence

$$\pi_1(S \otimes_k k_s, \alpha) \to \pi_1(S, \beta) \to \pi_1(\operatorname{Spec} k, \gamma)$$

is exact.

Let $X' \to S \otimes_k k_s$ be a connected etale covering space. By 1.10.9, 1.10.10 and 2.3.7, there exist a finite separable extension K of k contained in k_s and an etale covering space $X_1 \to S \otimes_k K$ inducing $X' \to S \otimes_k k_s$ by the base extension $K \to k_s$. Since the projection $X' \to X_1$ is surjective, X_1 is connected. The composite

$$X_1 \to S \otimes_k K \to S$$

is an etale covering space of S. The graph

$$\Gamma : X_1 \to X_1 \times_S (S \otimes_k K) \cong X_1 \otimes_k K$$

of the S-morphism $X_1 \to S \otimes_k K$ is a section of the projection $X_1 \otimes_k K \to X_1$. It induces an isomorphism of X_1 with a connected component of $X_1 \otimes_k K$. The inverse of this isomorphism is an $(S \otimes_k K)$-isomorphism from a connected component of $X_1 \otimes_k K$ to X_1. By the base extension $K \to k_s$, we get an $(S \otimes_k k_s)$-morphism from a connected component of $X_1 \otimes_k k_s$ to X'. By 3.3.4 (ii), $\pi_1(S \otimes_k k_s, \alpha) \to \pi_1(S, \beta)$ is injective.

$$\begin{array}{ccccc}
X_1 & \xrightarrow{\Gamma} & X_1 \otimes_k K & \to & X_1 \\
 & & & & \downarrow \\
 & \searrow & \downarrow & & S \otimes_k K \\
 & & & & \downarrow \\
 & & S \otimes_k K & \to & S
\end{array}$$

□

Chapter 4

Group Cohomology and Galois Cohomology

4.1 Group Cohomology

([Serre (1979)] VII, VIII.)

Let G be a group. An abelian group with a left G-action can be identified with a $\mathbb{Z}[G]$-module. We also call a $\mathbb{Z}[G]$-module a G-module. A homomorphism of G-modules is defined to be a homomorphism of $\mathbb{Z}[G]$-modules. It is a homomorphism of abelian groups compatible with the G-actions. For any G-module A, define

$$A^G = \{a \in A | ga = a \text{ for all } g \in G\},$$

$$A_G = A/\text{the subgroup generated by } \{ga - a | g \in G, a \in A\}.$$

We call A^G (resp. A_G) the group of G-invariants (resp. G-coinvariants) of A. It is the maximal subgroup (resp. quotient group) of A on which G acts trivially. The functor $A \mapsto A^G$ on the category of G-modules is denoted by Γ^G. It is left exact, and its i-th derived functor is denoted by $H^i(G, -)$ or $R^i\Gamma^G$. Obviously we have

$$A^G = \mathrm{Hom}_{\mathbb{Z}[G]}(\mathbb{Z}, A),$$

where $\mathbb{Z}$ is the group of integers with the trivial G-action. We thus have

$$H^i(G, A) \cong \mathrm{Ext}^i_{\mathbb{Z}[G]}(\mathbb{Z}, A).$$

So if $I^\cdot$ (resp. $P_\cdot$) is an injective (resp. projective) resolution of A (resp. $\mathbb{Z}$) in the category of G-modules, then we have

$$H^i(G, A) \cong H^i(I^{\cdot G}) \cong H^i(\mathrm{Hom}_G(P_\cdot, A)).$$

Proposition 4.1.1. *Let $G = \mathbb{Z}$. For any G-module A, we have*

$$H^i(G, A) \cong \begin{cases} A^G & \text{if } i = 0, \\ A_G & \text{if } i = 1, \\ 0 & \text{if } i \geq 2. \end{cases}$$

Proof. Let T correspond to the canonical generator 1 of the group $G = \mathbb{Z}$. We can identify $\mathbb{Z}[G]$ with $\mathbb{Z}[T, T^{-1}]$. Consider the exact sequence

$$0 \to \mathbb{Z}[T, T^{-1}] \stackrel{T-1}{\to} \mathbb{Z}[T, T^{-1}] \stackrel{\epsilon}{\to} \mathbb{Z} \to 0,$$

where $T - 1 : \mathbb{Z}[T, T^{-1}] \to \mathbb{Z}[T, T^{-1}]$ is the multiplication by $T - 1$, and $\epsilon(\sum_i a_i T^i) = \sum_i a_i$. This gives rise to a resolution of $\mathbb{Z}$ by free $\mathbb{Z}[G]$-modules. So $H^i(G, A)$ is the i-th cohomology group of the complex

$$0 \to \mathrm{Hom}_{\mathbb{Z}[G]}(\mathbb{Z}[G], A) \stackrel{T-1}{\to} \mathrm{Hom}_{\mathbb{Z}[G]}(\mathbb{Z}[G], A) \to 0.$$

This complex can be identified with

$$0 \to A \stackrel{T-1}{\to} A \to 0,$$

where $T - 1 : A \to A$ is the homomorphism defined by $(T - 1)(a) = Ta - a$. Our assertion follows. □

Proposition 4.1.2. *Let $G = \mathbb{Z}/n$ for some $n \in \mathbb{N}$. For any G-module A, let $T : A \to A$ be the homomorphism defined by the action of the canonical generator $\bar{1} \in \mathbb{Z}/n$. We have*

$$H^i(G, A) \cong \begin{cases} A^G & \text{if } i = 0, \\ \ker(T^{n-1} + \cdots + T + 1)/\mathrm{im}(T - 1) & \text{if } i = 1, 3, 5, \ldots, \\ \ker(T - 1)/\mathrm{im}(T^{n-1} + \cdots + T + 1) & \text{if } i = 2, 4, 6, \ldots. \end{cases}$$

Proof. We can identify $\mathbb{Z}[G]$ with $Z[T]/(T^n - 1)$. We have an exact sequence

$$\cdots \stackrel{T-1}{\to} \mathbb{Z}[G] \stackrel{T^{n-1}+\cdots+T+1}{\to} \mathbb{Z}[G] \stackrel{T-1}{\to} \mathbb{Z}[G] \stackrel{T^{n-1}+\cdots+T+1}{\to} \mathbb{Z}[G] \stackrel{T-1}{\to} \mathbb{Z}[G] \stackrel{\epsilon}{\to} \mathbb{Z} \to 0,$$

where $T - 1 : \mathbb{Z}[G] \to \mathbb{Z}[G]$ (resp. $T^{n-1} + \cdots + T + 1 : \mathbb{Z}[G] \to \mathbb{Z}[G]$) is the multiplication by $T - 1$ (resp. $T^{n-1} + \cdots + T + 1$), and $\epsilon(\sum_g a_g g) = \sum_g a_g$. This gives rise to a resolution of $\mathbb{Z}$ by free $\mathbb{Z}[G]$-modules. We use this resolution to calculate $H^i(G, A) \cong \mathrm{Ext}^i_{\mathbb{Z}[G]}(\mathbb{Z}, A)$. □

Let L_i be the free abelian group generated by elements $(g_0, \ldots, g_i)$ in G^{i+1} $(i \geq 0)$. Define a left G-action on L_i by

$$g(g_0, \ldots, g_i) = (gg_0, \ldots, gg_i).$$

Then L_i is a free $\mathbb{Z}[G]$-module with a basis consisting of elements of the form $(1, g_1, g_1 g_2, \ldots, g_1 g_2 \cdots g_i)$. Define a G-homomorphism $\partial_i : L_i \to L_{i-1}$ for each $i \geq 1$ by

$$\partial_i(g_0, \ldots, g_i) = \sum_{j=0}^{i} (-1)^j (g_0, \ldots, \hat{g}_j, \ldots, g_i),$$

and a G-homomorphism $\epsilon : L_0 \to \mathbb{Z}$ by

$$\epsilon(g_0) = 1.$$

Then the sequence

$$\cdots \to L_i \overset{\partial_i}{\to} L_{i-1} \to \cdots \to L_1 \overset{\partial_1}{\to} L_0 \overset{\epsilon}{\to} \mathbb{Z} \to 0$$

is exact. Indeed, fix an element $g \in G$, then the homomorphisms

$$\begin{aligned} h_i : L_i \to L_{i+1}, &\quad h_i(g_0, \ldots, g_i) = (g, g_0, \ldots, g_i), \\ h_{-1} : \mathbb{Z} \to L_0, &\quad h_{-1}(1) = (g) \end{aligned}$$

define a homotopy between the identity morphism and the zero morphism of the above complex. Thus $L.$ is a resolution of $\mathbb{Z}$ by free $\mathbb{Z}[G]$-modules, and for any $\mathbb{Z}[G]$-module A, $H^i(G, A)$ is the i-th cohomology group of the complex

$$\begin{aligned} 0 \to \mathrm{Hom}_{\mathbb{Z}[G]}(L_0, A) &\overset{d_0}{\to} \mathrm{Hom}_{\mathbb{Z}[G]}(L_1, A) \overset{d_1}{\to} \cdots \\ &\to \mathrm{Hom}_{\mathbb{Z}[G]}(L_i, A) \overset{d_i}{\to} \mathrm{Hom}_{\mathbb{Z}[G]}(L_{i+1}, A) \to \cdots . \end{aligned}$$

Let $C^i(G, A)$ be the set of all maps from G^i to A. We have isomorphisms

$$F_i : \mathrm{Hom}_{\mathbb{Z}[G]}(L_i, A) \to C^i(G, A)$$

defined by

$$F_i(\phi)(g_1, \ldots, g_i) = \phi(1, g_1, g_1g_2, \ldots, g_1g_2 \cdots g_i)$$

for any $\phi \in \mathrm{Hom}_{\mathbb{Z}[G]}(L_i, A)$ and $(g_1, \ldots, g_i) \in G^i$. Through these isomorphisms, the homomorphisms $d_i : \mathrm{Hom}_{\mathbb{Z}[G]}(L_i, A) \to \mathrm{Hom}_{\mathbb{Z}[G]}(L_{i+1}, A)$ are identified with the homomorphisms

$$d_i : C^i(G, A) \to C^{i+1}(G, A)$$

defined by

$$\begin{aligned} &(d_i f)(g_1, \ldots, g_{i+1}) \\ &= g_1 f(g_2, \ldots, g_{i+1}) + \sum_{j=1}^{i} (-1)^j f(g_1, \ldots, g_{j-1}, g_j g_{j+1}, g_{j+2}, \ldots, g_{i+1}) \\ &\quad + (-1)^{i+1} f(g_1, \ldots, g_i) \end{aligned}$$

for any $f \in C^i(G, A)$ and $(g_1, \ldots, g_{i+1}) \in G^{i+1}$. So $H^i(G, A)$ is the i-th cohomology group of the complex

$$0 \to C^0(G, A) \overset{d_0}{\to} C^1(G, A) \to \cdots \to C^i(G, A) \overset{d_i}{\to} C^{i+1}(G, A) \to \cdots .$$

A 1-cocycle of the complex $C^{\cdot}(G, A)$ is a map $f : G \to A$ satisfying the condition

$$f(g_1 g_2) = g_1 f(g_2) + f(g_1)$$

for any $g_1, g_2 \in G$. A 1-coboundary of the complex $C^{\cdot}(G, A)$ is a map $f : G \to A$ such that there exists $a \in A$ satisfying the condition

$$f(g) = ga - a$$

for any $g \in G$. If G acts trivially on A, we have $H^1(G, A) = \mathrm{Hom}(G, A)$.

Let $f : G' \to G$ be a homomorphism of groups, A a G-module, A' a G'-module, and $\phi : A \to A'$ an additive map compatible with the group actions, that is, we have

$$\phi(f(g')a) = g'\phi(a)$$

for any $g' \in G'$ and $a \in A$. Then ϕ induces a homomorphism $A^G \to A'^{G'}$. By the universal property of derived functors, it extends canonically to a family of homomorphisms $H^i(G, A) \to H^i(G', A')$.

In particular, if H is a subgroup of G and A is a G-module, we can take $f : G' \to G$ to be the inclusion $H \hookrightarrow G$, and $\phi : A \to A'$ to be $\mathrm{id} : A \to A$. We thus get the *restriction homomorphisms*

$$\mathrm{Res} : H^i(G, A) \to H^i(H, A).$$

If H is normal in G, we can take $f : G' \to G$ to be the projection $G \to G/H$, and $\phi : A \to A'$ to be the inclusion $A^H \hookrightarrow A$. We thus get the *inflation homomorphisms*

$$\mathrm{Inf} : H^i(G/H, A^H) \to H^i(G, A).$$

Let H be a subgroup of G and let B be an H-module. Define

$$\mathrm{Ind}_H^G B = \mathrm{Hom}_{\mathbb{Z}[H]}(\mathbb{Z}[G], B).$$

Note that $\mathrm{Ind}_H^G B$ can be identified with the set of maps $\phi : G \to B$ satisfying

$$\phi(hg) = h\phi(g)$$

for any $h \in H$ and $g \in G$. Define a G-action on $\mathrm{Ind}_H^G B$ by

$$(g\phi)(g') = \phi(g'g)$$

for any $\phi \in \mathrm{Ind}_H^G B$ and $g, g' \in G$. Let

$$\theta : \mathrm{Ind}_H^G B \to B$$

be the homomorphism defined by $\theta(\phi) = \phi(1)$ for any $\phi \in \mathrm{Ind}_H^G B$. It is compatible with the group actions.

Theorem 4.1.3 (Shapiro). *For any H-module B, θ induces isomorphisms $H^i(G, \mathrm{Ind}_H^G B) \overset{\cong}{\to} H^i(H, B)$ for all i.*

Proof. For any G-module A, one can check that the map

$$\mathrm{Hom}_G(A, \mathrm{Ind}_H^G B) \to \mathrm{Hom}_H(A, B), \quad f \mapsto \theta \circ f$$

is bijective. In other words, the functor $B \mapsto \mathrm{Ind}_H^G B$ from the category of H-modules to the category of G-modules is right adjoint to the canonical forgetful functor $A \mapsto A$ from the category of G-modules to the category of H-modules. As the forgetful functor is exact, for any injective H-module B, $\mathrm{Ind}_H^G B$ is an injective G-module. But the functor $B \mapsto \mathrm{Ind}_H^G B$ is exact and θ induces an isomorphism $(\mathrm{Ind}_H^G B)^G \cong B^H$. It follows that θ induces isomorphisms $H^i(G, \mathrm{Ind}_H^G B) \cong H^i(H, B)$ for all i. □

Consider the case where $H = \{1\}$. Any abelian group M can be considered as an H-module. We can form the G-module

$$\mathrm{Ind}_{\{1\}}^G M = \mathrm{Hom}(\mathbb{Z}[G], M).$$

By 4.1.3, we have

$$H^i(G, \mathrm{Ind}_{\{1\}}^G M) = H^i(\{1\}, M).$$

But $H^i(\{1\}, M) = 0$ for any $i \geq 1$. So we have $H^i(G, \mathrm{Ind}_{\{1\}}^G M) = 0$ for any $i \geq 1$. A G-module is called *induced* if it is isomorphic to $\mathrm{Ind}_{\{1\}}^G M$ for some abelian group M. A G-module is called *weakly injective* if it is a direct factor of an induced G-module. For any weakly injective G-module I, we have $H^i(G, I) = 0$ for any $i \geq 1$. Any G-module A can be embedded into an induced G-module. Indeed, the homomorphism

$$\iota : A \to \mathrm{Ind}_{\{1\}}^G A = \mathrm{Hom}(\mathbb{Z}[G], A), \quad \iota(a)(g) = ga$$

is a G-monomorphism. Injective G-modules are weakly injective. We can use resolutions of A by weakly injective G-modules to calculate $H^i(G, A)$.

Let H be a subgroup of G. Then any induced G-module is also an induced H-module. Indeed, for any abelian group M, we have

$$\begin{aligned} \mathrm{Ind}_{\{1\}}^G M &= \mathrm{Hom}(\mathbb{Z}[G], M) = \mathrm{Hom}\Big(\bigoplus_{gH \in G/H} \mathbb{Z}[gH], M\Big) \\ &\cong \prod_{gH \in G/H} \mathrm{Hom}(\mathbb{Z}[gH], M) \cong \mathrm{Hom}\Big(\mathbb{Z}[H], \prod_{gH \in G/H} M\Big). \end{aligned}$$

So any weakly injective G-module is also a weakly injective H-module. In particular, any injective G-module I is a weakly injective H-module, and we have $H^i(H, I) = 0$ for any $i \geq 1$. If

$$0 \to I^0 \to I^1 \to \cdots$$

is a resolution of a G-module A by injective G-modules, then

$$H^i(H, A) = H^i(I^{\cdot H}).$$

Suppose H is normal in G. Then for any G-module A, A^H is a G/H-module and we have

$$(A^H)^{G/H} = A^G.$$

If I is an injective G-module, then I^H is an injective G/H-module. Using these facts, one can construct a biregular spectral sequence

$$E_2^{ij} = H^i(G/H, H^j(H, A)) \Rightarrow H^{i+j}(G, A),$$

which is called the *Hochschild–Serre spectral sequence.* Note that for any $g \in G$, the action of $gH \in G/H$ on $H^j(H, A)$ is induced by the group isomorphism

$$H \to H, \quad h \mapsto g^{-1}hg$$

and the additive map

$$A \to A, \quad a \mapsto ga$$

which is compatible with the group actions.

Again let H be a subgroup of G. For any G-module A, the G-homomorphism

$$\iota : A \to \mathrm{Ind}_H^G A = \mathrm{Hom}_{\mathbb{Z}[H]}(\mathbb{Z}[G], A), \quad \iota(a)(g) = ga$$

induces homomorphisms

$$H^i(G, A) \to H^i(G, \mathrm{Ind}_H^G A).$$

Composed with the isomorphisms $H^i(G, \mathrm{Ind}_H^G A) \cong H^i(H, A)$, the resulting homomorphisms

$$H^i(G, A) \to H^i(H, A)$$

coincide with the restriction homomorphisms.

Suppose H has finite index in G. We have a G-homomorphism

$$\pi : \mathrm{Ind}_H^G A = \mathrm{Hom}_{\mathbb{Z}[H]}(\mathbb{Z}[G], A) \to A, \quad f \mapsto \sum_{gH \in G/H} gf(g^{-1}).$$

(Note that $gf(g^{-1})$ depends on the left coset gH, not on the choice of g.) It induces homomorphisms

$$H^i(G, \mathrm{Ind}_H^G A) \to H^i(G, A).$$

Composed with the isomorphisms $H^i(G, \mathrm{Ind}_H^G A) \cong H^i(H, A)$, the resulting homomorphisms

$$\mathrm{Cor} : H^i(H, A) \to H^i(G, A)$$

are called the *corestriction homomorphisms.*

Proposition 4.1.4. *Suppose H is a subgroup of G with finite index n. Then we have* $\mathrm{Cor} \circ \mathrm{Res} = n$.

Proof. Notation as above. The composite

$$A \xrightarrow{\iota} \mathrm{Ind}_H^G A \xrightarrow{\pi} A$$

coincides with multiplication by n. Our assertion follows. □

Corollary 4.1.5. *Let G be a finite group, p a prime number, G_p a Sylow p-subgroup of G, and A a G-module. Then* $\mathrm{Res} : H^i(G, A) \to H^i(G_p, A)$ *is injective on the p-primary part of $H^i(G, A)$ for any i. In particular, if $H^i(G_p, A) = 0$ for each prime number p, then $H^i(G, A) = 0$.*

Proof. Let $x \in H^i(G, A)$ be an element in the kernel of Res. Then we have

$$[G : G_p]x = \mathrm{Cor} \circ \mathrm{Res}(x) = 0.$$

If x lies in the p-primary part of $H^i(G, A)$, then $p^k x = 0$ for some natural number k. As $[G : G_p]$ is relatively prime to p, this implies that $x = 0$. □

Corollary 4.1.6. *Let G be a finite group of order n. Then for any G-module A and any $i \geq 1$, $H^i(G, A)$ is annihilated by n.*

Proof. Apply 4.1.4 to the subgroup $H = \{1\}$. □

Corollary 4.1.7. *Let G be a finite group and let A be a G-module finitely generated as an abelian group. Then $H^i(G, A)$ are finite for all $i \geq 1$.*

Proof. Since $H^i(G, A)$ are the cohomology groups of the complex $C^\cdot(G, A)$, they are finitely generated abelian groups. By 4.1.6, $H^i(G, A)$ are torsion groups for all $i \geq 1$. So they are finite. □

4.2 Profinite Groups

([Serre (1964)] I 1.)

Lemma 4.2.1. *Let X be a compact Hausdorff topological space and let $x \in X$. The connected component of X containing x is $\cap_\lambda U_\lambda$, where $\{U_\lambda\}$ is the family of compact open neighborhoods of x.*

Proof. Let C be the connected component of X containing x. Any compact open neighborhood U_λ of x is both open and closed, and hence contains C. So we have $C \subset \cap_\lambda U_\lambda$. To prove our assertion, it suffices to show that $\cap_\lambda U_\lambda$ is connected. Let $A \subset \cap_\lambda U_\lambda$ be open and closed in $\cap_\lambda U_\lambda$. Then A and $\cap_\lambda U_\lambda - A$ are closed in X. We can find disjoint open subsets U and V in X such that

$$A \subset U, \quad \cap_\lambda U_\lambda - A \subset V.$$

We have $\cap_\lambda U_\lambda \subset U \cup V$ and hence $X - U \cup V \subset \cup_\lambda (X - U_\lambda)$. But $X - U \cup V$ is compact. So $X - U \cup V \subset (X - U_{\lambda_1}) \cup \cdots \cup (X - U_{\lambda_n})$ for finitely many compact open neighborhoods U_{λ_i} $(i = 1, \ldots, n)$ of x. Then we have $U_{\lambda_1} \cap \cdots \cap U_{\lambda_n} \subset U \cup V$. So $U_{\lambda_1} \cap \cdots \cap U_{\lambda_n}$ is the disjoint union of the open subsets $U_{\lambda_1} \cap \cdots \cap U_{\lambda_n} \cap U$ and $U_{\lambda_1} \cap \cdots \cap U_{\lambda_n} \cap V$. Hence $U_{\lambda_1} \cap \cdots \cap U_{\lambda_n} \cap U$ and $U_{\lambda_1} \cap \cdots \cap U_{\lambda_n} \cap V$ are open and closed. As X is compact, $U_{\lambda_1} \cap \cdots \cap U_{\lambda_n} \cap U$ and $U_{\lambda_1} \cap \cdots \cap U_{\lambda_n} \cap V$ are also compact. Only one of them contains x. If $x \in U_{\lambda_1} \cap \cdots \cap U_{\lambda_n} \cap U$, then $U_{\lambda_1} \cap \cdots \cap U_{\lambda_n} \cap U$ is a compact open neighborhood of x, and hence $\cap_\lambda U_\lambda \subset U_{\lambda_1} \cap \cdots \cap U_{\lambda_n} \cap U$. This implies that $\cap_\lambda U_\lambda - A \subset U$. But $\cap_\lambda U_\lambda - A \subset V$, and U and V are disjoint. So we must have $\cap_\lambda U_\lambda - A = \emptyset$. If $x \in U_{\lambda_1} \cap \cdots \cap U_{\lambda_n} \cap V$, then $\cap_\lambda U_\lambda \subset U_{\lambda_1} \cap \cdots \cap U_{\lambda_n} \cap V$. This implies that $A \subset V$. But $A \subset U$, and U and V are disjoint. So we must have $A = \emptyset$. This shows that $\cap_\lambda U_\lambda$ is connected. □

Lemma 4.2.2. *Let G be a compact topological group. Then G is totally disconnected if and only if $\cap_\lambda U_\lambda = \{1\}$, where $\{U_\lambda\}$ is the family of compact open neighborhoods of* 1.

Recall that a topological space X is called *totally disconnected* if any connected subset of X contains only one point.

Proof. Suppose that G is compact and totally disconnected. Since $\overline{\{1\}}$ is connected, we have $\overline{\{1\}} = \{1\}$. Hence $\{1\}$ is a closed subset of G. Since the diagonal $\Delta = \{(x,x)|x \in G\}$ is the inverse image of $\{1\}$ under the continuous map

$$G \times G \to G, \quad (x,y) \mapsto xy^{-1},$$

it is closed and hence G is Hausdorff. By 4.2.1, we must have $\cap_\lambda U_\lambda = \{1\}$.

Conversely, suppose $\cap_\lambda U_\lambda = \{1\}$. For any $g \in \overline{\{1\}}$, as gU_λ^{-1} is an open neighborhood of g, we have $1 \in gU_\lambda^{-1}$, and hence $g \in U_\lambda$. So we have $g \in \cap_\lambda U_\lambda = \{1\}$. This shows that $\{1\}$ is closed. As above, this implies that G is Hausdorff. By 4.2.1, G is totally disconnected. □

Lemma 4.2.3. *Let G be a Hausdorff topological group and let U be a compact open neighborhood of $\{1\}$. Then U contains a compact open subgroup of G.*

Proof. Let $F = (G-U)\cap U^2$. Then F is closed. We have $U\cdot 1 \subset G-F$. Since U is compact, we can find an open neighborhood V of 1 contained in U such that $UV \subset G-F$. Replacing V by $V \cap V^{-1}$, we may assume $V = V^{-1}$. We have $UV \subset (G-F)\cap U^2 \subset U$. From this, we get $UV^n \subset U$ for all $n \geq 1$. Let $H = \bigcup_{n=1}^{\infty} V^n$. Then H is an open subgroup of G. Any open subgroup is necessarily closed. As H is contained in U, H must be compact. □

Lemma 4.2.4. *Let G be a totally disconnected group and let U be a compact neighborhood (not necessarily open) of 1. Then U contains a compact open subgroup of G.*

Proof. Let V be an open neighborhood of 1 contained in U. Since G is totally disconnected, it is Hausdorff by the proof of 4.2.2. By 4.2.1, we have $\{1\} = \cap_\lambda U_\lambda$, where $\{U_\lambda\}$ is the family of compact open neighborhoods of 1 in U. We have $\cap_\lambda U_\lambda = \{1\} \subset V$. Hence $U - V \subset \cup_\lambda (U - U_\lambda)$. But $U - V$ is compact. So $U - V \subset (U - U_{\lambda_1}) \cup \cdots \cup (U - U_{\lambda_n})$ for finitely many compact open neighborhoods U_{λ_i} $(i = 1, \ldots, n)$ of 1 in U. Then we have $U_{\lambda_1} \cap \cdots \cap U_{\lambda_n} \subset V$. Note that $U_{\lambda_1} \cap \cdots \cap U_{\lambda_n}$ is open in V and hence in G, and it is compact. By 4.2.3, $U_{\lambda_1} \cap \cdots \cap U_{\lambda_n}$ contains a compact open subgroup of G. □

Lemma 4.2.5. *Let G be a topological group, K a compact subset of G, and U a neighborhood of 1. Then there exists a neighborhood V of 1 such that $xVx^{-1} \subset U$ for any $x \in K$.*

Proof. The continuous map

$$G \times G \to G, \quad (x, y) \mapsto xyx^{-1}$$

maps $(x, 1)$ to 1. So for any $x \in K$, there exists a neighborhood W_x of x and a neighborhood W'_x of 1 such that $x'yx'^{-1} \in U$ for any $x' \in W_x$ and $y \in W'_x$. Let $\{W_{x_1}, \ldots, W_{x_n}\}$ be a finite covering of K. We can take $V = W'_{x_1} \cap \cdots \cap W'_{x_n}$. □

Proposition 4.2.6. *Let G be a compact totally disconnected group and let U be a neighborhood of 1. Then U contains a normal open subgroup of G.*

Proof. Since G is compact and Hausdorff, U contains a compact neighborhood (not necessarily open) of 1. By 4.2.4, U contains an open subgroup of V. By 4.2.5, there exists a neighborhood V' of 1 such that $xV'x^{-1} \subset V$ for any $x \in G$. Then $V' \subset \bigcap_{x \in G} x^{-1}Vx$. Note that $\bigcap_{x \in G} x^{-1}Vx$ is an open normal subgroup of G contained in U. □

Recall that a topological group is called profinite if it is isomorphic to the inverse limit of an inverse system of finite discrete groups.

Corollary 4.2.7.

(i) *A topological group is profinite if and only if it is compact and totally disconnected.*

(ii) *A closed subgroup of a profinite group is profinite.*

(iii) *Direct products of profinite groups are profinite.*

(iv) *The inverse limit of any inverse system of profinite groups is profinite.*

(v) *If H is a closed subgroup of a profinite group, then the space of left cosets G/H with the quotient topology is totally disconnected. If H is a normal closed subgroup of G, then G/H is profinite.*

Proof.

(i) Let $G = \varprojlim_i G_i$ be a profinite group, where $\{G_i\}$ is an inverse system of finite discrete groups. Note that $\varprojlim_i G_i$ is a closed subset of $\prod_i G_i$. By Tychonoff's Theorem, G is compact. Let C be a connected subset of G. Then for each i, the image of C under the projection $G \to G_i$ is connected, and hence has only one element since G_i is discrete. It follows that C has only one element. So G is totally disconnected.

Conversely, let G be a compact totally disconnected group. By 4.2.6, open normal subgroups of G form a base of neighborhoods of 1. For any open normal subgroup N of G, the group G/N is a compact discrete group and hence finite. Consider the profinite group $\varprojlim_N G/N$, where N goes over the family of open normal subgroups of G. The canonical homomorphism $G \to \varprojlim_N G/N$ is continuous and has dense image. As G is compact and $\varprojlim_N G/N$ is Hausdorff, this homomorphism is a closed map. By 4.2.2, this homomorphism is injective. These facts imply that $G \to \varprojlim_N G/N$ is an isomorphism of topological groups. So G is profinite.

(ii) and (iii) follow from (i). (iv) follows from (ii) and (iii) since the inverse limit of an inverse system of topological groups is isomorphic to a closed subgroup of the product of these topological groups.

(v) By 4.2.1, to show that G/H is totally disconnected, it suffices to show that for any $g \in G$, we have $\{gH\} = \cap_\lambda U_\lambda$, where $\{U_\lambda\}$ is the family of compact open neighborhoods of gH in G/H. If $xH \notin \{gH\}$, that is, if $1 \notin g^{-1}xH$, then there exists an open subgroup U disjoint from $g^{-1}xH$. The canonical homomorphism $\pi : G \to G/H$ is continuous and open. Since U is an open subgroup, it is necessarily closed and hence compact. It follows that $\pi(gU)$ is a compact open neighborhood of gH in G/H. We have $xH \notin \pi(gU)$. So $xH \notin \cap_\lambda U_\lambda$. Our assertion follows. □

A *surnatural number* is a formal product $\prod_p p^{n_p}$, where p goes over the set of prime numbers, and $n_p \in \mathbb{N} \cup \{0, \infty\}$. One defines the product, the greatest common divisor, and the least common multiple of a family of surnatural numbers in the obvious way.

Let G be a profinite group and let H be a closed subgroup of G. The index $[G : H]$ of H in G is defined to be the least common multiple of $[G/U : H/H \cap U]$, where U goes over the family of open normal subgroups of G. It is a surnatural number. It is also the least common multiple of $[G : V]$, where V goes over the family of open subgroups of G containing H. Indeed, if U is an open normal subgroup, then HU is an open subgroup containing H, and

$$[G/U : H/H \cap U] = [G : HU];$$

if V is an open subgroup containing $H = H \cdot 1$, then $HU \subset V$ for some open normal subgroup U and

$$[G : V] \bigm| [G : HU] = [G/U : H/H \cap U].$$

Proposition 4.2.8. *Let G be a profinite group.*

(i) *For all closed subgroups K and H of G such that $K \subset H$, we have $[G : K] = [G : H][H : K]$.*

(ii) *For all closed subgroups K and H of G such that $K \subset H$ and K is normal in G, we have $[G/K : H/K] = [G : H]$.*

(iii) *A closed subgroup H in G is open if and only if $[G:H]$ is finite.*

(iv) *If $\{H_i\}$ is a decreasing filtered family of closed subgroups in G and $H=\cap_i H_i$, then $[G:H]$ is the least common multiple of $[G:H_i]$.*

Proof.

(i) For any open normal subgroup U of G, we have

$[G/U : K/K\cap U] = [G/U : H/H\cap U][H/H\cap U : K/K\cap U] \mid [G:H][H:K]$.

So $[G:K] \mid [G:H][H:K]$. Given an open normal subgroup U of G and an open normal subgroup V of H, let U' be an open normal subgroup of G such that $H\cap U' \subset V$, and let $U''=U\cap U'$. We have

$$[G:HU][H:KV] \mid [G:HU''][H:K(H\cap U'')].$$

But

$$\begin{aligned}[G:HU][H:KV] &= [G/U : H/H\cap U][H/V : K/K\cap V],\\ [G:HU''][H:K(H\cap U'')] &= [G/U'' : H/H\cap U''][H/H\cap U'' : K/K\cap U'']\\ &= [G/U'' : K/K\cap U''].\end{aligned}$$

It follows that

$$[G/U : H/H\cap U][H/V : K/K\cap V] \mid [G/U'' : K/K\cap U''],$$

and hence $[G:H][H:K] \mid [G:K]$.

(ii) Use the fact that the set of open subgroups of G/K containing H/K are in one-to-one correspondence with the set of open subgroups of G containing H.

(iii) Suppose H is an open subgroup of G. Then $\{gH\}_{g\in G}$ is an open covering of G. Since G is compact, we have $G\subset g_1H\cup\cdots\cup g_nH$ for finitely many $g_1,\ldots,g_n\in G$. It follows that $[G:H]$ is finite.

Conversely, suppose $[G:H]$ is finite. We can find an open subgroup V of G containing H such that $[G:V]=[G:H]$. Suppose $V\neq H$. Take $x\in V-H$. We have $1\notin xH$. As xH is closed, there exists an open normal subgroup U of G such that $U\cap xH=\emptyset$. Then $x\notin UH$. Hence $V\cap UH$ is strictly contained in V. But $V\cap UH$ is an open subgroup containing H. So we have

$$[G:H]\geq [G:V\cap UH] > [G:V].$$

This contradicts $[G:V]=[G:H]$. So we must have $V=H$.

(iv) For any open subgroup V of G, using the fact that $G-V$ is compact, one can show that $\cap_i H_i\subset V$ if and only if $H_i\subset V$ for some i. It follows that $[G:H]$ is the least common multiple of $[G:V]$ for open subgroups V containing some H_i. Hence $[G:H]$ is the least common multiple of $[G:H_i]$. □

Fix a prime number p. A profinite group G is called a *pro-p-group* if $[G : 1]$ is a power of p, or equivalently, G is the inverse limit of an inverse system of finite p-groups. Given a profinite group G, a closed subgroup H of G is called a *Sylow p-subgroup* if H is a pro-p-group and $[G : H]$ is prime to p.

Proposition 4.2.9.

(i) *Let G be a profinite group. Any pro-p-subgroup H of G is contained in a Sylow p-subgroup. In particular, Sylow p-subgroups exist. Any two Sylow p-subgroups are conjugate in G.*

(ii) *Let $G \to G'$ be an epimorphism of profinite groups. Then the image of a Sylow p-subgroup of G in G' is a Sylow p-subgroup.*

Proof.

(i) For any open normal subgroup U of G, let $\mathscr{S}_U$ be the set of Sylow p-subgroups of G/U containing the p-subgroup $H/H \cap U$. For any two normal subgroups $V \subset U$ of G, the canonical homomorphism $G/V \to G/U$ induces a map $\mathscr{S}_V \to \mathscr{S}_U$. Since $\mathscr{S}_U$ is finite and nonempty for each open normal subgroup U, we have $\varprojlim_U \mathscr{S}_U \neq \emptyset$. Let $(P_U) \in \varprojlim_U \mathscr{S}_U$ and let $P = \varprojlim_U P_U$. Then P is a pro-p-subgroup of G containing H. $[G : P]$ is the least common multiple of $[G/U : P_U]$. Since $\frac{|G/U|}{|P_U|}$ is prime to p, $[G : P]$ is prime to p. So P is a Sylow p-subgroup.

Let P and P' be Sylow p-subgroups of G. For any open normal subgroup U of G, let A_U be the set of elements $x \in G/U$ such that $x^{-1}(PU/U)x = P'U/U$. Here we regard PU/U and $P'U/U$ as subgroups of G/U. As they are Sylow p-subgroups of G/U (confer the proof of (ii)), A_U is nonempty. Moreover A_U is finite. Hence $\varprojlim_U A_U \neq \emptyset$. Let $(x_U) \in \varprojlim_U A_U$ and choose $x \in G$ so that its image in G/U is x_U for each U. Then $x^{-1}Px = P'$.

(ii) Let N be the kernel of $G \to G'$. We have $G' \cong G/N$. Let P be a Sylow p-subgroup of G. Its image in G' is identified with the subgroup PN/N of G/N. We have

$$[G/N : PN/N] = [G : PN] \mid [G : P].$$

So $[G/N : PN/N]$ is prime to p. On the other hand, we have

$$[PN/N : 1] = [P/P \cap N : 1] = [P : P \cap N] \mid [P : 1].$$

So $[PN/N : 1]$ is a power of p. Therefore PN/N is a Sylow p-subgroup. □

4.3 Cohomology of Profinite Groups

([Serre (1964)] I 2.)

Let G be a profinite group and let A be an abelian group on which G acts on the left. Put the discrete topology on A. Then the following conditions are equivalent:

(i) G acts continuously on A, that is, the map

$$G \times A \to A, \quad (g, x) \mapsto gx$$

is continuous.

(ii) For any $x \in A$, the stabilizer

$$G_x = \{g \in G | gx = x\}$$

is open in G.

(iii) We have $A = \bigcup_U A^U$, where U goes over the set of open subgroups of G.

From now on, we simply call a discrete abelian group on which G acts left continuously a G-module. For any G-module A, and any nonnegative integer n, let $C^n(G, A)$ be the set of continuous maps from G^n to A. Note that $C^n(G, -)$ is an exact functor on the category of G-modules. Define

$$d_n : C^n(G, A) \to C^{n+1}(G, A)$$

by

$$\begin{aligned} &(d_n f)(g_1, \ldots, g_{n+1}) \\ &= g_1 f(g_2, \ldots, g_{n+1}) + \sum_{j=1}^{n} (-1)^j f(g_1, \ldots, g_{j-1}, g_j g_{j+1}, g_{j+2}, \ldots, g_{n+1}) \\ &\quad + (-1)^{n+1} f(g_1, \ldots, g_n) \end{aligned}$$

for any $f \in C^n(G, A)$ and $(g_1, \ldots, g_{n+1}) \in G^{n+1}$. One can verify $d \circ d = 0$. We define $H^n(G, A)$ to be the n-th cohomology group of the complex $C^\cdot(G, A)$. We also denote $H^n(G, A)$ by $R^n\Gamma^G(A)$.

Theorem 4.3.1. *Let $\{G_i\}_{i \in I}$ be an inverse system of profinite groups, and let $\{A_i\}_{i \in I}$ be a direct system of abelian groups such that each A_i is a G_i-module. For each pair $i \leq j$, suppose that the homomorphism $A_i \to A_j$ is compatible with the homomorphism $G_j \to G_i$ and the group actions of G_i and G_j on A_i and A_j, respectively. Set $G = \varprojlim_i G_i$ and $A = \varinjlim_i A_i$. Then the canonical chain map*

$$\varinjlim_i C^\cdot(G_i, A_i) \to C^\cdot(G, A)$$

is an isomorphism, and hence

$$H^{\cdot}(G, A) \cong \varinjlim_i H^{\cdot}(G_i, A_i).$$

Proof. Let $f_1, f_2 : G_i^n \to A_i$ be two continuous maps. Assume they induce the same map from G^n to A. Let $f_1', f_2' : G_i^n \to A$ be the maps induced by f_1 and f_2, respectively, and let

$$X = \{x \in G_i^n | f_1'(x) = f_2'(x)\}.$$

Since A has the discrete topology and f_1' and f_2' are continuous, X is open in G_i^n. Let

$$\pi_i : G^n = \varprojlim_i G_i^n \to G_i^n, \quad \pi_{ji} : G_j^n \to G_i^n \quad (j \geq i)$$

be the canonical homomorphisms. Note that

$$\operatorname{im} \pi_i = \bigcap_{j \geq i} \operatorname{im} \pi_{ji}.$$

Indeed, given any $x \in \bigcap_{j \geq i} \operatorname{im} \pi_{ji}$ and any $j_1 \geq j_2 \geq i$, let

$$A_{j_1 j_2} = \{(x_j) \in \prod_{j \geq i} G_j^n | \pi_{j_1 j_2}(x_{j_1}) = x_{j_2} \text{ and } \pi_{j_2 i}(x_{j_2}) = x\}.$$

Then $A_{j_1 j_2}$ are closed in $\prod_{j \geq i} G_j^n$, and any intersection of finitely many of them is nonempty. Since $\prod_{j \geq i} G_j^n$ is compact, we have $\bigcap_{j_1 \geq j_2 \geq i} A_{j_1 j_2} \neq \emptyset$. For any $y \in \bigcap_{j_1 \geq j_2 \geq i} A_{j_1 j_2}$, we have $y \in G^n = \varprojlim_i G_i^n$ and $\pi_i(y) = x$. So $x \in \operatorname{im} \pi_i$. By our assumption, we have $\operatorname{im} \pi_i \subset X$. It follows that $\bigcap_{j \geq i} \operatorname{im} \pi_{ji} \subset X$. As $G_i^n - X$ is compact and $G_i^n - \operatorname{im} \pi_{ji}$ $(j \geq i)$ form an open covering of $G_i^n - X$, we have $\operatorname{im} \pi_{ji} \subset X$ for a sufficiently large j. Then $f_1' \circ \pi_{ji} = f_2' \circ \pi_{ji}$. Since G_j^n is compact and A is discrete, $\operatorname{im}(f_1' \circ \pi_{ji}) = \operatorname{im}(f_2' \circ \pi_{ji})$ is finite. Taking j sufficiently large, we see that f_1 and f_2 induce the same map from G_j^n to A_j. This proves the canonical map $\varinjlim_i C^n(G_i, A_i) \to C^n(G, A)$ is injective.

Given a continuous map $f : G^n \to A$, let $\operatorname{im} f = \{a_1, \ldots, a_k\}$. For each a_λ, $f^{-1}(a_\lambda)$ is open and closed. So each $f^{-1}(a_\lambda)$ is a finite union of sets of the form

$$\left(\prod_i U_i\right) \bigcap \left(\varprojlim_i G_i^n\right),$$

where $U_i \subset G_i^n$ are open and $U_i = G_i^n$ for all but finitely many i. Taking j sufficiently large so that for any $i \geq j$, all U_i appeared above are G_i^n. Then for any $x, x' \in G^n$ such that $\pi_j(x) = \pi_j(x')$, we have $f(x) = f(x')$. So f can be factorized as a composite

$$G^n \xrightarrow{\pi_j} \pi_j(G^n) \xrightarrow{f'} A$$

for some continuous map $f' : \pi_j(G^n) \to A$. We can find disjoint open and closed subsets V_λ of G_j^n such that $f'^{-1}(a_\lambda) \subset V_\lambda$. Taking j sufficiently large, we may assume that there exist $a_{\lambda j} \in A_j$ with images a_λ in A, respectively. Define $f_j : G_j^n \to A_j$ by $f_j(V_\lambda) = \{a_{\lambda j}\}$ and $f_j(G_j^n - \cup_{\lambda=1}^k V_\lambda) = \{0\}$. Then f_j is continuous and induces f. This proves the canonical map $\varinjlim_i C^n(G_i, A_i) \to C^n(G, A)$ is surjective. □

Corollary 4.3.2. *Let G be a profinite group and let A be a G-module. Then we have*

$$H^i(G, A) \cong \varinjlim_U H^i(G/U, A^U),$$

where U goes over the set of open normal subgroups of G. For all $i \geq 1$, $H^i(G, A)$ are torsion groups.

Proof. Use 4.3.1 and 4.1.6. □

Corollary 4.3.3. *For each i, $H^i(G, -)$ is the i-th derived functor of the functor $A \mapsto A^G$ on the abelian category of G-modules.*

Proof. One can check $H^0(G, A) = A^G$. Using the fact that $C^i(G, -)$ are exact functors, one can show that $H^i(G, -)$ form a δ-functor. The abelian category of G-modules has enough injective objects. For any injective G-module I, one can check that I^U is an injective G/U-module for any open normal subgroup U of G. It follows that

$$H^i(G, I) = \varinjlim_U H^i(G/U, I^U) = 0$$

for any $i \geq 1$. Our assertion follows. □

Let $f : G' \to G$ be a continuous homomorphism of profinite groups, A a G-module, A' a G'-module, $\phi : A \to A'$ an additive map compatible with the group actions. Then ϕ induces a homomorphism $A^G \to A'^{G'}$. It extends canonically to a family of homomorphisms $H^i(G, A) \to H^i(G', A')$. These homomorphisms are induced by the chain map $C^\cdot(G, A) \to C^\cdot(G', A')$ induced by f and ϕ.

In particular, if H is a closed subgroup of G and A is a G-module, then we have the restriction homomorphisms

$$\mathrm{Res} : H^i(G, A) \to H^i(H, A).$$

Let H be a closed subgroup of G, let B be an H-module, and let $\mathrm{Ind}_H^G B$ be the group of continuous maps $\phi : G \to B$ satisfying $\phi(hg) = h\phi(g)$ for any $g \in G$ and $h \in H$. Define a G-action on $\mathrm{Ind}_H^G B$ by $(g\phi)(g') = \phi(g'g)$ for any $\phi \in \mathrm{Ind}_H^G B$ and $g, g' \in G$. Note that this action is continuous if we put the discrete topology on $\mathrm{Ind}_H^G B$. Let

$$\theta : \mathrm{Ind}_H^G B \to B$$

be the homomorphism defined by $\theta(\phi) = \phi(1)$ for any $\phi \in \mathrm{Ind}_H^G B$. It is compatible with the group actions. We again have the following theorem:

Theorem 4.3.4 (Shapiro). *For any H-module B, θ induces isomorphisms $H^i(G, \mathrm{Ind}_H^G B) \xrightarrow{\cong} H^i(H, B)$ for all i.*

Consider the case $H = \{1\}$. Any abelian group M can be considered as an H-module. We can form the G-module $\mathrm{Ind}_{\{1\}}^G M$ of continuous maps from G to M. A G-module is called *induced* if it is isomorphic to $\mathrm{Ind}_{\{1\}}^G M$ for some abelian group M. A G-module is called *weakly injective* if it is a direct factor of an induced G-module. For any weakly injective G-module I, we have $H^i(G, I) = 0$ for any $i \geq 1$. Any G-module A can be embedded into an induced G-module. Injective G-modules are weakly injective. We can use resolutions of A by weakly injective G-modules to calculate $H^i(G, A)$.

Let H be a closed subgroup of G. Then for any abelian group M, we have $H^i(H, \mathrm{Ind}_{\{1\}}^G M) = 0$ for any $i \geq 1$. Indeed, we have

$$H^i(H, \mathrm{Ind}_{\{1\}}^G M) = \varinjlim_U H^i(H/H \cap U, (\mathrm{Ind}_{\{1\}}^G M)^U),$$

where U goes over the set of open normal subgroups of G. We have

$$(\mathrm{Ind}_{\{1\}}^G M)^U \cong \mathrm{Hom}(\mathbb{Z}[G/U], M).$$

So $(\mathrm{Ind}_{\{1\}}^G M)^U$ is an induced G/U-module and thus an induced $(H/H\cap U)$-module. It follows that $H^i(H/H \cap U, (\mathrm{Ind}_{\{1\}}^G M)^U) = 0$ and hence $H^i(H, \mathrm{Ind}_{\{1\}}^G M) = 0$ for any $i \geq 1$. For any weakly injective G-module A, we also have $H^i(H, A) = 0$ for any $i \geq 1$. Using this fact, we can show that if H is a normal closed subgroup of G and A is a G-module, then we have a biregular spectral sequence

$$E_2^{ij} = H^i(G/H, H^j(H, A)) \Rightarrow H^{i+j}(G, A),$$

which we call the *Hochschild–Serre spectral sequence*.

For any open subgroup H of G and any G-module A, we can define the *corestriction homomorphisms*

$$\mathrm{Cor} : H^i(H, A) \to H^i(G, A).$$

Proposition 4.3.5. *Let H be an open subgroup of G with index n. Then* $\mathrm{Cor} \circ \mathrm{Res} = n$.

Proposition 4.3.6. *Let H be a closed subgroup of a profinite group G such that $[G:H]$ is relatively prime to a prime number p. Then for any G-module A and any i, the restriction homomorphism $\mathrm{Res}: H^i(G,A) \to H^i(H,A)$ is injective on the p-primary part of $H^i(G,A)$.*

Proof. When $[G:H] < \infty$, this follows from 4.3.5. In general, we have $H^i(H,A) = \varinjlim_U H^i(U,A)$ by 4.3.1, where U goes over the set of open subgroups of G containing H. For each U, $[G:U]$ is prime to p. By 4.3.5, $\mathrm{Res}: H^i(G,A) \to H^i(U,A)$ is injective on the p-primary part of $H^i(G,A)$. It follows that $\mathrm{Res}: H^i(G,A) \to H^i(H,A)$ is injective on the p-primary part of $H^i(G,A)$. □

Proposition 4.3.7. *Let $\hat{\mathbb{Z}} = \varprojlim_{n\in\mathbb{N}} \mathbb{Z}/n$, where the partial order on the direct set $\mathbb{N}$ is defined by $n \leq m$ if and only if $n \mid m$ and the transition homomorphism $\mathbb{Z}/m \to \mathbb{Z}/n$ is the canonical one. For any torsion abelian group A with continuous $\hat{\mathbb{Z}}$ action, we have*

$$H^i(\hat{\mathbb{Z}}, A) \cong \begin{cases} A^{\hat{\mathbb{Z}}} & \text{if } i = 0, \\ A_{\hat{\mathbb{Z}}} & \text{if } i = 1, \\ 0 & \text{if } i \neq 0, 1. \end{cases}$$

Proof. Let $T = (\bar{1}) \in \varprojlim_{n\in\mathbb{N}} \mathbb{Z}/n$ be the canonical topological generator of $\hat{\mathbb{Z}}$. For any $n \in \mathbb{N}$, let

$$A^{T^n} = \{x \in A | T^n(x) = x\}.$$

We have

$$H^i(\hat{\mathbb{Z}}, A) \cong \varinjlim_n H^i(\mathbb{Z}/n, A^{T^n}).$$

For any $m, n \in \mathbb{N}$, consider the following commutative diagram:

$$\begin{array}{ccccccccc} 0 \to & A^{T^n} & \overset{T-1}{\to} & A^{T^n} & \overset{T^{n-1}+\cdots+T+1}{\to} & A^{T^n} & \overset{T-1}{\to} & A^{T^n} & \to \\ & \downarrow 1 & & \downarrow 1 & & \downarrow m & & \downarrow m & \\ 0 \to & A^{T^{mn}} & \overset{T-1}{\to} & A^{T^{mn}} & \overset{T^{mn-1}+\cdots+T+1}{\to} & A^{T^{mn}} & \overset{T-1}{\to} & A^{T^{mn}} & \to \end{array}$$

$$\begin{array}{cccccccc} \overset{T^{n-1}+\cdots+T+1}{\to} & A^{T^n} & \overset{T-1}{\to} & A^{T^n} & \overset{T^{n-1}+\cdots+T+1}{\to} & A^{T^n} & \to \cdots \\ & \downarrow m^2 & & \downarrow m^2 & & \downarrow m^3 & \\ \overset{T^{mn-1}+\cdots+T+1}{\to} & A^{T^{mn}} & \overset{T-1}{\to} & A^{T^{mn}} & \overset{T^{mn-1}+\cdots+T+1}{\to} & A^{T^{mn}} & \to \cdots, \end{array}$$

where the vertical arrows are multiplication by powers of m. By the proof of 4.1.2, the cohomology groups of the two horizontal complexes are

$H^{\cdot}(\mathbb{Z}/n, A^{T^n})$ and $H^{\cdot}(\mathbb{Z}/mn, A^{T^{mn}})$, respectively. Moreover, the vertical arrows in the above commutative diagram induce the transition homomorphisms for the direct limit $H^i(\hat{\mathbb{Z}}, A) \cong \varinjlim_n H^i(\mathbb{Z}/n, A^{T^n})$.

We have

$$H^0(\hat{\mathbb{Z}}, A) \cong \varinjlim_n \ker(T - 1|_{A^{T^n}}),$$

where the transition homomorphism of the direct system are inclusions

$$\ker(T - 1|_{A^{T^n}}) \hookrightarrow \ker(T - 1|_{A^{T^{mn}}}).$$

It follows that

$$H^0(\hat{\mathbb{Z}}, A) \cong A^{\hat{\mathbb{Z}}}.$$

Of course, this also follows from the definition of $H^0(\hat{\mathbb{Z}}, A)$.

Set

$$A' = \{x \in A | T^n(x) = x \text{ and } (T^{n-1} + \cdots + T + 1)(x) = 0 \text{ for some } n \in \mathbb{N}\}.$$

Using the fact that $A = \bigcup_{n \in \mathbb{N}} A^{T^n}$, we see that

$$\begin{aligned} H^1(\hat{\mathbb{Z}}, A) &\cong \varinjlim_n H^1(\mathbb{Z}/n, A^{T^n}) \\ &\cong \varinjlim_n \ker(T^{n-1} + \cdots + T + 1|_{A^{T^n}})/\mathrm{im}(T - 1|_{A^{T^n}}) \\ &\cong A'/(T-1)A. \end{aligned}$$

Since A is a torsion $\hat{\mathbb{Z}}$-module, for any $x \in A$, there exist $m, n \in \mathbb{N}$ such that $mx = 0$ and $x \in A^{T^n}$. We then have

$$(T^{mn-1} + \cdots + T + 1)(x) = m(T^{n-1} + \cdots + T + 1)(x) = 0.$$

Hence $x \in A'$. Therefore $A = A'$ and

$$H^1(\hat{\mathbb{Z}}, A) \cong A'/(T-1)A = A_{\hat{\mathbb{Z}}}.$$

Using the fact that A is a torsion abelian group, one sees that for any $i \geq 2$ and any $x \in H^i(\mathbb{Z}/n, A^{T^n})$, there exists $m \in \mathbb{N}$ such that the transition homomorphism

$$H^i(\mathbb{Z}/n, A^{T^n}) \to H^i(\mathbb{Z}/mn, A^{T^{mn}})$$

for the direct limit $H^i(\hat{\mathbb{Z}}, A) = \varinjlim_n H^i(\mathbb{Z}/n, A^{T^n})$ maps x to 0. It follows that $H^i(\hat{\mathbb{Z}}, A) = 0$ for any $i \geq 2$. □

Fix a prime number p. Put a topology on $\mathbb{Z}$ so that subgroups $n\mathbb{Z}$ with n relatively prime to p form a base of neighborhoods of 0. The completion of $\mathbb{Z}$ with respect to this topology is denoted by $\hat{\mathbb{Z}}^{(p)}$. We have

$$\hat{\mathbb{Z}}^{(p)} = \varprojlim_{(n,p)=1} \mathbb{Z}/n.$$

Let q be a power of p. Note that multiplication by q induces a continuous isomorphism of groups

$$q : \hat{\mathbb{Z}}^{(p)} \to \hat{\mathbb{Z}}^{(p)}.$$

Proposition 4.3.8. *Let A be a $\hat{\mathbb{Z}}^{(p)}$-module. Suppose that every element in A is annihilated by some integer relatively prime to p. Then we have*

$$H^i(\hat{\mathbb{Z}}^{(p)}, A) \cong \begin{cases} A^{\hat{\mathbb{Z}}^{(p)}} & \text{if } i = 0, \\ A_{\hat{\mathbb{Z}}^{(p)}} & \text{if } i = 1, \\ 0 & \text{if } i \neq 0, 1. \end{cases}$$

Let $\phi : A \to A$ be a homomorphism compatible with the homomorphism $q : \hat{\mathbb{Z}}^{(p)} \to \hat{\mathbb{Z}}^{(p)}$ and the group actions, that is,

$$\phi((qg)x) = g\phi(x)$$

for any $g \in \hat{\mathbb{Z}}^{(p)}$ and $x \in A$. Then the homomorphism induced by (q, ϕ) on $H^0(\hat{\mathbb{Z}}^{(p)}, A) \cong A^{\hat{\mathbb{Z}}^{(p)}}$ is the homomorphism

$$\phi : A^{\hat{\mathbb{Z}}^{(p)}} \to A^{\hat{\mathbb{Z}}^{(p)}}$$

induced by ϕ, and the homomorphism induced by (q, ϕ) on $H^1(\hat{\mathbb{Z}}^{(p)}, A) \cong A_{\hat{\mathbb{Z}}^{(p)}}$ is the homomorphism

$$q\phi : A_{\hat{\mathbb{Z}}^{(p)}} \to A_{\hat{\mathbb{Z}}^{(p)}}$$

induced by $q\phi$.

Proof. The statement about the structure of $H^i(\hat{\mathbb{Z}}^{(p)}, A)$ can be proved by the same method as in the proof of 4.3.7. The determination of the homomorphism induced by (q, ϕ) on $H^0(\hat{\mathbb{Z}}^{(p)}, A)$ can be done using the definition of $H^0(\hat{\mathbb{Z}}^{(p)}, A)$.

To determine the homomorphism induced by (q, ϕ) on $H^1(\hat{\mathbb{Z}}^{(p)}, A)$, we first calculate $H^1(\hat{\mathbb{Z}}^{(p)}, A)$ using its definition. A 1-cocycle is a continuous map $f : \hat{\mathbb{Z}}^{(p)} \to A$ satisfying

$$f(g_1 + g_2) = g_1 f(g_2) + f(g_1)$$

for any $g_1, g_2 \in \hat{\mathbb{Z}}^{(p)}$. Let $T = (\bar{1}) \in \hat{\mathbb{Z}}^{(p)} = \varprojlim_{(n,p)=1} \mathbb{Z}/n$ be the canonical topological generator of $\hat{\mathbb{Z}}^{(p)}$. Then a 1-cocycle f is completely determined by the value $f(T) \in A$. Conversely, given $a \in A$, define $f : \mathbb{Z} \to A$ by

$$\begin{aligned} f(0) &= 0, \\ f(1) &= a, \\ f(n) &= T^{n-1}f(1) + f(n-1), \\ f(-n) &= -T^{-n}f(n) \end{aligned}$$

for any $n \in \mathbb{N}$. Note that we have

$$f(g_1 + g_2) = g_1 f(g_2) + f(g_1)$$

for any $g_1, g_2 \in \mathbb{Z}$. We claim that $f : \mathbb{Z} \to A$ is continuous. Indeed, we can find $m, n \in \mathbb{N}$ which are relatively prime to p such that $ma = 0$ and $a \in A^{T^n}$. We then have

$$(T^{mn-1} + \cdots + T + 1)(a) = m(T^{n-1} + \cdots T + 1)(a) = 0.$$

It follows that $f(mn) = 0$. This implies that $f(\delta) = 0$ for any $\delta \in mn\mathbb{Z}$. For any $g \in \mathbb{Z}$ and any $\delta \in mn\mathbb{Z}$, we then have

$$f(g + \delta) = f(g).$$

Thus f is continuous. We can then extend f to a 1-cocycle $f : \hat{\mathbb{Z}}^{(p)} \to A$. Therefore the group of 1-cocycles can be identified with A. A 1-cocycle $f : \hat{\mathbb{Z}}^{(p)} \to A$ is a 1-coboundary if there exists $a \in A$ such that $f(g) = ga - a$ for all $g \in \hat{\mathbb{Z}}^{(p)}$. Under the above identification, the group of 1-coboundaries is identified with the subgroup $\{Ta - a | a \in A\}$ of A. We thus have

$$H^1(\hat{\mathbb{Z}}^{(p)}, A) \cong A/\{Ta - a | a \in A\} = A_{\hat{\mathbb{Z}}^{(p)}}.$$

Note that ϕ induces a homomorphism on $A_{\hat{\mathbb{Z}}}$. To see this, it suffices to show that the subset $\{gx - x | g \in \hat{\mathbb{Z}}^{(p)},\ x \in A\}$ is invariant under ϕ. Given $g \in \hat{\mathbb{Z}}^{(p)}$, there exists $g' \in \hat{\mathbb{Z}}^{(p)}$ such that $qg' = g$. We then have

$$\phi(gx - x) = \phi((qg')x) - \phi(x) = g'\phi(x) - \phi(x).$$

This proves our assertion. Let $f : \hat{\mathbb{Z}}^{(p)} \to A$ be a 1-cocyle and let $a = f(T)$. The homomorphism on $H^1(\hat{\mathbb{Z}}^{(p)}, A)$ induced by (ϕ, q) maps the cohomology class of f to the cohomology class of $\phi \circ f \circ q$. We have

$$\phi \circ f \circ q(T) = \phi(T^{q-1}(a) + \cdots + T(a) + a) = q\phi(a) + \phi\left(\sum_{i=0}^{q-1}(T^i(a) - a)\right).$$

As $\phi\left(\sum_{i=0}^{q-1}(T^i(a) - a)\right)$ lies in the subgroup generated by $gx - x$ ($g \in \hat{\mathbb{Z}}^{(p)}$, $x \in A$), $\phi \circ f \circ q(T)$ has the same image in $A_{\hat{\mathbb{Z}}^{(p)}}$ as $q\phi(a)$. Therefore the homomorphism induced by (q, ϕ) on $H^1(\hat{\mathbb{Z}}^{(p)}, A) \cong A_{\hat{\mathbb{Z}}^{(p)}}$ is the homomorphism $q\phi : A_{\hat{\mathbb{Z}}^{(p)}} \to A_{\hat{\mathbb{Z}}^{(p)}}$ induced by $q\phi$. □

Proposition 4.3.9. *Let k be a separably closed field of characteristic p. For any positive integer n relatively prime to p, let*

$$\mu_n(k) = \{\zeta \in k | \zeta^n = 1\}$$

be the group of n-th roots of unity in k, let $G = \varprojlim_{(n,p)=1} \mu_n(k)$, and let A be a G-module such that every element in A is annihilated by some integer relatively prime to p. Then we have

$$H^i(G, A) \cong \begin{cases} A^G & \text{if } i = 0, \\ \mathrm{Hom}(\varprojlim_{(n,p)=1} \mu_n, A_G) & \text{if } i = 1, \\ 0 & \text{if } i \neq 0, 1. \end{cases}$$

Let q be a power of p, let $q : G \to G$ be the isomorphism of raising to q-th power, and let $\phi : A \to A$ be a homomorphism compatible with the homomorphism $q : G \to G$ and the group actions, that is,

$$\phi((qg)x) = g\phi(x)$$

for any $g \in G$ and $x \in A$. Then the homomorphism induced by (q, ϕ) on $H^0(G, A) \cong A^G$ is the homomorphism $\phi : A^G \to A^G$ induced by ϕ, and the homomorphism induced by (q, ϕ) on $H^1(G, A) \cong \mathrm{Hom}(\varprojlim_{(n,p)=1} \mu_n, A_G)$ is the homomorphism

$$\mathrm{Hom}(\varprojlim_{(n,p)=1} \mu_n, A_G) \to \mathrm{Hom}(\varprojlim_{(n,p)=1} \mu_n, A_G)$$

that maps any $\psi \in \mathrm{Hom}(\varprojlim_{(n,p)=1} \mu_n, A_G)$ to $\phi \circ \psi \circ q$.

Proof. For each positive integer n relatively prime to p, choose a primitive n-th root of unity ζ_n in k such that $\zeta_n^{\frac{n}{m}} = \zeta_m$ whenever $m|n$. Then $\zeta = (\zeta_n)$ is a topological generator of $G = \varprojlim_{(n,p)=1} \mu_n(k)$. We have an isomorphism

$$\varprojlim_{(n,p)=1} \mu_n(k) \cong \varprojlim_{(n,p)=1} \mathbb{Z}/n$$

that maps ζ to the canonical topological generator $1 = (\bar{1})$ of $\varprojlim_{(n,p)=1} \mathbb{Z}/n$. By 4.3.8, we have

$$H^i(G, A) \cong \begin{cases} A^G & \text{if } i = 0, \\ A_G & \text{if } i = 1, \\ 0 & \text{if } i \neq 0, 1. \end{cases}$$

By the proof of 4.3.8, in the case $i = 1$, the isomorphism $H^1(G, A) \cong A_G$ is induced by the map that maps each 1-cocycle $f : G \to A$ to $f(\zeta)$. This isomorphism depends on the choice of ζ. Consider the map

$$H^1(G, A) \otimes_{\mathbb{Z}^{(p)}} (\varprojlim_{(n,p)=1} \mu_n(k)) \to A_G$$

defined by

$$[f] \otimes g \to f(g),$$

where $[f]$ denotes the cohomology class of any 1-cocycle f, and g is any element in $G = \varprojlim_{(n,p)=1} \mu_n(k)$. One can show that it is a well-defined homomorphism. Through the isomorphism $\varprojlim_{(n,p)=1} \mu_n(k) \cong \varprojlim_{(n,p)=1} \mathbb{Z}/n$ defined above, this homomorphism is identified with the isomorphism $H^1(G, A) \cong A_G$ defined in the proof of 4.3.8. So we have

$$H^1(G, A) \otimes_{\mathbb{Z}^{(p)}} (\varprojlim_{(n,p)=1} \mu_n(k)) \cong A_G,$$

and hence we have a canonical isomorphism

$$H^1(G, A) \cong \mathrm{Hom}(\varprojlim_{(n,p)=1} \mu_n(k), A_G).$$

The other statements of the proposition follows directly from 4.3.8. □

4.4 Cohomological Dimensions

([Serre (1964)] I 3.)

Let G be a profinite group. For each prime number p, we define the *p-cohomological dimension* $\mathrm{cd}_p(G)$ (resp. the *strict p-cohomological dimension* $\mathrm{scd}_p(G)$) of G to be the smallest integer n such that for any torsion G-module (resp. any G-module) A and any $i > n$, the p-primary part of $H^i(G, A)$ vanishes. We define the *cohomological dimension* $\mathrm{cd}(G)$ (resp. the *strict cohomological dimension* $\mathrm{scd}(G)$) to be $\sup_p(\mathrm{cd}_p(G))$ (resp. $\sup_p(\mathrm{scd}_p(G))$).

Proposition 4.4.1. *Let G be a profinite group. The following conditions are equivalent:*

(i) $\mathrm{cd}_p(G) \leq n$.

(ii) $H^i(G, A) = 0$ *for any $i > n$ and any p-torsion G-module A.*

(iii) $H^{n+1}(G, A) = 0$ *for any simple p-torsion G-module A.*

Proof. For any torsion G-module A, we have $A = \bigoplus_p A(p)$, where for each prime p, $A(p)$ is the p-primary part of A. Note that $H^i(G, A(p))$ is the p-primary part of $H^i(G, A)$ for each i. So (i)⇔(ii). (i)⇒(iii) is clear. Suppose that (iii) holds. Any finite p-torsion G-module has a finite filtration with simple successive quotients, and any torsion G-module is the direct limit of its finite G-submodules. So $H^{n+1}(G,A)=0$ for any p-torsion G-module A. A can be embedded into the induced G-module $\mathrm{Ind}_{\{1\}}^G A$. Since $H^i(G,\mathrm{Ind}_{\{1\}}^G A)=0$ for all $i\geq 1$, we have $H^{i+1}(G,A)\cong H^i(G,\mathrm{Ind}_{\{1\}}^G A/A)$ for any $i\geq 1$. $\mathrm{Ind}_{\{1\}}^G A/A$ is also a p-torsion G-module. By induction on i, we see that $H^i(G,A)=0$ for any $i>n$ and any p-torsion G-module A. So (ii) holds. □

Proposition 4.4.2. *For any profinite group G, we have*

$$\mathrm{cd}_p(G)\leq \mathrm{scd}_p(G)\leq \mathrm{cd}_p(G)+1.$$

Proof. The first inequality follows from the definition of $\mathrm{cd}_p(G)$ and $\mathrm{scd}_p(G)$. For any G-module A, let $A_p=\ker(p:A\to A)$. We have exact sequences

$$0\to A_p\to A\stackrel{p}{\to} pA\to 0,$$
$$0\to pA\to A\to A/pA\to 0.$$

Note that A_p and A/pA are p-torsion G-modules. So $H^i(G,A_p)$ and $H^{i-1}(G,A/pA)$ vanish for any $i>\mathrm{cd}_p(G)+1$. Hence

$$H^i(G,A)\stackrel{p}{\to} H^i(G,pA),\quad H^i(G,pA)\to H^i(G,A)$$

are injective for any $i>\mathrm{cd}_p(G)+1$. It follows that their composite

$$H^i(G,A)\stackrel{p}{\to} H^i(G,A)$$

is injective. This implies that the p-primary part of $H^i(G,A)$ vanishes for any $i>\mathrm{cd}_p(G)+1$. So $\mathrm{scd}_p(G)\leq \mathrm{cd}_p(G)+1$. □

Proposition 4.4.3. *Let H be a closed subgroup of a profinite group G. Then*

$$\mathrm{cd}_p(H)\leq \mathrm{cd}_p(G),\quad \mathrm{scd}_p(H)\leq \mathrm{scd}_p(G).$$

If $[G:H]$ is relatively prime to p, then

$$\mathrm{cd}_p(H)=\mathrm{cd}_p(G),\quad \mathrm{scd}_p(H)=\mathrm{scd}_p(G).$$

Proof. The first statement follows from 4.3.4. The second statement follows from 4.3.6. □

Corollary 4.4.4. *Let G_p be a Sylow p-subgroup of a profinite group G. Then*

$$\mathrm{cd}_p(G)=\mathrm{cd}_p(G_p)=\mathrm{cd}(G_p),$$
$$\mathrm{scd}_p(G)=\mathrm{scd}_p(G_p)=\mathrm{scd}(G_p).$$

Proof. By 4.4.3, we have $\mathrm{cd}_p(G) = \mathrm{cd}_p(G_p)$ and $\mathrm{scd}_p(G) = \mathrm{scd}_p(G_p)$. Moreover, for any prime number ℓ distinct form p, we have

$$\mathrm{cd}_\ell(G_p) = \mathrm{cd}_\ell(\{1\}) = 0,$$
$$\mathrm{scd}_\ell(G_p) = \mathrm{scd}_\ell(\{1\}) = 0.$$

It follows that $\mathrm{cd}_p(G_p) = \mathrm{cd}(G_p)$ and $\mathrm{scd}_p(G_p) = \mathrm{scd}(G_p)$. □

Proposition 4.4.5. *Let H be a closed normal subgroup of G. We have*

$$\mathrm{cd}_p(G) \leq \mathrm{cd}_p(H) + \mathrm{cd}_p(G/H).$$

Proof. Use the Hochschild–Serre spectral sequence. □

Lemma 4.4.6. *Let G be a finite p-group and let A be a nonzero p-torsion G-module. Then $A^G \neq 0$.*

Proof. Replacing A by a subgroup generated by an orbit of a nonzero element of A, we may assume that A is finite. Since A is a nonzero p-torsion abelian group, we have $|A| = p^n$ for some positive integer n. Take $x_1, \ldots, x_n$ so that A is the disjoint union of the orbits $Gx_1, \ldots, Gx_n$. For any x_i, we have a bijection

$$G/G_{x_i} \to Gx_i, \quad gG_{x_i} \mapsto gx_i,$$

where G_{x_i} is the stabilizer of x_i. Hence $|Gx_i| = |G/G_{x_i}|$. Since G is a finite p-group, we have $|G/G_{x_i}| = p^{n_i}$ for some nonnegative integer n_i, and we have $n_i = 0$ if and only if $x_i \in A^G$. So we have

$$p^n = |A| = \sum_{i=1}^{k} |Gx_i| = \sum_{n_i \geq 1} p^{n_i} + |A^G|.$$

This implies that $p \mid |A^G|$. But $|A^G| \neq 0$ since $0 \in A^G$. So A^G contains at least p elements. □

Lemma 4.4.7. *Let G be a pro-p-group. Any nonzero simple p-torsion G-module is isomorphic to $\mathbb{Z}/p$ with the trivial G-action.*

Proof. Let A be a nonzero simple p-torsion G-module. It is necessarily finite. The action of G on A factors through a finite quotient group. By 4.4.6, we have $A^G \neq 0$. Since A is simple, we must have $A = A^G$. So G acts trivially on A. A has no nontrivial subgroup. So we have $A = \mathbb{Z}/p$. □

Proposition 4.4.8. *Let G be a pro-p-group. Then $\mathrm{cd}(G) \leq n$ if and only if $H^{n+1}(G, \mathbb{Z}/p) = 0$.*

Proof. By 4.4.4, we have $\mathrm{cd}(G) = \mathrm{cd}_p(G)$. We then apply 4.4.1 (iii) and 4.4.7. □

4.5 Galois Cohomology

([Serre (1964)] II 1–4, [SGA $4\frac{1}{2}$] Arcata III 2.)

In this section, K is a field and we fix a separable closure K_s of K. We study the cohomology of the profinite group

$$\mathrm{Gal}(K_s/K) = \varprojlim_L \mathrm{Gal}(L/K),$$

where L goes over the family of finite galois extensions of K in K_s. For any $\mathrm{Gal}(K_s/K)$-module A, by 4.3.2, we have

$$H^i(\mathrm{Gal}(K_s/K), A) = \varinjlim_L H^i(\mathrm{Gal}(L/K), A^{\mathrm{Gal}(K_s/L)}).$$

Lemma 4.5.1. *Let L be an extension of K and let $\sigma_1, \ldots, \sigma_n$ be a finite family of distinct K-automorphisms of L. Then $\{\sigma_1, \ldots, \sigma_n\}$ is linearly independent over L, that is, if $\{a_1, \ldots, a_n\}$ is a family of elements in L such that $\sum\limits_{i=1}^{n} a_i\sigma_i(x) = 0$ for all $x \in L$, then $a_i = 0$ for all i.*

Proof. Suppose $\{\sigma_1, \ldots, \sigma_n\}$ is not linearly independent over L. Let m be the smallest positive integer such that there exist distinct $i_1, \ldots, i_m \in \{1, \ldots, n\}$ and nonzero $a_1, \ldots, a_m \in L - \{0\}$ satisfying

$$a_1\sigma_{i_1} + \cdots + a_m\sigma_{i_m} = 0.$$

Note that we must have $m \geq 2$. For any $x, y \in L$, we have

$$a_1\sigma_{i_1}(xy) + \cdots + a_m\sigma_{i_m}(xy) = 0,$$

that is,

$$a_1\sigma_{i_1}(x)\sigma_{i_1} + \cdots + a_m\sigma_{i_m}(x)\sigma_{i_m} = 0.$$

From these equations, we get

$$a_2(\sigma_{i_1}(x) - \sigma_{i_2}(x))\sigma_{i_2} + \cdots + a_m(\sigma_{i_1}(x) - \sigma_{i_m}(x))\sigma_{i_m} = 0.$$

Since $\sigma_{i_1} \neq \sigma_{i_2}$, there exists $x \in L$ such that $\sigma_{i_1}(x) \neq \sigma_{i_2}(x)$. But then the last equation above contradicts the minimality of m. So $\{\sigma_1, \ldots, \sigma_n\}$ is linearly independent over L. □

Lemma 4.5.2. *Let L/K be a finite galois extension and let* $\mathrm{Gal}(L/K) = \{\sigma_1, \ldots, \sigma_n\}$. *Then there exists $x \in L$ such that $\{\sigma_1(x), \ldots, \sigma_n(x)\}$ is a basis of L over K.*

A basis of L over K of the above form is called a *normal basis.*

Proof. Note that $\{\sigma_1(x), \ldots, \sigma_n(x)\}$ is a basis if and only if $\det(\sigma_i\sigma_j(x)) \neq 0$. Indeed, suppose that $\det(\sigma_i\sigma_j(x)) \neq 0$ and $\sum_j a_j\sigma_j(x) = 0$ for some $a_j \in K$. Then $\sum_j a_j\sigma_i\sigma_j(x) = 0$ for all i. This implies that $a_j = 0$. Hence $\{\sigma_1(x), \ldots, \sigma_n(x)\}$ is linearly independent. As $n = |\mathrm{Gal}(L/K)| = [L : K]$, $\{\sigma_1(x), \ldots, \sigma_n(x)\}$ is a basis of L over K. Suppose $\det(\sigma_i\sigma_j(x)) = 0$. Then there exist $b_1, \ldots, b_n \in L$, not all equal to 0, such that $\sum_j b_j\sigma_i\sigma_j(x) = 0$ for all i. Without loss of generality, assume $b_1 \neq 0$. By 3.2.2, there exists $b \in L$ such that $\mathrm{Tr}_{L/K}(b) \neq 0$. Replacing b_j by $bb_1^{-1}b_j$, we may assume $\mathrm{Tr}_{L/K}(b_1) \neq 0$. We have $\sum_j \sigma_i^{-1}(b_j)\sigma_j(x) = 0$. Summing over i, we get $\sum_j \mathrm{Tr}_{L/K}(b_j)\sigma_j(x) = 0$. So $\{\sigma_1(x), \ldots, \sigma_n(x)\}$ is not linearly independent over K.

Assume K is infinite and let us prove the existence of a normal basis. Define $p(i,j) \in \{1, \ldots, n\}$ by $\sigma_i\sigma_j = \sigma_{p(i,j)}$. Consider the polynomial

$$f(X_1, \ldots, X_n) = \det(X_{p(i,j)}).$$

We claim $f(X_1, \ldots, X_n) \neq 0$. Indeed, if $p(i,j) = p(i',j)$ for some i, i', j, then $i = i'$; if $p(i,j) = p(i,j')$ for some i, j, j', then $j = j'$. So $f(1, 0, \ldots, 0)$ is the determinant of a matrix with only one nonzero entry on each row and each column. We have $f(1, 0, \ldots, 0) = \pm 1$ and hence $f(X_1, \ldots, X_n) \neq 0$. We have

$$\det(\sigma_i\sigma_j(x)) = f(\sigma_1(x), \ldots, \sigma_n(x)).$$

Let $\{x_1, \ldots, x_n\}$ be a basis of L over K. Any $x \in L$ can be written as $x = \sum_j a_jx_j$ $(a_j \in K)$. We have $\sigma_i(x) = \sum_j a_j\sigma_i(x_j)$. Let

$$g(Y_1, \ldots, Y_n) = f\Big(\sum_j \sigma_1(x_j)Y_j, \ldots, \sum_j \sigma_n(x_j)Y_j\Big).$$

We claim that $(\sigma_i(x_j))$ is an invertible matrix. Indeed, if $\sum_i b_i\sigma_i(x_j) = 0$ for some $b_1, \ldots, b_n \in L$ and all j, then $\sum_i b_i\sigma_i = 0$. By 4.5.1, we have $b_i = 0$ for all i. So $(\sigma_i(x_j))$ is an invertible matrix. Let (b_{ij}) be its inverse. Then

$$f(X_1, \ldots, X_n) = g\Big(\sum_j b_{1j}X_j, \ldots, \sum_j b_{nj}X_j\Big).$$

As $f(X_1, \ldots, X_n) \neq 0$, we have $g(Y_1, \ldots, Y_n) \neq 0$. Since K is an infinite field, there exist $a_1, \ldots, a_n \in K$ such that $g(a_1, \ldots, a_n) \neq 0$, that is,

$$f\Big(\sigma_1\Big(\sum_j a_jx_j\Big), \ldots, \sigma_n\Big(\sum_j a_jx_j\Big)\Big) \neq 0.$$

Let $x = \sum_j a_j x_j$. Then $\det(\sigma_i\sigma_j(x)) \neq 0$. Hence $\{\sigma_1(x), \ldots, \sigma_n(x)\}$ is a normal basis.

Assume $\mathrm{Gal}(L/K)$ is cyclic. (This holds if K is finite.) Let us prove the existence of a normal basis. Choose a generator σ of $\mathrm{Gal}(L/K)$. Define a $K[T]$-module structure on L by

$$\left(\sum_i a_i T^i\right) \cdot x = \sum_i a_i \sigma^i(x)$$

for any $\sum_i a_i T^i \in K[T]$ and $x \in L$. Note that σ satisfies the polynomial equation $T^n - 1 = 0$. By 4.5.1, σ does not satisfy any polynomial equation of degree $< n$. So $T^n - 1$ is the minimal polynomial of σ. Since $[L : K] = n$, we have $L \cong K[T]/(T^n - 1)$ as $K[T]$-modules. (We leave the proof of this fact to the reader as an exercise in linear algebra.) Let x be the element in L corresponding to 1 in $K[T]/(T^n - 1)$. Then $\{x, \sigma(x), \ldots, \sigma^{n-1}(x)\}$ is a normal basis. □

Theorem 4.5.3. *For any finite galois extension L/K, we have*

$$H^i(\mathrm{Gal}(L/K), L) = \begin{cases} K & \text{if } i = 0, \\ 0 & \text{if } i \geq 1, \end{cases}$$

and hence

$$H^i(\mathrm{Gal}(K_s/K), K_s) = \begin{cases} K & \text{if } i = 0, \\ 0 & \text{if } i \geq 1. \end{cases}$$

Proof. By 4.5.2, L is an induced $\mathrm{Gal}(L/K)$-module. The first statement follows. The second statement then follows from the fact that

$$H^i(\mathrm{Gal}(K_s/K), K_s) \cong \varinjlim_L H^i(\mathrm{Gal}(L/K), L),$$

where L goes over the family of finite galois extensions of K contained in K_s. □

Theorem 4.5.4 (Hilbert 90). *For any finite galois extension L/K, we have $H^1(\mathrm{Gal}(L/K), L^*) = 0$, and hence $H^1(\mathrm{Gal}(K_s/K), K_s^*) = 0$.*

Proof. Let $f : \mathrm{Gal}(L/K) \to L^*$ be a 1-cocycle. By Theorem 4.5.1, there exists $x \in L$ such that

$$\sum_{\sigma \in \mathrm{Gal}(L/K)} f(\sigma)\sigma(x) \neq 0.$$

Let $y = \sum_{\sigma \in \mathrm{Gal}(L/K)} f(\sigma)\sigma(x) \in L^*$. For any $\tau \in \mathrm{Gal}(L/K)$, we have

$$f(\tau\sigma) = \tau(f(\sigma)) \cdot f(\tau),$$

and

$$\begin{aligned}\tau(y) &= \sum_{\sigma\in \mathrm{Gal}(L/K)} \tau(f(\sigma))\tau\sigma(x)\\ &= \sum_{\sigma\in \mathrm{Gal}(L/K)} f(\tau)^{-1}f(\tau\sigma)\tau\sigma(x)\\ &= f(\tau)^{-1}y.\end{aligned}$$

So we have $f(\tau) = y\tau(y)^{-1}$. This shows that f is a 1-coboundary. Therefore $H^1(\mathrm{Gal}(K_s/K), K_s^*) = 0$. □

In 5.7.9, we will give another proof of the Hilbert 90 using the descent theory and the etale cohomology.

Theorem 4.5.5.

(i) *If K is a field of characteristic p, then* $\mathrm{cd}_p(\mathrm{Gal}(K_s/K)) \leq 1$.

(ii) *Assume* $H^2(\mathrm{Gal}(K_s/L), K_s^*) = 0$ *for any finite extension L of K contained in K_s. Then* $\mathrm{cd}(\mathrm{Gal}(K_s/K)) \leq 1$, *that is, for any torsion* $\mathrm{Gal}(K_s/K)$*-module A, we have*

$$H^i(\mathrm{Gal}(K_s/K), A) = 0$$

for any $i \geq 2$. Moreover, we have

$$H^i(\mathrm{Gal}(K_s/K), K_s^*) = 0$$

for any $i \geq 1$.

Proof.

(i) Let G_p be the Sylow p-subgroup of $\mathrm{Gal}(K_s/K)$ and let $M = K_s^{G_p}$. We have an exact sequence of $\mathrm{Gal}(K_s/K)$-modules

$$0 \to \mathbb{Z}/p \to K_s \xrightarrow{\wp} K_s \to 0,$$

where $\wp(x) = x^p - x$ for any $x \in K_s$. By 4.5.3, we have

$$H^i(G_p, K_s) = H^i(\mathrm{Gal}(K_s/M), K_s) = 0$$

for any $i \geq 1$. The long exact sequence of cohomology groups associated to the above short exact sequence shows that $H^i(G_p, \mathbb{Z}/p) = 0$ for any $i \geq 2$. By 4.4.8, we have $\mathrm{cd}(G_p) \leq 1$. By 4.4.4, we have $\mathrm{cd}_p(\mathrm{Gal}(K_s/K)) \leq 1$.

(ii) Let ℓ be a prime number distinct from the characteristic of K, let G_ℓ be the Sylow ℓ-subgroup of $\mathrm{Gal}(K_s/K)$, and let $M = K_s^{G_\ell}$. We have $M = \varinjlim_L L$, where L goes over the set of finite extensions of K contained in M. We have an exact sequence of $\mathrm{Gal}(K_s/K)$-modules

$$0 \to \boldsymbol{\mu}_\ell(K_s) \to K_s^* \stackrel{x\mapsto x^\ell}{\to} K_s^* \to 0,$$

where $\mu_\ell(K_s)$ is the group of ℓ-th roots of unity in K_s^*. For any finite extension L of K contained in M, we have

$$H^i(\mathrm{Gal}(K_s/L), K_s^*) = 0$$

for $i = 1, 2$ by 4.5.4 and the assumption. The long exact sequence of cohomology groups associated to the above short exact sequence then shows that $H^2(\mathrm{Gal}(K_s/L), \mu_\ell(K_s)) = 0$. By 4.3.1, we have

$$H^2(G_\ell, \mu_\ell(K_s)) = \varinjlim_{L} H^2(\mathrm{Gal}(K_s/L), \mu_\ell(K_s)) = 0.$$

Note that $\mu_\ell(K_s)$ has no nontrivial subgroup. So as a G_ℓ-module, it is simple and isomorphic to $\mathbb{Z}/\ell$ with the trivial G_ℓ-action. By 4.4.8, we have $\mathrm{cd}(G_\ell) \leq 1$. By 4.4.4, we have $\mathrm{cd}_\ell(\mathrm{Gal}(K_s/K)) \leq 1$. Combined with (i), we get $\mathrm{cd}(\mathrm{Gal}(K_s/K)) \leq 1$. By 4.4.2, we have $\mathrm{scd}(\mathrm{Gal}(K_s/K)) \leq 2$. It follows that

$$H^i(\mathrm{Gal}(K_s/K), K_s^*) = 0$$

for any $i \geq 3$. □

A field K is called *quasi-algebraically closed* if for any positive integer n and any n-variable homogeneous polynomial $f(X_1, \ldots, X_n)$ of degree $< n$ with coefficients in K, $f(X_1, \ldots, X_n) = 0$ has nonzero solutions in K^n. In 5.7.15, we will prove the following using the descent theory and the etale cohomology:

Proposition 4.5.6. *Suppose K is quasi-algebraically closed. We have* $H^2(\mathrm{Gal}(K_s/K), K_s^*) = 0$.

Lemma 4.5.7. *If K is quasi-algebraically closed and L is algebraic over K, then L is quasi-algebraically closed.*

Proof. Let $F(X_1, \ldots, X_n)$ be an n-variable homogeneous polynomial of degree $< n$ with coefficients in L. Then the coefficients of $F(X_1, \ldots, X_n)$ lies in a finite extension of K contained in L. So to prove that $F(X_1, \ldots, X_n) = 0$ has a nonzero solution, we may assume that $[L : K] < \infty$. Let $\{e_1, \ldots, e_m\}$ be a basis of L over K. Consider the function

$$\begin{aligned} &f(x_{11}, \ldots, x_{1m}, \ldots, x_{n1}, \ldots, x_{nm}) \\ &= \mathrm{Norm}_{L/K}(F(x_{11}e_1 + \cdots + x_{1m}e_m, \ldots, x_{n1}e_1 + \cdots + x_{nm}e_m)) \end{aligned}$$

defined on K^{nm}. It is a homogeneous polynomial of degree $< nm$ with coefficients in K. Since K is quasi-algebraically closed, $f(x_{11}, \ldots, x_{1m}, \ldots, x_{n1}, \ldots, x_{nm})$ has a nonzero solution in K^{nm}. Hence $F(X_1, \ldots, X_n)$ has a nonzero solution in L^n. □

Theorem 4.5.8 (Tsen). *Let k be an algebraically closed field and let K/k be an extension of transcendental degree 1. Then K is quasi-algebraically closed.*

Proof. By 4.5.7, it suffices to consider the case where $K = k(t)$ is purely transcendental. Let

$$f(X_1, \ldots, X_n) = \sum_{i_1+\cdots+i_n=d} a_{i_1\ldots i_n}(t) X_1^{i_1} \cdots X_n^{i_n}$$

be an n-variable homogeneous polynomial of degree $d < n$ with coefficients $a_{i_1\ldots i_n}(t) \in k(t)$. We need to show that $f(X_1, \ldots, X_n)$ has a nonzero solution in $k(t)^n$. It suffices to treat the case where $a_{i_1\ldots i_n}(t)$ are polynomials. Let

$$\delta = \max\{\deg(a_{i_1\ldots i_n}(t)) | i_1 + \cdots + i_n = d\}.$$

Substituting $X_i = \sum_{j=0}^N a_{ij} t^j$ $(i = 1, \ldots, n)$ into $f(X_1, \ldots, X_n)$, and setting the coefficients of t^k $(k = 0, 1, \ldots, \delta + dN)$ to zero, we get a system of $\delta + dN + 1$ homogeneous equations with $n(N+1)$ unknowns a_{ij}. Since $d < n$, we have $\delta + dN + 1 < n(N+1)$ for sufficiently large N. In this case, the system of homogeneous equations has a nonzero solution by [Hartshorne (1977)] I 7.2. So $f(X_1, \ldots, X_n) = 0$ has a nonzero solution. □

Theorem 4.5.9. *Let K be a quasi-algebraically closed field. Then for any torsion $\mathrm{Gal}(K_s/K)$-module A, we have $H^i(\mathrm{Gal}(K_s/K), A) = 0$ for any $i \geq 2$. Moreover we have $H^i(\mathrm{Gal}(K_s/K), K_s^*) = 0$ for any $i \geq 1$.*

Proof. This follows from 4.5.5 (ii), 4.5.6 and 4.5.7. □

Theorem 4.5.10. *Let L/K be an extension of fields, and let K_s and L_s be their separable closures, respectively. For any prime number p, we have*

$$\mathrm{cd}_p(\mathrm{Gal}(L_s/L)) \leq \mathrm{cd}_p(\mathrm{Gal}(K_s/K)) + \mathrm{tr.deg}(L/K),$$

where $\mathrm{tr.deg}(L/K)$ denotes the transcendental degree of L over K.

Proof. It suffices to treat the case where $\mathrm{tr.deg}(L/K) < \infty$. There exists a subextension L'/K of L/K such that $L' \cong K(t_1, \ldots, t_n)$ is purely transcendental, and L/L' is algebraic. Note that $\mathrm{Gal}(L_s/L)$ can be regarded as a closed subgroup of $\mathrm{Gal}(L'_s/L')$, where L'_s is a separable closure of L'. By 4.4.3, we have $\mathrm{cd}_p(\mathrm{Gal}(L_s/L)) \leq \mathrm{cd}_p(\mathrm{Gal}(L'_s/L'))$. It suffices to prove

$$\mathrm{cd}_p(\mathrm{Gal}(L'_s/L')) \leq \mathrm{cd}_p(\mathrm{Gal}(K_s/K)) + \mathrm{tr.deg}(L/K).$$

So we may assume that $L = K(t_1, \ldots, t_n)$. By induction on n, we are reduced to the case where $L = K(t)$. Let $K(t)_s$ be a separable closure of $K(t)$. By 4.4.5, we have

$$\mathrm{cd}_p\Big(\mathrm{Gal}(K(t)_s/K(t))\Big) \leq \mathrm{cd}_p\Big(\mathrm{Gal}(K(t)_s/K_s(t))\Big) + \mathrm{cd}_p\Big(\mathrm{Gal}(K_s(t)/K(t))\Big).$$

Let $\overline{K}$ be an algebraic closure of K and let $\overline{K}(t)_s$ be a separable closure of $\overline{K}(t)$. We have

$$\mathrm{Gal}(K(t)_s/K_s(t)) \cong \mathrm{Gal}(\overline{K}(t)_s/\overline{K}(t)).$$

By 4.5.8 and 4.5.9, we have

$$\mathrm{cd}_p\Big(\mathrm{Gal}(\overline{K}(t)_s/\overline{K}(t))\Big) \leq 1.$$

On the other hand, we have

$$\mathrm{Gal}(K_s(t)/K(t)) \cong \mathrm{Gal}(K_s/K).$$

It follows that

$$\mathrm{cd}_p\Big(\mathrm{Gal}(K(t)_s/K(t))\Big) \leq 1 + \mathrm{cd}_p(\mathrm{Gal}(K_s/K)).$$

This proves our assertion. □

Theorem 4.5.11. *Let k be a separably closed field and let K/k be an extension of transcendental degree 1.*

(i) *We have $H^i(\mathrm{Gal}(K_s/K), A) = 0$ for any $i \geq 2$ and any torsion $\mathrm{Gal}(K_s/K)$-module A.*

(ii) *We have $H^i(\mathrm{Gal}(K_s/K), K_s^*) = 0$ for $i = 1$ and any $i \geq 3$. Let p be the characteristic of K. Then $H^2(\mathrm{Gal}(K_s/K), K_s^*)$ is a p-torsion group.*

Proof.

(i) By 4.5.10, for any prime number p, we have

$$\mathrm{cd}_p(\mathrm{Gal}(K_s/K)) \leq \mathrm{cd}_p(\mathrm{Gal}(k_s/k)) + 1 = 1.$$

(ii) By 4.5.4, we have $H^1(\mathrm{Gal}(K_s/K), K_s^*) = 0$. By 4.4.2, we have

$$\mathrm{scd}(\mathrm{Gal}(K_s/K)) \leq \mathrm{cd}(\mathrm{Gal}(K_s/K)) + 1 \leq 2.$$

So $H^i(\mathrm{Gal}(K_s/K), K_s^*) = 0$ for any $i \geq 3$. For any prime number ℓ distinct from p, we have an exact sequence of $\mathrm{Gal}(K_s/K)$-modules

$$0 \to \boldsymbol{\mu}_\ell(K_s) \to K_s^* \stackrel{x \mapsto x^\ell}{\to} K_s^* \to 0.$$

By (i), we have $H^2(\mathrm{Gal}(K_s/K), \boldsymbol{\mu}_\ell(K_s)) = 0$. So

$$H^2(\mathrm{Gal}(K_s/K), K_s^*) \stackrel{\ell}{\to} H^2(\mathrm{Gal}(K_s/K), K_s^*)$$

is injective. Hence $H^2(\mathrm{Gal}(K_s/K), K_s^*)$ has no ℓ-torsion, and must be a p-torsion group. □

Chapter 5

Etale Cohomology

5.1 Presheaves and Čech Cohomology

Let $\mathscr{C}$ be a category. A *presheaf* of sets (resp. abelian groups, rings, etc.) is a contravariant functor from $\mathscr{C}$ to the category of sets (resp. abelian groups, rings, etc.). Morphisms of presheaves are defined to be morphisms of functors. Given a presheaf $\mathscr{F}$ on $\mathscr{C}$, an object U in $\mathscr{C}$ and a morphism $f : U \to V$ in $\mathscr{C}$, elements in $\mathscr{F}(U)$ are called sections of $\mathscr{F}$ on U, and the map $\mathscr{F}(f) : \mathscr{F}(V) \to \mathscr{F}(U)$ is called the *restriction* induced by f. For any $s \in \mathscr{F}(V)$, we often denote $\mathscr{F}(f)(s)$ by $s|_U$.

In the following, we mainly discuss presheaves of abelian groups, and leave it for the reader to treat presheaves with other structures. We simply call a presheaf of abelian groups a presheaf.

Denote the category of presheaves on $\mathscr{C}$ by $\mathscr{P}_{\mathscr{C}}$. It is an abelian category. A sequence

$$\mathscr{F} \to \mathscr{G} \to \mathscr{H}$$

of presheaves is exact if and only if for any object U in $\mathscr{C}$, the sequence

$$\mathscr{F}(U) \to \mathscr{G}(U) \to \mathscr{H}(U)$$

is exact.

Let $f : \mathscr{C} \to \mathscr{C}'$ be a covariant functor. It induces a covariant functor

$$f^{\mathscr{P}} : \mathscr{P}_{\mathscr{C}'} \to \mathscr{P}_{\mathscr{C}}, \quad f^{\mathscr{P}}(\mathscr{F}')(U) = \mathscr{F}'(f(U)),$$

where $\mathscr{F}'$ is any object in $\mathscr{P}_{\mathscr{C}'}$ and U is any object in $\mathscr{C}$. For any object U' in $\mathscr{C}'$, define a category $I_{U'}$ as follows: Objects in $I_{U'}$ are pairs (U, ϕ) such that U are objects in $\mathscr{C}$ and $\phi : U' \to f(U)$ are morphisms in $\mathscr{C}'$. Given two objects (U_i, ϕ_i) $(i = 1, 2)$ in $I_{U'}$, a morphism $\xi : (U_1, \phi_1) \to (U_2, \phi_2)$ in $I_{U'}$ is a morphism $\xi : U_1 \to U_2$ in $\mathscr{C}$ such that $f(\xi)\phi_1 = \phi_2$. Let $I^\circ_{U'}$

be the opposite category of $I_{U'}$. For any presheaf $\mathscr{F}$ on $\mathscr{C}$, the assignment $(U,\phi)\mapsto \mathscr{F}(U)$ defines a contravariant functor on $I_{U'}$ and hence a covariant functor on $I_{U'}^{\circ}$. Suppose for any presheaf $\mathscr{F}$ on $\mathscr{C}$, and any object U' of $\mathscr{C}'$, that $\varinjlim_{(U,\phi)\in \mathrm{ob}\, I_{U'}^{\circ}} \mathscr{F}(U)$ exists. We define

$$(f_{\mathscr{P}}\mathscr{F})(U') = \varinjlim_{(U,\phi)\in \mathrm{ob}\, I_{U'}^{\circ}} \mathscr{F}(U).$$

Then $f_{\mathscr{P}}\mathscr{F}$ is a presheaf on $\mathscr{C}'$. One can show that $f_{\mathscr{P}}$ is left adjoint to the functor $f^{\mathscr{P}}$. That is, for any presheaf $\mathscr{F}$ on $\mathscr{C}$ and any presheaf $\mathscr{F}'$ on $\mathscr{C}'$, we have a one-to-one correspondence

$$\mathrm{Hom}(f_{\mathscr{P}}\mathscr{F},\mathscr{F}')\cong \mathrm{Hom}(\mathscr{F}', f^{\mathscr{P}}\mathscr{F})$$

functorial in $\mathscr{F}$ and $\mathscr{F}'$.

Suppose fiber products exist in $\mathscr{C}$. Let U be an object in $\mathscr{C}$, and let $\mathfrak{U}=\{U_\alpha\to U\}_{\alpha\in I}$ be a set of morphisms in $\mathscr{C}$ with a common target U. For any $\alpha_0,\ldots,\alpha_n$, let

$$U_{\alpha_0\ldots\alpha_n} = U_{\alpha_0}\times_U\cdots\times_U U_{\alpha_n}.$$

For any presheaf $\mathscr{F}$ on $\mathscr{C}$ and any nonnegative integer n, let

$$C^n(\mathfrak{U},\mathscr{F}) = \prod_{\alpha_0,\ldots,\alpha_n}\mathscr{F}(U_{\alpha_0\ldots\alpha_n}).$$

Define a homomorphism

$$d: C^n(\mathfrak{U},\mathscr{F})\to C^{n+1}(\mathfrak{U},\mathscr{F})$$

as follows: For any $s=(s_{\alpha_0\ldots\alpha_n})\in C^n(\mathfrak{U},\mathscr{F})$, define $ds=((ds)_{\alpha_0\ldots\alpha_{n+1}})\in \mathscr{C}^{n+1}(\mathfrak{U},\mathscr{F})$ by

$$(ds)_{\alpha_0\ldots\alpha_{n+1}} = \sum_{i=1}^{n+1}(-1)^i s_{\alpha_0\ldots\hat{\alpha}_i\ldots\alpha_{n+1}}|_{U_{\alpha_0\cdots\alpha_{n+1}}},$$

where $s_{\alpha_0\ldots\hat{\alpha}_i\ldots\alpha_{n+1}}|_{U_{\alpha_0\cdots\alpha_{n+1}}}$ is the image of $s_{\alpha_0\ldots\hat{\alpha}_i\ldots\alpha_{n+1}}$ under the restriction

$$\mathscr{F}(U_{\alpha_0\ldots\hat{\alpha}_i\ldots\alpha_{n+1}})\to\mathscr{F}(U_{\alpha_0\ldots\alpha_{n+1}})$$

induced by the projection

$$U_{\alpha_0}\times_U\cdots\times_U U_{\alpha_{n+1}}\to U_{\alpha_0}\times_U\cdots\times_U \widehat{U}_{\alpha_i}\times_U\cdots\times_U U_{\alpha_{n+1}}.$$

One can check that $dd=0$. The complex $(C^{\cdot}(\mathfrak{U},\mathscr{F}),d)$ is called the *Čech complex*, and its cohomology groups are called the *Čech cohomology groups* of $\mathscr{F}$ with respect to the family $\mathfrak{U}$, and are denoted by $\check{H}^{\cdot}(\mathfrak{U},\mathscr{F})$.

Unlike the Čech cohomology for ordinary topological spaces, we cannot use alternative cochains to calculate the Čech cohomology for presheaves on categories. For example, if we take $\mathscr{C}$ to be the category of open subsets of a topological space X, $\mathfrak{U} = \{U_\alpha \to U\}_{\alpha\in I}$ an open covering, $\mathfrak{U}' = \{\coprod_{\alpha\in I} U_\alpha \to U\}$, and $\mathscr{F}$ a sheaf on X, then

$$C^\cdot(\mathfrak{U}, \mathscr{F}) \cong C^\cdot(\mathfrak{U}', \mathscr{F}).$$

Since $\mathfrak{U}'$ has only one element, alternative n-cochains for $\mathfrak{U}'$ and $\mathscr{F}$ are trivial for any $n \geq 1$, whereas in general, $\check{H}^n(\mathfrak{U}, \mathscr{F})$ is not trivial.

It is clear that $\mathscr{F} \to C^n(\mathfrak{U}, \mathscr{F})$ is an exact functor on $\mathscr{P}_{\mathscr{C}}$. Using this fact, one checks that $\check{H}^\cdot(\mathfrak{U}, -)$ is a δ-functor. The following proposition shows that it is effaceable under mild conditions.

Proposition 5.1.1. *Assume that there exists a set Λ of objects in $\mathscr{C}$ such that for any presheaf $\mathscr{A}$ and any proper sub-presheaf $\mathscr{B}$ of $\mathscr{A}$, there exists $X \in \Lambda$ such that $\mathscr{A}(X) \neq \mathscr{B}(X)$. Then the category $\mathscr{P}_{\mathscr{C}}$ has enough injective objects, and $\check{H}^n(\mathfrak{U}, \mathscr{I}) = 0$ for any injective presheaf $\mathscr{I}$ and any $n \geq 1$.*

Proof. For any object X in $\mathscr{C}$, consider the category $\{X\}$ with only one object X and only one morphism id_X. Let $i : \{X\} \to \mathscr{C}$ be the inclusion functor. For any abelian group M, denote the presheaf $X \mapsto M$ on $\{X\}$ by M, and denote the presheaf $i_{\mathscr{P}}M$ on $\mathscr{C}$ by M_X. For any object U in $\mathscr{C}$, we have

$$M_X(U) = \bigoplus_{\mathrm{Hom}_{\mathscr{C}}(U,X)} M.$$

For any presheaf $\mathscr{F}$ on $\mathscr{C}$, we have

$$\mathrm{Hom}(\mathbb{Z}_X, \mathscr{F}) \cong \mathrm{Hom}(\mathbb{Z}, i^{\mathscr{P}}\mathscr{F}) \cong \mathrm{Hom}(\mathbb{Z}, \mathscr{F}(X)) \cong \mathscr{F}(X).$$

Taking into account of our assumption, we see that for any presheaf $\mathscr{A}$ and any proper sub-presheaf $\mathscr{B}$ of $\mathscr{A}$, there exists $X \in \Lambda$ such that there exists a morphism $\mathbb{Z}_X \to \mathscr{A}$ which does not factor through $\mathscr{B}$. So $\{\mathbb{Z}_X | X \in \Lambda\}$ is a set of generators for $\mathscr{P}_{\mathscr{C}}$. By the same argument as in the proof [Fu (2006)] 2.1.6 or [Grothendieck (1957)] 1.10.1, one can show that $\mathscr{P}_{\mathscr{C}}$ has enough injective objects.

Let us prove $\check{H}^n(\mathscr{U}, \mathscr{I}) = 0$ for any injective presheaf $\mathscr{I}$ and any $n \geq 1$. We need to show that the sequence

$$\prod_{\alpha_0\in I} \mathscr{I}(U_{\alpha_0}) \xrightarrow{d} \prod_{\alpha_0,\alpha_1\in I} \mathscr{I}(U_{\alpha_0\alpha_1}) \xrightarrow{d} \prod_{\alpha_0,\alpha_1,\alpha_2\in I} \mathscr{I}(U_{\alpha_0\alpha_1\alpha_2}) \xrightarrow{d} \cdots$$

is exact. This sequence can be identified with

$$\mathrm{Hom}(\bigoplus_{\alpha_0\in I}\mathbb{Z}_{U_{\alpha_0}},\mathscr{I})\to\mathrm{Hom}(\bigoplus_{\alpha_0,\alpha_1\in I}\mathbb{Z}_{U_{\alpha_0\alpha_1}},\mathscr{I})\to\mathrm{Hom}(\bigoplus_{\alpha_0,\alpha_1,\alpha_2\in I}\mathbb{Z}_{U_{\alpha_0\alpha_1\alpha_2}},\mathscr{I})\to\cdots.$$

Since $\mathscr{I}$ is an injective presheaf, it suffices to show that the sequence

$$\bigoplus_{\alpha_0\in I}\mathbb{Z}_{U_{\alpha_0}}\stackrel{\delta_0}{\leftarrow}\bigoplus_{\alpha_0,\alpha_1\in I}\mathbb{Z}_{U_{\alpha_0\alpha_1}}\stackrel{\delta_1}{\leftarrow}\bigoplus_{\alpha_0,\alpha_1,\alpha_2\in I}\mathbb{Z}_{U_{\alpha_0\alpha_1\alpha_2}}\stackrel{\delta_2}{\leftarrow}\cdots$$

is an exact sequence of presheaves. We need to show that for any object X in $\mathscr{C}$, the sequence

$$\bigoplus_{\alpha_0\in I}\mathbb{Z}_{U_{\alpha_0}}(X)\stackrel{\delta_0}{\leftarrow}\bigoplus_{\alpha_0,\alpha_1\in I}\mathbb{Z}_{U_{\alpha_0\alpha_1}}(X)\stackrel{\delta_1}{\leftarrow}\bigoplus_{\alpha_0,\alpha_1,\alpha_2\in I}\mathbb{Z}_{U_{\alpha_0\alpha_1\alpha_2}}(X)\stackrel{\delta_2}{\leftarrow}\cdots$$

is exact, that is, the sequence

$$\bigoplus_{\alpha_0\in I}\bigoplus_{\mathrm{Hom}(X,U_{\alpha_0})}\mathbb{Z}\stackrel{\delta_0}{\leftarrow}\bigoplus_{\alpha_0,\alpha_1\in I}\bigoplus_{\mathrm{Hom}(X,U_{\alpha_0\alpha_1})}\mathbb{Z}\stackrel{\delta_1}{\leftarrow}\bigoplus_{\alpha_0,\alpha_1,\alpha_2\in I}\bigoplus_{\mathrm{Hom}(X,U_{\alpha_0\alpha_1\alpha_2})}\mathbb{Z}\stackrel{\delta_2}{\leftarrow}\cdots$$

is exact. Let us describe the homomorphisms δ_n. For any $f\in\mathrm{Hom}(X,U_{\alpha_0\ldots\alpha_n})$, let e_f be the element in $\bigoplus_{\alpha_0,\ldots,\alpha_n\in I}\bigoplus_{\mathrm{Hom}(X,U_{\alpha_0\ldots\alpha_n})}\mathbb{Z}$ whose component corresponding to f is 1 and whose other components are 0. Let

$$p_i:U_{\alpha_0}\times_U\cdots\times_U U_{\alpha_{n+1}}\to U_{\alpha_0}\times_U\cdots\times_U\widehat{U}_{\alpha_i}\times_U\cdots\times_U U_{\alpha_{n+1}}$$

be the projections. Then

$$\delta_n:\bigoplus_{\alpha_0,\ldots,\alpha_{n+1}\in I}\bigoplus_{\mathrm{Hom}(X,U_{\alpha_0\ldots\alpha_{n+1}})}\mathbb{Z}\to\bigoplus_{\alpha_0,\ldots,\alpha_n\in I}\bigoplus_{\mathrm{Hom}(X,U_{\alpha_0\ldots\alpha_n})}\mathbb{Z}$$

is defined by

$$\delta_n(e_f)=\sum_{i=0}^{n+1}(-1)^i e_{p_if}.$$

Let π_α $(\alpha\in I)$ be the morphisms $U_\alpha\to U$. For any $\phi\in\mathrm{Hom}(X,U)$, let

$$S(\phi)=\coprod_\alpha\{g\in\mathrm{Hom}(X,U_\alpha)|\pi_\alpha g=\phi\}.$$

The the above sequence can be written as

$$\bigoplus_{\phi\in\mathrm{Hom}(X,U)}\bigoplus_{S(\phi)}\mathbb{Z}\stackrel{\delta_0}{\leftarrow}\bigoplus_{\phi\in\mathrm{Hom}(X,U)}\bigoplus_{S(\phi)\times S(\phi)}\mathbb{Z}\stackrel{\delta_1}{\leftarrow}\bigoplus_{\phi\in\mathrm{Hom}(X,U)}\bigoplus_{S(\phi)\times S(\phi)\times S(\phi)}\mathbb{Z}\stackrel{\delta_2}{\leftarrow}\cdots.$$

To prove that this sequence is exact, it suffices to show that for any ϕ, the sequence

$$\bigoplus_{S(\phi)}\mathbb{Z}\stackrel{\delta_0}{\leftarrow}\bigoplus_{S(\phi)\times S(\phi)}\mathbb{Z}\stackrel{\delta_1}{\leftarrow}\bigoplus_{S(\phi)\times S(\phi)\times S(\phi)}\mathbb{Z}\stackrel{\delta_2}{\leftarrow}\cdots$$

is exact. Here δ_n can be described as follows: For any $(f_0, \ldots, f_{n+1}) \in S(\phi)^{n+2}$, let $e_{(f_0,\ldots,f_{n+1})}$ be the element in $\bigoplus_{S(\phi)^{n+2}} \mathbb{Z}$ whose component corresponding to $(f_0, \ldots, f_{n+1})$ is 1 and whose other components are 0. Then

$$\delta_n : \bigoplus_{S(\phi)^{n+2}} \mathbb{Z} \to \bigoplus_{S(\phi)^{n+1}} \mathbb{Z}$$

is defined by

$$\delta_n(e_{(f_0,\ldots,f_{n+1})}) = \sum_{i=0}^{n+1} (-1)^i e_{(f_0,\ldots,\hat{f}_i,\ldots,f_{n+1})}.$$

Let

$$\delta_{-1} : \bigoplus_{S(\phi)} \mathbb{Z} \to \mathbb{Z}$$

be the homomorphism defined by $\delta_0(e_f) = 1$ for any $f \in S(\phi)$. Fix an element g in $S(\phi)$. Consider the homomorphisms

$$D_n : \bigoplus_{S(\phi)^{n+1}} \mathbb{Z} \to \bigoplus_{S(\phi)^{n+2}} \mathbb{Z}, \qquad D_n(e_{(f_0,\ldots,f_{n+1})}) = e_{(g,f_0,\ldots,f_{n+1})} \quad (n \geq 0),$$

$$D_{-1} : \mathbb{Z} \to \bigoplus_{S(\phi)} \mathbb{Z}, \qquad D_{-1}(1) = e_g.$$

Then D_n define a homotopy between the identity and the zero morphism of the complex

$$0 \leftarrow \mathbb{Z} \overset{\delta_{-1}}{\leftarrow} \bigoplus_{S(\phi)} \mathbb{Z} \overset{\delta_0}{\leftarrow} \bigoplus_{S(\phi)\times S(\phi)} \mathbb{Z} \overset{\delta_1}{\leftarrow} \bigoplus_{S(\phi)\times S(\phi)\times S(\phi)} \mathbb{Z} \overset{\delta_2}{\leftarrow} \cdots.$$

Hence this complex is acyclic. □

Let $\mathfrak{U} = \{U_\alpha \to U\}_{\alpha\in I}$ and $\mathfrak{U}' = \{U'_{\alpha'} \to U\}_{\alpha'\in I'}$ be two sets of morphisms in $\mathscr{C}$ with a common target U. A morphism $f : \mathfrak{U} \to \mathfrak{U}'$ consists of a map $\epsilon : I \to I'$ and a morphism $f_\alpha : U_\alpha \to U'_{\epsilon(\alpha)}$ in $\mathscr{C}$ for each $\alpha \in I$. Such a morphism induces a morphism of complexes

$$f^* : C^\cdot(\mathfrak{U}', \mathscr{F}) \to C^\cdot(\mathfrak{U}, \mathscr{F})$$

as follows: Given $s = (s_{\alpha'_0\ldots\alpha'_n}) \in C^n(\mathfrak{U}', \mathscr{F})$, define $f^*s = ((f^*s)_{\alpha_0\ldots\alpha_n}) \in C^n(\mathfrak{U}, \mathscr{F})$ by

$$(f^*s)_{\alpha_0\ldots\alpha_n} = s_{\epsilon(\alpha_0)\ldots\epsilon(\alpha_n)}|_{U_{\alpha_0\cdots\alpha_n}},$$

where $s_{\epsilon(\alpha_0)\ldots\epsilon(\alpha_n)}|_{U_{\alpha_0\cdots\alpha_n}}$ is the image of $s_{\epsilon(\alpha_0)\ldots\epsilon(\alpha_n)}$ under the restriction induced by the morphism

$$f_{\alpha_0} \times \cdots \times f_{\alpha_n} : U_{\alpha_0} \times_U \cdots \times_U U_{\alpha_n} \to U'_{\epsilon(\alpha_0)} \times_U \cdots \times_U U'_{\epsilon(\alpha_n)}.$$

Let $g : \mathfrak{U} \to \mathfrak{U}'$ be another morphism consisting of a map $\delta : I \to I'$ and morphisms $g_\alpha : U_\alpha \to U'_{\delta(\alpha)}$ $(\alpha \in I)$. Define homomorphisms

$$H : C^n(\mathfrak{U}', \mathscr{F}) \to C^{n-1}(\mathfrak{U}, \mathscr{F})$$

as follows: Given $s = (s_{\alpha'_0 \ldots \alpha'_n}) \in C^n(\mathfrak{U}', \mathscr{F})$, define $Hs = ((Hs)_{\alpha_0 \ldots \alpha_{n-1}}) \in C^{n-1}(\mathfrak{U}, \mathscr{F})$ by

$$(Hs)_{\alpha_0 \ldots \alpha_{n-1}} = \sum_{i=0}^{n} (-1)^i s_{\epsilon(\alpha_0) \ldots \epsilon(\alpha_i)\delta(\alpha_i) \ldots \delta(\alpha_{n-1})}|_{U_{\alpha_0 \ldots \alpha_{n-1}}},$$

where $s_{\epsilon(\alpha_0) \ldots \epsilon(\alpha_i)\delta(\alpha_i) \ldots \delta(\alpha_{n-1})}|_{U_{\alpha_0 \ldots \alpha_{n-1}}}$ is the image of $s_{\epsilon(\alpha_0) \ldots \epsilon(\alpha_i)\delta(\alpha_i) \ldots \delta(\alpha_{n-1})}$ under the restriction induced by the morphism

$$f_{\alpha_0} \times \cdots \times f_{\alpha_{i-1}} \times (f_{\alpha_i}, g_{\alpha_i}) \times g_{\alpha_{i+1}} \times \cdots \times g_{\alpha_{n-1}} :$$
$$U_{\alpha_0} \times_U \cdots \times_U U_{\alpha_{n-1}} \to U'_{\epsilon(\alpha_0)} \times_U \cdots \times_U U'_{\epsilon(\alpha_i)} \times_U U'_{\delta(\alpha_i)} \times_U \cdots \times_U U'_{\delta(\alpha_{n-1})}.$$

Then we have

$$Hd + dH = g^* - f^*.$$

So f^* and g^* are homotopic to each other, and hence induce the same homomorphisms $\check{H}^{\cdot}(\mathfrak{U}', \mathscr{F}) \to \check{H}^{\cdot}(\mathfrak{U}, \mathscr{F})$ on Čech cohomology groups. We say that $\mathfrak{U}$ is a refinement of $\mathfrak{U}'$ if there exists a morphism from $\mathfrak{U}$ to $\mathfrak{U}'$. So we have canonical homomorphisms $\check{H}^{\cdot}(\mathfrak{U}', \mathscr{F}) \to \check{H}^{\cdot}(\mathfrak{U}, \mathscr{F})$ on Čech cohomology groups whenever $\mathfrak{U}$ is a refinement of $\mathfrak{U}'$.

5.2 Etale Sheaves

([SGA 4] VII 1–2, [SGA $4\frac{1}{2}$] Arcata I, II.)

Let X be a scheme. Define a category X_{et} as follows: Objects in X_{et} are X-schemes U so that the structure morphisms $U \to X$ are etale. For any two etale X-schemes U_1 and U_2, morphisms from U_1 to U_2 in X_{et} are X-morphisms from U_1 to U_2. Note that morphisms in X_{et} are etale, and fiber products exist in X_{et}. An *etale presheaf* on X is a presheaf on the category X_{et}. The category of etale presheaves on X is denoted by $\mathscr{P}_X$. It is an abelian category.

Let $f : X' \to X$ be a morphism of schemes. It induces a functor

$$f_{\text{et}} : X_{\text{et}} \to X'_{\text{et}}, \quad U \mapsto U \times_X X'.$$

For any etale presheaf $\mathscr{F}'$ on X', denote the etale presheaf $(f_{\text{et}})^{\mathscr{P}}(\mathscr{F}')$ by $f_{\mathscr{P}}\mathscr{F}'$. We have

$$(f_{\mathscr{P}}\mathscr{F}')(U) = \mathscr{F}'(U \times_X X')$$

for any $U \in \operatorname{ob} X_{\mathrm{et}}$. For any object U' in X'_{et}, define a category $I_{U'}$ as follows: Object in $I_{U'}$ are pairs (U, ϕ), where U is an object in X_{et}, and $\phi : U' \to U$ is an X-morphism. For any two such pairs (U_i, ϕ_i) $(i = 1, 2)$, a morphism $\xi : (U_1, \phi_1) \to (U_2, \phi_2)$ in $I_{U'}$ is an X-morphism $\xi : U_1 \to U_2$ such that $\xi\phi_1 = \phi_2$. Note that this category is equivalent to the category defined in 5.1 for the functor $f_{\mathrm{et}} : X_{\mathrm{et}} \to X'_{\mathrm{et}}$. We will show shortly that the opposite category $I^{\circ}_{U'}$ of $I_{U'}$ satisfies the conditions (I2) and (I3) in 2.7, and that there exists a cofinal subcategory of $I^{\circ}_{U'}$ whose objects form a set. For any etale presheaf $\mathscr{F}$ on X, denote the etale presheaf $(f_{\mathrm{et}})_{\mathscr{P}}(\mathscr{F})$ on X' by $f^{\mathscr{P}}\mathscr{F}$. We have

$$(f^{\mathscr{P}}\mathscr{F})(U') = \varinjlim_{(U,\phi)\in \operatorname{ob} I^{\circ}_{U'}} \mathscr{F}(U)$$

for any $U' \in \operatorname{ob} X'_{\mathrm{et}}$. The functor $f^{\mathscr{P}} : \mathscr{P}_X \to \mathscr{P}_{X'}$ is exact, and is left adjoint to the functor $f_{\mathscr{P}} : \mathscr{P}_{X'} \to \mathscr{P}_X$, that is, for any etale presheaf $\mathscr{F}$ on X and any etale presheaf $\mathscr{F}'$ on X', we have a canonical one-to-one correspondence

$$\operatorname{Hom}(f^{\mathscr{P}}\mathscr{F}, \mathscr{F}') \cong \operatorname{Hom}(\mathscr{F}, f_{\mathscr{P}}\mathscr{F}')$$

functorial in $\mathscr{F}$ and $\mathscr{F}'$.

Let us show that $I^{\circ}_{U'}$ satisfies the conditions (I2) and (I3). Given two morphisms $\alpha, \beta : (U_2, \phi_2) \to (U_1, \phi_1)$ in $I_{U'}$, consider the Cartesian diagram

$$\begin{array}{ccc} K = U_2 \times_{(U_2 \times_X U_1)} U_2 & \stackrel{p_2}{\to} & U_2 \\ p_1 \downarrow & & \downarrow \Gamma_\beta \\ U_2 & \stackrel{\Gamma_\alpha}{\to} & U_2 \times_X U_1, \end{array}$$

where Γ_α and Γ_β are the graphs of α and β, respectively. Note that the schemes in this diagram are etale X-schemes. Let $\pi_1 : U_2 \times_X U_1 \to U_2$ be the projection. We have

$$p_1 = \pi_1\Gamma_\alpha p_1 = \pi_1\Gamma_\beta p_2 = p_2.$$

Denote $p_1 = p_2$ by p. One can show that the sequence

$$K \stackrel{p}{\to} U_2 \underset{\beta}{\overset{\alpha}{\rightrightarrows}} U_1$$

is exact, that is, $\alpha p = \beta p$, and that for any morphism $\psi : V \to U_2$ of schemes with the property $\alpha\psi = \beta\psi$, there exists a unique morphism $\psi' : V \to K$ such that $p\psi' = \psi$. We have

$$\alpha\phi_2 = \beta\phi_2 = \phi_1.$$

So there exists a morphism $\phi : U' \to K$ such that $p\phi = \phi_2$. The morphism p defines a morphism $p : (K, \phi) \to (U_2, \phi_2)$ in $I_{U'}$ and we have $\alpha p = \beta p$. So $I^\circ_{U'}$ satisfies (I2). Given two objects (U_i, ϕ_i) $(i = 1, 2)$ in $I_{U'}$, let p_i : $U_1 \times_X U_2 \to U_i$ be the projections, and let $(\phi_1, \phi_2) : U' \to U_1 \times_X U_2$ be the morphism with the property

$$p_i \circ (\phi_1, \phi_2) = \phi_i.$$

Then the pair $(U_1 \times_X U_2, (\phi_1, \phi_2))$ is an object in $I_{U'}$, and p_i define morphisms $p_i : (U_1 \times_X U_2, (\phi_1, \phi_2)) \to (U_i, \phi_i)$ in $I_{U'}$. So $I^\circ_{U'}$ satisfies (I3).

Next we show $I^\circ_{U'}$ has a cofinal subcategory whose objects form a set. For any $u' \in U'$, choose an affine open subset $X_{u'}$ of X containing the image of u' in X. We will construct a set $\mathscr{S}$ of X-schemes Y with the following property (*) so that any X-scheme with the property (*) is isomorphic to an X-scheme in $\mathscr{S}$:

(*) Y can be covered by affine open subsets $Y_{u'}$ $(u' \in U')$ so that for each u', the image of $Y_{u'}$ in X is contained in $X_{u'}$ and that $\Gamma(Y_{u'}, \mathscr{O}_Y)$ is a $\Gamma(X_{u'}, \mathscr{O}_X)$-algebra of finite presentation.

Then objects (U, ϕ) in $I^\circ_{U'}$ with $U \in \mathscr{S}$ (and $\phi \in \mathrm{Hom}_X(U', U)$) form a set, and the full subcategory of $I^\circ_{U'}$ consisting of these objects is cofinal in $I^\circ_{U'}$. Let $\mathscr{S}_{u'}$ be the family of affine schemes $\mathrm{Spec}\,(\Gamma(X_{u'}, \mathscr{O}_X)[t_1, \ldots, t_n]/I)$ with $n \in \mathbb{N}$ and with I being finitely generated ideals of $\Gamma(X_{u'}, \mathscr{O}_X)[t_1, \ldots, t_n]$. Then $\mathscr{S}_{u'}$ is a set. So

$$\mathscr{T}_{u'} = \{(Z, W) | Z \in \mathscr{S}_{u'}, W \text{ is an open subset of } Z\}$$

is a set. Hence

$$\begin{aligned} \{((Z_{u'}, W_{u'v'}, \psi_{u'v'})_{u',v' \in U'}) \mid\ & (Z_{u'}, W_{u'v'}) \in \mathscr{T}_{u'}, \\ & \psi_{u'v'} \in \mathrm{Hom}_X(W_{u'v'}, W_{v'u'}), \\ & W_{u'u'} = Z_{u'},\ \psi_{u'u'} = \mathrm{id}, \\ & \psi_{u'v'}(W_{u'w'} \cap W_{u'v'}) \subset W_{v'w'} \cap W_{v'u'}, \\ & \psi_{v'w'}\psi_{u'v'} = \psi_{u'w'} \text{ on } W_{u'v'} \cap W_{u'w'}\} \end{aligned}$$

is a set. This set is nothing but the set of gluing data for X-schemes with the property (*). We take $\mathscr{S}$ to be the set of X-schemes defined by gluing data in this set.

Let X be a scheme. A set $\mathfrak{U} = \{U_\alpha \to U\}_{\alpha \in I}$ of morphisms in X_{et} with a common target U is called an *etale covering* of U if U is the union of images of U_α. An etale presheaf $\mathscr{F}$ on X is called an *etale sheaf* if for any etale covering $\mathfrak{U} = \{U_\alpha \to U\}_{\alpha\in I}$, the sequence

$$0 \to \mathscr{F}(U) \to \prod_{\alpha\in I} \mathscr{F}(U_\alpha) \to \prod_{\alpha,\beta\in I} \mathscr{F}(U_\alpha \times_U U_\beta)$$

is exact, where the homomorphisms in this sequence are defined by

$$\mathscr{F}(U) \to \prod_{\alpha\in I} \mathscr{F}(U_\alpha), \qquad s \mapsto (s|_{U_\alpha}),$$

$$\prod_{\alpha\in I} \mathscr{F}(U_\alpha) \to \prod_{\alpha,\beta\in I} \mathscr{F}(U_\alpha \times_U U_\beta), \qquad (s_\alpha) \mapsto (s_\beta|_{U_\alpha\times_U U_\beta} - s_\alpha|_{U_\alpha\times_U U_\beta}).$$

Morphisms between etale sheaves are defined to be morphisms between them that are considered as presheaves. Denote the category of etale sheaves on X by $\mathscr{S}_X$. It is a full subcategory of the category $\mathscr{P}_X$ of etale presheaves on X. Denote by $i : \mathscr{S}_X \to \mathscr{P}_X$ be the inclusion.

¿From now on, we call etale sheaves and etale presheaves simply sheaves and presheaves.

Proposition 5.2.1. *There exists a functor* $\# : \mathscr{P}_X \to \mathscr{S}_X$ *that is left adjoint to* $i : \mathscr{S}_X \to \mathscr{P}_X$*, that is, for any presheaf* $\mathscr{G}$ *on* X*, there exists a pair* $(\mathscr{G}^\#, \theta)$ *consisting of a sheaf* $\mathscr{G}^\#$ *on* X *and a morphism* $\theta : \mathscr{G} \to \mathscr{G}^\#$ *of presheaves such that for any sheaf* $\mathscr{F}$ *on* X *and any morphism of presheaves* $\phi : \mathscr{G} \to \mathscr{F}$*, there exists one and only one morphism* $\psi : \mathscr{G}^\# \to \mathscr{F}$ *such that* $\psi\theta = \phi$*. The pair* $(\mathscr{G}^\#, \theta)$ *is unique up to a unique isomorphism.*

We call $\mathscr{G}^\#$ the *sheaf associated to the presheaf* $\mathscr{G}$.

To prove the proposition, we first define a functor $+ : \mathscr{P}_X \to \mathscr{P}_X$. For every object U in X_{et}, let J_U be the category defined as follows: Objects in J_U are etale coverings of U. Morphisms in J_U are defined in the same way as in the end of 5.1. We say that an etale covering $\mathfrak{U}$ is a *refinement* of an etale covering $\mathfrak{U}'$ and write $\mathfrak{U}' \leq \mathfrak{U}$ if there exists a morphism $\mathfrak{U} \to \mathfrak{U}'$ in J_U. One can show that $(J_U, \leq)$ is a directed family, and that it has a cofinal subfamily whose objects form a set. By the discussion at the end of 5.1, we have a well-defined homomorphism

$$\check{H}^0(\mathfrak{U}', \mathscr{G}) \to \check{H}^0(\mathfrak{U}, \mathscr{G})$$

for any presheaf $\mathscr{G}$ on X. We define

$$\mathscr{G}^+(U) = \varinjlim_{\mathfrak{U}\in \mathrm{ob}\, J_U} \check{H}^0(\mathfrak{U}, \mathscr{G}).$$

Let $V \to U$ be a morphism in X_{et}, and let $\mathfrak{U} = \{U_\alpha \to U\}_{\alpha \in I}$ be an etale covering of U. Then $\mathfrak{V} = \{U_\alpha \times_U V \to V\}_{\alpha \in I}$ is an etale covering of V. We have a canonical morphism of Čech complexes

$$C^{\cdot}(\mathfrak{U}, \mathscr{G}) \to C^{\cdot}(\mathfrak{V}, \mathscr{G}).$$

It induces a homomorphism

$$\check{H}^0(\mathfrak{U}, \mathscr{G}) \to \check{H}^0(\mathfrak{V}, \mathscr{G})$$

and hence a homomorphism

$$\mathscr{G}^+(U) = \varinjlim_{\mathfrak{U} \in \text{ob}\, J_U} \check{H}^0(\mathfrak{U}, \mathscr{G}) \to \mathscr{G}^+(V) = \varinjlim_{\mathfrak{V} \in \text{ob}\, J_V} \check{H}^0(\mathfrak{V}, \mathscr{G}).$$

We thus get a presheaf $\mathscr{G}^+$ on X. The functor

$$+ : \mathscr{P}_X \to \mathscr{P}_X, \quad \mathscr{G} \mapsto \mathscr{G}^+$$

is leaf exact. Note that $\mathfrak{U}_0 = \{\text{id} : U \to U\}$ is an etale covering, and that any etale covering $\mathfrak{U}$ of U is a refinement of $\mathfrak{U}_0$. We have

$$\check{H}^0(\mathfrak{U}_0, \mathscr{G}) = \mathscr{G}(U).$$

So we have a canonical homomorphism

$$\mathscr{G}(U) \to \varinjlim_{\mathfrak{U} \in \text{ob}\, J_U} \check{H}^0(\mathfrak{U}, \mathscr{G}) = \mathscr{G}^+(U).$$

We thus have a morphism of presheaves $\mathscr{G} \to \mathscr{G}^+$. If $\mathscr{G}$ is a sheaf, then this morphism is an isomorphism.

Lemma 5.2.2. *For any presheaf $\mathscr{G}$ and any sheaf $\mathscr{F}$ on X, the canonical morphism $\mathscr{G} \to \mathscr{G}^+$ induces a one-to-one correspondence*

$$\Phi : \text{Hom}(\mathscr{G}^+, \mathscr{F}) \cong \text{Hom}(\mathscr{G}, \mathscr{F}).$$

Proof. Any homomorphism $\phi : \mathscr{G} \to \mathscr{F}$ induces a morphism

$$\phi^+ : \mathscr{G}^+ \to \mathscr{F}^+ \cong \mathscr{F}.$$

We thus have a map

$$\Psi : \text{Hom}(\mathscr{G}, \mathscr{F}) \to \text{Hom}(\mathscr{G}^+, \mathscr{F}), \quad \phi \mapsto \phi^+.$$

It is clear that $\Phi\Psi = \text{id}$. Let $\mathfrak{U} = \{U_\alpha \to U\}_{\alpha \in I}$ be an etale covering. For any $\alpha_0 \in I$ and any morphism $\psi : \mathscr{G}^+ \to \mathscr{F}$, we have a commutative diagram

$$\begin{array}{ccccc} \mathscr{G}(U) & \to & \mathscr{G}^+(U) & \stackrel{\psi}{\to} & \mathscr{F}(U) \\ \downarrow & & \downarrow & & \downarrow \\ \mathscr{G}(U_{\alpha_0}) & \to & \mathscr{G}^+(U_{\alpha_0}) & \stackrel{\psi}{\to} & \mathscr{F}(U_{\alpha_0}). \end{array}$$

Note that the etale covering $\mathfrak{V}_0 = \{\mathrm{id} : U_{\alpha_0} \to U_{\alpha_0}\}$ is a refinement of the etale covering $\mathfrak{V} = \{U_\alpha \times_U U_{\alpha_0} \to U_{\alpha_0}\}_{\alpha \in I}$. Indeed, the diagonal morphism $\Delta : U_{\alpha_0} \to U_{\alpha_0} \times_U U_{\alpha_0}$ defines a morphism $\mathfrak{V}_0 \to \mathfrak{V}$. Given

$$\xi = (\xi_\alpha) \in \check{H}^0(\mathfrak{U}, \mathscr{G}) = \ker\Big(\prod_\alpha \mathscr{G}(U_\alpha) \to \prod_{\alpha,\beta} \mathscr{G}(U_\alpha \times_U U_\beta)\Big),$$

one can check the image of ξ under the composite

$$\check{H}^0(\mathfrak{U}, \mathscr{G}) \to \check{H}^0(\mathfrak{V}, \mathscr{G}) \to \check{H}^0(\mathfrak{V}_0, \mathscr{G}) = \mathscr{G}(U_{\alpha_0})$$

is ξ_{α_0}. Let $\bar{\xi}$ be the image of $\xi \in \check{H}^0(\mathfrak{U}, \mathscr{G})$ in $\mathscr{G}^+(U) = \varinjlim_{\mathfrak{U} \in \mathrm{ob}\, J_U} \check{H}^0(\mathfrak{U}, \mathscr{G})$. Then the image of $\bar{\xi}$ under the homomorphism $\mathscr{G}^+(U) \to \mathscr{G}^+(U_{\alpha_0})$ is the same as the image of ξ_{α_0} under the homomorphism $\mathscr{G}(U_{\alpha_0}) \to \mathscr{G}^+(U_{\alpha_0})$. So $\psi(\bar{\xi})|_{U_{\alpha_0}} \in \mathscr{F}(U_{\alpha_0})$ coincides with the image of ξ_{α_0} under the composite

$$\mathscr{G}(U_{\alpha_0}) \to \mathscr{G}^+(U_{\alpha_0}) \to \mathscr{F}(U_{\alpha_0}).$$

This is true for any $\alpha_0 \in I$. From this fact, one deduces $\Psi\Phi(\psi) = \psi$. □

Lemma 5.2.3. *We say that a presheaf $\mathscr{G}$ on X has the property (+) if for any etale covering $\{U_\alpha \to U\}_{\alpha \in I}$ in X_{et}, the canonical homomorphism*

$$\mathscr{G}(U) \to \prod_{\alpha \in I} \mathscr{G}(U_\alpha)$$

is injective.

(i) *For any presheaf $\mathscr{G}$, $\mathscr{G}^+$ has the property (+).*

(ii) *If a presheaf $\mathscr{G}$ has the property (+), then $\mathscr{G}^+$ is a sheaf.*

Proof.

(i) Let $\{U_\alpha \to U\}_{\alpha \in I}$ be an etale covering, and let $\bar{\xi}_i \in \mathscr{G}^+(U)$ $(i = 1, 2)$ such that they have the same image under the homomorphism

$$\mathscr{G}^+(U) \to \prod_{\alpha \in I} \mathscr{G}^+(U_\alpha).$$

Choose an etale covering $\mathfrak{V} = \{V_\beta \to U\}_{\beta \in J}$ such that $\bar{\xi}_i$ are the images of

$$\xi_i \in \check{H}^0(\mathfrak{V}, \mathscr{G}) = \ker\Big(\prod_{\beta \in J} \mathscr{G}(V_\beta) \to \prod_{\beta_1, \beta_2 \in J} \mathscr{G}(V_{\beta_1} \times_U V_{\beta_2})\Big)$$

under the homomorphism

$$\check{H}^0(\mathfrak{V}, \mathscr{G}) \to \varinjlim_{\mathfrak{U} \in \mathrm{ob}\, J_U} \check{H}^0(\mathfrak{U}, \mathscr{G}) = \mathscr{G}^+(U).$$

For each $i \in \{1, 2\}$ and each $\alpha \in I$, the image of $\bar{\xi}_i$ under the homomorphism $\mathscr{G}^+(U) \to \mathscr{G}^+(U_\alpha)$ is the image of ξ_i under the composite

$$\check{H}^0(\mathfrak{V}, \mathscr{G}) \to \check{H}^0(\mathfrak{V} \times_U U_\alpha, \mathscr{G}) \to \mathscr{G}^+(U_\alpha),$$

where $\mathfrak{V} \times_U U_\alpha = \{V_\beta \times_U U_\alpha \to U_\alpha\}_{\beta \in J}$. Since $\bar{\xi}_i$ $(i = 1, 2)$ have the same image in $\mathscr{G}^+(U_\alpha)$, there exists an etale covering $\mathfrak{W}_\alpha = \{W_{\alpha\gamma} \to U_\alpha\}_{\gamma \in K_\alpha}$ which is a refinement of $\mathfrak{V} \times_U U_\alpha$ such that the images of ξ_i $(i = 1, 2)$ under the composite

$$\check{H}^0(\mathfrak{V}, \mathscr{G}) \to \check{H}^0(\mathfrak{V} \times_U U_\alpha, \mathscr{G}) \to \check{H}^0(\mathfrak{W}_\alpha, \mathscr{G})$$

are the same, that is, ξ_i $(i = 1, 2)$ have the same image in

$$\check{H}^0(\mathfrak{W}_\alpha, \mathscr{G}) = \ker\Big(\prod_{\gamma \in K_\alpha} \mathscr{G}(W_{\alpha\gamma}) \to \prod_{\gamma_1, \gamma_2 \in K_\alpha} \mathscr{G}(W_{\alpha\gamma_1} \times_{U_\alpha} W_{\alpha\gamma_2})\Big).$$

But $\{W_{\alpha\gamma} \to U_\alpha \to U\}_{\alpha \in I, \gamma \in K_\alpha}$ is an etale covering of U, which is a refinement of $\mathfrak{V}$. Denote this etale covering by $\mathfrak{W}$. Then ξ_i $(i = 1, 2)$ have the same image under the map

$$\begin{aligned} \check{H}^0(\mathfrak{V}, \mathscr{G}) &\to \check{H}^0(\mathfrak{W}, \mathscr{G}) \\ &= \ker\Big(\prod_{\alpha \in I, \gamma \in K_\alpha} \mathscr{G}(W_{\alpha\gamma}) \to \prod_{\substack{\alpha_1, \alpha_2 \in I, \\ \gamma_1 \in K_{\alpha_1}, \\ \gamma_2 \in K_{\alpha_2}}} \mathscr{G}(W_{\alpha_1\gamma_1} \times_U W_{\alpha_2\gamma_2})\Big). \end{aligned}$$

So ξ_i $(i = 1, 2)$ have the same image in $\mathscr{G}^+(U)$, that is, $\bar{\xi}_1 = \bar{\xi}_2$. This proves that the homomorphism $\mathscr{G}^+(U) \to \prod_{\alpha \in I} \mathscr{G}^+(U_\alpha)$ is injective.

(ii) We claim that if $\mathscr{G}$ has the property $(+)$, then for any etale covering $\mathfrak{U} = \{U_\alpha \to U\}_{\alpha \in I}$ of $U \in \mathrm{ob}\, X_{\mathrm{et}}$ and any refinement $\mathfrak{V} = \{V_\beta \to U\}_{\beta \in J}$ of $\mathfrak{U}$, the canonical homomorphism $\check{H}^0(\mathfrak{U}, \mathscr{G}) \to \check{H}^0(\mathfrak{V}, \mathscr{G})$ is injective.

Let $f : \mathfrak{V} \to \mathfrak{U}$ be a morphism. Then $\mathfrak{W} = \{U_\alpha \times_U V_\beta \to U\}_{\alpha \in I, \beta \in J}$ is an etale covering U and the projections $U_\alpha \times_U V_\beta \to U_\alpha$ and $U_\alpha \times_U V_\beta \to V_\beta$ define morphisms $p_1 : \mathfrak{W} \to \mathfrak{U}$ and $p_2 : \mathfrak{W} \to \mathfrak{V}$. For each fixed $\alpha \in I$, $\{U_\alpha \times_U V_\beta \to U_\alpha\}_{\beta \in J}$ is an etale covering of U_α. Since $\mathscr{G}$ has the property $(+)$, the canonical homomorphism

$$\mathscr{G}(U_\alpha) \to \prod_{\beta \in J} \mathscr{G}(U_\alpha \times_U V_\beta)$$

is injective. So

$$\prod_{\alpha \in I} \mathscr{G}(U_\alpha) \to \prod_{\alpha \in I,\, \beta \in J} \mathscr{G}(U_\alpha \times_U V_\beta)$$

is injective, and hence

$$p_1^* : \check{H}^0(\mathfrak{U}, \mathscr{G}) \to \check{H}^0(\mathfrak{W}, \mathscr{G})$$

is injective. This last homomorphism coincides with the composite

$$\check{H}^0(\mathfrak{U},\mathscr{G}) \stackrel{f^*}{\to} \check{H}^0(\mathfrak{V},\mathscr{G}) \stackrel{p_2^*}{\to} \check{H}^0(\mathfrak{W},\mathscr{G}).$$

It follows that $\check{H}^0(\mathfrak{U},\mathscr{G}) \to \check{H}^0(\mathfrak{V},\mathscr{G})$ is injective. This prove our claim.

To prove that $\mathscr{G}^+$ is a sheaf, we need to show that for any etale covering $\{U_\alpha \to U\}_{\alpha\in I}$, the sequence

$$0 \to \mathscr{G}^+(U) \to \prod_{\alpha\in I} \mathscr{G}^+(U_\alpha) \to \prod_{\alpha,\beta\in I} \mathscr{G}^+(U_\alpha \times_U U_\beta)$$

is exact. Since $\mathscr{G}^+$ has the property (+), we only need to show that any

$$\bar{\xi} = (\bar{\xi}_\alpha) \in \ker\Big(\prod_{\alpha\in I} \mathscr{G}^+(U_\alpha) \to \prod_{\alpha,\beta\in I} \mathscr{G}^+(U_\alpha \times_U U_\beta)\Big)$$

lies in the image of the homomorphism $\mathscr{G}^+(U) \to \prod_{\alpha\in I} \mathscr{G}^+(U_\alpha)$. For each $\alpha \in I$, choose an etale covering $\{U_{\alpha\lambda} \to U_\alpha\}_{\lambda\in I_\alpha}$ such that $\bar{\xi}_\alpha$ is the image in $\mathscr{G}^+(U_\alpha)$ of

$$\begin{aligned}\xi_\alpha = (\xi_{\alpha\lambda}) &\in \check{H}^0(\{U_{\alpha\lambda} \to U_\alpha\},\mathscr{G}) \\ &= \ker\Big(\prod_{\lambda\in I_\alpha} \mathscr{G}(U_{\alpha\lambda}) \to \prod_{\lambda_1,\lambda_2\in I_\alpha} \mathscr{G}(U_{\alpha\lambda_1} \times_{U_\alpha} U_{\alpha\lambda_2})\Big).\end{aligned}$$

For each pair $\alpha, \beta \in I$, by the base change $U_\alpha \times_U U_\beta \to U_\alpha$, ξ_α induces an element

$$\begin{aligned}&\xi^1_{\alpha\beta} = (\xi^1_{\alpha\beta\lambda}) \\ &\in \ker\Big(\textstyle\prod_{\lambda\in I_\alpha} \mathscr{G}(U_{\alpha\lambda}\times_U U_\beta) \to \prod_{\lambda_1,\lambda_2\in I_\alpha} \mathscr{G}((U_{\alpha\lambda_1}\times_U U_\beta)\times_{(U_\alpha\times_U U_\beta)}(U_{\alpha\lambda_2}\times_U U_\beta))\Big),\end{aligned}$$

and by the base change $U_\alpha \times_U U_\beta \to U_\beta$, ξ_β induces an element

$$\begin{aligned}&\xi^2_{\alpha\beta} = (\xi^2_{\alpha\beta\mu}) \\ &\in \ker\Big(\textstyle\prod_{\mu\in I_\beta} \mathscr{G}(U_\alpha\times_U U_{\beta\mu}) \to \prod_{\mu_1,\mu_2\in I_\beta} \mathscr{G}((U_\alpha\times_U U_{\beta\mu_1})\times_{(U_\alpha\times_U U_\beta)}(U_\alpha\times_U U_{\beta\mu_2}))\Big),\end{aligned}$$

where $\xi^1_{\alpha\beta\lambda}$ are the restrictions of $\xi_{\alpha\lambda} \in \mathscr{G}(U_{\alpha\lambda})$ to $U_{\alpha\lambda} \times_U U_\beta$, and $\xi^2_{\alpha\beta\mu}$ are the restrictions of $\xi_{\beta\mu} \in \mathscr{G}(U_{\beta\mu})$ to $U_\alpha \times_U U_{\beta\mu}$.

$$\begin{array}{ccccc}\{U_{\alpha\lambda}\times_U U_{\beta\mu}\}_{\lambda\in I_\alpha,\mu\in I_\beta} & \to & \{U_{\alpha\lambda}\times_U U_\beta\}_{\lambda\in I_\alpha} & \to & \{U_{\alpha\lambda}\}_{\lambda\in I_\alpha} \\ \downarrow & & \downarrow & & \downarrow \\ \{U_\alpha\times_U U_{\beta\mu}\}_{\mu\in I_\beta} & \to & U_\alpha\times_U U_\beta & \to & U_\alpha \\ \downarrow & & \downarrow & & \downarrow \\ \{U_{\beta\mu}\}_{\mu\in I_\beta} & \to & U_\beta & \to & U\end{array}$$

Since we have

$$\bar{\xi} = (\bar{\xi}_\alpha) \in \ker\Big(\prod_{\alpha\in I} \mathscr{G}^+(U_\alpha) \to \prod_{\alpha,\beta\in I} \mathscr{G}^+(U_\alpha \times_U U_\beta)\Big),$$

the images of $\xi^1_{\alpha\beta}$ and $\xi^2_{\alpha\beta}$ in $\mathscr{G}^+(U_\alpha \times_U U_\beta)$ are the same. So there exists an etale covering $\mathfrak{V}$ of $U_\alpha \times_U U_\beta$ which is a common refinement of the etale coverings $\{U_{\alpha\lambda} \times_U U_\beta \to U_\alpha \times_U U_\beta\}_{\lambda\in I_\alpha}$ and $\{U_\alpha \times_U U_{\beta\mu} \to U_\alpha \times_U U_\beta\}_{\mu\in I_\beta}$ such that $\xi^1_{\alpha\beta}$ and $\xi^2_{\alpha\beta}$ have the same image in $\check{H}^0(\mathfrak{V}, \mathscr{G})$. By the claim at the beginning of the proof, $\xi^1_{\alpha\beta}$ and $\xi^2_{\alpha\beta}$ have the same image in $\check{H}^0(\mathfrak{V}, \mathscr{G})$ for any common refinement $\mathfrak{V}$ of the above two etale coverings. Taking

$$\mathfrak{V} = \{U_{\alpha\lambda} \times_U U_{\beta\mu} \to U_\alpha \times_U U_\beta\}_{\lambda\in I_\alpha,\, \mu\in I_\beta},$$

we see that $\xi^1_{\alpha\beta\lambda}$ and $\xi^2_{\alpha\beta\mu}$ have the same image in $\mathscr{G}(U_{\alpha\lambda} \times_U U_{\beta\mu})$. So $\xi_{\alpha\lambda} \in \mathscr{G}(U_{\alpha\lambda})$ and $\xi_{\beta\mu} \in \mathscr{G}(U_{\beta\mu})$ have the same image in $\mathscr{G}(U_{\alpha\lambda} \times_U U_{\beta\mu})$, and hence we have

$$\begin{aligned}(\xi_{\alpha\lambda}) &\in \ker\Big(\prod_{\alpha\in I, \lambda\in I_\alpha} \mathscr{G}(U_{\alpha\lambda}) \to \prod_{\substack{\alpha,\beta\in I,\\ \lambda\in I_\alpha,\ \mu\in I_\beta}} \mathscr{G}(U_{\alpha\lambda} \times_U U_{\beta\mu})\Big)\\ &= \check{H}^0(\{U_{\alpha\lambda} \to U\}_{\alpha\in I, \lambda\in I_\alpha}, \mathscr{G}).\end{aligned}$$

One then checks that the image of $(\xi_{\alpha\lambda})$ under the homomorphism

$$\check{H}^0(\{U_{\alpha\lambda} \to U\}_{\alpha\in I, \lambda\in I_\alpha}, \mathscr{G}) \to \mathscr{G}^+(U)$$

is mapped to $\bar{\xi} = (\bar{\xi}_\alpha)$ under $\mathscr{G}^+(U) \to \prod_{\alpha\in I} \mathscr{G}^+(U_\alpha)$. □

Proof of 5.2.1. For any presheaf $\mathscr{G}$ on X, the presheaf $\mathscr{G}^{++}$ is a sheaf by 5.2.3. We have a canonical morphism $\mathscr{G} \to \mathscr{G}^{++}$. By 5.2.2, it induces a one-to-one correspondence

$$\mathrm{Hom}(\mathscr{G}^{++}, \mathscr{F}) \cong \mathrm{Hom}(\mathscr{G}, \mathscr{F})$$

for any sheaf $\mathscr{F}$ on X. It suffices to take $\mathscr{G}^\# = \mathscr{G}^{++}$. □

Proposition 5.2.4.

(i) *$\mathscr{S}_X$ is an abelian category with enough injective objects. Let $\phi : \mathscr{F} \to \mathscr{G}$ be a morphism of sheaves on X. Then* $\ker\phi$ *is the sheaf defined by*

$$U \mapsto \ker(\phi(U) : \mathscr{F}(U) \to \mathscr{G}(U))$$

for any $U \in \mathrm{ob}\, X_{\mathrm{et}}$, and $\mathrm{coker}\,\phi$ *(resp.* $\mathrm{im}\,\phi$*, resp.* $\mathrm{coim}\,\phi$*) is the sheaf associated to the presheaf defined by*

$$U \mapsto \mathrm{coker}(\phi(U)) \quad (\textit{resp. } U \mapsto \mathrm{im}(\phi(U)), \quad \textit{resp. } U \mapsto \mathrm{coim}(\phi(U))).$$

(ii) *The inclusion $i : \mathscr{S}_X \to \mathscr{P}_X$ is left exact, and $\# : \mathscr{P}_X \to \mathscr{S}_X$ is exact.*

Proof.

(i) Let $\phi : \mathscr{F} \to \mathscr{G}$ be a morphism of sheaves. The kernel of ϕ in the category of presheaves is a sheaf, and it is the kernel of ϕ in the category of sheaves. Let $\mathscr{C}$ be the cokernel of ϕ in the category of presheaves. One can verify that $\mathscr{C}^\#$ is the cokernel of ϕ in the category of sheaves. Let $\mathscr{P}$ be the cokernel of the canonical morphism $\ker\phi \to \mathscr{F}$ in the category of presheaves. We have an exact sequence

$$0 \to \mathscr{P} \to \mathscr{G} \to \mathscr{C} \to 0$$

in the category of presheaves. Since the functor $+ : \mathscr{P}_X \to \mathscr{P}_X$ is left exact, and $\# = + \circ +$, the sequence

$$0 \to \mathscr{P}^\# \to \mathscr{G}^\# \to \mathscr{C}^\#$$

is exact in the category of presheaves. It follows that

$$\mathscr{P}^\# \cong \ker(\mathscr{G} \to \mathscr{C}^\#).$$

Note that $\mathscr{P}^\#$ is the coimage of ϕ and $\ker(\mathscr{G} \to \mathscr{C}^\#)$ is the image of ϕ in the category of sheaves. So $\mathscr{S}_X$ is an abelian category.

For every object U in X_{et}, we have the presheaf $\mathbb{Z}_U$ on X defined in the proof of 5.1.1. Denote the sheaf $(\mathbb{Z}_U)^\#$ also by $\mathbb{Z}_U$. We can find a set of objects in X_{et} such that for any sheaf $\mathscr{A}$ and any proper subsheaf $\mathscr{B}$ of $\mathscr{A}$, there exists an object U in this set such that $\mathscr{A}(U) \neq \mathscr{B}(U)$. Then $\{\mathbb{Z}_U\}$ is a set of generators for $\mathscr{S}_X$. By the same argument as in the proof of [Fu (2006)] 2.1.6 or [Grothendieck (1957)] 1.10.1, one can show that $\mathscr{S}_X$ has enough injective objects.

(ii) We have seen that i and $\#$ is left exact in the proof of (i). Given an exact sequence

$$0 \to \mathscr{G}' \to \mathscr{G} \to \mathscr{G}'' \to 0$$

in the category of presheaves, for any sheaf $\mathscr{F}$, the sequence

$$0 \to \mathrm{Hom}(\mathscr{G}'', \mathscr{F}) \to \mathrm{Hom}(\mathscr{G}, \mathscr{F}) \to \mathrm{Hom}(\mathscr{G}, \mathscr{F})$$

is exact. Hence the sequence

$$0 \to \mathrm{Hom}(\mathscr{G}''^\#, \mathscr{F}) \to \mathrm{Hom}(\mathscr{G}^\#, \mathscr{F}) \to \mathrm{Hom}(\mathscr{G}^\#, \mathscr{F})$$

is exact. It follows that

$$\mathscr{G}'^\# \to \mathscr{G}^\# \to \mathscr{G}''^\# \to 0$$

is exact in the category of sheaves. So $\#$ is exact.

Let $f : X' \to X$ be a morphism of schemes. For any sheaf $\mathscr{F}'$ on X', the presheaf $f_{\mathscr{P}}\mathscr{F}'$ is a sheaf. We denote it by $f_*\mathscr{F}'$ and call it the *direct image* of $\mathscr{F}'$. For any object U in X_{et}, we have

$$f_*(\mathscr{F}') = \mathscr{F}'(U \times_X X').$$

For any sheaf $\mathscr{F}$ on X, define the *inverse image* $f^*\mathscr{F}$ of $\mathscr{F}$ on X' to be the sheaf $(f^{\mathscr{P}}\mathscr{F})^\#$. We often denote $f^*\mathscr{F}$ by $\mathscr{F}|_{X'}$. The functor f^* is left adjoint to f_*, that is, we have a one-to-one correspondence

$$\mathrm{Hom}(f^*\mathscr{F}, \mathscr{F}') \cong \mathrm{Hom}(\mathscr{F}, f_*\mathscr{F}')$$

functorial in $\mathscr{F}$ and $\mathscr{F}'$.

Proposition 5.2.5.

(i) *For any presheaf $\mathscr{G}$ on X, we have $f^*(\mathscr{G}^\#) \cong (f^{\mathscr{P}}\mathscr{G})^\#$.*

(ii) *$f_* : \mathscr{S}_{X'} \to \mathscr{S}_X$ is left exact and $f^* : \mathscr{S}_X \to \mathscr{S}_{X'}$ is exact.*

(iii) *Suppose $f : X' \to X$ is etale. Then for any sheaf $\mathscr{F}$ on X, $f^*\mathscr{F}$ is the sheaf defined by*

$$(f^*\mathscr{F})(U') = \mathscr{F}(U')$$

for any object U' in X'_{et}, where on the right-hand side, we regard U' as an etale X-scheme.

(iv) *Let $f : X' \to X$ and $g : X'' \to X'$ be two morphisms of schemes. We have $(fg)_* \cong f_*g_*$ and $(fg)^* \cong g^*f^*$.*

Proof.

(i) For any sheaf $\mathscr{F}'$ on X', we have one-to-one correspondences

$$\begin{aligned}\mathrm{Hom}(f^*(\mathscr{G}^\#), \mathscr{F}') &\cong \mathrm{Hom}(\mathscr{G}^\#, f_*\mathscr{F}')\\ &\cong \mathrm{Hom}(\mathscr{G}, f_{\mathscr{P}}\mathscr{F}')\\ &\cong \mathrm{Hom}(f^{\mathscr{P}}\mathscr{G}, \mathscr{F}')\\ &\cong \mathrm{Hom}((f^{\mathscr{P}}\mathscr{G})^\#, \mathscr{F}').\end{aligned}$$

So we have $(f^{\mathscr{P}}\mathscr{G})^\# \cong f^*(\mathscr{G}^\#)$.

(ii) The left exactness of f_* follows from the definition. Let

$$0 \to \mathscr{F}_1 \to \mathscr{F}_2 \to \mathscr{F}_3 \to 0$$

be an exact sequence of sheaves on X. Let $\mathscr{C}$ be cokernel of $\mathscr{F}_2 \to \mathscr{F}_3$ in the category of presheaves. Then the sequence

$$0 \to \mathscr{F}_1 \to \mathscr{F}_2 \to \mathscr{F}_3 \to \mathscr{C} \to 0$$

is exact in the category of presheaves and $\mathscr{C}^\# = 0$. Since $f^{\mathscr{P}}$ is exact, the sequence

$$0 \to f^{\mathscr{P}}\mathscr{F}_1 \to f^{\mathscr{P}}\mathscr{F}_2 \to f^{\mathscr{P}}\mathscr{F}_3 \to f^{\mathscr{P}}\mathscr{C} \to 0$$

is exact in the category of presheaves. Since $\#$ is exact, the sequence

$$0 \to (f^{\mathscr{P}}\mathscr{F}_1)^\# \to (f^{\mathscr{P}}\mathscr{F}_2)^\# \to (f^{\mathscr{P}}\mathscr{F}_3)^\# \to (f^{\mathscr{P}}\mathscr{C})^\# \to 0$$

is exact in the category of sheaves. By (i), we have

$$(f^{\mathscr{P}}\mathscr{C})^\# \cong f^*(\mathscr{C}^\#) = 0.$$

It follows that

$$0 \to (f^{\mathscr{P}}\mathscr{F}_1)^\# \to (f^{\mathscr{P}}\mathscr{F}_2)^\# \to (f^{\mathscr{P}}\mathscr{F}_3)^\# \to 0$$

is exact in the category of sheaves. Hence f^* is exact.

(iii) We have

$$(f^{\mathscr{P}}\mathscr{F})(U') = \varinjlim_{(U,\phi)\in \mathrm{ob}\, I_{U'}^\circ} \mathscr{F}(U).$$

One can show that (U', id) is a final object in the category $I_{U'}^\circ$. So we have

$$(f^{\mathscr{P}}\mathscr{F})(U') = \mathscr{F}(U').$$

One easily shows that $f^{\mathscr{P}}\mathscr{F}$, given by this formula, is a sheaf. So we have

$$(f^*\mathscr{F})(U') = \mathscr{F}(U').$$

(iv) We have $(fg)_* \cong f_*g_*$ by definition. This implies that $(fg)^* \cong g^*f^*$ by adjunction. □

Proposition 5.2.6. *Let $\mathscr{F}$ be a presheaf on a scheme X. Suppose for any etale covering $\{U_\alpha \to U\}_{\alpha\in I}$ with the property* (a) *or* (b) *below, that the sequence*

$$0 \to \mathscr{F}(U) \to \prod_{\alpha\in I} \mathscr{F}(U_\alpha) \to \prod_{\alpha,\beta\in I} \mathscr{F}(U_\alpha \times_U U_\beta)$$

is exact:

(a) *U_α and U are affine and I is finite.*

(b) *$U_\alpha \to U$ are open immersions for all α.*

Then $\mathscr{F}$ is a sheaf.

Proof. Let $\mathfrak{U} = \{U_\alpha \to U\}_{\alpha \in I}$ be an etale covering of U, let $\{V_\lambda\}_{\lambda \in J}$ be a covering of U by affine open subsets, and let $\mathfrak{U}_\lambda = \{U_\alpha \times_U V_\lambda \to V_\lambda\}_{\alpha \in I}$ for each $\lambda \in J$. There exists a refinement $\mathfrak{U}'_\lambda$ of $\mathfrak{U}_\lambda$ satisfying the condition (a). By our assumption, we have

$$\mathscr{F}(V_\lambda) \cong \check{H}^0(\mathfrak{U}'_\lambda, \mathscr{F}).$$

This isomorphism coincides with the composite

$$\mathscr{F}(V_\lambda) \to \check{H}^0(\mathfrak{U}_\lambda, \mathscr{F}) \to \check{H}^0(\mathfrak{U}'_\lambda, \mathscr{F}).$$

So

$$\mathscr{F}(V_\lambda) \to \check{H}^0(\mathfrak{U}_\lambda, \mathscr{F})$$

is injective. We have a commutative diagram

$$\begin{array}{ccc} \mathscr{F}(U) & \to & \prod_{\alpha \in I} \mathscr{F}(U_\alpha) \\ \downarrow & & \downarrow \\ \prod_{\lambda \in I} \mathscr{F}(V_\lambda) & \to & \prod_{\alpha \in I} \prod_{\lambda \in I} \mathscr{F}(U_\alpha \times_U V_\lambda). \end{array}$$

The vertical arrows are injective by (b). We have just shown that the bottom horizontal arrow is injective. So

$$\mathscr{F}(U) \to \prod_{\alpha \in I} \mathscr{F}(U_\alpha)$$

is injective. Hence $\mathscr{F}$ has the property (+). By the claim at the beginning of the proof of 5.2.3 (ii),

$$\check{H}^0(\mathfrak{U}_\lambda, \mathscr{F}) \to \check{H}^0(\mathfrak{U}'_\lambda, \mathscr{F})$$

is injective. Since the composite

$$\mathscr{F}(V_\lambda) \to \check{H}^0(\mathfrak{U}_\lambda, \mathscr{F}) \to \check{H}^0(\mathfrak{U}'_\lambda, \mathscr{F})$$

is an isomorphism, we have

$$\mathscr{F}(V_\lambda) \cong \check{H}^0(\mathfrak{U}_\lambda, \mathscr{F}).$$

Consider the commutative diagram

$$\begin{array}{cccccc} & 0 & & 0 & & 0 \\ & \downarrow & & \downarrow & & \downarrow \\ 0 \to & \mathscr{F}(U) & \to & \prod\limits_{\alpha \in I} \mathscr{F}(U_\alpha) & \to & \prod\limits_{\alpha, \beta \in I} \mathscr{F}(U_\alpha \times_U U_\beta) \\ & \downarrow & & \downarrow & & \downarrow \\ 0 \to & \prod\limits_{\lambda \in J} \mathscr{F}(V_\lambda) & \to & \prod\limits_{\lambda \in J} \prod\limits_{\alpha \in I} \mathscr{F}(U_\alpha \times_U V_\lambda) & \to & \prod\limits_{\lambda \in J} \prod\limits_{\alpha, \beta \in I} \mathscr{F}(U_\alpha \times_U U_\beta \times_U V_\lambda) \\ & \downarrow & & \downarrow & & \downarrow \\ 0 \to & \prod\limits_{\lambda, \mu \in J} \mathscr{F}(V_\lambda \cap V_\mu) & \to & \prod\limits_{\lambda, \mu \in J} \prod\limits_{\alpha \in I} \mathscr{F}(U_\alpha \times_U (V_\lambda \cap V_\mu)) & \to & \prod\limits_{\lambda, \mu \in J} \prod\limits_{\alpha, \beta \in I} \mathscr{F}(U_\alpha \times_U U_\beta \times_U (V_\lambda \cap V_\mu)). \end{array}$$

By (b), the vertical sequences are exact. We have just shown that the second horizontal sequence is exact. By a diagram chasing, one can show the first horizontal sequence is exact. So $\mathscr{F}$ is a sheaf. □

Let X be a scheme and let Y be an X-scheme. Consider the presheaf $\widetilde{Y}$ on X defined by

$$\widetilde{Y}(U) = \mathrm{Hom}_X(U, Y)$$

for any object U in X_{et}. Let $\{U_\alpha \to U\}_{\alpha\in I}$ be an etale covering. If U_α and U are affine and I is finite, then $\coprod_{\alpha\in I} U_\alpha \to U$ is a quasi-compact faithfully flat morphism. By 1.8.3, the sequence

$$\mathrm{Hom}_X(U, Y) \to \prod_{\alpha\in I} \mathrm{Hom}_X(U_\alpha, Y) \rightrightarrows \prod_{\alpha,\beta\in I} \mathrm{Hom}_X(U_\alpha \times_U U_\beta, Y)$$

is exact. If each $U_\alpha \to U$ is an open immersion, then the above sequence is also exact. By 5.2.6, $\widetilde{Y}$ is a sheaf of sets on X. We call it the *sheaf represented by the X-scheme Y*.

Proposition 5.2.7. *Let Y be an etale X-scheme.*

(i) *For any sheaf $\mathscr{F}$ of sets on X, we have a one-to-one correspondence*

$$\mathrm{Hom}(\widetilde{Y}, \mathscr{F}) \cong \mathscr{F}(Y).$$

(ii) *Any subsheaf of $\widetilde{Y}$ is of the form $\widetilde{U}$ for some open subset U of Y.*

(iii) *Let $f : X' \to X$ be a morphism. We have $f^*\widetilde{Y} \cong (Y \times_X X')^\sim$.*

Proof.

(i) Given a morphism of sheaves $\phi : \widetilde{Y} \to \mathscr{F}$, let $S(\phi) \in \mathscr{F}(Y)$ be the image of id_Y under the map

$$\phi(Y) : \mathrm{Hom}(Y, Y) = \widetilde{Y}(Y) \to \mathscr{F}(Y).$$

Given a section $s \in \mathscr{F}(Y)$, define a morphism $T(s) : \widetilde{Y} \to \mathscr{F}$ so that for any object U in X_{et} and any $t \in \widetilde{Y}(U)$, the image of t under the map

$$T(s)(U) : \widetilde{Y}(U) \to \mathscr{F}(U)$$

is the image of s under the restriction $\mathscr{F}(Y) \to \mathscr{F}(U)$ induced by the morphism $t : U \to Y$. Then $S : \mathrm{Hom}(\widetilde{Y}, \mathscr{F}) \to \mathscr{F}(Y)$ and $T : \mathscr{F}(Y) \to \mathrm{Hom}(\widetilde{Y}, \mathscr{F})$ are inverses to each other.

(ii) Let $\mathscr{F}$ be a subsheaf of $\widetilde{Y}$. Objects in Y_{et} can be considered as objects in X_{et}. Let $\{U_\alpha\}_{\alpha\in I}$ be the family of all objects in Y_{et} such that the structure morphisms $U_\alpha \to Y$ lie in the subset $\mathscr{F}(U_\alpha)$ of $\mathrm{Hom}_X(U_\alpha, Y)$. Let U be the union of the images of U_α in Y. Then for any object V in X_{et}, we have

$$\mathscr{F}(V) \subset \mathrm{Hom}_X(V, U).$$

So $\mathscr{F}$ is a subsheaf of $\widetilde{U}$. By our construction, $\{U_\alpha \to U\}_{\alpha\in I}$ is a covering and the image of the inclusion $i_U : U \hookrightarrow Y$ under the maps $\tilde{Y}(U) \to \tilde{Y}(U_\alpha)$

lie in $\mathscr{F}(U_\alpha)$ for all $\alpha \in I$. So we have $i_U \in \mathscr{F}(U)$. It follows that $\mathrm{id}_U \in \widetilde{U}(U)$ lies in $\mathscr{F}(U)$. For any object V in X_{et}, if $\widetilde{U}(V) = \emptyset$, then we have $\mathscr{F}(V) = \emptyset$ since $\mathscr{F}$ is a subsheaf of $\widetilde{U}$. If $\widetilde{U}(V) \neq \emptyset$, then an arbitrary section $s \in \widetilde{U}(V)$ is the image of id_U under the restriction $\widetilde{U}(U) \to \widetilde{U}(V)$ induced by $s : V \to U$. As $\mathrm{id}_U \in \mathscr{F}(U)$, we have $s \in \mathscr{F}(V)$. We thus have $\mathscr{F}(V) = \widetilde{U}(V)$, and hence $\mathscr{F} = \widetilde{U}$.

A flaw in the above proof is that the family $\{U_\alpha\}_{\alpha \in I}$ may not be a set. To avoid this, one fixes an affine open covering $\{V_j\}_{j \in J}$ of Y with indices j lying in a set J. For each $j \in J$, one can find a set of objects $\{U_{j\beta}\}_{\beta \in I_j}$ in Y_{et} such that each $U_{j\beta}$ is affine, its image in Y is contained in V_j, and any affine object in $(V_j)_{\mathrm{et}}$ is Y-isomorphic to some $U_{j\beta}$. Then the above argument works if we take $\{U_\alpha\}_{\alpha \in I}$ to be the set consisting of those $U_{j\beta}$ $(j \in J,\ \beta \in I_j)$ such that the structure morphisms $U_{j\beta} \to Y$ lie in $\mathscr{F}(U_{j\beta})$.

(iii) For any sheaf $\mathscr{F}'$ on X', we have

$$\begin{aligned}\mathrm{Hom}(f^*\widetilde{Y}, \mathscr{F}') &\cong \mathrm{Hom}(\widetilde{Y}, f_*\mathscr{F}') \\ &\cong (f_*\mathscr{F}')(Y) \\ &\cong \mathscr{F}'(Y \times_X X') \\ &\cong \mathrm{Hom}((Y \times_X X')^\sim, \mathscr{F}').\end{aligned}$$

So we have $f^*\widetilde{Y} \cong (Y \times_X X')^\sim$. □

Let X be a scheme and let $\mathscr{M}$ be a quasi-coherent $\mathscr{O}_X$-module. By 1.6.2 and 5.2.6, the presheaf on X defined by

$$U \mapsto \Gamma(U, \pi^*\mathscr{M})$$

for any object $\pi : U \to X$ in X_{et} is a sheaf. We denote the sheaf by $\mathscr{M}_{\mathrm{et}}$. Taking $\mathscr{M} = \mathscr{O}_X$, we get the sheaf $\mathscr{O}_{X_{\mathrm{et}}}$. We have

$$\mathscr{O}_{X_{\mathrm{et}}}(U) = \Gamma(U, \mathscr{O}_U).$$

The subsheaf of $\mathscr{O}_{X_{\mathrm{et}}}$ defined by

$$\mathscr{O}_{X_{\mathrm{et}}}(U) = \Gamma(U, \mathscr{O}_U^*)$$

is denoted by $\mathscr{O}^*_{X_{\mathrm{et}}}$.

Suppose that X is quasi-compact and quasi-separated. Let X^f_{et} be the full subcategory of X_{et} whose objects are etale X-schemes with finite presentation. Note that fiber products exist in X^f_{et}. An etale covering in X^f_{et} is defined to be a family of morphisms $\{U_\alpha \to U\}_{\alpha \in I}$ such that I is finite and U is the union of the images of U_α. A presheaf $\mathscr{F}$ on X^f_{et} is called a sheaf if for any etale covering $\{U_\alpha \to U\}_{\alpha \in I}$ as above, the canonical sequence

$$0 \to \mathscr{F}(U) \to \prod_{\alpha \in I} \mathscr{F}(U_\alpha) \to \prod_{\alpha, \beta \in I} \mathscr{F}(U_\alpha \times_U U_\beta)$$

is exact. Denote the category of presheaves and the category of sheaves on X^f_{et} by $\mathscr{P}_{X^f}$ and $\mathscr{S}_{X^f}$, respectively. Let $i : X^f_{\text{et}} \to X_{\text{et}}$ be the inclusion. It induces functors

$$i^{\mathscr{P}} : \mathscr{P}_X \to \mathscr{P}_{X^f}, \quad i_{\mathscr{P}} : \mathscr{P}_{X^f} \to \mathscr{P}_X.$$

If $\mathscr{F}$ is a sheaf on X, then $i^{\mathscr{P}}\mathscr{F}$ is a sheaf on X^f_{et}. Let $r : \mathscr{S}_X \to \mathscr{S}_{X^f}$ be the restriction of $i^{\mathscr{P}}$ to the category of sheaves.

Proposition 5.2.8. *Suppose that X is quasi-compact and quasi-separated. Then the functor $r : \mathscr{S}_X \to \mathscr{S}_{X^f}$ defines an equivalence of categories.*

Proof. Let $\mathscr{F}$ and $\mathscr{G}$ be two sheaves on X. Consider the map

$$\text{Hom}_{\mathscr{S}_X}(\mathscr{F}, \mathscr{G}) \to \text{Hom}_{\mathscr{S}_{X^f}}(\mathscr{F}, \mathscr{G}).$$

Suppose that $\phi_i : \mathscr{F} \to \mathscr{G}$ $(i = 1, 2)$ are two morphisms which have the same image in $\text{Hom}_{\mathscr{S}_{X^f}}(\mathscr{F}, \mathscr{G})$. Then for any object U in X_{et} with finite presentation, ϕ_i $(i = 1, 2)$ induce the same homomorphism $\mathscr{F}(U) \to \mathscr{G}(V)$. For a general object U in X_{et}, let $\{U_\alpha\}_{\alpha \in I}$ be a covering of U by affine open subsets. Then each U_α has finite presentation over X. Since

$$\mathscr{F}(U) \to \prod_{\alpha \in I} \mathscr{F}(U_\alpha), \quad \mathscr{G}(U) \to \prod_{\alpha \in I} \mathscr{G}(U_\alpha)$$

are injective and ϕ_i $(i = 1, 2)$ induce the same homomorphism $\mathscr{F}(U_\alpha) \to \mathscr{G}(U_\alpha)$ for each α, ϕ_i induce the same homomorphism $\mathscr{F}(U) \to \mathscr{G}(U)$. So we have $\phi_1 = \phi_2$. This proves that the above map is injective. We leave it for the reader to prove the map is surjective.

Let $\mathscr{F}$ be a sheaf on X^f_{et}. We claim that

$$r((i_{\mathscr{P}}\mathscr{F})^\#) \cong \mathscr{F}.$$

This proves that the functor r is essentially surjective. Let U be an object in X^f_{et}. It can also be regarded as an object in X_{et}. Using the definition of the functor $i_{\mathscr{P}}$, one can verify that $(i_{\mathscr{P}}\mathscr{F})(U) = \mathscr{F}(U)$. For any etale covering of U in X_{et}, there exists an etale covering of U in X^f_{et} refining the given etale covering. As $\mathscr{F}$ is a sheaf on X^f_{et}, we have $(i_{\mathscr{P}}\mathscr{F})^+(U) = \mathscr{F}(U)$ and hence $(i_{\mathscr{P}}\mathscr{F})^{++}(U) = \mathscr{F}(U)$. We thus have $(i_{\mathscr{P}}\mathscr{F})^\#(U) = \mathscr{F}(U)$. So $r((i_{\mathscr{P}}\mathscr{F})^\#) \cong \mathscr{F}$. □

Proposition 5.2.9. *Let X be a quasi-compact quasi-separated scheme, and let Λ be a directed set.*

(i) *For any direct system $(\mathscr{F}_\lambda, \phi_{\lambda\mu})_{\lambda \in \Lambda}$ of sheaves on X^f_{et}, the presheaf defined by*

$$U \mapsto \varinjlim_{\lambda \in \Lambda} \mathscr{F}_\lambda(U)$$

on X^f_{et} is a sheaf.

(ii) *For any set of sheaves* $\{\mathscr{F}_i\}_{i\in I}$ *on* X^f_{et}, *the presheaf*

$$U \mapsto \bigoplus_{i\in I} \mathscr{F}_\lambda(U)$$

on X^f_{et} *is a sheaf.*

Proof.

(i) Let $\{U_\alpha \to U\}_{\alpha\in I}$ be a covering in X^f_{et}. We have an exact sequence

$$0 \to \mathscr{F}_\lambda(U) \to \prod_{\alpha\in I} \mathscr{F}_\lambda(U_\alpha) \to \prod_{\alpha,\beta\in I} \mathscr{F}_\lambda(U_\alpha \times_U U_\beta).$$

So the sequence

$$0 \to \varinjlim_{\lambda\in\Lambda} \mathscr{F}_\lambda(U) \to \varinjlim_{\lambda\in\Lambda} \prod_{\alpha\in I} \mathscr{F}_\lambda(U_\alpha) \to \varinjlim_{\lambda\in\Lambda} \prod_{\alpha,\beta\in I} \mathscr{F}_\lambda(U_\alpha \times_U U_\beta)$$

is exact. Since I is finite, the above sequence can be identified with the exact sequence

$$0 \to \varinjlim_{\lambda\in\Lambda} \mathscr{F}_\lambda(U) \to \prod_{\alpha\in I} \varinjlim_{\lambda\in\Lambda} \mathscr{F}_\lambda(U_\alpha) \to \prod_{\alpha,\beta\in I} \varinjlim_{\lambda\in\Lambda} \mathscr{F}_\lambda(U_\alpha \times_U U_\beta).$$

Hence $U \mapsto \varinjlim_{\lambda\in\Lambda} \mathscr{F}_\lambda(U)$ is a sheaf on X^f_{et}.

(ii) follows from (i) and the fact that a direct sum over a set I is the direct limit of directs sums over finite subsets of I. □

Let $\mathscr{C}$ be a category on which fiber products exist. For each object U in $\mathscr{C}$, suppose we specify a family $\mathscr{T}_U$ whose elements are sets of morphisms in $\mathscr{C}$ of the form $\{U_\alpha \to U\}_{\alpha\in I}$. We call the elements in $\mathscr{T}_U$ the coverings of U. We say that $\mathscr{T}$ is a *Grothendieck topology* on $\mathscr{C}$ if the following axioms hold:

(GT1) For any isomorphism $V \xrightarrow{\cong} U$ in $\mathscr{C}$, the set $\{V \to U\}$ is a covering in $\mathscr{T}_U$.

(GT2) Let $\{U_\alpha \to U\}_{\alpha\in I}$ be a covering in $\mathscr{T}_U$. For any morphism $V \to U$ in $\mathscr{C}$, the set $\{U_\alpha \times_U V \to V\}_{\alpha\in I}$ defined by base change is a covering in $\mathscr{T}_V$.

(GT3) Let $\{U_\alpha \to U\}_{\alpha\in I}$ be a covering in $\mathscr{T}_U$, and for each $\alpha \in I$, let $\{U_{\alpha\beta} \to U_\alpha\}_{\beta\in I_\alpha}$ be a covering in $\mathscr{T}_{U_\alpha}$. Then the set $\{U_{\alpha\beta} \to U\}_{\alpha\in I,\beta\in I_\alpha}$ defined by taking composites is a covering in $\mathscr{T}_U$.

A presheaf $\mathscr{F}$ on $\mathscr{C}$ is called a sheaf if for any $U \in \operatorname{ob}\mathscr{C}$ and any covering $\{U_\alpha \to U\}_{\alpha\in I}$ in $\mathscr{T}_U$, the sequence

$$0 \to \mathscr{F}(U) \to \prod_{\alpha\in I} \mathscr{F}(U_\alpha) \to \prod_{\alpha,\beta\in I} \mathscr{F}(U_\alpha \times_U U_\beta)$$

is exact, where the homomorphisms in this sequence are defined by

$$\mathscr{F}(U) \to \prod_{\alpha \in I} \mathscr{F}(U_\alpha), \qquad s \mapsto (s|_{U_\alpha}),$$

$$\prod_{\alpha \in I} \mathscr{F}(U_\alpha) \to \prod_{\alpha,\beta \in I} \mathscr{F}(U_\alpha \times_U U_\beta), \qquad (s_\alpha) \mapsto (s_\beta|_{U_\alpha \times_U U_\beta} - s_\alpha|_{U_\alpha \times_U U_\beta}).$$

Many results in this section hold for sheaves on general Grothendieck topologies.

5.3 Stalks of Sheaves

([SGA 4] VIII 3, 4, 7.)

Let Ω be a separably closed field. Schemes etale over $\operatorname{Spec}\Omega$ are disjoint unions of copies of $\operatorname{Spec}\Omega$. So the functor

$$\mathscr{F} \to \Gamma(\operatorname{Spec}\Omega, \mathscr{F})$$

defines an equivalence between the category of sheaves of sets on $\operatorname{Spec}\Omega$ and the category of sets. Let Ω' be a separably closed field containing Ω, and let $u : \operatorname{Spec}\Omega' \to \operatorname{Spec}\Omega$ be the canonical morphism. Then the functor $U \mapsto U \otimes_\Omega \Omega'$ defines an equivalence between the category of etale Ω-schemes and the category of etale Ω'-schemes. So the functor $\mathscr{F} \mapsto u^*\mathscr{F}$ defines an equivalence between the category of sheaves on $\operatorname{Spec}\Omega$ and the category of sheaves on $\operatorname{Spec}\Omega'$. We have

$$\mathscr{F}(\operatorname{Spec}\Omega) \cong (u^*\mathscr{F})(\operatorname{Spec}\Omega').$$

Let X be a scheme, let $\mathscr{F}$ be a sheaf on X, and let $s \to X$ be a geometric point of X, where $s = \operatorname{Spec} k(s)$ is the spectrum of a separably closed field $k(s)$. For convenience, we often denote the geometric point simply by s. Let $\mathscr{F}|_s$ be the inverse image of $\mathscr{F}$ on s. We define the *stalk* of $\mathscr{F}$ at s to be $\Gamma(s, \mathscr{F}|_s)$ and denote it by $\mathscr{F}_s$. Let $x \in X$. Denote by $\bar{x}$ the geometric point $\operatorname{Spec}\overline{k(x)} \to X$, where $\overline{k(x)}$ is a separable closure of $k(x)$. If x is the image of s, then we have

$$\mathscr{F}_{\bar{x}} \cong \mathscr{F}_s.$$

Let $f : X' \to X$ be a morphism, let s' be a geometric point of X', and let $f(s')$ be the geometric point defined by the composite

$$s' \to X' \xrightarrow{f} X.$$

Then for any sheaf $\mathscr{F}$ on X, we have

$$(f^*\mathscr{F})_{s'} \cong \mathscr{F}_{f(s')}.$$

Let X be a scheme and let s be a geometric point of X. Define a category I_s as follows: Objects in I_s are pairs (U, s_U), where U are objects in X_{et}, and $s_U : s \to U$ are X-morphisms. For any two objects (U_1, s_{U_1}) and (U_2, s_{U_2}) in I_s, a morphism $f : (U_1, s_{U_1}) \to (U_2, s_{U_2})$ in I_s is an X-morphism $f : U_1 \to U_2$ such that $f s_{U_1} = s_{U_2}$. We often denote an object (U, s_U) in I_s simply by U. Objects in I_s are called *etale neighborhoods* of s in X.

Proposition 5.3.1. *Let X be a scheme and let $s \to X$ be a geometric point of X. For any sheaf (resp. presheaf) $\mathscr{F}$ on X, we have*

$$\mathscr{F}_s \cong \varinjlim_{(U,s_U)\in \text{ob}\, I_s^\circ} \mathscr{F}(U) \quad (\text{resp. } \mathscr{F}_s^\# \cong \varinjlim_{(U,s_U)\in \text{ob}\, I_s^\circ} \mathscr{F}(U)).$$

Proof. Note that for any presheaf $\mathscr{G}$ on s, we have

$$\Gamma(s, \mathscr{G}) \cong \Gamma(s, \mathscr{G}^\#).$$

Let $\gamma : s \to X$ be the structure morphism. If $\mathscr{F}$ is a sheaf on X, we have

$$\mathscr{F}_s = \Gamma(s, \gamma^*\mathscr{F}) \cong \Gamma(s, \gamma^{\mathscr{P}}\mathscr{F}) = \varinjlim_{(U,s_U)\in \text{ob}\, I_s^\circ} \mathscr{F}(U).$$

If $\mathscr{F}$ is a presheaf on X, we have

$$\mathscr{F}_s^\# = \Gamma(s, \gamma^*(\mathscr{F}^\#)) \cong \Gamma(s, (\gamma^{\mathscr{P}}\mathscr{F})^\#) \cong \Gamma(s, \gamma^{\mathscr{P}}\mathscr{F}) = \varinjlim_{(U,s_U)\in \text{ob}\, I_s^\circ} \mathscr{F}(U).$$ □

Notation as above. Given an etale neighborhood (U, s_U) of s and any section $t \in \mathscr{F}(U)$, we call the image of t in $\mathscr{F}_s \cong \varinjlim_{(U,s_U)\in \text{ob}\, I_S^\circ} \mathscr{F}(U)$ the *germ* of t at s.

Lemma 5.3.2. *Let X be a scheme and let $\mathscr{F}$ and $\mathscr{G}$ be sheaves on X.*

(i) *Given a morphism $u : \mathscr{F} \to \mathscr{G}$, if for any point $x \in X$, the map $u_{\bar{x}} : \mathscr{F}_{\bar{x}} \to \mathscr{G}_{\bar{x}}$ induced by u is an isomorphism, then u is an isomorphism.*

(ii) *Let $u_i : \mathscr{F} \to \mathscr{G}$ $(i = 1, 2)$ be two morphisms of sheaves. If for any $x \in X$, u_i induce the same map $\mathscr{F}_{\bar{x}} \to \mathscr{G}_{\bar{x}}$, then $u_1 = u_2$.*

Proof. We give a proof of (i) and leave it for the reader to prove (ii). Let U be an object in X_{et}, and let $s_i \in \mathscr{F}(U)$ $(i = 1, 2)$ be two sections such that $u(s_1) = u(s_2)$ in $\mathscr{G}(U)$. By our assumption, for any point $x' \in U$, s_1 and s_2 have the same image in $\mathscr{F}_{\bar{x}}$, where x is the image of x' in X. By 5.3.1, there exists an etale neighborhood $U_{x'}$ of $\bar{x}'$ in U such that $s_1|_{U_{x'}} = s_2|_{U_{x'}}$. When x' goes over the set of points of U, $U_{x'}$ form an etale covering of U. It follows that $s_1 = s_2$. So $u : \mathscr{F}(U) \to \mathscr{G}(U)$ is injective.

Let $t \in \mathscr{G}(U)$. By our assumption, for any point $x' \in U$, the image of t in $\mathscr{G}_{\bar{x}}$ lies in the image of $\mathscr{F}_{\bar{x}} \to \mathscr{G}_{\bar{x}}$. So we can find an etale neighborhood $U_{x'}$ of $\bar{x}'$ and a section $s_{x'} \in \mathscr{F}(U_{x'})$ such that $u(s_{x'}) = t|_{U_{x'}}$. By the injectivity that we have shown, for any two points $x'_1, x'_2 \in U'$, we have

$$s_{x'_1}|_{U_{x'_1} \times_U U_{x'_2}} = s_{x'_2}|_{U_{x'_1} \times_U U_{x'_2}}.$$

So there exists a section $s \in \mathscr{F}(U)$ such that $s|_{U_{x'}} = s_{x'}$ for all $x' \in U$. We then have $u(s) = t$. So $u : \mathscr{F}(U) \to \mathscr{G}(U)$ is surjective. □

Proposition 5.3.3. *Let X be a scheme. A sequence*

$$\mathscr{F} \to \mathscr{G} \to \mathscr{H}$$

of morphisms of sheaves on X is exact if and only if for any geometric point s of X, the sequence

$$\mathscr{F}_s \to \mathscr{G}_s \to \mathscr{H}_s$$

is exact.

Proof. By 5.3.1, we have

$$(\ker(\mathscr{G} \to \mathscr{H}))_s \cong \ker(\mathscr{G}_s \to \mathscr{H}_s), \quad (\mathrm{im}(\mathscr{F} \to \mathscr{G}))_s \cong \mathrm{im}(\mathscr{F}_s \to \mathscr{G}_s).$$

This implies the "only if" part. If $\mathscr{F}_s \to \mathscr{G}_s \to \mathscr{H}_s$ is exact for all geometric points s in X, then by 5.3.2, the composite $\mathscr{F} \to \mathscr{G} \to \mathscr{H}$ is 0, and we have

$$\mathrm{im}(\mathscr{F} \to \mathscr{G}) \cong \ker(\mathscr{G} \to \mathscr{H}).$$

So $\mathscr{F} \to \mathscr{G} \to \mathscr{H}$ is exact. □

Let A be a strictly henselian local ring and let s be the closed point of $\mathrm{Spec}\, A$. By 2.3.10 (i) and 2.8.3 (vii), $\mathrm{Spec}\, A$ is a final object in I_s°. So we have

$$\mathscr{F}_s \cong \mathscr{F}(\mathrm{Spec}\, A)$$

for any sheaf $\mathscr{F}$ on $\mathrm{Spec}\, A$.

Let X be a scheme and let $s \to X$ be a geometric point of X. Consider the sheaf $\mathscr{O}_{X_{\mathrm{et}}}$ on X. We have $\mathscr{O}_{X_{\mathrm{et}}}(U) = \mathscr{O}_U(U)$ for any object U of X_{et}. So we have

$$\mathscr{O}_{X_{\mathrm{et}},s} \cong \varinjlim_{(U,s_U)\in I_s^\circ} \mathscr{O}_U(U).$$

Let x be the image of s in X, and let $\widetilde{\mathscr{O}}_{X,s}$ be the strict henselization of the local ring $\mathscr{O}_{X,x}$ with respect to the separable closure of $k(x)$ in $k(s)$. Then we have

$$\mathscr{O}_{X_{\mathrm{et}},s} \cong \widetilde{\mathscr{O}}_{X,s}.$$

We call $\operatorname{Spec} \widetilde{\mathscr{O}}_{X,s}$ the *strict localization* of X at s and denote it by $\widetilde{X}_s$. We have a canonical morphism $\widetilde{X}_s \to X$, and $s \to X$ can be factorized as

$$s \to \widetilde{X}_s \to X.$$

For any sheaf $\mathscr{F}$ on X, we have

$$\mathscr{F}_s \cong \Gamma(\widetilde{X}_s, \mathscr{F}|_{\widetilde{X}_s}).$$

Let s and s' be two geometric points of X. A *specialization morphism* is an X-morphism $\widetilde{X}_{s'} \to \widetilde{X}_s$. When such a morphism exists, we say that s' is a *generalization* of s, and that s is a *specialization* of s'.

Proposition 5.3.4. *Let X be a scheme.*

(i) *For any geometric point s' in X and any etale X-scheme U, the canonical map*

$$\operatorname{Hom}_X(\widetilde{X}_{s'}, U) \to \operatorname{Hom}_X(s', U)$$

is bijective.

(ii) *Let $s \to X$ be another geometric point of X. Then the canonical map*

$$\operatorname{Hom}_X(\widetilde{X}_{s'}, \widetilde{X}_s) \to \operatorname{Hom}_X(s', \widetilde{X}_s)$$

is bijective.

Proof.

(i) Taking graphs of X-morphisms, we get the following one-to-one correspondences:

$$\operatorname{Hom}_X(\widetilde{X}_{s'}, U) \cong \operatorname{Hom}_{\widetilde{X}_{s'}}(\widetilde{X}_{s'}, \widetilde{X}_{s'} \times_X U),$$
$$\operatorname{Hom}_X(s', U) \cong \operatorname{Hom}_{s'}(s', s' \times_X U).$$

To prove our assertion, it suffices to show for any etale $\widetilde{X}_{s'}$-scheme W, that the canonical map

$$\operatorname{Hom}_{\widetilde{X}_{s'}}(\widetilde{X}_{s'}, W) \to \operatorname{Hom}_{s'}(s', s' \times_{\widetilde{X}_{s'}} W)$$

is bijective. This follows from 2.3.10 (i) and 2.8.3 (vii).

(ii) Let U_0 be an affine open neighborhood of the image of s in X. Consider the full subcategory J_s of I_s consisting of those etale neighborhoods

(U, s_U) of s such that U are affine and that the images of U in X are contained in U_0. Then J_s is cofinal in I_s. Moreover, J_s is equivalent to a category whose objects form a set. We have

$$\widetilde{\mathscr{O}}_{X,s} \cong \varinjlim_{(U,s_U)\in \mathrm{ob}\, J_s^\circ} \mathscr{O}_U(U).$$

By 1.10.1 (i), we have

$$\widetilde{X}_s \cong \varprojlim_{(U,s_U)\in \mathrm{ob}\, J_s^\circ} U$$

in the category of schemes. So we have one-on-one correspondences

$$\mathrm{Hom}_X(\widetilde{X}_{s'}, \widetilde{X}_s) \cong \varprojlim_{(U,s_U)\in \mathrm{ob}\, J_s^\circ} \mathrm{Hom}_X(\widetilde{X}_{s'}, U),$$

$$\mathrm{Hom}_X(s', \widetilde{X}_s) \cong \varprojlim_{(U,s_U)\in \mathrm{ob}\, J_s^\circ} \mathrm{Hom}_X(s', U).$$

By (i), we have one-to-one correspondences

$$\mathrm{Hom}_X(\widetilde{X}_{s'}, U) \cong \mathrm{Hom}_X(s', U).$$

Our assertion follows. □

Proposition 5.3.5. *Let s and s' be two geometric points in a scheme X, and let x and x' be their images in X, respectively. Then s is a specialization of s' if and only if $x \in \overline{\{x'\}}$.*

Proof. If s is a specialization of s', then by 5.3.4, $\mathrm{Hom}_X(s', \widetilde{X}_s)$ is not empty. Let y' be the image of an X-morphism $s' \to \widetilde{X}_s$, and let y be the closed point of $\widetilde{X}_s$. We have $y \in \overline{\{y'\}}$. The images of y and y' in X are x and x', respectively. So we have $x \in \overline{\{x'\}}$.

Conversely, suppose $x \in \overline{\{x'\}}$. Let (U, s_U) be an object in the category J_s defined in the proof of 5.3.4 (ii). The image of U in X is open and contains x, and hence contains x'. It follows that the set $\mathrm{Hom}_X(s', U)$ of geometric points in U above s' is not empty. This set is finite. So $\varprojlim_{(U,s_U)\in \mathrm{ob}\, J_s} \mathrm{Hom}_X(s', U)$ is not empty. Hence $\mathrm{Hom}_X(s', \widetilde{X}_s)$ is not empty. By 5.3.4 (ii), there exists an X-morphism $\widetilde{X}_{s'} \to \widetilde{X}_s$. So s is a specialization of s'. □

Let s and s' be two geometric points in a scheme X. Any specialization morphism $\widetilde{X}_{s'} \to \widetilde{X}_s$ induces a homomorphism

$$\mathscr{F}_s \to \mathscr{F}_{s'}$$

for any sheaf $\mathscr{F}$ on X. We call it the *specialization map*. This homomorphism can be described as follows: We have

$$\mathscr{F}_s \cong \varinjlim_{(U,s_U)\in \mathrm{ob}\, I_s^\circ} \mathscr{F}(U),$$

$$\mathscr{F}_{s'} \cong \varinjlim_{(U',s'_{U'})\in \mathrm{ob}\, I_{s'}^\circ} \mathscr{F}(U').$$

Given an object (V, s_V) in I_s, the morphism $s_V : s \to V$ induces a morphism $\widetilde{X}_s \to V$. Denote the composite

$$s' \to \widetilde{X}_{s'} \to \widetilde{X}_s \to V$$

by s'_V. Then (V, s'_V) is an object in $I_{s'}$. So we have canonical homomorphisms

$$\mathscr{F}(V) \stackrel{\phi_{(V,s_V)}}{\to} \varinjlim_{(U,s_U)\in \mathrm{ob}\, I_s^\circ} \mathscr{F}(U),$$

$$\mathscr{F}(V) \stackrel{\phi_{(V,s'_V)}}{\to} \varinjlim_{(U',s'_{U'})\in \mathrm{ob}\, I_{s'}^\circ} \mathscr{F}(U).$$

For any section $t \in \mathscr{F}(V)$, the homomorphism $\mathscr{F}_s \to \mathscr{F}_{s'}$ maps $\phi_{(V,s_V)}(t)$ to $\phi_{(V,s'_V)}(t)$.

Proposition 5.3.6. *Let X be a scheme, s a geometric point in X, and s' a geometric point of $\widetilde{X}_s$. Denote the image of s' in X also by s'. Then $\widetilde{X}_{s'}$ is X-isomorphic to the strict localization $(\widetilde{X}_s)^{\sim}_{s'}$ of $\widetilde{X}_s$ at s'.*

Proof. By 5.3.4, we have a canonical X-morphism $\widetilde{X}_{s'} \to \widetilde{X}_s$. From its construction, one can show this is a flat morphism. Taking strict localizations at s' of the following commutative diagram,

$$\begin{array}{ll} \widetilde{X}_{s'} \to \widetilde{X}_s, \\ \downarrow \ \swarrow & \\ X & \end{array}$$

we get the commutative diagram

$$\begin{array}{ll} \widetilde{X}_{s'} \to (\widetilde{X}_s)^{\sim}_{s'}, \\ \mathrm{id} \downarrow \ \swarrow & \\ \widetilde{X}_{s'} & \end{array}$$

and hence a commutative diagram

$$\begin{array}{ll} \widetilde{\mathscr{O}}_{X,s'} \leftarrow \widetilde{\mathscr{O}}_{\widetilde{X}_s,s'}. \\ \mathrm{id} \uparrow \ \nearrow & \\ \widetilde{\mathscr{O}}_{X,s'} & \end{array}$$

The horizontal arrow in the last diagram is thus surjective. It is also faithfully flat and hence injective. So it is an isomorphism. □

Proposition 5.3.7. *Let $f : X \to Y$ be a finite morphism, y a point in Y, $\overline{k(y)}$ a separable closure of $k(y)$, and $\mathscr{F}$ a sheaf on X. Then*

$$(f_*\mathscr{F})_{\bar{y}} \cong \bigoplus_{x \in X \otimes_{\mathscr{O}_Y} \overline{k(y)}} \mathscr{F}_{\bar{x}}.$$

The functor f_* is exact and faithful. In particular, a sequence

$$\mathscr{F} \to \mathscr{G} \to \mathscr{H}$$

is exact if and only if

$$f_*\mathscr{F} \to f_*\mathscr{G} \to f_*\mathscr{H}$$

is exact.

Proof. Let x_i $(i = 1, \ldots, n)$ be all the points of $X \otimes_{\mathscr{O}_Y} \overline{k(y)}$. By 2.8.20, we have

$$X \times_Y \widetilde{Y}_{\bar{y}} \cong \coprod_{i=1}^{n} \widetilde{X}_{\bar{x}_i}.$$

The argument in the proof of 2.8.20 shows that there exists an etale morphism $U \to Y$ such that U is affine, $X \times_Y U \cong \coprod_{i=1}^n V_i$ for some finite U-schemes V_i, $\widetilde{Y}_{\bar{y}} \to Y$ factors through U, and $V_i \times_U \widetilde{Y}_{\bar{y}} \cong \widetilde{X}_{\bar{x}_i}$. Replacing Y by U and X by each V_i, we are reduced to the case where Y is affine, $X \otimes_{\mathscr{O}_Y} \overline{k(y)}$ contains only one point x and $X \times_Y \widetilde{Y}_{\bar{y}} \cong \widetilde{X}_{\bar{x}}$.

Fix an affine open neighborhood U_0 of y in Y. By 5.3.4 (i), for any affine etale neighborhood $(V, \bar{x}_V)$ of $\bar{x}$ so that the image of V in Y is contained in U_0, the X-morphism $\bar{x}_V : \operatorname{Spec} \overline{k(x)} \to V$ defines an X-morphism

$$g : \widetilde{X}_{\bar{x}} \to V$$

so that the composite of g with $\operatorname{Spec} \overline{k(x)} \to \widetilde{X}_{\bar{x}}$ coincides with $\bar{x}_V$. Let $J_{\bar{y}}$ be the category of affine etale neighborhoods of $\bar{y}$ in U_0. We have

$$\widetilde{Y}_{\bar{y}} \cong \operatorname{Spec} \big(\varinjlim_{(U, s_U) \in \operatorname{ob} J_{\bar{y}}^{\circ}} \mathscr{O}_U(U) \big).$$

By 1.10.9, there exist an affine etale neighborhood (U, s_U) of $\bar{y}$ in Y and an X-morphism

$$g_U : X \times_Y U \to V$$

so that the composite

$$\widetilde{X}_{\bar{x}} \cong X \times_Y \widetilde{Y}_{\bar{y}} \to X \times_Y U \stackrel{g_U}{\to} V$$

coincides with $g : \widetilde{X}_{\bar{x}} \to V$. The composite

$$\mathrm{Spec}\,\overline{k(x)} \to \widetilde{X}_{\bar{x}} \cong X \times_Y \widetilde{Y}_{\bar{y}} \to X \times_Y U$$

makes $X \times_Y U$ an etale neighborhood of $\bar{x}$ in X that dominates the etale neighborhood $(V, \bar{x}_V)$. It follows that

$$\mathscr{F}_{\bar{x}} \cong \varinjlim_{(U,s_U)\in \mathrm{ob}\, J_{\bar{y}}^{\circ}} \mathscr{F}(X \times_Y U).$$

But we have

$$(f_*\mathscr{F})_{\bar{y}} \cong \varinjlim_{(U,s_U)\in \mathrm{ob}\, J_{\bar{y}}^{\circ}} (f_*\mathscr{F})(U) \cong \varinjlim_{(U,s_U)\in \mathrm{ob}\, J_{\bar{y}}^{\circ}} \mathscr{F}(X \times_Y U).$$

So we have $(f_*\mathscr{F})_{\bar{y}} \cong \mathscr{F}_{\bar{x}}$. This proves the first statement of the proposition. It implies that f_* is exact, and that if $f_*\mathscr{F} = 0$, then $\mathscr{F} = 0$. So f_* is exact and faithful. □

Corollary 5.3.8. *Let $f : X \to Y$ be an immersion. Then for any sheaf $\mathscr{F}$ on X, the canonical morphism $f^* f_*\mathscr{F} \to \mathscr{F}$ is an isomorphism.*

Proof. The assertion is clear for any open immersion. If f is a closed immersion, we have

$$(f^* f_*\mathscr{F})_{\bar{x}} \cong (f_*\mathscr{F})_{f(\bar{x})} \cong \mathscr{F}_{\bar{x}}$$

for any $x \in X$ by 5.3.7. So $f^* f_*\mathscr{F} \cong \mathscr{F}$ for any immersion f. □

Corollary 5.3.9. *Consider a Cartesian diagram*

$$\begin{array}{ccc} X \times_Y Y' & \stackrel{g'}{\to} & X \\ f' \downarrow & & \downarrow f \\ Y' & \stackrel{g}{\to} & Y. \end{array}$$

Suppose f is finite. Then for any sheaf $\mathscr{F}$ on X, we have a canonical isomorphism

$$g^* f_*\mathscr{F} \stackrel{\cong}{\to} f'_* g'^* \mathscr{F}.$$

Proof. The composite of the canonical morphisms

$$f_*\mathscr{F} \stackrel{f_*(\mathrm{adj})}{\to} f_* g'_* g'^* \mathscr{F} \cong g_* f'_* g'^* \mathscr{F}$$

induces a morphism

$$g^* f_*\mathscr{F} \to f'_* g'^* \mathscr{F}$$

by adjunction. Using 5.3.7, one proves that this morphism induces isomorphisms on stalks. So it is an isomorphism. □

Corollary 5.3.10. *Let $f : X \to Y$ be a finite surjective radiciel morphism. Then for any sheaf $\mathscr{F}$ on X and any sheaf $\mathscr{G}$ on Y, the canonical morphisms $f^*f_*\mathscr{F} \to \mathscr{F}$ and $\mathscr{G} \to f_*f^*\mathscr{G}$ are isomorphisms. So the functors f_* and f^* define an equivalence between the category of sheaves on X and the category of sheaves on Y.*

Proof. By our assumption, for any $y \in Y$, $X \otimes_{\mathscr{O}_Y} \overline{k(y)}$ consists of only one point. Using 5.3.7, one proves that the canonical morphisms $f^*f_*\mathscr{F} \to \mathscr{F}$ and $\mathscr{G} \to f_*f^*\mathscr{G}$ induce isomorphisms on stalks. □

5.4 Recollement of Sheaves

Proposition 5.4.1. *Let $i : Y \to X$ be a closed immersion, and let $j : X - Y \hookrightarrow X$ be the complement. The functor $i_* : \mathscr{S}_Y \to \mathscr{S}_X$ is fully faithful, and a sheaf $\mathscr{F}$ on X is isomorphic to a sheaf of the form $i_*\mathscr{G}$ for some sheaf $\mathscr{G}$ on Y if and only if $j^*\mathscr{F} = 0$.*

Proof. Let $\mathscr{G}_k$ $(k = 1, 2)$ be two sheaves on Y. The canonical morphisms

$$i^*i_*\mathscr{G}_k \to \mathscr{G}_k$$

are isomorphisms. One can check the map

$$\mathrm{Hom}(\mathscr{G}_1, \mathscr{G}_2) \to \mathrm{Hom}(i_*\mathscr{G}_1, i_*\mathscr{G}_2)$$

and the composite

$$\mathrm{Hom}(i_*\mathscr{G}_1, i_*\mathscr{G}_2) \to \mathrm{Hom}(i^*(i_*\mathscr{G}_1), i^*(i_*\mathscr{G}_2)) \cong \mathrm{Hom}(\mathscr{G}_1, \mathscr{G}_2)$$

are inverses to each other. So i_* is fully faithful. It is clear that $j^*i_*\mathscr{G} = 0$ for any sheaf $\mathscr{G}$ on Y. If $\mathscr{F}$ is a sheaf on X with the property $j^*\mathscr{F} = 0$, then the canonical morphism $\mathscr{F} \to i_*i^*\mathscr{F}$ induces isomorphisms on stalks, and hence is an isomorphism. □

Let $\mathscr{F}$ be a sheaf on a scheme X. The smallest closed subset F of X such that $\mathscr{F}|_{X-F} = 0$ is called the *support* of $\mathscr{F}$. Let U be an object in X_{et}, and let $s \in \mathscr{F}(U)$ be a section. The *support* of s is the smallest closed subset F of U such that $s|_{U-F} = 0$. Let $V \to U$ be an etale X-morphism. Then the support of $s|_V$ is the inverse image of the support of s in V.

Let $i : Y \to X$ be a closed immersion, $j : X - Y \hookrightarrow X$ the complement, $\mathscr{F}$ a sheaf on X, and $\mathscr{K}$ the kernel of the canonical morphism $\mathscr{F} \to j_*j^*\mathscr{F}$. Since j^* is exact, $j^*\mathscr{K}$ is the kernel of the morphism $j^*\mathscr{F} \to j^*(j_*j^*\mathscr{F})$. But this morphism is clearly an isomorphism. So $j^*\mathscr{K} = 0$. By 5.4.1, there exists a sheaf $i^!\mathscr{F}$ on Y such that $\mathscr{K} \cong i_*i^!\mathscr{F}$. For any etale X-scheme U, sections in $(i_*i^!\mathscr{F})(U)$ are those sections in $\mathscr{F}(U)$ whose supports are contained in the inverse image of Y in U. For any sheaf $\mathscr{G}$ on Y, we have $i^!i_*\mathscr{G} \cong \mathscr{G}$. For any morphism $\phi : i_*\mathscr{G} \to \mathscr{F}$, we have a commutative diagram

$$\begin{array}{ccc} i_*\mathscr{G} & \to & \mathscr{F} \\ \downarrow & & \downarrow \\ j_*j^*(i_*\mathscr{G}) & \to & j_*j^*\mathscr{F}. \end{array}$$

As $j_*j^*(i_*\mathscr{G}) = 0$, the image of ϕ is contained in the kernel of $\mathscr{F} \to j_*j^*\mathscr{F}$. So we have

$$\mathrm{Hom}(i_*\mathscr{G}, \mathscr{F}) \cong \mathrm{Hom}(i_*\mathscr{G}, i_*i^!\mathscr{F}) \cong \mathrm{Hom}(\mathscr{G}, i^!\mathscr{F}).$$

Hence i_* is left adjoint to $i^!$.

Let $\mathscr{H}$ be a sheaf on $X - Y$. Define a presheaf $\mathscr{P}$ on X as follows: For any object V in X_{et}, if the image of V in X is not contained in $X - Y$, we let $\mathscr{P}(V) = 0$; otherwise, we regard V as an object of $(X - Y)_{\mathrm{et}}$ and let $\mathscr{P}(V) = \mathscr{H}(V)$. Define $j_!\mathscr{H} = \mathscr{P}^\#$. We have $j^*j_!\mathscr{H} \cong \mathscr{H}$. For any sheaf $\mathscr{F}$ on X, we have

$$\mathrm{Hom}(j_!\mathscr{H}, \mathscr{F}) \cong \mathrm{Hom}(\mathscr{P}, \mathscr{F}) \cong \mathrm{Hom}(\mathscr{H}, j^*\mathscr{F}).$$

So $j_!$ is left adjoint to j^*.

We summarize the above results as follows:

Proposition 5.4.2. *Let $i : Y \to X$ be a closed immersion, and let $j : X - Y \hookrightarrow X$ be the complement.*

(i) *(i^*, i_*), $(j_!, j^*)$, $(i_*, i^!)$ and (j^*, j_*) are pairs of adjoint functors.*

(ii) *$i_*, i^*, j^*, j_!$ are exact, and $j_*, i^!$ are left exact.*

(iii)

$$\begin{aligned} &i^!i_* \cong i^*i_* \cong \mathrm{id}, \\ &j^*j_! \cong j^*j_* \cong \mathrm{id}, \\ &i^*j_! \cong i^!j_! \cong i^!j_* \cong 0, \\ &j^*i_* \cong 0. \end{aligned}$$

(iv) *For any sheaf $\mathscr{F}$ on X, the following sequences are exact:*

$$\begin{aligned} &0 \to i_*i^!\mathscr{F} \to \mathscr{F} \to j_*j^*\mathscr{F}, \\ &0 \to j_!j^*\mathscr{F} \to \mathscr{F} \to i_*i^*\mathscr{F} \to 0. \end{aligned}$$

Proposition 5.4.3. *Let $i : Y \to X$ be a closed immersion, and let $j : X - Y \hookrightarrow X$ be the complement. Define a category as follows: Objects in $\mathscr{C}$ are triples $(\mathscr{G}, \mathscr{H}, \phi)$, where $\mathscr{G}$ is a sheaf on Y, $\mathscr{H}$ is a sheaf on $X - Y$, and $\phi : \mathscr{G} \to i^*j_*\mathscr{H}$ is a morphism. Given two objects $(\mathscr{G}_k, \mathscr{H}_k, \phi_k)$ $(k = 1, 2)$ in $\mathscr{C}$, a morphism $(\xi, \eta) : (\mathscr{G}_1, \mathscr{H}_1, \phi_1) \to (\mathscr{G}_2, \mathscr{H}_2, \phi_2)$ in $\mathscr{C}$ consists of morphisms of sheaves $\xi : \mathscr{G}_1 \to \mathscr{G}_2$ and $\eta : \mathscr{H}_1 \to \mathscr{H}_2$ such that the following diagram commutes:*

$$\begin{array}{ccc} \mathscr{G}_1 & \stackrel{\phi_1}{\to} & i^*j_*\mathscr{H}_1 \\ {\scriptstyle\xi}\downarrow & & \downarrow{\scriptstyle i^*j_*(\eta)} \\ \mathscr{G}_2 & \stackrel{\phi_2}{\to} & i^*j_*\mathscr{H}_2. \end{array}$$

Then the functor

$$\mathscr{S}_X \to \mathscr{C}, \quad \mathscr{F} \mapsto (i^*\mathscr{F}, j^*\mathscr{F}, i^*\mathscr{F} \to i^*(j_*j^*\mathscr{F}))$$

defines an equivalence of categories.

Proof. For any sheaf $\mathscr{F}$ on X, we have a commutative diagram

$$\begin{array}{ccc} \mathscr{F} & \to & i_*i^*\mathscr{F} \\ \downarrow & & \downarrow \\ j_*j^*\mathscr{F} & \to & i_*i^*j_*j^*\mathscr{F}. \end{array}$$

We claim that it is Cartesian. There exists a morphism

$$\mathscr{F} \to i_*i^*\mathscr{F} \times_{(i_*i^*j_*j^*\mathscr{F})} j_*j^*\mathscr{F}$$

whose composite with the projections are the canonical morphisms $\mathscr{F} \to i_*i^*\mathscr{F}$ and $\mathscr{F} \to j_*j^*\mathscr{F}$. Since i^* and j^* are exact, we have

$$\begin{aligned} i^*(i_*i^*\mathscr{F} \times_{(i_*i^*j_*j^*\mathscr{F})} j_*j^*\mathscr{F}) &\cong i^*i_*i^*\mathscr{F} \times_{i^*i_*i^*j_*j^*\mathscr{F}} i^*j_*j^*\mathscr{F} \\ &\cong i^*\mathscr{F} \times_{i^*j_*j^*\mathscr{F}} i^*j_*j^*\mathscr{F} \\ &\cong i^*\mathscr{F}, \\ j^*(i_*i^*\mathscr{F} \times_{(i_*i^*j_*j^*\mathscr{F})} j_*j^*\mathscr{F}) &\cong j^*i_*i^*\mathscr{F} \times_{j^*i_*i^*j_*j^*\mathscr{F}} j^*j_*j^*\mathscr{F} \\ &\cong 0 \times_0 j^*\mathscr{F} \\ &\cong j^*\mathscr{F}. \end{aligned}$$

It follows that the restriction of the morphism

$$\mathscr{F} \to i_*i^*\mathscr{F} \times_{(i_*i^*j_*j^*\mathscr{F})} j_*j^*\mathscr{F}$$

to Y and to $X - Y$ are isomorphisms, and hence the morphism itself is an isomorphism. This proves our claim.

Given two sheaves $\mathscr{F}_i$ $(i = 1, 2)$ on X, the map

$$\begin{aligned}&\mathrm{Hom}(\mathscr{F}_1, \mathscr{F}_2)\\ \to\ &\mathrm{Hom}((i^*\mathscr{F}_1, j^*\mathscr{F}_1, i^*\mathscr{F}_1 \to i^*(j_*j^*\mathscr{F}_1)), (i^*\mathscr{F}_2, j^*\mathscr{F}_2, i^*\mathscr{F}_2 \to i^*(j_*j^*\mathscr{F}_2)))\end{aligned}$$

is injective by 5.3.2 (ii). Let

$$(\xi, \eta) : (i^*\mathscr{F}_1, j^*\mathscr{F}_1, i^*\mathscr{F}_1 \to i^*(j_*j^*\mathscr{F}_1)) \to (i^*\mathscr{F}_2, j^*\mathscr{F}_2, i^*\mathscr{F}_2 \to i^*(j_*j^*\mathscr{F}_2))$$

be a morphism in $\mathscr{C}$. By the claim above, there exists a morphism $\varphi : \mathscr{F}_1 \to \mathscr{F}_2$ such that the following diagram commutes:

$$\begin{array}{ccccc} i_*i^*\mathscr{F}_1 & \leftarrow & \mathscr{F}_1 & \to & j_*j^*\mathscr{F}_1 \\ {\scriptstyle i_*(\xi)}\downarrow & & {\scriptstyle\varphi}\downarrow & & \downarrow{\scriptstyle j_*(\eta)} \\ i_*i^*\mathscr{F}_2 & \leftarrow & \mathscr{F}_2 & \to & j_*j^*\mathscr{F}_2. \end{array}$$

The morphism $\varphi \in \mathrm{Hom}(\mathscr{F}_1, \mathscr{F}_2)$ is thus mapped to (ξ, η). This proves that our functor is fully faithful.

For any object $(\mathscr{G}, \mathscr{H}, \phi)$ in $\mathscr{C}$, define a sheaf $\mathscr{F}$ on X by the Cartesian diagram

$$\begin{array}{ccc} \mathscr{F} & \to & i_*\mathscr{G} \\ \downarrow & & \downarrow{\scriptstyle i_*\phi} \\ j_*\mathscr{H} & \to & i_*i^*j_*\mathscr{H}. \end{array}$$

Applying i^* and j^* to this diagram, we get Cartesian diagrams

$$\begin{array}{ccc} i^*\mathscr{F} & \to & \mathscr{G} \\ \downarrow & & \downarrow{\scriptstyle \phi} \\ i^*j_*\mathscr{H} & \stackrel{\mathrm{id}}{\to} & i^*j_*\mathscr{H}, \end{array} \qquad \begin{array}{ccc} j^*\mathscr{F} & \to & 0 \\ \downarrow & & \downarrow \\ \mathscr{H} & \to & 0. \end{array}$$

It follows that $i^*\mathscr{F} \cong \mathscr{G}$ and $j^*\mathscr{F} \cong \mathscr{H}$. Moreover, the diagram

$$\begin{array}{ccccc} & & i^*\mathscr{F} & \cong & \mathscr{G} \\ & \swarrow & \downarrow & & \downarrow{\scriptstyle \phi} \\ i^*j_*j^*\mathscr{F} & \to & i^*j_*\mathscr{H} & \stackrel{\mathrm{id}}{\to} & i^*j_*\mathscr{H}. \end{array}$$

commutes. So

$$(i^*\mathscr{F}, j^*\mathscr{F}, i^*\mathscr{F} \to i^*j_*j^*\mathscr{F}) \cong (\mathscr{G}, \mathscr{H}, \phi).$$

This proves that our functor is essentially surjective. □

5.5 The Functor $f_!$

([SGA 4] XVII 6.1.)

Let $f : X' \to X$ be an etale morphism. We have a functor $i : X'_{\text{et}} \to X_{\text{et}}$ by regarding etale X'-schemes as etale X-schemes. For any sheaf $\mathscr{F}$ on X, we have $i^{\mathscr{P}}\mathscr{F} \cong f^*\mathscr{F}$ by 5.2.5 (iii). The functor $i^{\mathscr{P}}$ has a left adjoint $i_{\mathscr{P}}$. For any object U in X_{et}, define a category K_U as follows: Objects of K_U are pairs (U', ϕ), where U' is an object in X'_{et}, and $\phi : U \to U'$ is an X-morphism. Given objects (U'_i, ϕ_i) $(i = 1, 2)$ in K_U, a morphism $\xi : (U'_1, \phi_1) \to (U'_2, \phi_2)$ in K_U is an X'-morphism $\xi : U'_1 \to U'_2$ such that $\xi\phi_1 = \phi_2$. The category K_U° satisfies the conditions (I1) and (I2) in 2.7. For any presheaf $\mathscr{F}'$ on X', we have

$$(i_{\mathscr{P}}\mathscr{F}')(U) = \varinjlim_{(U',\phi)\in \text{ob}\, K_U^\circ} \mathscr{F}'(U').$$

Given an X-morphism $\psi : U \to X'$, which is necessarily etale, denote by U_ψ the etale X'-scheme U with the structure morphism ψ. The subcategory of K_U° consisting of objects (U_ψ, id_U) $(\psi \in \text{Hom}_X(U, X'))$ and the identity morphisms of these objects is a full cofinal subcategory. It follows that

$$(i_{\mathscr{P}}\mathscr{F}')(U) = \bigoplus_{\psi\in \text{Hom}_X(U,X')} \mathscr{F}'(U_\psi).$$

When $\mathscr{F}'$ is a sheaf on X', we denote the sheaf $(i_{\mathscr{P}}\mathscr{F}')^\#$ by $f_!\mathscr{F}'$. If f is an open immersion j, then $f_!$ coincides with the functor $j_!$ defined in 5.4.

Proposition 5.5.1. *Let $f : X' \to X$ be an etale morphism.*

(i) *$f_!$ is left adjoint to f^*.*

(ii) *Let*

$$\begin{array}{ccc} X' \times_X Y & \stackrel{g'}{\to} & X' \\ {\scriptstyle f'}\downarrow & & \downarrow {\scriptstyle f} \\ Y & \stackrel{g}{\to} & X \end{array}$$

be a Cartesian diagram. For any sheaf $\mathscr{F}'$ on X', we have a canonical isomorphism

$$f'_! g'^* \mathscr{F}' \stackrel{\cong}{\to} g^* f_! \mathscr{F}'.$$

(iii) *Let $h : X'' \to X'$ be another etale morphism. We have $(fh)_! \cong f_! h_!$.*

(iv) *For any point x in X, let $\overline{k(x)}$ be a separable closure of $k(x)$. We have*

$$(f_!\mathscr{F}')_{\bar{x}} \cong \bigoplus_{x' \in X' \otimes_{\mathscr{O}_X} \overline{k(x)}} \mathscr{F}'_{\bar{x}'}.$$

In particular, the functor $f_!$ is exact and faithful.

Proof.

(i) For sheaves $\mathscr{F}'$ on X' and $\mathscr{F}$ on X, we have

$$\begin{aligned}\mathrm{Hom}(f_!\mathscr{F}', \mathscr{F}) &= \mathrm{Hom}((i_{\mathscr{P}}\mathscr{F}')^{\#}, \mathscr{F}) \\ &\cong \mathrm{Hom}(i_{\mathscr{P}}\mathscr{F}', \mathscr{F}) \\ &\cong \mathrm{Hom}(\mathscr{F}', i^{\mathscr{P}}\mathscr{F}) \\ &\cong \mathrm{Hom}(\mathscr{F}', f^*\mathscr{F}).\end{aligned}$$

(ii) For any sheaf $\mathscr{G}$ on Y, since f is etale, we have a canonical isomorphism

$$f^*g_*\mathscr{G} \stackrel{\cong}{\to} g'_*f'^*\mathscr{G}.$$

So we have

$$\begin{aligned}\mathrm{Hom}(g^*f_!\mathscr{F}', \mathscr{G}) &\cong \mathrm{Hom}(\mathscr{F}', f^*g_*\mathscr{G}) \\ &\cong \mathrm{Hom}(\mathscr{F}', g'_*f'^*\mathscr{G}) \\ &\cong \mathrm{Hom}(f'_!g'^*\mathscr{F}', \mathscr{G}).\end{aligned}$$

It follows that $f'_!g'^*\mathscr{F}' \cong g^*f_!\mathscr{F}'$.

(iii) follows from $(fh)^* \cong h^*f^*$ by adjunction.

(iv) follows from (ii) by taking g to be the canonical morphism $\mathrm{Spec}\,\overline{k(x)} \to X$. □

Suppose that $f : X' \to X$ is a separated etale morphism. Given a sheaf $\mathscr{F}'$ on X, we construct a morphism

$$f_!\mathscr{F}' \to f_*\mathscr{F}'$$

by adjunction from a morphism

$$\mathscr{F}' \to f^*f_*\mathscr{F}'$$

as follows. Let $\phi : U' \to X'$ be an etale morphism. We have

$$(f^*f_*\mathscr{F}')(U') \cong \mathscr{F}'(U' \times_X X').$$

Since f is separated, the graph

$$\Gamma_\phi : U' \to U' \times_X X'$$

of ϕ is a closed immersion. Since the projection $U' \times_X X' \to U'$ is etale, Γ_ϕ is etale. So Γ_ϕ is an open and closed immersion by 2.3.8. We define a homomorphism

$$\mathscr{F}'(U') \to \mathscr{F}'(U' \times_X X')$$

by mapping each section $s \in \mathscr{F}'(U')$ to the section of $\mathscr{F}'(U' \times_X X')$ whose restriction via Γ_ϕ is s, and whose restriction to the complement of Γ_ϕ is 0.

Recall that $f_!\mathscr{F}'$ is the sheaf associated to the presheaf $i_{\mathscr{P}}\mathscr{F}'$ defined by

$$(i_{\mathscr{P}}\mathscr{F}')(U) = \bigoplus_{\psi \in \mathrm{Hom}_X(U,X')} \mathscr{F}'(U_\psi)$$

for any etale X-scheme U. The morphism $f_!\mathscr{F}' \to f_*\mathscr{F}'$ is induced from the morphism of presheaves

$$i_{\mathscr{P}}\mathscr{F}' \to f_*\mathscr{F}'$$

defined by the homomorphism

$$\bigoplus_{\psi \in \mathrm{Hom}_X(U,X')} \mathscr{F}'(U_\psi) \to \mathscr{F}'(U \times_X X')$$

that maps each section $s \in \mathscr{F}'(U_\psi)$ to the section of $\mathscr{F}'(U \times_X X')$ whose restriction via Γ_ψ is s, and whose restriction to the complement of Γ_ψ is 0.

Proposition 5.5.2. *Suppose that $f : X' \to X$ is an etale separated morphism of finite type. Then for any sheaf $\mathscr{F}'$ on X', the canonical morphism*

$$f_!\mathscr{F}' \to f_*\mathscr{F}'$$

is injective. For any object U in X_{et}, a section $s \in (f_*\mathscr{F}')(U)$ lies in the image of this morphism if any only if the support of s, considered as a section in $\mathscr{F}'(U \times_X X')$, is proper over U.

Here we say that a closed subset of $U \times_X X'$ is proper over U if any closed subscheme of $U \times_X X'$, whose underlying topological space is the given closed subset, is proper over U.

Proof. Note that the map $\psi \mapsto \mathrm{im}\,\Gamma_\psi$ defines a one-to-one correspondence from the set $\mathrm{Hom}_X(U, X')$ to the set of open and closed subsets of $U \times_X X'$ so that the restriction of the projection $U \times_X X' \to U$ to them are isomorphisms. (Confer 2.3.10 (ii).) From the above description of the morphism $i_{\mathscr{P}}\mathscr{F}' \to f_*\mathscr{F}'$, one can verify that the morphism $f_!\mathscr{F}' \to f_*\mathscr{F}'$ is injective. It is clear that any section $s \in \mathscr{F}'(U \times_X X')$ lying in the image of the morphism $i_{\mathscr{P}}\mathscr{F}' \to f_*\mathscr{F}'$ of presheaves has proper support over U. Using 1.7.12, one deduces that any section $s \in \mathscr{F}'(U \times_X X')$ lying in the image of the morphism $f_!\mathscr{F}' \to f_*\mathscr{F}'$ of sheaves has proper support over U.

Let F be a closed subscheme of $U \times_X X'$ that is proper over U. Since f is necessarily quasi-finite, F is finite over U by [Fu (2006)] 2.5.12 or

[EGA] III 4.4.2. For any $x \in U$, let $\widetilde{U}_{\bar{x}}$ be the strict localization of U at $\bar{x}$. Then $\widetilde{U}_{\bar{x}} \times_U F$ is finite over $\widetilde{U}_{\bar{x}}$. Regard $\widetilde{U}_{\bar{x}} \times_U F$ as a closed subscheme of $\widetilde{U}_{\bar{x}} \times_X X'$. Let $\bar{x}'_1, \ldots, \bar{x}'_k$ be all the points in $\widetilde{U}_{\bar{x}} \times_U F$ that lie above the closed point of $\widetilde{U}_{\bar{x}}$. By 2.8.14 (iii), there exist sections

$$\Gamma_{\psi'_i} : \widetilde{U}_{\bar{x}} \to \widetilde{U}_{\bar{x}} \times_X X' \quad (i = 1, \ldots, k)$$

of the projection $\widetilde{U}_{\bar{x}} \times_X X' \to \widetilde{U}_{\bar{x}}$ such that $\Gamma_{\psi'_i}$ map the closed point of $\widetilde{U}_{\bar{x}}$ to $\bar{x}'_i$. The images of $\Gamma_{\psi'_i}$ are open and closed connected subsets of $\widetilde{U}_{\bar{x}} \times_X X'$, and they are disjoint. Moreover, we have

$$\widetilde{U}_{\bar{x}} \times_U F \subset \bigcup_{i=1}^{k} \operatorname{im} \Gamma_{\psi'_i}.$$

Indeed, $\widetilde{U}_{\bar{x}} \times_U F - \bigcup_{i=1}^{k} \operatorname{im} \Gamma_{\psi'_i}$ is a closed subset of $\widetilde{U}_{\bar{x}} \times_U F$. Since F is proper over U, the image of $\widetilde{U}_{\bar{x}} \times_U F - \bigcup_{i=1}^{k} \operatorname{im} \Gamma_{\psi'_i}$ in $\widetilde{U}_{\bar{x}}$ is closed. If it is nonempty, then this image contains the closed point of $\widetilde{U}_{\bar{x}}$. This is impossible since all the preimages x'_i of the closed point of $\widetilde{U}_{\bar{x}}$ in $\widetilde{U}_{\bar{x}} \times_U F$ lie in $\bigcup_{i=1}^{k} \operatorname{im} \Gamma_{\psi'_i}$. Hence $\widetilde{U}_{\bar{x}} \times_U F - \bigcup_{i=1}^{k} \operatorname{im} \Gamma_{\psi'_i}$ is empty. By 1.10.1 (iv) and 1.10.9, there exist an etale neighborhood $U_{\bar{x}}$ of $\bar{x}$ in U and sections

$$\Gamma_{\psi_i} : U_{\bar{x}} \to U_{\bar{x}} \times_X X' \quad (i = 1, \ldots, k)$$

of the projection $U_{\bar{x}} \times_X X' \to U_{\bar{x}}$ such that $\operatorname{im} \Gamma_{\psi_i}$ are disjoint open and closed subsets of $U_{\bar{x}} \times_X X'$, and

$$U_{\bar{x}} \times_U F \subset \bigcup_{i=1}^{k} \operatorname{im} \Gamma_{\psi_i}.$$

Let $\psi_i : U_{\bar{x}} \to X'$ be the composite

$$U_{\bar{x}} \stackrel{\Gamma_{\psi_i}}{\to} U_{\bar{x}} \times_X X' \to X'.$$

For any section $s \in \mathscr{F}'(U \times_X X')$ with the property $s|_{U \times_X X' - F} = 0$, note that $s|_{U_{\bar{x}} \times_X X'}$ lies in the image of the homomorphism

$$\bigoplus_{\psi \in \operatorname{Hom}_X(U_{\bar{x}}, X')} \mathscr{F}'(U_{\bar{x},\psi}) \to \mathscr{F}'(U_{\bar{x}} \times_X X')$$

described before 5.5.2. As $U_{\bar{x}}$ $(x \in U)$ form an etale covering of U, s lies in the image of the morphism of sheaves $f_! \mathscr{F}' \to f_* \mathscr{F}'$. $\square$

Let $f : X' \to X$ be a separated morphism of finite type and let $\mathscr{F}'$ be a sheaf on X'. Define a presheaf $f_!\mathscr{F}'$ on X so that for any object U in X_{et}, we have
$(f_!\mathscr{F}')(U) = \{s | s \in \mathscr{F}'(U \times_X X')$ and the support of s is proper over $U\}$. Using 1.7.12, one can show that $f_!\mathscr{F}'$ is a sheaf. We have a canonical monomorphism

$$f_!\mathscr{F} \hookrightarrow f_*\mathscr{F}.$$

If f is etale, then $f_!$ coincides with the functor defined at the beginning of this section by 5.5.2. If f is proper, we have $f_! = f_*$.

Proposition 5.5.3. *Let $f : X' \to X$ and $f' : X'' \to X'$ be two separated morphisms of finite type and let $g = ff'$. Then we have an isomorphism of functors*

$$g_! \cong f_! f'_!$$

such that the following diagram commutes:

$$\begin{array}{ccc} g_! & \cong & f_! f'_! \\ \downarrow & & \downarrow \\ g_* & \cong & f_* f'_*. \end{array}$$

Proof. Let U be an object in X_{et} and let $\mathscr{F}$ be a sheaf on X''. We need to show that under the identification

$$(g_*\mathscr{F})(U) \cong (f_* f'_*\mathscr{F})(U),$$

the subset $(g_!\mathscr{F})(U)$ of $(g_*\mathscr{F})(U)$ is identified with the subset $(f_! f'_!\mathscr{F})(U)$ of $(f_* f'_*\mathscr{F})(U)$. Let $s \in \mathscr{F}(U \times_X X'')$ be a section lying in $(f_! f'_!\mathscr{F})(U)$. Then s lies in

$$(f_* f'_!\mathscr{F})(U) = (f'_!\mathscr{F})(U \times_X X').$$

So the support of s is proper over $U \times_X X'$. On the other hand, s lies in $(f_! f'_*\mathscr{F})(U)$. So if we consider s as a section of $(f'_*\mathscr{F})(U \times_X X')$, its support is proper over U. The support of s considered as a section of $(f'_*\mathscr{F})(U \times_X X')$ is the closure of the image of the support of s under the morphism $U \times_X X'' \to U \times_X X'$. One deduces from these facts that the support of s is proper over U. So $s \in (g_!\mathscr{F})(U)$.

Let $s \in \mathscr{F}(U \times_X X'')$ be a section lying in $(g_!\mathscr{F})(U)$. Then the support of s is proper over U. Since $U \times_X X' \to U$ is separated, the support of s is also proper over $U \times_X X'$. Hence s lies in $(f'_!\mathscr{F})(U \times_X X')$. Moreover, the image of the support of s in $U \times_X X'$ is closed. So the support of s considered as a section of $(f'_!\mathscr{F})(U \times_X X')$ coincides with image of the support of s under the morphism $U \times_X X'' \to U \times_X X'$. Since the support of s is proper over U, its image in $U \times_X X'$ is also proper over U. It follows that $s \in (f_! f'_!\mathscr{F})(U)$. □

5.6 Etale Cohomology

([SGA 4] XVII 4.2.2–4.2.6.)

Let X be a scheme, and let $F : \mathscr{S}_X \to \mathscr{A}$ be a left exact functor from the category $\mathscr{S}_X$ of sheaves of abelian groups on X to an abelian category $\mathscr{A}$. By 5.2.4 (i), we can define the right derived functors R^iF $(i \geq 0)$ of F.

For any object U in X_{et}, the functor $\mathscr{F} \mapsto \mathscr{F}(U)$ is left exact. Its right derived functors are denoted by $\mathscr{F} \mapsto H^i(U, \mathscr{F})$, or $\mathscr{F} \mapsto R^i\Gamma(U, \mathscr{F})$. We call $H^i(X, \mathscr{F})$ the *etale cohomology groups* of $\mathscr{F}$. Note that we have $H^i(U, \mathscr{F}) \cong H^i(U, \mathscr{F}|_U)$.

Let $f : X' \to X$ be a morphism of schemes. The functor $\mathscr{F}' \mapsto f_*\mathscr{F}'$ is left exact. Its right derived functors are denoted by $\mathscr{F}' \mapsto R^if_*\mathscr{F}'$. We call $R^if_*\mathscr{F}'$ the *higher direct images* of $\mathscr{F}'$. Each $R^if_*\mathscr{F}'$ is the sheaf on X associated to the presheaf $U \mapsto H^i(U \times_X X', \mathscr{F}')$ for any $U \in \mathrm{ob}\, X_{\mathrm{et}}$.

Let A be a ring. The category of sheaves of A-modules on X has enough injective objects. For any sheaf of A-modules $\mathscr{G}$ on X. The functors

$$\mathscr{F} \mapsto \mathrm{Hom}_A(\mathscr{G}, \mathscr{F}), \quad \mathscr{F} \mapsto \mathscr{H}om_A(\mathscr{G}, \mathscr{F})$$

are left exact. Their derived functors are denoted by

$$\mathscr{F} \mapsto \mathrm{Ext}^i_A(\mathscr{G}, \mathscr{F}), \quad \mathscr{F} \mapsto \mathscr{E}xt^i_A(\mathscr{G}, \mathscr{F})$$

respectively. We often denote $\mathrm{Ext}^i_A(\mathscr{F}, \mathscr{G})$ and $\mathscr{E}xt^i_A(\mathscr{F}, \mathscr{G})$ by $\mathrm{Ext}^i(\mathscr{F}, \mathscr{G})$ and $\mathscr{E}xt^i(\mathscr{F}, \mathscr{G})$, respectively.

Lemma 5.6.1. *Let $\mathscr{F}$ be a sheaf on a scheme X and let $\mathscr{H}^i(\mathscr{F})$ be the presheaf defined by $\mathscr{H}^i(\mathscr{F})(U) = H^i(U, \mathscr{F})$ for any object U in X_{et}. Then $\mathscr{H}^i(\mathscr{F})^+ = 0$ for all $i \geq 1$.*

Proof. Let $\mathscr{I}^\cdot$ be an injective resolution of $\mathscr{F}$. Then $\mathscr{H}^i(\mathscr{F})(U)$ is the i-th cohomology group of the complex $\mathscr{I}^\cdot(U)$. So $\mathscr{H}^i(\mathscr{F})^\#$ is the i-th cohomology sheaf of the complex $\mathscr{I}^\cdot$. Hence $\mathscr{H}^i(\mathscr{F})^\# = 0$ for all $i \geq 1$. Since $\mathscr{H}^i(\mathscr{F})^+$ are subsheaves of $\mathscr{H}^i(\mathscr{F})^\#$, we have $\mathscr{H}^i(\mathscr{F})^+ = 0$ for all $i \geq 1$. □

Proposition 5.6.2. *Let X be a scheme and let $\mathfrak{U} = \{U_\alpha \to U\}_{\alpha \in I}$ be an etale covering in X_{et}. For any sheaf $\mathscr{F}$ on X, we have biregular spectral sequences*

$$E_2^{ij} = \check{H}^i(\mathfrak{U}, \mathscr{H}^j(\mathscr{F})) \Rightarrow H^{i+j}(U, \mathscr{F}),$$
$$E_2^{ij} = \check{H}^i(U, \mathscr{H}^j(\mathscr{F})) \Rightarrow H^{i+j}(U, \mathscr{F}),$$

where

$$\check{H}^i(U, -) = \varinjlim_{\mathfrak{U}\in \mathrm{ob}\, J_U} \check{H}^i(\mathfrak{U}, -).$$

(See the discussion after 5.2.1 for the definition of J_U).

Proof. If $\mathscr{I}$ is an injective sheaf, then one can show that it is an injective object in the category of presheaves. So we have $\check{H}^j(\mathfrak{U}, \mathscr{I}) = 0$ for all $j \geq 1$ by 5.1.1. Let $\mathscr{I}^\cdot$ be an injective resolution of $\mathscr{F}$. Consider the bicomplex $C^\cdot(\mathfrak{U}, \mathscr{I}^\cdot)$. We have

$$H^j(C^\cdot(\mathfrak{U}, \mathscr{I}^i)) = \check{H}^j(\mathfrak{U}, \mathscr{I}^i) = \begin{cases} \Gamma(U, \mathscr{I}^i) & \text{if } j = 0, \\ 0 & \text{if } j \geq 1. \end{cases}$$

So the spectral sequence

$$E_2^{ij} = H^i_{II} H^j_I (C^\cdot(\mathfrak{U}, \mathscr{I}^\cdot)) \Rightarrow H^{i+j}(C^\cdot(\mathfrak{U}, \mathscr{I}^\cdot))$$

degenerates and we have

$$H^i(C^\cdot(\mathfrak{U}, \mathscr{I}^\cdot)) \cong H^i(\Gamma(U, \mathscr{I}^\cdot)) \cong H^i(U, \mathscr{F}).$$

The spectral sequence

$$E_2^{ij} = H^i_I H^j_{II} (C^\cdot(\mathfrak{U}, \mathscr{I}^\cdot)) \Rightarrow H^{i+j}(C^\cdot(\mathfrak{U}, \mathscr{I}^\cdot))$$

can be identified with

$$E_2^{ij} = \check{H}^i(\mathfrak{U}, \mathscr{H}^j(\mathscr{F})) \Rightarrow H^{i+j}(U, \mathscr{F}).$$

Taking direct limit, we get the spectral sequence

$$E_2^{ij} = \check{H}^i(U, \mathscr{H}^j(\mathscr{F})) \Rightarrow H^{i+j}(U, \mathscr{F}).$$

□

Corollary 5.6.3. *We have a canonical isomorphism $\check{H}^1(U, \mathscr{F}) \cong H^1(U, \mathscr{F})$ and a canonical monomorphism $\check{H}^2(U, \mathscr{F}) \hookrightarrow H^2(U, \mathscr{F})$.*

Proof. We have

$$\check{H}^0(U, \mathscr{H}^j(\mathscr{F})) \cong \mathscr{H}^j(\mathscr{F})^+(U) = 0$$

for all $j \geq 1$ by 5.6.1. We then use the exact sequence

$$0 \to E_2^{10} \to H^1(U, \mathscr{F}) \to E_2^{01} \to E_2^{20} \to H^2(U, \mathscr{F})$$

in [Fu (2006)] 2.2.3 deduced from the second spectral sequence in 5.6.2. □

A sheaf $\mathscr{F}$ on a scheme X is called *flasque* if for any etale covering $\mathfrak{U} = \{U_\alpha \to U\}_{\alpha \in I}$ in X_{et}, we have $\check{H}^1(\mathfrak{U}, \mathscr{F}) = 0$. By 5.6.3, we have

$$H^1(U, \mathscr{F}) \cong \check{H}^1(U \mathscr{F}) \cong \varinjlim_{\mathfrak{U} \in \text{ob}\, J_U} \check{H}^1(\mathfrak{U}, \mathscr{F}) = 0$$

for any $U \in \text{ob}\, X_{\text{et}}$ and any flasque sheaf $\mathscr{F}$ on X.

Proposition 5.6.4. *Let*

$$0 \to \mathscr{F}' \to \mathscr{F} \to \mathscr{F}'' \to 0$$

be an exact sequence of sheaves on a scheme X.

(i) *If $\mathscr{F}'$ is flasque, then the sequence is exact in the category of presheaves.*

(ii) *If $\mathscr{F}'$ and $\mathscr{F}$ are flasque, then $\mathscr{F}''$ is flasque.*

(iii) *If the sequence splits and $\mathscr{F}$ is flasque, then $\mathscr{F}'$ and $\mathscr{F}''$ are flasque.*

(iv) *Injective sheaves are flasque.*

Proof. We prove (i) and leave the rest to the reader. For any object U in X_{et}, we have a long exact sequence

$$0 \to \mathscr{F}'(U) \to \mathscr{F}(U) \to \mathscr{F}''(U) \to H^1(U, \mathscr{F}') \to \cdots.$$

If $\mathscr{F}'$ is flasque, then $H^1(U, \mathscr{F}') = 0$ and hence the sequence

$$0 \to \mathscr{F}'(U) \to \mathscr{F}(U) \to \mathscr{F}''(U) \to 0$$

is exact. □

Proposition 5.6.5. *Let $f : X' \to X$ be a morphism of schemes.*

(i) *If $\mathscr{F}'$ is a flasque sheaf on X', then $f_*\mathscr{F}'$ is flasque.*

(ii) *Let*

$$0 \to \mathscr{F}' \to \mathscr{F} \to \mathscr{F}'' \to 0$$

be an exact sequence of sheaves on X'. If $\mathscr{F}'$ is flasque, then

$$0 \to f_*\mathscr{F}' \to f_*\mathscr{F} \to f_*\mathscr{F}'' \to 0$$

is exact.

Proof.

(i) Let $\mathfrak{U} = \{U_\alpha \to U\}$ be an etale covering in X_{et}. Then $f^*\mathfrak{U} = \{U_\alpha \times_X X' \to U \times_X X'\}$ is an etale covering in X'_{et}. We have

$$\check{H}^i(\mathfrak{U}, f_*\mathscr{F}') \cong \check{H}^i(f^*\mathfrak{U}, \mathscr{F}').$$

If $\mathscr{F}'$ is flasque, we have $\check{H}^1(f^*\mathfrak{U}, \mathscr{F}') = 0$. So $\check{H}^1(\mathfrak{U}, f_*\mathscr{F}') = 0$. Hence $f_*\mathscr{F}'$ is flasque.

(ii) By 5.6.4 (i), the sequence

$$0 \to \mathscr{F}' \to \mathscr{F} \to \mathscr{F}'' \to 0$$

is exact in the category of presheaves. So

$$0 \to f_*\mathscr{F}' \to f_*\mathscr{F} \to f_*\mathscr{F}'' \to 0$$

is exact in the category of presheaves and hence exact in the category of sheaves. □

Corollary 5.6.6.

(i) *If $\mathscr{F}$ is a flasque sheaf on a scheme X, then $H^i(U,\mathscr{F}) = 0$ for all $i \geq 1$ and all objects U in X_{et}. So we can use flasque resolutions to calculate $H^i(U,-)$.*

(ii) *Let $f : X' \to X$ be a morphism of schemes. For all flasque sheaves $\mathscr{F}'$ on X', we have $R^i f_*\mathscr{F}' = 0$ for all $i \geq 1$. So we can use flasque resolutions to calculate $R^i f_*$.*

Proof. We prove (i) and leave it for the reader to prove (ii). Let $\mathscr{I}^\cdot$ be an injective resolution of $\mathscr{F}$, and let

$$\mathscr{Z}^i = \ker(\mathscr{I}^i \to \mathscr{I}^{i+1}).$$

Then we have short exact sequences

$$\begin{aligned} 0 \to \mathscr{F} \to \mathscr{I}^0 \to \mathscr{Z}^1 \to 0, \\ 0 \to \mathscr{Z}^1 \to \mathscr{I}^1 \to \mathscr{Z}^2 \to 0, \\ \vdots \end{aligned}$$

Using 5.6.4 (ii) and (iv), one can show that $\mathscr{F} = \mathscr{Z}^0, \mathscr{Z}^1, \cdots$ are flasque. By 5.6.4 (i), the following sequences are exact:

$$\begin{aligned} 0 \to \mathscr{F}(U) \to \mathscr{I}^0(U) \to \mathscr{Z}^1(U) \to 0, \\ 0 \to \mathscr{Z}^1(U) \to \mathscr{I}^1(U) \to \mathscr{Z}^2(U) \to 0, \\ \vdots \end{aligned}$$

So the sequence

$$\mathscr{I}^0(U) \to \mathscr{I}^1(U) \to \cdots$$

is exact. Hence $H^i(U,\mathscr{F}) = 0$ for all $i \geq 1$. □

Proposition 5.6.7. *Let $\mathscr{F}$ be a sheaf on a scheme X. The following conditions are equivalent:*

(i) *For any object U in X_{et}, we have $H^i(U,\mathscr{F}) = 0$ for all $i \geq 1$.*

(ii) *For any object U in X_{et}, we have $\check{H}^i(U,\mathscr{F}) = 0$ for all $i \geq 1$.*

(iii) *For any etale covering $\mathfrak{U} = \{U_\alpha \to U\}_{\alpha\in I}$ in X_{et}, we have $\check{H}^i(\mathfrak{U},\mathscr{F}) = 0$ for all $i \geq 1$.*

(iv) *$\mathscr{F}$ is flasque.*

Proof.

(i)⇒(iii) Suppose $H^j(U,\mathscr{F})=0$ for all $U\in \mathrm{ob}\, X_{\mathrm{et}}$ and all $j\geq 1$. Then $\mathscr{H}^j(\mathscr{F})=0$ for all $j\geq 1$. The spectral sequence

$$E_2^{ij}=\check{H}^i(\mathfrak{U},\mathscr{H}^j(\mathscr{F}))\Rightarrow H^{i+j}(U,\mathscr{F})$$

degenerates and we have

$$\check{H}^i(\mathfrak{U},\mathscr{F})\cong \check{H}^i(\mathfrak{U},\mathscr{H}^0(\mathscr{F}))\cong H^i(U,\mathscr{F})=0$$

for all $i\geq 1$.

(iii)⇒(ii) is clear.

(ii)⇒(i) Let us prove $\mathscr{H}^j(\mathscr{F})=0$ for all $j\geq 1$ by induction on j. For any object V in X_{et}, by 5.6.3, we have

$$\mathscr{H}^1(\mathscr{F})(V)=H^1(V,\mathscr{F})\cong \check{H}^1(V,\mathscr{F})=0.$$

Suppose we have shown $\mathscr{H}^j(\mathscr{F})=0$ for all $1\leq j\leq n$. Consider the spectral sequence

$$E_2^{ij}=\check{H}^i(U,\mathscr{H}^j(\mathscr{F}))\Rightarrow H^{i+j}(U,\mathscr{F}).$$

We have

$$E_2^{0j}=\mathscr{H}^j(\mathscr{F})^+(U)=0$$

for all $j\geq 1$ by 5.6.1 and $E_2^{ij}=0$ for all $1\leq j\leq n$. This implies that

$$\check{H}^{n+1}(U,\mathscr{F})\cong H^{n+1}(U,\mathscr{F})$$

by [Fu (2006)] 2.2.3. So $H^{n+1}(U,\mathscr{F})=0$ and hence $\mathscr{H}^{n+1}(\mathscr{F})=0$.

(iii)⇒(iv) is clear.

(iv)⇒(i) follows from 5.6.6 (i). □

Let X be a scheme and let A be a ring. Injective objects in the category of sheaves of A-modules on X are called injective sheaves of A-modules. Using the same argument as in the proof of 5.1.1 by replacing $\mathbb{Z}$ by A, one can show that injective sheaves of A-modules are flasque. So for any sheaf $\mathscr{F}$ of A-modules on X, we can also use resolutions of $\mathscr{F}$ by injective sheaves of A-modules to calculate $H^i(U,\mathscr{F})$ and $R^if_*\mathscr{F}$ for any $U\in\mathrm{ob}\,X_{\mathrm{et}}$ and any morphism $f:X\to Y$. We can define A-module structures on the groups $H^i(U,\mathscr{F})$ and the sheaves $R^if_*\mathscr{F}$.

We say that a sheaf $\mathscr{F}$ of A-modules on X is *flat* if the stalks of $\mathscr{F}$ at all geometric points of X are flat A-modules. This is equivalent to saying that the functor $\mathscr{F}\otimes_A -$ is exact in the category of sheaves of A-modules.

Lemma 5.6.8. *Let X be a scheme, A a ring, $\mathscr{F}$ a sheaf of A-modules, and $\mathscr{I}$ an injective sheaf of A-modules. If $\mathscr{F}$ is flat, then $\mathscr{H}om_A(\mathscr{F},\mathscr{I})$ is injective. In general, $\mathscr{H}om_A(\mathscr{F},\mathscr{I})$ is flasque.*

Proof. We have

$$\mathrm{Hom}_A(-, \mathscr{H}om_A(\mathscr{F}, \mathscr{I})) \cong \mathrm{Hom}_A(\mathscr{F} \otimes_A -, \mathscr{I}).$$

If $\mathscr{F}$ is flat and $\mathscr{I}$ is injective, then the functor $\mathrm{Hom}_A(\mathscr{F} \otimes_A -, \mathscr{I})$ is exact. So the functor $\mathrm{Hom}_A(-, \mathscr{H}om_A(\mathscr{F}, \mathscr{I}))$ is exact, and hence $\mathscr{H}om_A(\mathscr{F}, \mathscr{I})$ is an injective sheaf of A-modules. This implies that $\mathscr{H}om_A(\mathscr{F}, \mathscr{I})$ is flasque.

For any etale morphism $f : U \to X$, let $A_U = f_! A$, where A on the right-hand side denotes the constant sheaf on U associated to A, that is, the sheaf associated to the constant presheaf $V \mapsto A$ for any etale U-scheme V. Let $\mathfrak{U} = \{U_\alpha \to U\}$ be an etale covering of U. As in the proof of 5.1.1, we have an exact sequence

$$0 \leftarrow A_U \overset{\delta_{-1}}{\leftarrow} \bigoplus_{\alpha_0} A_{U_{\alpha_0}} \overset{\delta_0}{\leftarrow} \bigoplus_{\alpha_0,\alpha_1} A_{U_{\alpha_0\alpha_1}} \overset{\delta_1}{\leftarrow} \cdots.$$

For convenience, we write the above sequence as

$$A_{\mathfrak{U}}^{\cdot} \to A_U \to 0.$$

Since $\mathscr{I}$ is an injective sheaf of A-modules, the sequence

$$0 \to \mathscr{H}om_A(A_U, \mathscr{I}) \to \mathscr{H}om_A(A_{\mathfrak{U}}^{\cdot}, \mathscr{I})$$

is exact. Since A_U and $A_{U_{\alpha_0\cdots\alpha_i}}$ are flat sheaves, $\mathscr{H}om_A(A_{\mathfrak{U}}^{\cdot}, \mathscr{I})$ is an injective resolution of the injective sheaf $\mathscr{H}om_A(A_U, \mathscr{I})$ in the category of sheaves of A-modules. So we have

$$H^i(\mathrm{Hom}_A(\mathscr{F}, \mathscr{H}om_A(A_{\mathfrak{U}}^{\cdot}, \mathscr{I}))) \cong \mathrm{Ext}^i(\mathscr{F}, \mathscr{H}om_A(A_U, \mathscr{I})) = 0$$

for all $i \geq 1$. On the other hand, we have

$$\begin{aligned} H^i(\mathrm{Hom}_A(\mathscr{F}, \mathscr{H}om_A(A_{\mathfrak{U}}^{\cdot}, \mathscr{I}))) &\cong H^i(\mathrm{Hom}_A(A_{\mathfrak{U}}^{\cdot}, \mathscr{H}om_A(\mathscr{F}, \mathscr{I}))) \\ &\cong H^i(C^{\cdot}(\mathfrak{U}, \mathscr{H}om_A(\mathscr{F}, \mathscr{I}))) \\ &\cong \check{H}^i(\mathfrak{U}, \mathscr{H}om_A(\mathscr{F}, \mathscr{I})). \end{aligned}$$

So we have $\check{H}^i(\mathfrak{U}, \mathscr{H}om_A(\mathscr{F}, \mathscr{I})) = 0$ for all $i \geq 1$. Hence $\mathscr{H}om_A(\mathscr{F}, \mathscr{I})$ is flasque. □

Corollary 5.6.9.

(i) *Let $f : X' \to X$ be a morphism of schemes and let $\mathscr{F}'$ be a sheaf on X'. We have a biregular spectral sequence*

$$E_2^{pq} = H^i(X, R^j f_* \mathscr{F}') \Rightarrow H^{i+j}(X', \mathscr{F}').$$

(ii) *Let $f' : X'' \to X'$ and $f : X' \to X$ be morphisms of schemes and let $\mathscr{F}''$ be a sheaf on X''. We have a biregular spectral sequence*

$$E_2^{ij} = R^i f_* R^j f'_* \mathscr{F}'' \Rightarrow R^{i+j}(ff')_* \mathscr{F}''.$$

(iii) *Let $\mathscr{F}$ and $\mathscr{G}$ be sheaves of A-modules on a scheme X. We have a biregular spectral sequence*

$$E_2^{ij} = H^i(X, \mathscr{E}xt_A^i(\mathscr{F}, \mathscr{G})) \Rightarrow \mathrm{Ext}_A^{i+j}(\mathscr{F}, \mathscr{G}).$$

Proof. This follows from 5.6.5, 5.6.6, 5.6.8, [Fu (2006)] 2.2.10 or [Grothendieck (1957)] 2.4.1. □

Let X be a scheme and let $\{U_\alpha \to X\}$ be an etale covering of X. If $\mathscr{I}$ is an injective sheaf, the proof of 5.6.8 shows that

$$0 \to \mathscr{H}om(\bigoplus_{\alpha_0} \mathbb{Z}_{U_{\alpha_0}}, \mathscr{I}) \to \mathscr{H}om(\bigoplus_{\alpha_0,\alpha_1} \mathbb{Z}_{U_{\alpha_0\alpha_1}}, \mathscr{I}) \to \cdots$$

is an injective resolution of $\mathscr{H}om(\mathbb{Z}_X, \mathscr{I})$. Denote by $\pi_{\alpha_0\ldots\alpha_n}$ the canonical morphism

$$U_{\alpha_0\ldots\alpha_n} = U_{\alpha_0} \times_X \cdots \times_X U_{\alpha_n} \to X.$$

This shows that

$$0 \to \prod_{\alpha_0} \pi_{\alpha_0 *}\pi_{\alpha_0}^* \mathscr{I} \to \prod_{\alpha_0,\alpha_1} \pi_{\alpha_0\alpha_1 *}\pi_{\alpha_0\alpha_1}^* \mathscr{I} \to \cdots$$

is an injective resolution of $\mathscr{I}$.

Let $f : X \to Y$ be a morphism, let $\mathscr{F}$ be a sheaf on X, a let $\mathscr{I}^\cdot$ be an injective resolution of $\mathscr{F}$. Consider the bicomplex

$$\begin{array}{ccccc}
 & & \vdots & & \vdots \\
 & & \uparrow & & \uparrow \\
0 \to & \prod_{\alpha_0} \pi_{\alpha_0 *}\pi_{\alpha_0}^* \mathscr{I}_1 & \to & \prod_{\alpha_0,\alpha_1} \pi_{\alpha_0\alpha_1 *}\pi_{\alpha_0\alpha_1}^* \mathscr{I}_1 & \to \cdots \\
 & \uparrow & & \uparrow & \\
0 \to & \prod_{\alpha_0} \pi_{\alpha_0 *}\pi_{\alpha_0}^* \mathscr{I}_0 & \to & \prod_{\alpha_0,\alpha_1} \pi_{\alpha_0\alpha_1 *}\pi_{\alpha_0\alpha_1}^* \mathscr{I}_0 & \to \cdots \\
 & \uparrow & & \uparrow & \\
 & 0 & & 0 &
\end{array}$$

Applying f_* to this bicomplex and analyzing the spectral sequences associated to the resulting bicomplex, we get the following.

Proposition 5.6.10. *Notation as above. We have a biregular spectral sequence*

$$E_2^{ij} \Rightarrow R^{i+j} f_* \mathscr{F},$$

where E_2^{ij} is the i-th cohomology sheaf of the complex

$$0 \to \prod_{\alpha_0} R^j(f\pi_{\alpha_0})_* \pi_{\alpha_0}^* \mathscr{F} \to \prod_{\alpha_0,\alpha_1} R^j(f\pi_{\alpha_0\alpha_1})_* \pi_{\alpha_0\alpha_1}^* \mathscr{F} \cdots.$$

Let $i : Y \to X$ be a closed immersion and let $j : U = X - Y \hookrightarrow X$ be its complement. For any sheaf $\mathscr{F}$ on X, let

$$\Gamma_Y(X, \mathscr{F}) = \{s \in \Gamma(X, \mathscr{F}) | \text{ the support of } s \text{ is contained in } Y\}.$$

The functor $\Gamma_Y(X, -)$ is left exact. Denote its right derived functors by $H^q_Y(X, -)$ or $R^q\Gamma_Y(X, -)$. Let $R^q i^!$ be the right derived functors of $i^!$.

Proposition 5.6.11. *Notation as above.*

(i) *We have a long exact sequence*

$$\cdots \to H^q_Y(X, \mathscr{F}) \to H^q(X, \mathscr{F}) \to H^q(U, \mathscr{F}) \to \cdots.$$

(ii) *We have an exact sequence*

$$0 \to i_*i^!\mathscr{F} \to \mathscr{F} \to j_*j^*\mathscr{F} \to i_*R^1i^!\mathscr{F} \to 0$$

and isomorphisms

$$R^q j_*(j^*\mathscr{F}) \cong i_*(R^{q+1}i^!\mathscr{F}) \quad (q \geq 1).$$

(iii) *We have a biregular spectral sequence*

$$E_2^{pq} = H^p(Y, R^q i^! \mathscr{F}) \Rightarrow H^{p+q}_Y(X, \mathscr{F}).$$

Proof. Denote the constant sheaves on U and on X associated to $\mathbb{Z}$ both by $\mathbb{Z}$. Note that $j_!\mathbb{Z}$ is a subsheaf of $\mathbb{Z}$. For any injective sheaf $\mathscr{I}$ on X, the canonical homomorphism

$$\mathrm{Hom}(\mathbb{Z}, \mathscr{I}) \to \mathrm{Hom}(j_!\mathbb{Z}, \mathscr{I})$$

is surjective. So the restriction

$$\Gamma(X, \mathscr{I}) \to \Gamma(U, \mathscr{I})$$

is surjective. Similarly, for any etale X-scheme V, the restriction

$$\mathscr{I}(V) \to \mathscr{I}(U \times_X V)$$

is surjective. So the morphism

$$\mathscr{I} \to j_*j^*\mathscr{I}$$

is surjective. Let $\mathscr{I}^\cdot$ be an injective resolution of $\mathscr{F}$. We have a short exact sequence of complexes

$$0 \to \Gamma_Y(X, \mathscr{I}^\cdot) \to \Gamma(X, \mathscr{I}^\cdot) \to \Gamma(U, \mathscr{I}^\cdot) \to 0.$$

Taking the long exact sequence of cohomology groups associated to this short exact sequence, we get (i). We have a short exact sequence of complexes

$$0 \to i_*i^!\mathscr{I}^\cdot \to \mathscr{I}^\cdot \to j_*j^*\mathscr{I}^\cdot \to 0.$$

Taking the long exact sequence of cohomology sheaves associated to this short exact sequence, we get (ii). Since the functor i_* is exact and is left adjoint to $i^!$, the functor $i^!$ maps injective sheaves to injective sheaves. Moreover, we have

$$\Gamma(Y, i^! -) = \Gamma_Y(X, -).$$

By [Fu (2006)] 2.2.10 or [Grothendieck (1957)] 2.4.1, we have a biregular spectral sequence

$$E_2^{pq} = H^p(Y, R^q i^! \mathscr{F}) \Rightarrow H_Y^{p+q}(X, \mathscr{F}). \qquad \square$$

Proposition 5.6.12 (Excision Theorem). *Let $f : X' \to X$ be an etale morphism, $Y \to X$ a closed immersion, $Y' = X' \times_X Y$, and $\mathscr{F}$ a sheaf on X. Suppose that the projection $Y' \to Y$ is an isomorphism. Then we have*

$$H_Y^q(X, \mathscr{F}) \cong H_{Y'}^q(X', f^*\mathscr{F})$$

for all q.

Proof. Since $f_!$ is left adjoint to f^* and is exact, f^* maps injective sheaves to injective sheaves. Moreover f^* is exact. To prove our assertion, it suffices to prove that the canonical homomorphism

$$\Gamma_Y(X, \mathscr{F}) \to \Gamma_{Y'}(X', f^*\mathscr{F})$$

is bijective for any sheaf $\mathscr{F}$ on X. Suppose $s \in \Gamma_Y(X, \mathscr{F})$ and $s|_{X'} = 0$. We have $s|_{X-Y} = 0$. Since $X - Y \hookrightarrow X$ and $X' \to X$ form an etale covering of X, we have $s = 0$ and hence the above homomorphism is injective. Suppose $s' \in \Gamma_{Y'}(X', f^*\mathscr{F})$. Let us prove that $(s', 0)$ lies in the kernel of the homomorphism

$$\begin{aligned}\mathscr{F}(X') \times \mathscr{F}(X - Y) \stackrel{d_0}{\to} \mathscr{F}(X' \times_X X') \times \mathscr{F}(X' \times_X (X - Y)) \\ \times \mathscr{F}((X - Y) \times_X X') \times \mathscr{F}((X - Y) \times_X (X - Y))\end{aligned}$$

in the Čech complex for the etale covering $\{X - Y \hookrightarrow X,\ X' \to X\}$, and hence there exists $s \in \Gamma(X, \mathscr{F})$ such that $s|_{X'} = s'$ and $s|_{X-Y} = 0$. This shows that the above homomorphism is surjective. The nontrivial part is to verify that s' is mapped to the same section under the restrictions

$$p_i^* : \mathscr{F}(X') \to \mathscr{F}(X' \times_X X') \quad (i = 1, 2)$$

defined by the projections $p_i : X' \times_X X' \to X'$. Note that $X' \times_X X'$ can be covered by the open subscheme $(X' - Y') \times_{(X-Y)} (X' - Y')$ and the closed subscheme $Y' \times_Y Y'$. It is clear that $p_i^*(s')$ vanish on $(X' - Y') \times_{(X-Y)} (X' - Y')$. Since $Y' \to Y$ is an isomorphism, the two projections $Y' \times_Y Y' \to Y'$ are the same. So $p_1^*(s')$ and $p_2^*(s')$ have the same germ at every geometric point of $Y' \times_Y Y'$. We thus have $p_1^*(s') = p_2^*(s')$. $\square$

Let X be a scheme and let $P = \{\gamma_s : s \to X\}$ be a set of geometric points of X such that for every point x in X, there exists a geometric point γ_s in P with image x, where s are spectra of separably closed fields. For any sheaf $\mathscr{F}$ on X, let

$$\mathscr{C}^0(\mathscr{F}) = \prod_{s\in P} \gamma_{s*}\gamma_s^*\mathscr{F}.$$

The canonical morphism $\mathscr{F} \to \mathscr{C}^0(\mathscr{F})$ is injective. Let $\mathscr{C}^{-1}(\mathscr{F}) = \mathscr{F}$ and let $d^{-1} : \mathscr{C}^{-1}(\mathscr{F}) \to \mathscr{C}^0(\mathscr{F})$ be this monomorphism. Suppose we have defined an exact sequence

$$0 \to \mathscr{F} \stackrel{d^{-1}}{\to} \mathscr{C}^0(\mathscr{F}) \stackrel{d^0}{\to} \cdots \stackrel{d^{n-1}}{\to} \mathscr{C}^n(\mathscr{F})$$

for some $n \geq 0$. We define

$$\mathscr{C}^{n+1}(\mathscr{F}) = \mathscr{C}^0(\operatorname{coker}(\mathscr{C}^{n-1}(\mathscr{F}) \stackrel{d^{n-1}}{\to} \mathscr{C}^n(\mathscr{F})))$$

and define $d^{n+1} : \mathscr{C}^n(\mathscr{F}) \to \mathscr{C}^{n+1}(\mathscr{F})$ to be the composite

$$\mathscr{C}^n(\mathscr{F}) \to \operatorname{coker}(\mathscr{C}^{n-1}(\mathscr{F}) \stackrel{d^{n-1}}{\to} \mathscr{C}^n(\mathscr{F})) \to \mathscr{C}^0(\operatorname{coker}(\mathscr{C}^{n-1}(\mathscr{F}) \stackrel{d^{n-1}}{\to} \mathscr{C}^n(\mathscr{F}))).$$

Then we get a resolution

$$0 \to \mathscr{C}^0(\mathscr{F}) \stackrel{d^0}{\to} \cdots \stackrel{d^{n-1}}{\to} \mathscr{C}^n(\mathscr{F}) \stackrel{d^n}{\to} \mathscr{C}^{n+1}(\mathscr{F}) \stackrel{d^{n+1}}{\to} \cdots$$

of $\mathscr{F}$. We denote it by $\mathscr{C}^\cdot(\mathscr{F})$ and call it the *Godement resolution* of $\mathscr{F}$.

Let A be a ring,

$$K^\cdot = (\cdots \to K^i \stackrel{d^i}{\to} K^{i+1} \to \cdots)$$

an acyclic complex of A-modules, and $Z^i = \ker d^i$. Then we have short exact sequences

$$0 \to Z^i \to K^i \to Z^{i+1} \to 0.$$

We say $K^\cdot$ is *split* if all these short exact sequences are split.

Proposition 5.6.13.

(i) *The functors* $\mathscr{F} \mapsto \mathscr{C}^n(\mathscr{F})$ *and* $\mathscr{F} \mapsto \operatorname{im}(\mathscr{C}^{n-1}(\mathscr{F}) \stackrel{d^{n-1}}{\to} \mathscr{C}^n(\mathscr{F}))$ *are exact for all* $n \geq 0$.

(ii) $\mathscr{C}^n(\mathscr{F})$ $(n \geq 0)$ *are flasque.*

(iii) *The stalk of the complex*

$$0 \to \mathscr{F} \to \mathscr{C}^\cdot(\mathscr{F})$$

at every geometric point s *in* P *is split.*

Proof.

(i) For any object U in X_{et} and any geometric point s in P, $s \times_X U$ is a disjoint union of copies of s, and we have

$$(\gamma_{s*}\gamma_s^*\mathscr{F})(U) = (\gamma_s^*\mathscr{F})(s \times_X U).$$

It follows that $\gamma_{s*}\gamma_s^*$ is an exact functor and hence $\mathscr{C}^0(-)$ is an exact functor. Suppose that

$$0 \to \mathscr{F} \to \mathscr{G} \to \mathscr{H} \to 0$$

is an exact sequence. Applying the snake lemma to the commutative diagram

$$\begin{array}{ccccccccc} 0 \to & \mathscr{F} & \to & \mathscr{G} & \to & \mathscr{H} & \to 0 \\ & \downarrow & & \downarrow & & \downarrow & \\ 0 \to & \mathscr{C}^0(\mathscr{F}) & \to & \mathscr{C}^0(\mathscr{G}) & \to & \mathscr{C}^0(\mathscr{H}) & \to 0, \end{array}$$

and using the fact that the vertical arrows are monomorphisms, we get an exact sequence

$$0 \to \mathrm{coker}\,(\mathscr{F} \to \mathscr{C}^0(\mathscr{F})) \to \mathrm{coker}\,(\mathscr{G} \to \mathscr{C}^0(\mathscr{G})) \to \mathrm{coker}\,(\mathscr{H} \to \mathscr{C}^0(\mathscr{H})) \to 0.$$

Suppose we have proved that

$$0 \to \mathscr{C}^n(\mathscr{F}) \to \mathscr{C}^n(\mathscr{G}) \to \mathscr{C}^n(\mathscr{H}) \to 0,$$

$$0 \to \mathrm{coker}\,(\mathscr{C}^{n-1}(\mathscr{F}) \to \mathscr{C}^n(\mathscr{F})) \to \mathrm{coker}\,(\mathscr{C}^{n-1}(\mathscr{G}) \to \mathscr{C}^n(\mathscr{G})) \to \mathrm{coker}\,(\mathscr{C}^{n-1}(\mathscr{H}) \to \mathscr{C}^n(\mathscr{H})) \to 0$$

are exact. Applying $\mathscr{C}^0(-)$ to the second exact sequence, we get an exact sequence

$$0 \to \mathscr{C}^{n+1}(\mathscr{F}) \to \mathscr{C}^{n+1}(\mathscr{G}) \to \mathscr{C}^{n+1}(\mathscr{H}) \to 0.$$

Applying the snake lemma to the commutative diagram

$$\begin{array}{ccccccccc} 0 \to & \mathrm{coker}(\mathscr{C}^{n-1}(\mathscr{F}) \to \mathscr{C}^n(\mathscr{F})) & \to & \mathrm{coker}(\mathscr{C}^{n-1}(\mathscr{G}) \to \mathscr{C}^n(\mathscr{G})) & \to & \mathrm{coker}(\mathscr{C}^{n-1}(\mathscr{H}) \to \mathscr{C}^n(\mathscr{H})) & \to 0 \\ & \downarrow & & \downarrow & & \downarrow & \\ 0 \to & \mathscr{C}^{n+1}(\mathscr{F}) & \to & \mathscr{C}^{n+1}(\mathscr{G}) & \to & \mathscr{C}^{n+1}(\mathscr{H}) & \to 0 \end{array}$$

and using the fact that vertical arrows are monomorphisms, we get an exact sequence

$$0 \to \mathrm{coker}\,(\mathscr{C}^n(\mathscr{F}) \to \mathscr{C}^{n+1}(\mathscr{F})) \to \mathrm{coker}\,(\mathscr{C}^n(\mathscr{G}) \to \mathscr{C}^{n+1}(\mathscr{G})) \to \mathrm{coker}\,(\mathscr{C}^n(\mathscr{H}) \to \mathscr{C}^{n+1}(\mathscr{H})) \to 0.$$

So the functors $\mathscr{F} \mapsto \mathscr{C}^n(\mathscr{F})$ and $\mathscr{F} \mapsto \mathrm{coker}\,(\mathscr{C}^{n-1}(\mathscr{F}) \to \mathscr{C}^n(\mathscr{F}))$ are exact for all $n \geq 0$. We have

$$\mathrm{coker}\,(\mathscr{C}^{n-1}(\mathscr{F}) \to \mathscr{C}^n(\mathscr{F})) \cong \mathrm{im}\,(\mathscr{C}^n(\mathscr{F}) \to \mathscr{C}^{n+1}(\mathscr{F})).$$

So the functors $\mathscr{F} \mapsto \operatorname{im}(\mathscr{C}^n(\mathscr{F}) \to \mathscr{C}^{n+1}(\mathscr{F}))$ are exact for all $n \geq 0$. For $n = -1$, this functor can be identified with the identity functor and hence is exact.

(ii) It is enough to show that $\mathscr{C}^0(\mathscr{F})$ is flasque. For any $s \in P$, the functor $\Gamma(s, -)$ is exact and hence $H^i(s, -) = 0$ for all $i \geq 1$. So any sheaf on s is flasque. By 5.6.5 (i), $\gamma_{s*}\gamma_s^*\mathscr{F}$ is flasque. This implies that $\mathscr{C}^0(\mathscr{F}) = \prod_{s\in P} \gamma_{s*}\gamma_s^*\mathscr{F}$ is flasque.

(iii) We first prove the following fact: For any morphism $f : X \to Y$ and any sheaf $\mathscr{F}$ on Y, the diagram

$$\begin{array}{ccc} f^*\mathscr{F} & \to & f^*(f_*f^*\mathscr{F}) \\ \scriptstyle\text{id}\| & & \|\scriptstyle\text{id} \\ f^*\mathscr{F} & \leftarrow & (f^*f_*)f^*\mathscr{F} \end{array}$$

commutes, where the horizontal arrows are induced by the canonical morphisms $\text{id} \stackrel{\text{adj}}{\to} f_*f^*$ and $f^*f_* \stackrel{\text{adj}}{\to} \text{id}$, respectively. Indeed, the composite

$$f^*\mathscr{F} \to f^*(f_*f^*\mathscr{F}) = (f^*f_*)f^*\mathscr{F} \to f^*\mathscr{F}$$

is induced by the composite

$$\mathscr{F} \stackrel{\text{adj}}{\to} f_*f^*\mathscr{F} \stackrel{\text{id}}{\to} f_*f^*\mathscr{F}$$

by adjunction. It is clear that $\text{id} : f^*\mathscr{F} \to f^*\mathscr{F}$ is also induced by the morphism $\mathscr{F} \stackrel{\text{adj}}{\to} f_*f^*\mathscr{F}$ by adjunction. So the above diagram commutes.

Let $t \in P$. Consider the morphism $\gamma_t^*\mathscr{C}^0(\mathscr{F}) \to \gamma_t^*\mathscr{F}$ defined by composing the projection

$$\gamma_t^*\mathscr{C}^0(\mathscr{F}) = \gamma_t^*\Big(\prod_{s\in P} \gamma_{s*}\gamma_s^*\mathscr{F}\Big) \to \gamma_t^*(\gamma_{t*}\gamma_t^*\mathscr{F})$$

with the canonical morphism

$$\gamma_t^*(\gamma_{t*}\gamma_t^*\mathscr{F}) = (\gamma_t^*\gamma_{t*})\gamma_t^*\mathscr{F} \to \gamma_t^*\mathscr{F}.$$

By the above discussion, this morphism is a left inverse of the stalk at t of the morphism $\mathscr{F} \to \mathscr{C}^0(\mathscr{F})$. It follows that the short exact sequence

$$0 \to \mathscr{F}_t \to \mathscr{C}^0(\mathscr{F})_t \to \ker(\mathscr{C}^1(\mathscr{F}) \to \mathscr{C}^2(\mathscr{F}))_t \to 0$$

is split. Applying this result to $\operatorname{coker}(\mathscr{C}^{n-2}(\mathscr{F}) \to \mathscr{C}^{n-1}(\mathscr{F}))$, we see that the short exact sequences

$$0\to\ker(\mathscr{C}^n(\mathscr{F})\to\mathscr{C}^{n+1}(\mathscr{F}))_t\to\mathscr{C}^n(\mathscr{F})_t\to\ker(\mathscr{C}^{n+1}(\mathscr{F})\to\mathscr{C}^{n+2}(\mathscr{F}))_t\to 0$$

are split for all n. □

Let S be a scheme. For every $s \in S$, fix an algebraic closure $\overline{k(s)}$ of the residue field $k(s)$. For every S-scheme X locally of finite type, let

$$P_X = \bigcup_{s\in S} \mathrm{Hom}_S(\mathrm{Spec}\,\overline{k(s)}, X)$$

be the set of geometric points above the geometric points $\mathrm{Spec}\,\overline{k(s)} \to S$ in S. For every sheaf $\mathscr{F}$ on X, we define a complex of sheaves $\mathscr{C}^{\cdot}(\mathscr{F})$ and a morphism $\mathscr{F} \to \mathscr{C}^0(\mathscr{F})$ as before using P_X instead of P. Then $\mathscr{C}^{\cdot}(\mathscr{F})$ is resolution of $\mathscr{F}$ (see the discussion below) and 5.6.13 still holds. We also call $\mathscr{C}^{\cdot}(\mathscr{F})$ the Godement resolution of $\mathscr{F}$. This resolution has the advantage that it is functorial, that is, if $f : X' \to X$ is an S-morphism between S-schemes locally of finite type, then for any sheaf $\mathscr{F}$ on X, we have a canonical morphism of complexes

$$\mathscr{C}^{\cdot}(\mathscr{F}) \to f_*\mathscr{C}^{\cdot}(f^*\mathscr{F}).$$

Let us prove that the morphism $\mathscr{F} \to \mathscr{C}^0(\mathscr{F})$ is injective. Let $\pi : U \to X$ be an object in X_{et}, and let $\xi \in \mathscr{F}(U)$ such that its image in $\mathscr{C}^0(\mathscr{F})(U)$ is 0. We need to show $\xi = 0$. Let $f : X \to S$ be the structure morphism. For every $u \in U$ such that u is closed in the fiber $(f\pi)^{-1}(f\pi(u))$, $k(u)$ is a finite extension of $k(f(\pi(u)))$. So there exists an S-morphism $\mathrm{Spec}\,\overline{k(f\pi(u))} \to U$ whose image is u. Composing this morphism with π, we get a geometric point in P_X. Since the image of ξ in

$$\mathscr{C}^0(\mathscr{F})(U) = \prod_{t\in P_X} (\gamma_{t*}\gamma_t^*\mathscr{F})(U)$$

is 0, the germ of ξ at the geometric point $\bar{u}$ is 0. So there exists an open neighborhood V_u of u in U such that $\xi|_{V_u} = 0$. Let $V = \bigcup_{u\in U} V_u$, where u goes over all points in U which are closed in the fibers $(f\pi)^{-1}(f\pi(u))$. Then V is an open subset of U and $\xi|_V = 0$. To prove $\xi = 0$, it suffices to show $V = U$. For any $s \in S$, $(f\pi)^{-1}(s)$ is a $k(s)$-scheme locally of finite type, and $V \cap (f\pi)^{-1}(s)$ is an open subset containing all the closed points of $(f\pi)^{-1}(s)$. It follows that $V \cap (f\pi)^{-1}(s) = (f\pi)^{-1}(s)$.

5.7 Calculation of Etale Cohomology

([SGA 4] VII 4, VIII 1, 2, [SGA $4\frac{1}{2}$] Arcata I 5, III 1.)

Proposition 5.7.1. *Let $f : X \to Y$ be a finite surjective radiciel morphism. For any sheaf $\mathscr{F}$ on Y, we have $H^i(Y, \mathscr{F}) \cong H^i(X, f^*\mathscr{F})$.*

Proof. Use 5.3.10. □

Corollary 5.7.2.

(i) *Let X be a scheme. Then $H^i(X, \mathscr{F}) \cong H^i(X_{\text{red}}, \mathscr{F}|_{X_{\text{red}}})$ for any sheaf $\mathscr{F}$ on X.*

(ii) *Let X be a scheme over a field K and let L be a finite purely inseparable extension of K. Then $H^i(X, \mathscr{F}) \cong H^i(X \otimes_K L, \mathscr{F}|_{X\otimes_K L})$ for any sheaf $\mathscr{F}$ on X.*

Proposition 5.7.3. *Let A be a strict henselian local ring. For any sheaf $\mathscr{F}$ on $X = \operatorname{Spec} A$, we have $H^i(X, \mathscr{F}) = 0$ for all $i \geq 1$.*

Proof. Let s be the closed point of X. We have $\Gamma(X, \mathscr{F}) = \mathscr{F}_s$. Hence $\Gamma(X, -)$ is an exact functor. Our assertion follows. □

Proposition 5.7.4. *Let $f : X \to Y$ be a finite morphism and let $\mathscr{F}$ be a sheaf on X. We have $R^i f_* \mathscr{F} = 0$ for all $i \geq 1$, and $H^i(Y, f_*\mathscr{F}) \cong H^i(X, \mathscr{F})$ for all i.*

Proof. The first assertion follows from the fact that f_* is exact (5.3.7), and the second assertion follows from the first one and 5.6.9 (i). □

Let X be a scheme. For any etale sheaf $\mathscr{F}$ on X, define a sheaf $i_*\mathscr{F}$ with respect to the Zariski topology by

$$(i_*\mathscr{F})(V) = \mathscr{F}(V)$$

for any open subset V of X. It is clear that i_* is left exact. For every Zariski presheaf $\mathscr{G}$ on X, define an etale presheaf $i^{\mathscr{P}}\mathscr{G}$ by

$$(i^{\mathscr{P}}\mathscr{G})(U) = \mathscr{G}(\operatorname{im}(U \to X))$$

for any object U in X_{et}. Let $i^*\mathscr{G} = (i^{\mathscr{P}}\mathscr{G})^{\#}$ be the sheaf associated to $i^{\mathscr{P}}\mathscr{G}$. We have a one-to-one correspondence

$$\operatorname{Hom}(\mathscr{G}, i_*\mathscr{F}) \cong \operatorname{Hom}(i^*\mathscr{G}, \mathscr{F}).$$

So i^* is left adjoint to i_*. Let us prove i^* is exact. Given an exact sequence of Zariski sheaves

$$0 \to \mathscr{G}' \to \mathscr{G} \to \mathscr{G}'' \to 0,$$

let $\mathscr{C}$ be the Zariski presheaf defined by

$$\mathscr{C}(U) = \operatorname{coker}(\mathscr{G}(U) \to \mathscr{G}''(U)).$$

Then

$$0 \to i^{\mathscr{P}}\mathscr{G}' \to i^{\mathscr{P}}\mathscr{G} \to i^{\mathscr{P}}\mathscr{G}'' \to i^{\mathscr{P}}\mathscr{C} \to 0$$

is an exact sequence of presheaves. So

$$0 \to i^*\mathscr{G}' \to i^*\mathscr{G} \to i^*\mathscr{G}'' \to i^*\mathscr{C} \to 0$$

is an exact sequence of sheaves. To prove that i^* is exact, it suffices to show $(i^{\mathscr{P}}\mathscr{C})^{\#} = 0$. This follows from the fact that the Zariski sheaf associated to the Zariski presheaf $\mathscr{C}$ is 0. So i_* maps injective etale sheaves to injective Zariski sheaves. For any etale sheaf $\mathscr{F}$ on X, we have a biregular spectral sequence

$$E_2^{pq} = H^p_{\mathrm{Zar}}(X, R^q i_*\mathscr{F}) \Rightarrow H^{p+q}_{\mathrm{et}}(X, \mathscr{F}).$$

Proposition 5.7.5. *Let X be a scheme, $\mathscr{M}$ a quasi-coherent $\mathscr{O}_X$-module, and $\mathscr{M}_{\mathrm{et}}$ the etale sheaf defined by $U \mapsto \Gamma(U, \pi^*M)$ for any object $\pi : U \to X$ in X_{et}. We have $H^q_{\mathrm{Zar}}(X, \mathscr{M}) \cong H^q_{\mathrm{et}}(X, \mathscr{M}_{\mathrm{et}})$ for all q.*

Proof. It suffices to show that $R^q i_*\mathscr{M}_{\mathrm{et}} = 0$ for any $q \geq 1$. Note that $R^q i_*\mathscr{M}_{\mathrm{et}}$ is the Zariski sheaf associated to the Zariski presheaf defined by $U \mapsto H^q_{\mathrm{et}}(U, \mathscr{M}_{\mathrm{et}})$ for open subsets U of X. So it suffices to show $H^q_{\mathrm{et}}(U, \mathscr{M}_{\mathrm{et}}) = 0$ for any $q \geq 1$, any affine scheme U, and any quasi-coherent $\mathscr{O}_U$-module $\mathscr{M}$. By 5.7.6 below, for any etale covering $\mathfrak{U} = \{U_\alpha \to U\}_{\alpha \in I}$ with I being finite and with U and each U_α being affine, we have $\check{H}^q(\mathfrak{U}, \mathscr{M}_{\mathrm{et}}) = 0$ for all $q \geq 1$. So we have

$$\check{H}^q(U, \mathscr{M}_{\mathrm{et}}) \cong \varinjlim_{\mathfrak{U}} \check{H}^q(\mathfrak{U}, \mathscr{M}_{\mathrm{et}}) = 0$$

for all $q \geq 1$. We have

$$H^1_{\mathrm{et}}(U, \mathscr{M}_{\mathrm{et}}) \cong \check{H}^1(U, \mathscr{M}_{\mathrm{et}})$$

by 5.6.3. So we have $H^1_{\mathrm{et}}(U, \mathscr{M}_{\mathrm{et}}) = 0$. Suppose we have shown that $H^q_{\mathrm{et}}(U, \mathscr{M}_{\mathrm{et}}) = 0$ for any $1 \leq q \leq n$. Consider the biregular spectral sequence

$$E_2^{pq} = \check{H}^p_{\mathrm{et}}(U, \mathscr{H}^q(\mathscr{M}_{\mathrm{et}})) \Rightarrow H^{p+q}_{\mathrm{et}}(U, \mathscr{M}_{\mathrm{et}})$$

in 5.6.2. We have

$$\check{H}^p_{\mathrm{et}}(U, \mathscr{H}^q(\mathscr{M}_{\mathrm{et}})) = \varinjlim_{\mathfrak{U}} \check{H}^p(\mathfrak{U}, \mathscr{H}^q(\mathscr{M}_{\mathrm{et}})),$$

where $\mathfrak{U}$ goes over etale coverings $\mathfrak{U} = \{U_\alpha \to U\}_{\alpha \in I}$ of U with I being finite and with each U_α being affine. Each $U_{\alpha_0 \ldots \alpha_n}$ is affine. So

$$\mathscr{H}^q(\mathscr{M}_{\mathrm{et}})(U_{\alpha_0 \ldots \alpha_n}) = H^q_{\mathrm{et}}(U_{\alpha_0 \ldots \alpha_n}, \mathscr{M}_{\mathrm{et}}) = 0$$

for any $1 \leq q \leq n$. Hence $\check{H}^p(\mathfrak{U}, \mathscr{H}^q(\mathscr{M}_{\text{et}})) = 0$ and $E_2^{pq} = 0$ for any p and $1 \leq q \leq n$. We have an exact sequence

$$0 \to E_2^{n+1,0} \to H_{\text{et}}^{n+1}(U, \mathscr{M}_{\text{et}}) \to E_2^{0,n+1}$$

by [Fu (2006)] 2.2.3. Moreover, we have

$$E_2^{0,n+1} = \check{H}_{\text{et}}^0(U, \mathscr{H}^{n+1}(\mathscr{M}_{\text{et}})) = \mathscr{H}^{n+1}(\mathscr{M}_{\text{et}})^+(U) = 0$$

by 5.6.1. So we have

$$H_{\text{et}}^{n+1}(U, \mathscr{M}_{\text{et}}) \cong E_2^{n+1,0} = \check{H}_{\text{et}}^{n+1}(U, \mathscr{M}_{\text{et}}) = 0.$$

□

Lemma 5.7.6. *Let $A \to B$ be a faithfully flat homomorphism of rings and let M be an A-module. Then the sequence*

$$0 \to M \stackrel{d_{-1}}{\to} M \otimes_A B \stackrel{d_0}{\to} M \otimes_A B \otimes_A B \stackrel{d_1}{\to} M \otimes_A B \otimes_A B \otimes_A B \stackrel{d_2}{\to} \cdots$$

is exact, where

$$d_n(x \otimes b_0 \otimes \cdots \otimes b_n) = \sum_{i=0}^{n+1} (-1)^i x \otimes b_0 \otimes \cdots \otimes b_{i-1} \otimes 1 \otimes b_i \otimes \cdots \otimes b_n.$$

Proof. Since B is faithfully flat over A, it suffices to prove that the sequence is exact after tensoring with B, that is, the sequence

$$0 \to M \otimes_A B \stackrel{d_{-1}}{\to} M \otimes_A B \otimes_A B \stackrel{d_0}{\to} M \otimes_A B \otimes_A B \otimes_A B \stackrel{d_1}{\to} M \otimes_A B \otimes_A B \otimes_A B \otimes_A B \stackrel{d_2}{\to} \cdots$$

is exact, where

$$d_n(x \otimes b_0 \otimes \cdots \otimes b_n \otimes b) = \sum_{i=0}^{n+1} (-1)^i x \otimes b_0 \otimes \cdots \otimes b_{i-1} \otimes 1 \otimes b_i \otimes \cdots \otimes b_n \otimes b.$$

Define

$$D_n : M \otimes_A B^{\otimes(n+2)} \to M \otimes_A B^{\otimes(n+1)}$$

by

$$D_n(x \otimes b_0 \otimes \cdots \otimes b_n \otimes b) = x \otimes b_1 \otimes \cdots \otimes b_n \otimes b_0 b.$$

Then

$$d_{n-1} D_n + D_{n+1} d_n = \text{id}.$$

Our assertion follows. □

Proposition 5.7.7. *Let X be a scheme and let $\mathscr{O}_{X_{\text{et}}}^*$ be the etale sheaf defined by $U \mapsto \mathscr{O}_U(U)^*$ for any object U in X_{et}. We have*

$$H_{\text{et}}^1(X, \mathscr{O}_{X_{\text{et}}}^*) \cong H_{\text{Zar}}^1(X, \mathscr{O}_X^*) \cong \text{Pic}\,(X).$$

Proof. It is known that $H^1_{\text{Zar}}(X, \mathscr{O}_X^*) \cong \text{Pic}(X)$ ([Fu (2006)] 2.3.12). By [Fu (2006)] 2.2.3 applied to the spectral sequence

$$E_2^{pq} = H^p_{\text{Zar}}(X, R^q i_* \mathscr{O}^*_{X_{\text{et}}}) \Rightarrow H^{p+q}_{\text{et}}(X, \mathscr{O}^*_{X_{\text{et}}}),$$

we have an exact sequence

$$0 \to H^1_{\text{Zar}}(X, \mathscr{O}^*_X) \to H^1_{\text{et}}(X, \mathscr{O}^*_{X_{\text{et}}}) \to H^0_{\text{Zar}}(X, R^1 i_* \mathscr{O}^*_{X_{\text{et}}}).$$

To prove our assertion, it suffices to show that the homomorphism

$$H^1_{\text{et}}(X, \mathscr{O}^*_{X_{\text{et}}}) \to H^0_{\text{Zar}}(X, R^1 i_* \mathscr{O}^*_{X_{\text{et}}})$$

is 0. Since $R^1 i_* \mathscr{O}^*_{X_{\text{et}}}$ is the Zariski sheaf associated to the Zariski presheaf $V \mapsto H^1_{\text{et}}(V, \mathscr{O}^*_{X_{\text{et}}})$ for any open subset V of X, it suffices to show that for any element $\xi \in H^1_{\text{et}}(X, \mathscr{O}^*_{X_{\text{et}}})$, there exists an open covering $\{V_\lambda\}$ of X such that the image of ξ under the canonical homomorphism

$$H^1_{\text{et}}(X, \mathscr{O}^*_{X_{\text{et}}}) \to H^1_{\text{et}}(V_\lambda, \mathscr{O}^*_{X_{\text{et}}})$$

is 0 for every λ. The problem is local. So we may assume that X is affine. We have

$$H^1_{\text{et}}(X, \mathscr{O}^*_{X_{\text{et}}}) \cong \check{H}^1_{\text{et}}(X, \mathscr{O}^*_{X_{\text{et}}}) \cong \varinjlim_{\mathfrak{U}} \check{H}^1_{\text{et}}(\mathfrak{U}, \mathscr{O}^*_{X_{\text{et}}}),$$

where $\mathfrak{U}$ goes over the family of etale coverings of the form $\mathfrak{U} = \{U_\alpha \to X\}_{\alpha \in I}$ such that I is finite and U_α are affine. Suppose that an element $\xi \in H^1_{\text{et}}(X, \mathscr{O}^*_{X_{\text{et}}})$ is represented by the image of an element

$$(\xi_{\alpha\beta}) \in \ker\Big(\prod_{\alpha,\beta} \mathscr{O}^*(U_{\alpha\beta}) \to \prod_{\alpha,\beta,\gamma} \mathscr{O}^*(U_{\alpha\beta\gamma})\Big) = \check{H}^1(\mathfrak{U}, \mathscr{O}^*_{X_{\text{et}}}).$$

The X-scheme $U = \coprod_{\alpha \in I} U_\alpha$ is quasi-compact and faithfully flat over X. Let

$$\begin{aligned} &p : U \to X, \quad p_i : U \times_X U \to U \quad (i = 1, 2), \\ &p_{ij} : U \times_X U \times_X U \to U \times_X U \quad (1 \leq i < j \leq 3) \end{aligned}$$

be the projections. We have

$$U \times_X U = \coprod_{\alpha,\beta} U_{\alpha\beta}, \quad U \times_X U \times_X U = \coprod_{\alpha,\beta,\gamma} U_{\alpha\beta\gamma}.$$

Define an isomorphism $\sigma : p_1^* \mathscr{O}_U \stackrel{\cong}{\to} p_2^* \mathscr{O}_U$ so that on each $U_{\alpha\beta}$, the following diagram commutes:

$$\begin{array}{ccc} p_1^* \mathscr{O}_U|_{U_{\alpha\beta}} & \stackrel{\sigma}{\to} & p_2^* \mathscr{O}_U|_{U_{\alpha\beta}} \\ \cong\downarrow & & \downarrow\cong \\ \mathscr{O}_{U_{\alpha\beta}} & \stackrel{\xi_{\alpha\beta}}{\to} & \mathscr{O}_{U_{\alpha\beta}}, \end{array}$$

where the vertical arrows are the canonical isomorphisms, and the bottom horizontal arrow is the multiplication by $\xi_{\alpha\beta} \in \mathscr{O}^*(U_{\alpha\beta})$. Since $(\xi_{\alpha\beta})$ lies in the kernel of the homomorphism $\prod_{\alpha,\beta} \mathscr{O}^*(U_{\alpha\beta}) \to \prod_{\alpha,\beta,\gamma} \mathscr{O}^*(U_{\alpha\beta\gamma})$ in the Čech complex, we have

$$p_{13}^*(\sigma) \cong p_{23}^*(\sigma) \circ p_{12}^*(\sigma).$$

By 1.6.1 (ii), there exist a quasi-coherent $\mathscr{O}_X$-module $\mathscr{L}$ and an isomorphism $\tau : p^*\mathscr{L} \xrightarrow{\cong} \mathscr{O}_U$ such that $\sigma \circ p_1^*\tau = p_2^*\tau$. By 1.6.6, $\mathscr{L}$ is an invertible $\mathscr{O}_X$-module. So there exists an open covering $\{V_\lambda\}$ of X such that $\mathscr{L}|_{V_\lambda}$ is isomorphic to the trivial invertible $\mathscr{O}_{V_\lambda}$-module for each λ. One can check that the image of ξ in each $\check{H}^1_{\mathrm{et}}(V_\lambda, \mathscr{O}^*_{X_{\mathrm{et}}})$ is 0. □

Let K be a field. Choose a separable closure K_s of K. It defines a geometric point $s : \operatorname{Spec} K_s \to \operatorname{Spec} K$. The galois group $\mathrm{Gal}(K_s/K)$ acts on K_s on the left, and hence on $\operatorname{Spec} K_s$ on the right. For any sheaf $\mathscr{F}$ on $\operatorname{Spec} K$, $\mathrm{Gal}(K_s/K)$ acts on the stalk $\mathscr{F}_s$ on the left. The action is continuous with respect to the discrete topology on $\mathscr{F}_s$. Indeed, we have

$$\mathscr{F}_s = \varinjlim_{K'} \mathscr{F}(\operatorname{Spec} K'),$$

where K' goes over the set of finite extensions of K contained in K_s, and the open subgroup $\mathrm{Gal}(K_s/K')$ of $\mathrm{Gal}(K_s/K)$ acts trivially on $\mathscr{F}(\operatorname{Spec} K')$.

Proposition 5.7.8. *Notation as above. The functor $\mathscr{F} \mapsto \mathscr{F}_s$ defines an equivalence between the category of etale sheaves on* $\operatorname{Spec} K$ *and the category of* $\mathrm{Gal}(K_s/K)$*-modules. We have*

$$H^q(\operatorname{Spec} K, \mathscr{F}) \cong H^q(\mathrm{Gal}(K_s/K), \mathscr{F}_s)$$

for all q.

Proof. Let K' be a finite separable extension of K contained in K_s. We claim that

$$\mathscr{F}(\operatorname{Spec} K') \cong \mathscr{F}_s^{\mathrm{Gal}(K_s/K')}$$

for any sheaf $\mathscr{F}$ on $\operatorname{Spec} K$. We have

$$\mathscr{F}_s = \varinjlim_{L} \mathscr{F}(\operatorname{Spec} L), \quad \mathrm{Gal}(K_s/K') = \varprojlim_{L} \mathrm{Gal}(L/K'),$$

where L goes over the set of finite galois extensions of K' contained in K_s. Given $\alpha \in \mathscr{F}_s^{\mathrm{Gal}(K_s/K')}$, let L be a finite galois extension of K' contained in K_s such that $\alpha \in \mathscr{F}(\operatorname{Spec} L)$. Note that $\mathrm{Gal}(K_s/K')$ acts on $\mathscr{F}(\operatorname{Spec} L)$

via the finite quotient $\mathrm{Gal}(L/K')$ and α is invariant under the action of $\mathrm{Gal}(L/K')$. It follows that α lies in the kernel of the homomorphism

$$\mathscr{F}(\mathrm{Spec}\, L) \stackrel{d_0}{\to} \mathscr{F}(\mathrm{Spec}\,(L\otimes_{K'} L))$$

in the Čech complex. Since the sequence

$$0 \to \mathscr{F}(\mathrm{Spec}\, K') \to \mathscr{F}(\mathrm{Spec}\, L) \stackrel{d_0}{\to} \mathscr{F}(\mathrm{Spec}\,(L\otimes_{K'} L))$$

is exact, we have $\alpha \in \mathscr{F}(\mathrm{Spec}\, K')$. So we have $\mathscr{F}_s^{\mathrm{Gal}(K_s/K')} \subset \mathscr{F}(\mathrm{Spec}\, K')$. It is clear that $\mathscr{F}(\mathrm{Spec}\, K') \subset \mathscr{F}_s^{\mathrm{Gal}(K_s/K')}$. So $\mathscr{F}(\mathrm{Spec}\, K') = \mathscr{F}_s^{\mathrm{Gal}(K_s/K')}$. Using this fact, one can show that for all sheaves $\mathscr{F}_1$ and $\mathscr{F}_2$ on $\mathrm{Spec}\, K$, the canonical map

$$\mathrm{Hom}(\mathscr{F}_1, \mathscr{F}_2) \to \mathrm{Hom}_{\mathrm{Gal}(K_s/K)}(\mathscr{F}_{1s}, \mathscr{F}_{2s})$$

is bijective.

Let M be a $\mathrm{Gal}(K_s/K)$-module. For any finite extension K' of K contained in K_s, define $\mathscr{F}(\mathrm{Spec}\, K') = M^{\mathrm{Gal}(K_s/K')}$. Every etale K-scheme is isomorphic to a disjoint union of K-schemes of the form $\mathrm{Spec}\, K'$. We define

$$\mathscr{F}(\coprod_i \mathrm{Spec}\, K_i) = \prod_i \mathscr{F}(\mathrm{Spec}\, K_i)$$

for all finite extensions K_i of K contained in K_s. Let K_1 and K_2 be two finite extensions of K contained in K_s, and let

$$f : \mathrm{Spec}\, K_1 \to \mathrm{Spec}\, K_2$$

be a K-morphism. Then f is induced by a K-homomorphism $\sigma : K_2 \to K_1$. We can extend σ to an element $\bar{\sigma} : K_s \to K_s$ in $\mathrm{Gal}(K_s/K)$. Define the restriction

$$\mathscr{F}(\mathrm{Spec}\, K_2) \to \mathscr{F}(\mathrm{Spec}\, K_1)$$

to be the homomorphism

$$M^{\mathrm{Gal}(K_s/K_2)} \to M^{\mathrm{Gal}(K_s/K_1)}, \quad x \mapsto \bar{\sigma}(x).$$

This definition is independent of the choice of $\bar{\sigma}$. In this way, we get a presheaf $\mathscr{F}$ on $\mathrm{Spec}\, K$. For any finite galois extension L of K' contained in K_s, the sequence

$$0 \to \mathscr{F}(\mathrm{Spec}\, K') \to \mathscr{F}(\mathrm{Spec}\, L) \stackrel{d_0}{\to} \mathscr{F}(\mathrm{Spec}\,(L\otimes_{K'} L))$$

is exact by our construction. So we have

$$\mathscr{F}(\mathrm{Spec}\, K') \cong \check{H}^0(\{\mathrm{Spec}\, L \to \mathrm{Spec}\, K'\}, \mathscr{F}).$$

Any etale covering $\mathfrak{U}$ of $\operatorname{Spec} K'$ has a refinement of the form $\{\operatorname{Spec} L \to \operatorname{Spec} K'\}$, and the above isomorphism coincides with the composite

$$\mathscr{F}(\operatorname{Spec} K') \to \check{H}^0(\mathfrak{U}, \mathscr{F}) \to \check{H}^0(\{\operatorname{Spec} L \to \operatorname{Spec} K'\}, \mathscr{F}).$$

So the homomorphism

$$\mathscr{F}(\operatorname{Spec} K') \to \check{H}^0(\mathfrak{U}, \mathscr{F})$$

is injective, that is, $\mathscr{F}$ has the property (+) in 5.2.3. By the claim in the proof of 5.2.3 (ii), the homomorphism

$$\check{H}^0(\mathfrak{U}, \mathscr{F}) \to \check{H}^0(\{\operatorname{Spec} L \to \operatorname{Spec} K'\}, \mathscr{F})$$

is injective. This implies that the homomorphism

$$\mathscr{F}(\operatorname{Spec} K') \to \check{H}^0(\mathfrak{U}, \mathscr{F})$$

is bijective. So $\mathscr{F}$ is a sheaf. We have $\mathscr{F}_s \cong M$. □

The following corollary also follows from 4.5.4.

Corollary 5.7.9 (Hilbert 90). *We have $H^1(\operatorname{Gal}(K_s/K), K_s^*) = 0$ for any field K.*

Proof. Use 5.7.7, 5.7.8, and the fact that $\operatorname{Pic}(\operatorname{Spec} K) = 0$. □

Let X be a scheme and let $\mathscr{G}$ be a sheaf of groups (not necessarily commutative) on X. Given an etale covering $\mathfrak{U} = \{U_\alpha \to X\}_{\alpha \in I}$ of X, an element $(g_{\alpha_0\alpha_1}) \in \prod_{\alpha_0, \alpha_1 \in I} \mathscr{G}(U_{\alpha_0\alpha_1})$ is called a 1-cocycle if

$$g_{\alpha_1\alpha_2}|_{U_{\alpha_0\alpha_1\alpha_2}} \cdot g_{\alpha_0\alpha_1}|_{U_{\alpha_0\alpha_1\alpha_2}} = g_{\alpha_0\alpha_2}|_{U_{\alpha_0\alpha_1\alpha_2}}$$

for any $\alpha_j \in I$ $(j = 0, 1, 2)$. Two 1-cocycles $(g_{\alpha_0\alpha_1})$ and $(g'_{\alpha_0\alpha_1})$ are called equivalent if there exists $(h_{\alpha_0}) \in \prod_{\alpha_0 \in I} \mathscr{G}(U_{\alpha_0})$ such that

$$g'_{\alpha_0\alpha_1} = h_{\alpha_1}|_{U_{\alpha_0\alpha_1}} \cdot g_{\alpha_0\alpha_1} \cdot h_{\alpha_0}^{-1}|_{U_{\alpha_0\alpha_1}}.$$

One checks this defines an equivalence relation on the set of 1-cocyles. We define $\check{H}^1(\mathfrak{U}, \mathscr{G})$ to be the set of equivalent classes of 1-cocyles. In general, $\check{H}^1(\mathfrak{U}, \mathscr{G})$ is not a group. But it has a distinguished element corresponding to the equivalent class containing the 1-cocycle $(g_{\alpha_0\alpha_1})$ defined by $g_{\alpha_0\alpha_1} = 1$ for all $\alpha_j \in I$ $(j = 0, 1)$. We denote this element also by 1. Define

$$\check{H}^1(X, \mathscr{G}) = \varinjlim_{\mathfrak{U} \in J_X} \check{H}^1(\mathfrak{U}, \mathscr{G}).$$

(See the discussion after 5.2.1 for the definition of J_X). Again $\check{H}^1(X, \mathscr{G})$ is not necessarily a group. It has a distinguished element 1.

Let (A, a) and (A', a') be two sets with a distinguished element. A morphism $\phi : (A, a) \to (A', a')$ between them is a map $\phi : A \to A'$ such that $\phi(a) = a'$. We define its kernel $\ker \phi$ to be the subset $\phi^{-1}(a')$ of A. A sequence

$$(A'', a'') \xrightarrow{\psi} (A, a) \xrightarrow{\phi} (A', a')$$

is called exact if $\ker \phi = \operatorname{im} \psi$.

Proposition 5.7.10. *Let*

$$1 \to \mathscr{G}_1 \to \mathscr{G}_2 \to \mathscr{G}_3 \to 1$$

be an exact sequence of sheaves of groups (not necessarily commutative) on a scheme X. Then we have an exact sequence of distinguished sets

$$\begin{aligned} 0 &\to \Gamma(X, \mathscr{G}_1) \to \Gamma(X, \mathscr{G}_2) \to \Gamma(X, \mathscr{G}_3) \\ &\xrightarrow{\delta_0} \check{H}^1(X, \mathscr{G}_1) \to \check{H}^1(X, \mathscr{G}_2) \to \check{H}^1(X, \mathscr{G}_3). \end{aligned}$$

Suppose for any $U \in \operatorname{ob} X_{\text{et}}$, $\mathscr{G}_1(U)$ is mapped to the center of $\mathscr{G}_2(U)$. Then $\mathscr{G}_1$ is a sheaf of abelian groups and we have an exact sequence of distinguished sets

$$\begin{aligned} 0 &\to \Gamma(X, \mathscr{G}_1) \to \Gamma(X, \mathscr{G}_2) \to \Gamma(X, \mathscr{G}_3) \\ &\xrightarrow{\delta_0} \check{H}^1(X, \mathscr{G}_1) \to \check{H}^1(X, \mathscr{G}_2) \to \check{H}^1(X, \mathscr{G}_3) \xrightarrow{\delta_1} H^2(X, \mathscr{G}_1). \end{aligned}$$

Proof. We define maps of distinguished sets

$$\delta_0 : \Gamma(X, \mathscr{G}_3) \to \check{H}^1(X, \mathscr{G}_1), \quad \delta_1 : \check{H}^1(X, \mathscr{G}_3) \to \check{H}^2(X, \mathscr{G}_1)$$

and leave it for the reader to verify that the above sequences are exact.

Given a section $s \in \Gamma(X, \mathscr{G}_3)$, there exist an etale covering $\{U_i \to X\}_{i \in I}$ and sections $t_i \in \mathscr{G}_2(U_i)$ lifting $s|_{U_i}$. The sections $t_j|_{U_{ij}} \cdot t_i^{-1}|_{U_{ij}}$ in $\mathscr{G}_2(U_{ij})$ are liftings of 1. So they can be lifted to sections $t_{ij} \in \mathscr{G}_1(U_{ij})$. Note that (t_{ij}) is a 1-cocycle. We define $\delta_0(s)$ to be its image in $\check{H}^1(X, \mathscr{G}_1)$. One can check δ_0 is well-defined.

Regard each $\mathscr{G}_1(U)$ as a subgroup of $\mathscr{G}_2(U)$, and suppose that it lies in the center. Given an element $s \in \check{H}^1(X, \mathscr{G}_3)$, let $\mathfrak{U} = \{U_i \to X\}_{i \in I}$ be an etale covering of X such that s is represented by a 1-cocycle

$$(s_{ij}) \in \ker \Big(\prod_{i,j \in I} \mathscr{G}_3(U_{ij}) \to \prod_{i,j,k \in I} \mathscr{G}_3(U_{ijk}) \Big).$$

Replacing the etale covering by a refinement, we may assume that s_{ij} can be lifted to sections $t_{ij} \in \mathscr{G}_2(U_{ij})$. Since

$$s_{jk}|_{U_{ijk}} \cdot s_{ij}|_{U_{ijk}} = s_{ik}|_{U_{ijk}},$$

there exist $t_{ijk} \in \mathscr{G}_1(U_{ijk})$ such that

$$t_{jk}|_{U_{ijk}} \cdot t_{ij}|_{U_{ijk}} \cdot t_{ik}^{-1}|_{U_{ijk}} = t_{ijk}.$$

We claim that

$$t_{jkl}|_{U_{ijkl}} \cdot t_{ikl}^{-1}|_{U_{ijkl}} \cdot t_{ijl}|_{U_{ijkl}} \cdot t_{ijk}^{-1}|_{U_{ijkl}} = 1.$$

For convenience, we omit mentioning the restriction to U_{ijkl}. We have

$$\begin{aligned}
t_{jkl}t_{ikl}^{-1}t_{ijl}t_{ijk}^{-1} &= t_{ikl}^{-1}t_{jkl}t_{ijl}t_{ijk}^{-1} \\
&= (t_{il}t_{ik}^{-1}t_{kl}^{-1})(t_{kl}t_{jk}t_{jl}^{-1})(t_{jl}t_{ij}t_{il}^{-1})t_{ijk}^{-1} \\
&= (t_{il}t_{ik}^{-1})(t_{jk}t_{ij}t_{il}^{-1})t_{ijk}^{-1} \\
&= (t_{il}t_{ik}^{-1})t_{ijk}^{-1}(t_{jk}t_{ij}t_{il}^{-1}) \\
&= (t_{il}t_{ik}^{-1})(t_{ik}t_{ij}^{-1}t_{jk}^{-1})(t_{jk}t_{ij}t_{il}^{-1}) \\
&= 1.
\end{aligned}$$

Here we use the fact that t_{ijk}^{-1} lies in the center. This proves our claim. So (t_{ijk}) defines an element in $\check{H}^2(\mathfrak{U}, \mathscr{G}_1)$ and hence an element in $\check{H}^2(X, \mathscr{G}_1)$. By 5.6.3, we have a monomorphism $\check{H}^2(X, \mathscr{G}_1) \hookrightarrow H^2(X, \mathscr{G}_2)$. We define $\delta_1(s)$ to be the image of (t_{ijk}) in $H^2(X, \mathscr{G}_2)$. □

Let K be a field. Define an etale sheaf of groups $\mathrm{GL}(n, \mathscr{O}_{\mathrm{Spec}\, K,\mathrm{et}})$ on $\mathrm{Spec}\, K$ by setting

$$\mathrm{GL}(n, \mathscr{O}_{\mathrm{Spec}\, K,\mathrm{et}})(\mathrm{Spec}\, L) = \mathrm{GL}(n, L)$$

for any finite separable extension L of K. We have a monomorphism

$$\mathscr{O}^*_{\mathrm{Spec}\, K,\mathrm{et}} \to \mathrm{GL}(n, \mathscr{O}_{\mathrm{Spec}\, K,\mathrm{et}})$$

which maps each section $\alpha \in L^* = \mathscr{O}^*_{\mathrm{Spec}\, K,\mathrm{et}}(\mathrm{Spec}\, L)$ to the diagonal matrix αI in $\mathrm{GL}(n, L)$. Let $\mathrm{PGL}(n, \mathscr{O}_{\mathrm{Spec}\, K,\mathrm{et}})$ be the cokernel of this monomorphism. We have a short exact sequence

$$1 \to \mathscr{O}^*_{\mathrm{Spec}\, K,\mathrm{et}} \to \mathrm{GL}(n, \mathscr{O}_{\mathrm{Spec}\, K,\mathrm{et}}) \to \mathrm{PGL}(n, \mathscr{O}_{\mathrm{Spec}\, K,\mathrm{et}}) \to 0.$$

By 5.7.10, we have a canonical map

$$\delta_1 : \check{H}^1(\mathrm{Spec}\, K, \mathrm{PGL}(n, \mathscr{O}_{\mathrm{Spec}\, K,\mathrm{et}})) \to H^2(\mathrm{Spec}\, K, \mathscr{O}^*_{\mathrm{Spec}\, K,\mathrm{et}}).$$

Lemma 5.7.11. *Let K'/K be a finite separable extension of degree n, $f : \mathrm{Spec}\, K' \to \mathrm{Spec}\, K$ the corresponding morphism of schemes, and α an element in the kernel of the homomorphism*

$$H^2(\mathrm{Spec}\, K, \mathscr{O}^*_{\mathrm{Spec}\, K,\mathrm{et}}) \to H^2(\mathrm{Spec}\, K', f^*\mathscr{O}^*_{\mathrm{Spec}\, K,\mathrm{et}}).$$

Then α lies in the image of the map

$$\delta_1 : \check{H}^1(\mathrm{Spec}\, K, \mathrm{PGL}(n, \mathscr{O}_{\mathrm{Spec}\, K,\mathrm{et}})) \to H^2(\mathrm{Spec}\, K, \mathscr{O}^*_{\mathrm{Spec}\, K,\mathrm{et}}).$$

Proof. Note that f is a finite etale morphism. Fix a basis $\{e_1, \ldots, e_n\}$ of K' over K. Define a morphism

$$f_*f^*\mathcal{O}^*_{\operatorname{Spec} K,\mathrm{et}} \to \mathrm{GL}(n, \mathcal{O}_{\operatorname{Spec} K,\mathrm{et}})$$

by mapping each section

$$s \in (f_*f^*\mathcal{O}^*_{\operatorname{Spec} K,\mathrm{et}})(\operatorname{Spec} L) = (K' \otimes_K L)^*$$

to the section

$$(a_{ij}) \in \mathrm{GL}(n, L) = \mathrm{GL}(n, \mathcal{O}_{\operatorname{Spec} K,\mathrm{et}})(\operatorname{Spec} L)$$

defined by

$$s \cdot (e_i \otimes 1) = \sum_{j=1}^{n} e_j \otimes a_{ij}.$$

We have a commutative diagram

$$\begin{array}{ccccccc}
0 \to \mathcal{O}^*_{\operatorname{Spec} K,\mathrm{et}} & \to & f_*f^*\mathcal{O}^*_{\operatorname{Spec} K,\mathrm{et}} & \to & f_*f^*\mathcal{O}^*_{\operatorname{Spec} K,\mathrm{et}}/\mathcal{O}^*_{\operatorname{Spec} K,\mathrm{et}} & \to 0 \\
\| & & \downarrow & & \downarrow & \\
0 \to \mathcal{O}^*_{\operatorname{Spec} K,\mathrm{et}} & \to & \mathrm{GL}(n, \mathcal{O}_{\operatorname{Spec} K,\mathrm{et}}) & \to & \mathrm{PGL}(n, \mathcal{O}_{\operatorname{Spec} K,\mathrm{et}}) & \to 0.
\end{array}$$

It induces a commutative diagram

$$\begin{array}{ccccc}
\check{H}^1(\operatorname{Spec} K, f_*f^*\mathcal{O}^*_{\operatorname{Spec} K,\mathrm{et}}/\mathcal{O}^*_{\operatorname{Spec} K,\mathrm{et}}) & \to & H^2(\operatorname{Spec} K, \mathcal{O}^*_{\operatorname{Spec} K,\mathrm{et}}) & \to & H^2(\operatorname{Spec} K, f_*f^*\mathcal{O}^*_{\operatorname{Spec} K,\mathrm{et}}) \\
\downarrow & & \| & & \\
\check{H}^1(\operatorname{Spec} K, \mathrm{PGL}(n, \mathcal{O}_{\operatorname{Spec} K,\mathrm{et}})) & \stackrel{\delta_1}{\to} & H^2(\operatorname{Spec} K, \mathcal{O}^*_{\operatorname{Spec} K,\mathrm{et}}). & &
\end{array}$$

Our assertion follows by a diagram chasing. □

Let K be a field and A a finite dimensional algebra (not necessarily commutative) such that K is contained in the center of A. A is called a *division algebra* if any nonzero element in A is invertible. The opposite algebra A° of A has the same underlying addition group structure as A and its multiplication is defined by $(a_1, a_2) \mapsto a_2a_1$. We require that for any left A-module M, multiplication by 1 is the identity. A left A-module M is called *simple* if it is nonzero and it has no nontrivial submodule. M is called semisimple if it is a direct sum of simple modules. A is called *simple* if A has no nontrivial two-sided ideals. A is called *a central K-algebra* if K is the center of A.

Lemma 5.7.12. *Notation as above. Suppose M is semisimple. Then for any submodule N of M, there exists a submodule N' of M such that $M = N \oplus N'$.*

Proof. Suppose $M = \bigoplus_{i\in I} M_i$ for some simple submodules M_i of M. We have either $N \cap M_i = M_i$ or $N \cap M_i = 0$. If $N \cap M_i = M_i$ for all i, then $M = N \oplus 0$. Otherwise, we can use Zorn's lemma to find a nonempty subset J of I which is maximal with respect to the property $N \cap \left(\bigoplus_{j\in J} M_j\right) = 0$. We claim that $M = N \bigoplus \left(\bigoplus_{j\in J} M_j\right)$. It suffices to show $M_i \subset N \bigoplus \left(\bigoplus_{j\in J} M_j\right)$ for all $i \in I$. If $i \in J$, this is obvious. Suppose $i \notin J$ but $M_i \not\subset N \bigoplus \left(\bigoplus_{j\in J} M_j\right)$. Since M_i is simple, we have $M_i \bigcap \left(N \bigoplus \left(\bigoplus_{j\in J} M_j\right)\right) = 0$. This implies that $N \cap \left(M_i \bigoplus \left(\bigoplus_{j\in J} M_j\right)\right) = 0$ which contradicts the maximality of J. □

Lemma 5.7.13 (Jacobson density theorem). *Notation as above. Suppose M is a semisimple A-module. Let $D = \mathrm{End}_A(M)$. Then M is a D-module. For any D-linear homomorphism $f : M \to M$ and any finitely many elements $x_1, \ldots, x_n \in M$, there exists $a \in A$ such that $f(x_i) = ax_i$ for all i.*

Proof. First consider the case $n = 1$. By 5.7.12, there exists a submodule N of M such that $M = Ax_1 \oplus N$. Let $\pi : M \to Ax_1$ be the projection. Then $\pi \in D$. Since f is D-linear, we have

$$f(x_1) = f(\pi \cdot x_1) = \pi \cdot f(x_1) = \pi(f(x_1)) \in Ax_1.$$

So $f(x_1) = ax_1$ for some $a \in A$.

In general, consider the homomorphism

$$f^n : M^n \to M^n, \quad (y_1, \ldots, y_n) \mapsto (f(y_1), \ldots, f(y_n)).$$

Let $D' = \mathrm{End}_A(M^n)$. One can check f^n is D'-linear. By the $n = 1$ case treated above, there exists $a \in A$ such that $f^n(x_1, \ldots, x_n) = a(x_1, \ldots, x_n)$. Then $f(x_i) = ax_i$ for all i. □

Lemma 5.7.14. *Let K be a field and let A be a finite dimensional K-algebra (not necessarily commutative) such that K is contained in the center of A.*

(i) *Let M be a simple left A-module such that $\mathrm{Ann}_A(M) = 0$. Then $D = \mathrm{End}_A(M)$ is a finite dimensional division K-algebra and A is isomorphic to the algebra $M_n(D^\circ)$ of $n \times n$ matrices with entries in D° for $n = \frac{\dim_K M}{\dim_K D}$.*

(ii) *A is a simple K-algebra if and only if A is K-isomorphic to some matrix algebra $M_n(D)$ for some finite dimensional division K-algebra D. The integer n is uniquely determined, and D is uniquely determined up to isomorphism. D and A have the same center.*

(iii) *Suppose A is a central K-algebra. Consider the K-linear map*

$$F : A \otimes_K A^\circ \to \mathrm{End}_K(A)$$

defined by

$$F(a_1 \otimes a_2)(x) = a_1 x a_2.$$

Then A is simple if and only if F is an isomorphism.

(iv) *Let L be a field containing K. Then K is the center of A if and only if L is the center of $A \otimes_K L$. A is a central simple K-algebra if and only if $A \otimes_K L$ is a central simple L-algebra.*

(v) *Suppose A is a central simple K-algebra. Let A^* be the group of units of A and let $\mathrm{Aut}_K(A)$ be the group of K-algebra automorphisms of A. Then the homomorphism*

$$A^* \to \mathrm{Aut}_K(A)$$

defined by mapping each $a \in A^*$ to the automorphism $x \mapsto axa^{-1}$ induces an isomorphism $A^*/K^* \cong \mathrm{Aut}_K(A)$.

(vi) *Suppose K is separably closed. Then any finite dimensional central division K-algebra is isomorphic to K.*

(vii) *Suppose K is quasi-algebraically closed. Then any finite dimensional central K-division algebra is isomorphic to K.*

Proof.

(i) Since M is simple, for any nonzero element $x_1 \in M$, we have $M = Ax_1$. In particular, M is finite dimensional over K. Let $f \in D = \mathrm{End}_A(M)$ be a nonzero element. We have $\ker f \neq M$. Since M is simple, we must have $\ker f = 0$. As M is finite dimensional, f must be an isomorphism. So D is a finite dimensional division K-algebra. Consider the homomorphism $A \to \mathrm{End}_D(M)$ that maps each $a \in A$ to the endomorphism of M defined by the multiplication by a. By 5.7.13, this homomorphism is surjective. Since $\mathrm{Ann}_A(M) = 0$, it is injective. So $A \cong \mathrm{End}_D(M)$. Let $n = \dim_D M = \frac{\dim_K M}{\dim_K D}$. Then $\mathrm{End}_D(M)$ is isomorphic to the matrix algebra $M_n(D^\circ)$.

(ii) Suppose A is a simple algebra. Note that simple A-modules exist. In fact, any nonzero left ideal of A with minimal dimension is a simple A-module. For any simple A-module M, $\mathrm{Ann}_A(M)$ is a two-sided ideal of A and hence is 0. By (i), A is isomorphic to a matrix algebra $M_n(D)$ for some finite dimensional division K-algebra D.

Let I be a nonzero two-sided ideal of $M_n(D)$. For any $1 \leq i, j \leq n$, denote by E_{ij} the matrix whose (i,j)-entry is 1 and whose other entries are 0. Let (a_{ij}) be a nonzero element in I and suppose $a_{i_0j_0} \neq 0$. Then we have

$$E_{ij} = a_{i_0j_0}^{-1} E_{ii_0} \cdot (a_{ij}) \cdot E_{j_0j} \in I$$

for any $1 \leq i, j \leq n$. It follows that $I = M_n(D)$ and hence $M_n(D)$ is a simple algebra.

For each $1 \leq j \leq n$, let J_j be the left ideal of $M_n(D)$ consisting of matrices whose nonzero entries are on the j-th column. Note that J_j are isomorphic to each other as left $M_n(D)$-modules since they are all isomorphic to the space of column vectors with n-entries. Let (a_{ij}) be an arbitrary nonzero element in J_1. Suppose $a_{i_01} \neq 0$. Then we have

$$E_{i1} = a_{i_01}^{-1} E_{ii_0} \cdot (a_{ij}) \in J_1$$

for any i. It follows that J_1 is generated by any nonzero element in it, and hence is a simple left $M_n(D)$-module. Any simple left $M_n(D)$-module M is isomorphic to J_1. Indeed, let x be a nonzero element in M. For each $1 \leq j \leq n$, $\mathrm{Ann}(x) \cap J_j$ is a left-submodule of J_j. We have either $\mathrm{Ann}(x) \cap J_j = 0$ or $\mathrm{Ann}(x) \cap J_j = J_j$. If $\mathrm{Ann}(x) \cap J_j = J_j$ for all j, then

$$M_n(D) = J_1 + \cdots + J_n \subset \mathrm{Ann}(x).$$

This is impossible since $x \neq 0$. So we have $\mathrm{Ann}(x) \cap J_j = 0$ for some j. Let us prove that M is then isomorphic to J_j, and hence isomorphic to J_1. Consider the homomorphism

$$\phi : J_j \to M, \quad a \mapsto ax.$$

We have $\ker \phi = 0$. The image of ϕ is a nonzero submodule of M. Since M is simple, we have $\mathrm{im}\, \phi = M$. So ϕ is an isomorphism.

We claim that $D^\circ \cong \mathrm{End}_{M_n(D)}(M)$ for any simple left $M_n(D)$-module M. So D is uniquely determined up to isomorphism by the algebra $M_n(D)$. We can take $M = J_1$. For any matrix X in J_1, we have $X = XE_{11}$. For any $f \in \mathrm{End}_{M_n(D)}(J_1)$, we have

$$f(X) = f(XE_{11}) = Xf(E_{11}) = X(E_{11}f(E_{11})).$$

Note that the only nonzero entry of $E_{11}f(E_{11})$ is the $(1,1)$-entry. We get an isomorphism between $\mathrm{End}_{M_n(D)}(J_1)$ and D° by mapping f to the $(1,1)$-entry of $E_{11}f(E_{11})$.

We leave it for the reader to prove that $M_n(D)$ and D have the same center.

(iii) If A has a nontrivial two-sided ideal I, then for any element $\alpha \in A \otimes_K A^\circ$, we have $F(\alpha)(I) \subset I$. But there are K-endomorphisms of A not mapping I to I. So F cannot be an isomorphism.

Suppose A is simple. Let M be the left $(A \otimes_K A^\circ)$-module whose underlying addition group is A and whose scalar multiplication is given by

$$\alpha \cdot x = F(\alpha)(x)$$

for any $\alpha \in A \otimes_K A^\circ$ and $x \in M$. Note that submodules of M are two-sided ideals of A. So M is a simple left $(A \otimes_K A^\circ)$-module. Let $\phi : M \to M$ be an $(A \otimes_K A^\circ)$-module homomorphism, that is, $\phi : A \to A$ is a K-linear map such that

$$\phi(axb) = a\phi(x)b$$

for any $a, b, x \in A$. Taking $x = b = 1$, we get

$$\phi(a) = a\phi(1).$$

Taking $a = x = 1$, we get

$$\phi(b) = \phi(1)b.$$

So we have

$$a\phi(1) = \phi(1)a,$$

and hence $\phi(1)$ lies in the center K of A. We can thus identify $\mathrm{End}_{A\otimes_K A^\circ}(M)$ with K. So M is a simple left module with trivial annihilator over $(A \otimes_K A^\circ)/\mathrm{Ann}_{A\otimes_K A^\circ}(M)$, and we have

$$\mathrm{End}_{(A\otimes_K A^\circ)/\mathrm{Ann}_{A\otimes_K A^\circ}(M)}(M) \cong \mathrm{End}_{A\otimes_K A^\circ}(M) \cong K.$$

By (i), we have

$$(A \otimes_K A^\circ)/\mathrm{Ann}_{A\otimes_K A^\circ}(M) \cong M_n(K)$$

for $n = \dim_K M = \dim_K A$. So $\dim_K((A \otimes_K A^\circ)/\mathrm{Ann}_{A\otimes_K A^\circ}(M)) = n^2$. But $\dim_K(A \otimes_K A^\circ) = n^2$. It follows that $\mathrm{Ann}_{A\otimes_K A^\circ}(M) = 0$, that is, $\ker F = 0$. As $A \otimes_K A^\circ$ and $\mathrm{End}_K(A)$ have the same dimension, F is an isomorphism.

(iv) Let C be the center of A. Fix a basis $\{a_1, \ldots, a_n\}$ of A over K. Consider the K-linear map

$$\phi : A \to A^n, \quad \phi(a) = (aa_1 - a_1 a, \ldots, aa_n - a_n a).$$

Then $C = \ker \phi$ and hence $C \otimes_K L \cong \ker(\phi \otimes \mathrm{id}_L)$. As $\ker(\phi \otimes \mathrm{id}_L)$ is the center of $A \otimes_K L$, this shows that $C \otimes_K L$ can be identified with the center of $A \otimes_K L$. So K is the center of A if and only if L is the center of $A \otimes_K L$. Using (iii), one can show A is a central simple K-algebra if and only if $A \otimes_K L$ is a central simple L-algebra.

(v) Fix a basis $\{a_1, \ldots, a_n\}$ of A over K. Let σ be an element in $\mathrm{Aut}_K(A)$. By (iii), σ is of the form

$$x \mapsto \sum_{i=1}^{n} a_i x b_i$$

for some $b_i \in A$. From $\sigma(xy) = \sigma(x)\sigma(y)$, we get

$$\sum_{i=1}^{n} a_i x(yb_i - b_i\sigma(y)) = 0.$$

By (iii) again, this implies that

$$yb_i - b_i\sigma(y) = 0$$

for all i and all $y \in A$. So each b_iA is a two-sided ideal of A. Hence $b_iA = 0$ or $b_iA = A$, that is, $b_i = 0$ or b_i is invertible. If some b_i is invertible, then we have

$$\sigma(y) = b_i^{-1} y b_i.$$

If no b_i is invertible, all b_i vanish and $\sigma = 0$. This is impossible since σ is an automorphism. So σ is of the form $x \mapsto axa^{-1}$ for some unit a in A.

If a is a unit in A such that $axa^{-1} = x$ for all $x \in A$, then a lies in the center of A. Hence $a \in K^*$. We thus have $A^*/K^* \cong \mathrm{Aut}_K(A)$.

(vi) Let D be a finite dimensional central division K-algebra. Suppose K is algebraically closed. For any $\alpha \in D$, the subalgebra $K[\alpha]$ of D generated by K and α is commutative and finite dimensional over K. It is a field. Since K is algebraically closed, we have $K[\alpha] = K$. So $\alpha \in K$. We thus have $K = D$. By (ii), any finite dimensional central simple K-algebra is isomorphic to the matrix algebra $M_n(K)$ for some n.

Now suppose K is separably closed. For any element $\alpha \in D$, $K[\alpha]$ is field finite and purely inseparable over K. Let $p = \mathrm{char}\, K$. Then the degree of α over K is p^m for some positive integer m and $\alpha^{p^m} \in K$. Let q be the largest power of p dividing $\dim_K D$. Then $p^m | q$ and hence $\alpha^q \in K$. Fix a decomposition $D = K \oplus E$ as K-vector spaces. Let $\overline{K}$ be an algebraic closure of K. This decomposition induces a decomposition

$$D \otimes_K \overline{K} = \overline{K} \oplus (E \otimes_K \overline{K}).$$

Let $\pi : D \otimes_K \overline{K} \to E \otimes_K \overline{K}$ be the projection. The map

$$\phi : D \otimes_K \overline{K} \to E \otimes_K \overline{K},$$
$$x \mapsto \pi(x^q)$$

is a polynomial map defined over K, that is, if we fix bases for D and E over K and express this map in terms of coordinates, then we get a polynomial map defined by polynomials with coefficients in K. By the above discussion, we have $\phi(x) = 0$ for any $x \in D$. As K is necessarily an infinite field, this implies that $\phi(x) = 0$ for any $x \in D \otimes_K \overline{K}$, that is, $x^q \in \overline{K}$ for any $x \in D \otimes_K \overline{K}$. On the other hand, by the discussion for the algebraically closed field case, we have $D \otimes_K \overline{K} \cong M_n(\overline{K})$ for some $n \geq 1$. For the matrix E_{11} in $M_n(\overline{K})$, E_{11}^q does not lie in the image of $\overline{K}$ in $M_n(\overline{K})$ if $n \geq 2$. So we have $n = 1$ and hence $D = K$.

(vii) We first introduce the reduced norm. Let K be a field, let K_s be a separable closure of K, and let A be a finite dimensional central simple K-algebra. By (ii), (iv) and (vi), we have an isomorphism of K_s-algebras

$$\phi : A \otimes_K K_s \cong M_n(K_s)$$

for some integer n. In particular, $\dim_K A$ is the square of an integer. For any $a \in A$, we defined the *reduced norm* $\mathrm{Nrd}(a)$ of a to be the determinant of $\phi(a \otimes 1)$. By (v), if

$$\phi' : A \otimes_K K_s \cong M_n(K_s)$$

is another isomorphism of K_s-algebras, then there exists an invertible matrix P such that $\phi' = P^{-1}\phi P$. It follows that

$$\det(\phi(a \otimes 1)) = \det(\phi'(a \otimes 1)).$$

So $\mathrm{Nrd}(a)$ is independent of the choice of the isomorphism ϕ. We will prove shortly that $\mathrm{Nrd}(a) \in K$ for any $a \in A$. As $\mathrm{Nrd} : A \to K$ is the restriction to A of the map

$$A \otimes_K K_s \cong M_n(K_s) \stackrel{\det}{\to} K_s,$$

it is given by a homogeneous polynomial of degree $\sqrt{\dim_K A}$.

Let us prove that Nrd takes values in K. There exists a finite galois extension K' of K contained in K_s such that we have an isomorphism of K'-algebras.

$$\phi : A \otimes_K K' \cong M_n(K').$$

For any $\sigma \in \mathrm{Gal}(K'/K)$, let $\mathrm{id} \otimes \sigma$ and $M_n(\sigma)$ be the automorphisms induced by σ on $A \otimes_K K'$ and $M_n(K')$ respectively. Then

$$M_n(\sigma) \circ \phi \circ (\mathrm{id} \otimes \sigma^{-1}) : A \otimes_K K' \to M_n(K')$$

is also an isomorphism of K'-algebras. By (v), there is an invertible matrix P in $M_n(K')$ such that

$$M_n(\sigma) \circ \phi \circ (\mathrm{id} \otimes \sigma^{-1}) = P\phi P^{-1}.$$

For any $a \in A$, we have

$$\det(M_n(\sigma) \circ \phi \circ (\mathrm{id} \otimes \sigma^{-1})(a \otimes 1)) = \sigma(\det(\phi(a \otimes 1))).$$

It follows that

$$\sigma(\det(\phi(a \otimes 1))) = \det(P\phi(a \otimes 1)P^{-1}) = \det(\phi(a \otimes 1)),$$

that is, $\sigma(\mathrm{Nrd}(a)) = \mathrm{Nrd}(a)$. So we have $\mathrm{Nrd}(a) \in K$.

Now let us prove (vii). Suppose K is quasi-algebraically closed. Let D be a finite dimensional central division K-algebra, and let $n = \sqrt{\dim_K D}$. The reduced norm $\mathrm{Nrd} : D \to K$ is given by a homogeneous polynomial of degree n with n^2 variables. First consider the case where K is infinite. Then the coefficients of the polynomial expressing the map Nrd lie in K. For any $x \in D - \{0\}$, we have

$$\mathrm{Nrd}(x) \cdot \mathrm{Nrd}(x^{-1}) = 1.$$

So $\mathrm{Nrd}(x)$ has no nontrivial zero in D. Since K is quasi-algebraically closed, the number of variables of the polynomial expressing Nrd does not exceed the degree, that is, $n^2 \leq n$. So $n = 1$ and hence $D = K$. In the case where K is finite, D is a finite division algebra. A theorem of Wedderburn says that D must be a field. Since K is the center of D, we have $K = D$. □

The following is 4.5.6.

Proposition 5.7.15. *Suppose that K is a quasi-algebraically closed field. We have $H^2(\mathrm{Gal}(K_s/K), K_s^*) = 0$.*

Proof. By 5.7.8, we need to show

$$H^2(\mathrm{Spec}\, K, \mathscr{O}^*_{\mathrm{Spec}\, K,\mathrm{et}}) = 0.$$

Let $\alpha \in H^2(\mathrm{Spec}\, K, \mathscr{O}^*_{\mathrm{Spec}\, K,\mathrm{et}})$. Choose an injective resolution $\mathscr{I}^\cdot$ of $\mathscr{O}^*_{\mathrm{Spec}\, K,\mathrm{et}}$ and suppose that α is given by the cohomology class of a section $s \in \Gamma(\mathrm{Spec}\, K, \mathscr{I}^2)$ whose image in $\Gamma(\mathrm{Spec}\, K, \mathscr{I}^3)$ is 0. There exists a finite separable extension K' of K such that the restriction of s to $\mathrm{Spec}\, K'$ can be lifted to a section in $\Gamma(\mathrm{Spec}\, K', \mathscr{I}^1)$. The image of α under the canonical homomorphism

$$H^2(\mathrm{Spec}\, K, \mathscr{O}^*_{\mathrm{Spec}\, K,\mathrm{et}}) \to H^2(\mathrm{Spec}\, K', \mathscr{O}^*_{\mathrm{Spec}\, K',\mathrm{et}})$$

is then 0. By 5.7.11, α lies in the image of the canonical homomorphism

$$\check{H}^1(\mathrm{Spec}\, K, \mathrm{PGL}(n, \mathscr{O}_{\mathrm{Spec}\, K,\mathrm{et}})) \to H^2(\mathrm{Spec}\, K, \mathscr{O}^*_{\mathrm{Spec}\, K,\mathrm{et}}),$$

where $n = [K' : K]$. To prove our assertion, it suffices to show

$$\check{H}^1(\mathrm{Spec}\, K, \mathrm{PGL}(n, \mathscr{O}_{\mathrm{Spec}\, K,\mathrm{et}})) = 0.$$

Let L be a finite separable extension of K. Then $\{\mathrm{Spec}\, L \to \mathrm{Spec}\, K\}$ is an etale covering of $\mathrm{Spec}\, K$, and any etale covering of $\mathrm{Spec}\, K$ has a refinement of this form. Let $\sigma \in \mathrm{PGL}(n, L \otimes_K L)$ be a 1-cocyle for this etale covering and the sheaf $\mathrm{PGL}(n, \mathscr{O}_{\mathrm{Spec\,K,et}})$. Consider the sheaf of matrix algebra $M_n(\mathscr{O}_{\mathrm{Spec}\, L})$ over $\mathscr{O}_{\mathrm{Spec}\, L}$. Let

$$p_i : \mathrm{Spec}\,(L \otimes_K L) \to \mathrm{Spec}\, L \quad (i = 1, 2),$$
$$p_{ij} : \mathrm{Spec}\,(L \otimes_K \otimes_K L) \to \mathrm{Spec}\,(L \otimes_K L) \quad (1 \leq i < j \leq 3)$$

be the projections. Then the conjugation by $\sigma \in \mathrm{PGL}(n, L \otimes_K L)$ defines an automorphism

$$\sigma : p_1^* M_n(\mathscr{O}_{\mathrm{Spec}\, L}) \to p_2^* M_n(\mathscr{O}_{\mathrm{Spec}\, L})$$

such that

$$p_{13}^*(\sigma) = p_{23}^*(\sigma) \circ p_{12}^*(\sigma).$$

So σ defines a descent datum for the matrix algebra $M_n(L)$ over L. By 1.6.1, 5.7.14 (iv) and (v), this matrix algebra can be descended down to a central simple K-algebra A. By 5.7.14 (ii) and (vii), A is isomorphic to $M_n(K)$. So the descent datum is isomorphic to the trivial one and the 1-cocycle is equivalent to the distinguished 1-cocycle. Hence $\check{H}^1(\mathrm{Spec}\, K, \mathrm{PGL}(n, \mathscr{O}_{\mathrm{Spec}\, K,\mathrm{et}})) = 0$. □

Let X be a scheme. An X-scheme G is called a *group scheme* if for any X-scheme Y, the set $\mathrm{Hom}_X(Y, G)$ has a group structure, and for any X-morphism of X-schemes $f : Y_1 \to Y_2$, the map

$$\mathrm{Hom}_X(Y_2, G) \to \mathrm{Hom}_X(Y_1, G), \quad g \mapsto gf$$

is a homomorphism of groups. A right action of G on an X-scheme P is an X-morphism $P \times_X G \to P$ such that for any X-scheme Y, this X-morphism induces a group action

$$\mathrm{Hom}_X(Y, P) \times \mathrm{Hom}_X(Y, G) \to \mathrm{Hom}_X(Y, P).$$

We often denote this X-morphism by

$$P \times_X G \to P, \quad (p, g) \mapsto pg.$$

We say that G is an etale (resp. flat, resp. smooth) group scheme if the structure morphism $G \to X$ is etale (resp. flat, resp. smooth).

Proposition 5.7.16. *Let G be an etale group scheme over X, let P be an X-scheme on which G acts on the right, and let*

$$F : P \times_X G \to P \times_X P, \quad (p, g) \mapsto (p, pg)$$

be the morphism so that $\pi_1 F$ is the projection

$$P \times_X G \to P, \quad (p, g) \mapsto p$$

to the first factor, and $\pi_2 F$ is the group action

$$P \times_X G \to P, \quad (p, g) \mapsto pg,$$

where

$$\pi_1, \pi_2 : P \times_X P \to P$$

are the two projections. The following conditions are equivalent:

(i) *The structure morphism $P \to X$ is surjective and etale, and the morphism $F : P \times_X G \to P \times_X P$ is an isomorphism.*

(ii) *There exists an etale covering $\{U_i \to X\}_{i \in I}$ such that for any $i \in I$, there exists a U_i-isomorphism*

$$U_i \times_X G \cong U_i \times_X P$$

which is compatible with the canonical right actions of $U_i \times_X G$ on itself and on $U_i \times_X P$.

(iii) *The structure morphism $P \to X$ is surjective and etale, and for any X-scheme Y, the action of the group $\mathrm{Hom}_X(Y, G)$ on $\mathrm{Hom}_X(Y, P)$ is transitive, and the stabilizer of any element in $\mathrm{Hom}_X(Y, P)$ is trivial.*

Proof.

(i)$\Rightarrow$(ii) Use the etale covering $\{P \to X\}$.

(ii)$\Rightarrow$(i) The problem is local and we may assume that X is affine, I is finite, and each $U_i \to X$ has finite presentation. Let $U = \coprod_{i \in i} U_i$. Then $U \to X$ is faithfully flat and quasi-compact. The unit in the group $\mathrm{Hom}_X(U, G)$ defines a section for the projection $U \times_X G \to U$. In particular, this projection is surjective. Since G is etale over X, this projection is etale. But $U \times_X G \cong U \times_X P$ as U-schemes. So $U \times_X P \to U$ is surjective and etale. By 1.7.2 and 2.5.10, $P \to X$ is surjective and etale. Its clear that

$$G \times_X G \to G \times_X G, \quad (g', g) \mapsto (g', g'g)$$

is an isomorphism. Since $U \times_X G \cong U \times_X P$ as U-schemes with group actions, the base change of the morphism

$$P \times_X G \to P \times_X P, \quad (p, g) \mapsto (p, pg)$$

by $U \to X$ is an isomorphism. So this morphism is an isomorphism by 1.8.4.

(i)⇔(iii) The condition that the action of the group $\mathrm{Hom}_X(Y, G)$ on $\mathrm{Hom}_X(Y, P)$ is transitive, and that the stabilizer of any element in $\mathrm{Hom}_X(Y, P)$ is trivial is equivalent to the condition that the map

$$\mathrm{Hom}_X(Y, P) \times \mathrm{Hom}_X(Y, G) \to \mathrm{Hom}_X(Y, P) \times \mathrm{Hom}_X(Y, P)$$

induced by the morphism

$$P \times_X G \to P \times_X P, \quad (p, g) \mapsto (p, pg)$$

is an isomorphism. □

When P satisfies the equivalent conditions of 5.7.16, we say that P is a *G-torsor*. We say that P is a trivial G-torsor if P is X-isomorphic to G.

Proposition 5.7.17. *Let G be an etale group scheme over X. A G-torsor P is trivial if and only if the structure morphism $P \to X$ has a section.*

Proof. The unit in the group $\mathrm{Hom}_X(X, G)$ defines a section for the structure morphism $G \to X$. Suppose $P \to X$ has a section s. Consider the composite

$$G \cong X \times_X G \stackrel{s\times \mathrm{id}}{\to} P \times_X G \to P,$$

where the second morphism is the group action. Using condition (iii) in 5.7.16, one verifies that this morphism induces a bijection

$$\mathrm{Hom}_X(Y, G) \to \mathrm{Hom}_X(Y, P)$$

for any X-scheme Y. It follows that it is an isomorphism. □

Let G be an etale group scheme over X, and let $\widetilde{G}$ be the sheaf on X represented by G. Let $\mathscr{F}$ be a sheaf of sets on X. A right action of $\widetilde{G}$ on $\mathscr{F}$ is a morphism

$$\mathscr{F} \times \widetilde{G} \to \mathscr{F}$$

such that for any object U in X_{et}, the map

$$\mathscr{F}(U) \times \widetilde{G}(U) \to \mathscr{F}(U)$$

defines a right action of $\widetilde{G}(U)$ on $\mathscr{F}(U)$. We say that $\mathscr{F}$ is a $\widetilde{G}$-torsor if there exists an etale covering $\{U_i \to X\}$ of X such that for each i, we have an isomorphism $\widetilde{G}|_{U_i} \cong \mathscr{F}|_{U_i}$ compatible with the actions of $\widetilde{G}|_{U_i}$ on itself and on $\mathscr{F}|_{U_i}$. Note that if P is a G-torsor, then the sheaf $\widetilde{P}$ represented by P is a $\widetilde{G}$-torsor.

Proposition 5.7.18. *Suppose G is a separated etale group scheme with finite presentation over a scheme X. Then the functor $P \mapsto \widetilde{P}$ from the category of G-torsors to the category of $\widetilde{G}$-torsors is an equivalence of categories.*

Proof. Let P_1 and P_2 be G-torsors. Define a map

$$T : \mathrm{Hom}(\widetilde{P}_1, \widetilde{P}_2) \to \mathrm{Hom}_X(P_1, P_2)$$

as follows. Let $\phi : \widetilde{P}_1 \to \widetilde{P}_2$ be a morphism of sheaves. id_{P_1} is a section of $\widetilde{P}_1$ over $P_1 \in \mathrm{ob}\, X_{\mathrm{et}}$. We define the X-morphism $T(\phi) : P_1 \to P_2$ to be the image of id_{P_1} under the map $\phi(P_1) : \widetilde{P}_1(P_1) \to \widetilde{P}_2(P_1)$. One checks that T is the inverse of the canonical map

$$\mathrm{Hom}_X(P_1, P_2) \to \mathrm{Hom}(\widetilde{P}_1, \widetilde{P}_2).$$

Hence the functor $P \mapsto \widetilde{P}$ from the category of G-torsors to the category of $\widetilde{G}$-torsors is fully faithful.

Let $\mathscr{F}$ be a $\widetilde{G}$-torsor. Let us prove there exists a G-torsor P such that $\mathscr{F} \cong \widetilde{P}$. Let $\{V_i\}_{i\in I}$ be an affine open covering of X. If we can find G-torsors P_i on V_i such that we have isomorphisms $\phi_i : \mathscr{F}|_{V_i} \stackrel{\cong}{\to} \widetilde{P}_i$, then by the discussion above, the isomorphisms $\phi_j \circ \phi_i^{-1}$ define isomorphisms

$$\phi_{ij} : P_i|_{V_i \cap V_j} \cong P_j|_{V_i \cap V_j}$$

such that

$$\phi_{jk}\phi_{ij} \cong \phi_{ik}$$

on $V_i \cap V_j \cap V_k$. We can glue P_i together to get a G-torsor P on X such that $\mathscr{F} \cong \widetilde{P}$. We are thus reduced to the case where X is affine. By 1.10.12 and 2.3.7, G is quasi-affine over X. We can find an etale covering $\{U_j \to X\}_{j\in J}$ such that J is finite, each $U_j \to X$ has finite presentation and $\mathscr{F}|_{U_j}$ is isomorphic to the trivial $\widetilde{G}|_{U_j}$-torsor. Let $U = \coprod_{j\in J} U_j$. Then $U \to X$ is quasi-compact and faithfully flat, and $\mathscr{F}|_U$ is represented by the trivial $G|_U$-torsor P'. The descent datum on $\mathscr{F}|_U$ defining the sheaf $\mathscr{F}$ defines a descent datum for the G-torsor P'. By 1.8.11, P' can be descended down to a G-torsor P on X. We then have $\mathscr{F} \cong \widetilde{P}$. □

Proposition 5.7.19. *Let G be a separated etale group scheme (not necessarily commutative) with finite presentation over X. Then the set of isomorphic classes of G-torsors is isomorphic to $\check{H}^1(X, \widetilde{G})$.*

Proof. By 5.7.18, it suffices to show that the isomorphic classes of $\widetilde{G}$-torsors is isomorphic to $\check{H}^1(X, \widetilde{G})$. Let $\mathscr{F}$ be a $\widetilde{G}$-torsor and let $\{U_i \to X\}_{i\in i}$ be an etale covering such that we have isomorphisms

$$\phi_i : \mathscr{F}|_{U_i} \stackrel{\cong}{\to} \widetilde{G}|_{U_i}$$

of $\widetilde{G}|_{U_i}$-torsors. For each pair (i, j), $\phi_j \circ \phi_i^{-1}|_{U_i\times_X U_j}$ is an isomorphism of the trivial $\widetilde{G}|_{U_i\times_X U_j}$-torsor. So there exists a section $s_{ij} \in \widetilde{G}(U_i \times_X U_j)$

such that $\phi_j \circ \phi_i^{-1}|_{U_i \times_X U_j}$ is induced by left multiplication by s_{ij}. One can verify that (s_{ij}) is a 1-cocycle. We define a map from the set of isomorphic classes of $\widetilde{G}$-torsors to the set $\check{H}^1(X, \widetilde{G})$ by assigning the image of the 1-cocyle (s_{ij}) in $\check{H}^1(X, \widetilde{G})$ to $\mathscr{F}$. One checks that this map is bijective. □

Let G be a finite group (not necessarily commutative). For any scheme X, let $G_X = \coprod_{g \in G} X_g$, where each X_g is a copy of X. For any X-scheme Y, we have

$$\mathrm{Hom}_X(Y, G_X) = \coprod_{g \in G} \mathrm{Hom}_X(Y, X_g).$$

But $\mathrm{Hom}_X(Y, X_g)$ has only one element. So $\coprod_{g \in G} \mathrm{Hom}_X(Y, X_g)$ can be identified with G and G_X is a group scheme over X.

Assume that X is a connected noetherian scheme and $\gamma : s \to X$ is a geometric point of X. For every X-scheme X', let $X'(\gamma)$ be the set of geometric points of X' lying above γ. By 3.2.12, the functor $X' \mapsto X'(\gamma)$ is an equivalence from the category of etale covering spaces of X to the category of finite sets on which $\pi_1(X, \gamma)$ acts continuously on the left.

Let P be a G_X-torsor. Then P is an etale covering spaces of X. So $\pi_1(X, \gamma)$ acts continuously on $P(\gamma)$ on the left. On the other hand, G acts on $P(\gamma)$ on the right. Note that these two actions on $P(\gamma)$ commute with each other, and the map

$$P(\gamma) \times G \to P(\gamma) \times P(\gamma), \quad (p, g) \mapsto (p, pg)$$

is bijective, that is, the action of G on $P(\gamma)$ is transitive and stabilizers of points in $P(\gamma)$ are trivial.

Conversely, let A be a finite set on which $\pi_1(X, \gamma)$ acts continuously on the left and G acts on the right such that these two actions commute with each other, G acts transitively, and stabilizers of points in A are trivial. Then there exists an etale covering space P of X such that $P(\gamma)$ is isomorphic to A as $\pi_1(X, \gamma)$-sets. The action of G on A induces a homomorphism $G \to \mathrm{Aut}(P/X)^\circ$, and hence a morphism

$$P \times_X G_X \to P.$$

Since the map

$$A \times G \to A \times A, \quad (a, g) \mapsto (a, ag)$$

is bijective, the morphism

$$P \times_X G_X \to P \times_X P, \quad (p, g) \mapsto (p, pg)$$

is an isomorphism. So P is a G_X-torsor.

The set of isomorphic classes of G_X-torsors is thus isomorphic to the isomorphic classes of finite sets on which $\pi_1(X, \gamma)$ acts continuously on the left and G acts on the right, such that these two actions commute with each other, G acts transitively, and stabilizers of points in A are trivial.

Any finite set A on which G acts transitively on the right such that stabilizers of points in A are trivial is isomorphic to G with G acting on it by right multiplication. Suppose $\pi_1(X, \gamma)$ acts on G on the left such that this action commutes with the right multiplication on G. It is completely determined by the map

$$\pi_1(X, \gamma) \to G, \quad \sigma \mapsto \sigma e.$$

Indeed, for any $g \in G$ and $\sigma \in \pi_1(X, \gamma)$, we have

$$\sigma(g) = \sigma(eg) = (\sigma e)g.$$

This map is a continuous homomorphism. Indeed, for any $\sigma_1, \sigma_2 \in \pi_1(X, \gamma)$, we have

$$(\sigma_1\sigma_2)e = \sigma_1(\sigma_2 e) = (\sigma_1 e)(\sigma_2 e).$$

Conversely, any continuous homomorphism from $\pi_1(X, \gamma)$ to G defines a continuous left action of $\pi_1(X, \gamma)$ on G which commutes with the right multiplication on G. We thus prove the following.

Proposition 5.7.20. *Let X be a connected noetherian scheme, γ a geometric point in X, and G a finite group. Then we have*

$$\check{H}^1(X, \widetilde{G}_X) \cong \text{cont.Hom}(\pi_1(X, \gamma), G),$$

where $\text{cont.Hom}(\pi_1(X, \gamma), G)$ *is the set of continuous homomorphisms from* $\pi_1(X, \gamma)$ *to* G.

5.8 Constructible Sheaves

([SGA 4] IX 2.)

Let X be a scheme. For any set G, the *constant sheaf* on X associated to G is the sheaf associated to the constant presheaf $U \mapsto G$ for any $U \in \mathrm{ob}\, X_{\mathrm{et}}$. We denote this sheaf also by G. For any connected object U in X_{et}, we have $G(U) = G$. The sheaf G is represented by the X-scheme $G_X = \coprod_{g \in G} X_g$, where each X_g is a copy of X. A sheaf $\mathscr{F}$ is called *locally constant* if there exists an etale covering $\{U_i \to X\}_{i \in I}$ such that $\mathscr{F}|_{U_i}$ are constant sheaves.

Proposition 5.8.1. *Let X be a scheme and let $\mathscr{F}$ be a sheaf of sets on X.*

(i) *Suppose that the stalks of $\mathscr{F}$ are finite. Then $\mathscr{F}$ is locally constant if and only if it is represented by an etale covering space of X. Moreover, if $\mathscr{F}$ is locally constant and has finite stalks, then there exists a surjective finite etale morphism $\pi : Y \to X$ such that $\pi^*\mathscr{F}$ is constant.*

(ii) *If $\mathscr{F}$ is locally constant and X has finitely many irreducible components, then there exist a dense open subset V of X and a surjective finite etale morphism $\pi : V' \to V$ such that $\pi^*(\mathscr{F}|_V)$ is constant.*

Proof.

(i) Let $\mathscr{F}$ be a locally constant sheaf with finite stalks. Let us prove that there exists an etale covering space $X' \to X$ such that $\mathscr{F} \cong \widetilde{X}'$. Let $\{V_i\}_{i\in I}$ be an affine open covering of X. If we can find etale covering spaces $\pi_i : V_i' \to V_i$ such that we have isomorphisms

$$\phi_i : \mathscr{F}|_{V_i} \cong \widetilde{V}_i',$$

then $\phi_j \circ \phi_i^{-1}|_{V_i\cap V_j}$ induce isomorphisms

$$\phi_{ij} : \pi_i^{-1}(V_i \cap V_j) \stackrel{\cong}{\to} \pi_j^{-1}(V_i \cap V_j)$$

and $\phi_{jk}\phi_{ij} \cong \phi_{ik}$ over $V_i \cap V_j \cap V_k$. We can glue V_i' together to get an etale covering space X' of X such that $\mathscr{F} \cong \widetilde{X}'$. We are thus reduced to the case where X is affine. We can find an etale covering $\{U_j \to X\}_{j\in J}$ such that J is finite, each $U_j \to X$ has finite presentation and each $\mathscr{F}|_{U_j}$ is a constant sheaf with finite stalks. Let $U = \coprod_{j\in J} U_j$. Then $U \to X$ is quasi-compact and faithfully flat, and $\mathscr{F}|_U$ is represented by a trivial etale covering space U' of U. The descent datum on $\mathscr{F}|_U$ defining the sheaf $\mathscr{F}$ defines a descent datum for U'. By 1.8.6, U' can be descended down to an X-scheme X'. By 1.8.8 and 2.5.10, X' is an etale covering space of X. We then have $\mathscr{F} \cong \widetilde{X}'$.

Conversely, let $f : X' \to X$ be an etale covering space. By 5.2.7 (iii), the stalk of $\widetilde{X}'$ at a geometric point $\gamma : s \to X$ can be identified with the set $X'(\gamma)$ of geometric points of X' lying above γ. So it is finite. Let us prove that $\widetilde{X}'$ is locally constant, and that there exists a finite surjective etale morphism $\pi : Y \to X$ such that $\pi^*\widetilde{X}'$ is constant. We may assume that X' and X are connected. It suffices to show that there exists a surjective finite etale morphism $\pi : Y \to X$ such that as a Y-scheme, $X' \times_X Y$ is a disjoint union of copies of Y. Note that $\operatorname{im} f$ is an open and closed subset of X and hence coincides with X. So f is surjective. The diagonal morphism $\Delta : X' \to X' \times_X X'$ is an open and closed immersion. By induction on the number of elements of the set $X'(\gamma)$ and applying the induction hypothesis to the etale covering space

$$X' \times_X X' - \operatorname{im} \Delta \to X'$$

induced by the projection $X' \times_X X' \to X'$ to the first factor, we may assume that there exists a surjective finite etale morphism $Y \to X'$ such that $(X' \times_X X' - \operatorname{im} \Delta) \times_{X'} Y$ is a disjoint union of copies of Y. Then the composite $Y \to X' \to X$ is a finite surjective etale morphism with the required property.

(ii) Let $\eta_1, \ldots, \eta_m$ be all the generic points of X. For each η_i, let V_i be an affine open neighborhood of η_i in X such that $V_i \cap V_j = \emptyset$ for $i \neq j$. Choose an etale morphism $\pi_i : V_i' \to V_i$ of finite presentation such that $\pi_i^*(\mathscr{F}|_{V_i})$ is constant. The generic fiber $\pi_i^{-1}(\eta_i) \to \eta_i$ of π_i is finite, surjective, and etale. By 1.10.10 (iv) and (vi) and 2.3.7, after shrinking V_i, we may assume that π_i is finite, surjective, and etale. Let π be the morphism $\coprod \pi_i : V' = \coprod V_i' \to V = \coprod V_i$. Then π is a surjective finite etale morphism and $\pi^*(\mathscr{F}|_V)$ is constant. □

Let X be a noetherian scheme and A a noetherian ring. A sheaf $\mathscr{F}$ of sets (resp. A-modules) is called *constructible* if there exists a decomposition $X = \bigcup_{i=1}^n X_i$ of X into finitely many locally closed subsets X_i such that each $\mathscr{F}|_{X_i}$ is locally constant and the stalks of $\mathscr{F}$ are finite (resp. finitely generated A-modules). A sheaf of abelian groups $\mathscr{F}$ on X is called a *torsion sheaf* if $\mathscr{F}(U)$ are torsion abelian groups of all $U \in \operatorname{ob} X_{\mathrm{et}}$. A sheaf of abelian groups on X is called constructible if it is constructible as a sheaf of sets. Any constructible sheaf of abelian group $\mathscr{F}$ on X is a torsion sheaf, and there exists a nonzero integer n such that $n\mathscr{F} = 0$. Note that a sheaf of $\mathbb{Z}$-modules is constructible as a sheaf of abelian groups if and only if it is a torsion sheaf and is constructible as a sheaf of $\mathbb{Z}$-modules. The constant sheaf $\mathbb{Z}$ is a constructible sheaf of $\mathbb{Z}$-modules, but not a constructible sheaf of abelian groups.

Proposition 5.8.2. *Let $f : X' \to X$ be a morphism of noetherian schemes, A a noetherian ring, and $\mathscr{F}$ a sheaf of sets (resp. abelian groups, resp. A-modules). If $\mathscr{F}$ is constructible, then so is $f^*\mathscr{F}$.*

Proposition 5.8.3. *Let X be a noetherian scheme, A a noetherian ring, and $\mathscr{F}$ a sheaf of sets (resp. abelian groups, resp. A-modules). $\mathscr{F}$ is constructible if and only if for any irreducible closed subset Y of X, there exists a nonempty open subset V of Y such that $\mathscr{F}|_V$ is locally constant with finite (resp. finite, resp. finitely generated) stalks.*

Proof. Suppose $\mathscr{F}$ is constructible. For any irreducible closed subset Y, $\mathscr{F}|_Y$ is constructible. So there exists a decomposition $Y = \bigcup_{i=1}^n Y_i$ of Y into finitely many locally closed subsets Y_i such that $\mathscr{F}|_{Y_i}$ are locally constant. Let Y_{i_0} be a locally closed subset containing the generic point of Y. Then Y_{i_0} is a nonempty open subset of Y and $\mathscr{F}|_{Y_{i_0}}$ is locally constant.

Conversely, suppose for any irreducible closed subset Y of X, that there exists a nonempty open subset V of Y such that $\mathscr{F}|_V$ is locally constant with finite (resp. finite, resp. finitely generated) stalks. Let

$$\mathscr{S} = \{Y | Y \subset X \text{ is nonemtpy and closed and } \mathscr{F}|_Y \text{ is not constructible}\}.$$

If $\mathscr{S} \neq \emptyset$, then since X is noetherian, $\mathscr{S}$ has a minimal element Y. If Y is not irreducible, then $Y = Y_1 \cup Y_2$ for two proper closed subsets Y_i $(i = 1, 2)$. By the minimality of Y, $\mathscr{F}|_{Y_i}$ are constructible, and hence $\mathscr{F}|_Y$ is constructible. This contradicts the fact that $Y \in \mathscr{S}$. So Y is irreducible. By our assumption, there exists a nonempty open subset V of Y such that $\mathscr{F}|_V$ is constructible. By the minimality of Y, $\mathscr{F}|_{Y-V}$ is constructible. It follows that $\mathscr{F}|_Y$ is constructible. Contradiction. So $\mathscr{S} = \emptyset$. Hence $\mathscr{F}$ is constructible. □

Proposition 5.8.4. *Let X be a noetherian scheme, $f : U \to X$ an etale morphism of finite type, and A a noetherian ring.*

(i) *The sheaf $\widetilde{U}$ represented by U is a constructible sheaf of sets.*

(ii) *The sheaf $A_U = f_!A$ is a constructible sheaf of A-modules.*

Proof. Let Y be an integral closed subscheme of X and let K be its function field. Then

$$U \times_X \operatorname{Spec} K \cong \coprod_{i=1}^n \operatorname{Spec} L_i$$

for some finite separable extensions L_i of K. By 1.10.10 (iv) and 2.3.7, there exists a nonempty open subset V of Y such that $U \times_X V$ is finite etale over V. Note that $\widetilde{U}|_V$ is represented by the V-scheme $U \times_X V$ by 5.2.7 (iii). By 5.8.1 (i), $\widetilde{U}|_V$ is locally constant with finite stalks. So by 5.8.3, $\widetilde{U}$ is constructible. Let $f' : U \times_X V \to V$ be the base change of f. By 5.5.1 (ii), we have

$$(f_!A)|_V \cong f'_!A.$$

By the proof of 5.8.1 (i), there exists an etale covering $\{V_i \to V\}_{i \in i}$ such that V_i are connected and $U \times_X V_i$ are disjoint union of copies of V_i. It follows that $(f'_!A)|_{V_i}$ are constant sheaves of free A-modules with finite rank, and hence $f'_!A$ is locally constant. By 5.8.3, $f_!A$ is constructible. □

Let X be a noetherian scheme and A a noetherian ring. A sheaf $\mathscr{F}$ of sets (resp. abelian groups, resp. A-modules) is called *noetherian* if any ascending chain of subsheaves of sets (resp. abelian groups, resp. A-modules)

$$\mathscr{F}_1 \subset \mathscr{F}_2 \subset \cdots \subset \mathscr{F}$$

in $\mathscr{F}$ is stationary, that is, $\mathscr{F}_i = \mathscr{F}_{i+1} = \cdots$ for large i.

Proposition 5.8.5. *Let X be a noetherian scheme and A a noetherian ring.*

(i) *For any sheaf of A-modules $\mathscr{F}$ on X, the following conditions are equivalent:*

(a) *$\mathscr{F}$ is a constructible sheaf of A-modules.*

(b) *$\mathscr{F}$ is noetherian.*

(c) *There exists an exact sequence of the form*

$$A_V \to A_U \to \mathscr{F} \to 0,$$

where $A_U = f_!A$ and $A_V = g_!A$ for some etale morphisms $f: U \to X$ and $g: V \to X$ of finite type.

(ii) *For any sheaf of abelian groups $\mathscr{F}$ on X, the following conditions are equivalent:*

(a) *$\mathscr{F}$ is a constructible sheaf of abelian groups.*

(b) *$\mathscr{F}$ is a torsion sheaf and noetherian.*

(c) *$\mathscr{F}$ is a constructible sheaf of $\mathbb{Z}/n$-modules for some nonzero integer n.*

Proof.

(i) (a)$\Rightarrow$(b) Let

$$\mathscr{F}_1 \subset \mathscr{F}_2 \subset \cdots$$

be an ascending chain of subsheaves of $\mathscr{F}$ and let

$$\mathscr{S} = \{Y|\ Y \subset X \text{ is nonempty and closed and } \mathscr{F}_1|_Y \subset \mathscr{F}_2|_Y \subset \cdots \text{ is not stationary}\}.$$

If $\mathscr{S} \neq \emptyset$, then $\mathscr{S}$ has a minimal element Y, and Y is necessarily irreducible. By 5.8.3, there exists a nonempty open subset V of Y such that $\mathscr{F}|_V$ is locally constant with finitely generated stalks. Let $\eta \in V$ be the generic point of Y and let $s_{1\bar{\eta}}, \ldots, s_{n\bar{\eta}}$ be a finite family of generators for the A-module $\bigcup_{i=1}^{\infty} \mathscr{F}_{i\bar{\eta}}$. Choose i_0 sufficiently large, we may assume $s_{1\bar{\eta}}, \ldots, s_{n\bar{\eta}} \in \mathscr{F}_{i_0\bar{\eta}}$. We may find an etale neighborhood U of $\bar{\eta}$ in V such that $s_{1\bar{\eta}}, \ldots, s_{n\bar{\eta}}$ are germs at $\bar{\eta}$ of sections $s_1, \ldots, s_n \in \mathscr{F}_{i_0}(U)$, respectively. It follows that if x lies in the image of U in V and $i \geq i_0$, then $s_{1\bar{\eta}}, \ldots, s_{n\bar{\eta}}$ lie in the image of specialization homomorphism $\mathscr{F}_{i\bar{x}} \to \mathscr{F}_{i\bar{\eta}}$, and hence the specialization homomorphism is surjective for any $i \geq i_0$ and any $x \in \mathrm{im}\,(U \to V)$. Since $\mathscr{F}|_V$ is locally constant, the specialization homomorphism $\mathscr{F}_{\bar{x}} \to \mathscr{F}_{\bar{\eta}}$ is bijective, and hence the specialization homomorphism $\mathscr{F}_{i\bar{x}} \to \mathscr{F}_{i\bar{\eta}}$ is injective for any i. So the specialization homomorphism $\mathscr{F}_{i\bar{x}} \to \mathscr{F}_{i\bar{\eta}}$ is bijective for any $i \geq i_0$ and any $x \in \mathrm{im}\,(U \to V)$. We have

$$(\mathscr{F}_{i_0})_{\bar{\eta}} = (\mathscr{F}_{i_0+1})_{\bar{\eta}} = \cdots .$$

So we have

$$(\mathscr{F}_{i_0})_{\bar{x}} = (\mathscr{F}_{i_0+1})_{\bar{x}} = \cdots$$

for any $x \in \mathrm{im}\,(U \to V)$. By the minimality of Y, the chain $\mathscr{F}_1 \subset \mathscr{F}_2 \subset \cdots$ is stationary on $Y - \mathrm{im}\,(U \to V)$. Hence the chain is stationary on Y. This contradicts the fact that $Y \in \mathscr{S}$. So $\mathscr{S} = \emptyset$ and the chain is stationary.

(b)⇒(c) For any etale X-scheme U, we have canonical one-to-one correspondences

$$\mathrm{Hom}(A_U, \mathscr{F}) \cong \mathrm{Hom}(A, \mathscr{F}|_U) \cong \mathscr{F}(U).$$

For any section $s \in \mathscr{F}(U)$, let $\phi_s : A_U \to \mathscr{F}$ be the morphism corresponding to s. It maps the section 1 in $A_U(U)$ to s. We can find a set $\{U_\alpha\}$ of objects U_α in X_{et} with the property that U_α are affine and the images of U_α in X are contained in affine open subsets of X such that any object in X_{et} with this property is isomorphic to some U_α. Then we have an epimorphism

$$\bigoplus_{\alpha, s \in \mathscr{F}(U_\alpha)} A_{U_\alpha} \stackrel{\oplus\phi_s}{\to} \mathscr{F}.$$

Since $\mathscr{F}$ is noetherian, we can find finitely many objects U_i $(i = 1, \ldots, n)$ in $\{U_\alpha\}$ and sections $s_i \in \mathscr{F}(U_i)$ such that

$$\bigoplus_{i=1}^{n} A_{U_i} \stackrel{\oplus\phi_{s_i}}{\to} \mathscr{F}$$

is an epimorphism. Let $U = \coprod_{i=1}^n U_i$. We then have an epimorphism $\phi : A_U \to \mathscr{F}$. By 5.8.4, A_U is constructible and hence noetherian. So $\ker\phi$ is noetherian. The above argument then shows that there exists an epimorphism $A_V \to \ker\phi$ for some etale X-scheme V of finite type. We then have an exact sequence

$$A_V \to A_U \to \mathscr{F} \to 0.$$

(c)⇒(a) By 5.8.4, A_U and A_V are constructible. So there exists a decomposition $X = \bigcup_{i=1}^n X_i$ of X into finitely many locally closed subsets such that $A_U|_{X_i}$ and $A_V|_{X_i}$ are locally constant. Let $X_i' \to X_i$ be surjective etale morphisms such that $A_U|_{X_i'}$ and $A_V|_{X_i'}$ are constant. Then the inverse image of $\mathscr{F}$ on each connected component of X_i' is constant. So $\mathscr{F}$ is constructible.

(ii) (a)⇒(c) Let $X = \bigcup_{i=1}^n X_i$ be a decomposition of X into finitely many locally closed subsets such that $\mathscr{F}|_{X_i}$ are locally constant sheaves with finite stalks. We can find a nonzero integer n such that each $\mathscr{F}|_{X_i}$ is a sheaf of $\mathbb{Z}/n$-modules. Then $\mathscr{F}$ is a constructible sheaf of $\mathbb{Z}/n$-modules.

(c)⇒(b) follows from (i) (a)⇒(b).

(b)⇒(a) Since $\mathscr{F}$ is a torsion sheaf, we have

$$\mathscr{F} = \bigcup_{n\in\mathbb{N}} \ker(n : \mathscr{F} \to \mathscr{F}).$$

Since $\mathscr{F}$ is noetherian, there exists a nonzero integer n such that

$$\mathscr{F} = \ker(n : \mathscr{F} \to \mathscr{F}).$$

Then $\mathscr{F}$ is a sheaf of $\mathbb{Z}/n$-modules. By (i) (b)⇒(a), $\mathscr{F}$ is a constructible sheaf of $\mathbb{Z}/n$-modules. This implies that $\mathscr{F}$ is a constructible sheaf of abelian groups. □

Proposition 5.8.6. *Let X be a noetherian scheme and A a noetherian ring.*

(i) *Subsheaves and quotient sheaves of a constructible sheaf of abelian groups (resp. A-modules) are constructible.*

(ii) *Let $\phi : \mathscr{F} \to \mathscr{G}$ be a morphism of constructible sheaves of abelian groups (resp. A-modules). Then* $\ker\phi$, $\operatorname{coker}\phi$ *and* $\operatorname{im}\phi$ *are constructible.*

(iii) *The category of constructible sheaves of abelian groups (resp. A-modules) is an abelian category.*

Proof. Use 5.8.5 and the fact that subsheaves and quotient sheaves of a noetherian sheaf are noetherian. □

Proposition 5.8.7. *Let X be a noetherian scheme, A a noetherian ring, and*

$$0 \to \mathscr{F}' \to \mathscr{F} \to \mathscr{F}'' \to 0$$

an exact sequence of sheaves of abelian groups or A-modules. If $\mathscr{F}'$ and $\mathscr{F}''$ are constructible (resp. locally constant), then so is $\mathscr{F}$.

Proof. By 5.8.5, it suffices to treat the case of sheaves of A-modules. Suppose that $\mathscr{F}'$ and $\mathscr{F}''$ are constructible sheaves of A-modules. We can find a decomposition $X = \bigcup_{i=1}^n X_i$ of X into finitely many locally closed subsets X_i and surjective etale morphisms $X_i' \to X_i$ such that $\mathscr{F}'|_{X_i'}$ and $\mathscr{F}''|_{X_i'}$ are constant sheaves associated to A-modules M_i' and M_i'', respectively. It suffices to show that each $\mathscr{F}|_{X_i'}$ is locally constant. Let $s_1'', \ldots, s_n'' \in M_i''$ be a finite family of generators for M_i''. We can find an etale covering $\{U_{i\alpha} \to X_i'\}_{\alpha \in I_\alpha}$ of X_i' such that each $U_{i\alpha}$ is connected and each s_j'' can be lifted to a section in $\mathscr{F}(U_{i\alpha})$. We claim that $\mathscr{F}|_{U_{i\alpha}}$ is a constant sheaf. For any connected etale $U_{i\alpha}$-scheme V, consider the commutative diagram

$$\begin{array}{ccccccccc} 0 \to & \mathscr{F}'(U_{i\alpha}) & \to & \mathscr{F}(U_{i\alpha}) & \to & \mathscr{F}''(U_{i\alpha}) & \to 0 \\ & \downarrow & & \downarrow & & \downarrow & \\ 0 \to & \mathscr{F}'(V) & \to & \mathscr{F}(V) & \to & \mathscr{F}''(V) & \to 0. \end{array}$$

The horizontal lines are exact. Since $\mathscr{F}'|_{X_i'}$ and $\mathscr{F}''|_{X_i'}$ are constant, the first and the last vertical arrows are isomorphisms. It follows that $\mathscr{F}(U_{i\alpha}) \to \mathscr{F}(V)$ is an isomorphism. So $\mathscr{F}|_{U_{i\alpha}}$ is constant. Therefore $\mathscr{F}|_{X_i'}$ is locally constant and $\mathscr{F}$ is constructible. □

Proposition 5.8.8. *Let X be a noetherian scheme, A a noetherian ring, and $\mathscr{F}$ a torsion sheaf of abelian groups (resp. a sheaf of A-modules). Then constructible subsheaves of abelian groups (resp. A-modules) of $\mathscr{F}$ form a direct system and $\mathscr{F}$ is the direct limit of this system.*

Proof. If $\mathscr{F}_1$ and $\mathscr{F}_2$ are two constructible subsheaves of $\mathscr{F}$, then so is $\mathscr{F}_1 + \mathscr{F}_2$ by 5.8.6 (ii). So constructible subsheaves of $\mathscr{F}$ form a direct system. If $\mathscr{F}$ is a torsion sheaf, then $\mathscr{F} = \bigcup_{n \in \mathbb{N}} \ker(n : \mathscr{F} \to \mathscr{F})$. Since $\ker(n : \mathscr{F} \to \mathscr{F})$ are sheaves of $\mathbb{Z}/n$-modules, to prove the proposition, it suffices to treat sheaves of A-modules. For any etale X-scheme U of finite type and any section $s \in \mathscr{F}(U)$, we have a canonical morphism $A_U \to \mathscr{F}$ which maps the section $1 \in A_U(U)$ to s. The image of this morphism is a constructible subsheaf of $\mathscr{F}$ by 5.8.6 (i), and s is a section of it. So $\mathscr{F}$ is the direct limit of its constructible subsheaves. □

Proposition 5.8.9. *Let X be a noetherian scheme, A a noetherian ring, $\mathscr{F}$ a sheaf of sets (resp. abelian groups, resp. A-modules) with finite (resp. finite, resp. finitely generated) stalks. Then $\mathscr{F}$ is locally constant if and only if for any $x, y \in X$ with $x \in \overline{\{y\}}$ and any specialization morphism $\widetilde{X}_{\bar{y}} \to \widetilde{X}_{\bar{x}}$, the specialization map $\mathscr{F}_{\bar{x}} \to \mathscr{F}_{\bar{y}}$ is bijective.*

Proof. The "only if" part is clear. Let us prove the "if" part. We say that two points $x, y \in X$ can be connected by a path if there exists a sequence of points $x_1 = x, x_2, \ldots, x_n = y$ such that for any $i \in \{1, \ldots, n-1\}$, we have either $x_i \in \overline{\{x_{i+1}\}}$, or $x_{i+1} \in \overline{\{x_i\}}$. A subset Y of X is called path connected if any two points in Y can be connected by a path formed by a sequence of points in Y. Any irreducible component of X is path connected since any point in it can be connected to the generic point. If X_1 and X_2 are path connected subsets of X with nonempty intersection, then $X_1 \cup X_2$ is path connected. It follows that connected components of X are path connected.

Let $x \in X$, let $s_{1\bar{x}}, \ldots, s_{n\bar{x}} \in \mathscr{F}_{\bar{x}}$ be all the elements of $\mathscr{F}_{\bar{x}}$ (resp. all elements of $\mathscr{F}_{\bar{x}}$, resp. a finite family of generators of $\mathscr{F}_{\bar{x}}$), and let U be a connected quasi-compact etale neighborhood of $\bar{x}$ so that there exist sections $s_1, \ldots, s_n \in \mathscr{F}(U)$ whose germs at $\bar{x}$ are $s_{1\bar{x}}, \ldots, s_{n\bar{x}}$, respectively. Let G be the set $\{s_1, \ldots, s_n\}$ (resp. the subgroup of $\mathscr{F}(U)$ generated by $\{s_1, \ldots, s_n\}$, resp. the submodule of $\mathscr{F}(U)$ generated by $\{s_1, \ldots, s_n\}$). We claim that when U is sufficiently small, the canonical map $G \to \mathscr{F}_{\bar{x}}$ is bijective. In the case of sheaf of sets, this is clear. In the case of sheaf of abelian groups, $\{s_{1\bar{x}}, \ldots, s_{n\bar{x}}\} = \mathscr{F}_{\bar{x}}$ is a group. So for any pair $i, j \in \{1, \ldots, n\}$, there exists $k(i,j) \in \{1, \ldots, n\}$ such that

$$s_{i\bar{x}} - s_{j\bar{x}} = s_{k(i,j),\bar{x}}.$$

Replacing U by a sufficiently small connected quasi-compact etale neighborhood of $\bar{x}$ in U, we may assume

$$s_i - s_j = s_{k(i,j)}.$$

Then $\{s_1, \ldots, s_n\}$ is a group. We have $G = \{s_1, \ldots, s_n\}$ and $G \to \mathscr{F}_{\bar{x}}$ is bijective. In the case of sheaf of A-modules, let $(a_{i1}, \ldots, a_{in})$ $(i = 1, \ldots, m)$ be a finite family of generators for the kernel of the homomorphism

$$A^n \to \mathscr{F}_{\bar{x}}, \quad (a_1, \ldots, a_n) \mapsto a_1 s_{1\bar{x}} + \cdots + a_n s_{n\bar{x}}.$$

We have

$$a_{i1} s_{1\bar{x}} + \cdots + a_{in} s_{n\bar{x}} = 0 \quad (i = 1, \ldots, m).$$

Replacing U by a sufficiently small connected quasi-compact etale neighborhood of $\bar{x}$ in U, we may assume

$$a_{i1} s_1 + \cdots + a_{in} s_n = 0 \quad (i = 1, \ldots, m).$$

Then the kernel of the homomorphism

$$A^n \to \mathscr{F}(U), \quad (a_1, \ldots, a_n) \mapsto a_1 s_1 + \cdots + a_n s_n$$

is also generated by $(a_{i1}, \ldots, a_{in})$ $(i = 1, \ldots, m)$. This implies that $G \cong \mathscr{F}_{\bar{x}}$.

We have a canonical morphism of sheaves $G \to \mathscr{F}|_U$. By our choice of G, the map $G_{\bar{x}} \to (\mathscr{F}|_U)_{\bar{x}}$ is bijective. Specialization maps for G are bijective. By assumption, specialization maps for $\mathscr{F}$ are also bijective. Moreover U is connected and hence path connected. These facts imply that $G_{\bar{y}} \to (\mathscr{F}|_U)_{\bar{y}}$ is bijective for any $y \in U$. So $G \cong \mathscr{F}|_U$. Hence $\mathscr{F}$ is locally constant. □

Corollary 5.8.10. *Let $f : X \to Y$ be a finite surjective radiciel morphism between noetherian schemes. Then the functor $Y' \mapsto X \times_Y Y'$ defines an equivalence from the category of etale covering spaces of Y to the category of etale covering spaces of X. In particular, if X is connected, and $\gamma : s \to X$ is a geometric point of X, then f induces an isomorphism*

$$f_* : \pi_1(X, \gamma) \stackrel{\cong}{\to} \pi_1(Y, f \circ \gamma).$$

Proof. By 5.8.1 (i) and 5.2.7 (iii), it suffices to show that the functor $\mathscr{F} \to f^*\mathscr{F}$ defines an equivalence from the category of locally constant sheaves of sets with finite stalks on Y to that on X. By 5.3.10, f^* defines an equivalence from the category of sheaves of sets on Y to that on X. Using 5.8.9, one shows that $\mathscr{F}$ is locally constant with finite stalks if and only if $f^*\mathscr{F}$ has the same property. Our assertions follows. □

Proposition 5.8.11.

(i) *Let $f : X \to Y$ be a finite morphism between noetherian schemes, A a noetherian ring, and $\mathscr{F}$ a sheaf of A-modules on X. Then $\mathscr{F}$ is constructible if and only if $f_*\mathscr{F}$ is constructible.*

(ii) *Suppose that X is a scheme of finite type over a field or over $\mathbb{Z}$. Let A be a noetherian ring and let $\mathscr{F}$ be a constructible sheaf of A-modules on X. There exist noetherian integral schemes X_i $(i = 1, \ldots, m)$, finite morphisms $p_i : X_i \to X$, and finitely generated A-modules M_i such that we have a monomorphism*

$$\mathscr{F} \hookrightarrow \bigoplus_{i=1}^{m} p_{i*} M_i.$$

When $A = \mathbb{Z}/n$, we can take $M_i = \mathbb{Z}/n$.

Proof.

(i) Using 5.3.7, one can check that the canonical morphism $f^*f_*\mathscr{F} \to \mathscr{F}$ is surjective. If $f_*\mathscr{F}$ is constructible, then by 5.8.2 and 5.8.6 (i), $\mathscr{F}$ is constructible.

Suppose $\mathscr{F}$ is constructible. By 5.8.3, to prove that $f_*\mathscr{F}$ is constructible, it suffices to show that for any irreducible closed subset F of Y, there exists a nonempty open subset V of F such that $(f_*\mathscr{F})|_V$ is locally constant. By 5.3.9, we have

$$(f_*\mathscr{F})|_V \cong \big((f_F)_*(\mathscr{F}|_{f^{-1}(F)})\big)|_V,$$

where we put a reduced closed subscheme structure on F and $f_F : f^{-1}(F) \to F$ is the base change of f. Replacing f by f_F, we may assume that Y is an integral scheme and prove that there exists a nonempty open subset V of Y such that $(f_*\mathscr{F})|_V$ is locally constant. First consider the case where f is finite surjective and radiciel. Then f is a homeomorphism. There exists a nonempty open subset V of Y such that $\mathscr{F}|_{f^{-1}(V)}$ is locally constant. Using 5.8.9, one can check that $(f_*\mathscr{F})|_V$ is locally constant. This proves our assertion under the assumption that f is finite surjective and radiciel. Now suppose that f is a finite morphism. Let K be the function field of Y. Then the base change $X \times_Y \operatorname{Spec} K \to \operatorname{Spec} K$ of f is finite. So $X \times_Y \operatorname{Spec} K$ is artinian, and hence

$$(X \times_Y \operatorname{Spec} K)_{\mathrm{red}} \cong \coprod_{i=1}^n \operatorname{Spec} L_i,$$

where L_i are the residue fields of $X \times_Y \operatorname{Spec} K$ at its generic points. For each i, let K_i be the separable closure of K in L_i. We have a commutative diagram

$$\begin{array}{ccc} \coprod_{i=1}^n \operatorname{Spec} L_i \cong (X \times_Y \operatorname{Spec} K)_{\mathrm{red}} & \to & X \times_Y \operatorname{Spec} K \\ \downarrow & & \downarrow \\ \coprod_{i=1}^n \operatorname{Spec} K_i & \to & \operatorname{Spec} K. \end{array}$$

By 1.10.10 (iv), (vi), (vii) and 2.3.7, there exists a nonempty open subset W in Y such that we have a commutative diagram

$$\begin{array}{ccc} U' & \stackrel{p}{\to} & f^{-1}(W) \\ {\scriptstyle q}\downarrow & & \downarrow {\scriptstyle f_W} \\ W' & \stackrel{r}{\to} & W \end{array}$$

with p and q being finite surjective and radiciel, and r being finite etale, where f_W is the base change of f. We have

$$(f_*\mathscr{F})|_W \cong f_{W*}(\mathscr{F}|_{f^{-1}(W)}) \cong f_{W*}p_*(\mathscr{F}|_{U'}) = r_*q_*(\mathscr{F}|_{U'}) \cong r_!q_*(\mathscr{F}|_{U'}).$$

We have shown that $q_*(\mathscr{F}|_{U'})$ is constructible. Using 5.8.5 (i) (c) and the fact that $r_!$ is an exact functor, one can show that $r_!q_*(\mathscr{F}|_{U'})$ is constructible. So $(f_*\mathscr{F})|_W$ is constructible. Hence there exists a nonempty open subset V of W such that $(f_*\mathscr{F})|_V$ is locally constant.

(ii) Since $\mathscr{F}$ is constructible, there exists a decomposition $X = \bigcup_{i=1}^n X_i$ of X into finitely many locally closed subsets such that $\mathscr{F}|_{X_i}$ are locally constant. Let $k_i : X_i \to X$ be the immersions. For each i, there exists a surjective separated etale morphism of finite type $f_i : X_i' \to X_i$ such that $f_i^* k_i^* \mathscr{F}$ is the constant sheaf associated to some A-module M_i. Let X_{ij}' be the irreducible components of X_i' with the reduced subscheme structures, let $l_{ij} : X_{ij}' \to X_i'$ be the closed immersions, and let $\widetilde{X}_{ij}'$ be the normalizations of X_{ij}'. Since X is a scheme of finite type over a field or over $\mathbb{Z}$, the canonical morphisms $\pi_{ij} : \widetilde{X}_{ij}' \to X_{ij}'$ are finite ([Matsumura (1970)] (31 H) Theorem 72) and surjective. Note that $k_i f_i l_{ij} \pi_{ij}$ are quasi-finite separated morphisms and we have monomorphisms

$$\mathscr{F} \hookrightarrow \bigoplus_i k_{i*} f_{i*} f_i^* k_i^* \mathscr{F} \cong \bigoplus_i k_{i*} f_{i*} M_i \hookrightarrow \bigoplus_{i,j} (k_i f_i l_{ij} \pi_{ij})_* M_i.$$

By the Zariski Main Theorem 1.10.13, there exist finite morphisms $p_{ij} : X_{ij} \to X$ and open immersions $g_{ij} : \widetilde{X}_{ij}' \hookrightarrow X_{ij}$ such that $p_{ij} g_{ij} = k_i f_i l_{ij} \pi_{ij}$. Replacing X_{ij} by the closures of $\mathrm{im}\,(g_{ij})$ with the reduced closed subscheme structures, we may assume that X_{ij} are integral schemes. Let $\widetilde{X}_{ij}$ be the normalizations of X_{ij}. Then $\widetilde{X}_{ij}$ are finite over X_{ij}, and $\widetilde{X}_{ij}'$ can be regarded as open subschemes of $\widetilde{X}_{ij}$. Replacing X_{ij} by $\widetilde{X}_{ij}$, we may assume that X_{ij} are normal integral schemes. We claim that we then have $g_{ij*}(M_i) = M_i$. Indeed, any connected object U of $(X_{ij})_{\mathrm{et}}$ is normal and hence integral. $\widetilde{X}_{ij}' \times_{X_{ij}} U$ is a nonempty open subscheme of U and hence is irreducible and connected. It follows that

$$(g_{ij*}(M_i))(U) = M_i(\widetilde{X}_{ij}' \times_{X_{ij}} U) = M_i = M_i(U).$$

This proves our claim. We thus have

$$\mathscr{F} \hookrightarrow \bigoplus_{i,j} (k_i f_i l_{ij} \pi_{ij})_* M_i \cong \bigoplus_{i,j} p_{ij*} g_{ij*} M_i = \bigoplus_{i,j} p_{ij*} M_i.$$

So we have a monomorphism $\mathscr{F} \hookrightarrow \bigoplus_{i,j} p_{ij*} M_i$. When $A = \mathbb{Z}/n$, any finitely generated A-module can be embedded into a free $\mathbb{Z}/n$-module of finite rank. So we can take $M_i = \mathbb{Z}/n$ in our conclusion. □

One can also study constructible sheaves on non-noetherian schemes. Confer [SGA 4] IX 2.

5.9 Passage to Limit

([SGA 4] VII 5, IX 2.)

We use the same notation as in 1.10. We assume S_0 is quasi-compact and quasi-separated. Then S_λ and S are also quasi-compact and quasi-separated. By 5.2.8, we can work with sheaves on $(S_\lambda)^f_{\text{et}}$ and S^f_{et} instead of sheaves on $(S_\lambda)_{\text{et}}$ and S_{et}.

Define the category $\mathscr{S}$ of sheaves on (S_λ, u_λ) as follows: An object in $\mathscr{S}$ is a family $(\mathscr{F}_\lambda, \psi_{\lambda\mu})_{\lambda\in I}$, where $\mathscr{F}_\lambda$ is a sheaf of abelian groups on $(S_\lambda)^f_{\text{et}}$ for each $\lambda \in I$,

$$\psi_{\lambda\mu} : u^*_{\lambda\mu}(\mathscr{F}_\lambda) \to \mathscr{F}_\mu$$

is a morphism of sheaves for each pair $\lambda \leq \mu$, and

$$\psi_{\lambda\nu} = \psi_{\mu\nu} \circ u^*_{\mu\nu}(\psi_{\lambda\mu})$$

for each triple $\lambda \leq \mu \leq \nu$. We call such an object a sheaf on $(S_\lambda, u_{\lambda\mu})$. We often denote it by $(\mathscr{F}_\lambda)$ for simplicity. A morphism from a sheaf $(\mathscr{F}_\lambda, \psi_{\lambda\mu})$ to another sheaf $(\mathscr{F}'_\lambda, \psi'_{\lambda\mu})$ on $(S_\lambda, u_{\lambda\mu})$ is a family (θ_λ), where $\theta_\lambda : \mathscr{F}_\lambda \to \mathscr{F}'_\lambda$ are morphisms of sheaves such that for any pair $\lambda \leq \mu$, we have

$$\theta_\mu \circ \psi_{\lambda\mu} = \psi'_{\lambda\mu} \circ u^*_{\lambda\mu}(\theta_\lambda).$$

One can show that $\mathscr{S}$ is an abelian category.

We can also define the category $\mathscr{P}$ of presheaves on $(S_\lambda, u_{\lambda\mu})$. A presheaf on $(S_\lambda, u_{\lambda\mu})$ is a family $(\mathscr{F}_\lambda, \psi_{\lambda\mu})_{\lambda\in I}$, where $\mathscr{F}_\lambda$ is a presheaf of abelian groups on $(S_\lambda)^f_{\text{et}}$ for each $\lambda \in I$,

$$\psi_{\lambda\mu} : u^{\mathscr{P}}_{\lambda\mu}(\mathscr{F}_\lambda) \to \mathscr{F}_\mu$$

is a morphism of presheaves for each pair $\lambda \leq \mu$, and

$$\psi_{\lambda\nu} = \psi_{\mu\nu} \circ u^{\mathscr{P}}_{\mu\nu}(\psi_{\lambda\mu})$$

for each triple $\lambda \leq \mu \leq \nu$.

For any λ and any sheaf $\mathscr{F}_\lambda$ on X_λ, let $\overline{\mathscr{F}}_\lambda = ((\overline{\mathscr{F}}_\lambda)_\mu, \psi_{\mu\nu})_{\mu\in I}$ be the sheaf on $(S_\lambda, u_{\lambda\mu})$ defined as follows: If $\mu \geq \lambda$, we define the μ-component $(\overline{\mathscr{F}}_\lambda)_\mu$ of $\overline{\mathscr{F}}_\lambda$ to be $u^*_{\lambda\mu}\mathscr{F}_\lambda$. Otherwise, we define the $(\overline{\mathscr{F}}_\lambda)_\mu$ to be 0. For any $\mu \leq \nu$, we define $\psi_{\mu\nu} : u^*_{\mu\nu}(\overline{\mathscr{F}}_\lambda)_\mu \to (\overline{\mathscr{F}}_\lambda)_\nu$ to be 0 if $(\overline{\mathscr{F}}_\lambda)_\mu = 0$, and we define it to be the canonical isomorphism $u^*_{\mu\nu}(u^*_{\lambda\mu}\mathscr{F}_\lambda) \cong u^*_{\lambda\nu}\mathscr{F}_\lambda$ if $\mu \geq \lambda$. For any sheaf $\mathscr{G} = (\mathscr{G}_\lambda)$ on $(S_\lambda, u_{\lambda\mu})$, we have a one-to-one correspondence

$$\text{Hom}(\overline{\mathscr{F}}_\lambda, \mathscr{G}) = \text{Hom}(\mathscr{F}_\lambda, \mathscr{G}_\lambda).$$

So the functor

$$\mathscr{S}_{S_\lambda} \to \mathscr{S}, \quad \mathscr{F}_\lambda \mapsto \overline{\mathscr{F}}_\lambda$$

is left adjoint to the functor

$$\mathscr{S} \to \mathscr{S}_{S_\lambda}, \quad (\mathscr{G}_\lambda) \mapsto \mathscr{G}_\lambda.$$

For each $\lambda \in I$, let J_λ be a set of objects $\pi_\lambda : U_\lambda \to S_\lambda$ in $(S_\lambda)^f_{\text{et}}$ with the property that U_λ are affine and their images in S_λ are also affine, and any object in $(S_\lambda)^f_{\text{et}}$ with this property is isomorphic to an object in J_λ. When λ goes over I and π_λ goes over J_λ, $\overline{\pi_{\lambda!}\mathbb{Z}}$ form a set of generators for the category $\mathscr{S}$. By the same argument as the proof of [Fu (2006)] 2.1.6 or [Grothendieck (1957)] 1.10.1, $\mathscr{S}$ has enough injective objects. Since the functor $\mathscr{F}_\lambda \mapsto \overline{\mathscr{F}}_\lambda$ is exact, if $(\mathscr{I}_\lambda)$ is an injective sheaf on $(S_\lambda, u_{\lambda\mu})$, then $\mathscr{I}_\lambda$ is an injective sheaf on S_λ for each λ.

Given a presheaf $(\mathscr{F}_\lambda)$ on $(S_\lambda, u_{\lambda\mu})$, define a presheaf $\mathscr{F}$ on S as follows. For any object U in S^f_{et}, by 1.10.9 (iii) and 2.3.7, we can find $\lambda_U \in I$ and an object U_{λ_U} in $(S_{\lambda_U})^f_{\text{et}}$ such that we have an S-isomorphism $U \cong U_{\lambda_U} \times_{S_{\lambda_U}} S$. Define

$$\mathscr{F}(U) = \varinjlim_{\lambda \geq \lambda_U} \mathscr{F}_\lambda(U_\lambda),$$

where $U_\lambda = U_{\lambda_U} \times_{S_{\lambda_U}} S_\lambda$ for any $\lambda \geq \lambda_U$. Given a morphism $U \to V$ in S^f_{et}, we can find $\lambda_0 \geq \lambda_U, \lambda_V$ and an S_{λ_0}-morphism $U_{\lambda_0} \to V_{\lambda_0}$ inducing the given S-morphism $U \to V$. We define the restriction $\mathscr{F}(V) \to \mathscr{F}(U)$ to be the direct limit of the restrictions $\mathscr{F}_\lambda(V_\lambda) \to \mathscr{F}_\lambda(U_\lambda)$ for $\lambda \geq \lambda_0$. We thus get a presheaf $\mathscr{F}$. Suppose that $(\mathscr{F}_\lambda)$ is a sheaf on $(S_\lambda, u_{\lambda\mu})$, then $\mathscr{F}$ is a sheaf on S^f_{et}. Indeed, for any etale covering $\mathfrak{U} = \{U_i \to U\}_{i\in I}$ in S^f_{et}, we can find $\lambda_{\mathfrak{U}} \geq \lambda_U, \lambda_{U_i}$ and an etale covering $\{U_{i\lambda_{\mathfrak{U}}} \to U_{\lambda_{\mathfrak{U}}}\}_{i\in I}$ in $(S_{\lambda_{\mathfrak{U}}})^f_{\text{et}}$ inducing the etale covering $\mathfrak{U}$. For any $\lambda \geq \lambda_{\mathfrak{U}}$, we have the canonical exact sequence

$$0 \to \mathscr{F}_\lambda(U_\lambda) \to \prod_i \mathscr{F}_\lambda(U_{i\lambda}) \to \prod_{i,j} \mathscr{F}_\lambda(U_{i\lambda} \times_{U_\lambda} U_{j\lambda}).$$

Taking direct limit, we get an exact sequence

$$0 \to \mathscr{F}(U) \to \prod_i \mathscr{F}(U_i) \to \prod_{i,j} \mathscr{F}(U_i \times_U U_j).$$

So $\mathscr{F}$ is a sheaf.

Lemma 5.9.1. *Let $(\mathscr{P}_\lambda)$ be a presheaf on $(S_\lambda, u_{\lambda\mu})$. Then $(\mathscr{P}^\#_\lambda)$ is a sheaf on $(S_\lambda, u_{\lambda\mu})$. Let $\mathscr{P}$ (resp. $\mathscr{F}$) be the presheaf (resp. sheaf) on S corresponding to $(\mathscr{P}_\lambda)$ (resp. $(\mathscr{P}^\#_\lambda)$) defined as above. We have $\mathscr{P}^\# \cong \mathscr{F}$.*

Proof. For any sheaf $\mathscr{G}$ on S^f_{et}, we have

$$\begin{aligned}\mathrm{Hom}(\mathscr{P}^\#,\mathscr{G}) &\cong \mathrm{Hom}(\mathscr{P},\mathscr{G})\\ &\cong \varprojlim_\lambda \mathrm{Hom}(\mathscr{P}_\lambda, u_{\lambda *}\mathscr{G})\\ &\cong \varprojlim_\lambda \mathrm{Hom}(\mathscr{P}_\lambda^\#, u_{\lambda *}\mathscr{G})\\ &\cong \mathrm{Hom}(\mathscr{F},\mathscr{G}).\end{aligned}$$

□

Proposition 5.9.2. *Given a sheaf $(\mathscr{F}_\lambda)$ on $(S_\lambda, u_{\lambda\mu})$, define the sheaf $\mathscr{F}$ on S as above.*

(i) *We have $\mathscr{F} \cong \varinjlim_\lambda u_\lambda^* \mathscr{F}_\lambda$.*

(ii) *If each $\mathscr{F}_\lambda$ is flasque, then $\mathscr{F}$ is flasque.*

(iii) *We have $H^q(S,\mathscr{F}) \cong \varinjlim_\lambda H^q(S_\lambda, \mathscr{F}_\lambda)$ for all q.*

Proof.

(i) For any sheaf $\mathscr{G}$ on S^f_{et}, we have

$$\begin{aligned}\mathrm{Hom}(\mathscr{F},\mathscr{G}) &\cong \varprojlim_\lambda \mathrm{Hom}(\mathscr{F}_\lambda, u_{\lambda *}\mathscr{G})\\ &\cong \varprojlim_\lambda \mathrm{Hom}(u_\lambda^*\mathscr{F}_\lambda, \mathscr{G})\\ &\cong \mathrm{Hom}(\varinjlim_\lambda u_\lambda^*\mathscr{F}_\lambda, \mathscr{G}).\end{aligned}$$

(ii) Let $\mathfrak{U} = \{U_i \to U\}_{i\in I}$ be an etale covering in S^f_{et}. We can find $\lambda \geq \lambda_U, \lambda_{U_i}$ such that there exists an etale covering $\mathfrak{U}_\lambda = \{U_{i\lambda} \to U_\lambda\}_{i\in I}$ in $(S_\lambda)^f_{\mathrm{et}}$ inducing $\mathfrak{U}$. It is clear that

$$\check{H}^q(\mathfrak{U},\mathscr{F}) \cong \varinjlim_{\lambda\geq\lambda_{\mathfrak{U}}} \check{H}^q(\mathfrak{U}_\lambda,\mathscr{F}_\lambda).$$

Since $\mathscr{F}_\lambda$ are flasque, we have $\check{H}^q(\mathfrak{U}_\lambda,\mathscr{F}_\lambda) = 0$ for all $q \geq 1$. It follows that $\check{H}^q(\mathfrak{U},\mathscr{F}) = 0$ for all $q \geq 1$. So $\mathscr{F}$ is flasque.

(iii) Let

$$0 \to (\mathscr{I}^0_\lambda) \to (\mathscr{I}^1_\lambda) \to \cdots$$

be an injective resolution of $(\mathscr{F}_\lambda)$ in the category of sheaves on $(S_\lambda, u_{\lambda\mu})$. Then for each λ, $\mathscr{I}^\cdot_\lambda$ is an injective resolution of $\mathscr{F}_\lambda$. By (i) and (ii), $\varinjlim_\lambda u_\lambda^*\mathscr{I}^\cdot_\lambda$ is a flasque resolution of $\mathscr{F}$. So we have

$$\begin{aligned}H^q(S,\mathscr{F}) &\cong H^q(\Gamma(S, \varinjlim_\lambda u_\lambda^*\mathscr{I}^\cdot_\lambda))\\ &\cong H^q(\varinjlim_\lambda \Gamma(S_\lambda, \mathscr{I}^\cdot_\lambda))\\ &\cong \varinjlim_\lambda H^q(\Gamma(S_\lambda, \mathscr{I}^\cdot_\lambda))\\ &\cong \varinjlim_\lambda H^q(S_\lambda, \mathscr{F}_\lambda).\end{aligned}$$

□

Corollary 5.9.3. *Let $\mathscr{F}_0$ be a sheaf on S_0, and let $\mathscr{F}_\lambda$ and $\mathscr{F}$ be the inverse images of $\mathscr{F}_0$ on S_λ and S, respectively. We have $H^q(S, \mathscr{F}) \cong \varinjlim_\lambda H^q(S_\lambda, \mathscr{F}_\lambda)$ for all q.*

Corollary 5.9.4. *Let $\mathscr{F}$ be a sheaf on S and let $\mathscr{F}_\lambda = u_{\lambda *}\mathscr{F}$. We have $H^q(S, \mathscr{F}) \cong \varinjlim_\lambda H^q(S_\lambda, \mathscr{F}_\lambda)$ for all q.*

Corollary 5.9.5. *Let $f : X \to Y$ be a quasi-compact and quasi-separated morphism, $\mathscr{F}$ a sheaf on X, y a point on Y, and $\widetilde{Y}_{\bar{y}}$ the strict localization of Y at $\bar{y}$. Then for all q, we have*

$$(R^q f_*\mathscr{F})_{\bar{y}} \cong H^q(X \times_Y \widetilde{Y}_{\bar{y}}, \mathscr{F}|_{X\times_Y \widetilde{Y}_{\bar{y}}}).$$

Proof. Let $\{V_\lambda\}$ be the family of affine etale neighborhood of $\bar{y}$ in Y. We have

$$\widetilde{Y}_{\bar{y}} = \operatorname{Spec}(\varinjlim_\lambda \Gamma(V_\lambda, \mathscr{O}_{V_\lambda})).$$

For each q, $R^q f_*\mathscr{F}$ is the sheaf associated to the presheaf

$$V \mapsto H^q(X \times_Y V, \mathscr{F}|_{X\times_Y V})$$

for any $V \in \operatorname{ob} Y_{\text{et}}$. So we have

$$(R^q f_*\mathscr{F})_{\bar{y}} \cong \varinjlim_\lambda H^q(X \times_Y V_\lambda, \mathscr{F}|_{X\times_Y V_\lambda}).$$

By 5.9.3, we have

$$\varinjlim_\lambda H^q(X \times_Y V_\lambda, \mathscr{F}|_{X\times_Y V_\lambda}) \cong H^q(X \times_Y \widetilde{Y}_{\bar{y}}, \mathscr{F}|_{X\times_Y \widetilde{Y}_{\bar{y}}}).$$

Our assertion follows. □

Corollary 5.9.6. *Let X_0 be an S_0-scheme, $(X_\lambda, v_{\lambda\mu})$ another inverse system of X_0-schemes such that X_λ are affine over X_0 for all $\lambda \in I$, $X = \varprojlim_\lambda X_\lambda$, $v : X \to X_\lambda$ and $v_{\lambda\mu} : X_\mu \to X_\lambda$ $(\lambda \leq \mu)$ the projections, $f_\lambda : X_\lambda \to S_\lambda$ quasi-compact quasi-separated S_0-morphisms satisfying $f_\lambda v_{\lambda\mu} = u_{\lambda\mu} f_\mu$ for any $\lambda \leq \mu$, and $f : X \to S$ the morphism induced by (f_λ) by passing to limit.*

(i) *Let $(\mathscr{F}_\lambda)$ be a sheaf on $(X_\lambda, v_{\lambda\mu})$, and let $\mathscr{F} = \varinjlim_\lambda v_\lambda^*\mathscr{F}_\lambda$. Then we have*

$$R^q f_*\mathscr{F} \cong \varinjlim_\lambda u_\lambda^* R^q f_{\lambda *}\mathscr{F}_\lambda$$

for all q.

(ii) *Let $(\mathscr{G}_\lambda)$ be a sheaf on $(S_\lambda, v_{\lambda\mu})$, and let $\mathscr{G} = \varinjlim_\lambda u_\lambda^*\mathscr{G}_\lambda$. Then we have*

$$f^*\mathscr{G} \cong \varinjlim_\lambda v_\lambda^* f_\lambda^*\mathscr{G}_\lambda.$$

Proof.

(i) For each q, $R^q f_* \mathscr{F}$ (resp. $R^q f_{\lambda *} \mathscr{F}_\lambda$) is the sheaf on S_{et}^f (resp. $(S_\lambda)_{\mathrm{et}}^f$) associated to the presheaf

$$V \mapsto H^q(X \times_S V, \mathscr{F}) \quad (\text{resp. } V_\lambda \mapsto H^q(X_\lambda \times_{S_\lambda} V_\lambda, \mathscr{F}_\lambda))$$

for any $V \in \mathrm{ob}\, S_{\mathrm{et}}^f$ (resp. $V_\lambda \in \mathrm{ob}\,(S_\lambda)_{\mathrm{et}}^f$). By 5.9.2 (iii), we have

$$H^q(X \times_S V, \mathscr{F}) \cong \varinjlim_{\lambda \geq \lambda_V} H^q(X_\lambda \times_{S_\lambda} V_\lambda, \mathscr{F}_\lambda).$$

Our assertion then follows from 5.9.1.

(ii) We have

$$\begin{aligned} f^* \mathscr{G} &\cong f^*(\varinjlim_\lambda u_\lambda^* \mathscr{G}_\lambda) \\ &\cong \varinjlim_\lambda f^* u_\lambda^* \mathscr{G}_\lambda \\ &\cong \varinjlim_\lambda v_\lambda^* f_\lambda^* \mathscr{G}_\lambda. \end{aligned}$$

□

Corollary 5.9.7. *Let $f : X \to S$ be an integral morphism and let $\mathscr{F}$ be a sheaf on X. Then $R^q f_* \mathscr{F} = 0$ for any $q \geq 1$. Moreover, for any Cartesian diagram*

$$\begin{array}{ccc} X \times_S S' & \xrightarrow{g'} & X \\ f' \downarrow & & \downarrow f \\ S' & \xrightarrow{g} & S, \end{array}$$

we have a canonical isomorphism

$$g^* f_* \mathscr{F} \xrightarrow{\cong} f'_* g'^* \mathscr{F}.$$

Proof. The canonical morphism $g^* f_* \mathscr{F} \to f'_* g^* \mathscr{F}$ is obtained by adjunction from the composite of the canonical morphisms

$$f_* \mathscr{F} \xrightarrow{f_*(\mathrm{adj})} f_* g'_* g'^* \mathscr{F} \cong g_* f'_* g'^* \mathscr{F}.$$

By 5.3.9, we have $g^* f_* \mathscr{F} \cong f'_* g^* \mathscr{F}$ if f is a finite morphism.

To prove 5.9.7, the problem is local on S, and we may assume $S = \mathrm{Spec}\, A$ is affine. Since X is integral over S, it is also affine. Let $\{A_\lambda\}$ be the family subalgebras of $\Gamma(X, \mathscr{O}_X)$ finitely generated over A, let $S_\lambda = \mathrm{Spec}\, A_\lambda$, and let $u_\lambda : X \to S_\lambda$ and $f_\lambda : S_\lambda \to S$ be the canonical morphisms. Using 5.9.2 (i), one can show $\mathscr{F} \cong \varinjlim_\lambda u_\lambda^* u_{\lambda *} \mathscr{F}$. Applying 5.9.6 to the morphism (f_λ) from the inverse system (S_λ) to the constant inverse system (S), we get

$$R^q f_* \mathscr{F} \cong \varinjlim_\lambda R^q f_{\lambda *}(u_{\lambda *} \mathscr{F}).$$

Each f_λ is a finite morphism. By 5.7.4, we have

$$R^q f_{\lambda *}(u_{\lambda *}\mathscr{F}) = 0$$

for any $q \geq 1$. So we have

$$R^q f_* \mathscr{F} = 0$$

for any $q \geq 1$. In the case $q = 0$, we get

$$f_* \mathscr{F} \cong \varinjlim_{\lambda} f_{\lambda *} u_{\lambda *} \mathscr{F}.$$

Fix notation by the following commutative diagram:

$$\begin{array}{ccccc} X \times_S S' & \stackrel{u'_\lambda}{\rightarrow} & S_\lambda \times_S S' & \stackrel{f'_\lambda}{\rightarrow} & S' \\ {\scriptstyle g'} \downarrow & & {\scriptstyle g_\lambda} \downarrow & & {\scriptstyle g} \downarrow \\ X & \stackrel{u_\lambda}{\rightarrow} & S_\lambda & \stackrel{f_\lambda}{\rightarrow} & S. \end{array}$$

Since $\mathscr{F} \cong \varinjlim_\lambda u_\lambda^* u_{\lambda *} \mathscr{F}$, we have

$$g'^* \mathscr{F} \cong \varinjlim_{\lambda} u_\lambda'^* g_\lambda^* u_{\lambda *} \mathscr{F}.$$

Applying 5.9.6 to the morphism (f'_λ) from the inverse system $(S_\lambda \times_S S')$ to the constant inverse system (S'), we get

$$f'_* g'^* \mathscr{F} \cong \varinjlim_{\lambda} f'_{\lambda *} g_\lambda^* u_{\lambda *} \mathscr{F}.$$

So we have

$$\begin{aligned} g^* f_* \mathscr{F} &\cong g^*(\varinjlim_{\lambda} f_{\lambda *} u_{\lambda *} \mathscr{F}) \\ &\cong \varinjlim_{\lambda} g^* f_{\lambda *} u_{\lambda *} \mathscr{F} \\ &\cong \varinjlim_{\lambda} f'_{\lambda *} g_\lambda^* u_{\lambda *} \mathscr{F} \\ &\cong f'_* g'^* \mathscr{F}, \end{aligned}$$

where for the third isomorphism, we apply 5.3.9 to the finite morphisms f_λ. □

Lemma 5.9.8. *Suppose that S_0, S_λ, and S are noetherian schemes, and A is a noetherian ring.*

(i) *Let $\mathscr{F}_0$ be a sheaf of A-modules on S_0, $\mathscr{F}_\lambda$ and $\mathscr{F}$ its inverse images on S_λ and S, respectively. If $\mathscr{F}_0$ is a constructible sheaf of A-modules, then for any sheaf of A-modules $(\mathscr{G}_\lambda)$ on $(S_\lambda, u_{\lambda\mu})$, we have a one-to-one correspondence*

$$\varinjlim_{\lambda} \mathrm{Hom}_A(\mathscr{F}_\lambda, \mathscr{G}_\lambda) \to \mathrm{Hom}_A(\mathscr{F}, \mathscr{G}),$$

where $\mathscr{G} = \varinjlim_\lambda u_\lambda^* \mathscr{G}_\lambda$.

(ii) *Let* $\mathscr{F}_0$ *and* $\mathscr{G}_0$ *be constructible sheaves of A-modules on* S_0*,* $\mathscr{F}_\lambda$ *and* $\mathscr{G}_\lambda$ *(resp.* $\mathscr{F}$ *and* $\mathscr{G}$*) their inverse images on* S_λ *(resp.* S*),* $\phi_0 : \mathscr{F}_0 \to \mathscr{G}_0$ *a morphism, and* $\phi_\lambda : \mathscr{F}_\lambda \to \mathscr{G}_\lambda$ *and* $\phi : \mathscr{F} \to \mathscr{G}$ *the morphisms induced by* ϕ_0*. If* ϕ *is an isomorphism, then* ϕ_λ *are isomorphisms for sufficiently large* λ*.*

(iii) *For any constructible sheaf of A-modules* $\mathscr{F}$ *on* S*, we can find a sufficiently large* λ *and a constructible sheaf of A-modules* $\mathscr{F}_\lambda$ *on* S_λ *such that* $\mathscr{F} \cong u_\lambda^* \mathscr{F}$*.*

(iv) *Let*

$$\mathscr{F}_0 \stackrel{\phi_0}{\to} \mathscr{G}_0 \stackrel{\psi_0}{\to} \mathscr{H}_0$$

be a sequence of constructible sheaves on S_0*. If its inverse image*

$$\mathscr{F} \stackrel{\phi}{\to} \mathscr{G} \stackrel{\psi}{\to} \mathscr{H}$$

on S *is exact, then for sufficiently large* λ*, its inverse image*

$$\mathscr{F}_\lambda \stackrel{\phi_\lambda}{\to} \mathscr{G}_\lambda \stackrel{\psi_\lambda}{\to} \mathscr{H}_\lambda$$

on S_λ *is exact.*

Proof.

(i) Any morphism $\phi_\lambda : \mathscr{F}_\lambda \to \mathscr{G}_\lambda$ induces a morphism $\mathscr{F} \to \mathscr{G}$ by taking the composite

$$\mathscr{F} = u_\lambda^* \mathscr{F}_\lambda \stackrel{u_\lambda^*(\phi_\lambda)}{\to} u_\lambda^* \mathscr{G}_\lambda \to \varinjlim_\lambda u_\lambda^* \mathscr{G}_\lambda = \mathscr{G}.$$

We thus get maps

$$\mathrm{Hom}_A(\mathscr{F}_\lambda, \mathscr{G}_\lambda) \to \mathrm{Hom}_A(\mathscr{F}, \mathscr{G}).$$

They are compatible with each other and induce a map

$$\varinjlim_\lambda \mathrm{Hom}_A(\mathscr{F}_\lambda, \mathscr{G}_\lambda) \to \mathrm{Hom}_A(\mathscr{F}, \mathscr{G}).$$

Let us prove that the last map is bijective.

For any etale S_0-scheme U_0 of finite type, let $U_\lambda = U_0 \times_{S_0} S_\lambda$ and $U = U_0 \times_{S_0} S$. We have

$$\begin{aligned}\varinjlim_\lambda \mathrm{Hom}_A(A_{U_\lambda}, \mathscr{G}_\lambda) &\cong \varinjlim_\lambda \mathscr{G}_\lambda(U_\lambda) \\ &\cong \mathscr{G}(U) \\ &\cong \mathrm{Hom}_A(A_U, \mathscr{G}).\end{aligned}$$

In general, by 5.8.5, we have an exact sequence

$$A_{U_0} \to A_{V_0} \to \mathscr{F}_0 \to 0$$

for some etale S_0-schemes U_0 and V_0 of finite type. Consider the commutative diagram

$$\begin{array}{ccccccc} 0 \to & \varinjlim_\lambda \mathrm{Hom}_A(\mathscr{F}_\lambda, \mathscr{G}_\lambda) & \to & \varinjlim_\lambda \mathrm{Hom}_A(A_{V_\lambda}, \mathscr{G}_\lambda) & \to & \mathrm{Hom}_A(A_{U_\lambda}, \mathscr{G}_\lambda) \\ & \downarrow & & \downarrow & & \downarrow \\ 0 \to & \mathrm{Hom}_A(\mathscr{F}, \mathscr{G}) & \to & \mathrm{Hom}_A(A_V, \mathscr{G}) & \to & \mathrm{Hom}_A(A_U, \mathscr{G}). \end{array}$$

The horizontal lines are exact. The last two vertical arrows are bijective. It follows that the first vertical arrow is bijective.

(ii) follows from (i).

(iii) We can find an exact sequence

$$A_U \to A_V \to \mathscr{F} \to 0$$

for some etale S-schemes U and V of finite type. We can find λ and etale S_λ-schemes U_λ and V_λ of finite type such that $U \cong U_\lambda \times_{S_\lambda} S$ and $V \cong V_\lambda \times_{S_\lambda} S$. By (i), taking λ sufficiently large, we can find a morphism $A_{U_\lambda} \to A_{V_\lambda}$ inducing the morphism $A_U \to A_V$. Let $\mathscr{F}_\lambda$ be the cokernel of $A_{U_\lambda} \to A_{V_\lambda}$. Then $\mathscr{F} \cong u_\lambda^* \mathscr{F}_\lambda$.

(iv) We have $\psi\phi = 0$. By (i), we have $\psi_\lambda \phi_\lambda = 0$ for sufficiently large λ. Let $\mathscr{K}_\lambda$ (resp. $\mathscr{K}$) be the kernel of ψ_λ (resp. ψ), and let $\mathscr{L}_\lambda$ (resp. $\mathscr{L}$) be the image of ϕ_λ (resp. ϕ). We have $\mathscr{L} \cong \mathscr{K}$. By (ii), we have $\mathscr{L}_\lambda \cong \mathscr{K}_\lambda$ for sufficiently large λ. □

Lemma 5.9.9. *Suppose that S_0, S_λ and S are noetherian schemes, and A is a noetherian ring. Let $(\mathscr{I}_\lambda)$ be an injective sheaf of A-modules on $(S_\lambda, u_{\lambda\mu})$, and let $\mathscr{I} = \varinjlim_\lambda u_\lambda^* \mathscr{I}_\lambda$. Then $\mathscr{I}$ is an injective of A-modules on S.*

Proof. Sheaves of the form A_U with U being etale S-schemes of finite type form a family of generators for the category of sheaves of A-modules on S. To prove that $\mathscr{I}$ is injective, it suffices to show that for any subsheaf $\mathscr{F}$ of A_U, any morphism $\mathscr{F} \to \mathscr{I}$ can be extended to a morphism $A_U \to \mathscr{I}$. By 5.8.6, $\mathscr{F}$ is constructible. By 5.9.8, for sufficiently large λ, we can find an etale S_λ-scheme U_λ of finite type such that $U \cong U_\lambda \times_{S_\lambda} S$, a subsheaf $\mathscr{F}_\lambda$ of A_{U_λ} such that $\mathscr{F} \cong u_\lambda^* \mathscr{F}_\lambda$, and a morphism $\mathscr{F}_\lambda \to \mathscr{I}_\lambda$ inducing the given morphism $\mathscr{F} \to \mathscr{I}$. Since $\mathscr{I}_\lambda$ is necessarily injective, the morphism $\mathscr{F}_\lambda \to \mathscr{I}_\lambda$ can be extended to a morphism $A_{U_\lambda} \to \mathscr{I}_\lambda$. It follows that $\mathscr{F} \to \mathscr{I}$ can be extended to a morphism $A_U \to \mathscr{I}$. □

Proposition 5.9.10. *Suppose that* S_0, S_λ *and* S *are noetherian schemes, and* A *is a noetherian ring. Let* $\mathscr{F}_0$ *be a sheaf of* A*-modules on* S_0, $\mathscr{F}_\lambda$ *and* $\mathscr{F}$ *its inverse images on* S_λ *and* S, *respectively. If* $\mathscr{F}$ *is constructible, then for any sheaf of* A*-modules* $(\mathscr{G}_\lambda)$ *on* $(S_\lambda, u_{\lambda\mu})$, *we have*

$$\mathrm{Ext}^q_A(\mathscr{F}, \mathscr{G}) \cong \varinjlim_\lambda \mathrm{Ext}^q_A(\mathscr{F}_\lambda, \mathscr{G}_\lambda),$$
$$\mathscr{E}xt^q_A(\mathscr{F}, \mathscr{G}) \cong \varinjlim_\lambda u_\lambda^* \mathscr{E}xt^q_A(\mathscr{F}_\lambda, \mathscr{G}_\lambda)$$

for all q, *where* $\mathscr{G} = \varinjlim_\lambda u_\lambda^* \mathscr{G}_\lambda$.

Proof. Let

$$0 \to (\mathscr{I}_\lambda^0) \to (\mathscr{I}_\lambda^1) \to \cdots$$

be a resolution of $(\mathscr{G}_\lambda)$ by injective sheaves of A-modules on $(S_\lambda, u_{\lambda\mu})$. By 5.9.9, $\varinjlim_\lambda u_\lambda^* \mathscr{I}_\lambda^\cdot$ is an injective resolution of $\mathscr{G}$, and each $\mathscr{I}_\lambda^\cdot$ is an injective resolution of $\mathscr{G}_\lambda$. By 5.9.8, we have

$$\mathrm{Hom}_A(\mathscr{F}, \varinjlim_\lambda u_\lambda^* \mathscr{I}_\lambda^\cdot) \cong \varinjlim_\lambda \mathrm{Hom}_A(\mathscr{F}_\lambda, \mathscr{I}_\lambda^\cdot).$$

So we have

$$\begin{aligned}\mathrm{Ext}^q_A(\mathscr{F}, \mathscr{G}) &\cong H^q(\mathrm{Hom}_A(\mathscr{F}, \varinjlim_\lambda u_\lambda^* \mathscr{I}_\lambda^\cdot))\\ &\cong H^q(\varinjlim_\lambda \mathrm{Hom}_A(\mathscr{F}_\lambda, \mathscr{I}_\lambda^\cdot))\\ &\cong \varinjlim_\lambda H^q(\mathrm{Hom}_A(\mathscr{F}_\lambda, \mathscr{I}_\lambda^\cdot))\\ &\cong \varinjlim_\lambda \mathrm{Ext}^q_A(\mathscr{F}_\lambda, \mathscr{G}_\lambda).\end{aligned}$$

Note that $\mathscr{E}xt^q(\mathscr{F}, \mathscr{G})$ is the sheaf associated to the presheaf $U \mapsto \mathrm{Ext}^q_A(\mathscr{F}|_U, \mathscr{G}|_U)$ for any $U \in \mathrm{ob}\, S^f_{\mathrm{et}}$. We have

$$\mathrm{Ext}^q_A(\mathscr{F}|_U, \mathscr{G}|_U) \cong \varinjlim_{\lambda \geq \lambda_U} \mathrm{Ext}^q_A(\mathscr{F}_\lambda|_{U_\lambda}, \mathscr{G}_\lambda|_{U_\lambda}).$$

Since $\mathscr{E}xt^q(\mathscr{F}_\lambda, \mathscr{G}_\lambda)$ is the sheaf associated to the presheaf $U_\lambda \mapsto \mathrm{Ext}^q_A(\mathscr{F}_\lambda|_{U_\lambda}, \mathscr{G}_\lambda|_{U_\lambda})$ for any $U_\lambda \in \mathrm{ob}\,(S_\lambda)^f_{\mathrm{et}}$, we have

$$\mathscr{E}xt^q_A(\mathscr{F}, \mathscr{G}) \cong \varinjlim_\lambda u_\lambda^* \mathscr{E}xt^q_A(\mathscr{F}_\lambda, \mathscr{G}_\lambda)$$

by 5.9.1 and 5.9.2. □

Corollary 5.9.11. *Let S be a noetherian scheme, A a noetherian ring, $s \in S$, $\widetilde{S}_{\bar{s}}$ the strict henselization of S at $\bar{s}$, and $\mathscr{F}$ and $\mathscr{G}$ sheaves of A-modules on S. If $\mathscr{F}$ is constructible, then*

$$(\mathscr{E}xt^q_A(\mathscr{F},\mathscr{G}))_{\bar{s}} \cong \mathrm{Ext}^q_A(\mathscr{F}|_{\widetilde{S}_{\bar{s}}}, \mathscr{G}|_{\widetilde{S}_{\bar{s}}}).$$

Proof. Let (V_λ) be the family of affine etale neighborhoods of $\bar{s}$ in S. We have

$$\widetilde{S}_{\bar{s}} = \mathrm{Spec}\,(\varinjlim_{\lambda} \Gamma(V_\lambda, \mathscr{O}_{V_\lambda})).$$

Since $\mathscr{E}xt^q_A(\mathscr{F},\mathscr{G})$ is the sheaf associated to the presheaf $V \mapsto \mathrm{Ext}^q_A(\mathscr{F}|_V, \mathscr{G}|_V)$, we have

$$(\mathscr{E}xt^q_A(\mathscr{F},\mathscr{G}))_{\bar{s}} \cong \varinjlim_{\lambda} \mathrm{Ext}^q_A(\mathscr{F}|_{V_\lambda}, \mathscr{G}|_{V_\lambda}).$$

By 5.9.10, we have

$$\varinjlim_{\lambda} \mathrm{Ext}^q_A(\mathscr{F}|_{V_\lambda}, \mathscr{G}|_{V_\lambda}) \cong \mathrm{Ext}^q_A(\mathscr{F}|_{\widetilde{S}_{\bar{s}}}, \mathscr{G}|_{\widetilde{S}_{\bar{s}}}).$$

Our assertion follows. □

Chapter 6

Derived Categories and Derived Functors

In this chapter, covariant functors between categories are simply called functors.

6.1 Triangulated Categories

([Beilinson, Bernstein and Deligne (1982)] 1.1, [Hartshorne (1966)] I 0–2, [SGA 4] XVII 1.1, [SGA $4\frac{1}{2}$] C.D. I 1.)

A *triangulated category* $\mathscr{C}$ is an additive category provided with an automorphism $T : \mathscr{C} \to \mathscr{C}$, called the *translation functor*, and a collection of sextuples (X, Y, Z, u, v, w), called *distinguished triangles* and also denoted by

$$X \stackrel{u}{\to} Y \stackrel{v}{\to} Z \stackrel{w}{\to},$$

where X, Y and Z are objects in $\mathscr{C}$, $u : X \to Y$, $v : Y \to Z$ and $w : Z \to T(X)$ are morphisms in $\mathscr{C}$, such that the following axioms hold:

(TR1) Any morphism $u : X \to Y$ in $\mathscr{C}$ can be extended to a distinguished triangle

$$X \stackrel{u}{\to} Y \to Z \to .$$

The sextuple $(X, X, 0, \mathrm{id}_X, 0, 0)$ is a distinguished triangle for any object X in $\mathscr{C}$. Any sextuple isomorphic to a distinguished triangle is a distinguished triangle. Here a morphism from a sextuple (X, Y, Z, u, v, w) to another sextuple (X', Y', Z', u', v', w') is a triple $(X \stackrel{f}{\to} X', Y \stackrel{g}{\to} Y', Z \stackrel{h}{\to} Z')$ of morphisms in $\mathscr{C}$ such that the following diagram commutes:

$$\begin{array}{ccccccc} X & \stackrel{u}{\to} & Y & \stackrel{v}{\to} & Z & \stackrel{w}{\to} & T(X) \\ {\scriptstyle f}\downarrow & & {\scriptstyle g}\downarrow & & {\scriptstyle h}\downarrow & & \downarrow{\scriptstyle T(f)} \\ X' & \stackrel{u'}{\to} & Y' & \stackrel{v'}{\to} & Z' & \stackrel{w'}{\to} & T(X'). \end{array}$$

We define composites of morphisms of sextuples in the obvious way. An isomorphism of sextuples is a morphism with a two-sided inverse.

(TR2) A sextuple (X, Y, Z, u, v, w) is a distinguished triangle if and only if $(Y, Z, T(X), v, w, -T(u))$ is a distinguished triangle.

(TR3) Given two distinguished triangles

$$X \to Y \to Z \to \text{ and } X' \to Y' \to Z' \to,$$

and two morphisms $f : X \to X'$ and $g : Y \to Y'$ such that the diagram

$$\begin{array}{ccc} X & \to & Y \\ f\downarrow & & \downarrow g \\ X' & \to & Y' \end{array}$$

commutes, there exists a morphism $h : Z \to Z'$ such that (f, g, h) is a morphism of distinguished triangles.

(TR4) Given two morphisms $u : X \to Y$ and $v : Y \to Z$, and three triangles

$$X \xrightarrow{u} Y \xrightarrow{j} Z' \to, \quad Y \xrightarrow{v} Z \to X' \xrightarrow{i}, \quad X \xrightarrow{vu} Z \to Y' \to,$$

there exist morphisms $f : Z' \to Y'$ and $g : Y' \to X'$ such that (id_X, v, f) and (u, id_Z, g) are morphisms of distinguished triangles, and the sextuple $(Z', Y', X', f, g, T(j) \circ i)$ is a distinguished triangle.

Let $\mathscr{C}'$ and $\mathscr{C}$ be triangulated categories with the translation functors T' and T, respectively. An additive covariant (resp. contravariant) functor $F : \mathscr{C}' \to \mathscr{C}$ is called *exact* if there exists an isomorphism of functors $\phi : F \circ T' \xrightarrow{\cong} T \circ F$ (resp. $\phi : F \circ T'^{-1} \xrightarrow{\cong} T \circ F$) such that for any distinguished triangle (X', Y', Z', u', v', w') in $\mathscr{C}'$, the sextuple

$$\begin{array}{ll} & (F(X'), F(Y'), F(Z'), F(u'), F(v'), \phi_{X'} \circ F(w')) \\ (\text{resp.} & (F(Z'), F(Y'), F(X'), F(v'), F(u'), \phi_{Z'} \circ F(T^{-1}(w')))) \end{array}$$

is a distinguished triangle in $\mathscr{C}$.

Let I be a finite set, $\epsilon : I \to \{\pm 1\}$ a map, $\mathscr{C}$ a triangulated category with the translation functor T, and $\mathscr{C}_i$ $(i \in I)$ triangulated categories with the translation functors T_i, and $F : \prod_i \mathscr{C}_i \to \mathscr{C}$ an additive functor covariant (resp. contravariant) in the i-th component if $\epsilon(i) = 1$ (resp. $\epsilon(i) = -1$). For any $i \in I$, denote by $(T_i^{\epsilon(i)})$ the functor $\prod_i \mathscr{C}_i \to \prod_i \mathscr{C}_i$ induced by $T_i^{\epsilon(i)}$ and $\mathrm{id}_{\mathscr{C}_j}$ for $j \neq i$. We say that F is an *exact functor* if the following conditions hold:

(a) There exist isomorphisms of functors $\phi_i : F \circ (T_i^{\epsilon(i)}) \overset{\cong}{\to} T \circ F$ $(i \in I)$ such that for any $i \neq j$, the diagram

$$\begin{array}{ccc} F \circ (T_i^{\epsilon(i)}) \circ (T_j^{\epsilon(j)}) = F \circ (T_j^{\epsilon(j)}) \circ (T_i^{\epsilon(i)}) & \overset{\phi_j}{\to} & T \circ F \circ (T_i^{\epsilon(i)}) \\ \phi_i \downarrow & & \downarrow \phi_i \\ T \circ F \circ (T_j^{\epsilon(j)}) & \overset{\phi_j}{\to} & T \circ T \circ F \end{array}$$

is anti-commutative, that is, $\phi_i \phi_j = -\phi_j \phi_i$.

(b) With respect to each component of $\prod_i \mathscr{C}_i$, F is an exact functor.

Proposition 6.1.1. *Let $\mathscr{C}$ be a triangulated category.*

(i) *For any distinguished triangle $X \overset{u}{\to} Y \overset{v}{\to} Z \overset{w}{\to}$ in $\mathscr{C}$, we have $vu = 0$ and $wv = 0$.*

(ii) *For any distinguished triangle $X \overset{u}{\to} Y \overset{v}{\to} Z \overset{w}{\to}$ and any object W in $\mathscr{C}$, the following sequences are exact*

$$\cdots \to \mathrm{Hom}(W, T^i X) \to \mathrm{Hom}(W, T^i Y) \to \mathrm{Hom}(W, T^i Z) \to \mathrm{Hom}(W, T^{i+1} X) \to \cdots,$$
$$\cdots \leftarrow \mathrm{Hom}(T^i X, W) \leftarrow \mathrm{Hom}(T^i Y, W) \leftarrow \mathrm{Hom}(T^i Z, W) \leftarrow \mathrm{Hom}(T^{i+1} X, W) \leftarrow \cdots.$$

(iii) *In the axiom (TR3), if f and g are isomorphisms, then so is h.*

Proof.

(i) The pair of morphisms (id_X, u) gives rise to a commutative diagram

$$\begin{array}{ccc} X & \overset{\mathrm{id}_X}{\to} & X \\ \mathrm{id}_X \downarrow & & \downarrow u \\ X & \overset{u}{\to} & Y. \end{array}$$

By (TR3), it can be extended to a morphism from the distinguished triangle

$$X \overset{\mathrm{id}_X}{\to} X \to 0 \to$$

to the distinguished triangle

$$X \overset{u}{\to} Y \overset{v}{\to} Z \overset{w}{\to}.$$

So we have a commutative diagram

$$\begin{array}{ccccccc} X & \overset{\mathrm{id}_X}{\to} & X & \to & 0 & \to & T(X) \\ \mathrm{id}_X \downarrow & & \downarrow u & & \downarrow & & \downarrow \mathrm{id}_{T(X)} \\ X & \overset{u}{\to} & Y & \overset{v}{\to} & Z & \overset{w}{\to} & T(X). \end{array}$$

It follows that $vu = 0$. By (TR2),

$$Y \overset{v}{\to} Z \overset{w}{\to} T(X) \overset{-T(u)}{\to}$$

is also a distinguished triangle. It follows that $wv = 0$ from the discussion above.

(ii) It suffices to prove that

$$\begin{array}{c}\mathrm{Hom}(W,X) \to \mathrm{Hom}(W,Y) \to \mathrm{Hom}(W,Z),\\ \mathrm{Hom}(X,W) \leftarrow \mathrm{Hom}(Y,W) \leftarrow \mathrm{Hom}(Z,W)\end{array}$$

are exact and then apply (TR2) repeatedly. By (i), the composites of morphisms in these two sequences are zero. Suppose that f lies in the kernel of $\mathrm{Hom}(W,Y) \to \mathrm{Hom}(W,Z)$. We have a commutative diagram

$$\begin{array}{ccc} W & \to & 0 \\ {\scriptstyle f}\downarrow & & \downarrow \\ Y & \stackrel{v}{\to} & Z. \end{array}$$

By (TR2) and (TR3), there exists $h \in \mathrm{Hom}(W,X)$ making the following diagram commute:

$$\begin{array}{ccccccc} W & \stackrel{\mathrm{id}}{\to} & W & \to & 0 & \to & \\ {\scriptstyle h}\downarrow & & {\scriptstyle f}\downarrow & & \downarrow & & \\ X & \stackrel{u}{\to} & Y & \stackrel{v}{\to} & Z & \stackrel{w}{\to} & . \end{array}$$

We then have $f = uh$, and f lies in the image of $\mathrm{Hom}(W,X) \to \mathrm{Hom}(W,Y)$. This proves that the first sequence is exact. Similarly, one can prove that the second sequence is exact.

(iii) Consider the commutative diagram

$$\begin{array}{ccccccccc} \mathrm{Hom}(Z',X) & \to & \mathrm{Hom}(Z',Y) & \to & \mathrm{Hom}(Z',Z) & \to & \mathrm{Hom}(Z',T(X)) & \to & \mathrm{Hom}(Z',T(Y)) \\ \downarrow & & \downarrow & & \downarrow & & \downarrow & & \downarrow \\ \mathrm{Hom}(Z',X') & \to & \mathrm{Hom}(Z',Y') & \to & \mathrm{Hom}(Z',Z') & \to & \mathrm{Hom}(Z',T(X')) & \to & \mathrm{Hom}(Z',T(Y')), \end{array}$$

where the vertical arrows are induced by $f, g, h, T(f), T(g)$, respectively. By (ii), the horizontal lines are exact. If f and g are isomorphisms, then by the five lemma, the morphism

$$\mathrm{Hom}(Z', Z) \to \mathrm{Hom}(Z', Z')$$

induced by h is an isomorphism. So there exists $h' : Z' \to Z$ such that $hh' = \mathrm{id}_{Z'}$. Similarly one can prove that h has a left inverse. Hence h is an isomorphism. □

Let $\mathscr{A}$ be an additive category and let $K(\mathscr{A})$ be the category whose objects are complexes of objects in $\mathscr{A}$, and whose morphisms are homotopy classes of morphisms of complexes. A complex $X^\cdot$ is *bounded below* (resp. *bounded above*, resp. *bounded*) if $X^n = 0$ for $n \ll 0$ (resp. $n \gg 0$, resp.

$n \ll 0$ and $n \gg 0$). Denote by $K^+(\mathscr{A})$, $K^-(\mathscr{A})$ and $K^b(\mathscr{A})$ the full subcategories of $K(\mathscr{A})$ consisting of bounded below, bounded above, and bounded complexes, respectively.

Define the translation functor $T : K(\mathscr{A}) \to K(\mathscr{A})$ as follows. For any complex $X^\cdot \in \operatorname{ob} K(\mathscr{A})$, $T(X^\cdot)$ is the complex defined by

$$T(X^\cdot)^i = X^{i+1}, \quad d^i_{T(X^\cdot)} = -d^{i+1}_{X^\cdot}.$$

For any morphism $u^\cdot : X^\cdot \to Y^\cdot$, define $T(u^\cdot) : T(X^\cdot) \to T(Y^\cdot)$ by

$$T(u^\cdot)^i = u^{i+1}.$$

We also write $X^\cdot[1]$ for $T(X^\cdot)$ and write $X^\cdot[n]$ for $T^n(X)$ for any integer n.

For any morphism of complexes $u : X^\cdot \to Y^\cdot$ in $K(\mathscr{A})$, define the *mapping cone* $C(u)^\cdot$ of u to be the complex defined by

$$C(u)^p = X^{p+1} \oplus Y^p,$$
$$d : C(u)^p \to C(u)^{p+1}, \quad (x_{p+1}, y_p) \mapsto (-dx_{p+1}, u(x_{p+1}) + dy_p).$$

One can check that $dd = 0$. We have canonical morphisms $Y^\cdot \to C(u)^\cdot$ and $C(u)^\cdot \to X^\cdot[1]$. Any sextuple isomorphic to

$$X^\cdot \xrightarrow{u} Y^\cdot \to C(u)^\cdot \to$$

is called a distinguished triangle in $K(\mathscr{A})$. One can show that $K(\mathscr{A})$ is a triangulated category, and the subcategories $K^+(\mathscr{A})$, $K^-(\mathscr{A})$ and $K^b(\mathscr{A})$ are also triangulated categories. Moreover, we have the following.

Proposition 6.1.2. *Suppose that $\mathscr{A}$ is an abelian category. Let*

$$X^\cdot \to Y^\cdot \to Z^\cdot \to$$

be a distinguished triangle in $K(\mathscr{A})$. Then the sequence

$$\cdots \to H^i(X^\cdot) \to H^i(Y^\cdot) \to H^i(Z^\cdot) \to H^{i+1}(X^\cdot) \to \cdots$$

is exact.

Suppose that $\mathscr{A}$ is an abelian category. An abelian subcategory $\mathscr{A}'$ of $\mathscr{A}$ is called *thick* if for any exact sequence

$$0 \to X' \to X \to X'' \to 0$$

in $\mathscr{A}$ with $X', X'' \in \operatorname{ob}\mathscr{A}'$, we have $X \in \operatorname{ob}\mathscr{A}'$. For any thick abelian subcategory $\mathscr{A}'$ of $\mathscr{A}$, we defined $K_{\mathscr{A}'}(\mathscr{A})$ to be the full subcategory of $K(\mathscr{A})$ consisting of the complexes $X^\cdot$ in $K(\mathscr{A})$ such that $\mathscr{H}^q(X^\cdot)$ are objects in $\mathscr{A}'$ for all q. By 6.1.2, $K_{\mathscr{A}'}(\mathscr{A})$ is a triangulated subcategory of $K(\mathscr{A})$. Similarly, we can define triangulated subcategories $K^+_{\mathscr{A}'}(\mathscr{A})$, $K^-_{\mathscr{A}'}(\mathscr{A})$ and $K^b_{\mathscr{A}'}(\mathscr{A})$.

6.2 Derived Categories

([Hartshorne (1966)] I 3–4, [SGA $4\frac{1}{2}$] C.D. I 2, II 1.)

Let $\mathscr{C}$ be a category. A family S of morphisms in $\mathscr{C}$ is called a *multiplicative system* if it satisfies the following conditions:

(a) For any $f, g \in S$, we have $fg \in S$ whenever the composite fg is defined. Identity morphisms are in S.

(b) Any diagram

$$\begin{array}{ccc} X & \to & Y \\ {\scriptstyle s}\downarrow & & \\ Z & & \end{array} \qquad \left(\text{resp.} \quad \begin{array}{ccc} & & Y \\ & & \downarrow{\scriptstyle t} \\ Z & \to & W \end{array}\right)$$

with $s \in S$ (resp. $t \in S$) can be completed to a commutative diagram

$$\begin{array}{ccc} X & \to & Y \\ {\scriptstyle s}\downarrow & & \downarrow{\scriptstyle t} \\ Z & \to & W \end{array}$$

with $t \in S$ (resp. $s \in S$).

(c) Let $f, g : X \to Y$ be two morphisms. There exists $s \in S$ such that $sf = sg$ if and only if there exists $t \in S$ such that $ft = gt$.

Let $\mathscr{C}$ be a category and let S be a multiplicative system of morphisms in $\mathscr{C}$. For any $X \in \operatorname{ob}\mathscr{C}$, define a category X^- (resp. X^+) as follows: Objects of X^- (resp. X^+) are morphisms $X' \stackrel{s}{\to} X$ (resp. $X \stackrel{s}{\to} X'$) in S. We also denote these objects by X' for simplicity. Given two objects $X_i' \stackrel{s_i}{\to} X$ (resp. $X \stackrel{s_i}{\to} X_i'$) $(i = 1, 2)$ in X^- (resp. X^+), morphisms from s_1 to s_2 are morphisms $f : X_1' \to X_2'$ in $\mathscr{C}$ such that $s_2 f = s_1$ (resp. $fs_1 = s_2$). One can verify $(X^-)^\circ$ and X^+ satisfy the conditions (I2) and (I3) in 2.7, where $(X^-)^\circ$ is the opposite category of X^-.

Define a category $S^{-1}\mathscr{C}$ as follows: $S^{-1}\mathscr{C}$ has the same family of objects as $\mathscr{C}$. For any $X, Y \in \operatorname{ob} S^{-1}\mathscr{C}$, $\operatorname{Hom}_{\mathscr{C}}(-, Y)$ is a contravariant functor on X^-, and we define

$$\operatorname{Hom}_{S^{-1}\mathscr{C}}(X, Y) = \varinjlim_{X' \in \operatorname{ob}(X^-)^\circ} \operatorname{Hom}_{\mathscr{C}}(X', Y).$$

If an element in $\operatorname{Hom}_{S^{-1}\mathscr{C}}(X, Y)$ is represented by an element in $f : X' \to Y$ in $\operatorname{Hom}_{\mathscr{C}}(X', Y)$ for some object $s : X' \to X$ in X^-, we denote this element by the diagram

$$\begin{array}{ccc} X' & & \\ {\scriptstyle s}\downarrow & {\scriptstyle f}\searrow & \\ X & & Y. \end{array}$$

Given two morphisms

$$\begin{array}{ccc} X' & & \\ {\scriptstyle s}\downarrow{\scriptstyle f} & \searrow & \\ X & & Y, \end{array} \qquad \begin{array}{ccc} Y' & & \\ {\scriptstyle t}\downarrow{\scriptstyle g} & \searrow & \\ Y & & Z \end{array}$$

in $S^{-1}\mathscr{C}$, find a commutative diagram

$$\begin{array}{ccc} X'' & & \\ {\scriptstyle t'}\downarrow{\scriptstyle f'} & \searrow & \\ X' & & Y' \\ & {\scriptstyle f}\searrow & \downarrow{\scriptstyle t} \\ & & Y \end{array}$$

such that $t' : X'' \to X'$ lies in S. The composite of the above two morphisms is defined to be the morphism represented by the diagram

$$\begin{array}{ccc} X'' & & \\ {\scriptstyle st'}\downarrow{\scriptstyle gf'} & \searrow & \\ X & & Z. \end{array}$$

One can verify that the composite of morphisms in $S^{-1}\mathscr{C}$ is well-defined, that is, independent of the choice of diagrams representing morphisms. We have

$$\begin{aligned} \mathrm{Hom}_{S^{-1}\mathscr{C}}(X, Y) &= \varinjlim_{X' \in \mathrm{ob}\,(X^-)^\circ} \mathrm{Hom}_{\mathscr{C}}(X', Y) \\ &= \varinjlim_{Y' \in \mathrm{ob}\, Y^+} \mathrm{Hom}_{\mathscr{C}}(X, Y') \\ &= \varinjlim_{\substack{X' \in \mathrm{ob}\,(X^-)^\circ, \\ Y' \in \mathrm{ob}\, Y^+}} \mathrm{Hom}_{\mathscr{C}}(X', Y'). \end{aligned}$$

We call $S^{-1}\mathscr{C}$ the *localization* of $\mathscr{C}$ with respect to S. We have a canonical functor $Q : \mathscr{C} \to S^{-1}\mathscr{C}$. It transforms morphisms of $\mathscr{C}$ in S into isomorphisms in $S^{-1}\mathscr{C}$.

Proposition 6.2.1. *Let $\mathscr{C}$ and $\mathscr{D}$ be categories, S a multiplicative system of morphisms in $\mathscr{C}$, $F : \mathscr{C} \to \mathscr{D}$ a functor that transforms morphisms of $\mathscr{C}$ lying in S into isomorphisms in $\mathscr{D}$, and $Q : \mathscr{C} \to S^{-1}\mathscr{C}$ the canonical functor. Then there exists a unique functor $G : S^{-1}\mathscr{C} \to \mathscr{D}$ such that $F = G \circ Q$.*

Proposition 6.2.2. *Let $\mathscr{C}$ be a triangulated category with the translation functor T and let S be a multiplicative system of morphisms in $\mathscr{C}$. Suppose the following two conditions hold:*

(a) *For any* $s \in S$, *we have* $T(s) \in S$.

(b) *In the axiom (TR3), if* $f, g \in S$, *then* $h \in S$.

By (a), T *induces an automorphism on* $S^{-1}\mathscr{C}$. *A sextuple in* $S^{-1}\mathscr{C}$ *is called a distinguished triangle if it is isomorphic in* $S^{-1}\mathscr{C}$ *to a sextuple coming from a distinguished triangle in* $\mathscr{C}$. *Then* $S^{-1}\mathscr{C}$ *is a triangulated category, and the canonical functor* $Q : \mathscr{C} \to S^{-1}\mathscr{C}$ *is exact.*

Let $\mathscr{A}$ be an abelian category, and let $X^{\cdot}$ and $Y^{\cdot}$ be two complexes of objects in $\mathscr{A}$. A morphism $u : X^{\cdot} \to Y^{\cdot}$ in $K(\mathscr{A})$ is called a *quasi-isomorphism* if u induces isomorphisms $H^i(X^{\cdot}) \to H^i(Y^{\cdot})$ for all i. By 6.1.2, this is equivalent to saying that the mapping cone $C(u)^{\cdot}$ is acyclic, that is, $H^i(C(u)^{\cdot}) = 0$ for all i.

Proposition 6.2.3. *Let* $\mathscr{A}$ *be an abelian category, let* $\mathscr{A}'$ *be a thick abelian subcategory of* $\mathscr{A}$, *and let* S *be the family of quasi-isomorphisms in* $K(\mathscr{A})$ *(resp.* $K^+(\mathscr{A})$, $K^-(\mathscr{A})$, $K^b(\mathscr{A})$, $K_{\mathscr{A}'}(\mathscr{A})$, $K^+_{\mathscr{A}'}(\mathscr{A})$, $K^-_{\mathscr{A}'}(\mathscr{A})$, $K^b_{\mathscr{A}'}(\mathscr{A})$*). Then* S *is a multiplicative system satisfying the conditions* (a) *and* (b) *in 6.2.2.*

Proof. It is clear that if $s : Y \to Z$ and $t : X \to Y$ are quasi-isomorphisms, then so is st.

Given a diagram

$$\begin{array}{ccc} X & \xrightarrow{u} & Y \\ {\scriptstyle s}\downarrow & & \\ Z & & \end{array}$$

with $s \in S$, by (TR1) and (TR2), we can choose a distinguished triangle

$$C \xrightarrow{v} X \xrightarrow{s} Z \to .$$

Since s is a quasi-isomorphism, C is acyclic by 6.1.2. By (TR1), we can choose a distinguished triangle

$$C \xrightarrow{uv} Y \xrightarrow{t} W \to .$$

Then t is a quasi-isomorphism. By (TR3), the pair (id_C, u) can be extended to a morphism of triangles (id_C, u, g) from $C \xrightarrow{v} X \xrightarrow{s} Z \to$ to $C \xrightarrow{uv} Y \xrightarrow{t} W \to$. Then we have a commutative diagram

$$\begin{array}{ccc} X & \xrightarrow{u} & Y \\ {\scriptstyle s}\downarrow & & \downarrow{\scriptstyle t} \\ Z & \xrightarrow{g} & W \end{array}$$

with $t \in S$. Similarly any diagram

$$\begin{array}{ccc} & & Y \\ & & \downarrow{\scriptstyle t} \\ Z & \to & W \end{array}$$

with $t \in S$ can be completed to a commutative diagram

$$\begin{array}{ccc} X & \rightarrow & Y \\ {\scriptstyle s}\downarrow & & \downarrow{\scriptstyle t} \\ Z & \rightarrow & W \end{array}$$

with $s \in S$.

Let $f : X \to Y$ be a morphism and let $s : Y \to Y'$ be a quasi-isomorphism such that $sf = 0$. By (TR1) and (TR2), we can choose a distinguished triangle

$$C \stackrel{u}{\to} Y \stackrel{s}{\to} Y' \to .$$

Since s is a quasi-isomorphism, C is acyclic. Since $sf = 0$, by 6.1.1 (ii), there exists a morphism $v : X \to C$ such that $f = uv$. By (TR1) and (TR2), we can choose a distinguished triangle

$$X' \stackrel{t}{\to} X \stackrel{v}{\to} C \to .$$

Since C is acyclic, t is a quasi-isomorphism. We have $ft = uvt = 0$. Similarly, one can prove that if there exists a quasi-isomorphism t such that $ft = 0$, then there exists a quasi-isomorphism s such that $sf = 0$.

It is clear that S satisfies the condition (a) in 6.2.2. It follows from 6.1.2 and the five lemma that S satisfies the condition (b). □

Let $\mathscr{A}$ be an abelian category, $\mathscr{A}'$ a thick abelian subcategory of $\mathscr{A}$, and S the family of quasi-isomorphisms in

$$\begin{aligned} K(\mathscr{A}) \ \ (\text{resp. } & K^+(\mathscr{A}), K^-(\mathscr{A}), K^b(\mathscr{A}), \\ & K_{\mathscr{A}'}(\mathscr{A}), K^+_{\mathscr{A}'}(\mathscr{A}), K^-_{\mathscr{A}'}(\mathscr{A}), K^b_{\mathscr{A}'}(\mathscr{A})). \end{aligned}$$

The category

$$\begin{aligned} S^{-1}K(\mathscr{A}) \ \ (\text{resp. } & S^{-1}K^+(\mathscr{A}), S^{-1}K^-(\mathscr{A}), S^{-1}K^b(\mathscr{A}), \\ & S^{-1}K_{\mathscr{A}'}(\mathscr{A}), S^{-1}K^+_{\mathscr{A}'}(\mathscr{A}), S^{-1}K^-_{\mathscr{A}'}(\mathscr{A}), S^{-1}K^b_{\mathscr{A}'}(\mathscr{A})) \end{aligned}$$

is called a *derived category* of $\mathscr{A}$, and is denoted by

$$\begin{aligned} D(\mathscr{A}) \ \ (\text{resp. } & D^+(\mathscr{A}), D^-(\mathscr{A}), D^b(\mathscr{A}), \\ & D_{\mathscr{A}'}(\mathscr{A}), D^+_{\mathscr{A}'}(\mathscr{A}), D^-_{\mathscr{A}'}(\mathscr{A}), D^b_{\mathscr{A}'}(\mathscr{A})). \end{aligned}$$

We have a commutative diagram

$$\begin{array}{ccc} D^b_{\mathscr{A}'}(\mathscr{A}) & \rightarrow & D^b(\mathscr{A}) \\ \downarrow & & \downarrow \\ D^{\pm}_{\mathscr{A}'}(\mathscr{A}) & \rightarrow & D^{\pm}(\mathscr{A}) \\ \downarrow & & \downarrow \\ D_{\mathscr{A}'}(\mathscr{A}) & \rightarrow & D(\mathscr{A}), \end{array}$$

where all arrows are fully faithful exact functors. We have a canonical exact functor $D(\mathscr{A}') \to D_{\mathscr{A}'}(\mathscr{A})$ which in general is neither fully faithful nor essentially surjective.

Proposition 6.2.4. *Let $\mathscr{A}$ be an abelian category. For any object A in $\mathscr{A}$, denote also by A the complex*

$$\cdots \to 0 \to A \to 0 \to \cdots$$

whose 0-th component is A and whose other components are 0. The functor

$$\mathscr{A} \to D(\mathscr{A}), \quad A \mapsto A$$

defines an equivalence between the category $\mathscr{A}$ and the full subcategory of $D(\mathscr{A})$ consisting of complexes $X^\cdot$ satisfying $H^i(X^\cdot) = 0$ for all $i \neq 0$.

Proposition 6.2.5. *Let $\mathscr{A}$ be an abelian category, and let*

$$0 \to X^\cdot \stackrel{u}{\to} Y^\cdot \stackrel{v}{\to} Z^\cdot \to 0$$

be a short exact sequence of complexes of objects in $\mathscr{A}$.

(i) *There are canonical morphisms $\delta^i : H^i(Z^\cdot) \to H^{i+1}(X^\cdot)$ such that we have a long exact sequence*

$$\cdots \to H^i(X^\cdot) \stackrel{u}{\to} H^i(Y^\cdot) \stackrel{v}{\to} H^i(Z^\cdot) \stackrel{\delta}{\to} H^{i+1}(X^\cdot) \to \cdots.$$

(ii) *There exists a morphism $w : Z^\cdot \to X^\cdot[1]$ in $D(\mathscr{A})$ such that*

$$X^\cdot \stackrel{u}{\to} Y^\cdot \stackrel{v}{\to} Z^\cdot \stackrel{w}{\to}$$

is a distinguished triangle.

Proof. (i) is well-known. Let us prove (ii). We have a distinguished triangle

$$X^\cdot \stackrel{u}{\to} Y^\cdot \stackrel{v'}{\to} C(u)^\cdot \stackrel{w'}{\to}$$

in $K(\mathscr{A})$. Since $vu = 0$, there exists a morphism $f : C(u)^\cdot \to Z^\cdot$ in $K(\mathscr{A})$ such that $v = fv'$ by 6.1.1 (ii). One can check that the following diagram commutes:

$$\begin{array}{ccccccccc}
\cdots \to & H^i(X^\cdot) & \stackrel{u}{\to} & H^i(Y^\cdot) & \stackrel{v'}{\to} & H^i(C(u)^\cdot) & \stackrel{w'}{\to} & H^{i+1}(X^\cdot) & \to \cdots \\
& \| & & \| & & \downarrow f & & \| & \\
\cdots \to & H^i(X^\cdot) & \stackrel{u}{\to} & H^i(Y^\cdot) & \stackrel{v}{\to} & H^i(Z^\cdot) & \stackrel{-\delta}{\to} & H^{i+1}(X^\cdot) & \to \cdots.
\end{array}$$

By (i), 6.1.2, and the five lemma, $f : C(u)^\cdot \to Z^\cdot$ is a quasi-isomorphism. So f induces an isomorphism in $D(\mathscr{A})$. Define $w = w' \circ f^{-1}$. Then

$$X^\cdot \stackrel{u}{\to} Y^\cdot \stackrel{v}{\to} Z^\cdot \stackrel{w}{\to}$$

is a distinguished triangle in $D(\mathscr{A})$. □

Lemma 6.2.6. *Let $f : Z^{\cdot} \to I^{\cdot}$ be a morphism of complexes in an abelian category $\mathscr{A}$. Suppose that $Z^{\cdot}$ is acyclic, and $I^{\cdot}$ is a bounded below complex of injective objects in $\mathscr{A}$. Then f is homotopic to 0.*

Proof. We leave the proof to the reader. Confer [Fu (2006)] 2.1.1 (ii). □

Lemma 6.2.7. *Let $s : I^{\cdot} \to X^{\cdot}$ be a quasi-isomorphism of complexes in an abelian category $\mathscr{A}$. Suppose $I^{\cdot}$ is a bounded below complex of injective objects. Then there exists a quasi-isomorphism $t : X^{\cdot} \to I^{\cdot}$ such that ts is homotopic to $\mathrm{id}_{I^{\cdot}}$.*

Proof. Choose a distinguished triangle

$$C^{\cdot} \to I^{\cdot} \stackrel{s}{\to} X^{\cdot} \to .$$

Since s is a quasi-isomorphism, $C^{\cdot}$ is acyclic. By 6.2.6, $C^{\cdot} \to I^{\cdot}$ is homotopic to 0. So we have a commutative diagram

$$\begin{array}{ccc} C^{\cdot} & \to & I^{\cdot} \\ \downarrow & & \downarrow \mathrm{id}_{I^{\cdot}} \\ 0 & \to & I^{\cdot} \end{array}$$

in $K(\mathscr{A})$. By (TR3), there exists $t : X^{\cdot} \to I^{\cdot}$ such that $(0, \mathrm{id}_{I^{\cdot}}, t)$ is a morphism of triangles.

$$\begin{array}{ccccc} C^{\cdot} & \to & I^{\cdot} & \stackrel{s}{\to} & X^{\cdot} & \to \\ \downarrow & & \downarrow \mathrm{id}_{I^{\cdot}} & & \downarrow t & \\ 0 & \to & I^{\cdot} & \stackrel{\mathrm{id}_{I^{\cdot}}}{\to} & I^{\cdot} & \to . \end{array}$$

Then ts is homotopic to $\mathrm{id}_{I^{\cdot}}$, and t is a quasi-isomorphism. □

Lemma 6.2.8. *Let $\mathscr{A}$ be an abelian category, $X^{\cdot}, Y^{\cdot} \in \mathrm{ob}\, K(\mathscr{A})$, $I^{\cdot}$ a bounded below complex of injective objects in $\mathscr{A}$, and $s : Y^{\cdot} \to I^{\cdot}$ a quasi-isomorphism. Then we have*

$$\mathrm{Hom}_{K(\mathscr{A})}(X^{\cdot}, I^{\cdot}) \cong \mathrm{Hom}_{D(\mathscr{A})}(X^{\cdot}, Y^{\cdot}).$$

Proof. Let $Y^{\cdot +}$ be the category so that objects in $Y^{\cdot +}$ are quasi-isomorphisms $Y^{\cdot} \to Y'^{\cdot}$, and morphisms in $Y^{\cdot +}$ from an object $Y^{\cdot} \stackrel{s_1}{\to} Y_1'^{\cdot}$ to an object $Y^{\cdot} \stackrel{s_2}{\to} Y_2'^{\cdot}$ are morphisms $f : Y_1'^{\cdot} \to Y_2'^{\cdot}$ in $K(\mathscr{A})$ with the property $fs_1 = s_2$. We claim that $Y^{\cdot} \stackrel{s}{\to} I^{\cdot}$ is a final object in $Y^{\cdot +}$ and hence we have

$$\mathrm{Hom}_{D(\mathscr{A})}(X^{\cdot}, Y^{\cdot}) \cong \varinjlim_{Y'^{\cdot} \in \mathrm{ob}\, Y^{\cdot +}} \mathrm{Hom}_{K(\mathscr{A})}(X^{\cdot}, Y'^{\cdot}) \cong \mathrm{Hom}_{K(\mathscr{A})}(X^{\cdot}, I^{\cdot}).$$

Given an arbitrary object $Y^{\cdot} \stackrel{s'}{\to} Y'^{\cdot}$ in $Y^{\cdot +}$, by 6.2.3, we can find a commutative diagram

$$\begin{array}{ccc} Y^{\cdot} & \stackrel{s}{\to} & I^{\cdot} \\ s' \downarrow & & \downarrow s'' \\ Y'^{\cdot} & \stackrel{f}{\to} & J^{\cdot} \end{array}$$

in $K(\mathscr{A})$ such that s'' is a quasi-isomorphism. By 6.2.7, there exists a morphism $t : J^{\cdot} \to I^{\cdot}$ such that $ts'' = \mathrm{id}_{I^{\cdot}}$ in $K(\mathscr{A})$. Then tf is a morphism in $Y^{\cdot +}$ from $Y^{\cdot} \stackrel{s'}{\to} Y'^{\cdot}$ to $Y^{\cdot} \stackrel{s}{\to} I^{\cdot}$. Suppose $f_1, f_2 : Y'^{\cdot} \to I^{\cdot}$ are two morphisms in $Y^{\cdot +}$ from $Y^{\cdot} \stackrel{s'}{\to} Y'^{\cdot}$ to $Y^{\cdot} \stackrel{s}{\to} I^{\cdot}$. We then have

$$f_1 s' = s = f_2 s'.$$

By 6.2.3, there exists a quasi-isomorphism $s''' : I^{\cdot} \to J^{\cdot}$ such that $s''' f_1 = s''' f_2$. By 6.2.7, there exists a morphism $t' : J^{\cdot} \to I^{\cdot}$ such that $t's''' = \mathrm{id}_{I^{\cdot}}$ in $K(\mathscr{A})$. Then we have

$$f_1 = t's''' f_1 = t's''' f_2 = f_2.$$

So there exists one and only one morphism in $Y^{\cdot +}$ from $Y^{\cdot} \stackrel{s'}{\to} Y'^{\cdot}$ to $Y^{\cdot} \stackrel{s}{\to} I^{\cdot}$. □

Lemma 6.2.9. *Let $\mathscr{I}$ be a family of objects in an abelian category $\mathscr{A}$ such that every object in $\mathscr{A}$ can be embedded into an object in $\mathscr{I}$. Then for any $X^{\cdot} \in \mathrm{ob}\, K(\mathscr{A})$ such that $H^i(X) = 0$ for $i \ll 0$, there exists a quasi-isomorphism $X^{\cdot} \to I^{\cdot}$ such that $I^{\cdot}$ is a bounded below complex of objects in $\mathscr{I}$.*

Proof. Define $I^i = 0$ for $i \ll 0$. Suppose that we have defined I^i and morphisms $s^i : X^i \to I^i$ for all $i < n$ such that $H^i(X^{\cdot}) \cong H^i(I^{\cdot})$ for all $i < n-1$ and such that $H^{n-1}(X^{\cdot}) \to I^{n-1}/\mathrm{im}\, d_{I^{\cdot}}^{n-2}$ is injective. Consider the diagram

$$\begin{array}{ccccc} X^{n-1}/\mathrm{im}\, d_{X^{\cdot}}^{n-2} & \stackrel{d^{n-1}}{\to} & \ker d_{X^{\cdot}}^{n} & \stackrel{i}{\to} & X^n \\ s^{n-1} \downarrow & & \downarrow & & \downarrow \\ I^{n-1}/\mathrm{im}\, d_{I^{\cdot}}^{n-2} & \stackrel{\alpha}{\to} & A & \stackrel{\beta}{\to} & B, \end{array}$$

where on the first line the morphisms are uniquely determined by the property that i is the inclusion and $i \circ d^{n-1}$ is the morphism $X^{n-1}/\mathrm{im}\, d_{X^{\cdot}}^{n-2} \to X^n$ induced by $d_{X^{\cdot}}^{n-1}$, and on the second line, we have

$$A = \mathrm{coker}(X^{n-1}/\mathrm{im}\, d_{X^{\cdot}}^{n-2} \stackrel{(s^{n-1}, -d^{n-1})}{\to} I^{n-1}/\mathrm{im}\, d_{I^{\cdot}}^{n-2} \oplus \ker d_{X^{\cdot}}^{n}),$$

$$B = \mathrm{coker}(X^{n-1}/\mathrm{im}\, d_{X^{\cdot}}^{n-2} \stackrel{(s^{n-1}, -d_{X^{\cdot}}^{n-1})}{\to} I^{n-1}/\mathrm{im}\, d_{I^{\cdot}}^{n-2} \oplus X^{n}),$$

and all arrows in the diagram are the canonical morphisms. The first vertical arrow induces an epimorphism $H^{n-1}(X^\cdot) \to \ker \alpha$. Combined with the hypothesis that $H^{n-1}(X^\cdot) \to I^{n-1}/\mathrm{im}\, d_{I^\cdot}^{n-2}$ is injective, we see that it induces an isomorphism $H^{n-1}(X^\cdot) \cong \ker \alpha$. The middle vertical arrow induces an isomorphism $H^n(X^\cdot) \cong A/\mathrm{im}\, \alpha$, and β is injective. Embed B into an object I^n in $\mathscr{I}$. Define $d_{I^\cdot}^{n-1} : I^{n-1} \to I^n$ to be the composite

$$I^{n-1} \to I^{n-1}/\mathrm{im}\, d_{I^\cdot}^{n-2} \stackrel{\alpha}{\to} A \stackrel{\beta}{\to} B \to I^n,$$

and define $X^n \to I^n$ to be the composite

$$X^n \to B \to I^n.$$

Then we have $H^i(X^\cdot) \cong H^i(I^\cdot)$ for all $i < n$ and $H^n(X^\cdot) \to I^n/\mathrm{im}\, d_{I^\cdot}^{n-1}$ is injective. □

Corollary 6.2.10. *Let $\mathscr{A}$ be an abelian category and let $\mathscr{I}$ be the full subcategory consisting of injective objects. Then the canonical functor $K^+(\mathscr{I}) \to D^+(\mathscr{A})$ is fully faithful. If $\mathscr{A}$ has enough injective objects, then $K^+(\mathscr{I}) \to D^+(\mathscr{A})$ is an equivalence of categories.*

Proof. Let I and J be objects in $K^+(\mathscr{I})$. By 6.2.8, we have

$$\mathrm{Hom}_{K(\mathscr{A})}(I, J) \cong \mathrm{Hom}_{D(\mathscr{A})}(I, J).$$

So the functor $K^+(\mathscr{I}) \to D^+(\mathscr{A})$ is fully faithful. If $\mathscr{A}$ has enough injective objects, then by 6.2.9, any object in $D^+(\mathscr{A})$ is quasi-isomorphic to a complex of injective objects. The above functor is then essentially surjective. □

6.3 Derived Functors

([Hartshorne (1966)] I 5, [SGA 4] XVII 1.2, [SGA $4\frac{1}{2}$] C.D. II 2.)

Let $\mathscr{A}$ and $\mathscr{B}$ be abelian categories, $\mathscr{A}'$ a thick abelian subcategory of $\mathscr{A}$, $K^*(\mathscr{A})$ one of the categories

$$K(\mathscr{A}), K^+(\mathscr{A}), K^-(\mathscr{A}), K^b(\mathscr{A}), K_{\mathscr{A}'}(\mathscr{A}), K^+_{\mathscr{A}'}(\mathscr{A}), K^-_{\mathscr{A}'}(\mathscr{A}), K^b_{\mathscr{A}'}(\mathscr{A}),$$

$D^*(\mathscr{A})$ the corresponding derived category, and $F : K^*(\mathscr{A}) \to K(\mathscr{B})$ an exact functor. For any $X \in \mathrm{ob}\, K^*(\mathscr{A})$, let

$$F'(X) : D(\mathscr{B}) \to (\mathbf{Sets})$$

be the contravariant functor from $D(\mathscr{B})$ to the category of sets defined by

$$F'(X)(Y) = \varinjlim_{X' \in \mathrm{ob}\, X^+} \mathrm{Hom}_{D(\mathscr{B})}(Y, F(X'))$$

for any $Y \in \operatorname{ob} D(\mathscr{B})$, where X^+ is the category so that objects in X^+ are quasi-isomorphisms $X \to X'$, and morphisms in X^+ from an object $X \stackrel{s_1}{\to} X_1'$ to an object $X \stackrel{s_2}{\to} X_2'$ are morphisms $f : X_1' \to X_2'$ in $K^*(\mathscr{A})$ with the property $fs_1 = s_2$. Given a morphism $X_1 \to X_2$ in $K^*(\mathscr{A})$, define a morphism of functors $F'(X_1) \to F'(X_2)$ as follows: For any object $X_1 \to X_1'$ in X_1^+, we can find a commutative diagram

$$\begin{array}{ccc} X_1 & \to & X_2 \\ \downarrow & & \downarrow \\ X_1' & \to & X_2' \end{array}$$

such that $X_2 \to X_2'$ is an object in X_2^+. The morphism $X_1' \to X_2'$ induces a map

$$\operatorname{Hom}_{D(\mathscr{B})}(Y, F(X_1')) \to \operatorname{Hom}_{D(\mathscr{B})}(Y, F(X_2'))$$

for any $Y \in \operatorname{ob} D(\mathscr{B})$. We can take the limit of these maps and get a map $F'(X_1)(Y) \to F'(X_2)(Y)$. We thus get a functor

$$F' : K^*(\mathscr{A}) \to \mathbf{Hom}(D(\mathscr{B})^\circ, (\mathbf{Sets})),$$

where $\mathbf{Hom}(D(\mathscr{B})^\circ, (\mathbf{Sets}))$ denotes the category of contravariant functors from $D(\mathscr{B})$ to the category of sets. One can check that F' transforms quasi-isomorphisms in $K^*(\mathscr{A})$ to isomorphisms in $\mathbf{Hom}(D(\mathscr{B})^\circ, (\mathrm{Sets}))$. So there exists a unique functor

$$RF : D^*(\mathscr{A}) \to \mathbf{Hom}(D(\mathscr{B})^\circ, (\mathbf{Sets})),$$

such that $F' = RF \circ Q$, where $Q : K^*(\mathscr{A}) \to D^*(\mathscr{A})$ is the canonical functor. We call RF the *right derived functor* of F. We say that RF is defined at $X \in \operatorname{ob} D^*(\mathscr{A})$ if the functor $RF(X)$ is representable, that is,

$$F'(X) \cong \operatorname{Hom}_{D(\mathscr{B})}(-, Z)$$

for some object $Z \in D(\mathscr{B})$. Suppose that RF is defined everywhere on $D^*(\mathscr{A})$. Then there exists a functor $D^*(\mathscr{A}) \to D(\mathscr{B})$ that maps each $X \in \operatorname{ob} D^*(\mathscr{A})$ to an object in $D(\mathscr{B})$ representing $F'(X)$. We also call this functor the right derived functor of F and denote it by

$$RF : D^*(\mathscr{A}) \to D(\mathscr{B}).$$

It it uniquely determined up to isomorphism. For any integer i, we define

$$R^iF(X) = H^i(RF(X)).$$

Similarly, for any $X \in \operatorname{ob} K^*(\mathscr{A})$, let

$${}'F(X) : D(\mathscr{B}) \to (\mathbf{Sets})$$

be the covariant functor from $D(\mathscr{B})$ to the category of sets defined by

$$'F(X)(Y) = \varinjlim_{X' \in \mathrm{ob}\,(X^-)^\circ} \mathrm{Hom}_{D(\mathscr{B})}(F(X'), Y)$$

for any $Y \in \mathrm{ob}\, D(\mathscr{B})$, where X^- is the category such that objects in X^- are quasi-isomorphisms $X' \to X$, and morphisms in X^- from an object $X_1' \stackrel{s_1}{\to} X$ to an object $X_2' \stackrel{s_2}{\to} X$ are morphisms $f : X_1' \to X_2'$ in $K^*(\mathscr{A})$ with the property $s_2 f = s_1$. It defines a contravariant functor

$$'F : K^*(\mathscr{A}) \to \mathbf{Hom}(D(\mathscr{B}), (\mathbf{Sets})),$$

where $\mathbf{Hom}(D(\mathscr{B}), (\mathbf{Sets}))$ denotes the category of covariant functors from $D(\mathscr{B})$ to the category of sets. One can check that $'F$ transforms quasi-isomorphisms in $K^*(\mathscr{A})$ to isomorphisms in $\mathbf{Hom}(D(\mathscr{B}), (\mathrm{Sets}))$. So there exists a unique functor

$$LF : D^*(\mathscr{A}) \to \mathbf{Hom}(D(\mathscr{B}), (\mathbf{Sets})),$$

such that $'F = LF \circ Q$. We call LF the *left derived functor* of F. We say that LF is defined at $X \in \mathrm{ob}\, D^*(\mathscr{A})$ if the functor $LF(X)$ is representable, that is,

$$'F(X) \cong \mathrm{Hom}_{D(\mathscr{B})}(Z, -)$$

for some object $Z \in D(\mathscr{B})$. Suppose that LF is defined everywhere on $D^*(\mathscr{A})$. Then there exists a functor $D^*(\mathscr{A}) \to D(\mathscr{B})$ that maps each $X \in \mathrm{ob}\, D^*(\mathscr{A})$ to an object in $D(\mathscr{B})$ representing $'F(X)$. We also call this functor the left derived functor of F and denote it by

$$LF : D^*(\mathscr{A}) \to D(\mathscr{B}).$$

It is uniquely determined up to isomorphism. For any integer i, we define

$$L^i F(X) = H^i(LF(X)).$$

We leave it for the reader to define left derived functors and right derived functors for contravariant functors. In the following, we state results only for right derived functors of covariant functors. We leave it for the reader to extend them to other derived functors.

Proposition 6.3.1. *Notation as above. Suppose that there exists a family of objects $\mathscr{L}$ of $K^*(\mathscr{A})$ satisfying the following conditions:*

(a) *Any object X in $K^*(\mathscr{A})$ admits a quasi-isomorphism $X \to X'$ such that X' is an object in $\mathscr{L}$.*

(b) *The exact functor $F : K^*(\mathscr{A}) \to K(\mathscr{B})$ transforms quasi-isomorphisms between objects in $\mathscr{L}$ to quasi-isomorphisms in $K(\mathscr{B})$.*

(c) *For any morphism $X' \to X''$ in $K^*(\mathscr{A})$ such that X' and X'' are objects in $\mathscr{L}$, there exists a distinguished triangle*

$$X' \to X'' \to X''' \to$$

in $K^*(\mathscr{A})$ such that X''' is also an object in $\mathscr{L}$.
Then RF is everywhere defined and $RF : D^*(\mathscr{A}) \to D(\mathscr{B})$ is an exact functor. For any $X \in \operatorname{ob} D^*(\mathscr{A})$, let $X \to X'$ be a quasi-isomorphism such that X' is an object in $\mathscr{L}$. We have $RF(X) \cong F(X')$.

Proof. Let $X \to X'$ be a quasi-isomorphism such that X' is an object in $\mathscr{L}$. For any object $X \to X''$ in X^+, we can find a commutative diagram

$$\begin{array}{ccc} X & \to & X' \\ \downarrow & & \downarrow \\ X'' & \to & X''' \end{array}$$

such that all arrows are quasi-isomorphisms. By condition (a), we may assume that X''' is an object in $\mathscr{L}$. By condition (b), $X' \to X'''$ induces an isomorphism

$$F(X') \stackrel{\cong}{\to} F(X''')$$

in $D(\mathscr{B})$. So we have a one-to-one correspondence

$$\operatorname{Hom}_{D(\mathscr{B})}(Y, F(X')) \cong \operatorname{Hom}_{D(\mathscr{B})}(Y, F(X'''))$$

for any $Y \in \operatorname{ob} D(\mathscr{B})$. It follows that

$$F'(X)(Y) = \varinjlim_{X'' \in \operatorname{ob} X^+} \operatorname{Hom}_{D(\mathscr{B})}(Y, F(X'')) \cong \operatorname{Hom}_{D(\mathscr{B})}(Y, F(X')).$$

So the functor $F'(X)$ is represented by $F(X')$. It follows that RF is defined at X and $RF(X) \cong F(X')$. To prove that RF is an exact functor, we use the condition that F is exact, and that any distinguished triangle in $D^*(\mathscr{A})$ is isomorphic to a distinguished triangle in $K^*(\mathscr{A})$ whose vertices are objects in $\mathscr{L}$. □

Corollary 6.3.2. *Let $K^\star(\mathscr{A})$ be one of the categories*

$$K(\mathscr{A}), K^+(\mathscr{A}), K^-(\mathscr{A}), K^b(\mathscr{A}), K_{\mathscr{A}'}(\mathscr{A}), K^+_{\mathscr{A}'}(\mathscr{A}), K^-_{\mathscr{A}'}(\mathscr{A}), K^b_{\mathscr{A}'}(\mathscr{A})$$

contained in $K^*(\mathscr{A})$ and let $D^\star(\mathscr{A})$ be the corresponding derived category. Suppose that the conditions of 6.3.1 hold, and suppose furthermore that any object X in $K^\star(\mathscr{A})$ admits a quasi-isomorphism from X to an object in $\mathscr{L} \cap \operatorname{ob} K^\star(\mathscr{A})$. Then $R(F|_{K^\star(\mathscr{A})})$ is defined everywhere on $D^\star(\mathscr{A})$ and $R(F|_{K^\star(\mathscr{A})}) \cong (RF)|_{D^\star(\mathscr{A})}$.

Corollary 6.3.3. *Let $\mathscr{A}$, $\mathscr{B}$ and $\mathscr{C}$ be abelian categories, $\mathscr{A}'$ (resp. $\mathscr{B}'$) a thick abelian subcategory of $\mathscr{A}$ (resp. $\mathscr{B}$), $K^*(\mathscr{A})$ (resp. $K^\star(\mathscr{B})$) one of the categories*

$$K(\mathscr{A}), K^+(\mathscr{A}), K^-(\mathscr{A}), K^b(\mathscr{A}), K_{\mathscr{A}'}(\mathscr{A}), K^+_{\mathscr{A}'}(\mathscr{A}), K^-_{\mathscr{A}'}(\mathscr{A}), K^b_{\mathscr{A}'}(\mathscr{A})$$

$$(\textit{resp.}\, K(\mathscr{B}), K^+(\mathscr{B}), K^-(\mathscr{B}), K^b(\mathscr{B}), K_{\mathscr{B}'}(\mathscr{B}), K^+_{\mathscr{B}'}(\mathscr{B}), K^-_{\mathscr{B}'}(\mathscr{B}), K^b_{\mathscr{B}'}(\mathscr{B})),$$

$D^*(\mathscr{A})$ (resp. $D^\star(\mathscr{B})$) the corresponding derived category, $F : K^*(\mathscr{A}) \to K^\star(\mathscr{B})$ and $G : K^\star(\mathscr{B}) \to K(\mathscr{C})$ exact functors. Suppose that the following conditions hold:

(a) *There exists a family $\mathscr{L}$ (resp. $\mathscr{M}$) of objects in $K^*(\mathscr{A})$ (resp. $K^\star(\mathscr{B})$) such that any object X in $K^*(\mathscr{A})$ (resp. $K^\star(\mathscr{B})$) admits a quasi-isomorphism from X to an object X' in $\mathscr{L}$ (resp. $\mathscr{M}$), and for any morphism $X' \to X''$ in $K^*(\mathscr{A})$ (resp. $K^\star(\mathscr{B})$) such that X' and X'' are objects in $\mathscr{L}$ (resp. $\mathscr{M}$), there exists a distinguished triangle*

$$X' \to X'' \to X''' \to$$

in $K^*(\mathscr{A})$ (resp. $K^\star(\mathscr{B})$) such that X''' is also an object in $\mathscr{L}$ (resp. $\mathscr{M}$).

(b) *F (resp. G) transforms quasi-isomorphisms between objects in $\mathscr{L}$ (resp. $\mathscr{M}$) to quasi-isomorphisms in $K^\star(\mathscr{B})$ (resp. $K(\mathscr{C})$).*

(c) *$F(\mathscr{L}) \subset \mathscr{M}$.*

Then RF, RG and $R(GF)$ are everywhere defined, and $R(GF) \cong RG \circ RF$.

Proposition 6.3.4. *Let $\mathscr{A}$ and $\mathscr{B}$ be abelian categories. Suppose that $\mathscr{A}$ has enough injective objects.*

(i) *For any exact functor $F : K^+(\mathscr{A}) \to K(\mathscr{B})$, RF is everywhere defined on $D^+(\mathscr{A})$. For any object X in $K^+(\mathscr{A})$, we can find a quasi-isomorphism $X \to I$ such that I is a bounded below complex of injective objects in $\mathscr{A}$, and we have $RF(X) \cong F(I)$.*

(ii) *Let $F : \mathscr{A} \to \mathscr{B}$ be a left exact functor. Denote the functor $K(\mathscr{A}) \to K(\mathscr{B})$ induced by F also by F. By* (i), *RF is everywhere defined on $D^+(\mathscr{A})$. Suppose that RF has finite cohomological dimension. Then RF is everywhere defined on $D(\mathscr{A})$.*

Here we say RF has *finite cohomological dimension* if there exists a nonnegative integer n such that $R^iF(X) = 0$ for any $i > n$ and any $X \in \operatorname{ob} \mathscr{A}$.

Proof.

(i) Let $\mathscr{L}$ be the family of objects in $K^+(\mathscr{A})$ consisting of bounded below complexes of injective objects in $\mathscr{A}$. Since $\mathscr{A}$ has enough injective objects, by 6.2.9, for any object X in $K^+(\mathscr{A})$, we can find a quasi-isomorphism

$X \to I$ so that I is an object in $\mathscr{L}$. If $s : I_1 \to I_2$ is a quasi-isomorphism of objects in $\mathscr{L}$, then by 6.2.7, s is an isomorphism in $K^+(\mathscr{A})$. Hence $F(s)$ is an isomorphism in $K(\mathscr{B})$. Our assertion then follows from 6.3.1.

(ii) An object X in $\mathscr{A}$ is called *RF-acyclic* if $R^iF(X) = 0$ for all $i > 0$. Injective objects are RF-acyclic. If

$$0 \to X' \to X \to X'' \to 0$$

is an exact sequence of objects in $\mathscr{A}$, and X' and X'' are RF-acyclic, then X is also RF-acyclic, and the sequence

$$0 \to F(X') \to F(X) \to F(X'') \to 0$$

is exact. Take $\mathscr{L}$ to be the family of objects in $K(\mathscr{A})$ consisting of complexes of RF-acyclic objects. Let n be an integer such that $R^iF(X) = 0$ for any $i > n$ and any $X \in \operatorname{ob} \mathscr{A}$, and let

$$0 \to I^0 \stackrel{d^0}{\to} I^1 \stackrel{d^1}{\to} \cdots$$

be an injective resolution of an object X in $\mathscr{A}$. Then

$$0 \to I^n \stackrel{d^n}{\to} I^{n+1} \stackrel{d^{n+1}}{\to} \cdots$$

is an injective resolution of $\ker d^n$. For any $i > 0$, we have

$$R^iF(\ker d^n) \cong H^i(F(I^\cdot[n])) \cong R^{i+n}F(X) = 0.$$

So $\ker d^n$ is RF-acyclic. Hence

$$0 \to I^0 \to \cdots \to I^{n-1} \to \ker d^n_{I^\cdot} \to 0$$

is an RF-acyclic resolution of X. Let $X^\cdot$ be a complex in $K(\mathscr{A})$, and let $I^{\cdot\cdot}$ be the Cartan–Eilenberg resolution of $X^\cdot$. For each j,

$$0 \to I^{0j} \to I^{1j} \to \cdots$$

is an injective resolution of X^j. We have a quasi-isomorphism from $X^\cdot$ to the complex associated to the truncated bicomplex

$$\begin{array}{ccccccc}
 & & 0 & & 0 & & \\
 & & \uparrow & & \uparrow & & \\
\cdots & \to & \ker d_1^{n,j} & \to & \ker d_1^{n,j+1} & \to & \cdots \\
 & & \uparrow & & \uparrow & & \\
\cdots & \to & I^{n-1,j} & \to & I^{n-1,j+1} & \to & \cdots \\
 & & \uparrow & & \uparrow & & \\
 & & \vdots & & \vdots & & \\
 & & \uparrow & & \uparrow & & \\
\cdots & \to & I^{0,j} & \to & I^{0,j+1} & \to & \cdots \\
 & & \uparrow & & \uparrow & & \\
 & & 0 & & 0 & &
\end{array}$$

and this is a complex of RF-acyclic objects. Hence every object X in $K(\mathscr{A})$ admits a quasi-isomorphism from X to an object in $\mathscr{L}$. Next we show that F transforms quasi-isomorphisms of objects in $\mathscr{L}$ into quasi-isomorphisms in $K(\mathscr{B})$. Then (ii) follows from 6.3.1. It suffices to show that F transforms acyclic complexes in $\mathscr{L}$ to acyclic complexes. Let $X^\cdot$ be an acyclic complex of RF-acyclic objects, and let $Z^i = \ker d^i_{X^\cdot}$. We have short exact sequences

$$0 \to Z^{i-1} \to X^{i-1} \to Z^i \to 0.$$

Since X^i are RF-acyclic, for any $q > 0$, we have

$$R^qF(Z^i) \cong R^{q+1}F(Z^{i-1}) \cong \cdots \cong R^{q+n}F(Z^{i-n}) = 0.$$

So Z^i are RF-acyclic. It follows that

$$0 \to F(Z^{i-1}) \to F(X^{i-1}) \to F(Z^i) \to 0$$

are exact, and hence $F(X^\cdot)$ is acyclic. □

Proposition 6.3.5. *Let $\mathscr{A}$ and $\mathscr{B}$ be abelian categories and let $F : \mathscr{A} \to \mathscr{B}$ be a left exact functor. Suppose that $\mathscr{A}$ has enough injective objects. Then for any $K^\cdot \in \operatorname{ob} D^+(\mathscr{A})$, we have two biregular spectral sequences*

$$E_2^{pq} = R^pF(H^q(K^\cdot)) \Rightarrow R^{p+q}F(K^\cdot),$$
$$E_1^{pq} = R^qF(K^p) \Rightarrow R^{p+q}F(K^\cdot).$$

Suppose furthermore that RF has finite cohomological dimension. Then we have the same result for any $K^\cdot \in \operatorname{ob} D(\mathscr{A})$.

Proof. Let us prove the second part of the proposition. Let n be an integer such that $R^iF(X) = 0$ for any object X in $\mathscr{A}$ and any $i > n$. Given an object $X^\cdot$ in $K(\mathscr{A})$, let

$$Z^j = \ker d^j_{X^\cdot}, \quad B^j = \operatorname{im} d^{j-1}_{X^\cdot}, \quad H^j = Z^j/B^j.$$

Choose a Cartan–Eilenberg resolution $I^{\cdot\cdot}$ of $X^\cdot$. Then for each j,

$$(I^{\cdot,j}, d_1^{\cdot,j}),\ (\ker d_2^{\cdot,j}, d_1^{\cdot,j}),\ (\operatorname{im} d_2^{\cdot,j-1}, d_1^{\cdot,j}),\ (\ker d_2^{\cdot,j}/\operatorname{im} d_2^{\cdot,j-1}, d_1^{\cdot,j})$$

are injective resolutions of X^j, Z^j, B^j, H^j, respectively. For any complex $Y^\cdot$, let $\tau_{\leq n}Y^\cdot$ be the truncated complex

$$\cdots \to Y^{n-1} \to \ker d^n_{Y^\cdot} \to 0.$$

Then

$$(\tau_{\leq n}(I^{\cdot,j}), d_1^{\cdot,j}),\ (\tau_{\leq n}(\ker d_2^{\cdot,j}), d_1^{\cdot,j}),\ (\tau_{\leq n}(\operatorname{im} d_2^{\cdot,j-1}), d_1^{\cdot,j}),$$
$$(\tau_{\leq n}(\ker d_2^{\cdot,j}/\operatorname{im} d_2^{\cdot,j-1}), d_1^{\cdot,j})$$

are resolutions by RF-acyclic objects of X^j, Z^j, B^j, H^j, respectively. Moreover, we have short exact sequences of complexes

$$0 \to \tau_{\leq n}(\operatorname{im} d_2^{\cdot,j-1}) \to \tau_{\leq n}(\ker d_2^{\cdot,j}) \to \tau_{\leq n}(\ker d_2^{\cdot,j}/\operatorname{im} d_2^{\cdot,j-1}) \to 0,$$
$$0 \to \tau_{\leq n}(\ker d_2^{\cdot,j}) \to \tau_{\leq n}(I^{\cdot,j}) \to \tau_{\leq n}(\operatorname{im} d_2^{\cdot,j}) \to 0.$$

Let $I'^{\cdot\cdot}$ be the truncated bicomplex

$$\begin{array}{ccccccc} & & 0 & & 0 & & \\ & & \uparrow & & \uparrow & & \\ \cdots & \to & \ker d_1^{n,j} & \to & \ker d_1^{n,j+1} & \to & \cdots \\ & & \uparrow & & \uparrow & & \\ \cdots & \to & I^{n-1,j} & \to & I^{n-1,j+1} & \to & \cdots \\ & & \uparrow & & \uparrow & & \\ & & \vdots & & \vdots & & \\ & & \uparrow & & \uparrow & & \\ \cdots & \to & I^{0,j} & \to & I^{0,j+1} & \to & \cdots \\ & & \uparrow & & \uparrow & & \\ & & 0 & & 0 & & \end{array}$$

and denote objects associated to this complex by putting the supscript $'$. Then the complexes

$$(I'^{\cdot,j}, d_1'^{\cdot,j}),\ (\ker d_2'^{\cdot,j}, d_1'^{\cdot,j}),\ (\operatorname{im} d_2'^{\cdot,j-1}, d_1'^{\cdot,j}),\ (\ker d_2'^{\cdot,j}/\operatorname{im} d_2'^{\cdot,j-1}, d_1'^{\cdot,j})$$

can be identified with

$$(\tau_{\leq n}(I^{\cdot,j}), d_1^{\cdot,j}),\ (\tau_{\leq n}(\ker d_2^{\cdot,j}), d_1^{\cdot,j}),\ (\tau_{\leq n}(\operatorname{im} d_2^{\cdot,j-1}), d_1^{\cdot,j}),$$
$$(\tau_{\leq n}(\ker d_2^{\cdot,j}/\operatorname{im} d_2^{\cdot,j-1}), d_1^{\cdot,j})$$

respectively, and hence are resolutions by RF-acyclic objects of X^j, Z^j, B^j, H^j, respectively. The two spectral sequences are those associated to the bicomplex $F(I'^{\cdot\cdot})$. □

Let X be a scheme and let $\mathscr{A}$ be the abelian category of sheaves of abelian groups on X. Denote the derived category $D(\mathscr{A})$ by $D(X)$. Let $\mathscr{A}'$ be the thick abelian subcategory consisting of torsion sheaves (resp. constructible sheaves). Denote the derived category $D_{\mathscr{A}'}(\mathscr{A})$ by $D_{\mathrm{tor}}(X)$ (resp. $D_c(X)$). Let A be a ring, let $\mathscr{A}$ be the abelian category of sheaves of A-modules on X, and let $\mathscr{A}'$ be the thick abelian subcategory consisting of constructible sheaves of A-modules. Denote the derived category $D(\mathscr{A})$ (resp. $D_{\mathscr{A}'}(\mathscr{A})$) by $D(X, A)$ (resp. $D_c(X, A)$). We get the full subcategories $D^*(X), D^*_{\mathrm{tor}}(X), D^*_c(X), D^*(X, A), D^*_c(X, A)$ by taking $* = +, -, b$.

The category of sheaves of abelian groups has enough injective objects. We can define the right derived functor

$$R\Gamma(X,-) : D^+(X) \to D^+((\mathbf{Ab.\ gps}))$$

of the functor $\Gamma(X,-)$ of taking global sections, where $(\mathbf{Ab.\ gps})$ is the category of abelian groups. If $R\Gamma(X,-)$ has finite cohomological dimension, we can extend $R\Gamma(X,-)$ to

$$R\Gamma(X,-) : D(X) \to D((\mathbf{Ab.\ gps})).$$

For any sheaf of abelian groups $\mathscr{F}$ on X and any i, we have

$$H^i(R\Gamma(X,\mathscr{F})) = H^i(X,\mathscr{F}).$$

Let $f : X \to Y$ be a morphism of schemes. We can define the right derived functor

$$Rf_* : D^+(X) \to D^+(Y)$$

of f_*. If Rf_* has finite cohomological dimension, we can then define

$$Rf_* : D(X) \to D(Y).$$

For any sheaf of abelian groups $\mathscr{F}$ on X and any i, we have

$$\mathscr{H}^i(Rf_*\mathscr{F}) = R^if_*\mathscr{F}.$$

6.4 $R\mathbf{Hom}(-,-)$ and $-\otimes_A^L -$

([Hartshorne (1966)] I 6, [SGA 4] XVII 1.1, [SGA $4\frac{1}{2}$] C.D. II 3.)

Let $\mathscr{A}$ be an abelian category. For all complexes $X^\cdot$ and $Y^\cdot$ of objects in $\mathscr{A}$, define a complex $\mathrm{Hom}^\cdot(X^\cdot,Y^\cdot)$ of abelian groups as follows: Take

$$\mathrm{Hom}^n(X^\cdot,Y^\cdot) = \prod_i \mathrm{Hom}_{\mathscr{A}}(X^i,Y^{i+n}).$$

For any

$$f = (f_i) \in \mathrm{Hom}^n(X^\cdot,Y^\cdot) = \prod_i \mathrm{Hom}_{\mathscr{A}}(X^i,Y^{i+n}),$$

define

$$df = ((df)_i) \in \mathrm{Hom}^{n+1}(X^\cdot,Y^\cdot) = \prod_i \mathrm{Hom}_{\mathscr{A}}(X^i,Y^{i+n+1})$$

by

$$(df)_i = d_{Y^\cdot}^{i+n} \circ f_i - (-1)^n f_{i+1} \circ d_{X^\cdot}^i.$$

We have $dd = 0$. Note that the kernel of

$$d : \mathrm{Hom}^n(X^\cdot, Y^\cdot) \to \mathrm{Hom}^{n+1}(X^\cdot, Y^\cdot)$$

can be identified with the group of morphisms of complexes from $X^\cdot$ to $Y^\cdot[n]$, and the image of

$$d : \mathrm{Hom}^{n-1}(X^\cdot, Y^\cdot) \to \mathrm{Hom}^n(X^\cdot, Y^\cdot)$$

can be identified with the subgroup of morphisms of complexes from $X^\cdot$ to $Y^\cdot[n]$ homotopic to 0. So we have

$$H^n(\mathrm{Hom}^\cdot(X^\cdot, Y^\cdot)) \cong \mathrm{Hom}_{K(\mathscr{A})}(X^\cdot, Y^\cdot[n]).$$

Define

$$\mathrm{Hom}^\cdot(X^\cdot, Y^\cdot[1]) \stackrel{\cong}{\to} \mathrm{Hom}^\cdot(X^\cdot, Y^\cdot)[1]$$

so that it is the identity through the identification

$$\mathrm{Hom}^n(X^\cdot, Y^\cdot[1]) = \prod_i \mathrm{Hom}_{\mathscr{A}}(X^i, Y^{i+n+1}) = (\mathrm{Hom}^\cdot(X^\cdot, Y^\cdot)[1])^n.$$

Define

$$\mathrm{Hom}^\cdot(X^\cdot[-1], Y^\cdot) \stackrel{\cong}{\to} \mathrm{Hom}^\cdot(X^\cdot, Y^\cdot)[1]$$

so that it is $(-1)^{n+1}\mathrm{id}$ through the identification

$$\mathrm{Hom}^n(X^\cdot[-1], Y^\cdot) = \prod_i \mathrm{Hom}_{\mathscr{A}}(X^i, Y^{i+n+1}) = (\mathrm{Hom}^\cdot(X^\cdot, Y^\cdot)[1])^n.$$

One can check they are morphisms of complexes. Moreover, the following diagram is anti-commutative:

$$\begin{array}{ccc} \mathrm{Hom}^\cdot(X^\cdot[-1], Y^\cdot[1]) & \stackrel{\cong}{\to} & \mathrm{Hom}^\cdot(X^\cdot[-1], Y^\cdot)[1]. \\ \cong\downarrow & & \downarrow\cong \\ \mathrm{Hom}^\cdot(X^\cdot, Y^\cdot[1])[1] & \stackrel{\cong}{\to} & \mathrm{Hom}^\cdot(X^\cdot, Y^\cdot)[1][1]. \end{array}$$

Fix an object $X^\cdot$ (resp. $Y^\cdot$) in $K(\mathscr{A})$. Then $\mathrm{Hom}^\cdot(X^\cdot, -)$ (resp. $\mathrm{Hom}^\cdot(-, Y^\cdot)$) transforms distinguished triangles to distinguished triangles, and hence

$$\mathrm{Hom}^\cdot(-,-) : K(\mathscr{A}) \times K(\mathscr{A}) \to K((\mathbf{Ab.gps}))$$

is an exact functor contravariant in the first variable, and covariant in the second variable. As an example, given a distinguished triangle

$$X_1^\cdot \stackrel{u}{\to} X_2^\cdot \stackrel{v}{\to} C(u) \stackrel{w}{\to},$$

where $C(u)$ is the mapping cone of u, let us verify that

$$\mathrm{Hom}^{\cdot}(C(u), Y^{\cdot}) \xrightarrow{v^*} \mathrm{Hom}^{\cdot}(X_2^{\cdot}, Y^{\cdot}) \xrightarrow{u^*} \mathrm{Hom}^{\cdot}(X_1^{\cdot}, Y^{\cdot}) \xrightarrow{w'}$$

is a distinguished triangle, where w' is defined by the commutative diagram

$$\begin{array}{ccc} \mathrm{Hom}^{\cdot}(X_1^{\cdot}, Y^{\cdot}) & \xrightarrow{w'} & \mathrm{Hom}^{\cdot}(C(u), Y^{\cdot})[1] \\ \| & & \uparrow \\ \mathrm{Hom}^{\cdot}(X_1^{\cdot}, Y^{\cdot}) & \xrightarrow{(w[-1])^*} & \mathrm{Hom}^{\cdot}(C(u)[-1], Y^{\cdot}). \end{array}$$

It suffices to show that

$$\mathrm{Hom}^{\cdot}(X_2^{\cdot}, Y^{\cdot}) \xrightarrow{u^*} \mathrm{Hom}^{\cdot}(X_1^{\cdot}, Y^{\cdot}) \xrightarrow{w'} \mathrm{Hom}^{\cdot}(C(u), Y^{\cdot})[1] \xrightarrow{-v^*[1]}$$

is a distinguished triangle. We need to define an isomorphism

$$\mathrm{Hom}^{\cdot}(C(u), Y^{\cdot})[1] \cong C(u^*)$$

so that the following diagram commutes:

$$\begin{array}{ccccc} \mathrm{Hom}^{\cdot}(X_1^{\cdot}, Y^{\cdot}) & \xrightarrow{w'} & \mathrm{Hom}^{\cdot}(C(u), Y^{\cdot})[1] & \xrightarrow{-v^*[1]} & \mathrm{Hom}^{\cdot}(X_2^{\cdot}, Y^{\cdot})[1] \\ \| & & \downarrow \cong & & \| \\ \mathrm{Hom}^{\cdot}(X_1^{\cdot}, Y^{\cdot}) & \to & C(u^*) & \to & \mathrm{Hom}^{\cdot}(X_2^{\cdot}, Y^{\cdot})[1]. \end{array}$$

For each integer n, the isomorphism

$$(\mathrm{Hom}^{\cdot}(C(u), Y^{\cdot})[1])^n \to (C(u^*))^n$$

is defined to be the composite of the following isomorphisms:

$$\begin{aligned} & (\mathrm{Hom}^{\cdot}(C(u), Y^{\cdot})[1])^n \\ = \quad & \prod_i \mathrm{Hom}(C(u)^i, Y^{i+n+1}) \\ = \quad & \prod_i \mathrm{Hom}(X_1^{i+1} \oplus X_2^i, Y^{i+n+1}) \\ \cong \quad & \prod_i \mathrm{Hom}(X_1^{i+1}, Y^{i+n+1}) \bigoplus \prod_i \mathrm{Hom}(X_2^i, Y^{i+n+1}) \\ \xrightarrow{((-1)^{n+1}\mathrm{id}, -\mathrm{id})} \quad & \prod_i \mathrm{Hom}(X_1^{i+1}, Y^{i+n+1}) \bigoplus \prod_i \mathrm{Hom}(X_2^i, Y^{i+n+1}) \\ = \quad & \mathrm{Hom}^n(X_1^{\cdot}, Y^{\cdot}) \bigoplus \mathrm{Hom}^{n+1}(X_2^{\cdot}, Y^{\cdot}) \\ = \quad & (C(u^*))^n. \end{aligned}$$

Lemma 6.4.1. *Let $X^{\cdot}$ be an acyclic complex of objects in $\mathscr{A}$ and let $Y^{\cdot}$ be a bounded below complex of injective objects in $\mathscr{A}$. Then* $\mathrm{Hom}^{\cdot}(X^{\cdot}, Y^{\cdot})$ *is acyclic.*

Proof. We have

$$H^n(\mathrm{Hom}^{\cdot}(X^{\cdot},Y^{\cdot}))\cong \mathrm{Hom}_{K(\mathscr{A})}(X^{\cdot},Y^{\cdot}[n]).$$

By 6.2.6, we have $\mathrm{Hom}_{K(\mathscr{A})}(X^{\cdot},Y^{\cdot}[n])=0$. Our assertion follows. □

Suppose that $\mathscr{A}$ is an abelian category with enough injective objects. For any fixed $X^{\cdot}\in \mathrm{ob}\, K(\mathscr{A})$, the right derived functor of

$$\mathrm{Hom}^{\cdot}(X^{\cdot},-):K^+(\mathscr{A})\to K((\mathbf{Ab.\ gps}))$$

is defined everywhere on $D^+(\mathscr{A})$ by 6.3.4 (i). Denote it by

$$R_{II}\mathrm{Hom}^{\cdot}(X^{\cdot},-):D^+(\mathscr{A})\to D((\mathbf{Ab.\ gps})).$$

We thus get a functor

$$R_{II}\mathrm{Hom}^{\cdot}(-,-):K(\mathscr{A})\times D^+(\mathscr{A})\to D((\mathbf{Ab.\ gps})).$$

By 6.4.1, the functor $R_{II}\mathrm{Hom}^{\cdot}(-,Y)$ transforms quasi-isomorphisms in $K(\mathscr{A})$ to isomorphisms in $D((\mathbf{Ab.\ gps}))$. So it factors through $D(\mathscr{A})\times D^+(\mathscr{A})$ and defines a functor

$$D(\mathscr{A})\times D^+(\mathscr{A})\to D((\mathbf{Ab.\ gps})),$$

which we denote by $R_IR_{II}\mathrm{Hom}^{\cdot}(-,-)$, or simply $R\mathrm{Hom}(-,-)$. Define

$$\mathrm{Ext}^i(X^{\cdot},Y^{\cdot})=H^i(R\mathrm{Hom}^{\cdot}(X^{\cdot},Y^{\cdot}))$$

for any $X\in \mathrm{ob}\, D(\mathscr{A})$ and $Y\in \mathrm{ob}\, D^+(\mathscr{A})$.

Proposition 6.4.2. *Let $\mathscr{A}$ be an abelian category with enough injective objects. For any $X\in \mathrm{ob}\, D(\mathscr{A})$ and $Y\in \mathrm{ob}\, D^+(\mathscr{A})$, we have*

$$\mathrm{Ext}^i(X^{\cdot},Y^{\cdot})\cong \mathrm{Hom}_{D(\mathscr{A})}(X^{\cdot},Y^{\cdot}[i]).$$

Proof. Let $Y^{\cdot}\to I^{\cdot}$ be a quasi-isomorphism so that $I^{\cdot}$ is a bounded below complex of injective objects in $\mathscr{A}$. We have

$$\mathrm{Ext}^i(X^{\cdot},Y^{\cdot})=H^i(R\mathrm{Hom}^{\cdot}(X^{\cdot},Y^{\cdot}))\cong H^i(\mathrm{Hom}^{\cdot}(X^{\cdot},I^{\cdot}))\cong \mathrm{Hom}_{K(\mathscr{A})}(X^{\cdot},I^{\cdot}[i]).$$

Our assertion follows from 6.2.8. □

For any $X^{\cdot},Y^{\cdot}\in \mathrm{ob}\, D(\mathscr{A})$, define

$$\mathrm{Ext}^i(X^{\cdot},Y^{\cdot})\cong \mathrm{Hom}_{D(\mathscr{A})}(X^{\cdot},Y^{\cdot}[i]).$$

By 6.4.2, this coincides with our previous definition of $\mathrm{Ext}^i(X^{\cdot},Y^{\cdot})$ if $\mathscr{A}$ has enough injective objects, $X\in \mathrm{ob}\, D(\mathscr{A})$ and $Y\in \mathrm{ob}\, D^+(\mathscr{A})$. By 6.1.1 (ii) and 6.2.5 (ii), we have the following.

Proposition 6.4.3. *Let $\mathscr{A}$ be an abelian category and let*

$$0 \to X^{\cdot} \to Y^{\cdot} \to Z^{\cdot} \to 0$$

be an exact sequence of complexes of objects in $\mathscr{A}$. For any $W^{\cdot} \in \operatorname{ob} D(\mathscr{A})$, we have the following exact sequences

$$\cdots \to \operatorname{Ext}^i(W^{\cdot}, X^{\cdot}) \to \operatorname{Ext}^i(W^{\cdot}, Y^{\cdot}) \to \operatorname{Ext}^i(W^{\cdot}, Z^{\cdot}) \to \operatorname{Ext}^{i+1}(W^{\cdot}, X^{\cdot}) \to \cdots,$$
$$\cdots \leftarrow \operatorname{Ext}^i(X^{\cdot}, W^{\cdot}) \leftarrow \operatorname{Ext}^i(Y^{\cdot}, W^{\cdot}) \leftarrow \operatorname{Ext}^i(Z^{\cdot}, W^{\cdot}) \leftarrow \operatorname{Ext}^{i-1}(X^{\cdot}, W^{\cdot}) \leftarrow \cdots.$$

Let X be a scheme and let A be a ring. Taking $\mathscr{A}$ to be the abelian category of sheaves of A-modules on X, we get the functors

$$R\mathrm{Hom} : D(X, A) \times D^+(X, A) \to D(\mathbf{Ab\ gps})$$

and

$$\operatorname{Ext}^i_A(-,-) : D(X, A) \times D(X, A) \to (\mathbf{Ab.\ gps}).$$

Given two complexes of sheaves of A-modules $\mathscr{F}^{\cdot}$ and $\mathscr{G}^{\cdot}$, we can define a complex of sheaves of A-modules $\mathscr{H}om^{\cdot}(\mathscr{F}^{\cdot}, \mathscr{G}^{\cdot})$ in the same way as above by replacing Hom by the sheafified $\mathscr{H}om$. We can then define the functors

$$R\mathscr{H}om : D(X, A) \times D^+(X, A) \to D(X, A)$$

and

$$\mathscr{E}xt^i_A(-,-) = \mathscr{H}^i(R\mathscr{H}om(-,-)).$$

Given two complexes $\mathscr{F}^{\cdot}$ and $\mathscr{G}^{\cdot}$ of sheaves of A-modules on X, define a complex $\mathscr{F}^{\cdot} \otimes_A \mathscr{G}^{\cdot}$ as follows: Take

$$(\mathscr{F}^{\cdot} \otimes_A \mathscr{G}^{\cdot})^n = \bigoplus_i \mathscr{F}^i \otimes_A \mathscr{G}^{n-i}.$$

Define

$$d : (\mathscr{F}^{\cdot} \otimes_A \mathscr{G}^{\cdot})^n \to (\mathscr{F}^{\cdot} \otimes_A \mathscr{G}^{\cdot})^{n+1}$$

by

$$d|_{\mathscr{F}^i \otimes_A \mathscr{G}^{n-i}} = d^i_{\mathscr{F}^{\cdot}} \otimes \mathrm{id}_{\mathscr{G}^{n-i}} + (-1)^i \mathrm{id}_{\mathscr{F}^i} \otimes d^{n-i}_{\mathscr{G}^{\cdot}}.$$

Define

$$(\mathscr{F}^{\cdot}[1]) \otimes_A \mathscr{G}^{\cdot} \to (\mathscr{F}^{\cdot} \otimes_A \mathscr{G}^{\cdot})[1]$$

so that its restriction to $\mathscr{F}^i \otimes_A \mathscr{G}^{n-i+1}$ is the identity, and define

$$\mathscr{F}^{\cdot} \otimes_A (\mathscr{G}^{\cdot}[1]) \to (\mathscr{F}^{\cdot} \otimes_A \mathscr{G}^{\cdot})[1]$$

so that its restriction to $\mathscr{F}^i \otimes_A \mathscr{G}^{n-i+1}$ is $(-1)^i\mathrm{id}$. One checks that the diagram

$$\begin{array}{ccc} (\mathscr{F}^{\cdot}[1]) \otimes_A (\mathscr{G}^{\cdot}[1]) & \to & (\mathscr{F}^{\cdot} \otimes_A (\mathscr{G}^{\cdot}[1]))[1] \\ \downarrow & & \downarrow \\ ((\mathscr{F}^{\cdot}[1]) \otimes_A \mathscr{G}^{\cdot})[1] & \to & (\mathscr{F}^{\cdot} \otimes_A \mathscr{G}^{\cdot})[1][1] \end{array}$$

is anti-commutative, and

$$- \otimes_A - : K(X, A) \times K(X, A) \to K(X, A)$$

is an exact functor, where $K(X, A)$ is the triangulated category of complexes of sheaves of A-modules on X. We have an isomorphism of complexes

$$\mathscr{F}^{\cdot} \otimes_A \mathscr{G}^{\cdot} \to \mathscr{G}^{\cdot} \otimes_A \mathscr{F}^{\cdot}$$

defined by

$$x^i \otimes y^j \mapsto (-1)^{ij} y^j \otimes x^i$$

for any $x^i \in \mathscr{F}^i$ and $y^j \in \mathscr{G}^j$.

Lemma 6.4.4. *Let X be a scheme, A a ring, $\mathscr{F}^{\cdot}$ and $\mathscr{G}^{\cdot}$ complexes of sheaves of A-modules on X. Suppose that the following conditions hold:*

(a) *All components of $\mathscr{G}$ are flat sheaves of A-modules.*

(b) *Either $\mathscr{F}^{\cdot}$ or $\mathscr{G}^{\cdot}$ is acyclic.*

(c) *Either both $\mathscr{F}^{\cdot}$ and $\mathscr{G}^{\cdot}$ are bounded above, or $\mathscr{G}^{\cdot}$ is bounded.*

Then $\mathscr{F}^{\cdot} \otimes_A \mathscr{G}^{\cdot}$ is acyclic.

Proof. By the condition (c), the two spectral sequences

$$\begin{aligned} {}'E_2^{pq} &= H_I^p H_{II}^q(\mathscr{F}^{\cdot} \otimes_A \mathscr{G}^{\cdot}) \Rightarrow H^{p+q}(\mathscr{F}^{\cdot} \otimes_A \mathscr{G}^{\cdot}), \\ {}''E_2^{pq} &= H_{II}^p H_I^q(\mathscr{F}^{\cdot} \otimes_A \mathscr{G}^{\cdot}) \Rightarrow H^{p+q}(\mathscr{F}^{\cdot} \otimes_A \mathscr{G}^{\cdot}) \end{aligned}$$

associated to the bicomplex $\mathscr{F}^{\cdot} \otimes_A \mathscr{G}^{\cdot}$ are biregular. If $\mathscr{F}^{\cdot}$ is acyclic, then by condition (a), we have $H_I^q(\mathscr{F}^{\cdot} \otimes_A \mathscr{G}^{\cdot}) = 0$ for all q. So ${}''E_2^{pq} = 0$ for all p and q, and hence $H^n(\mathscr{F}^{\cdot} \otimes_A \mathscr{G}^{\cdot}) = 0$ for all n. Suppose $\mathscr{G}^{\cdot}$ is acyclic. By condition (c), $\mathscr{G}^{\cdot}$ is bounded above, say of the form

$$\cdots \stackrel{d^{m-2}}{\to} \mathscr{G}^{m-1} \stackrel{d^{m-1}}{\to} \mathscr{G}^m \to 0.$$

Then we have short exact sequences

$$\begin{array}{l} 0 \to \ker d_{\mathscr{G}^{\cdot}}^{m-1} \to \mathscr{G}^{m-1} \to \quad \mathscr{G}^m \quad \to 0, \\ 0 \to \ker d_{\mathscr{G}^{\cdot}}^{m-2} \to \mathscr{G}^{m-2} \to \ker d_{\mathscr{G}^{\cdot}}^{m-1} \to 0, \\ \qquad\qquad\qquad \vdots \end{array}$$

Since $\mathscr{G}^q$ are flat, it follows from the above exact sequences that $\ker d^q_{\mathscr{G}^\cdot}$ are all flat. Hence we have short exact sequences

$$\begin{array}{llll} 0 \to \mathscr{F}^p \otimes_A \ker d^{m-1}_{\mathscr{G}^\cdot} \to \mathscr{F}^p \otimes_A \mathscr{G}^{m-1} \to & \mathscr{F}^p \otimes_A \mathscr{G}^m & \to 0, \\ 0 \to \mathscr{F}^p \otimes_A \ker d^{m-2}_{\mathscr{G}^\cdot} \to \mathscr{F}^p \otimes_A \mathscr{G}^{m-2} \to & \mathscr{F}^p \otimes_A \ker d^{m-1}_{\mathscr{G}^\cdot} & \to 0, \\ & \vdots & \end{array}$$

It follows that

$$\cdots \to \mathscr{F}^p \otimes_A \mathscr{G}^q \to \mathscr{F}^p \otimes_A \mathscr{G}^{q+1} \to \cdots$$

are exact. So $'E_2^{pq} = 0$ for all p and q, and hence $H^n(\mathscr{F}^\cdot \otimes_A \mathscr{G}^\cdot) = 0$ for all n. □

Let $\mathscr{L}$ be the family of objects in $K^-(X, A)$ consisting of bounded above complexes of flat sheaves A-modules. For each fixed object $\mathscr{F}^\cdot$ in $K^-(X, A)$, by 6.4.4, the functor $\mathscr{F}^\cdot \otimes_A -$ transforms quasi-isomorphisms in $\mathscr{L}$ to quasi-isomorphisms. On the other hand, any sheaf of A-modules on X is a quotient of a flat sheaf. Indeed, any sheaf of A-modules is a quotient of a sheaf of the form $\bigoplus_U A_U$, where $U \stackrel{f_U}{\to} X$ are etale X-schemes and $A_U = f_{U!}A$. By the dual version of 6.2.9, for any $\mathscr{G}^\cdot \in \operatorname{ob} K^-(X, A)$, there exists a quasi-isomorphism from an object in $\mathscr{L}$ to $\mathscr{G}^\cdot$. (Confer 6.4.5 below.) By the dual version of 6.3.1, the left derived functor of $\mathscr{F}^\cdot \otimes_A -$ is defined everywhere on $D^-(X, A)$. Denote the derived functor by

$$L_{II}(\mathscr{F}^\cdot \otimes_A -) : D^-(X, A) \to D^-(X, A).$$

So we get a functor

$$L_{II}(- \otimes_A -) : K^-(X, A) \times D^-(X, A) \to D^-(X, A).$$

By 6.4.4, the functor $L_{II}(- \otimes_A \mathscr{G}^\cdot)$ transforms quasi-isomorphisms in $K^-(X, A)$ to isomorphisms in $D^-(X, A)$. So it factors through $D^-(X, A)$ and defines a functor

$$D^-(X, A) \times D^-(X, A) \to D^-(X, A),$$

which we denote by $L_I L_{II}(- \otimes_A -)$. Similarly we can define $L_{II} L_I(- \otimes_A -)$, and we have

$$L_{II} L_I(- \otimes_A -) \cong L_I L_{II}(- \otimes_A -).$$

We denote $L_I L_{II}(- \otimes_A -)$ by $- \otimes_A^L -$. For any $\mathscr{F}^\cdot, \mathscr{G}^\cdot \in \operatorname{ob} D^-(X, A)$, we define

$$\operatorname{Tor}_i(\mathscr{F}^\cdot, \mathscr{G}^\cdot) = \mathscr{H}^{-i}(\mathscr{F}^\cdot \otimes_A^L \mathscr{G}^\cdot).$$

Lemma 6.4.5. *Let X be a noetherian scheme, A a noetherian ring, and $\mathscr{F}^\cdot$ a complex of sheaves of A-modules on X such that $\mathscr{H}^i(\mathscr{F}^\cdot)$ are constructible sheaves of A-modules for all i and $\mathscr{H}^i(\mathscr{F}^\cdot)=0$ for $i\gg 0$. Then there exists a quasi-isomorphism $\mathscr{F}'^\cdot\to\mathscr{F}^\cdot$ such that $\mathscr{F}'^\cdot$ is a bounded above complex of flat constructible sheaves of A-modules.*

Proof. We use the dual argument of the proof of 6.2.9. Define $\mathscr{F}'^i=0$ for $i\gg 0$. Suppose that we have defined constructible flat sheaves of A-modules $\mathscr{F}'^i$ and morphisms $\mathscr{F}'^i\to\mathscr{F}^i$ for all $i>n$ such that $\mathscr{H}^i(\mathscr{F}'^\cdot)\cong\mathscr{H}^i(\mathscr{F}^\cdot)$ for all $i>n+1$ and such that $\ker d^{n+1}_{\mathscr{F}'^\cdot}\to\mathscr{H}^{n+1}(\mathscr{F}^\cdot)$ is surjective. Consider the diagram

$$\begin{array}{ccccc}\mathscr{A} & \stackrel{\alpha}{\to} & \mathscr{B} & \stackrel{\beta}{\to} & \ker d^{n+1}_{\mathscr{F}'^\cdot}\\ \downarrow & & \downarrow & & \downarrow\\ \mathscr{F}^n & \stackrel{p}{\to} & \mathscr{F}^n/\mathrm{im}\, d^{n-1}_{\mathscr{F}^\cdot} & \stackrel{d^n}{\to} & \ker d^{n+1}_{\mathscr{F}^\cdot},\end{array}$$

where on the second line the morphisms are uniquely determined by the property that p is the projection and $d^n\circ p$ is the morphism $\mathscr{F}^n\to\ker d^{n+1}_{\mathscr{F}^\cdot}$ induced by $d^n_{\mathscr{F}^\cdot}$, and on the first line, we have

$$\mathscr{A}=\mathscr{F}^n\times_{\ker d^{n+1}_{\mathscr{F}^\cdot}}\ker d^{n+1}_{\mathscr{F}'^\cdot},$$

$$\mathscr{B}=\mathscr{F}^n/\mathrm{im}\, d^{n-1}_{\mathscr{F}^\cdot}\times_{\ker d^{n+1}_{\mathscr{F}^\cdot}}\ker d^{n+1}_{\mathscr{F}'^\cdot},$$

and all arrows in the diagram are the canonical morphisms. The last vertical arrow induces a monomorphism $\mathrm{coker}\,\beta\to\mathscr{H}^{n+1}(\mathscr{F}^\cdot)$. Combined with the hypothesis that $\ker d^{n+1}_{\mathscr{F}'^\cdot}\to\mathscr{H}^{n+1}(\mathscr{F}^\cdot)$ is surjective, we see that it induces an isomorphism $\mathrm{coker}\,\beta\cong\mathscr{H}^{n+1}(\mathscr{F}^\cdot)$. The middle vertical arrow induces an isomorphism $\ker\beta\cong\mathscr{H}^n(\mathscr{F}^\cdot)$. So we have an exact sequence

$$0\to\mathscr{H}^n(\mathscr{F}^\cdot)\to\mathscr{B}\stackrel{\beta}{\to}\ker d^{n+1}_{\mathscr{F}'^\cdot}\to\mathscr{H}^{n+1}(\mathscr{F}^\cdot)\to 0.$$

Since $\mathscr{H}^n(\mathscr{F}^\cdot)$, $\mathscr{H}^{n+1}(\mathscr{F}^\cdot)$ and $\ker d^{n+1}_{\mathscr{F}'^\cdot}$ are constructible, $\mathscr{B}$ is also constructible. We will construct a constructible flat sheaf of A-modules $\mathscr{F}'^n$ and a morphism $\gamma:\mathscr{F}'^n\to\mathscr{A}$ such that $\alpha\gamma$ is surjective. Define $d^n_{\mathscr{F}'^\cdot}:\mathscr{F}'^n\to\mathscr{F}'^{n+1}$ to be the composite

$$\mathscr{F}'^n\stackrel{\gamma}{\to}\mathscr{A}\stackrel{\alpha}{\to}\mathscr{B}\stackrel{\beta}{\to}\ker d^{n+1}_{\mathscr{F}'^\cdot}\to\mathscr{F}'^{n+1},$$

and define $\mathscr{F}'^n\to\mathscr{F}^n$ to be the composite

$$\mathscr{F}'^n\stackrel{\alpha}{\to}\mathscr{A}\to\mathscr{F}^n.$$

Then $\mathscr{H}^i(\mathscr{F}'^\cdot)\cong\mathscr{H}^i(\mathscr{F}^\cdot)$ for all $i>n$ and $\ker d^n_{\mathscr{F}'^\cdot}\to\mathscr{H}^n(\mathscr{F}^\cdot)$ is surjective. To construct $\mathscr{F}'^n$, choose an epimorphism $\bigoplus_U A_U\to\mathscr{A}$, where

$U \xrightarrow{f_U} X$ are some etale X-schemes and $A_U = f_{U!}A$. Since α is necessarily surjective, the composite $\bigoplus_U A_U \to \mathscr{A} \to \mathscr{B}$ is surjective. Since $\mathscr{B}$ is constructible, there exists a finite family $U_1, \ldots, U_k$ such that $\bigoplus_{i=1}^k A_{U_i} \to \mathscr{B}$ is surjective. We then take $\mathscr{F}'^n = \bigoplus_{i=1}^k A_{U_i}$. □

Let $\mathscr{F}^\cdot$ be a bounded complex of sheaves of A-modules on X. We say that $\mathscr{F}^\cdot$ has *finite* Tor-*dimension* if there exists an integer n such that $\mathrm{Tor}_i(\mathscr{F}^\cdot, M) = 0$ for any $i > n$ and any constant sheaf of A-modules M on X.

Proposition 6.4.6. *Let X be a noetherian scheme, A a noetherian ring, and $\mathscr{F}^\cdot$ a bounded complex of sheaves of A-modules. The following conditions are equivalent:*

(i) *There exists a quasi-isomorphism $\mathscr{F}'^\cdot \to \mathscr{F}^\cdot$ such that $\mathscr{F}'^\cdot$ is a bounded complex of constructible flat sheaves of A-modules.*

(ii) *$\mathscr{F}^\cdot$ has finite* Tor-*dimension and $\mathscr{H}^i(\mathscr{F})$ are constructible for all i.*

Proof. (i)⇒(ii) is clear. Suppose (ii) holds. Let n be an integer such that $\mathrm{Tor}_i(\mathscr{F}^\cdot, M) = 0$ for any $i > n$ and any constant sheaf of A-modules M on X. Taking $M = A$, we get $\mathscr{H}^i(\mathscr{F}^\cdot) = 0$ for all $i < -n$. Let $\mathscr{F}'^\cdot \to \mathscr{F}$ be the quasi-isomorphism constructed in 6.4.5. Then

$$\cdots \to \mathscr{F}'^{-n-1} \to \mathscr{F}'^{-n} \to 0$$

is a flat resolution of $\mathrm{coker}\, d_{\mathscr{F}'^\cdot}^{-n-1}$. So we have

$$\begin{aligned}\mathrm{Tor}_i(\mathrm{coker}\, d_{\mathscr{F}'^\cdot}^{-n-1}, M) &\cong \mathscr{H}^{-i}(\mathscr{F}'^\cdot[-n] \otimes_A M) \cong \mathscr{H}^{-i-n}(\mathscr{F}'^\cdot \otimes_A M)\\ &\cong \mathrm{Tor}_{i+n}(\mathscr{F}^\cdot, M) = 0\end{aligned}$$

for all $i > 0$. Hence $\mathrm{coker}\, d_{\mathscr{F}'^\cdot}^{-n-1}$ is a flat sheaf of A-modules. It is also constructible. Let $\tau_{\geq -n}\mathscr{F}'^\cdot$ be the truncated complex

$$0 \to \mathrm{coker}\, d_{\mathscr{F}'^\cdot}^{-n-1} \to \mathscr{F}'^{-n+1} \to \mathscr{F}'^{-n+2} \to \cdots.$$

It is a complex of constructible flat sheaves of A-modules. Moreover, $\tau_{\geq -n}\mathscr{F}'^\cdot \to \tau_{\geq -n}\mathscr{F}^\cdot$ is a quasi-isomorphism. Since $\mathscr{F}^\cdot$ is bounded, we have $\tau_{\geq -n}\mathscr{F}^\cdot = \mathscr{F}^\cdot$ if we take n sufficiently large. □

Let X be a scheme and let A be a ring. Denote by $D^b_{\mathrm{tf}}(X, A)$ the full subcategory of $D^b(X, A)$ consisting of objects with finite Tor-dimension. Then $D^b_{\mathrm{tf}}(X, A)$ is a triangulated subcategory of $D^b(X, A)$. Suppose that X is a noetherian scheme and A a noetherian ring. Denote by $D^b_{\mathrm{ctf}}(X, A)$ the full subcategory of $D^b(X, A)$ consisting of objects $\mathscr{F}^\cdot$ with finite Tor-dimension, and such that $\mathscr{H}^i(\mathscr{F})$ are constructible for all i. Then $D^b_{\mathrm{ctf}}(X, A)$ is a triangulated subcategory of $D^b(X, A)$. By 6.4.6, any object in $D^b_{\mathrm{ctf}}(X, A)$ is isomorphic to a bounded complex of constructible flat sheaves of A-modules on X.

Let $\mathscr{L}$ be the family of objects in $K^b(X, A)$ consisting of bounded complexes of flat sheaves of A-modules. For each fixed object $\mathscr{F}^\cdot$ in $K(X, A)$, by 6.4.4, the functor $\mathscr{F}^\cdot \otimes_A -$ transforms quasi-isomorphisms in $\mathscr{L}$ to quasi-isomorphisms. On the other hand, using an argument similar to the proof of 6.4.6, one can show that for any $\mathscr{G}^\cdot \in \mathrm{ob}\, K^b(X, A)$ with finite Tor-dimension, there exists a quasi-isomorphism from an object in $\mathscr{L}$ to $\mathscr{G}^\cdot$. By the dual version of 6.3.1, the left derived functor of $\mathscr{F}^\cdot \otimes_A -$ is defined everywhere on $D^b_{\mathrm{tf}}(X, A)$. Denote the derived functor by

$$L_{II}(\mathscr{F}^\cdot \otimes_A -) : D^b_{\mathrm{tf}}(X, A) \to D(X, A).$$

So we get a functor

$$L_{II}(- \otimes_A -) : K(X, A) \times D^b_{\mathrm{tf}}(X, A) \to D(X, A).$$

By 6.4.4, the functor $L_{II}(- \otimes_A \mathscr{G}^\cdot)$ transforms quasi-isomorphisms in $K(X, A)$ to isomorphisms in $D(X, A)$. So it factors through $D(X, A)$ and defines a functor

$$- \otimes^L_A - : D(X, A) \times D^b_{\mathrm{tf}}(X, A) \to D(X, A).$$

For any $\mathscr{F}^\cdot, \mathscr{G}^\cdot \in \mathrm{ob}\, K(X, A)$, the canonical evaluation morphism

$$\mathscr{F}^\cdot \otimes_A \mathscr{H}om^\cdot(\mathscr{F}^\cdot, \mathscr{G}^\cdot) \to \mathscr{G}^\cdot$$

is not a morphism of complexes, but the evaluation morphism

$$\mathscr{H}om^\cdot(\mathscr{F}^\cdot, \mathscr{G}^\cdot) \otimes_A \mathscr{F}^\cdot \to \mathscr{G}^\cdot$$

is a morphism of complexes. If $\mathscr{F}^\cdot \in \mathrm{ob}\, D^b_{\mathrm{tf}}(X, A)$ and $\mathscr{G}^\cdot \in \mathrm{ob}\, D^+(X, A)$, then the second evaluation morphism induces a morphism

$$\mathrm{Ev} : R\mathscr{H}om(\mathscr{F}^\cdot, \mathscr{G}^\cdot) \otimes^L_A \mathscr{F}^\cdot \to \mathscr{G}^\cdot$$

in $D(X, A)$. Let $\mathscr{F}'^\cdot \to \mathscr{F}^\cdot$ and $\mathscr{G}^\cdot \to \mathscr{I}^\cdot$ be quasi-isomorphisms such that $\mathscr{F}'^\cdot$ is a bounded complex of flat sheaves of A-modules, and $\mathscr{I}^\cdot$ is a bounded below complex of injective sheaves of A-modules. Ev is the composite

$$\begin{aligned} R\mathscr{H}om(\mathscr{F}^\cdot, \mathscr{G}^\cdot) \otimes^L_A \mathscr{F}^\cdot &\cong \mathscr{H}om(\mathscr{F}^\cdot, \mathscr{I}^\cdot) \otimes_A \mathscr{F}'^\cdot \\ &\to \mathscr{H}om(\mathscr{F}^\cdot, \mathscr{I}^\cdot) \otimes_A \mathscr{F}^\cdot \\ &\stackrel{\mathrm{Ev}}{\to} \mathscr{I}^\cdot \cong \mathscr{G}^\cdot. \end{aligned}$$

Proposition 6.4.7. *Let X be a scheme, A a ring, $\mathscr{F}^\cdot \in \mathrm{ob}\, D^b_{\mathrm{tf}}(X,A)$, and $\mathscr{G}^\cdot \in \mathrm{ob}\, D^+(X,A)$. For any $\mathscr{H}^\cdot \in \mathrm{ob}\, D(X,A)$ and any morphism $\phi : \mathscr{H}^\cdot \to R\mathscr{H}om(\mathscr{F}^\cdot, \mathscr{G}^\cdot)$, let $F(\phi) : \mathscr{H}^\cdot \otimes^L_A \mathscr{F}^\cdot \to \mathscr{G}^\cdot$ be the composite*

$$\mathscr{H}^\cdot \otimes^L_A \mathscr{F}^\cdot \stackrel{\phi\otimes^L_A \mathrm{id}}{\to} R\mathscr{H}om(\mathscr{F}^\cdot, \mathscr{G}^\cdot) \otimes^L_A \mathscr{F}^\cdot \stackrel{\mathrm{Ev}}{\to} \mathscr{G}^\cdot.$$

Then F defines a one-to-one correspondence

$$F : \mathrm{Hom}_{D(X,A)}(\mathscr{H}^\cdot, R\mathscr{H}om(\mathscr{F}^\cdot, \mathscr{G}^\cdot)) \stackrel{\cong}{\to} \mathrm{Hom}_{D(X,A)}(\mathscr{H}^\cdot \otimes^L_A \mathscr{F}^\cdot, \mathscr{G}^\cdot).$$

Proof. We may assume that $\mathscr{F}^\cdot$ is a bounded complex of flat sheaves and $\mathscr{G}^\cdot$ is a bounded below complex of injective sheaves. Then $\mathrm{Hom}^\cdot(\mathscr{F}^\cdot, \mathscr{G}^\cdot)$ is a bounded below complex of injective sheaves by 5.6.8. We have

$$\begin{aligned}\mathrm{Hom}_{D(X,A)}(\mathscr{H}^\cdot \otimes^L_A \mathscr{F}^\cdot, \mathscr{G}^\cdot) &\cong H^0(\mathrm{Hom}^\cdot(\mathscr{H}^\cdot \otimes_A \mathscr{F}^\cdot, \mathscr{G}^\cdot)) \\ &\cong H^0(\mathrm{Hom}^\cdot(\mathscr{H}^\cdot, \mathscr{H}om(\mathscr{F}^\cdot, \mathscr{G}^\cdot)) \\ &\cong \mathrm{Hom}_{D(X,A)}(\mathscr{H}^\cdot, R\mathscr{H}om(\mathscr{F}^\cdot, \mathscr{G}^\cdot)).\end{aligned}$$

□

Lemma 6.4.8. *Let A be a ring, let*

$$\cdots \to K^i \stackrel{d^i}{\to} K^{i+1} \to \cdots$$

be a complex of A-modules, let

$$\begin{array}{ccccccc} & & \vdots & & \vdots & & \\ & & \uparrow & & \uparrow & & \\ \cdots \to & & C^{i,j+1} & \stackrel{d_1^{i,j+1}}{\to} & C^{i+1,j+1} & & \to \cdots \\ & & d_2^{i,j} \uparrow & & \uparrow d_2^{i+1,j} & & \\ \cdots \to & & C^{ij} & \stackrel{d_1^{ij}}{\to} & C^{i+1,j} & & \to \cdots \\ & & \uparrow & & \uparrow & & \\ & & \vdots & & \vdots & & \\ & & \uparrow & & \uparrow & & \\ \cdots \to & & C^{i0} & \stackrel{d_1^{i0}}{\to} & C^{i+1,0} & & \to \cdots \\ & & \uparrow & & \uparrow & & \\ & & 0 & & 0 & & \end{array}$$

be a bicomplex of A-modules, and let $f^\cdot : (K^\cdot, d^\cdot) \to (C^{\cdot 0}, d_1^{\cdot 0})$ be a morphism of complexes. Suppose for each i, the sequence

$$0 \to K^i \to C^{i0} \stackrel{d_2^{i0}}{\to} C^{i1} \to \cdots$$

is exact. Let

$$Z^{ij} = \ker(C^{ij} \stackrel{d_2^{ij}}{\to} C^{i,j+1}).$$

Suppose for each pair (i,j), *there exists a left inverse* $p^{ij} : C^{ij} \to Z^{ij}$ *of the inclusion* $Z^{ij} \hookrightarrow C^{ij}$ *such that the following diagram commutes:*

$$\begin{array}{ccc} C^{ij} & \stackrel{d_1^{ij}}{\to} & C^{i+1,j} \\ p^{ij} \downarrow & & \downarrow p^{i+1,j} \\ Z^{ij} & \stackrel{d_1^{ij}|_{Z^{ij}}}{\to} & Z^{i+1,j} \end{array}$$

Then $K^{\cdot}$ *is homotopically equivalent to the complex associated to the bicomplex* $C^{\cdot\cdot}$.

Proof. Let $D^{ij} = Z^{ij} \bigoplus Z^{i,j+1}$. Define

$$d_1^{ij} : D^{ij} \to D^{i+1,j}$$

to be the homomorphisms induced by the restrictions of $d_1^{ij} : C^{ij} \to C^{i+1,j}$ and $d_1^{i,j+1} : C^{i,j+1} \to C^{i+1,j+1}$ to Z^{ij} and $Z^{i,j+1}$, respectively, and define

$$d_2^{ij} : D^{ij} \to D^{i,j+1}$$

by

$$d_2^{ij}(x_1, x_2) = (x_2, 0)$$

for any $(x_1, x_2) \in Z^{ij} \bigoplus Z^{i,j+1}$. Then $D^{\cdot\cdot}$ is a bicomplex, and the homomorphisms

$$C^{ij} \to D^{ij}, \quad x \mapsto (p^{ij}(x), d_2^{ij}(x))$$

define an isomorphism of bicomplexes. For each n, let

$$D^n = \bigoplus_{i+j=n, j\geq 0} D^{ij} = \bigoplus_{i+j=n, j\geq 0} (Z^{ij} \oplus Z^{i,j+1})$$

and define

$$d^n : D^n \to D^{n+1}$$

to be $\bigoplus_{i+j=n,j\geq 0}(d_1^{ij} \bigoplus (-1)^i d_2^{ij})$. To prove the lemma, it suffices to prove that the complex $K^{\cdot}$ is homotopically equivalent to the complex $D^{\cdot}$. Note that for each n, the homomorphism $f^n : K^n \to C^{n0}$ induces an isomorphism of K^n with Z^{n0}. Define

$$\phi^n : K^n \to D^n$$

so that any element x in K^n is mapped to

$$\Big((f^n(x),0),(0,0),\dots\Big)\in D^n=(Z^{n0}\oplus Z^{n1})\bigoplus(Z^{n-1,1}\oplus Z^{n-1,2})\bigoplus\cdots,$$

and define

$$\psi^n : D^n\to K^n$$

so that any element

$$\begin{aligned}&\Big((x^{n0},y^{n1}),(x^{n-1,1},y^{n-1,2}),\dots\Big)\\ &\in D^n=(Z^{n0}\oplus Z^{n1})\bigoplus(Z^{n-1,1}\oplus Z^{n-1,2})\bigoplus\cdots\end{aligned}$$

is mapped to the preimage of $x^{n0}\in\operatorname{im} f^n$ in K^n. Then $\phi^\cdot : K^\cdot\to D^\cdot$ and $\psi^\cdot : D^\cdot\to K^\cdot$ are morphisms of complexes, and $\psi^\cdot\circ\phi^\cdot=\mathrm{id}_{K^\cdot}$. Define

$$H^n : D^n\to D^{n-1}$$

so that any element

$$\begin{aligned}&\Big((x^{n0},y^{n1}),(x^{n-1,1},y^{n-1,2}),\dots\Big)\\ &\in D^n=(Z^{n0}\oplus Z^{n1})\bigoplus(Z^{n-1,1}\oplus Z^{n-1,2})\bigoplus\cdots\end{aligned}$$

is mapped to

$$\begin{aligned}&\Big((0,(-1)^{n-1}x^{n-1,1}),(0,(-1)^{n-2}x^{n-2,2}),\dots\Big)\\ &\in D^{n-1}=(Z^{n-1,0}\oplus Z^{n-1,1})\bigoplus(Z^{n-2,1}\oplus Z^{n-2,2})\bigoplus\cdots.\end{aligned}$$

Then we have

$$d^{n-1}H^n+H^{n+1}d^n=\mathrm{id}_{D^n}-\phi^n\psi^n.$$

So $K^\cdot$ is homotopically equivalent to $D^\cdot$. □

Corollary 6.4.9. *Let X be a scheme, A a ring, $\mathscr{F}^\cdot$ a complex of sheaves of A-modules on X, $\mathscr{C}^\cdot(\mathscr{F}^i)$ the Godement resolution of $\mathscr{F}^i$ for any i, $\tau_{\leq n}\mathscr{C}^\cdot(\mathscr{F}^i)$ the truncated Godement resolution*

$$0\to\mathscr{C}^0(\mathscr{F}_i)\to\cdots\to\mathscr{C}^{n-1}(\mathscr{F}_i)\to\ker\Big(\mathscr{C}^n(\mathscr{F}^i)\to\mathscr{C}^{n+1}(\mathscr{F}^i)\Big)\to 0$$

for any i and any $n\geq 0$, and P the set of geometric points of X in the definition of the Godement resolution. Then for any geometric point $t\in P$, $\mathscr{F}^\cdot_t$ is homotopically equivalent to the complex associated to the bicomplex $(\mathscr{C}^\cdot(\mathscr{F}^\cdot))_t$, and to the complex associated to the bicomplex $(\tau_{\leq n}\mathscr{C}^\cdot(\mathscr{F}^\cdot))_t$.

Proof. This follows from 6.4.8, the fact that the complexes

$$0 \to \mathscr{F}_{it} \to (\mathscr{C}^\cdot(\mathscr{F}^i))_t,$$
$$0 \to \mathscr{F}_{it} \to (\tau_{\leq n}\mathscr{C}^\cdot(\mathscr{F}^i))_t$$

are split, and the splitting is functorial with respect to $\mathscr{F}_i$. Confer the proof of 5.6.13 (iii). □

Corollary 6.4.10. *Let X be a scheme, A a ring, $\mathscr{F}^\cdot$ and $\mathscr{G}^\cdot$ bounded above complexes of sheaves of A-modules on X, and $\mathscr{G}'^\cdot \to \mathscr{G}^\cdot$ a quasi-isomorphism such that $\mathscr{G}^\cdot$ is a bounded above complex of flat sheaves of A-modules on X. Denote the complex associated to the bicomplex $\mathscr{C}^\cdot(\mathscr{G}'^\cdot)$ (resp. $\tau_{\leq n}\mathscr{C}^\cdot(\mathscr{F}^\cdot)$) of Godement resolutions (resp. truncated Godement resolutions) by the same notation. Then we have*

$$\mathscr{F}^\cdot \otimes_A^L \mathscr{G}^\cdot \cong \mathscr{F}^\cdot \otimes_A \mathscr{C}^\cdot(\mathscr{G}'^\cdot),$$
$$\mathscr{F}^\cdot \otimes_A^L \mathscr{G}^\cdot \cong \mathscr{F}^\cdot \otimes_A \tau_{\leq n}\mathscr{C}^\cdot(\mathscr{G}'^\cdot).$$

Suppose that $\mathscr{G}^\cdot$ has finite Tor*-dimension, and $\mathscr{G}'^\cdot$ is a bounded complex of flat sheaves of A-modules quasi-isomorphic to $\mathscr{G}^\cdot$. Then the above results hold for any complex $\mathscr{F}^\cdot$ of sheaves of A-modules.*

Proof. Let P be the set of geometric points of X in the definition of the Godement resolution. By 6.4.9, for any $t \in P$, the canonical morphisms

$$\mathscr{G}'^\cdot_t \to (\mathscr{C}^\cdot(\mathscr{G}'^\cdot))_t,$$
$$\mathscr{G}'^\cdot_t \to (\tau_{\leq n}\mathscr{C}^\cdot(\mathscr{G}'^\cdot))_t$$

are homotopically invertible. It follows that

$$(\mathscr{F}^\cdot \otimes_A \mathscr{G}'^\cdot)_t \to (\mathscr{F}^\cdot \otimes_A \mathscr{C}^\cdot(\mathscr{G}'^\cdot))_t,$$
$$(\mathscr{F}^\cdot \otimes_A \mathscr{G}'^\cdot)_t \to (\mathscr{F}^\cdot \otimes_A \tau_{\leq n}\mathscr{C}^\cdot(\mathscr{G}'^\cdot))_t$$

are homotopically invertible. Hence

$$\mathscr{F}^\cdot \otimes_A \mathscr{G}'^\cdot \to \mathscr{F}^\cdot \otimes_A \mathscr{C}^\cdot(\mathscr{G}'^\cdot),$$
$$\mathscr{F}^\cdot \otimes_A \mathscr{G}'^\cdot \to \mathscr{F}^\cdot \otimes_A \tau_{\leq n}\mathscr{C}^\cdot(\mathscr{G}'^\cdot)$$

are quasi-isomorphisms. We have $\mathscr{F}^\cdot \otimes_A \mathscr{G}'^\cdot \cong \mathscr{F}^\cdot \otimes_A^L \mathscr{G}^\cdot$. Our assertion follows. □

Let A be a ring, let

$$\cdots \to C^i \stackrel{d^i}{\to} C^{i+1} \to \cdots$$

be a complex of A modules, and let

$$Z^i = \ker d^i, \quad B^i = \operatorname{im} d^{i-1}, \quad H^i = Z^i/B^i.$$

We have short exact sequences

$$0 \to Z^i \to C^i \to B^{i+1} \to 0,$$
$$0 \to B^i \to Z^i \to H^i \to 0.$$

We say that the complex $C^\cdot$ is *split* if the above short exact sequences are split for all i.

Lemma 6.4.11. *Let A be a ring, and let $C^\cdot$ be a split complex of A modules. Then $C^\cdot$ is homotopically equivalent to the complex*

$$\cdots \to H^i(C^\cdot) \xrightarrow{0} H^{i+1}(C^\cdot) \to \cdots.$$

Proof. Notation as above. We have isomorphisms

$$C^i \cong Z^i \oplus B^{i+1} \cong H^i \oplus B^i \oplus B^{i+1}.$$

Through these isomorphisms, $d^i : C^i \to C^{i+1}$ can be identified with the homomorphisms

$$d^i : H^i \oplus B^i \oplus B^{i+1} \to H^{i+1} \oplus B^{i+1} \oplus B^{i+2}, \quad (h_i, b_i, b_{i+1}) \mapsto (0, b_{i+1}, 0).$$

The homomorphisms

$$\phi^i : H^i \to H^i \oplus B^i \oplus B^{i+1}, \qquad h_i \mapsto (h_i, 0, 0),$$
$$\psi^i : H^i \oplus B^i \oplus B^{i+1} \to H^i, \qquad (h_i, b_i, b_{i+1}) \mapsto h_i$$

define morphisms of complexes $\phi^\cdot$ and $\psi^\cdot$ between the complex

$$\cdots \to H^i \xrightarrow{0} H^{i+1} \to \cdots$$

and the complex

$$\cdots \to H^i \oplus B^i \oplus B^{i+1} \xrightarrow{d^i} H^{i+1} \oplus B^{i+1} \oplus B^{i+2} \to \cdots.$$

We have $\psi^\cdot \phi^\cdot = \mathrm{id}$. Consider the homomorphisms

$$G^i : H^i \oplus B^i \oplus B^{i+1} \to H^{i-1} \oplus B^{i-1} \oplus B^i, \quad (h_i, b_i, b_{i+1}) \mapsto (0, 0, b_i).$$

We have

$$d^{i-1}G^i + G^{i+1}d^i = \mathrm{id} - \phi^i\psi^i.$$

Our assertion follows. □

Proposition 6.4.12. *Let A be a ring, and let $K^\cdot$ and $L^\cdot$ be two bounded above complexes of A-modules. Then we have a biregular spectral sequence*

$$E_2^{pq} = \bigoplus_{i+j=q} \mathrm{Tor}_{-p}^A(H^i(K^\cdot), H^j(L^\cdot)) \Rightarrow H^{p+q}(K^\cdot \otimes^L L^\cdot).$$

Proof. Choose a projective Cartan–Eilenberg resolution $P^{\cdot\cdot}$ of $K^\cdot$. For each i,

$$(P^{\cdot,i}, d_1^{\cdot,i}),\ (\ker d_2^{\cdot,i}, d_1^{\cdot,i}),\ (\mathrm{im}\, d_2^{\cdot,i-1}, d_1^{\cdot,i}),\ (\ker d_2^{\cdot,i}/\mathrm{im}\, d_2^{\cdot,i-1}, d_1^{\cdot,i})$$

are projective resolutions of

$$K^i,\ \ker d_{K^\cdot}^i, \mathrm{im}\, d_{K^\cdot}^{i-1},\ \ker d_{K^\cdot}^i/\mathrm{im}\, d_{K^\cdot}^{i-1},$$

respectively. Moreover, the short exact sequences

$$0 \to \ker d_2^{\cdot,i} \to P^{\cdot,i} \to \mathrm{im}\, d_2^{\cdot,i} \to 0,$$
$$0 \to \mathrm{im}\, d_2^{\cdot,i-1} \to \ker d_2^{\cdot,i} \to \ker d_2^{\cdot,i}/\mathrm{im}\, d_2^{\cdot,i-1} \to 0$$

are split. Choose a projective Cartan–Eilenberg resolution $Q^{\cdot\cdot}$ for $L^\cdot$. Consider the bicomplex

$$C^{pq} = \bigoplus_{k+l=p,\ i+j=q} P^{ki} \otimes_A Q^{lj}.$$

We have a biregular spectral sequence

$$E_2^{pq} = H_I^p H_{II}^q(C^{\cdot\cdot}) \Rightarrow H^{p+q}(K^\cdot \otimes^L L^\cdot).$$

By Lemma 6.4.11, for each fixed k, the complex

$$\cdots \to P^{ki} \to P^{k,i+1} \to \cdots$$

is homotopically equivalent to the complex

$$\cdots \to H_{II}^i(P^{k\cdot}) \xrightarrow{0} H_{II}^{i+1}(P^{k\cdot}) \to \cdots,$$

and for each fixed l, the complex

$$\cdots \to Q^{lj} \to Q^{l,j+1} \to \cdots$$

is homotopically equivalent to the complex

$$\cdots \to H_{II}^j(Q^{l\cdot}) \xrightarrow{0} H_{II}^{j+1}(Q^{l\cdot}) \to \cdots.$$

So we have

$$H^q(P^{k\cdot} \otimes_A Q^{l\cdot}) \cong \bigoplus_{i+j=q} H_{II}^i(P^{k\cdot}) \otimes_A H_{II}^j(Q^{l\cdot}).$$

But $H_{II}^i(P^{\cdot\cdot})$ and $H_{II}^j(Q^{\cdot\cdot})$ are projective resolutions of $H^i(K^\cdot)$ and $H^j(L^\cdot)$, respectively. It follows that

$$H_I^p H_{II}^q(C^{\cdot\cdot}) \cong \bigoplus_{i+j=q} \mathrm{Tor}_{-p}^A(H^i(K^\cdot), H^j(L^\cdot)).$$

Our assertion follows. □

6.5 Way-out Functors

([Hartshorne (1966)] I 7.)

Let $\mathscr{A}$ and $\mathscr{B}$ be abelian categories, $\mathscr{A}'$ a thick abelian subcategory of $\mathscr{A}$, $D^*(\mathscr{A})$ one of the following derived categories

$$D(\mathscr{A}), D^+(\mathscr{A}), D^-(\mathscr{A}), D^b(\mathscr{A}), D_{\mathscr{A}'}(\mathscr{A}), D^+_{\mathscr{A}'}(\mathscr{A}), D^-_{\mathscr{A}'}(\mathscr{A}), D^b_{\mathscr{A}'}(\mathscr{A}),$$

and $F : D^*(\mathscr{A}) \to D(\mathscr{B})$ a covariant (resp. contravariant) exact functor. We say F is *way-out right* if for any integer N, there exists an integer M such that for any $X^\cdot \in \operatorname{ob} D^*(\mathscr{A})$ with $H^i(X^\cdot) = 0$ for all $i < M$ (resp. $i > M$), we have $H^i(F(X^\cdot)) = 0$ for all $i < N$. We say F is *way-out left* if for any integer N, there exists an integer M such that for any $X^\cdot \in \operatorname{ob} D^*(\mathscr{A})$ with $H^i(X^\cdot) = 0$ for all $i > M$ (resp. $i < M$), we have $H^i(F(X^\cdot)) = 0$ for all $i > N$.

Proposition 6.5.1. *Let $F : \mathscr{A} \to \mathscr{B}$ be a left exact functor. Suppose that $\mathscr{A}$ has enough injective objects. Then $RF : D^+(\mathscr{A}) \to D(\mathscr{B})$ is way-out right. If RF has finite cohomological dimension, then RF is also way-out left.*

Proof. For any integer N, let $X^\cdot$ be an object in $D^+(\mathscr{A})$ such that $H^i(X^\cdot) = 0$ for all $i < N$, then we can find a complex $I^\cdot$ of injective objects in $\mathscr{A}$ with $I^i = 0$ for all $i < N$ such that $I^\cdot$ is isomorphic to $X^\cdot$ in $D^+(\mathscr{A})$. We have

$$R^iF(X^\cdot) \cong H^i(F(I^\cdot)) = 0$$

for all $i < N$. So RF is way-out right. Suppose that RF has finite cohomological dimension, and let n be a nonnegative integer such that $R^iF(X) = 0$ for any $i > n$ and any $X \in \operatorname{ob} \mathscr{A}$. For any object $X^\cdot$ in $D(\mathscr{A})$, let $X^\cdot \to X'^\cdot$ be a quasi-isomorphism such that $X'^\cdot$ is a complex of RF-acyclic objects in $\mathscr{A}$. Given an integer N, if $H^i(X^\cdot) = 0$ for all $i > N - n$, then

$$0 \to X'^{N-n} \to X'^{N-n+1} \to \cdots$$

is an RF-acyclic resolution of $\ker d_{X'^\cdot}^{N-n}$. So we have

$$R^iF(\ker d_{X'^\cdot}^{N-n}) \cong H^i(F(X'^\cdot)[N-n]) \cong H^{i+N-n}(F(X'^\cdot)) \cong R^{i+N-n}F(X^\cdot)$$

for all $i > 0$. Since $R^iF(\ker d_{X'^\cdot}^{N-n}) = 0$ for all $i > n$, we have $R^iF(X^\cdot) = 0$ for all $i > N$. Hence RF is way-out left. $\square$

Given a complex $X^\cdot$ of objects in $\mathscr{A}$, consider the following truncated complexes:

$$\begin{aligned}
\tau_{\leq n}X^\cdot &= (\cdots \to X^{n-2} \to X^{n-1} \to \ker d^n_{X^\cdot} \to 0),\\
\tau_{\geq n}X^\cdot &= (0 \to X^n/\mathrm{im}\, d^{n-1}_{X^\cdot} \to X^{n+1} \to X^{n+2} \to \cdots),\\
\tau'_{\leq n}X^\cdot &= (\cdots \to X^{n-1} \to X^n \to \mathrm{im}\, d^{n+1}_{X^\cdot} \to 0),\\
\tau'_{\geq n}X^\cdot &= (0 \to \mathrm{im}\, d^{n-1}_{X^\cdot} \to X^n \to X^{n+1} \to \cdots).
\end{aligned}$$

We have quasi-isomorphisms

$$\tau_{\leq n}X^\cdot \to \tau'_{\leq n}X^\cdot, \quad \tau'_{\geq n}X^\cdot \to \tau_{\geq n}X^\cdot,$$

and the following short exact sequences of complexes

$$\begin{aligned}
&0 \to \tau_{\leq n}X^\cdot \to X^\cdot \to \tau'_{\geq n+1}X^\cdot \to 0,\\
&0 \to H^n(X^\cdot)[-n] \to \tau_{\geq n}X^\cdot \to \tau'_{\geq n+1}X^\cdot \to 0,\\
&0 \to \tau'_{\leq n-1}X^\cdot \to \tau_{\leq n}X^\cdot \to H^n(X^\cdot)[-n] \to 0.
\end{aligned}$$

Note that $\tau_{\leq n}$ and $\tau_{\geq n}$ transform quasi-isomorphisms to quasi-isomorphisms. So they can be defined on $D(\mathscr{A})$. We have the following distinguished triangles in $D(\mathscr{A})$:

$$\begin{aligned}
&\tau_{\leq n}X^\cdot \to X^\cdot \to \tau_{\geq n+1}X^\cdot \to,\\
&H^n(X^\cdot)[-n] \to \tau_{\geq n}X^\cdot \to \tau_{\geq n+1}X^\cdot \to,\\
&\tau_{\leq n-1}X^\cdot \to \tau_{\leq n}X^\cdot \to H^n(X^\cdot)[-n] \to .
\end{aligned}$$

Sometimes we also consider the following truncated complexes:

$$\begin{aligned}
\sigma_{\leq n}X^\cdot &= (\cdots \to X^{n-2} \to X^{n-1} \to X^n \to 0),\\
\sigma_{\geq n}X^\cdot &= (0 \to X^n \to X^{n+1} \to X^{n+2} \to \cdots).
\end{aligned}$$

But $\sigma_{\leq n}$ and $\sigma_{\geq n}$ do not transform quasi-isomorphisms to quasi-isomorphisms, and are not defined on $D(\mathscr{A})$.

Proposition 6.5.2. *Let $\mathscr{A}$ and $\mathscr{B}$ be abelian categories, $\mathscr{A}'$ a thick abelian subcategory of $\mathscr{A}$, F and G exact functors from $D_{\mathscr{A}'}(\mathscr{A})$, or $D^+_{\mathscr{A}'}(\mathscr{A})$, or $D^-_{\mathscr{A}'}(\mathscr{A})$ to $\mathscr{D}(\mathscr{B})$, and $\xi : F \to G$ a morphism of functors.*

(i) *If $\xi(X) : F(X) \to G(X)$ is an isomorphism for any $X \in \mathrm{ob}\,\mathscr{A}'$, then $\xi(X^\cdot) : F(X^\cdot) \to G(X^\cdot)$ is an isomorphism for any $X^\cdot \in \mathrm{ob}\, D^b_{\mathscr{A}'}(\mathscr{A})$.*

(ii) *If $\xi(X) : F(X) \to G(X)$ is an isomorphism for any $X \in \mathrm{ob}\,\mathscr{A}'$, and F and G are way-out right (resp. left), then $\xi(X^\cdot) : F(X^\cdot) \to G(X^\cdot)$ is an isomorphism for any $X^\cdot \in \mathrm{ob}\, D^+_{\mathscr{A}'}(\mathscr{A})$ (resp. $\mathrm{ob}\, D^-_{\mathscr{A}'}(\mathscr{A})$).*

(iii) *If $\xi(X) : F(X) \to G(X)$ is an isomorphism for any $X \in \mathrm{ob}\,\mathscr{A}'$, and F and G are way-out in both directions, then $\xi(X^\cdot) : F(X^\cdot) \to G(X^\cdot)$ is an isomorphism for any $X^\cdot \in \mathrm{ob}\, D_{\mathscr{A}'}(\mathscr{A})$.*

Proof.

(i) For any $X^\cdot \in \operatorname{ob} D^b_{\mathscr{A}'}(\mathscr{A})$, we have $\tau_{\geq n}X^\cdot = 0$ for sufficiently large n. So $\xi(\tau_{\geq n}X^\cdot)$ is an isomorphism for large n. ξ induces a morphism of distinguished triangles

$$\begin{array}{ccccc} F(H^{n-1}(X^\cdot)[-(n-1)]) & \to & F(\tau_{\geq n-1}X^\cdot) & \to & F(\tau_{\geq n}X^\cdot) \to \\ \xi\downarrow & & \xi\downarrow & & \xi\downarrow \\ G(H^{n-1}(X^\cdot)[-(n-1)]) & \to & G(\tau_{\geq n-1}X^\cdot) & \to & G(\tau_{\geq n}X^\cdot) \to . \end{array}$$

Since $H^{n-1}(X^\cdot) \in \operatorname{ob}\mathscr{A}'$, $\xi(H^{n-1}(X^\cdot)[-(n-1)])$ is an isomorphism by our assumption. Suppose we have shown that $\xi(\tau_{\geq n}X^\cdot)$ is an isomorphism. Then by 6.1.1 (iii) and (TR 2), $\xi(\tau_{\geq n-1}X^\cdot)$ is also an isomorphism. It follows that $\xi(\tau_{\geq n}X^\cdot)$ are isomorphisms for all n. We have $\tau_{\geq n}X^\cdot \cong X^\cdot$ for $n \ll 0$. So $\xi(X^\cdot)$ is an isomorphism.

(ii) Suppose that F and G are way-out right. Given $X^\cdot \in \operatorname{ob} D^+_{\mathscr{A}'}(\mathscr{A})$ and an integer j, let us prove that

$$H^j(\xi(X^\cdot)) : H^j(F(X)) \to H^j(G(X))$$

is an isomorphism. There exists an integer M such that for any $Y^\cdot \in \operatorname{ob} D^+_{\mathscr{A}'}(\mathscr{A})$ with $H^i(Y^\cdot) = 0$ for all $i < M$, we have $H^i(F(Y^\cdot)) = H^i(G(Y^\cdot)) = 0$ for all $i < j+1$. In particular, we have

$$H^i(F(\tau_{\geq M}X^\cdot)) = H^i(G(\tau_{\geq M}X^\cdot)) = 0$$

for $i = j-1, j$. We have a distinguished triangle

$$F(\tau_{\leq M-1}X^\cdot) \to F(X^\cdot) \to F(\tau_{\geq M}X^\cdot) \to .$$

From the long exact sequence of cohomology objects associated to this distinguished triangle, we get

$$H^j(F(\tau_{\leq M-1}X^\cdot)) \cong H^j(F(X^\cdot)).$$

Similarly, we have

$$H^j(G(\tau_{\leq M-1}X^\cdot)) \cong H^j(G(X^\cdot)).$$

By (i), ξ induces an isomorphism $H^j(F(\tau_{\leq M-1}X^\cdot)) \cong H^j(G(\tau_{\leq M-1}X^\cdot))$. So ξ induces an isomorphism $H^j(F(X^\cdot)) \cong H^j(G(X^\cdot))$.

(iii) For any $X^\cdot \in \operatorname{ob} D_{\mathscr{A}'}(\mathscr{A})$, ξ induces a morphism of distinguished triangles

$$\begin{array}{ccccc} F(\tau_{\leq 0}X^\cdot) & \to & F(X^\cdot) & \to & F(\tau_{\geq 1}X^\cdot) \to \\ \xi\downarrow & & \xi\downarrow & & \xi\downarrow \\ G(\tau_{\leq 0}X^\cdot) & \to & G(X^\cdot) & \to & G(\tau_{\geq 1}X^\cdot) \to \end{array}$$

By (ii), $\xi(\tau_{\leq 0}X^\cdot)$ and $\xi(\tau_{\geq 1}X^\cdot)$ are isomorphisms. It follows that $\xi(X^\cdot)$ is an isomorphism. □

Proposition 6.5.3. *Let $\mathscr{A}$ and $\mathscr{B}$ be abelian categories, $\mathscr{A}'$ and $\mathscr{B}'$ thick abelian subcategories of $\mathscr{A}$ and $\mathscr{B}$, respectively, F an exact functor from $D_{\mathscr{A}'}(\mathscr{A})$, or $D^+_{\mathscr{A}'}(\mathscr{A})$, or $D^-_{\mathscr{A}'}(\mathscr{A})$ to $\mathscr{D}(\mathscr{B})$.*

(i) *If $F(X) \in \mathrm{ob}\, D_{\mathscr{B}'}(\mathscr{B})$ for any $X \in \mathrm{ob}\,\mathscr{A}'$, then $F(X^\cdot) \in \mathrm{ob}\, D_{\mathscr{B}'}(\mathscr{B})$ for any $X^\cdot \in \mathrm{ob}\, D^b_{\mathscr{A}'}(\mathscr{A})$.*

(ii) *If $F(X) \in \mathrm{ob}\, D_{\mathscr{B}'}(\mathscr{B})$ for any $X \in \mathrm{ob}\,\mathscr{A}'$ and F is way-out right (resp. left), then $F(X^\cdot) \in \mathrm{ob}\, D_{\mathscr{B}'}(\mathscr{B})$ for any $X^\cdot \in \mathrm{ob}\, D^+_{\mathscr{A}'}(\mathscr{A})$ (resp. $X^\cdot \in \mathrm{ob}\, D^-_{\mathscr{A}'}(\mathscr{A})$).*

(iii) *If $F(X) \in \mathrm{ob}\, D_{\mathscr{B}'}(\mathscr{B})$ for any $X \in \mathrm{ob}\,\mathscr{A}'$ and F is way-out in both directions, then $F(X^\cdot) \in \mathrm{ob}\, D_{\mathscr{B}'}(\mathscr{B})$ for any $X^\cdot \in \mathrm{ob}\, D_{\mathscr{A}'}(\mathscr{A})$.*

Proposition 6.5.4. *Let X be a scheme, A a ring, and $\mathscr{F}^\cdot$ a bounded complex of sheaves of A-modules on X. The following conditions are equivalent:*

(i) *There exists a quasi-isomorphism $\mathscr{F}'^\cdot \to \mathscr{F}^\cdot$ such that $\mathscr{F}'^\cdot$ is a bounded complex of flat sheaves of A-modules.*

(ii) *The functor $\mathscr{F}^\cdot \otimes^L_A - : D^-(X, A) \to D^-(X, A)$ is way-out in both directions.*

(iii) *There exists an integer n such that $\mathrm{Tor}_i(\mathscr{F}^\cdot, \mathscr{G}) = 0$ for any $i > n$ and any sheaf of A-modules $\mathscr{G}$ on X.*

(iv) *$\mathscr{F}^\cdot$ has finite* Tor-*dimension.*

Let $f : X \to Y$ be a morphism of schemes, $\mathscr{F}$ a sheaf of A-modules on X, and $\mathscr{G}$ a sheaf of A-modules on Y. We have a canonical morphism

$$\mathscr{G} \otimes_A f_*\mathscr{F} \to f_*(f^*\mathscr{G} \otimes \mathscr{F}).$$

Suppose one of the following conditions hold:

(a) Rf_* has finite cohomological dimension, $\mathscr{F}^\cdot \in \mathrm{ob}\, D^-(X, A)$ and $\mathscr{G}^\cdot \in \mathrm{ob}\, D^-(Y, A)$.

(b) $\mathscr{G}^\cdot \in \mathrm{ob}\, D^b(Y, A)$ has finite Tor-dimension, and $\mathscr{F}^\cdot \in \mathrm{ob}\, D^+(X, A)$.

(c) Rf_* has finite cohomological dimension, $\mathscr{F}^\cdot \in \mathrm{ob}\, D(X, A)$, and $\mathscr{G}^\cdot \in \mathrm{ob}\, D^b(Y, A)$ has finite Tor-dimension.

Then we have a canonical morphism

$$\mathscr{G}^\cdot \otimes^L_A Rf_*\mathscr{F}^\cdot \to Rf_*(f^*\mathscr{G}^\cdot \otimes^L_A \mathscr{F}^\cdot).$$

For example, under the condition (a), let $\mathscr{F}^\cdot \to \mathscr{I}^\cdot$ (resp. $\mathscr{P}^\cdot \to \mathscr{G}^\cdot$, resp. $f^*\mathscr{P}^\cdot \otimes_A \mathscr{I}^\cdot \to \mathscr{J}^\cdot$) be a quasi-isomorphism such that $\mathscr{I}^\cdot$ (resp. $\mathscr{P}^\cdot$, resp. $\mathscr{J}^\cdot$) is a bounded above complex of Rf_*-acyclic sheaves (resp. flat sheaves, resp. Rf_*-acyclic sheaves) of A-modules on X (resp. on Y, resp. on X).

The above canonical morphism is the composite

$$\begin{aligned}\mathscr{G}^{\cdot} \otimes_A^L Rf_*\mathscr{F}^{\cdot} &\cong \mathscr{P}^{\cdot} \otimes_A f_*\mathscr{I}^{\cdot} \\ &\to f_*(f^*\mathscr{P} \otimes_A \mathscr{I}^{\cdot}) \\ &\to f_*\mathscr{J}^{\cdot} \\ &\cong Rf_*(f^*\mathscr{G} \otimes_A^L \mathscr{F}^{\cdot}).\end{aligned}$$

Proposition 6.5.5. *Let A be a ring and let $f : X \to Y$ be a morphism of schemes. Suppose Rf_* has finite cohomological dimension. For any $\mathscr{F}^{\cdot} \in \operatorname{ob} D^-(X, A)$ and any $\mathscr{G}^{\cdot} \in \operatorname{ob} D^-(Y, A)$ such that $\mathscr{H}^i(\mathscr{G}^{\cdot})$ are locally constant sheaves, we have a canonical isomorphism*

$$\mathscr{G}^{\cdot} \otimes_A^L Rf_*\mathscr{F}^{\cdot} \stackrel{\cong}{\to} Rf_*(f^*\mathscr{G}^{\cdot} \otimes_A^L \mathscr{F}^{\cdot}).$$

Proof. One can verify $- \otimes_A^L Rf_*\mathscr{F}^{\cdot}$ and $Rf_*(f^* - \otimes_A^L \mathscr{F}^{\cdot})$ are way-out left functors. So it suffices to prove

$$\mathscr{G} \otimes_A^L Rf_*\mathscr{F}^{\cdot} \cong Rf_*(f^*\mathscr{G} \otimes_A^L \mathscr{F}^{\cdot})$$

for any locally constant sheaf $\mathscr{G}$ of A-modules. The problem is local with respect to the etale topology on Y. We may assume that $\mathscr{G}$ is a constant sheaf, say associated to an A-module M. Let

$$\cdots \to L^{-1} \to L^0 \to 0$$

be a resolution of M by free A-modules, and let $\mathscr{F}^{\cdot} \to \mathscr{F}'^{\cdot}$ be a quasi-isomorphism such that $\mathscr{F}'^{\cdot}$ is a bounded above complex of Rf_*-acyclic sheaves. We have

$$\begin{aligned}\mathscr{G} \otimes_A^L Rf_*\mathscr{F}^{\cdot} &\cong L^{\cdot} \otimes_A f_*\mathscr{F}'^{\cdot}, \\ f^*\mathscr{G} \otimes_A^L \mathscr{F}^{\cdot} &\cong f^*L^{\cdot} \otimes_A \mathscr{F}'^{\cdot}.\end{aligned}$$

Since each L^i is free and each $\mathscr{F}'^j$ is Rf_*-acyclic, each $f^*L^i \otimes_A \mathscr{F}'^j$ is Rf_*-acyclic. So we have

$$Rf_*(f^*\mathscr{G} \otimes_A^L \mathscr{F}^{\cdot}) \cong f_*(f^*L^{\cdot} \otimes_A \mathscr{F}'^{\cdot}).$$

Since L^i are free, the canonical morphism

$$L^{\cdot} \otimes_A f_*\mathscr{F}'^{\cdot} \to f_*(f^*L^{\cdot} \otimes_A \mathscr{F}'^{\cdot})$$

is an isomorphism. Hence

$$L^{\cdot} \otimes_A^L Rf_*\mathscr{F}^{\cdot} \cong Rf_*(f^*L^{\cdot} \otimes_A^L \mathscr{F}^{\cdot}).$$

□

Corollary 6.5.6. *Let A be a ring and let $f : X \to Y$ be a morphism of schemes such that Rf_* has finite cohomological dimension. Then for any complex $\mathscr{F}^\cdot$ of sheaves of A-modules on X with finite* Tor*-dimension, $Rf_*\mathscr{F}^\cdot$ also has finite* Tor*-dimension. Moreover, for any $\mathscr{G}^\cdot \in \operatorname{ob} D(Y, A)$ such that $\mathscr{H}^i(\mathscr{G}^\cdot)$ are locally constant sheaves, we have*

$$\mathscr{G}^\cdot \otimes_A^L Rf_*\mathscr{F}^\cdot \cong Rf_*(f^*\mathscr{G}^\cdot \otimes_A^L \mathscr{F}^\cdot).$$

Proof. Let $\mathscr{G}$ be a constant sheaf of A-modules on Y. Since $\mathscr{F}^\cdot$ has finite Tor-dimension, there exists an integer n such that $\operatorname{Tor}_i(f^*\mathscr{G}, \mathscr{F}^\cdot) = 0$ for all $i > n$, that is, $\mathscr{H}^j(f^*\mathscr{G} \otimes_A^L \mathscr{F}^\cdot) = 0$ for all $j < -n$. Thus $R^jf_*(f^*\mathscr{G} \otimes_A^L \mathscr{F}^\cdot) = 0$ for all $j < -n$. By 6.5.5, we have

$$R^jf_*(f^*\mathscr{G} \otimes_A^L \mathscr{F}^\cdot) \cong \mathscr{H}^j(\mathscr{G} \otimes^L Rf_*\mathscr{F}^\cdot) \cong \operatorname{Tor}_{-j}(\mathscr{G}, Rf_*\mathscr{F}^\cdot).$$

So $\operatorname{Tor}_i(\mathscr{G}, Rf_*\mathscr{F}^\cdot) = 0$ for all $i > n$. Hence $Rf_*\mathscr{F}^\cdot$ has finite Tor-dimension. The functors $- \otimes_A^L Rf_*\mathscr{F}^\cdot$ and $Rf_*(f^* - \otimes_A^L \mathscr{F}^\cdot)$ are way-out in both directions. By 6.5.2 (ii) and 6.5.5, we have

$$\mathscr{G}^\cdot \otimes_A^L Rf_*\mathscr{F}^\cdot \cong Rf_*(f^*\mathscr{G}^\cdot \otimes_A^L \mathscr{F}^\cdot)$$

for any $\mathscr{G}^\cdot \in \operatorname{ob} D(Y, A)$ such that $\mathscr{H}^i(\mathscr{G}^\cdot)$ are locally constant sheaves. □

Corollary 6.5.7. *Let $f : X \to Y$ be a morphism such that Rf_* has finite cohomological dimension, and let $A \to B$ be a homomorphism of rings. For any $\mathscr{F}^\cdot \in \operatorname{ob} D^-(X, A)$ and any object $M^\cdot$ in the derived category $D^-(B)$ of the category of B-modules, we have a canonical isomorphism*

$$M^\cdot \otimes_A^L Rf_*\mathscr{F}^\cdot \xrightarrow{\cong} Rf_*(M^\cdot \otimes_A^L \mathscr{F}^\cdot)$$

in $D^-(Y, B)$.

Proof. (We cannot apply 6.5.5 directly since it gives an isomorphism in $D^-(Y, A)$, and not in $D^-(Y, B)$.) Let $\mathscr{F}'^\cdot$ be a bounded above complex of flat sheaves of A-modules on X quasi-isomorphic to $\mathscr{F}^\cdot$ and let $\mathscr{C}^\cdot(\mathscr{F}'^i)$ be the Godement resolution of $\mathscr{F}'^i$. Choose an integer d such that $R^if_*\mathscr{F} = 0$ for all $i > d$ and all sheaves $\mathscr{F}$ on X. Then $\tau_{\leq d}\mathscr{C}^\cdot(\mathscr{F}'^i)$ is a resolution of $\mathscr{F}'^i$ by Rf_*-acyclic sheaves. The complex associated to the bicomplex $\tau_{\leq d}\mathscr{C}^\cdot(\mathscr{F}'^\cdot)$ is a bounded above complex of Rf_*-acyclic sheaves quasi-isomorphic to $\mathscr{F}^\cdot$. Denote this complex by $\mathscr{F}''^\cdot$. We define a morphism

$$M^\cdot \otimes_A^L Rf_*\mathscr{F}^\cdot \to Rf_*(M^\cdot \otimes_A^L \mathscr{F}^\cdot)$$

in $D^-(Y, B)$ to be the composite

$$\begin{aligned} M^\cdot \otimes_A^L Rf_*\mathscr{F}^\cdot &\cong M^\cdot \otimes_A^L f_*\mathscr{F}''^\cdot \\ &\to M^\cdot \otimes_A f_*\mathscr{F}''^\cdot \\ &\to f_*(f^*M^\cdot \otimes_A \mathscr{F}''^\cdot) \\ &\to Rf_*(f^*M^\cdot \otimes_A \mathscr{F}''^\cdot) \\ &\cong Rf_*(M^\cdot \otimes_A^L \mathscr{F}^\cdot). \end{aligned}$$

To prove that it is an isomorphism, it suffices to show that it coincides with the morphism defined in 6.5.5. This follows from the commutativity of the following diagram, where $L^\cdot$ is a bounded above complex of free A-modules such that we have a quasi-isomorphism $L^\cdot \to M^\cdot$:

$$\begin{array}{ccccccccc} M^\cdot\otimes_A^L Rf_*\mathscr{F}^\cdot & \cong & L^\cdot\otimes_A f_*\mathscr{F}''^\cdot & \overset{\cong}{\to} & f_*(f^*L^\cdot\otimes_A\mathscr{F}''^\cdot) & \cong & Rf_*(f^*L^\cdot\otimes_A\mathscr{F}''^\cdot) & \cong & Rf_*(M^\cdot\otimes_A^L\mathscr{F}^\cdot) \\ \| & & \downarrow & & \downarrow & & \downarrow & & \| \\ M^\cdot\otimes_A^L Rf_*\mathscr{F}^\cdot & \to & M^\cdot\otimes_A f_*\mathscr{F}''^\cdot & \to & f_*(f^*M^\cdot\otimes_A\mathscr{F}''^\cdot) & \to & Rf_*(f^*M^\cdot\otimes_A\mathscr{F}''^\cdot) & \cong & Rf_*(M^\cdot\otimes_A^L\mathscr{F}^\cdot). \end{array}$$

□

Chapter 7

Base Change Theorems

7.1 Divisors

Let X be a noetherian scheme. We define the *dimension* $\dim X$ of X to be the supremum of all integers n so that there exists a chain

$$Z_0 \subset Z_1 \subset \cdots \subset Z_n$$

of distinct irreducible closed subsets in X. If $X = \operatorname{Spec} A$ for a noetherian ring A, then $\dim X$ coincides with the Krull dimension $\dim A$ of A. If Z is an irreducible closed subset of X, we define the *codimension* $\operatorname{codim}(Z, X)$ of Z in X to be the supremum of all integers n so that there exists a chain

$$Z_0 \subset Z_1 \subset \cdots \subset Z_n$$

of distinct irreducible closed subsets in X with $Z_0 = Z$. For any closed subset Y in X, we define the codimension $\operatorname{codim}(Y, X)$ of Y in X to be

$$\operatorname{codim}(Y, X) = \inf_{Z \subset Y} \operatorname{codim}(Z, X),$$

where Z goes over the set of irreducible closed subset of Y. If $X = \operatorname{Spec} A$ and $Y = V(\mathfrak{p})$ for some prime ideal $\mathfrak{p}$ of A, then $\operatorname{codim}(Y, X)$ coincides with the height $\operatorname{ht} \mathfrak{p}$ of $\mathfrak{p}$.

A noetherian scheme X is called *regular in codimension one* if for any $x \in X$ with $\dim \mathscr{O}_{X,x} = 1$, $\mathscr{O}_{X,x}$ is regular. Recall that a local noetherian integral domain of dimension 1 is regular if and only if it is normal ([Atiyah and Macdonald (1969)] 9.2). So any normal noetherian scheme is regular in codimension 1.

Suppose that X is an integral noetherian scheme regular in codimension one. A *prime divisor* on X is an integral closed subscheme of X of codimension 1. A *Weil divisor* on X is an element in the free abelian group $\operatorname{Div}(X)$ generated by prime divisors. We write a divisor as $D = \sum_i n_i Y_i$, where Y_i are prime divisors, n_i are integers, and n_i are nonzero only for finitely many i. If $n_i \geq 0$ for all i, we say that D is an *effective divisor*.

Let Y be a prime divisor, and let η be its generic point. Then $\dim \mathscr{O}_{X,\eta} = 1$. Since X is regular in codimension 1, $\mathscr{O}_{X,\eta}$ is a discrete valuation ring. It defines a discrete valuation v_Y on the function field K of X.

Lemma 7.1.1. *Let X be an integral noetherian scheme regular in codimension one, and let K be the function field of X. For any $f \in K^*$, there are only finitely many prime divisors Y such that $v_Y(f) \neq 0$.*

Proof. Let $U = \operatorname{Spec} A$ be a nonempty affine open subscheme of X so that $f \in \mathscr{O}_X(U) = A$, let Y be a prime divisor of X, and let η be the generic point of Y. If $\eta \notin U$, then $Y \subset X - U$. Since Y has codimension 1, Y must be an irreducible component of $X - U$. Since $X - U$ has only finitely many irreducible components, there are only finitely many prime divisors Y with $\eta \notin U$. Suppose $\eta \in U$ and $v_Y(f) \neq 0$. Let $\mathfrak{p}$ be the prime ideal of A corresponding to η. Since $v_Y(f) \neq 0$ and $f \in A$, we have $f \in \mathfrak{p}$. It follows that $Y \cap U = V(\mathfrak{p}) \subset V(f)$. As $Y \cap U$ has codimension 1 in U, $Y \cap U$ must be an irreducible component of $V(f)$. Since $V(f)$ has only finitely many irreducible components, there are only finitely many prime divisors Y such that $v_Y(f) \neq 0$. □

Let X be an integral noetherian scheme regular in codimension one and let K be its function field. For any $f \in K^*$, we define the divisor (f) of f to be

$$(f) = \sum_Y v_Y(f) Y,$$

where Y goes over the set of prime divisors of X. For any $f, g \in K^*$, we have

$$(fg) = (f) + (g).$$

A divisor on X is called *principal* if it is of the form (f) for some $f \in K^*$. Principal divisors form a subgroup of the group of divisors $\operatorname{Div}(X)$. The quotient group $\operatorname{Cl}(X)$ is called the *divisor class group* of X.

Proposition 7.1.2. *Let A be a normal noetherian integral domain, and let $X = \operatorname{Spec} A$. Then $\operatorname{Cl}(X) = 0$ if and only if A is a unique factorization domain.*

Proof. Suppose A is a unique factorization domain. Let Y be a prime divisor of X, and let $\mathfrak{p}$ be the prime ideal of A corresponding to the generic point of Y. Then $\mathrm{ht}\,\mathfrak{p} = 1$. By [Matsumura (1970)] (19.A) Theorem 47, we have $\mathfrak{p} = fA$ for some $f \in A$. For any prime ideal $\mathfrak{q}$ of A with $\mathrm{ht}\,\mathfrak{q} = 1$, denote by $v_{\mathfrak{q}}$ the valuation defined by $\mathfrak{q}$. If $v_{\mathfrak{q}}(f) \neq 0$, then we have $f \in \mathfrak{q}$, and hence $\mathfrak{p} \subset \mathfrak{q}$. We then must have $\mathfrak{p} = \mathfrak{q}$. It follows that the principal divisor (f) coincides with the prime divisor Y. So every prime divisor of X is principal and $\mathrm{Cl}(X) = 0$.

Conversely, suppose $\mathrm{Cl}(X) = 0$. For any prime ideal $\mathfrak{p}$ with $\mathrm{ht}\,\mathfrak{p} = 1$, let Y be the prime divisor with generic point $\mathfrak{p}$. Then Y is a principal divisor, say $Y = (f)$ for some nonzero element f in the fraction field K of A. For any prime ideal $\mathfrak{q}$ with $\mathrm{ht}\,\mathfrak{q} = 1$, we have

$$v_{\mathfrak{q}}(f) = \begin{cases} 1 \text{ if } \mathfrak{q} = \mathfrak{p}, \\ 0 \text{ if } \mathfrak{q} \neq \mathfrak{p}. \end{cases}$$

In particular, we have

$$f \in \bigcap_{\mathrm{ht}\,\mathfrak{q}=1} A_{\mathfrak{q}},$$

and hence $f \in A$ by [Matsumura (1970)] (17.H) Theorem 38. As $v_{\mathfrak{p}}(f) = 1$, we have $f \in \mathfrak{p}$. For any $g \in \mathfrak{p}$, we have $v_{\mathfrak{p}}(g) \geq 1$ and $v_{\mathfrak{q}}(g) \geq 0$ for any $\mathfrak{q}$ with $\mathrm{ht}\,\mathfrak{q} = 1$. It follows that $v_{\mathfrak{q}}(\frac{g}{f}) \geq 0$ for any $\mathfrak{q}$ with $\mathrm{ht}\,\mathfrak{q} = 1$. So we have

$$\frac{g}{f} \in \bigcap_{\mathrm{ht}\,\mathfrak{q}=1} A_{\mathfrak{q}} = A.$$

Hence $g \in fA$. We thus have $\mathfrak{p} = fA$. So every prime ideal of height 1 is principal. By [Matsumura (1970)] (19.A) Theorem 47, A is a unique factorization domain. □

Let X be an integral scheme, and let K be its function field. Denote by $\mathscr{K}_X$ the constant Zariski sheaf on X associated to K, and by $\mathscr{K}_X^*$ the constant Zariski sheaf associated to K^*. Any nonempty open subset U of X is irreducible and hence connected. So we have $\mathscr{K}_X^*(U) = K^*$. In particular, $\mathscr{K}_X^*(X) \to \mathscr{K}_X^*(U)$ is surjective. Hence $\mathscr{K}_X^*$ is a flasque sheaf (with respect to the Zariski topology) and we have

$$H^q(X, \mathscr{K}_X^*) = 0$$

for all $q \geq 1$.

We have a monomorphism

$$\mathscr{O}_X^* \to \mathscr{K}_X^*.$$

A *Cartier divisor* on X is an element in $\Gamma(X, \mathscr{K}_X^*/\mathscr{O}_X^*)$. A Cartier divisor can be represented as (f_i, U_i), where $\{U_i\}$ is an open covering of X, $f_i \in \Gamma(U_i, \mathscr{K}_X^*)$, and $\frac{f_i}{f_j} \in \Gamma(U_i \cap U_j, \mathscr{O}_X^*)$ for all i and j. If f_i can be chosen to lie in $\Gamma(U_i, \mathscr{O}_X)$, then we say that the Cartier divisor is *effective.* A Cartier divisor is called *principal* if it lies in the image of the canonical homomorphism

$$\Gamma(X, \mathscr{K}_X^*) \to \Gamma(X, \mathscr{K}_X^*/\mathscr{O}_X^*).$$

The cokernel of this homomorphism is called the *Cartier divisor class group,* and is denoted by $\mathrm{CaCl}(X)$.

Proposition 7.1.3. *Let X be an integral noetherian scheme. Suppose $\mathscr{O}_{X,x}$ are unique factorization domains for all $x \in X$. Then we have an isomorphism*

$$\Gamma(X, \mathscr{K}_X^*/\mathscr{O}_X^*) \stackrel{\cong}{\to} \mathrm{Div}(X),$$

and it induces an isomorphism between the group of principal Cartier divisors and the group of principal Weil divisors. So it induces an isomorphism

$$\mathrm{CaCl}(X) \stackrel{\cong}{\to} \mathrm{Cl}(X)$$

between the Cartier divisor class group and the Weil divisor class group.

Proof. Let $D = \{(f_i, U_i)\}$ be a Cartier divisor, where $\{U_i\}$ is an open covering of X, $f_i \in \Gamma(U_i, \mathscr{K}_X^*)$, and $\frac{f_i}{f_j} \in \Gamma(U_i \cap U_j, \mathscr{O}_X^*)$ for all i and j. For any prime divisor Y, let U_i be an open subset such that $U_i \cap Y \neq \emptyset$. We define

$$v_Y(D) = v_Y(f_i).$$

Note that $v_Y(D)$ is independent of the choice of the open subset U_i with nonempty intersection with Y. Define the Weil divisor associated to D to be $\sum_Y v_Y(D)Y$. We thus get a homomorphism

$$\Gamma(X, \mathscr{K}_X^*/\mathscr{O}_X^*) \to \mathrm{Div}(X).$$

First let us prove that it is injective. Suppose $D = \{(f_i, U_i)\}$ is mapped to the trivial Weil divisor. We need to show that D is the trivial Cartier divisor. We may assume that each $U_i = \mathrm{Spec}\, A_i$ is affine. Then for any prime ideal $\mathfrak{p}$ of A_i with $\mathrm{ht}\,\mathfrak{p} = 1$, we have $v_{\mathfrak{p}}(f_i) = 0$. By [Matsumura (1970)] (17.H) Theorem 38, we have $f_i \in \mathscr{O}_X^*(U_i)$. Hence D is trivial. Next we prove that the homomorphism $\Gamma(X, \mathscr{K}_X^*/\mathscr{O}_X^*) \to \mathrm{Div}(X)$ is surjective under the assumption that $\mathscr{O}_{X,x}$ is a unique factorization domain for any $x \in X$. For any Weil divisor $D = \sum_Y n_Y Y$ on X, let

$$D_x = \sum_Y n_Y (Y \otimes_{\mathscr{O}_X} \mathscr{O}_{X,x})$$

be the divisor on Spec $\mathscr{O}_{X,x}$ induced by D. By 7.1.2, we have $D_x = (f_x)$ for some $f_x \in \mathscr{O}_{X,x}$. Let U_x be an open neighborhood of x so that f_x can be extended to a section in $\Gamma(U_x, \mathscr{O}_X)$, which we still denote by f_x. There are only finitely many prime divisors Y with nonempty intersection with U_x such that both $v_Y(f_x)$ and n_Y are nonzero. Shrinking U_x, we may assume

$$\sum_Y n_Y(U_x \cap Y) = (f_x).$$

For any prime divisor Y with nonempty intersection with $U_x \cap U_{x'}$, we have

$$v_Y\left(\frac{f_x}{f_{x'}}\right) = 0.$$

So we have

$$\frac{f_x}{f_{x'}} \in \mathscr{O}_X^*(U_x \cap U_{x'})$$

by [Matsumura (1970)] (17.H) Theorem 38. Thus $\{(f_x, U_x)\}$ is a Cartier divisor. This Cartier divisor is mapped to D under the homomorphism $\Gamma(X, \mathscr{K}_X^*/\mathscr{O}_X^*) \to \mathrm{Div}(X)$. Hence the homomorphism is surjective. □

Remark 7.1.4. Let X be an integral noetherian scheme. Suppose $\mathscr{O}_{X,x}$ are regular for all $x \in X$, that is, X is a regular scheme. Then $\mathscr{O}_{X,x}$ are unique factorization domains for all x by [Matsumura (1970)] (19.B) Theorem 48. By 7.1.3, we have an isomorphism $\mathrm{CaCl}(X) \cong \mathrm{Cl}(X)$.

Proposition 7.1.5. *Let X be an integral scheme. We have a canonical isomorphism*

$$\mathrm{CaCl}(X) \xrightarrow{\cong} \mathrm{Pic}(X).$$

Proof. We have an exact sequence

$$0 \to \mathscr{O}_X^* \to \mathscr{K}_X^* \to \mathscr{K}_X^*/\mathscr{O}_X^* \to 0.$$

Since $H^1(X, \mathscr{K}_X^*) = 0$, we have an exact sequence

$$\Gamma(X, \mathscr{K}_X^*) \to \Gamma(X, \mathscr{K}_X^*/\mathscr{O}_X^*) \to H^1(X, \mathscr{O}_X^*) \to 0.$$

So we have an isomorphism

$$\mathrm{CaCl}(X) \cong H^1(X, \mathscr{O}_X^*).$$

But

$$H^1(X, \mathscr{O}_X^*) \cong \mathrm{Pic}(X).$$

So we have

$$\mathrm{CaCl}(X) \cong \mathrm{Pic}(X).$$

A more direct proof is as follows. Let $D = \{(f_i, U_i)\}$ be a Cartier divisor, where $\{U_i\}$ is an open covering of X, $f_i \in \Gamma(U_i, \mathscr{K}_X^*)$, and $\frac{f_i}{f_j} \in \Gamma(U_i \cap U_j, \mathscr{O}_X^*)$ for all i and j. We have

$$f_i^{-1}\mathscr{O}_{U_i \cap U_j} = f_j^{-1}\mathscr{O}_{U_i \cap U_j}.$$

So we have an $\mathscr{O}_X$-submodule $\mathscr{L}(D)$ of $\mathscr{K}$ so that

$$\mathscr{L}(D)|_{U_i} = f_i^{-1}\mathscr{O}_{U_i}.$$

$\mathscr{L}(D)$ is an invertible $\mathscr{O}_X$-module, and

$$\mathscr{L}(D_1 + D_2) \cong \mathscr{L}(D_1) \otimes_{\mathscr{O}_X} \mathscr{L}(D_2)$$

for all Cartier divisors D_1 and D_2. So $D \to \mathscr{L}(D)$ defines a homomorphism

$$\Gamma(X, \mathscr{K}_X^*/\mathscr{O}_X^*) \to \mathrm{Pic}(X).$$

One can check that $\mathscr{L}(D)$ is isomorphic to $\mathscr{O}_X$ if and only if D is a principal Cartier divisor. We thus have a monomorphism

$$\mathrm{CaCl}(X) \hookrightarrow \mathrm{Pic}(X).$$

For any invertible $\mathscr{O}_X$-module $\mathscr{L}$, we can find an open covering $\{U_i\}$ of X and $e_i \in \Gamma(U_i, \mathscr{L})$ so that we have isomorphisms

$$\mathscr{O}_X|_{U_i} \stackrel{\cong}{\to} \mathscr{L}|_{U_i}, \quad s \mapsto se_i.$$

We have

$$e_i|_{U_i \cap U_j} = f_{ij} e_j|_{U_i \cap U_j}$$

for some $f_{ij} \in \Gamma(U_i \cap U_j, \mathscr{O}_X^*)$. Fix a nonempty open subset U_{i_0} in the open covering. Let D be the Cartier divisor $D = \{(f_i, U_i)\}$ defined by

$$f_i = f_{i_0 i}.$$

Here we regard $f_{i_0 i}$ as a section of $\mathscr{K}_X^*$ over U_i. Then $\mathscr{L} \cong \mathscr{L}(D)$. This proves that the homomorphism $\Gamma(X, \mathscr{K}_X^*/\mathscr{O}_X^*) \to \mathrm{Pic}(X)$ is surjective. □

Let X be a smooth algebraic curve over a field k. Then a prime Weil divisor on X is a Zariski closed point of X with the reduced closed subscheme structure. Let $|X|$ be the set of all Zariski closed points of X. For any $x \in X$, let $k(x)$ be the residue field of $\mathscr{O}_{X,x}$. We define the degree of x to be

$$\deg(x) = [k(x) : k].$$

For any Weil divisor $D = \sum_{x\in|X|} n_x x$, we define the *degree* of D to be

$$\deg(D) = \sum_{x\in|X|} n_x \deg(x).$$

At the end of 8.1, we give a proof of the following fact.

Proposition 7.1.6. *Let X be a smooth proper algebraic curve over a field k, and let $K(X)$ be the function field of X. For any nonzero $f \in K(X)$, we have*

$$\deg(f) = 0,$$

where (f) is the principal divisor defined by f.

Suppose that X is a smooth proper algebraic curve over a field k. Taking degree defines a homomorphism

$$\deg : \mathrm{Div}(X) \to \mathbb{Z},$$

and it induces a homomorphism

$$\deg : \mathrm{Cl}(X) \to \mathbb{Z}.$$

For any invertible $\mathscr{O}_X$-module $\mathscr{L}$, choose a Weil divisor D on X such that $\mathscr{L} \cong \mathscr{L}(D)$, where $\mathscr{L}(D)$ is the invertible $\mathscr{O}_X$-module defined in the proof of 7.1.5 using the Cartier divisor corresponding to the Weil divisor D. We define the *degree* of $\mathscr{L}$ to be

$$\deg(\mathscr{L}) = \deg(D).$$

Then taking degree defines a homomorphism

$$\deg : \mathrm{Pic}(X) \to \mathbb{Z}.$$

7.2 Cohomology of Curves

([SGA 4] IX 4.)

Proposition 7.2.1 (Kummer's theory). *Let X be a scheme and let n be a positive integer invertible on X. Then the morphism of etale sheaves*

$$n : \mathscr{O}^*_{X_{\mathrm{et}}} \to \mathscr{O}^*_{X_{\mathrm{et}}}, \quad s \mapsto s^n$$

is surjective. Let $\boldsymbol{\mu}_{n,X}$ be its kernel. We thus have a short exact sequence

$$0 \to \boldsymbol{\mu}_{n,X} \to \mathscr{O}^*_{X_{\mathrm{et}}} \xrightarrow{n} \mathscr{O}^*_{X_{\mathrm{et}}} \to 0.$$

Proof. Let $U = \operatorname{Spec} A$ be an affine etale X-scheme and let $a \in \mathscr{O}^*_{X_{\text{et}}}(U) = A^*$ be a section. The canonical morphism

$$V = \operatorname{Spec} A[t]/(t^n - a) \to U = \operatorname{Spec} A$$

is etale by 2.3.3 since $t^n - a$ and its derivative nt^{n-1} generate $A[t]$. One can show that the fibers of $V \to U$ are nonempty and hence $V \to U$ is surjective. So $\{V \to U\}$ is an etale covering of U. The restriction of the section a to V has an n-th root t. So $n : \mathscr{O}^*_{X_{\text{et}}} \to \mathscr{O}^*_{X_{\text{et}}}$ is surjective. □

Proposition 7.2.2. *Let X be a scheme and let n be a positive integer invertible on X. The sheaf $\boldsymbol{\mu}_{n,X}$ is locally constant. When X is a scheme over a strictly local ring A such that n is invertible in A, $\boldsymbol{\mu}_{n,X}$ is isomorphic to the constant sheaf $\mathbb{Z}/n$.*

Proof. Note that for any etale X-scheme U, we have a one-to-one correspondence

$$\{s \in \mathscr{O}_U(U)^* | s^n = 1\} \cong \operatorname{Hom}_X(U, \mathbf{Spec}\, \mathscr{O}_X[t]/(t^n - 1)).$$

So $\boldsymbol{\mu}_{n,X}$ is represented by the X-scheme $\mathbf{Spec}\, \mathscr{O}_X[t]/(t^n - 1)$. This scheme is finite and etale over X. So $\boldsymbol{\mu}_{n,X}$ is locally constant by 5.8.1 (i). If A is strictly henselian and n is invertible in A, then $t^n - 1$ splits into a product of linear polynomials in $A[t]$ by 2.8.3 (v). For any A-scheme X, $\mathbf{Spec}\, \mathscr{O}_X[t]/(t^n - 1)$ is then a trivial etale covering of X of degree n. Hence $\boldsymbol{\mu}_{n,X}$ is isomorphic to the constant sheaf $\mathbb{Z}/n$. □

Proposition 7.2.3 (Artin–Schreier's theory). *Let p be a prime number and let X be a scheme such that $p \cdot 1 = 0$ in $\Gamma(X, \mathscr{O}_X)$. Then the morphism of etale sheaves*

$$\wp : \mathscr{O}_{X_{\text{et}}} \to \mathscr{O}_{X_{\text{et}}}, \quad s \mapsto s^p - s$$

is surjective. Its kernel is isomorphic to $\mathbb{Z}/p$. We thus have a short exact sequence

$$0 \to \mathbb{Z}/p \to \mathscr{O}_{X_{\text{et}}} \xrightarrow{\wp} \mathscr{O}_{X_{\text{et}}} \to 0.$$

Proof. Let $U = \operatorname{Spec} A$ be an affine etale X-scheme and let $a \in \mathscr{O}_{X_{\text{et}}}(U) = A$ be a section. The canonical morphism

$$V = \operatorname{Spec} A[t]/(t^p - t - a) \to U = \operatorname{Spec} A$$

is etale by 2.3.3 since the derivative -1 of $t^p - t - a$ generate $A[t]$. One can show the fibers of $V \to U$ are nonempty and hence $V \to U$ is surjective.

So $\{V \to U\}$ is an etale covering of U. The restriction of the section a to V is the image of t under the homomorphism

$$\wp : O_{X_{\text{et}}}(V) = A[t]/(t^p - t - a) \to O_{X_{\text{et}}}(V) = A[t]/(t^p - t - a).$$

So $\wp : \mathscr{O}_{X_{\text{et}}} \to \mathscr{O}_{X_{\text{et}}}$ is surjective. Note that for any etale X-scheme U, we have a one-to-one correspondence

$$\{s \in \mathscr{O}_U(U) | s^p - s = 0\} \cong \operatorname{Hom}_X(U, \mathbf{Spec}\, \mathscr{O}_X[t]/(t^p - t)),$$

that is, $\ker \wp$ is represented by the X-scheme $\mathbf{Spec}\, \mathscr{O}_X[t]/(t^p - t)$. We have

$$t^p - t = \prod_{i \in \mathbb{Z}/p} (t - i)$$

on X. It follows that $\ker \wp \cong \mathbb{Z}/p$. □

Lemma 7.2.4. *Let X be a noetherian scheme and let $\mathscr{F}$ be a sheaf on X. The following conditions are equivalent:*

(i) *For every non-closed point y in X, we have $\mathscr{F}_{\bar{y}} = 0$.*

(ii) *The canonical morphism*

$$\mathscr{F} \to \prod_{x \in |X|} i_{x*} i_x^* \mathscr{F}$$

induces an isomorphism

$$\mathscr{F} \cong \bigoplus_{x \in |X|} i_{x*} i_x^* \mathscr{F},$$

where $|X|$ is the set of Zariski closed points in X, and $i_x : \operatorname{Spec} k(x) \to X$ are the closed immersions.

When $\mathscr{F}$ satisfies these conditions, we say that $\mathscr{F}$ is a *skyscraper sheaf.*

Proof. Suppose (i) holds. Let $f : U \to X$ be an etale morphism of finite type. For any $s \in \mathscr{F}(U)$, if $u \in U$ is a point in the support of s, then $f(u)$ must be a closed point of X. Since f is quasi-finite, u is closed in $f^{-1}(f(u))$. Hence u is a closed point of U. Since U is noetherian, the support of s consists of finitely many closed points of U. It follows that the image of the canonical morphism

$$\mathscr{F} \to \prod_{x \in |X|} i_{x*} i_x^* \mathscr{F}$$

is contained in $\bigoplus_{x \in |X|} i_{x*} i_x^* \mathscr{F}$. One shows that under the condition (i), the morphism

$$\mathscr{F} \to \bigoplus_{x \in |X|} i_{x*} i_x^* \mathscr{F}$$

induces an isomorphism on stalks at every point of X, and hence it is an isomorphism. Here we use the formula

$$(i_{x*}i_x^*\mathscr{F})_{\bar{y}} \cong \begin{cases} \mathscr{F}_{\bar{x}} & \text{if } y = x, \\ 0 & \text{if } y \neq x \end{cases}$$

for any $x \in |X|$ and any $y \in X$. This proves (i)⇒(ii). It is clear that (ii)⇒(i). □

Lemma 7.2.5. *Let X be a noetherian scheme with $\dim X \leq 1$, let $\eta_1, \ldots, \eta_m$ be all the points in X satisfying $\dim \overline{\{\eta_i\}} = 1$ $(i = 1, \ldots, m)$, let $R = \prod_{i=1}^m \mathscr{O}_{X,\eta_i}$, and let $j : \operatorname{Spec} R \to X$ be the canonical morphism.*

(i) *For any sheaf $\mathscr{G}$ on $\operatorname{Spec} R$ and any $q \geq 1$, $R^q j_* \mathscr{G}$ is a skyscraper sheaf.*

(ii) *For any sheaf $\mathscr{F}$ on X, the kernel and cokernel of the canonical morphism $\mathscr{F} \to j_*j^*\mathscr{F}$ are skyscraper sheaves.*

Proof.

(i) For each i, let $\tilde{\mathscr{O}}_{X,\bar{\eta}_i}$ be the strict henselization of $\mathscr{O}_{X,\eta_i}$. By 5.9.5, we have

$$(R^q j_* \mathscr{G})_{\bar{\eta}_i} \cong H^q(\operatorname{Spec}(R \otimes_{\mathscr{O}_X} \tilde{\mathscr{O}}_{X,\bar{\eta}_i}), \mathscr{G}).$$

One can show

$$R \otimes_{\mathscr{O}_X} \tilde{\mathscr{O}}_{X,\bar{\eta}_i} \cong \tilde{\mathscr{O}}_{X,\bar{\eta}_i}.$$

By 5.7.3, we have

$$H^q(\operatorname{Spec} \tilde{\mathscr{O}}_{X,\bar{\eta}_i}, \mathscr{G}) = 0$$

for any $q \geq 1$. So $(R^q j_* \mathscr{G})_{\bar{\eta}_i} = 0$ for any $q \geq 1$. But η_i $(i = 1, \ldots, m)$ are all the non-closed points of X. So $R^q j_* \mathscr{G}$ is a skyscraper sheaf for any $q \geq 1$.

(ii) We have

$$(j_*j^*\mathscr{F})_{\bar{\eta}_i} \cong H^0(\operatorname{Spec}(R \otimes_{\mathscr{O}_X} \tilde{\mathscr{O}}_{X,\bar{\eta}_i}), j^*\mathscr{F}) \cong H^0(\operatorname{Spec} \tilde{\mathscr{O}}_{X,\bar{\eta}_i}, j^*\mathscr{F}) \cong \mathscr{F}_{\bar{\eta}_i}.$$

So the canonical morphism $\mathscr{F} \to j_*j^*\mathscr{F}$ induces isomorphisms on stalks at $\bar{\eta}_i$. It follows that its kernel and cokernel are skyscraper sheaves. □

Lemma 7.2.6. *Let X be a scheme of finite type over a separably closed field k. Then for any skyscraper sheaf $\mathscr{F}$ on X, we have $H^q(X, \mathscr{F}) = 0$ for any $q \geq 1$.*

Proof. We have

$$\begin{aligned} H^q(X, \mathscr{F}) &\cong H^q(X, \bigoplus_{x\in|X|} i_{x*}i_x^*\mathscr{F}) \\ &\cong \bigoplus_{x\in|X|} H^q(X, i_{x*}i_x^*\mathscr{F}) \\ &\cong \bigoplus_{x\in|X|} H^q(\operatorname{Spec} k(x), i_x^*\mathscr{F}). \end{aligned}$$

For any closed point x, the residue field $k(x)$ is separably closed. So $H^q(\operatorname{Spec} k(x), i_x^*\mathscr{F}) = 0$ for any $q \geq 1$. Our assertion follows. □

Theorem 7.2.7. *Let X be a reduced scheme of finite type over a separably closed field k of characteristic p with* $\dim X \leq 1$.

(i) *For any torsion sheaf $\mathscr{F}$ on X, we have $H^q(X, \mathscr{F}) = 0$ for any $q \geq 3$.*

(ii) *$H^2(X, \mathscr{O}^*_{X_{\text{et}}})$ and $H^3(X, \mathscr{O}^*_{X_{\text{et}}})$ are p-torsion groups, and $H^q(X, \mathscr{O}^*_{X_{\text{et}}}) = 0$ for any $q \geq 4$. If k is algebraically closed, then $H^q(X, \mathscr{O}^*_{X_{\text{et}}}) = 0$ for any $q \geq 2$.*

Proof. Let $\eta_1, \ldots, \eta_m$ be all the points in X satisfying $\dim \overline{\{\eta_i\}} = 1$ $(i = 1, \ldots, m)$, $R = \prod_{i=1}^m \mathscr{O}_{X,\eta_i}$, $j : \operatorname{Spec} R \to X$ be the canonical morphism, $\mathscr{F}$ a sheaf on X, $\mathscr{F} \to Rj_*j^*\mathscr{F}$ the canonical morphism, and

$$\mathscr{F} \to Rj_*j^*\mathscr{F} \to \Delta \to$$

a distinguished triangle. We have an exact sequence

$$0 \to \mathscr{H}^{-1}(\Delta) \to \mathscr{F} \to j_*j^*\mathscr{F} \to \mathscr{H}^0(\Delta) \to 0$$

and

$$R^q j_*j^*\mathscr{F} \cong \mathscr{H}^q(\Delta)$$

for any $q \geq 1$. By 7.2.5, $\mathscr{H}^q(\Delta)$ are skyscraper sheaves for any q. By 7.2.6, the biregular spectral sequence

$$E_2^{pq} = H^p(X, \mathscr{H}^q(\Delta)) \Rightarrow H^{p+q}(X, \Delta)$$

degenerates and we have

$$H^q(X, \Delta) \cong H^0(X, \mathscr{H}^q(\Delta))$$

for any q. Moreover, we have

$$H^q(X, Rj_*j^*\mathscr{F}) \cong H^q(\operatorname{Spec} R, j^*\mathscr{F}).$$

The long exact sequence of $H^{\cdot}(X,-)$ associated to the above distinguished triangle is

$$\cdots \to H^q(X,\mathscr{F}) \to H^q(\operatorname{Spec} R, j^*\mathscr{F}) \to H^0(X,\mathscr{H}^q(\Delta)) \to H^{q+1}(X,\mathscr{F}) \to \cdots.$$

Since X is reduced, each $\mathscr{O}_{X,\eta_i}$ is a field of transcendental degree 1 over k.

Suppose that $\mathscr{F}$ is a torsion sheaf. We have $H^q(\operatorname{Spec} R, j^*\mathscr{F}) = 0$ for any $q \geq 2$ by 5.7.8 and 4.5.11 (i). So we have

$$H^q(X,\mathscr{F}) \cong H^0(X,\mathscr{H}^{q-1}(\Delta))$$

for any $q \geq 3$. On the other hand, $R^q j_* j^*\mathscr{F}$ is the sheaf associated to the presheaf

$$V \mapsto H^q((\operatorname{Spec} R) \times_X V, j^*\mathscr{F})$$

for any etale X-scheme V. Note that $(\operatorname{Spec} R) \times_X V$ is a disjoint union of the spectra of some fields finite separable over some $\mathscr{O}_{X,\eta_i}$. Again by 5.7.8 and 4.5.11 (i), we have

$$H^q((\operatorname{Spec} R) \times_X V, j^*\mathscr{F}) = 0$$

for any $q \geq 2$. So $R^q j_* j^*\mathscr{F} = 0$ for any $q \geq 2$, and hence $\mathscr{H}^q(\Delta) = 0$ for any $q \geq 2$. Therefore

$$H^q(X,\mathscr{F}) \cong H^0(X,\mathscr{H}^{q-1}(\Delta)) = 0$$

for any $q \geq 3$. This proves (i).

Suppose $\mathscr{F} = \mathscr{O}^*_{X_{\text{et}}}$ (resp. $\mathscr{F} = \mathscr{O}^*_{X_{\text{et}}}$ and k is algebraically closed). We have $j^*\mathscr{O}^*_{X_{\text{et}}} \cong \mathscr{O}^*_{(\operatorname{Spec} R)_{\text{et}}}$. By 5.7.8 and 4.5.11 (ii) (resp. 4.5.8 and 4.5.9), we have

$$H^q(\operatorname{Spec} R, j^*\mathscr{O}^*_{X_{\text{et}}}) = 0$$

for $q = 1$ and any $q \geq 3$ (resp. $q \geq 1$). So we have

$$H^q(X, \mathscr{O}^*_{X_{\text{et}}}) \cong H^0(X,\mathscr{H}^{q-1}(\Delta))$$

for any $q \geq 4$ (resp. $q \geq 2$.) On the other hand, $R^q j_* j^*\mathscr{O}^*_{X_{\text{et}}}$ is the sheaf associated to the presheaf

$$V \mapsto H^q((\operatorname{Spec} R) \times_X V, \mathscr{O}^*_{(\operatorname{Spec} R)_{\text{et}}})$$

for any etale X-scheme V. Again by 5.7.8 and 4.5.11 (ii) (resp. 4.5.8 and 4.5.9), we have

$$H^q((\operatorname{Spec} R) \times_X V, \mathscr{O}^*_{(\operatorname{Spec} R)_{\text{et}}}) = 0$$

for $q = 1$ and any $q \geq 3$ (resp. $q \geq 1$). So $R^q j_* j^* \mathscr{O}^*_{X_{\text{et}}} = 0$ for $q = 1$ and any $q \geq 3$ (resp. $q \geq 1$), and hence $\mathscr{H}^q(\Delta) = 0$ for $q = 1$ and any $q \geq 3$ (resp. $q \geq 1$). Therefore

$$H^q(X, \mathscr{O}^*_{X_{\text{et}}}) \cong H^0(X, \mathscr{H}^{q-1}(\Delta)) = 0$$

for any $q \geq 4$ (resp. $q \geq 2$).

Since $H^0(X, \mathscr{H}^1(\Delta)) = 0$ and $H^3(\operatorname{Spec} R, j^* \mathscr{O}^*_{X_{\text{et}}}) = 0$, part of the above long exact sequence is

$$0 \to H^2(X, \mathscr{O}^*_{X_{\text{et}}}) \to H^2(\operatorname{Spec} R, j^* \mathscr{O}^*_{X_{\text{et}}}) \to H^0(X, \mathscr{H}^2(\Delta)) \to H^3(X, \mathscr{O}^*_{X_{\text{et}}}) \to 0.$$

By 4.5.11 (ii), $H^2(\operatorname{Spec} R, j^* \mathscr{O}^*_{X_{\text{et}}})$ is a p-torsion group and $R^2 j_* j^* \mathscr{O}^*_{X_{\text{et}}}$ is a p-torsion sheaf. Hence $\mathscr{H}^2(\Delta)$ is a p-torsion sheaf. It follows that $H^2(X, \mathscr{O}^*_{X_{\text{et}}})$ and $H^3(X, \mathscr{O}^*_{X_{\text{et}}})$ are p-torsion groups. □

Remark 7.2.8. Let X be a regular integral scheme of dimension 1, η the generic points, $K = \mathscr{O}_{X,\eta}$ the function field of X, and $j : \operatorname{Spec} K \to X$ the canonical morphism. Define a morphism

$$j_* \mathscr{O}^*_{(\operatorname{Spec} K)_{\text{et}}} \to \bigoplus_{x \in |X|} i_{x*}\mathbb{Z}$$

as follows: For any connected etale X-scheme V of finite type, let η' be the generic point of V and let $K' = \mathscr{O}_{V,\eta'}$. We have

$$(j_* \mathscr{O}^*_{(\operatorname{Spec} K)_{\text{et}}})(V) = K'^*,$$

$$\left(\bigoplus_{x \in |X|} i_{x*}\mathbb{Z}\right)(V) = \bigoplus_{x \in |V|} \mathbb{Z}.$$

Define

$$(j_* \mathscr{O}^*_{(\operatorname{Spec} K)_{\text{et}}})(V) \to \left(\bigoplus_{x \in |X|} i_{x*}\mathbb{Z}\right)(V)$$

to be

$$K'^* \to \bigoplus_{x \in |V|} \mathbb{Z}, \quad f \mapsto (v_x(f)),$$

where $v_x : K'^* \to \mathbb{Z}$ is the valuation at x. We can identify the sheaf $\bigoplus_{x \in |X|} i_{x*}\mathbb{Z}$ with the sheaf of Weil divisors $\mathscr{D}$ defined by setting $\mathscr{D}(V)$ to be the group of Weil divisors on V. The morphism $j_* \mathscr{O}^*_{(\operatorname{Spec} K)_{\text{et}}} \to \bigoplus_{x \in |X|} i_{x*}\mathbb{Z}$ is simply the morphism mapping each rational function to the principle Weil divisor associated to it. This morphism is surjective and its kernel is exactly $\mathscr{O}_{X_{\text{et}}}$. We thus have an exact sequence

$$0 \to \mathscr{O}_{X_{\text{et}}} \to j_* \mathscr{O}^*_{(\operatorname{Spec} K)_{\text{et}}} \to \bigoplus_{x \in |X|} i_{x*}\mathbb{Z} \to 0.$$

Theorem 7.2.9.

(i) *Let X be a reduced scheme of finite type over a separably closed field k with* $\dim X \leq 1$, *and let n be a positive integer invertible in k. Then $H^q(X, \mu_{n,X}) = 0$ for any $q \geq 3$ and we have an exact sequence*

$$\begin{aligned}0 &\to H^0(X, \mu_{n,X}) \to \Gamma(X, \mathscr{O}_X^*) \stackrel{n}{\to} \Gamma(X, \mathscr{O}_X^*) \to \\ &\to H^1(X, \mu_{n,X}) \to \mathrm{Pic}(X) \stackrel{n}{\to} \mathrm{Pic}(X) \to H^2(X, \mu_{n,X}) \to 0.\end{aligned}$$

(ii) *Let X be a smooth projective irreducible curve of genus g over an algebraically closed field k, and let n be a positive integer invertible in k. Then we have*

$$H^q(X, \mathbb{Z}/n) \cong \begin{cases} \mathbb{Z}/n & \text{if } q = 0, \\ (\mathbb{Z}/n)^{2g} & \text{if } q = 1, \\ \mathbb{Z}/n & \text{if } q = 2, \\ 0 & \text{if } q \geq 3. \end{cases}$$

Moreover, we have canonical isomorphisms

$$H^q(X, \mu_{n,X}) \cong \begin{cases} \{x \in k | x^n = 1\} & \text{if } q = 0, \\ \{\mathscr{L} \in \mathrm{Pic}(X) | n\mathscr{L} = 0\} & \text{if } q = 1, \\ \mathbb{Z}/n & \text{if } q = 2, \\ 0 & \text{if } q \geq 3. \end{cases}$$

Proof.

(i) By Kummer's theory 7.2.1, we have a long exact sequence

$$\begin{aligned}0 \to H^0(X, \mu_{n,X}) &\to H^0(X, \mathscr{O}_X^*) \stackrel{n}{\to} H^0(X, \mathscr{O}_X^*) \to H^1(X, \mu_{n,X}) \\ &\to H^1(X, \mathscr{O}_X^*) \stackrel{n}{\to} H^1(X, \mathscr{O}_X^*) \to H^2(X, \mu_{n,X}) \to H^2(X, \mathscr{O}_{\mathrm{et}}^*) \to \cdots.\end{aligned}$$

We have

$$H^0(X, \mathscr{O}_X^*) = \Gamma(X, \mathscr{O}_X^*), \quad H^1(X, \mathscr{O}_X^*) = \mathrm{Pic}(X).$$

Moreover, if $p = \mathrm{char}\, k$, then $H^q(X, \mathscr{O}_X^*)$ are p-torsion groups for any $q \geq 2$ by 7.2.7, whereas $H^q(X, \mu_{n,X})$ are $\mathbb{Z}/n$-modules. Since n is invertible in k, n is relatively prime to p. Our assertion follows.

(ii) Since X is projective and integral, and k is algebraically closed, we have $\Gamma(X, \mathscr{O}_X^*) = k^*$, and the homomorphism

$$n : k^* \to k^*, \quad x \mapsto x^n$$

is surjective. It follows from the long exact sequence in (i) that

$$\begin{aligned}H^0(X, \mu_{n,X}) &\cong \{x \in k | x^n = 1\}, \\ H^1(X, \mu_{n,X}) &\cong \mathrm{Pic}(X)_n, \\ H^2(X, \mu_{n,X}) &\cong \mathrm{Pic}(X)/n\mathrm{Pic}(X),\end{aligned}$$

where for any abelian group A, we set $A_n = \ker(n : A \to A)$. Let

$$\deg : \mathrm{Pic}(X) \to \mathbb{Z}$$

be the homomorphism mapping an invertible $\mathscr{O}_X$-module to its degree. It is surjective. Let $\mathrm{Pic}^0(X)$ be its kernel. We have a commutative diagram of short exact sequences

$$\begin{array}{ccccccccc} 0 & \to & \mathrm{Pic}^0(X) & \to & \mathrm{Pic}(X) & \stackrel{\deg}{\to} & \mathbb{Z} & \to & 0 \\ & & n\downarrow & & n\downarrow & & n\downarrow & & \\ 0 & \to & \mathrm{Pic}^0(X) & \to & \mathrm{Pic}(X) & \stackrel{\deg}{\to} & \mathbb{Z} & \to & 0, \end{array}$$

where all the vertical arrows are multiplication by n. By the snake lemma, we have an exact sequence

$$\begin{aligned} 0 &\to \mathrm{Pic}^0(X)_n \to \mathrm{Pic}(X)_n \to 0 \\ &\to \mathrm{Pic}^0(X)/n\mathrm{Pic}^0(X) \to \mathrm{Pic}(X)/n\mathrm{Pic}(X) \to \mathbb{Z}/n \to 0. \end{aligned}$$

$\mathrm{Pic}^0(X)$ is isomorphic to the group of k-points $\mathrm{Jac}_X(k)$ of the Jacobian Jac_X of X. Moreover the homomorphism $n : \mathrm{Jac}_X(k) \to \mathrm{Jac}_X(k)$ is surjective, and its kernel is isomorphic to $(\mathbb{Z}/n)^{2g}$. So we have

$$H^1(X, \boldsymbol{\mu}_{n,X}) \cong \mathrm{Pic}(X)_n \cong \mathrm{Pic}^0(X)_n \cong \mathrm{Jac}_X(k)_n \cong (\mathbb{Z}/n)^{2g},$$

and

$$H^2(X, \boldsymbol{\mu}_{n,X}) \cong \mathrm{Pic}(X)/n\mathrm{Pic}(X) \cong \mathbb{Z}/n.$$

By (i), we have $H^q(X, \boldsymbol{\mu}_{n,X}) = 0$ for any $q \geq 3$. This proves the second part of (ii). By 7.2.2, we have $\mathbb{Z}/n \cong \boldsymbol{\mu}_{n,X}$. The first part of (ii) follows.□

Proposition 7.2.10. *Let X be a smooth irreducible curve over a separably closed field k, and let n be a positive integer invertible in k. Then $H^q(X, \mathbb{Z}/n)$ are finite for all q and vanish for all $q \geq 3$. If X is not projective, then $H^q(X, \mathbb{Z}/n)$ vanish for all $q \geq 2$.*

Proof. Let $\bar{k}$ be an algebraic closure of k. Then $\bar{k} = \varinjlim_i k_i$, where k_i goes over the family of finite purely inseparable extensions of k in $\bar{k}$. By 5.7.2, we have

$$H^q(X, \mathbb{Z}/n) \cong H^q(X \otimes_k k_i, \mathbb{Z}/n),$$

and by 5.9.3, we have

$$H^q(X \otimes_k \bar{k}, \mathbb{Z}/n) \cong \varinjlim_i H^q(X \otimes_k k_i, \mathbb{Z}/n).$$

It follows that

$$H^q(X, \mathbb{Z}/n) \cong H^q(X \otimes_k \bar{k}, \mathbb{Z}/n).$$

So to prove our assertion, we may assume that k is algebraically closed. We have $\mathbb{Z}/n \cong \boldsymbol{\mu}_{n,X}$ by 7.2.2. So we may replace $\mathbb{Z}/n$ by $\boldsymbol{\mu}_{n,X}$. By 7.2.9 (i), we have $H^q(X, \boldsymbol{\mu}_{n,X}) = 0$ for any $q \geq 3$.

Since X is irreducible and hence connected, we have $H^0(X, \mathbb{Z}/n) \cong \mathbb{Z}/n$. In particular, $H^0(X, \mathbb{Z}/n)$ is finite.

Let $\overline{X}$ be the smooth compactification of X and let $\overline{X} - X = \{x_1, \ldots, x_m\}$. Then $\Gamma(X, \mathscr{O}_X^*)$ consists of nonzero rational functions on $\overline{X}$ regular and invertible on X. Consider the homomorphism

$$\Gamma(X, \mathscr{O}_X^*) \to \mathbb{Z}^m, \quad f \mapsto (v_{x_1}(f), \ldots, v_{x_m}(f)),$$

where v_{x_i} $(i = 1, \ldots, m)$ denote the valuation at x_i. The kernel of this homomorphism consists of rational functions on $\overline{X}$ regular everywhere, and hence is isomorphic to k^*. Let A be the image of this homomorphism. We have an exact sequence

$$0 \to k^* \to \Gamma(X, \mathscr{O}_X^*) \to A \to 0.$$

By the snake lemma, we have an exact sequence

$$k^*/k^{*n} \to \Gamma(X, \mathscr{O}_X^*)/\Gamma(X, \mathscr{O}_X^*)^n \to A/nA \to 0.$$

Since A is a subgroup of $\mathbb{Z}^m$, A/nA is finite. Since k is algebraically closed, we have $k^*/k^{*n} = 0$. It follows that $\Gamma(X, \mathscr{O}_X^*)/\Gamma(X, \mathscr{O}_X^*)^n$ is finite.

The canonical restriction homomorphism

$$\mathrm{Pic}(\overline{X}) \to \mathrm{Pic}(X), \quad \mathscr{L} \to \mathscr{L}|_X$$

is surjective. Let B be the kernel of this homomorphism. By the snake lemma, we have an exact sequence

$$\mathrm{Pic}(\overline{X})_n \to \mathrm{Pic}(X)_n \to B/nB \to \mathrm{Pic}(\overline{X})/n\mathrm{Pic}(\overline{X}) \to \mathrm{Pic}(X)/n\mathrm{Pic}(X) \to 0.$$

In the proof of 7.2.9, we have seen

$$\mathrm{Pic}(\overline{X})_n \cong (\mathbb{Z}/n)^{2g}, \quad \mathrm{Pic}(\overline{X})/n\mathrm{Pic}(\overline{X}) \cong \mathbb{Z}/n,$$

where g is the genus of $\overline{X}$. In particular, $\mathrm{Pic}(\overline{X})_n$ and $\mathrm{Pic}(\overline{X})/n\mathrm{Pic}(\overline{X})$ are finite. We will show in a moment that B/nB is finite. The above exact sequence then shows that $\mathrm{Pic}(X)_n$ and $\mathrm{Pic}(X)/n\mathrm{Pic}(X) \cong H^2(X, \boldsymbol{\mu}_{n,X})$ are finite.

To prove that B/nB is finite, consider the homomorphism $\mathbb{Z}^m \to B$ mapping $(n_1, \ldots, n_m)$ to the isomorphic class of the invertible $\mathscr{O}_{\overline{X}}$-module associated to the Weil divisor $n_1x_1 + \cdots + n_mx_m$. It is surjective. Indeed, for any $\mathscr{L} \in B$, we have $\mathscr{L}|_X \cong \mathscr{O}_X$. Let D be a Weil divisor on $\overline{X}$ so that $\mathscr{L} \cong \mathscr{L}(D)$. Then $D|_X$ is a principle divisor, say $D|_X = (f)|_X$ for some rational function f. Then $D - (f)$ is a Weil divisor on $\overline{X}$ supported on $\{x_1, \ldots, x_m\}$ and $\mathscr{L} \cong \mathscr{L}(D-(f))$. So $\mathscr{L}$ lies in the image of $\mathbb{Z}^m \to B$. Thus the homomorphism $\mathbb{Z}^m \to B$ is surjective. This implies that B/nB is finite.

By 7.2.9 (i), we have an exact sequence

$$\Gamma(X, \mathscr{O}_X^*) \xrightarrow{n} \Gamma(X, \mathscr{O}_X^*) \to H^1(X, \boldsymbol{\mu}_{n,X}) \to \mathrm{Pic}(X) \xrightarrow{n} \mathrm{Pic}(X).$$

Combined with the finiteness of $\Gamma(X, \mathscr{O}_X^*)/\Gamma(X, \mathscr{O}_X^*)^n$ and $\mathrm{Pic}(X)_n$, we see that $H^1(X, \boldsymbol{\mu}_n)$ is finite.

If X is not projective, then $\overline{X} - X \neq \emptyset$, and every Weil divisor on X is the restriction of a Weil divisor of degree 0 on $\overline{X}$. So the restriction homomorphism

$$\mathrm{Pic}^0(\overline{X}) \to \mathrm{Pic}(X), \quad \mathscr{L} \to \mathscr{L}|_X$$

is surjective. We have seen in the proof of 7.2.9 that $n : \mathrm{Pic}^0(\overline{X}) \to \mathrm{Pic}^0(\overline{X})$ is surjective. It follows that $n : \mathrm{Pic}(X) \to \mathrm{Pic}(X)$ is surjective. So

$$H^2(X, \boldsymbol{\mu}_{n,X}) \cong \mathrm{Pic}(X)/n\mathrm{Pic}(X) = 0.$$

□

Theorem 7.2.11. *Let X be a scheme of finite type over a separably closed field k of characteristic p. If X is affine, then $H^q(X, \mathbb{Z}/p) = 0$ for any $q \geq 2$. If X is proper, then $H^q(X, \mathbb{Z}/p) = 0$ for any $q > \dim X$, and we have an exact sequence*

$$0 \to H^q(X, \mathbb{Z}/p) \to H^q_{\mathrm{Zar}}(X, \mathscr{O}_X) \xrightarrow{\wp} H^q_{\mathrm{Zar}}(X, \mathscr{O}_X) \to 0$$

for every q, where $\wp : H^q_{\mathrm{Zar}}(X, \mathscr{O}_X) \to H^q_{\mathrm{Zar}}(X, \mathscr{O}_X)$ is the homomorphism induced by

$$\wp : \mathscr{O}_X \to \mathscr{O}_X, \quad s \mapsto s^p - s.$$

Proof. We may assume that k is algebraically closed. By Artin–Schreier's theory 7.2.3, we have an exact sequence

$$\begin{aligned} 0 &\to H^0(X, \mathbb{Z}/p) \to H^0(X, \mathscr{O}_{X_{\mathrm{et}}}) \xrightarrow{\wp} H^0(X, \mathscr{O}_{X_{\mathrm{et}}}) \to \cdots \\ &\to H^q(X, \mathbb{Z}/p) \to H^q(X, \mathscr{O}_{X_{\mathrm{et}}}) \xrightarrow{\wp} H^q(X, \mathscr{O}_{X_{\mathrm{et}}}) \to \cdots \end{aligned}$$

By 5.7.5, we have

$$H^q(X, \mathscr{O}_{X_{\mathrm{et}}}) \cong H^q_{\mathrm{Zar}}(X, \mathscr{O}_X)$$

for all q. If X is affine, we have $H^q_{\text{Zar}}(X, \mathscr{O}_X) = 0$ for any $q \geq 1$. So $H^q(X, \mathbb{Z}/p) = 0$ for any $q \geq 2$. If X is proper, each $H^q_{\text{Zar}}(X, \mathscr{O}_X)$ is a finite dimensional k-vector space. The homomorphism

$$\wp : H^q_{\text{Zar}}(X, \mathscr{O}_X) \to H^q_{\text{Zar}}(X, \mathscr{O}_X)$$

is of the form $\phi - \text{id}$, where

$$\phi : H^q_{\text{Zar}}(X, \mathscr{O}_X) \to H^q_{\text{Zar}}(X, \mathscr{O}_X)$$

is the homomorphism induced by

$$\mathscr{O}_X \to \mathscr{O}_X, \quad s \mapsto s^p.$$

We have

$$\phi(x_1 + x_2) = \phi(x_1) + \phi(x_2), \quad \phi(\lambda x) = \lambda^p \phi(x)$$

for any $x_1, x_2, x \in H^q_{\text{Zar}}(X, \mathscr{O}_X)$ and $\lambda \in k$. By 7.2.12 below, the homomorphism $\wp : H^q_{\text{Zar}}(X, \mathscr{O}_X) \to H^q_{\text{Zar}}(X, \mathscr{O}_X)$ is surjective for each q. So we have an exact sequence

$$0 \to H^q(X, \mathbb{Z}/p) \to H^q_{\text{Zar}}(X, \mathscr{O}_X) \xrightarrow{\wp} H^q_{\text{Zar}}(X, \mathscr{O}_X) \to 0.$$

If $q > \dim X$, we have $H^q_{\text{Zar}}(X, \mathscr{O}_X) = 0$, and hence $H^q(X, \mathbb{Z}/p) = 0$. □

Lemma 7.2.12. *Let k be an algebraically closed field of characteristic p, let V be a finite dimensional vector space over k, and let $\phi : V \to V$ be a map satisfying*

$$\phi(x_1 + x_2) = \phi(x_1) + \phi(x_2), \quad \phi(\lambda x) = \lambda^p \phi(x)$$

for any $x_1, x_2, x \in V$ and $\lambda \in k$. Then $\phi - \text{id}$ is surjective.

Proof. We identify V with the vector space k^n. Then ϕ is of the form

$$\phi(\lambda_1, \ldots, \lambda_n) = (a_{11}\lambda_1^p + \cdots + a_{n1}\lambda_n^p, \ldots, a_{1n}\lambda_1^p + \cdots + a_{nn}\lambda_n^p)$$

for some $a_{ij} \in k$. So

$$(\phi - \text{id})(\lambda_1, \ldots, \lambda_n) = (a_{11}\lambda_1^p + \cdots + a_{n1}\lambda_n^p - \lambda_1, \ldots, a_{1n}\lambda_1^p + \cdots + a_{nn}\lambda_n^p - \lambda_n).$$

Consider the k-algebra homomorphism

$$\psi : k[x_1, \ldots, x_n] \to k[x_1, \ldots, x_n], \quad x_i \mapsto a_{1i}x_1^p + \cdots + a_{ni}x_n^p - x_i.$$

It induces a k-morphism

$$f : \operatorname{Spec} k[x_1, \ldots, x_n] \to \operatorname{Spec} k[x_1, \ldots, x_n].$$

The homomorphism

$$\Omega_{k[x_1,\ldots,x_n]/k} \to \Omega_{k[x_1,\ldots,x_n]/k}$$

induced by ψ maps dx_i to

$$d(a_{1i}x_1^p + \cdots + a_{ni}x_n^p - x_i) = -dx_i.$$

Since $\Omega_{k[x_1,\ldots,x_n]/k}$ is a free $k[x_1,\ldots,x_n]$-module with basis $\{dx_1,\ldots,dx_n\}$, the homomorphism of $k[x_1,\ldots,x_n]$-modules

$$\Omega_{k[x_1,\ldots,x_n]/k} \otimes_\psi k[x_1,\ldots,x_n] \to \Omega_{k[x_1,\ldots,x_n]/k}$$

induced by ψ is an isomorphism. By 2.5.3, f is etale. Hence $\mathrm{im}\,(f)$ is open in $\mathrm{Spec}\,k[x_1,\ldots,x_n]$. The set of closed points in $\mathrm{Spec}\,k[x_1,\ldots,x_n]$ can be identified with k^n, and the restriction of f to this set can be identified with $\phi - \mathrm{id} : k^n \to k^n$. So $\mathrm{im}\,(\phi - \mathrm{id})$ is Zariski open in k^n. But k^n is an algebraic group, and $\mathrm{im}\,(\phi - \mathrm{id})$ is an open subgroup. So we must have $\mathrm{im}\,(\phi - \mathrm{id}) = k^n$. □

Theorem 7.2.13. *Let X be a scheme of finite type over a separably closed field k with* $\dim X \leq 1$. *For any torsion sheaf $\mathscr{F}$ on X, we have $H^q(X,\mathscr{F}) = 0$ for all $q \geq 3$. If X is affine, we have $H^q(X,\mathscr{F}) = 0$ for all $q \geq 2$.*

Proof. By 5.7.2 and 5.9.3, we may assume that k is algebraically closed and X is reduced. By 7.2.7, we have $H^q(X,\mathscr{F}) = 0$ for any torsion sheaf $\mathscr{F}$ and any $q \geq 3$. Suppose X is affine. To prove $H^2(X,\mathscr{F}) = 0$ for a torsion sheaf $\mathscr{F}$, we may assume that $\mathscr{F}$ is a constructible sheaf by 5.8.8 and 5.9.2. By 5.8.5 (ii), $\mathscr{F}$ is a constructible sheaf of $\mathbb{Z}/n$-modules for some n. By 5.8.5 (i), there exists an epimorphism $f_!\mathbb{Z}/n \to \mathscr{F}$ for some separated etale morphism $f : U \to X$ of finite type. Let $\mathscr{K}$ be the kernel of this epimorphism. We have $H^3(X,\mathscr{K}) = 0$. It follows that

$$H^2(X, f_!\mathbb{Z}/n) \to H^2(X,\mathscr{F})$$

is onto. To prove $H^2(X,\mathscr{F}) = 0$, it suffices to prove $H^2(X, f_!\mathbb{Z}/n) = 0$. Let η_i $(i = 1,\ldots,m)$ be the generic points of X. Since f is etale,

$$f_{\eta_i} : U \otimes_{\mathscr{O}_X} \mathscr{O}_{X,\eta_i} \to \mathrm{Spec}\,\mathscr{O}_{X,\eta_i}$$

are finite morphisms. By 1.10.10 (iv), for each i, there exists an open neighborhood V_i of η_i such that $f_{V_i} : f^{-1}(V_i) \to V_i$ is finite. Let $V = \cup_i V_i$ and let $j : V \to X$ be the open immersion. Then the canonical morphism

$$f_!\mathbb{Z}/n \to j_*j^*f_!\mathbb{Z}/n$$

is an isomorphism when restricted to V. Since $X - V$ is finite over $\operatorname{Spec} k$, the q-th cohomology groups of the kernel and the cokernel of $f_!\mathbb{Z}/n \to j_*j^*f_!\mathbb{Z}/n$ vanish for all $q \geq 1$. So

$$H^2(X, f_!\mathbb{Z}/n) \cong H^2(X, j_*j^*f_!\mathbb{Z}/n).$$

We are thus reduced to prove $H^2(X, j_*j^*f_!\mathbb{Z}/n) = 0$. We have

$$j_*j^*f_!\mathbb{Z}/n \cong j_*f_{V!}\mathbb{Z}/n.$$

Since f_V is finite, we have

$$f_{V!}\mathbb{Z}/n \cong f_{V*}\mathbb{Z}/n$$

by 5.5.2. It follows that

$$j_*j^*f_!\mathbb{Z}/n \cong (j \circ f_V)_*\mathbb{Z}/n.$$

Note that $j \circ f_V$ is etale. We are thus reduced to prove that $H^2(X, f_*\mathbb{Z}/n) = 0$ for any separated etale morphism $f : U \to X$ of finite type. By the Zariski Main Theorem 1.10.13, we have a commutative diagram

$$\begin{array}{ccc} U & \stackrel{i}{\hookrightarrow} & \overline{U} \\ f\downarrow & \swarrow \overline{f} & \\ X & & \end{array}$$

such that i is a dominant open immersion and $\overline{f}$ is finite. We have

$$H^2(X, f_*\mathbb{Z}/n) \cong H^2(X, \overline{f}_*i_*\mathbb{Z}/n) \cong H^2(\overline{U}, i_*\mathbb{Z}/n).$$

Note that $\overline{U}$ is affine and $\dim \overline{U} \leq 1$. Since i is dominant, $\overline{U} - U$ is finite over $\operatorname{Spec} k$. The kernel and cokernel of the canonical morphism $\mathbb{Z}/n \to i_*\mathbb{Z}/n$ are supported on $\overline{U} - U$. So we have

$$H^2(\overline{U}, i_*\mathbb{Z}/n) \cong H^2(\overline{U}, \mathbb{Z}/n),$$

and we are reduced to prove $H^2(\overline{U}, \mathbb{Z}/n) = 0$. We may assume that $\overline{U}$ is reduced. Let $\widetilde{U}$ be the normalization of $\overline{U}$ and let $\pi : \widetilde{U} \to \overline{U}$ be the canonical morphism. Then π is finite. It induces isomorphisms above generic points of $\overline{U}$, and hence induces an isomorphism above a dense open subset of $\overline{U}$ by 1.10.9 (ii). So the kernel and cokernel of the canonical morphism $\mathbb{Z}/n \to \pi_*\mathbb{Z}/n$ are supported on a closed subscheme of $\overline{U}$ finite over $\operatorname{Spec} k$. It follows that

$$H^2(\overline{U}, \mathbb{Z}/n) \cong H^2(\overline{U}, \pi_*\mathbb{Z}/n) \cong H^2(\widetilde{U}, \mathbb{Z}/n).$$

But $\widetilde{U}$ is a smooth affine curve. By 7.2.10, we have $H^2(\widetilde{U}, \mathbb{Z}/n) = 0$ if n is relatively prime to $p = \operatorname{char} k$. By 7.2.11, we have $H^2(\widetilde{U}, \mathbb{Z}/p) = 0$. It follows that $H^2(\widetilde{U}, \mathbb{Z}/p^r) = 0$ for any $r \geq 1$ since $\mathbb{Z}/p^r$ has a finite filtration with successive quotients isomorphic to $\mathbb{Z}/p$. Therefore $H^2(\widetilde{U}, \mathbb{Z}/n) = 0$ for all n. □

7.3 Proper Base Change Theorem

([SGA 4] XII, XIII, [SGA $4\frac{1}{2}$] Arcata IV 1–4.)

Consider a Cartesian square

$$\begin{array}{ccc} X\times_Y Y' & \xrightarrow{g'} & X \\ {\scriptstyle f'}\downarrow & & \downarrow{\scriptstyle f} \\ Y' & \xrightarrow{g} & Y. \end{array}$$

For any sheaf $\mathscr{F}$ on X, applying f_* to the canonical morphism $\mathscr{F} \to g'_* g'^* \mathscr{F}$, we get

$$f_*\mathscr{F} \to f_* g'_* g'^* \mathscr{F} \cong g_* f'_* g'^* \mathscr{F}.$$

From the adjointness of the functors (g^*, g_*), we get a morphism

$$g^* f_* \mathscr{F} \to f'_* g'^* \mathscr{F}.$$

Another way to get such a morphism is as follows: Apply g'^* to the canonical morphism $f^* f_* \mathscr{F} \to \mathscr{F}$, we get

$$f'^* g^* f_* \mathscr{F} \cong g'^* f^* f_* \mathscr{F} \to g'^* \mathscr{F}.$$

From the adjointness of the functors (f'^*, f'_*), we get a morphism $g^* f_* \mathscr{F} \to f'_* g'^* \mathscr{F}$ again. At the end of this section, we show that these two ways define the same morphism $g^* f_* \mathscr{F} \to f'_* g'^* \mathscr{F}$.

For any bounded below complex $K^\cdot$ of sheaves on X, let $K^\cdot \to I^\cdot$ and $g'^* I^\cdot \to J^\cdot$ be quasi-isomorphisms so that $I^\cdot$ and $J^\cdot$ respectively are bounded below complexes of injective sheaves on X and $X \times_Y Y'$. We have morphisms

$$g^* R f_* K^\cdot \cong g^* f_* I^\cdot \to f'_* g'^* I^\cdot \to f'_* J^\cdot \cong R f'_* g'^* K^\cdot.$$

In this way, we get a morphism

$$g^* R f_* K^\cdot \to R f'_* g'^* K^\cdot$$

for any object $K^\cdot$ in $D^+(X)$. In particular, we have a morphism

$$g^* R^q f_* K^\cdot \to R^q f'_* g'^* K^\cdot$$

for each q.

Theorem 7.3.1 (Proper base change theorem). *Let*

$$\begin{array}{ccc} X\times_Y Y' & \xrightarrow{g'} & X \\ {\scriptstyle f'}\downarrow & & \downarrow{\scriptstyle f} \\ Y' & \xrightarrow{g} & Y \end{array}$$

be a Cartesian square. Suppose that f is a proper morphism. Then for any torsion sheaf $\mathscr{F}$ on X, the canonical morphisms

$$g^*R^qf_*\mathscr{F} \to R^qf'_*g'^*\mathscr{F}$$

are isomorphisms for all q. For any object $K^\cdot$ in $D^+_{\mathrm{tor}}(X)$, the canonical morphism

$$g^*Rf_*K^\cdot \to Rf'_*g'^*K^\cdot$$

is an isomorphism. If Rf_* and Rf'_* have finite cohomological dimensions, the same assertion holds for any object $K^\cdot$ in $D_{\mathrm{tor}}(X)$.

Corollary 7.3.2. *Let $f : X \to Y$ be a proper morphism, let $\mathscr{F}$ be a torsion sheaf on X, and let $s \to Y$ be a geometric point of Y, where s is the spectrum of a separably closed field. Then*

$$(R^qf_*\mathscr{F})_s \cong H^q(X \times_Y s, \mathscr{F}|_{X\times_Y s})$$

for all q.

Corollary 7.3.3. *Let A be a strictly local ring, $Y = \mathrm{Spec}\, A$, $f : X \to Y$ a proper morphism, and X_0 the fiber of f above the closed point of Y. For any torsion sheaf $\mathscr{F}$ on X, the canonical homomorphisms*

$$H^q(X, \mathscr{F}) \to H^q(X_0, \mathscr{F}|_{X_0})$$

are isomorphisms for all q.

7.3.2 is a special case of 7.3.1 by taking $Y' = s$.

7.3.2 implies 7.3.3. Indeed, in the notation of 7.3.3, taking s to be the closed point of Y, we have

$$(R^qf_*\mathscr{F})_s \cong H^q(X_0, \mathscr{F}|_{X_0})$$

by 7.3.2. But we have

$$(R^qf_*\mathscr{F})_s \cong H^q(X, \mathscr{F})$$

by 5.9.5. So $H^q(X, \mathscr{F}) \cong H^q(X_0, \mathscr{F}|_{X_0})$.

7.3.3 implies 7.3.1. Note that the first assertion of 7.3.1 implies the rest by 6.5.1 and 6.5.2. To prove the first statement, we may reduce to the case where $Y = \mathrm{Spec}\, A$ and $Y' = \mathrm{Spec}\, A'$ are affine. Writing A' as a direct limit of finitely generated A-algebras and applying 5.9.6, we may reduce to the case where Y' is of finite type over Y. To prove that $g^*R^qf_*\mathscr{F} \to R^qf'_*g'^*\mathscr{F}$ is an isomorphism for each q, it suffices to show that for any point y' in Y' which is a close point in the fiber $g^{-1}(g(y'))$, the homomorphism on stalks

$$(g^*R^qf_*\mathscr{F})_{\bar{y}'} \to (R^qf'_*g'^*\mathscr{F})_{\bar{y}'}$$

is an isomorphism. Indeed, if $\pi : U \to Y'$ is an etale morphism and s is a section over U of the kernel (resp. cokernel) of the morphism $g^*R^qf_*\mathscr{F} \to R^qf'_*g'^*\mathscr{F}$, then for any point $u \in U$ which is closed in the fiber $(g\pi)^{-1}(g\pi(u))$, there exists an etale neighborhood W_u of $\bar{u}$ over U such that $s|_{W_u} = 0$. One can show that as u goes over the set of points in U which is closed in $(g\pi)^{-1}(g\pi(u))$, W_u form an etale covering of U. It follows that $s = 0$. Hence $g^*R^qf_*\mathscr{F} \to R^qf'_*g'^*\mathscr{F}$ is an isomorphism. Let us use 7.3.3 to verify

$$(g^*R^qf_*\mathscr{F})_{\bar{y}'} \to (R^qf'_*g'^*\mathscr{F})_{\bar{y}'}$$

is an isomorphism for any q and any point y' in Y' which is closed in the fiber $g^{-1}(g(y'))$. Let A (resp. A') be the strict henselization of $\mathscr{O}_{Y,g(y')}$ (resp. $\mathscr{O}_{Y',y'}$). By 5.9.5, we have

$$\begin{aligned}(g^*R^qf_*\mathscr{F})_{\bar{y}'} &\cong (R^qf_*\mathscr{F})_{\overline{g(y')}} \cong H^q(X \otimes_{\mathscr{O}_Y} A, \mathscr{F}|_{X\otimes_{\mathscr{O}_Y} A}),\\ (R^qf'_*g'^*\mathscr{F})_{\bar{y}'} &\cong H^q(X \otimes_{\mathscr{O}_Y} A', \mathscr{F}|_{X\otimes_{\mathscr{O}_Y} A'}).\end{aligned}$$

By 7.3.3, we have

$$\begin{aligned}H^q(X \otimes_{\mathscr{O}_Y} A, \mathscr{F}|_{X\otimes_{\mathscr{O}_Y} A}) &\cong H^q(X \otimes_{\mathscr{O}_Y} \overline{k(g(y'))}, \mathscr{F}|_{X\otimes_{\mathscr{O}_Y} \overline{k(g(y'))}}),\\ H^q(X \otimes_{\mathscr{O}_Y} A', \mathscr{F}|_{X\otimes_{\mathscr{O}_Y} A'}) &\cong H^q(X \otimes_{\mathscr{O}_Y} \overline{k(y')}, \mathscr{F}|_{X\otimes_{\mathscr{O}_Y} \overline{k(y')}}),\end{aligned}$$

where $\overline{k(g(y'))}$ (resp. $\overline{k(y')}$) is a separable closure of $k(g(y'))$ (resp. $k(y')$). Since y' is closed in the fiber $g^{-1}(g(y'))$, $k(y')$ is algebraic over $k(y)$, and hence $k(y')$ and $k(y)$ have the same algebraic closure. It follows from 5.7.2 and 5.9.3 that we have

$$H^q(X \otimes_{\mathscr{O}_Y} \overline{k(g(y'))}, \mathscr{F}|_{X\otimes_{\mathscr{O}_Y} \overline{k(g(y'))}}) \cong H^q(X \otimes_{\mathscr{O}_Y} \overline{k(y')}, \mathscr{F}|_{X\otimes_{\mathscr{O}_Y} \overline{k(y')}}).$$

Thus

$$(g^*R^qf_*\mathscr{F})_{\bar{y}'} \cong (R^qf'_*g'^*\mathscr{F})_{\bar{y}'},$$

and 7.3.1 holds.

We will prove 7.3.3 under the extra condition that A is noetherian. It implies that 7.3.1 and 7.3.2 hold under the extra condition that Y and Y' are locally noetherian. This is enough for applications.

Lemma 7.3.4. *Let $X_0 \to X$ be a morphism. For every torsion sheaf $\mathscr{F}$ on X, assume that the canonical homomorphism*

$$H^i(X, \mathscr{F}) \to H^i(X_0, \mathscr{F}|_{X_0})$$

is an isomorphism for each $i < n$ and a monomorphism for $i = n$. Then for every complex $K^\cdot$ satisfying $H^i(K^\cdot) = 0$ for all $i < 0$, the canonical homomorphism

$$H^i(X, K^\cdot) \to H^i(X_0, K^\cdot|_{X_0})$$

is an isomorphism for each $i < n$ and a monomorphism for $i = n$.

Proof. Let $m \geq 0$ be an integer such that $\mathscr{H}^i(K^\cdot) = 0$ for all $i < m$. Such m exists since $\mathscr{H}^i(K^\cdot) = 0$ for all $i < 0$. We prove the lemma by descending induction on m. We have a biregular spectral sequence

$$E_2^{pq} = H^p(X, \mathscr{H}^q(K^\cdot)) \Rightarrow H^{p+q}(X, K^\cdot).$$

If $p + q < m$, we have either $p < 0$ or $q < m$, and hence we have $H^p(X, \mathscr{H}^q(K^\cdot)) = 0$. It follows that $H^i(X, K^\cdot) = 0$ for any $i < m$. Similarly, we have $H^i(X_0, K^\cdot|_{X_0}) = 0$ for any $i < m$. Hence the lemma holds for those $K^\cdot$ with $\mathscr{H}^i(K^\cdot) = 0$ for all $i < n + 1$. Suppose that the lemma holds for those $K^\cdot$ with $\mathscr{H}^i(K^\cdot) = 0$ for all $i < m+1$. Let K be a complex with $\mathscr{H}^i(K^\cdot) = 0$ for all $i < m$. We have a distinguished triangle

$$\mathscr{H}^m(K^\cdot)[-m] \to K^\cdot \to \tau_{\geq m+1}K^\cdot \to,$$

where $\tau_{\geq m+1}K^\cdot$ is the complex

$$\cdots \to 0 \to K^{m+1}/\operatorname{im} d_m \to K^{m+2} \to \cdots .$$

Consider the commutative diagram

$$\begin{array}{ccccccc} H^{i-1}(X, \tau_{\geq m+1}K^\cdot) & \to & H^{i-m}(X, \mathscr{H}^m(K^\cdot)) & \to & H^i(X, K^\cdot) & \to \\ \downarrow & & \downarrow & & \downarrow & \\ H^{i-1}(X_0, \tau_{\geq m+1}K^\cdot|_{X_0}) & \to & H^{i-m}(X_0, \mathscr{H}^m(K^\cdot)|_{X_0}) & \to & H^i(X_0, K^\cdot|_{X_0}) & \to \end{array}$$

$$\begin{array}{ccccc} \to & H^i(X, \tau_{\geq m+1}K^\cdot) & \to & H^{i+1-m}(X, \mathscr{H}^m(K^\cdot)) \\ & \downarrow & & \downarrow \\ \to & H^i(X_0, \tau_{\geq m+1}K^\cdot|_{X_0}) & \to & H^{i+1-m}(X_0, \mathscr{H}^m(K^\cdot)|_{X_0}). \end{array}$$

If $i < n$, the second vertical arrow is bijective and the fifth is injective by assumption, and the first and the fourth are bijective by the induction hypothesis. By the five lemma, the third vertical arrow is bijective. If $i = n$, the second vertical arrow is injective by assumption, the first is bijective and the fourth is injective by the induction hypothesis. By the five lemma, the third vertical arrow is injective. This proves our assertion. □

Lemma 7.3.5. *Consider a Cartesian square*

$$\begin{array}{ccc} X_0' & \stackrel{i'}{\to} & X' \\ {\scriptstyle f'}\downarrow & & \downarrow{\scriptstyle f} \\ X_0 & \stackrel{i}{\to} & X \end{array}.$$

Assume that f is surjective, and the canonical morphism

$$i^* R^q f_* \mathscr{F}' \to R^q f'_* i'^* \mathscr{F}'$$

is an isomorphism for any torsion sheaf $\mathscr{F}'$ on X' and any q. Then the following conditions are equivalent:

(i) *For any torsion sheaf $\mathscr{F}'$ on X' and any q, the canonical homomorphism*

$$H^q(X', \mathscr{F}') \to H^q(X_0', i'^* \mathscr{F}')$$

is an isomorphism.

(ii) *For any torsion sheaf $\mathscr{F}$ on X and any q, the canonical homomorphism*

$$H^q(X, \mathscr{F}) \to H^q(X_0, i^* \mathscr{F})$$

is an isomorphism.

Proof.

(i)⇒(ii) Let $\mathscr{F} \to Rf_* f^* \mathscr{F}$ be the canonical morphism, and let

$$\mathscr{F} \to Rf_* f^* \mathscr{F} \to \Delta \to$$

be a distinguished triangle. Since f is surjective, the canonical morphism

$$\mathscr{F} \to f_* f^* \mathscr{F}$$

is injective. It follows that $\mathscr{H}^i(\Delta) = 0$ for all $i < 0$. We have

$$\begin{aligned} H^q(X, Rf_* f^* \mathscr{F}) &\cong H^q(X', f^* \mathscr{F}) \\ &\cong H^q(X_0', i'^* f^* \mathscr{F}) \\ &\cong H^q(X_0, Rf'_* i'^* f^* \mathscr{F}) \\ &\cong H^q(X_0, i^* Rf_* f^* \mathscr{F}). \end{aligned}$$

Hence the canonical homomorphism

$$H^q(X, Rf_* f^* \mathscr{F}) \to H^q(X_0, i^* Rf_* f^* \mathscr{F})$$

is an isomorphism for any q. Consider the commutative diagram

$$\begin{array}{ccccccc} 0 \to & H^0(X, \mathscr{F}) & \to & H^0(X, Rf_* f^* \mathscr{F}) & \to & H^0(X, \Delta) \\ & \downarrow & & \downarrow & & \downarrow \\ 0 \to & H^0(X_0, i^* \mathscr{F}) & \to & H^0(X_0, i^* Rf_* f^* \mathscr{F}) & \to & H^0(X_0, i^* \Delta). \end{array}$$

The second vertical arrow is an isomorphism. It follows that the first vertical arrow is injective. Then by 7.3.4, the third vertical arrow is injective. This then implies that the first vertical arrow is an isomorphism. Suppose that we have shown that the canonical homomorphism $H^q(X,\mathscr{F}) \to H^q(X_0, i^*\mathscr{F})$ is an isomorphism for any torsion sheaf $\mathscr{F}$ and any $q < n$. Consider the commutative diagram

$$\begin{array}{ccccccc} H^{n-1}(X, Rf_*f^*\mathscr{F}) & \to & H^{n-1}(X,\Delta) & \to & H^n(X,\mathscr{F}) & \to \\ \downarrow & & \downarrow & & \downarrow & \\ H^{n-1}(X_0, Rf_*f^*\mathscr{F}|_{X_0}) & \to & H^{n-1}(X_0,\Delta|_{X_0}) & \to & H^n(X_0,\mathscr{F}|_{X_0}) & \to \end{array}$$

$$\begin{array}{ccccc} \to & H^n(X, Rf_*f^*\mathscr{F}) & \to & H^n(X,\Delta) \\ & \downarrow & & \downarrow \\ \to & H^n(X_0, Rf_*f^*\mathscr{F}|_{X_0}) & \to & H^n(X_0,\Delta|_{X_0}). \end{array}$$

We have shown that the first and the fourth vertical arrows are bijective. By the induction hypothesis and 7.3.4, the second vertical arrow is injective. By the five lemma, the third vertical arrow is injective. Then by 7.3.4, the fifth vertical arrow is injective and the second is bijective. By the five lemma, the third vertical arrow is bijective.

(ii)⇒(i) We have

$$\begin{aligned} H^q(X',\mathscr{F}') &\cong H^q(X, Rf_*\mathscr{F}') \\ &\cong H^q(X_0, i^*Rf_*\mathscr{F}') \\ &\cong H^q(X_0, Rf'_*i'^*\mathscr{F}') \\ &\cong H^q(X'_0, i'^*\mathscr{F}'). \end{aligned}$$

□

Lemma 7.3.6. *Suppose that A is a strictly local noetherian ring. If 7.3.3 holds for any projective morphism f, then it holds for any proper morphism f.*

Proof. By Chow's lemma([Fu (2006)] 1.4.18, [EGA] II 5.6.1, [Hartshorne (1977)] Exer. II 4.10), for any proper morphism $f : X \to Y$, there exists a surjective projective morphism $g : X' \to X$ such that fg is projective. We then apply 7.3.5 and the equivalence of 7.3.1 and 7.3.3. □

Lemma 7.3.7. *Let A be a strictly local noetherian ring and let k be the residue field of A. Suppose for any nonnegative integer n and any torsion sheaf $\mathscr{F}$ on $\mathbb{P}^n_A$, that the canonical homomorphisms*

$$H^q(\mathbb{P}^n_A, \mathscr{F}) \to H^q(\mathbb{P}^n_k, \mathscr{F}|_{\mathbb{P}^n_k})$$

are isomorphisms for all q. Then 7.3.3 holds.

Proof. By 7.3.6, it suffices to prove 7.3.3 for any projective morphism f. Let

$$\begin{array}{ccc} X & \xrightarrow{i} & \mathbb{P}^n_A \\ {\scriptstyle f}\downarrow & \swarrow & \\ \operatorname{Spec} A & & \end{array}$$

be a commutative diagram such that i is a closed immersion, and let $i_0 : X_0 \to \mathbb{P}^n_k$ be the morphism induced from i by base change. For any torsion sheaf $\mathscr{F}$ on X, the canonical morphism

$$(i_*\mathscr{F})|_{\mathbb{P}^n_k} \to i_{0*}(\mathscr{F}|_{X_0})$$

is an isomorphism by 5.3.9. We have canonical isomorphisms

$$H^q(X, \mathscr{F}) \cong H^q(\mathbb{P}^n_A, i_*\mathscr{F}),$$
$$H^q(X_0, \mathscr{F}|_{X_0}) \cong H^q(\mathbb{P}^n_k, i_{0*}(\mathscr{F}|_{X_0})).$$

By our assumption, we have

$$H^q(\mathbb{P}^n_A, i_*\mathscr{F}) \cong H^q(\mathbb{P}^n_k, (i_*\mathscr{F})|_{\mathbb{P}^n_k}).$$

It follows that

$$H^q(X, \mathscr{F}) \cong H^q(X_0, \mathscr{F}|_{X_0}).$$

□

Lemma 7.3.8. *Suppose that A is a strictly local noetherian ring. If 7.3.3 holds in the case where* $\dim X_0 \leq 1$, *then it holds in general.*

Proof. By 7.3.7, it suffices to prove 7.3.3 for $X = \mathbb{P}^n_A$. We use induction on n. When $n = 0$, this is clear. When $n = 1$, this follows from our assumption. Suppose that 7.3.3 holds for $X = \mathbb{P}^{n-1}_A$. Let P be the closed subscheme of $\mathbb{P}^n_A \times_A \mathbb{P}^1_A$ defined by

$$P = \{([x_0 : \ldots : x_n], [t_0 : t_1]) \in \mathbb{P}^n_A \times_A \mathbb{P}^1_A | t_0x_0 + t_1x_1 = 0\}.$$

Let

$$i : P \to \mathbb{P}^n_A \times_A \mathbb{P}^1_A, \quad p_1 : \mathbb{P}^n_A \times_A \mathbb{P}^1_A \to \mathbb{P}^n_A, \quad p_2 : \mathbb{P}^n_A \times_A \mathbb{P}^1_A \to \mathbb{P}^1_A$$

be the closed immersion and the projections, respectively. Note that $p_1 i : P \to \mathbb{P}^n_A$ is surjective and its fibers have dimensions ≤ 1. By our assumption, the equivalence of 7.3.3 and 7.3.1, and 7.3.5, to prove that 7.3.3 holds for $X = \mathbb{P}^n_A$, it suffices to show that it holds for $X = P$. The fibers of $p_2 i : P \to \mathbb{P}^1_A$ are $(n-1)$-dimensional projective spaces. 7.3.3 follows from 7.3.5 applied to the morphism $p_2 i$. The conditions of 7.3.5 hold by our assumption, the induction hypothesis, and the equivalence of 7.3.1 and 7.3.3.

□

Lemma 7.3.9. *Suppose that A is a strictly local noetherian ring. In the notation of 7.3.3, if for any finite morphism $X' \to X$ and any positive integer n, the canonical homomorphism*

$$H^q(X', \mathbb{Z}/n) \to H^q(X' \times_X X_0, \mathbb{Z}/n)$$

is bijective for $q = 0$ and surjective for any $q \geq 1$, then for any torsion sheaf $\mathscr{F}$ on X, the canonical homomorphism

$$H^q(X, \mathscr{F}) \to H^q(X_0, \mathscr{F}|_{X_0})$$

is an isomorphism for any q.

Proof. Suppose that $\mathscr{F}$ is a constructible sheaf of $\mathbb{Z}/n$-modules. By 5.8.11 (ii), we can find a monomorphism $\mathscr{F} \hookrightarrow f_*\mathbb{Z}/n$ for some finite morphism $f : X' \to X$. (In 5.8.11 (ii), the scheme is assumed to be of finite type over a field or over $\mathbb{Z}$. To make it adapted to our case, we write $A = \varinjlim_i A_i$, where A_i are subalgebras of A finitely generated over $\mathbb{Z}$. By 1.10.9 and 5.9.8 (iii), we can find an A_i-scheme X_i and a constructible sheaf of $\mathbb{Z}/n$-modules $\mathscr{F}_i$ on X_i for some large i such that the pair $(X, \mathscr{F})$ is induced from the pair $(X_i, \mathscr{F}_i)$ by base change. We can apply 5.8.11 (ii) to $\mathscr{F}_i$ and then take base change.) Let $f_0 : X_0 \times_X X' \to X_0$ be the base change of f. We have

$$\begin{aligned} &H^q(X, f_*\mathbb{Z}/n) \cong H^q(X', \mathbb{Z}/n), \\ &H^q(X_0, (f_*\mathbb{Z}/n)|_{X_0}) \cong H^q(X_0, f_{0*}\mathbb{Z}/n) \cong H^q(X_0 \times_X X', \mathbb{Z}/n). \end{aligned}$$

So the canonical homomorphism

$$H^q(X, f_*\mathbb{Z}/n) \to H^q(X_0, (f_*\mathbb{Z}/n)|_{X_0})$$

is bijective for $q = 0$ and surjective for any $q > 0$. Let $\mathscr{C}$ be the cokernel of the monomorphism $\mathscr{F} \hookrightarrow f_*\mathbb{Z}/n$. Consider the commutative diagram

$$\begin{array}{ccccc} 0 \to & H^0(X, \mathscr{F}) & \to & H^0(X, f_*\mathbb{Z}/n) & \to & H^0(X, \mathscr{C}) \\ & \downarrow & & \downarrow & & \downarrow \\ 0 \to & H^0(X_0, \mathscr{F}|_{X_0}) & \to & H^0(X_0, (f_*\mathbb{Z}/n)|_{X_0}) & \to & H^0(X_0, \mathscr{C}|_{X_0}). \end{array}$$

We have seen that the second vertical arrow is bijective. It follows that the first vertical arrow is injective. This is true for any constructible sheaf $\mathscr{F}$ of $\mathbb{Z}/n$-modules. In particular, the last vertical arrow is injective. This implies that the first vertical arrow is bijective for any constructible sheaf $\mathscr{F}$ of $\mathbb{Z}/n$-modules. By 5.8.5 (ii), 5.8.8 and 5.9.2, the canonical homomorphism

$$H^0(X, \mathscr{F}) \to H^0(X_0, \mathscr{F}|_{X_0})$$

is bijective for any torsion sheaf $\mathscr{F}$.

Suppose that $H^i(X, \mathscr{F}) \to H^i(X_0, \mathscr{F}|_{X_0})$ is bijective for any $i < q$ ($q \geq 1$) and any torsion sheaf $\mathscr{F}$. Let $\mathscr{F}$ be a constructible sheaf of $\mathbb{Z}/n$-modules. Embed $\mathscr{F}$ into $f_*\mathbb{Z}/n$ for some finite morphism $f : X' \to X$, and embed $f_*\mathbb{Z}/n$ into a flasque sheaf $\mathscr{G}$ of $\mathbb{Z}/n$-modules. Let $\mathscr{C}_1$ be the cokernel of $f_*\mathbb{Z}/n \hookrightarrow \mathscr{G}$. Consider the commutative diagram

$$\begin{array}{ccccccc} H^{q-1}(X, \mathscr{G}) & \to & H^{q-1}(X, \mathscr{C}_1) & \to & H^q(X, f_*\mathbb{Z}/n) & \to & 0 \\ \downarrow & & \downarrow & & \downarrow & & \downarrow \\ H^{q-1}(X_0, \mathscr{G}|_{X_0}) & \to & H^{q-1}(X_0, \mathscr{C}_1|_{X_0}) & \to & H^q(X_0, (f_*\mathbb{Z}/n)|_{X_0}) & \to & H^q(X_0, \mathscr{G}|_{X_0}). \end{array}$$

By the induction hypothesis, the first two vertical arrows are bijective. It follows that the third vertical arrow is injective. We have shown that it is surjective at the beginning. So it is bijective. Let $\mathscr{C}_2$ be the cokernel of $\mathscr{F} \hookrightarrow f_*\mathbb{Z}/n$. Consider the commutative diagram

$$\begin{array}{cccccc} H^{q-1}(X, f_*\mathbb{Z}/n) & \to & H^{q-1}(X, \mathscr{C}_2) & \to & H^q(X, \mathscr{F}) & \to \\ \downarrow & & \downarrow & & \downarrow & \\ H^{q-1}(X_0, (f_*\mathbb{Z}/n)|_{X_0}) & \to & H^{q-1}(X_0, \mathscr{C}_2|_{X_0}) & \to & H^q(X_0, \mathscr{F}|_{X_0}) & \to \end{array}$$

$$\begin{array}{cccc} \to & H^q(X, f_*\mathbb{Z}/n) & \to & H^q(X, \mathscr{C}_2) \\ & \downarrow & & \downarrow \\ \to & H^q(X_0, (f_*\mathbb{Z}/n)|_{X_0}) & \to & H^q(X_0, \mathscr{C}_2|_{X_0}). \end{array}$$

By the induction hypothesis, the first two vertical arrows are bijective. We have just shown the fourth vertical arrow is bijective. So the third vertical arrow is injective. This is true for any constructible sheaf of $\mathbb{Z}/n$-modules $\mathscr{F}$. In particular, the last vertical arrow is injective. It then follows that the third vertical arrow is bijective for any constructible sheaf of $\mathbb{Z}/n$-modules $\mathscr{F}$. By 5.8.5 (ii), 5.8.8 and 5.9.2, the canonical homomorphism $H^q(X, \mathscr{F}) \to H^q(X_0, \mathscr{F}|_{X_0})$ is bijective for any torsion sheaf $\mathscr{F}$. □

By 7.3.8 and 7.3.9, to prove 7.3.3 in the case where A is noetherian, it suffices to prove the following:

Lemma 7.3.10. *Let A be a strictly local noetherian ring, $f : X \to \operatorname{Spec} A$ a proper morphism, and X_0 the fiber of f over the closed point of* $\operatorname{Spec} A$. *Suppose* $\dim X_0 \leq 1$. *Then the canonical homomorphism*

$$H^q(X, \mathbb{Z}/n) \to H^q(X_0, \mathbb{Z}/n)$$

is bijective for $q = 0$ and surjective for any $q \geq 1$.

Proof. By 7.2.13, we have $H^q(X_0, \mathbb{Z}/n) = 0$ for any $q \geq 3$. So it suffices to treat the cases where $q = 0, 1, 2$.

To prove that $H^0(X, \mathbb{Z}/n) \to H^0(X_0, \mathbb{Z}/n)$ is bijective, it suffices to show that the set of connected components of X is in one-to-one correspondence with the set of connected components of X_0. Connected components of a noetherian scheme are minimal open and closed subsets. It suffices to show that the set of open and closed subsets of X is in one-to-one correspondence with the set of open and closed subsets of X_0. But the set of open and closed subsets of X (resp. X_0) is in one-to-one correspondence with the set $\mathrm{Idem}\,\Gamma(X, \mathscr{O}_X)$ (resp. $\mathrm{Idem}\,\Gamma(X_0, \mathscr{O}_{X_0})$) of idempotent elements in $\Gamma(X, \mathscr{O}_X)$ (resp. $\Gamma(X_0, \mathscr{O}_{X_0})$). So it suffices to show that the canonical map

$$\mathrm{Idem}\,\Gamma(X, \mathscr{O}_X) \to \mathrm{Idem}\,\Gamma(X_0, \mathscr{O}_{X_0})$$

is bijective. Since X is proper over $\mathrm{Spec}\, A$, $\Gamma(X, \mathscr{O}_X)$ is a finite A-algebra by [Fu (2006)] 2.5.3 or [EGA] III 3.2.3. By 2.8.3 (iii), the canonical map

$$\mathrm{Idem}\,\Gamma(X, \mathscr{O}_X) \to \mathrm{Idem}\,\Gamma(X, \mathscr{O}_X)/\mathfrak{m}\Gamma(X, \mathscr{O}_X)$$

is bijective, where $\mathfrak{m}$ is the maximal ideal of A. For any positive integer n, since $\mathrm{Spec}\,(\Gamma(X, \mathscr{O}_X)/\mathfrak{m}^n\Gamma(X, \mathscr{O}_X))$ has the same underlying topological space as $\mathrm{Spec}\,(\Gamma(X, \mathscr{O}_X)/\mathfrak{m}\Gamma(X, \mathscr{O}_X))$, the canonical map

$$\mathrm{Idem}\,\Gamma(X, \mathscr{O}_X)/\mathfrak{m}^n\Gamma(X, \mathscr{O}_X) \to \mathrm{Idem}\,\Gamma(X, \mathscr{O}_X)/\mathfrak{m}\Gamma(X, \mathscr{O}_X)$$

is bijective. It follows that the canonical map

$$\mathrm{Idem}\,\Gamma(X, \mathscr{O}_X) \to \mathrm{Idem}\,\Gamma(X, \mathscr{O}_X)^\wedge$$

is bijective, where

$$\Gamma(X, \mathscr{O}_X)^\wedge = \varprojlim_n \Gamma(X, \mathscr{O}_X)/\mathfrak{m}^n\Gamma(X, \mathscr{O}_X).$$

By [Fu (2006)] 2.5.6 or [EGA] III 4.1.7, we have

$$\Gamma(X, \mathscr{O}_X)^\wedge \cong \varprojlim_n \Gamma(X_n, \mathscr{O}_{X_n}),$$

where $X_n = X \otimes_A A/\mathfrak{m}^n$ for any n. It follows that

$$\mathrm{Idem}\,\Gamma(X, \mathscr{O}_X)^\wedge \cong \varprojlim_n \mathrm{Idem}\,\Gamma(X_n, \mathscr{O}_{X_n}).$$

But X_n has the same underlying topological space as X_0. So we have

$$\mathrm{Idem}\,\Gamma(X_n, \mathscr{O}_{X_n}) \cong \mathrm{Idem}\,\Gamma(X_0, \mathscr{O}_{X_0})$$

and hence

$$\varprojlim_n \mathrm{Idem}\,\Gamma(X_n, \mathscr{O}_{X_n}) \cong \mathrm{Idem}\,\Gamma(X_0, \mathscr{O}_{X_0}).$$

It follows that

$$\operatorname{Idem}\Gamma(X, \mathscr{O}_X) \cong \operatorname{Idem}\Gamma(X_0, \mathscr{O}_{X_0}).$$

This proves our assertion.

To prove that $H^1(X, \mathbb{Z}/n) \to H^1(X_0, \mathbb{Z}/n)$ is surjective, we may assume that X and X_0 are connected by the above discussion. By 5.7.20, we have

$$\begin{aligned} H^1(X, \mathbb{Z}/n) &\cong \text{cont.Hom}(\pi_1(X), \mathbb{Z}/n), \\ H^1(X_0, \mathbb{Z}/n) &\cong \text{cont.Hom}(\pi_1(X_0), \mathbb{Z}/n). \end{aligned}$$

So it suffices to show that the canonical homomorphism $\pi_1(X_0) \to \pi_1(X)$ is bijective. This follows from 7.3.13 below.

Finally we prove that $H^2(X, \mathbb{Z}/n) \to H^2(X_0, \mathbb{Z}/n)$ is surjective. Let p be the characteristic of the residue field of A. By 7.2.11, we have $H^2(X_0, \mathbb{Z}/p) = 0$. For any positive integer r, $\mathbb{Z}/p^r$ has a finite filtration with successive quotients $\mathbb{Z}/p$. It follows that $H^2(X_0, \mathbb{Z}/p^r) = 0$ for any r. So to prove that $H^2(X, \mathbb{Z}/n) \to H^2(X_0, \mathbb{Z}/n)$ is surjective, we may assume that n is relatively prime to p, and prove that $H^2(X, \boldsymbol{\mu}_{n,X}) \to H^2(X_0, \boldsymbol{\mu}_{n,X_0})$ is surjective. We have a commutative diagram

$$\begin{array}{ccccccc} \cdots \to \operatorname{Pic}(X) & \to & H^2(X, \boldsymbol{\mu}_{n,X}) & \to & H^2(X, \mathscr{O}^*_{X_{\text{et}}}) & \to & \cdots \\ \downarrow & & \downarrow & & \downarrow & & \\ \cdots \to \operatorname{Pic}(X_0) & \to & H^2(X_0, \boldsymbol{\mu}_{n,X_0}) & \to & H^2(X_0, \mathscr{O}^*_{X_{0,\text{et}}}) & \to & \cdots . \end{array}$$

So it suffices to show the canonical homomorphisms

$$\operatorname{Pic}(X) \to \operatorname{Pic}(X_0), \quad \operatorname{Pic}(X_0) \to H^2(X_0, \boldsymbol{\mu}_{n,X_0})$$

are surjective. The first homomorphism is surjective by 7.3.15 below. By 7.2.9 (i), $\operatorname{Pic}(X_{0\text{red}}) \to H^2(X_{0\text{red}}, \boldsymbol{\mu}_{X_{n,0\text{red}}})$ is surjective. By 5.7.2 (i), we have

$$H^2(X_0, \boldsymbol{\mu}_{n,X_0}) \cong H^2(X_{0\text{red}}, \boldsymbol{\mu}_{n,X_{0\text{red}}}).$$

So to prove that $\operatorname{Pic}(X_0) \to H^2(X_0, \boldsymbol{\mu}_{n,X_0})$ is surjective, it suffices to prove that $\operatorname{Pic}(X_0) \to \operatorname{Pic}(X_{0,\text{red}})$ is surjective. This follows from 7.3.14 below.□

Proposition 7.3.11. *Let A be a noetherian ring, I an ideal of A such that A is complete with respect to the I-adic topology, $A_0 = A/I$, $S = \operatorname{Spec} A$, $S_0 = \operatorname{Spec} A_0$, X a proper S-scheme, and $X_0 = X \times_S S_0$. Then the functor $Y \mapsto Y \times_X X_0$ from the category of etale covering spaces of X to the category of etale covering spaces of X_0 is an equivalence of categories.*

Proof. Let Y and Y' be two etale covering spaces of X, and let

$$S_n = \mathrm{Spec}\, A/I^{n+1},\ X_n = X \times_S S_n,\ Y_n = Y \times_S S_n,\ Y'_n = Y' \times_S S_n.$$

By [Fu (2006)] 2.7.20 or [EGA] III 5.4.1, we have

$$\mathrm{Hom}_X(Y, Y') \cong \varprojlim_n \mathrm{Hom}_{X_n}(Y_n, Y'_n).$$

By 2.3.12, the canonical maps

$$\mathrm{Hom}_{X_{n+1}}(Y_{n+1}, Y'_{n+1}) \to \mathrm{Hom}_{X_n}(Y_n, Y'_n)$$

are bijective for all n. So

$$\mathrm{Hom}_X(Y, Y') \cong \mathrm{Hom}_{X_0}(Y_0, Y'_0)$$

and the functor $Y \mapsto Y \times_X X_0$ is fully faithful. Let $f_0 : Y_0 \to X_0$ be an etale covering space. By 2.3.12, there exist etale morphisms $f_n : Y_n \to X_n$ such that

$$Y_n \cong Y_{n+1} \times_{X_{n+1}} X_n$$

for all n. Since X_0 and X_n have the same underlying topological spaces and f_0 is proper, f_n is also proper. Since f_n is etale, it is quasi-finite. By [Fu (2006)] 2.5.12 or [EGA] III 4.4.2, f_n is finite. Let $\mathscr{B}_n = f_{n*}\mathscr{O}_{X_n}$. Then $\mathscr{B}_n$ is a coherent $\mathscr{O}_{X_n}$-algebra and $Y_n = \mathbf{Spec}\, \mathscr{B}_n$. We have

$$\mathscr{B}_{n+1} \otimes_{\mathscr{O}_{X_{n+1}}} \mathscr{O}_{X_n} \cong \mathscr{B}_n.$$

By Grothendieck Existence Theorem ([Fu (2006)] 2.7.11 or [EGA] III 5.1.4), there exists a coherent $\mathscr{O}_X$-module $\mathscr{B}$ such that

$$\mathscr{B}_n \cong \mathscr{B} \otimes_{\mathscr{O}_X} \mathscr{O}_{X_n}.$$

Defining an $\mathscr{O}_X$-algebra structure on $\mathscr{B}$ is equivalent to defining a morphism

$$\mathscr{B} \otimes_{\mathscr{O}_X} \mathscr{B} \to \mathscr{B}$$

which makes various diagrams expressing the associative law, the commutative law and the distributive law to commute, and that these diagrams involve only tensor products of $\mathscr{B}$. Using [Fu (2006)] 1.5.19 and 2.7.9, or [EGA] I 10.11.4 and III 5.1.3, one can show that $\mathscr{B}$ has an $\mathscr{O}_X$-algebra structure such that $\mathscr{B}_n \cong \mathscr{B} \otimes_{\mathscr{O}_X} \mathscr{O}_{X_n}$ are isomorphisms of $\mathscr{O}_{X_n}$-algebras. For any closed point x in X, since X is proper over S, x is above the closed point of S. Since $f_0 : \mathbf{Spec}\, \mathscr{B}_0 \to X_0$ is etale, $\mathscr{B}_0 \otimes_{\mathscr{O}_{X_0}} k(x)$ is a direct product of finite separable extensions of $k(x)$. We have

$$\mathscr{B} \otimes_{\mathscr{O}_X} k(x) \cong \mathscr{B}_0 \otimes_{\mathscr{O}_{X_0}} k(x).$$

We will prove in a moment that $\mathscr{B}$ is locally free as an $\mathscr{O}_X$-module. So the morphism $\mathbf{Spec}\,\mathscr{B} \to X$ is etale. It is an etale covering space of X inducing $f_0 : Y_0 \to X_0$. So the functor $Y \mapsto Y \times_X X_0$ is essentially surjective.

Finally let us prove that $\mathscr{B}$ is locally free as an $\mathscr{O}_X$-module. It suffices to show that the functor

$$\mathscr{G} \mapsto \mathscr{H}om_{\mathscr{O}_X}(\mathscr{B}, \mathscr{G})$$

is exact on the category of coherent $\mathscr{O}_X$-modules. Let $\widehat{X}$ be the formal completion of X with respect to $I^\sim \mathscr{O}_X$. For any coherent $\mathscr{O}_X$-module $\mathscr{G}$, let $\widehat{\mathscr{G}}$ be the formal completion of $\mathscr{G}$ with respect to $I^\sim \mathscr{O}_X$. By [Fu (2006)] 1.5.13 (i) and 2.7.9, or [EGA] I 10.8.8 (i) and III 5.1.3, the functor

$$\mathscr{G} \mapsto \widehat{\mathscr{G}}$$

is exact and faithful. So to prove that the functor $\mathscr{G} \mapsto \mathscr{H}om_{\mathscr{O}_X}(\mathscr{B}, \mathscr{G})$ is exact, it suffices to show that the functor

$$\mathscr{G} \mapsto \mathscr{H}om_{\mathscr{O}_X}(\mathscr{B}, \mathscr{G})^\wedge$$

is exact. By [Fu (2006)] 1.5.13 (iii) and 1.5.19, or [EGA] I 10.8.10 and 10.11.7, we have

$$\begin{aligned}\mathscr{H}om_{\mathscr{O}_X}(\mathscr{B}, \mathscr{G})^\wedge &\cong \mathscr{H}om_{\mathscr{O}_{\widehat{X}}}(\widehat{\mathscr{B}}, \widehat{\mathscr{G}}) \\ &\cong \varprojlim_n \mathscr{H}om_{\mathscr{O}_{X_n}}(\mathscr{B}_n, \mathscr{G}_n),\end{aligned}$$

where $\mathscr{G}_n$ is the inverse image of $\mathscr{G}$ on X_n. By our construction, each $\mathscr{B}_n$ is locally free. Moreover, we can find an open covering $\{U_\lambda\}$ of X_0 such that $\mathscr{B}_n|_{U_\lambda}$ is free for each λ. Here we regard each U_λ as an open subset of X_n. Using this fact, one can prove that the functor

$$\mathscr{G} \mapsto \varprojlim_n \mathscr{H}om_{\mathscr{O}_{X_n}}(\mathscr{B}_n, \mathscr{G}_n)$$

is exact. This proves our assertion. □

Theorem 7.3.12 (Artin's Approximation Theorem). *Let A be the henselization at a prime ideal of a finitely generated algebra over a field or over an excellent discrete valuation ring, let $\mathfrak{m}$ be the maximal ideal of A, let $\widehat{A}$ be the $\mathfrak{m}$-adic completion of A, and let*

$$f_1(Y_1, \ldots, Y_n), \ldots, f_m(Y_1, \ldots, Y_n) \in A[Y_1, \ldots, Y_n].$$

If there exist $\hat{y}_1, \ldots, \hat{y}_n \in \widehat{A}$ such that

$$f_1(\hat{y}_1, \ldots, \hat{y}_n) = \cdots = f_m(\hat{y}_1, \ldots, \hat{y}_n) = 0,$$

then for any given positive integer N, there exist $y_1, \ldots, y_n \in A$ such that

$$\begin{aligned}&y_i \equiv \hat{y}_i \bmod \mathfrak{m}^N \widehat{A}, \\ &f_1(y_1, \ldots, y_n) = \cdots = f_m(y_1, \ldots, y_n) = 0.\end{aligned}$$

For a proof of Artin's approximation theorem, confer [Artin (1969)] or [Bosch, Lütkebohmert and Raynaud (1990)] 3.6.

Lemma 7.3.13. *Let A be a noetherian henselian local ring, $f : X \to \operatorname{Spec} A$ a proper morphism, and X_0 the fiber of f over the closed point of $\operatorname{Spec} A$. The functor $Y \mapsto Y \times_X X_0$ from the category of etale covering spaces of X to the category of etale covering spaces of X_0 is an equivalence of categories.*

Proof. We can write $A = \varinjlim_\lambda B_\lambda$, where each B_λ is a henselization at a prime ideal of an algebra finitely generated over $\mathbb{Z}$. By 1.10.9 and 1.10.10 (xi), we may assume that $f : X \to \operatorname{Spec} A$ can be descended down to an inverse system of proper morphisms $f_\lambda : X_\lambda \to \operatorname{Spec} B_\lambda$. For each λ, let $X_{\lambda 0}$ be the fiber of f_λ over the closed point of $\operatorname{Spec} B_\lambda$. By 2.3.7 and 1.10.10 (iv), any etale covering space of X (resp. X_0) can be descended down to an etale covering space of X_λ (resp. $X_{\lambda 0}$) for sufficiently large λ. Moreover, if Y and Y' are two etale covering spaces of X which can be descended down to etale covering spaces Y_λ and Y'_λ of X_λ, respectively, we have

$$\operatorname{Hom}_X(Y, Y') \cong \varinjlim_{\mu \geq \lambda} \operatorname{Hom}_{X_\mu}(Y_\mu, Y'_\mu),$$

$$\operatorname{Hom}_{X_0}(Y_0, Y'_0) \cong \varinjlim_{\mu \geq \lambda} \operatorname{Hom}_{X_{\mu 0}}(Y_{\mu 0}, Y'_{\mu 0}),$$

where Y_μ (resp. Y'_μ) are induced by Y_λ (resp. Y'_λ) by the base changes $X_\mu \to X_\lambda$, Y_0 and Y'_0 (resp. $Y_{\mu 0}$ and $Y'_{\mu 0}$) are the fibers of Y and Y' (resp. Y_μ and Y'_μ) over the closed point of $\operatorname{Spec} A$ (resp. $\operatorname{Spec} B_\mu$). It follows that to prove the lemma for A, it suffices to prove the lemma for each B_λ. We may thus assume that A is the henselization at a prime ideal of a finitely generated algebra over $\mathbb{Z}$ so that Artin's Approximation Theorem is applicable.

Let Y and Y' be two etale covering spaces of X. Let us prove that

$$\operatorname{Hom}_X(Y, Y') \to \operatorname{Hom}_{X_0}(Y_0, Y'_0)$$

is injective. Supppose $g_1, g_2 \in \operatorname{Hom}_X(Y, Y')$ induce the same element in $\operatorname{Hom}_{X_0}(Y_0, Y'_0)$. Denote by $\mathfrak{m}$ the maximal ideal of A, and by $\widehat{A}$ the $\mathfrak{m}$-adic completion of A. By 7.3.11, g_1 and g_2 induce the same element in $\operatorname{Hom}_{\overline{X}}(\overline{Y}, \overline{Y}')$, where

$$\overline{X} = X \otimes_A \widehat{A}, \quad \overline{Y} = Y \otimes_A \widehat{A}, \quad \overline{Y}' = Y' \otimes_A \widehat{A}.$$

Write $\widehat{A} = \varinjlim_\alpha A_\alpha$, where $\{A_\alpha\}$ is the family of subalgebras of $\widehat{A}$ finitely generated over A. Then g_1 and g_2 induce the same element in

$\mathrm{Hom}_{X_\alpha}(Y_\alpha, Y'_\alpha)$ for a sufficiently large A_α, where

$$X_\alpha = X \otimes_A A_\alpha, \quad Y_\alpha = Y \otimes_A A_\alpha, \quad Y'_\alpha = Y' \otimes_A A_\alpha.$$

Fix an isomorphism of A-algebras

$$A[Y_1, \ldots, Y_n]/(f_1, \ldots, f_m) \cong A_\alpha$$

for some $f_1, \ldots, f_m \in A[Y_1, \ldots, Y_n]$. The inclusion $A_\alpha \hookrightarrow \widehat{A}$ induces an A-homomorphism

$$A[Y_1, \ldots, Y_n]/(f_1, \ldots, f_m) \to \widehat{A}.$$

Let $\hat{y}_1, \ldots, \hat{y}_n \in \widehat{A}$ be the images of $Y_1, \ldots, Y_n$ under this homomorphism, respectively. Then

$$f_1(\hat{y}_1, \ldots, \hat{y}_n) = \cdots = f_m(\hat{y}_1, \ldots, \hat{y}_n) = 0.$$

By Artin's Approximation Theorem, there exist $y_1, \ldots, y_n \in A$ such that

$$\begin{aligned} &y_i \equiv \hat{y}_i \operatorname{mod} \mathfrak{m}\widehat{A}, \\ &f_1(y_1, \ldots, y_n) = \cdots = f_m(y_1, \ldots, y_n) = 0. \end{aligned}$$

We have a homomorphism of A-algebras

$$A[Y_1, \ldots, Y_n]/(f_1, \ldots, f_m) \to A$$

which maps Y_i to y_i. It induces a homomorphism of A-algebras

$$\phi : A_\alpha \to A.$$

We have

$$X \cong X_\alpha \otimes_{A_\alpha, \phi} A, \quad Y \cong Y_\alpha \otimes_{A_\alpha, \phi} A, \quad Y' \cong Y'_\alpha \otimes_{A_\alpha, \phi} A.$$

Since g_1 and g_2 induce the same element in $\mathrm{Hom}_{X_\alpha}(Y_\alpha, Y'_\alpha)$, it follows that g_1 and g_2 are the same in $\mathrm{Hom}_X(Y, Y')$.

Next we show that $\mathrm{Hom}_X(Y, Y') \to \mathrm{Hom}_{X_0}(Y_0, Y'_0)$ is surjective. Let $g_0 \in \mathrm{Hom}_{X_0}(Y_0, Y'_0)$. By 7.3.11, there exists $\bar{g} \in \mathrm{Hom}_{\overline{X}}(\overline{Y}, \overline{Y}')$ inducing g_0. By 1.10.9, $\bar{g}$ can be descended down to a morphism $g_\alpha \in \mathrm{Hom}_{X_\alpha}(Y_\alpha, Y'_\alpha)$ for a sufficiently large A_α. Let $g \in \mathrm{Hom}_X(Y, Y')$ be the morphism induced by g_α by the base change $\phi : A_\alpha \to A$. Then g is the preimage of g_0.

Finally let Y_0 be an etale covering space of X_0. By 7.3.11, it is induced by an etale covering space $\overline{Y}$ of $\overline{X}$. By 1.10.9, 1.10.10 (iv) and 2.3.7, we may descend $\overline{Y}$ down to an etale covering space Y_α of X_α for a sufficiently large A_α. Let Y be the etale covering space of X induced by Y_α by base change $\phi : A_\alpha \to A$. Then Y_0 is induced by Y. □

Lemma 7.3.14. *Let X be a noetherian scheme of dimension ≤ 1, and let $\mathscr{I}$ be a coherent $\mathscr{O}_X$-ideal satisfying $\mathscr{I}^n = 0$ for some nonnegative integer n. Then the canonical homomorphism* $\mathrm{Pic}\,(X) \to \mathrm{Pic}\,(\mathbf{Spec}\,(\mathscr{O}_X/\mathscr{I}))$ *is surjective.*

Proof. First consider the case where $\mathscr{I}^2 = 0$. We have a short exact sequence

$$0 \to \mathscr{I} \stackrel{1+\mathrm{id}}{\to} \mathscr{O}_X^* \to (\mathscr{O}_X/\mathscr{I})^* \to 0,$$

where

$$1 + \mathrm{id} : \mathscr{I} \to \mathscr{O}_X^*$$

is defined by $(1+\mathrm{id})(s) = 1 + s$ for any section s of $\mathscr{I}$. This is a morphism of sheaves of abelian groups since $\mathscr{I}^2 = 0$. The above short exact sequence induces a long exact sequence

$$\cdots \to H^1_{\mathrm{Zar}}(X, \mathscr{O}_X^*) \to H^1_{\mathrm{Zar}}(X, (\mathscr{O}_X/\mathscr{I})^*) \to H^2_{\mathrm{Zar}}(X, \mathscr{I}) \to \cdots.$$

We have

$$H^1_{\mathrm{Zar}}(X, \mathscr{O}_X^*) \cong \mathrm{Pic}\,(X), \quad H^1_{\mathrm{Zar}}(X, (\mathscr{O}_X/\mathscr{I})^*) \cong \mathrm{Pic}\,(\mathbf{Spec}\,(\mathscr{O}_X/\mathscr{I})).$$

Since $\dim X \leq 1$, we have $H^2_{\mathrm{Zar}}(X, \mathscr{I}) = 0$ by [Hartshorne (1977)] III 2.7. So $\mathrm{Pic}\,(X) \to \mathrm{Pic}\,(\mathbf{Spec}\,(\mathscr{O}_X/\mathscr{I}))$ is surjective.

In general, $\mathrm{Pic}\,(\mathbf{Spec}\,(\mathscr{O}_X/\mathscr{I}^{k+1})) \to \mathrm{Pic}\,(\mathbf{Spec}\,(\mathscr{O}_X/\mathscr{I}^k))$ is surjective for all k. It follows that $\mathrm{Pic}\,(X) \to \mathrm{Pic}\,(\mathbf{Spec}\,(\mathscr{O}_X/\mathscr{I}))$ is surjective if $\mathscr{I}^n = 0$ for some nonnegative integer n. □

Lemma 7.3.15. *Let A be a noetherian henselian local ring, $f : X \to$ $\mathrm{Spec}\,A$ a proper morphism, and X_0 the fiber of f over the closed point of* $\mathrm{Spec}\,A$. *Suppose* $\dim X_0 \leq 1$. *Then the canonical homomorphism* $\mathrm{Pic}(X) \to \mathrm{Pic}(X_0)$ *is surjective.*

Proof. As in the proof of 7.3.13, we may assume that A is the henselization at a prime ideal of an algebra finitely generated over $\mathbb{Z}$ so that Artin's Approximation Theorem is applicable. Let $\mathfrak{m}$ be the maximal ideal of A, let $\widehat{A}$ be the $\mathfrak{m}$-adic completion of A, let $X_n = X \otimes_A A/\mathfrak{m}^{n+1}$, and let $\overline{X} = X \otimes_A \widehat{A}$. Given an invertible $\mathscr{O}_{X_0}$-module $\mathscr{L}_0$, by 7.3.14, for each n, we may find an invertible $\mathscr{O}_{X_n}$-module $\mathscr{L}_n$ such that its inverse image in X_{n+1} is isomorphic to $\mathscr{L}_{n+1}$. By Grothendieck's Existence Theorem ([Fu (2006)] 2.7.11 or [EGA] III 5.1.4), there exists a coherent $\mathscr{O}_{\overline{X}}$-module $\overline{\mathscr{L}}$ such that its inverse image on X_n is isomorphic to $\mathscr{L}_n$ for each n. As in the proof of 7.3.11, one can show the functor $\mathscr{H}om_{\mathscr{O}_{\overline{X}}}(\overline{\mathscr{L}}, -)$ is exact on the category of coherent $\mathscr{O}_{\overline{X}}$-modules. So $\overline{\mathscr{L}}$ is an invertible $\mathscr{O}_{\overline{X}}$-module. As in the proof of 7.3.13, write $\widehat{A} = \varinjlim_\alpha A_\alpha$, where $\{A_\alpha\}$ is the family of subalgebras of $\widehat{A}$ finitely generated over A. Let $X_\alpha = X \otimes_A A_\alpha$. By 1.10.2, we may descend $\overline{\mathscr{L}}$ to an invertible $\mathscr{O}_{X_\alpha}$-module $\mathscr{L}_\alpha$ for a sufficiently large A_α. Let $\mathscr{L}$ be the invertible $\mathscr{O}_X$-module induced by $\mathscr{L}_\alpha$ by the base change $\phi : A_\alpha \to A$ constructed in the proof of 7.3.13. Then $\mathscr{L}$ is a preimage of $\mathscr{L}_0$ under the canonical homomorphism $\mathrm{Pic}(X) \to \mathrm{Pic}(X_0)$. □

Finally, we prove that the two ways of defining the base change morphism at the beginning of this section are the same.

Proposition 7.3.16. *Let*

$$\begin{array}{ccc} X' & \stackrel{g'}{\to} & X \\ {\scriptstyle f'}\downarrow & & \downarrow{\scriptstyle f} \\ Y' & \stackrel{g}{\to} & Y \end{array}$$

be a commutative diagram, and let $\mathscr{F}$ be a sheaf on X. Then the morphisms

$$g^* f_* \mathscr{F} \to f'_* g'^* \mathscr{F}$$

defined by the following two ways are the same:

(i) *Take the composite*

$$f_* \mathscr{F} \stackrel{f_*(\mathrm{adj})}{\to} f_* g'_* g'^* \mathscr{F} \cong g_* f'_* g'^* \mathscr{F},$$

and define $g^* f_* \mathscr{F} \to f'_* g'^* \mathscr{F}$ to be the morphism induced by the composite by adjunction.

(ii) *Take the composite*

$$f'^* g^* f_* \mathscr{F} \cong g'^* f^* f_* \mathscr{F} \stackrel{g'^*(\mathrm{adj})}{\to} g'^* \mathscr{F},$$

and define $g^* f_* \mathscr{F} \to f'_* g'^* \mathscr{F}$ to be the morphism induced by the composite by adjunction.

Proof. Define a category $\mathscr{C}$ as follows: Objects in $\mathscr{C}$ are pairs $(Z, \mathscr{G})$ such that Z is a scheme and $\mathscr{G}$ is a sheaf on Z. Given two objects $(Z_1, \mathscr{G}_1)$ and $(Z_2, \mathscr{G}_2)$ in $\mathscr{C}$, a morphism $(Z_1, \mathscr{G}_1) \to (Z_2, \mathscr{G}_2)$ in $\mathscr{C}$ is a pair (h, ϕ) such that $h : Z_1 \to Z_2$ is a morphism of schemes and $\phi : \mathscr{G}_2 \to h_* \mathscr{G}_1$ is a morphism of sheaves. We define composites of morphisms in $\mathscr{C}$ in the obvious way. Note that we can also define a morphism $(Z_1, \mathscr{G}_1) \to (Z_2, \mathscr{G}_2)$ in $\mathscr{C}$ to be a pair (h, ψ) such that $h : Z_1 \to Z_2$ is a morphism of schemes and $\psi : h^* \mathscr{G}_2 \to \mathscr{G}_1$ is a morphism of sheaves. Given a morphism $h : Z_1 \to Z_2$ of schemes and sheaves $\mathscr{G}_1$ on Z_1 and $\mathscr{G}_2$ on Z_2, denote by $\mathrm{Hom}_h((Z_1, \mathscr{G}_1), (Z_2, \mathscr{G}_2))$ the set of morphisms $(h,\phi):(Z_1,\mathscr{G}_1)\to(Z_2,\mathscr{G}_2)$ in $\mathscr{C}$ whose first components are h. Given two morphism

$$Z_1\stackrel{h_1}{\to}Z_2\stackrel{h_2}{\to}Z_3$$

of schemes and sheaves $\mathscr{G}_i$ $(i=1,2,3)$ on Z_i, we have one-to-one correspondences

$$\begin{aligned}\mathrm{Hom}_{h_1}((Z_1,\mathscr{G}_1),(Z_2,h_2^*\mathscr{G}_3))&\cong\mathrm{Hom}_{h_2h_1}((Z_1,\mathscr{G}_1),(Z_3,\mathscr{G}_3)),\\ \mathrm{Hom}_{h_2}((Z_2,h_{1*}\mathscr{G}_1),(Z_3,\mathscr{G}_3))&\cong\mathrm{Hom}_{h_2h_1}((Z_1,\mathscr{G}_1),(Z_3,\mathscr{G}_3)).\end{aligned}$$

In the notation of the proposition, we have one-to-one correspondences

$$\begin{aligned}&\mathrm{Hom}(g^*f_*\mathscr{F},f'_*g'^*\mathscr{F})\\ \cong\ &\mathrm{Hom}_{\mathrm{id}_{Y'}}((Y',f'_*g'^*\mathscr{F}),(Y',g^*f_*\mathscr{F}))\\ \cong\ &\mathrm{Hom}_g((Y',f'_*g'^*\mathscr{F}),(Y,f_*\mathscr{F}))\\ \cong\ &\mathrm{Hom}_{gf'}((X',g'^*\mathscr{F}),(Y,f_*\mathscr{F}))\\ \cong\ &\mathrm{Hom}_{fg'}((X',g'^*\mathscr{F}),(Y,f_*\mathscr{F})).\end{aligned}$$

Let α be the morphism

$$(g',\mathrm{id}_{g'^*\mathscr{F}})\in\mathrm{Hom}_{g'}((X',g'^*\mathscr{F}),(X,\mathscr{F})),$$

and let β be the morphism

$$(f,\mathrm{id}_{f_*\mathscr{F}})\in\mathrm{Hom}_f((X,\mathscr{F}),(Y,f_*\mathscr{F})).$$

Let us prove that the two morphisms $g^*f_*\mathscr{F}\to f'_*g'^*\mathscr{F}$ defined in the proposition both correspond to the morphism

$$\beta\alpha\in\mathrm{Hom}_{fg'}((X',g'^*\mathscr{F}),(Y,f_*\mathscr{F}))$$

through the above one-to-one correspondences.

For any scheme Z and any morphism $\phi:\mathscr{G}_1\to\mathscr{G}_2$ of sheaves on Z, denote the morphism $(\mathrm{id},\phi):(Z,\mathscr{G}_2)\to(Z,\mathscr{G}_1)$ in $\mathscr{C}$ by $\phi^\star$. Let u be the composite

$$f_*\mathscr{F}\stackrel{f_*(\mathrm{adj})}{\to}f_*g'_*g'^*\mathscr{F}\cong g_*f'_*g'^*\mathscr{F},$$

and let $\theta_1:g^*f_*\mathscr{F}\to f'_*g'^*\mathscr{F}$ be the morphism defined in (i) of the proposition. We have a commutative diagram

$$\begin{array}{ccccccccc}(Y',f'_*g'^*\mathscr{F})&\to&(Y,g_*f'_*g'^*\mathscr{F})&\cong&(Y,f_*g'_*g'^*\mathscr{F})&\leftarrow&(X,g'_*g'^*\mathscr{F})&\leftarrow&(X',g'^*\mathscr{F})\\ \theta_1^\star\downarrow&&u^\star\downarrow&&\downarrow f_*(\mathrm{adj})^\star&&\downarrow\mathrm{adj}^\star&&\downarrow\alpha\\ (Y',g^*f_*\mathscr{F})&\to&(Y,f_*\mathscr{F})&=&(Y,f_*\mathscr{F})&\stackrel{\beta}{\leftarrow}&(X,\mathscr{F})&=&(X,\mathscr{F}),\end{array}$$

where the horizontal arrows are the canonical morphisms. Note that the composite of canonical morphisms

$$(X', g'^*\mathscr{F}) \to (X, g'_*g'^*\mathscr{F}) \to (Y, f_*g'_*g'^*\mathscr{F}) \cong (Y, g_*f'_*g'^*\mathscr{F})$$

coincides with the composite of canonical morphisms

$$(X', g'^*\mathscr{F}) \to (Y', f'_*g'^*\mathscr{F}) \to (Y, g_*f'_*g'^*\mathscr{F}).$$

So we have a commutative diagram

$$\begin{array}{ccc} (Y', f'_*g'^*\mathscr{F}) & \leftarrow & (X', g'^*\mathscr{F}) \\ \theta_1^\star \downarrow & & \downarrow \alpha \\ (Y', g^*f_*\mathscr{F}) \to (Y, f_*\mathscr{F}) & \stackrel{\beta}{\leftarrow} & (X, \mathscr{F}). \end{array}$$

This shows that θ_1 corresponds to $\beta\alpha$. Similarly, let v be the composite

$$f'^*g^*f_*\mathscr{F} \cong g'^*f^*f_*\mathscr{F} \stackrel{g'^*(\mathrm{adj})}{\to} g'^*\mathscr{F},$$

and let $\theta_2 : g^*f_*\mathscr{F} \to f'_*g'^*\mathscr{F}$ be the morphism defined in (ii) of the proposition. We have a commutative diagram

$$\begin{array}{ccccccccc} (Y', f'_*g'^*\mathscr{F}) & \leftarrow & (X', g'^*\mathscr{F}) & = & (X', g'^*\mathscr{F}) & \stackrel{\alpha}{\to} & (X, \mathscr{F}) & = & (X, \mathscr{F}) \\ \theta_2^\star \downarrow & & v^\star \downarrow & & \downarrow g'^*(\mathrm{adj})^\star & & \downarrow \mathrm{adj}^\star & & \downarrow \beta \\ (Y', g^*f_*\mathscr{F}) & \leftarrow & (X', f'^*g^*f_*\mathscr{F}) & \cong & (X', g'^*f^*f_*\mathscr{F}) & \to & (X, f^*f_*\mathscr{F}) & \to & (Y, f_*\mathscr{F}), \end{array}$$

where the horizontal arrows are the canonical morphisms. The composite of canonical morphisms

$$(X', f'^*g^*f_*\mathscr{F}) \cong (X', g'^*f^*f_*\mathscr{F}) \to (X, f^*f_*\mathscr{F}) \to (Y, f_*\mathscr{F})$$

coincides with the composite of canonical morphisms

$$(X', f'^*g^*f_*\mathscr{F}) \to (Y', g^*f_*\mathscr{F}) \to (Y, f_*\mathscr{F}).$$

So we have a commutative diagram

$$\begin{array}{ccc} (Y', f'_*g'^*\mathscr{F}) \leftarrow (X', g'^*\mathscr{F}) & \stackrel{\alpha}{\to} & (X, \mathscr{F}) \\ \theta_2^\star \downarrow & & \downarrow \beta \\ (Y', g^*f_*\mathscr{F}) & \to & (Y, f_*\mathscr{F}). \end{array}$$

This shows that θ_2 corresponds to $\beta\alpha$. $\square$

7.4 Cohomology with Proper Support

([SGA 4] XVII 3–6, [SGA $4\frac{1}{2}$] Arcata IV 5, 6.)

Throughout this section, we fix a scheme S. Let X and Y be S-schemes. An S-morphism $f : X \to Y$ is called *S-compactifiable* if Y is quasi-compact and quasi-separated and there exists a proper S-scheme P such that f is the composite of a separated quasi-finite morphism $X \to Y \times_S P$ and the projection $p_1 : Y \times_S P \to Y$. Note that f is then a quasi-compact separated morphism, and X is a quasi-compact quasi-separated scheme. We say that an S-scheme X is *compactifiable* if the structure morphism $X \to S$ is S-compactifiable.

$$\begin{array}{ccccc} X & \to & Y \times_S P & \stackrel{p_2}{\to} & P \\ & {\scriptstyle f}\searrow & {\scriptstyle p_1}\downarrow & & \downarrow \\ & & Y & \to & S \end{array}$$

A theorem of Nagata says that any separated morphism of finite type between noetherian schemes is compactifiable.

Proposition 7.4.1.

(i) *S-compactifiable morphisms are separated and of finite type.*

(ii) *Let $f : X \to Y$ be a separated quasi-finite S-morphism such that Y is quasi-compact and quasi-separated. Then f is S-compactifiable.*

(iii) *Let $f : X \to Y$ and $g : Y \to Z$ be two S-compactifiable morphisms. Then gf is S-compactifiable.*

(iv) *Let $f : X \to Y$ be an S-compactifiable morphism. For any morphism $Y' \to Y$ such that Y' is quasi-compact and quasi-separated, the base change $f' : X \times_Y Y' \to Y'$ is S-compactifiable.*

(v) *Let $f : X \to Y$ and $g : Y \to Z$ be two S-morphisms. Suppose g is quasi-compact and quasi-separated and gf is S-compactifiable. Then f is S-compactifiable.*

Proof. (i), (ii) and (iv) are clear. (v) follows from (iii) and (iv).

(iii) We can find proper S-schemes P and Q, and separated quasi-finite morphisms $X \to Y \times_S P$ and $Y \to Z \times_S Q$, so that their composites with the projections to the first components are f and g, respectively. Consider the commutative diagram

$$\begin{array}{ccccc} X & \to & Y \times_S P & \to & Z \times_S P \times_S Q \\ & \searrow & \downarrow & & \downarrow \\ & & Y & \to & Z \times_S Q \\ & & & \searrow & \downarrow \\ & & & & Z. \end{array}$$

Note that the square in the above diagram is Cartesian. All horizontal arrows are separated and quasi-finite, and all vertical arrows are projections. As $P \times_S Q$ is proper over S, gf is S-compactifiable. $\square$

Let $f : X \to Y$ be an S-compactifiable morphism. Then we can find a proper S-scheme P and a separated quasi-finite morphism $g : X \to Y \times_S P$ such that $f = p_1 g$. By the Zariski Main Theorem in [EGA] IV 18.12.13, (we treat the noetherian case in 1.10.13), we can find a finite morphism $\bar{g}$ and an open immersion j such that $g = \bar{g}j$. By 7.4.1 (ii), $\bar{g}$ is S-compactifiable, and by 7.4.1 (iii), $p_1\bar{g}$ is S-compactifiable. We have $f = (p_1\bar{g})j$. So any S-compactifiable morphism is the composite of an open immersion and a proper S-compactifiable morphism.

Let $X \xrightarrow{f} Y$ be a compactifiable S-morphism. A *compactification* of f is a factorization of f as a composite

$$X \overset{j}{\hookrightarrow} \overline{X} \xrightarrow{\bar{f}} Y$$

such that j is an open immersion, and $\bar{f}$ is a proper S-compactifiable morphism. Note that j is necessarily an S-compactifiable morphism. A morphism from the above compactification of f to another compactification

$$X \overset{j'}{\hookrightarrow} \overline{X}' \xrightarrow{\bar{f}'} Y$$

is a morphism $\phi : \overline{X} \to \overline{X}'$ such that $\phi j = j'$ and $\bar{f}'\phi = \bar{f}$. If such a morphism exists, we say the first compactification dominates the second one.

Let $X \xrightarrow{f} Y \xrightarrow{g} Z$ be S-compactifiable morphisms. A compactification of (f, g) is a commutative diagram

$$\begin{array}{ccccc} X & \hookrightarrow & \overline{X} & \hookrightarrow & \overline{X}' \\ & \searrow & \downarrow & & \downarrow \\ & & Y & \hookrightarrow & \overline{Y} \\ & & & \searrow & \downarrow \\ & & & & Z \end{array}$$

such that horizontal arrows are open immersions, and vertical arrows are proper S-compactifiable morphisms. Note that all arrows in the above diagram are necessarily S-compactifiable morphisms. We leave it for the reader to define morphisms between compactifications of (f, g).

Proposition 7.4.2. *Let $X \xrightarrow{f} Y$ (resp. $X \xrightarrow{f} Y \xrightarrow{g} Z$) be S-compactifiable morphism(s).*

(i) *There exists a compactification for f (resp. (f, g)).*

(ii) *Given two compactifications of f, there exists a compactification dominating both of them.*

(iii) *Given two compactifications*

$$X \overset{j_i}{\hookrightarrow} \overline{X}_i \overset{\bar{f}_i}{\to} Y \quad (i = 1, 2)$$

of f, let $\phi', \phi'' : \overline{X}_1 \to \overline{X}_2$ be two morphisms between these compactifications. There exists a compactification

$$X \hookrightarrow \overline{X} \to Y$$

and a morphism $\phi : \overline{X} \to \overline{X}_1$ of compactifications such that $\phi' \circ \phi = \phi'' \circ \phi$.

(iv) *Given two compactifications of (f, g), there exists a compactification dominating both of them.*

Proof.

(i) We have seen that f has a compactification. Choose compactifications

$$X \hookrightarrow \overline{X} \to Y \text{ and } Y \hookrightarrow \overline{Y} \to Z$$

of f and g, respectively. By 7.4.1 (iii), the composite $\overline{X} \to Y \hookrightarrow Y$ is S-compactifiable. Choose a compactification $X \hookrightarrow \overline{X}' \to \overline{Y}$ for this composite. We then get a commutative diagram

$$\begin{array}{ccccc} X & \hookrightarrow & \overline{X} & \hookrightarrow & \overline{X}' \\ & \searrow & \downarrow & & \downarrow \\ & & Y & \hookrightarrow & \overline{Y} \\ & & & \searrow & \downarrow \\ & & & & Z, \end{array}$$

which is a compactification of (f, g).

(ii) Given two compactifications

$$X \hookrightarrow \overline{X}_i \to Y \quad (i = 1, 2)$$

for f, the morphism $\overline{X}_1 \times_Y \overline{X}_2 \to Y$ is S-compactifiable by 7.4.1 (iii) and (iv). The morphism $X \to \overline{X}_1 \times_Y \overline{X}_2$ defined by $X \hookrightarrow \overline{X}_i$ $(i = 1, 2)$ is an immersion. It is S-compactifiable by 7.4.1 (v). So we can factorize it as the composite of an open immersion $X \hookrightarrow \overline{X}$ and a closed immersion $\overline{X} \to \overline{X}_1 \times_Y \overline{X}_2$. The composite $\overline{X} \to \overline{X}_1 \times_Y \overline{X}_2 \to Y$ is S-compactifiable by 7.4.1 (ii) and (iii), and it is proper. So $X \hookrightarrow \overline{X} \to Y$ is a compactification of f, and it dominates the given two compactifications.

(iii) Define K by the Cartesian square

$$\begin{array}{ccc} K & \overset{p_2}{\to} & \overline{X}_1 \\ {\scriptstyle p_1} \downarrow & & \downarrow {\scriptstyle \Gamma_{\phi''}} \\ \overline{X}_1 & \overset{\Gamma_{\phi'}}{\to} & \overline{X}_1 \times_Y \overline{X}_2, \end{array}$$

where $\Gamma_{\phi'}, \Gamma_{\phi''} : \overline{X}_1 \to \overline{X}_1 \times_Y \overline{X}_2$ are the graphs of ϕ' and ϕ'', respectively. Let $\pi_i : \overline{X}_1 \times_Y \overline{X}_2 \to \overline{X}_i$ $(i = 1, 2)$ be the projections. We have

$$p_1 = \mathrm{id} \circ p_1 = \pi_1 \Gamma_{\phi'} p_1 = \pi_1 \Gamma_{\phi''} p_2 = \mathrm{id} \circ p_2 = p_2.$$

Denote $p_1 = p_2$ by p. Note that p is a closed immersion. One can show that the sequence

$$K \xrightarrow{p} \overline{X}_1 \overset{\phi'}{\underset{\phi''}{\rightrightarrows}} \overline{X}_2$$

is exact, that is, $\phi' p = \phi'' p$, and for any morphism $\psi : K' \to \overline{X}_1$ with the property $\phi'\psi = \phi''\psi$, there exists a unique morphism $\psi' : K' \to K$ such that $p\psi' = \psi$. We have

$$\phi' j_1 = j_2 = \phi'' j_1.$$

So there exists a morphism $j : X \to K$ such that $pj = j_1$. j is necessarily an open immersion. The morphism $\bar{f}_1 p : K \to Y$ is S-compactifiable by 7.4.1 (ii) and (iii), and it is proper. So

$$X \overset{j}{\hookrightarrow} K \xrightarrow{\bar{f}_1 p} Y$$

is a compactification of f, and p is a morphism between compactifications such that $\phi'\phi = \phi''\phi$.

$$\begin{array}{ccccc}
X & = & X & = & X \\
j\downarrow & & j_1\downarrow & & \downarrow j_2 \\
K & \xrightarrow{p} & \overline{X}_1 & \overset{\phi'}{\underset{\phi''}{\rightrightarrows}} & \overline{X}_2 \\
 & & \bar{f}_1\downarrow & & \downarrow \bar{f}_2 \\
 & & Y & = & Y
\end{array}$$

(iv) Given two compactifications

$$\begin{array}{ccccc}
X & \hookrightarrow & \overline{X}_i & \hookrightarrow & \overline{X}'_i \\
 & \searrow & \downarrow & & \downarrow \\
 & & Y & \hookrightarrow & \overline{Y}_i \\
 & & & \searrow & \downarrow \\
 & & & & Z
\end{array} \quad (i = 1, 2)$$

of (f, g), find a compactification

$$X \hookrightarrow \overline{X} \to Y \quad (\text{resp. } Y \hookrightarrow \overline{Y} \to Z)$$

dominating

$$X \hookrightarrow \overline{X}_i \to Y \quad (\text{resp. } Y \hookrightarrow \overline{Y}_i \to Z)$$

for $i = 1, 2$. For each i, we have a commutative diagram

$$\begin{array}{ccccc} & & \overline{X}_i & \hookrightarrow & \overline{X}'_i \\ & \nearrow & \downarrow & & \downarrow \\ \overline{X} & & Y & \hookrightarrow & \overline{Y}_i. \\ \downarrow & \nearrow & & \nearrow & \\ Y & \hookrightarrow & \overline{Y} & & \end{array}$$

Let

$$\overline{X} \to \overline{X}'_i \times_{\overline{Y}_i} \overline{Y}$$

be the $\overline{Y}_i$-morphism with components given by the composites

$$\overline{X} \to \overline{X}_i \to \overline{X}'_i, \quad \overline{X} \to Y \to \overline{Y}.$$

It is S-compactifiable by 7.4.1 (v). Let

$$\overline{X} \hookrightarrow \overline{T}_i \to \overline{X}'_i \times_{\overline{Y}_i} \overline{Y}$$

be a compactification. Then the diagram

$$\begin{array}{ccc} \overline{X} & \hookrightarrow & \overline{T}_i \\ \downarrow & & \downarrow \\ Y & \hookrightarrow & \overline{Y} \end{array}$$

dominates the diagram

$$\begin{array}{ccc} \overline{X}_i & \hookrightarrow & \overline{X}'_i \\ \downarrow & & \downarrow \\ Y & \hookrightarrow & \overline{Y}_i. \end{array}$$

The morphism $\overline{T}_i \to \overline{Y}$ is S-compactifiable and proper. It follows that

$$\overline{X} \hookrightarrow \overline{T}_i \to \overline{Y}$$

is a compactification of the composite $\overline{X} \to Y \to \overline{Y}$. Choose a compactification

$$\overline{X} \hookrightarrow \overline{X}' \to \overline{Y}$$

of $\overline{X} \to Y \to \overline{Y}$ dominating $\overline{X} \hookrightarrow \overline{T}_i \to \overline{Y}$ for $i = 1, 2$. Then the diagram

$$\begin{array}{ccccc} X & \hookrightarrow & \overline{X} & \hookrightarrow & \overline{X}' \\ & \searrow & \downarrow & & \downarrow \\ & & Y & \hookrightarrow & \overline{Y} \\ & & & \searrow & \downarrow \\ & & & & Z \end{array}$$

is a compactification of (f, g) dominating the given compactifications. □

Lemma 7.4.3. *Consider a commutative diagram*

$$\begin{array}{ccc} X & \stackrel{k}{\hookrightarrow} & \overline{X} \\ f\downarrow & & \downarrow\bar{f} \\ Y & \stackrel{j}{\hookrightarrow} & \overline{Y}, \end{array}$$

where k and j are open immersions, and f and $\bar{f}$ are proper morphisms. Then for any $K \in \operatorname{ob} D^+_{\mathrm{tor}}(X)$, we have a canonical isomorphism

$$j_!Rf_*K \stackrel{\cong}{\to} R\bar{f}_*k_!K.$$

Proof. Fix notation by the commutative diagram

$$\begin{array}{ccccccc} X & \stackrel{l}{\hookrightarrow} & \overline{X}\times_{\overline{Y}}Y & \stackrel{j'}{\hookrightarrow} & \overline{X} & \stackrel{i'}{\hookleftarrow} & \overline{X}\times_{\overline{Y}}(\overline{Y}-Y) \\ & f\searrow & \bar{f}'\downarrow & & \downarrow\bar{f} & & \downarrow\bar{f}'' \\ & & Y & \stackrel{j}{\hookrightarrow} & \overline{Y} & \stackrel{i}{\hookleftarrow} & \overline{Y}-Y, \end{array}$$

where the squares are Cartesian, $i : \overline{Y} - Y \to \overline{Y}$ is a closed immersion, and l is the morphism such that $j'\iota = k$ and $\bar{f}'l = f$. Since j is an open immersion, we have a canonical isomorphism

$$j^*R\bar{f}_*k_!K \stackrel{\cong}{\to} R\bar{f}'_*l_!K.$$

Since f and $\bar{f}'$ are proper, l is also proper. Since j' and k are open immersions, l is also an open immersion. It follows that l is an open and closed immersion and hence $l_!K \cong Rl_*K$. So we have

$$R\bar{f}'_*l_!K \cong R\bar{f}'_*Rl_*K \cong Rf_*K.$$

We thus have an isomorphism

$$j^*R\bar{f}_*k_!K \cong Rf_*K.$$

The inverse of this isomorphism induces a morphism

$$j_!Rf_*K \to R\bar{f}_*k_!K,$$

and its restriction to Y is an isomorphism. To show that it is an isomorphism, it suffices to show $i^*R\bar{f}_*k_!K = 0$. We have $i'^*k_!K = 0$, and by the proper base change theorem 7.3.1, we have

$$i^*R\bar{f}_*k_!K \cong R\bar{f}''_*i'^*k_!K = 0.$$

□

Let $f : X \to Y$ be an S-compactifiable morphism, and let

$$X \overset{j_i}{\hookrightarrow} \overline{X}_i \overset{\bar{f}_i}{\to} Y \quad (i = 1, 2)$$

be two compactifications of f. Choose a compactification

$$X \overset{j_3}{\hookrightarrow} \overline{X}_3 \overset{\bar{f}_3}{\to} Y$$

dominating both of them, and let $\phi_i : \overline{X}_3 \to \overline{X}_i$ $(i = 1, 2)$ be morphisms making the following diagrams commute:

$$\begin{array}{ccccc} X & \overset{j_3}{\hookrightarrow} & \overline{X}_3 & \overset{\bar{f}_3}{\to} & Y \\ \| & & \phi_i \downarrow & & \| \\ X & \overset{j_i}{\hookrightarrow} & \overline{X}_i & \overset{\bar{f}_i}{\to} & Y. \end{array} \quad (i = 1, 2)$$

Then ϕ_i are proper, and by 7.4.3, we have isomorphisms

$$j_{i!}K \overset{\cong}{\to} R\phi_{i*}j_{3!}K$$

for any $K \in \mathrm{ob}\, D^+_{\mathrm{tor}}(X)$. So we have isomorphisms

$$R\bar{f}_{i*}j_{i!}K \overset{\cong}{\to} R\bar{f}_{i*}R\phi_{i*}j_{3!}K \cong R\bar{f}_{3*}j_{3!}K.$$

We thus have an isomorphism

$$R\bar{f}_{1*}j_{1!}K \cong R\bar{f}_{2*}j_{2!}K.$$

This isomorphism is independent of the choice of the third compactification of f dominating the first two compactifications. Indeed, given a fourth compactification dominating the first two, by 7.4.2 (ii) and (iii), we can find a fifth compactification dominating the third and the fourth such that for each $i = 1, 2$, the composite of the morphism from the fifth to the third and the morphism from the third to the i-th coincides with the composite of the morphism from the fifth to the fourth and the morphism from the fourth to the i-th. It is clear that the isomorphisms $R\bar{f}_{1*}j_{1!}K \cong R\bar{f}_{2*}j_{2!}K$ constructed from both the third and the fourth compactifications coincide with the isomorphism constructed from the fifth compactification. So the isomorphism $R\bar{f}_{1*}j_{1!}K \cong R\bar{f}_{2*}j_{2!}K$ is independent of the choice of the compactification dominating the first two compactifications.

For any S-compactifiable morphism $f : X \to Y$, choose a compactification

$$X \overset{j}{\hookrightarrow} \overline{X} \overset{\bar{f}}{\to} Y.$$

For any $K \in \mathrm{ob}\, D^+_{\mathrm{tor}}(X)$, we define

$$Rf_!K = R\bar{f}_*j_!K.$$

By the above discussion, the isomorphic class of the functor $Rf_!$ is independent of the choice of the compactification. Define

$$R^q f_! K = \mathscr{H}^q(Rf_! K).$$

Let $\mathscr{F}$ be a torsion sheaf on X, we define

$$f_!\mathscr{F} = R^0 f_!\mathscr{F} = \bar{f}_* j_!\mathscr{F}.$$

Note that $f_!$ coincides with the functor defined in 5.5.

Suppose that $f : X \to Y$ is a separated quasi-finite S-morphism such that Y is quasi-compact and quasi-separated. By the Zariski Main Theorem, we have a compactification

$$X \overset{j}{\hookrightarrow} \overline{X} \overset{\bar{f}}{\to} Y$$

such that $\bar{f}$ is finite. We then have

$$Rf_! = R\bar{f}_* j_! = \bar{f}_* j_! = f_!.$$

Suppose that $S = \operatorname{Spec} k$ is the spectrum of a separably closed field k. Let $f : X \to S$ be an S-compactifiable morphism, and let

$$X \overset{j}{\hookrightarrow} \overline{X} \overset{\bar{f}}{\to} \operatorname{Spec} k$$

be a compactification. For any $K \in \operatorname{ob} D^+_{\mathrm{tor}}(X)$, we define

$$R\Gamma_c(X, K) = \Gamma(\operatorname{Spec} k, Rf_! K) \cong R\Gamma(\overline{X}, j_! K),$$
$$H^q_c(X, K) = \mathscr{H}^q(R\Gamma_c(X, K)) \cong H^q(\overline{X}, j_! K).$$

For any torsion sheaf $\mathscr{F}$ on X, we define

$$\Gamma_c(X, \mathscr{F}) = H^0_c(X, \mathscr{F}) \cong \Gamma(\overline{X}, j_!\mathscr{F}).$$

Note that $\Gamma_c(X, \mathscr{F})$ consists of those sections in $\Gamma(X, \mathscr{F})$ with support proper over $\operatorname{Spec} k$.

In general, $Rf_!$ is not the right derived functor of $f_!$. For example, in the case where $S = \operatorname{Spec} k$ is the spectrum of a separably closed field k and X is a smooth affine algebraic curve over k, we have

$$\Gamma_c(X, \mathscr{F}) \cong \bigoplus_{x \in |X|} \Gamma_x(X, \mathscr{F}),$$

where $|X|$ is the set of Zariski closed points in X. It follows that the right derived functor of $\Gamma_c(X, -)$ is isomorphic to $\bigoplus_{x\in|X|} R\Gamma_x(X, -)$. Later we will show $H^1_x(X, \mathbb{Z}/n) \cong \mathbb{Z}/n$. Thus $\bigoplus_{x\in|X|} H^1_x(X, \mathbb{Z}/n)$ is infinite. But we will see that $H^1_c(X, \mathbb{Z}/n)$ is finite.

Theorem 7.4.4.

(i) *Consider a Cartesian diagram*

$$\begin{array}{ccc} X\times_Y Y' & \xrightarrow{g'} & X \\ {\scriptstyle f'}\downarrow & & \downarrow{\scriptstyle f} \\ Y' & \xrightarrow{g} & Y. \end{array}$$

Suppose that f is S-compactifiable, and Y' is quasi-compact and quasi-separated. Then for any $K\in \operatorname{ob} D^+_{\mathrm{tor}}(X)$, we have a canonical isomorphism

$$g^*Rf_!K \xrightarrow{\cong} Rf'_!g'^*K.$$

(ii) *Let $X \xrightarrow{f} Y \xrightarrow{g} Z$ be two S-compactifiable morphisms. For any $K\in \operatorname{ob} D^+_{\mathrm{tor}}(X)$, we have a canonical isomorphism*

$$Rg_!Rf_!K \xrightarrow{\cong} R(gf)_!K,$$

and we have a biregular spectral sequence

$$E_2^{pq} = R^pg_!R^qf_!K \Rightarrow R^{p+q}(gf)_!K.$$

(iii) *Let $f: X\to Y$ be an S-compactifiable morphism, $j: U\hookrightarrow X$ an S-compactifiable open immersion, and $i: A\to X$ a closed immersion with $A = X - U$. For any $K\in \operatorname{ob} D^+_{\mathrm{tor}}(X)$, we have a distinguished triangle*

$$R(fj)_!j^*K \to Rf_!K \to R(fi)_!i^*K \to$$

and a long exact sequence

$$\cdots \to R^q(fj)_!j^*K \to R^qf_!K \to R^q(fi)_!i^*K \to R^{p+1}(fj)_!j^*K \to \cdots.$$

Proof.

(i) Let

$$X \overset{j}{\hookrightarrow} \overline{X} \xrightarrow{\bar f} Y$$

be a compactification of f. Fix notation by the following commutative diagram

$$\begin{array}{ccc} X\times_Y Y' & \xrightarrow{g'} & X \\ {\scriptstyle j'}\downarrow & & \downarrow{\scriptstyle j} \\ \overline{X}\times_Y Y' & \xrightarrow{\bar g} & \overline{X} \\ {\scriptstyle \bar f'}\downarrow & & \downarrow{\scriptstyle \bar f} \\ Y' & \xrightarrow{g} & Y. \end{array}$$

We have

$$j'_!g'^*K \cong \bar g^*j_!K,$$

and by the proper base change theorem 7.3.1, we have

$$g^* R\bar{f}_* j_! K \cong R\bar{f}'_* \bar{g}^* j_! K.$$

It follows that

$$g^* R\bar{f}_* j_! K \cong R\bar{f}'_* j'_! g'^* K,$$

that is,

$$g^* Rf_! K \cong Rf'_! g'^* K.$$

One can verify that this isomorphism is independent of the choice of the compactification of f.

(ii) Let

$$\begin{array}{ccccc} X & \stackrel{j_1}{\hookrightarrow} & \overline{X} & \stackrel{j_2}{\hookrightarrow} & \overline{X}' \\ & {\scriptstyle f}\searrow & {\scriptstyle \bar{f}}\downarrow & & \downarrow{\scriptstyle \bar{f}'} \\ & & Y & \stackrel{k}{\hookrightarrow} & \overline{Y} \\ & & & {\scriptstyle g}\searrow & \downarrow{\scriptstyle \bar{g}} \\ & & & & Z \end{array}$$

be a compactification of (f, g). By 7.4.3, we have

$$k_! R\bar{f}_* j_{1!} K \cong R\bar{f}'_* j_{2!} j_{1!} K.$$

So we have

$$Rg_! Rf_! K \cong R\bar{g}_* k_! R\bar{f}_* j_{1!} K \cong R\bar{g}_* R\bar{f}'_* j_{2!} j_{1!} K \cong R(\bar{g}\bar{f}')_* (j_2 j_1)_! K \cong R(gf)_! K,$$

that is, $Rg_! Rf_! K \cong R(gf)_! K$. This isomorphism is independent of the choice of the compactification of (f, g).

For any $L \in \operatorname{ob} D^+_{\mathrm{tor}}(Y)$, we have a biregular spectral sequence

$$E_2^{pq} = R^p \bar{g}_* (\mathscr{H}^q(k_! L)) \Rightarrow R^{p+q} \bar{g}_* k_! L.$$

Taking $L = Rf_! K$, we get the spectral sequence

$$E_2^{pq} = R^p g_! R^q f_! K \Rightarrow R^{p+q} (gf)_! K.$$

(iii) By 5.4.2 (iv), we have a distinguished triangle

$$j_! j^* K \to K \to i_* i^* K \to,$$

and hence a distinguished triangle

$$Rf_! j_! j^* K \to Rf_! K \to Rf_! i_* i^* K \to .$$

We have $Rf_! j_! j^* K \cong R(fj)_! j^* K$ and $Rf_! i_* i^* K \cong R(fi)_! i^* K$. So we have a distinguished triangle

$$R(fj)_! j^* K \to Rf_! K \to R(fi)_! i^* K \to .$$

□

Theorem 7.4.5. *Let $f : X \to Y$ be an S-compactifiable morphism and let n be the supremum of dimensions of fibers of f. Then for any torsion sheaf $\mathscr{F}$ on X, we have $R^q f_! \mathscr{F} = 0$ for all $q > 2n$.*

Proof. By 7.4.4 (i), for any $y \in Y$, we have

$$(R^q f_! \mathscr{F})_{\bar{y}} \cong H^q_c(X \otimes_{\mathscr{O}_Y} \overline{k(y)}, \mathscr{F}|_{X \otimes_{\mathscr{O}_Y} \overline{k(y)}}),$$

where $\overline{k(y)}$ is a separable closure of the residue field $k(y)$. So it suffices to prove that for any separably closed field k and any S-compactifiable morphism $X \to \operatorname{Spec} k$, we have $H^q_c(X, \mathscr{F}) = 0$ for any $q > 2 \dim X$ and any torsion sheaf $\mathscr{F}$ on X. We prove this by induction on $\dim X$. If $\dim X = 0$, $\Gamma(X, -)$ is an exact functor on the category of sheaves on X, our assertion is clear. Suppose that $\dim X = n \geq 1$ and our assertion holds for those schemes over k of dimensions $< n$. We may assume that X is reduced. Let $\eta_1, \ldots, \eta_n$ be those generic points of X so that $\mathrm{tr.deg}(k(\eta_i)/k) = n$, where tr.deg denotes the transcendental degree. Since $n \geq 1$, for each i, we can choose $t_i \in k(\eta_i)$ which is transcendental over k. Choose an irreducible open affine neighborhood U_i of η_i such that t_i can be extended to a section in $\mathscr{O}_X(U_i)$, which we still denote by t_i. The k-algebra homomorphism

$$k[t] \to \mathscr{O}_X(U_i), \quad t \mapsto t_i$$

defines a k-morphism

$$U_i \to \mathbb{A}^1_k.$$

By 1.5.6 (i), we may assume that the morphism is flat by shrinking U_i. Since $\mathrm{tr.deg}(k(\eta_i)/k(t_i)) = n - 1$, the generic fiber of $U_i \to \mathbb{A}^1_k$ has dimension $\leq n - 1$. If $x \in U_i$ lies above a closed point y in $\mathbb{A}^1_k$, then by [Matsumura (1970)] (13.B) Theorem 19 (2), we have

$$\dim(\mathscr{O}_{U_i,x} \otimes_{\mathscr{O}_{\mathbb{A}^1_k,y}} k(y)) = \dim \mathscr{O}_{U_i,x} - \dim \mathscr{O}_{\mathbb{A}^1_k,y} \leq n - 1.$$

So the fibers of $U_i \to \mathbb{A}^1_k$ above closed points of $\mathbb{A}^1_k$ have dimensions $\leq n-1$. Shrinking U_i again, we may assume that they are disjoint. Let $U = \cup_i U_i$. Then we have a morphism $f : U \to \mathbb{A}^1_k$ whose fibers have dimensions $\leq n-1$. By the induction hypothesis, we have

$$R^q f_!(\mathscr{F}|_U) = 0$$

for any $q > 2n - 2$. By our construction, we have $\dim(X - U) \leq n - 1$. So by the induction hypothesis, we have

$$H^q_c(X - U, \mathscr{F}|_{X-U}) = 0$$

for any $q > 2n-2$. We have a long exact sequence

$$\cdots \to H_c^q(U, \mathscr{F}|_U) \to H_c^q(X, \mathscr{F}) \to H_c^q(X-U, \mathscr{F}|_{X-U}) \to \cdots.$$

To prove $H_c^q(X, \mathscr{F}) = 0$ for all $q > 2n$, it suffices to prove

$$H_c^q(U, \mathscr{F}|_U) = 0$$

for any $q > 2n$. We have a biregular spectral sequence

$$E_2^{pq} = H_c^p(\mathbb{A}_k^1, R^q f_!(\mathscr{F}|_U)) \Rightarrow H_c^{p+q}(U, \mathscr{F}|_U).$$

Since $R^q f_!(\mathscr{F}|_U) = 0$ for any $q > 2n-2$, to prove $H_c^q(U, \mathscr{F}|_U) = 0$ for any $q > 2n$, it suffices to prove

$$H_c^p(\mathbb{A}_k^1, \mathscr{G}) = 0$$

for any $p > 2$ and any torsion sheaf $\mathscr{G}$ on $\mathbb{A}_k^1$. Let $j : \mathbb{A}_k^1 \hookrightarrow \mathbb{P}_k^1$ be the open immersion. We have

$$H_c^p(\mathbb{A}_k^1, \mathscr{G}) \cong H^p(\mathbb{P}_k^1, j_!\mathscr{G}).$$

Our assertion follows from 7.2.7. □

Remark 7.4.6. Suppose that $f : X \to Y$ is an S-compactifiable morphism. Then the fibers of f have bounded dimensions. By 7.4.5, Rf_* has finite cohomological dimension if f is proper. So we can define $Rf_* : D_{\mathrm{tor}}(X) \to D_{\mathrm{tor}}(Y)$ if f is proper, and $Rf_! : D_{\mathrm{tor}}(X) \to D_{\mathrm{tor}}(Y)$ in general.

Theorem 7.4.7 (Projection formula). *Let $f : X \to Y$ be an S-compactifiable morphism, and let A be a torsion ring.*

(i) *For any $K \in \mathrm{ob}\, D^-(X, A)$ and $L \in \mathrm{ob}\, D^-(Y, A)$, we have a canonical isomorphism*

$$L \otimes_A^L Rf_!K \xrightarrow{\cong} Rf_!(f^*L \otimes_A^L K).$$

(ii) *For any $K \in \mathrm{ob}\, D_{\mathrm{tf}}^b(X, A)$, we have $Rf_!K \in \mathrm{ob}\, D_{\mathrm{tf}}^b(Y, A)$, and we have a canonical isomorphism*

$$L \otimes_A^L Rf_!K \xrightarrow{\cong} Rf_!(f^*L \otimes_A^L K)$$

for any $L \in \mathrm{ob}\, D(Y, A)$.

(iii) *Let $A \to B$ be a homomorphism of torsion rings. For any $K \in \mathrm{ob}\, D^-(X, A)$ and $M \in \mathrm{ob}\, D^-(Y, B)$, we have a canonical isomorphism*

$$M \otimes_A^L Rf_!K \xrightarrow{\cong} Rf_!(f^*M \otimes_A^L K)$$

in $D^-(Y, B)$.

Proof. We prove (i) and leave it for the reader to prove the rest. (Confer 6.5.6 and 6.6.7.) Let

$$X \stackrel{j}{\hookrightarrow} \overline{X} \stackrel{\bar{f}}{\to} Y$$

be a compactification of f. For any complexes of sheaves of A-modules $M^\cdot$ on X and $N^\cdot$ on $\overline{X}$, we have a canonical isomorphism

$$j_!(j^*N^\cdot \otimes_A M^\cdot) \cong N^\cdot \otimes_A j_!M^\cdot.$$

Taking $M^\cdot$ to be a complex of flat sheaves of A-modules quasi-isomorphic to K and $N^\cdot$ a complex quasi-isomorphic to $\bar{f}^*L$, we get an isomorphism

$$j_!(j^*\bar{f}^*L \otimes_A^L K) \cong \bar{f}^*L \otimes_A^L j_!K.$$

We have

$$\begin{aligned}&L \otimes_A^L Rf_!K \cong L \otimes_A^L R\bar{f}_*j_!K,\\ &Rf_!(f^*L \otimes_A^L K) \cong R\bar{f}_*j_!(j^*\bar{f}^*L \otimes_A^L K) \cong R\bar{f}_*(\bar{f}^*L \otimes_A^L j_!K).\end{aligned}$$

To prove (i), it suffices to show that the canonical morphism

$$L \otimes_A^L R\bar{f}_*j_!K \to R\bar{f}_*(\bar{f}^*L \otimes_A^L j_!K)$$

is an isomorphism. For any geometric point $s \to Y$, where s is the spectrum of a separably closed field, we need to show that the induced morphism

$$(L \otimes_A^L R\bar{f}_*j_!K)_s \to (R\bar{f}_*(\bar{f}^*L \otimes_A^L j_!K))_s$$

on stalks is an isomorphism. Let

$$\bar{f}_s : \overline{X}_s = \overline{X} \times_Y s \to s$$

be the base change of $\bar{f}$. By the proper base change theorem 7.3.2, we have

$$\begin{aligned}(L \otimes_A^L R\bar{f}_*j_!K)_s &\cong L_s \otimes_A^L R\bar{f}_{s*}((j_!K)|_{\overline{X}_s}),\\ (R\bar{f}_*(\bar{f}^*L \otimes_A^L j_!K))_s &\cong R\bar{f}_{s*}(\bar{f}_s^*L_s \otimes_A^L (j_!K)|_{\overline{X}_s}).\end{aligned}$$

So it suffices to prove that the canonical morphism

$$L_s \otimes_A^L R\bar{f}_{s*}((j_!K)|_{\overline{X}_s}) \to R\bar{f}_{s*}(\bar{f}_s^*L_s \otimes_A^L (j_!K)|_{\overline{X}_s})$$

is an isomorphism. This follows from 6.5.5 since $\mathscr{H}^i(L_s)$ are constant sheaves on s for all i. □

Corollary 7.4.8. *Consider a Cartesian diagram of S-schemes*

$$\begin{array}{ccc} X \times_Z Y & \stackrel{g'}{\to} & X \\ f' \downarrow & & \downarrow f \\ Y & \stackrel{g}{\to} & Z. \end{array}$$

Suppose that g is S-compactifiable and X is quasi-compact and quasi-separated. Let A be a torsion ring. For any $K \in \mathrm{ob}\, D^-(X, A)$ and $L \in \mathrm{ob}\, D^-(Y, A)$, we have a canonical isomorphism

$$K \otimes_A^L f^*Rg_!L \stackrel{\cong}{\to} Rg'_!(g'^*K \otimes_A^L f'^*L).$$

Proof. By 7.4.4 (i) and 7.4.7 (i), we have

$$K \otimes_A^L f^* Rg_! L \cong K \otimes_A^L Rg'_! f'^* L \cong Rg'_!(g'^* K \otimes_A^L f'^* L).$$

□

Corollary 7.4.9 (Künneth formula). *Let X_i, Y_i $(i = 1, 2)$ and Z be S-schemes, let $f_i : X_i \to Y_i$ and $Y_i \to Z$ be S-compactifiable morphisms, and let $p_i : X_1 \times_Z X_2 \to X_i$ and $q_i : Y_1 \times_Z Y_2 \to Y_i$ be the projections. For any $K_i \in \operatorname{ob} D^-(X_i, A)$, we have*

$$q_1^* Rf_{1!} K_1 \otimes_A^L q_2^* Rf_{2!} K_2 \cong R(f_1 \times f_2)_!(p_1^* K_1 \otimes_A^L p_2^* K_2).$$

Proof. Fix notation by the following commutative diagram of Cartesian squares:

$$\begin{array}{ccccc} X_1 \times_Z X_2 & \stackrel{f_1''}{\to} & Y_1 \times_Z X_2 & \stackrel{b_1''}{\to} & X_2 \\ f_2'' \downarrow & & f_2' \downarrow & & \downarrow f_2 \\ X_1 \times_Z Y_2 & \stackrel{f_1'}{\to} & Y_1 \times_Z Y_2 & \stackrel{q_2}{\to} & Y_2 \\ b_2'' \downarrow & & q_1 \downarrow & & \downarrow b_2 \\ X_1 & \stackrel{f_1}{\to} & Y_1 & \stackrel{b_1}{\to} & Z. \end{array}$$

By 7.4.8, we have isomorphisms

$$\begin{aligned} q_1^* Rf_{1!} K_1 \otimes_A^L q_2^* Rf_{2!} K_2 &\cong Rf'_{2!}(f_2'^* q_1^* Rf_{1!} K_1 \otimes_A^L b_1''^* K_2) \\ &\cong Rf'_{2!} Rf''_{1!}(f_2''^* b_2''^* K_1 \otimes_A^L f_1''^* b_1''^* K_2). \end{aligned}$$

Our assertion follows. □

Let X be a scheme, A a ring, $\mathscr{F}$ and $\mathscr{G}$ sheaves of A-modules on X, $\mathscr{C}^\cdot(\mathscr{F})$ and $\mathscr{C}^\cdot(\mathscr{G})$ the Godement resolutions of $\mathscr{F}$ and $\mathscr{G}$, respectively. By 6.4.9, we have a quasi-isomorphism

$$\mathscr{F} \otimes_A \mathscr{G} \to \mathscr{C}^\cdot(\mathscr{F}) \otimes_A \mathscr{C}^\cdot(\mathscr{G}).$$

Let

$$\mathscr{C}^\cdot(\mathscr{F}) \otimes_A \mathscr{C}^\cdot(\mathscr{G}) \to \mathscr{I}^\cdot$$

be a quasi-isomorphism so that $\mathscr{I}^\cdot$ is a complex of injective sheaves of A-modules. Suppose that $R\Gamma(X, -)$ has finite cohomological dimension. We define the *cup product*

$$R\Gamma(X, \mathscr{F}) \otimes_A^L R\Gamma(X, \mathscr{G}) \to R\Gamma(X, \mathscr{F} \otimes_A \mathscr{G})$$

to be the composite of the following morphisms

$$\begin{aligned} R\Gamma(X,\mathscr{F})\otimes_A^L R\Gamma(X,\mathscr{G}) &\cong \Gamma(X,\mathscr{C}^{\cdot}(\mathscr{F}))\otimes_A^L \Gamma(X,\mathscr{C}^{\cdot}(\mathscr{G}))\\ &\to \Gamma(X,\mathscr{C}^{\cdot}(\mathscr{F}))\otimes_A \Gamma(X,\mathscr{C}^{\cdot}(\mathscr{G}))\\ &\to \Gamma(X,\mathscr{C}^{\cdot}(\mathscr{F})\otimes_A \mathscr{C}^{\cdot}(\mathscr{G}))\\ &\to \Gamma(X,\mathscr{I}^{\cdot})\\ &\cong R\Gamma(X,\mathscr{F}\otimes_A\mathscr{G}). \end{aligned}$$

The cup product induces homomorphisms

$$H^i(X,\mathscr{F})\otimes_A H^j(X,\mathscr{G})\to H^{i+j}(X,\mathscr{F}\otimes_A\mathscr{G}).$$

For any $s\in H^i(X,\mathscr{F})$ and $t\in H^j(X,\mathscr{G})$, denote by $s\cup t\in H^{i+j}(X,\mathscr{F}\otimes_A\mathscr{G})$ the image of $s\otimes t$ under the above homomorphism.

Proposition 7.4.10. *Let X be a scheme, let A be a ring, and let $\mathscr{F}$ a sheaf of A-modules on X. Denote by*

$$\tau: H^n(X,\mathscr{F}\otimes_A\mathscr{F})\to H^n(X,\mathscr{F}\otimes_A\mathscr{F})$$

the homomorphisms induced by

$$\tau:\mathscr{F}\otimes_A\mathscr{F}\to\mathscr{F}\otimes_A\mathscr{F},\quad s_1\otimes s_2\mapsto s_2\otimes s_1$$

for all n. For any $s\in H^i(X,\mathscr{F})$ and $t\in H^j(X,\mathscr{F})$, we have

$$s\cup t=(-1)^{ij}\tau(t\cup s)$$

in $H^{i+j}(X,\mathscr{F}\otimes_A\mathscr{F})$.

Proof. First note that if $K^{\cdot}$ and $L^{\cdot}$ are two complexes of sheaves of A-modules, then the morphisms

$$\sigma: K^i\otimes_A L^j\to L^j\otimes_A K^i,\quad x\otimes y\mapsto(-1)^{ij}y\otimes x$$

induce a morphism between the complexes associated to the bicomplexes $K^{\cdot}\otimes_A L^{\cdot}$ and $L^{\cdot}\otimes_A K^{\cdot}$. Let $\mathscr{C}^{\cdot}(\mathscr{F})$ be the Godement resolution of $\mathscr{F}$ and let $\mathscr{I}^{\cdot}$ be an injective resolution of $\mathscr{C}^{\cdot}(\mathscr{F})\otimes_A\mathscr{C}^{\cdot}(\mathscr{F})$. Consider the following diagram

$$\begin{array}{ccccc} \mathscr{F}\otimes_A\mathscr{F} & \to & \mathscr{C}^{\cdot}(\mathscr{F})\otimes_A\mathscr{C}^{\cdot}(\mathscr{F}) & \to & \mathscr{I}^{\cdot}\\ {\scriptstyle\tau}\downarrow & & \downarrow{\scriptstyle\sigma} & & {\scriptstyle\rho}\downarrow\\ \mathscr{F}\otimes_A\mathscr{F} & \to & \mathscr{C}^{\cdot}(\mathscr{F})\otimes_A\mathscr{C}^{\cdot}(\mathscr{F}) & \to & \mathscr{I}^{\cdot}, \end{array}$$

where the middle vertical arrow σ is the morphism defined above by taking $K^{\cdot}=L^{\cdot}=\mathscr{C}^{\cdot}(\mathscr{F})$, and $\rho:\mathscr{I}\to\mathscr{I}$ is chosen to make the square on the right commute (up to homotopy). One can show that such ρ exists. (Confer the proof of 6.2.8.) So we have a commutative diagram

$$\begin{array}{ccc} H^{i+j}(\Gamma(X,\mathscr{C}^{\cdot}(\mathscr{F})\otimes_A \mathscr{C}^{\cdot}(\mathscr{F}))) & \rightarrow & H^{i+j}(\Gamma(X,\mathscr{I}^{\cdot})) \\ \sigma\downarrow & & \rho\downarrow \\ H^{i+j}(\Gamma(X,\mathscr{C}^{\cdot}(\mathscr{F})\otimes_A \mathscr{C}^{\cdot}(\mathscr{F}))) & \rightarrow & H^{i+j}(\Gamma(X,\mathscr{I}^{\cdot})). \end{array}$$

On the other hand, one can check the following diagram commutes

$$\begin{array}{ccc} H^i(\Gamma(X,\mathscr{C}^{\cdot}(\mathscr{F})))\otimes_A H^j(\Gamma(X,\mathscr{C}^{\cdot}(\mathscr{F}))) & \rightarrow & H^{i+j}(\Gamma(X,\mathscr{C}^{\cdot}(\mathscr{F})\otimes_A \mathscr{C}^{\cdot}(\mathscr{F}))) \\ \sigma\downarrow & & \downarrow\sigma \\ H^j(\Gamma(X,\mathscr{C}^{\cdot}(\mathscr{F})))\otimes_A H^i(\Gamma(X,\mathscr{C}^{\cdot}(\mathscr{F}))) & \rightarrow & H^{i+j}(\Gamma(X,\mathscr{C}^{\cdot}(\mathscr{F})\otimes_A \mathscr{C}^{\cdot}(\mathscr{F}))), \end{array}$$

where the first vertical arrow is the homomorphism $s\otimes t\mapsto(-1)^{ij}t\otimes s$. So we have a commutative diagram

$$\begin{array}{ccc} H^i(\Gamma(X,\mathscr{C}^{\cdot}(\mathscr{F})))\otimes_A H^j(\Gamma(X,\mathscr{C}^{\cdot}(\mathscr{F}))) & \rightarrow & H^{i+j}(\Gamma(X,\mathscr{I}^{\cdot})) \\ \sigma\downarrow & & \rho\downarrow \\ H^j(\Gamma(X,\mathscr{C}^{\cdot}(\mathscr{F})))\otimes_A H^i(\Gamma(X,\mathscr{C}^{\cdot}(\mathscr{F}))) & \rightarrow & H^{i+j}(\Gamma(X,\mathscr{I}^{\cdot})). \end{array}$$

This diagram can be identified with

$$\begin{array}{ccc} H^i(X,\mathscr{F})\otimes_A H^j(X,\mathscr{F}) & \rightarrow & H^{i+j}(X,\mathscr{F}\otimes_A\mathscr{F}) \\ \sigma\downarrow & & \tau\downarrow \\ H^j(X,\mathscr{F})\otimes_A H^i(X,\mathscr{F}) & \rightarrow & H^{i+j}(X,\mathscr{F}\otimes_A\mathscr{F}). \end{array}$$

Our assertion follows. □

Proposition 7.4.11. *Let X and Y be proper schemes over a separably closed field k, A a torsion ring, $\mathscr{F}$ a flat sheaf of A-modules on X, and $\mathscr{G}$ a sheaf of A-modules on Y. Denote by $p: X\times_k Y\to X$ and $q: X\times_k Y\to Y$ the projections. Then the homomorphism*

$$H^{2\dim X}(X,\mathscr{F})\otimes_A H^{2\dim Y}(Y,\mathscr{G})\rightarrow H^{2(\dim X+\dim Y)}(X\times_k Y,p^*\mathscr{F}\otimes_A q^*\mathscr{G}),$$
$$s\otimes t\mapsto p^*s\cup q^*t$$

is an isomorphism. Suppose that either $H^i(X,\mathscr{F})$ are flat A-modules for all i, or $H^i(Y,\mathscr{G})$ are flat A-modules for all i. Then the homomorphisms

$$\bigoplus_{i+j=v}(H^i(X,\mathscr{F})\otimes_A H^j(Y,\mathscr{G}))\rightarrow H^v(X\times_k Y,p^*\mathscr{F}\otimes_A q^*\mathscr{G}),$$
$$s\otimes t\mapsto p^*s\cup q^*t$$

are isomorphisms for all v.

Proof. By the Künneth formula 7.4.9, we have

$$R\Gamma(X,\mathscr{F})\otimes_A^L R\Gamma(Y,\mathscr{G})\cong R\Gamma(X\times_k Y,p^*\mathscr{F}\otimes_A^L q^*\mathscr{G}).$$

Since $\mathscr{F}$ is flat, we have

$$p^*\mathscr{F}\otimes_A^L q^*\mathscr{G}\cong p^*\mathscr{F}\otimes_A q^*\mathscr{G}.$$

By 6.4.12, we have a biregular spectral sequence

$$E_2^{uv}=\bigoplus_{i+j=v}\mathrm{Tor}_{-u}(H^i(X,\mathscr{F}),H^j(Y,\mathscr{G}))\Rightarrow H^{u+v}(R\Gamma(X,\mathscr{F})\otimes_A^L R\Gamma(Y,\mathscr{G})).$$

By 7.4.5, we have $H^i(X,\mathscr{F})=0$ for any $i>2\dim X$, and $H^j(Y,\mathscr{G})=0$ for any $j>2\dim Y$. It follows that $E_2^{uv}=0$ for $u>0$ or $v>2(\dim X+\dim Y)$. So we have

$$H^{2\dim X}(X,\mathscr{F})\otimes_A H^{2\dim Y}(Y,\mathscr{G})\cong H^{2(\dim X+\dim Y)}(R\Gamma(X,\mathscr{F})\otimes_A^L R\Gamma(Y,\mathscr{G})).$$

Hence

$$H^{2\dim X}(X,\mathscr{F})\otimes_A H^{2\dim Y}(Y,\mathscr{G})\cong H^{2(\dim X+\dim Y)}(X\times_k Y,p^*\mathscr{F}\otimes_A q^*\mathscr{G}).$$

If either $H^i(X,\mathscr{F})$ are flat A-modules for all i, or $H^i(Y,\mathscr{G})$ are flat A-modules for all i, then the above spectral sequence degenerates, and we have

$$\begin{aligned}\bigoplus_{i+j=v}(H^i(X,\mathscr{F})\otimes_A H^j(Y,\mathscr{G}))&\cong H^v(R\Gamma(X,\mathscr{F})\otimes_A^L R\Gamma(Y,\mathscr{G}))\\&\cong H^v(X\times_k Y,p^*\mathscr{F}\otimes_A q^*\mathscr{G}).\end{aligned}$$

To prove our assertion, it suffices to show that these isomorphisms are induced by cup products.

By 6.4.10, we have quasi-isomorphisms

$$p^*\mathscr{F}\otimes_A q^*\mathscr{G}\to p^*\mathscr{C}^\cdot(\mathscr{F})\otimes_A q^*\mathscr{C}^\cdot(\mathscr{G})\to\mathscr{C}^\cdot(p^*\mathscr{F})\otimes_A\mathscr{C}^\cdot(q^*\mathscr{G}).$$

Here we use the functorial Godement resolutions constructed in the end of 5.6. Let $f:X\to\operatorname{Spec}k$ and $g:Y\to\operatorname{Spec}k$ be the structure morphisms. The isomorphism

$$R\Gamma(X,\mathscr{F})\otimes_A^L R\Gamma(Y,\mathscr{G})\cong R\Gamma(X\times_k Y,p^*\mathscr{F}\otimes_A^L q^*\mathscr{G})$$

defined by the Künneth formula is the composite of the following morphisms:

$$\begin{aligned} R\Gamma(X, \mathscr{F}) \otimes_A^L R\Gamma(Y, \mathscr{G}) &\cong \Gamma(X, \mathscr{C}^\cdot(\mathscr{F})) \otimes_A^L \Gamma(Y, \mathscr{C}^\cdot(\mathscr{G})) \\ &\to \Gamma(X, \mathscr{C}^\cdot(\mathscr{F})) \otimes_A \Gamma(Y, \mathscr{C}^\cdot(\mathscr{G})) \\ &\to \Gamma(Y, g^* f_* \mathscr{C}^\cdot(\mathscr{F})) \otimes_A \Gamma(Y, \mathscr{C}^\cdot(\mathscr{G})) \\ &\to \Gamma(Y, g^* f_* \mathscr{C}^\cdot(\mathscr{F}) \otimes_A \mathscr{C}^\cdot(\mathscr{G})) \\ &\to \Gamma(Y, q_* p^* \mathscr{C}^\cdot(\mathscr{F}) \otimes_A \mathscr{C}^\cdot(\mathscr{G})) \\ &\to \Gamma(Y, q_*(p^* \mathscr{C}^\cdot(\mathscr{F}) \otimes_A q^* \mathscr{C}^\cdot(\mathscr{G}))) \\ &= \Gamma(X \times_k Y, p^* \mathscr{C}^\cdot(\mathscr{F}) \otimes_A q^* \mathscr{C}^\cdot(\mathscr{G})) \\ &\to R\Gamma(X \times_k Y, p^* \mathscr{C}^\cdot(\mathscr{F}) \otimes_A q^* \mathscr{C}^\cdot(\mathscr{G})) \\ &\cong R\Gamma(X \times_k Y, p^* \mathscr{F} \otimes_A^L q^* \mathscr{G}). \end{aligned}$$

The composite of the above arrows from the third to the seventh coincide with the composite

$$\begin{aligned} &\Gamma(X, \mathscr{C}^\cdot(\mathscr{F})) \otimes_A \Gamma(Y, \mathscr{C}^\cdot(\mathscr{G})) \\ &\to \Gamma(X \times_k Y, p^* \mathscr{C}^\cdot(\mathscr{F})) \otimes_A \Gamma(X \times_k Y, q^* \mathscr{C}^\cdot(\mathscr{G})) \\ &\to \Gamma(X \times_k Y, p^* \mathscr{C}^\cdot(\mathscr{F}) \otimes_A q^* \mathscr{C}^\cdot(\mathscr{G})). \end{aligned}$$

So the isomorphism defined by the Künneth formula is the composite of the following morphisms:

$$\begin{aligned} R\Gamma(X, \mathscr{F}) \otimes_A^L R\Gamma(Y, \mathscr{G}) &\cong \Gamma(X, \mathscr{C}^\cdot(\mathscr{F})) \otimes_A^L \Gamma(Y, \mathscr{C}^\cdot(\mathscr{G})) \\ &\to \Gamma(X, \mathscr{C}^\cdot(\mathscr{F})) \otimes_A \Gamma(Y, \mathscr{C}^\cdot(\mathscr{G})) \\ &\to \Gamma(X \times_k Y, p^* \mathscr{C}^\cdot(\mathscr{F})) \otimes_A \Gamma(X \times_k Y, q^* \mathscr{C}^\cdot(\mathscr{G})) \\ &\to \Gamma(X \times_k Y, p^* \mathscr{C}^\cdot(\mathscr{F}) \otimes_A q^* \mathscr{C}^\cdot(\mathscr{G})) \\ &\to R\Gamma(X \times_k Y, p^* \mathscr{C}^\cdot(\mathscr{F}) \otimes_A q^* \mathscr{C}^\cdot(\mathscr{G})) \\ &\cong R\Gamma(X \times_k Y, p^* \mathscr{F} \otimes_A^L q^* \mathscr{G}). \end{aligned}$$

The morphism defined via the cup product

$$\begin{aligned} R\Gamma(X, \mathscr{F}) \otimes_A^L R\Gamma(Y, \mathscr{G}) &\overset{p^* \otimes q^*}{\to} R\Gamma(X \times_k Y, p^* \mathscr{F}) \otimes_A^L R\Gamma(X \times_k Y, p^* \mathscr{G}) \\ &\overset{\cup}{\to} R\Gamma(X \times_k Y, p^* \mathscr{F} \otimes_A^L q^* \mathscr{G}) \end{aligned}$$

is the composite of the following morphisms:

$$\begin{aligned} R\Gamma(X,\mathscr{F})\otimes^L_A R\Gamma(Y,\mathscr{G}) &\cong \Gamma(X,\mathscr{C}^{\cdot}(\mathscr{F}))\otimes^L_A \Gamma(Y,\mathscr{C}^{\cdot}(\mathscr{G}))\\ &\to \Gamma(X,\mathscr{C}^{\cdot}(\mathscr{F}))\otimes_A \Gamma(Y,\mathscr{C}^{\cdot}(\mathscr{G}))\\ &\to \Gamma(X\times_k Y,p^*\mathscr{C}^{\cdot}(\mathscr{F}))\otimes_A \Gamma(X\times_k Y,q^*\mathscr{C}^{\cdot}(\mathscr{G}))\\ &\to \Gamma(X\times_k Y,\mathscr{C}^{\cdot}(p^*\mathscr{F}))\otimes_A \Gamma(X\times_k Y,\mathscr{C}^{\cdot}(q^*\mathscr{G}))\\ &\to \Gamma(X\times_k Y,\mathscr{C}^{\cdot}(p^*\mathscr{F})\otimes_A \mathscr{C}^{\cdot}(q^*\mathscr{G}))\\ &\to R\Gamma(X\times_k Y,\mathscr{C}^{\cdot}(p^*\mathscr{F})\otimes_A \mathscr{C}^{\cdot}(q^*\mathscr{G}))\\ &\cong R\Gamma(X\times_k Y,p^*\mathscr{F}\otimes^L_A q^*\mathscr{G}). \end{aligned}$$

The coincidence of the isomorphism defined by the Künneth formula and the morphism defined via the cup product follows from the commutativity of the following diagram:

$$\begin{array}{ccc} \Gamma(X\times_k Y,p^*\mathscr{C}^{\cdot}(\mathscr{F}))\otimes_A\Gamma(X\times_k Y,q^*\mathscr{C}^{\cdot}(\mathscr{G})) & \to & \Gamma(X\times_k Y,p^*\mathscr{C}^{\cdot}(\mathscr{F})\otimes_A q^*\mathscr{C}^{\cdot}(\mathscr{G}))\\ \downarrow & & \downarrow\\ \Gamma(X\times_k Y,\mathscr{C}^{\cdot}(p^*\mathscr{F}))\otimes_A \Gamma(X\times_k Y,\mathscr{C}^{\cdot}(q^*\mathscr{G})) & \to & \Gamma(X\times_k Y,\mathscr{C}^{\cdot}(p^*\mathscr{F})\otimes_A \mathscr{C}^{\cdot}(q^*\mathscr{G})). \end{array}$$

□

7.5 Cohomological Dimension of Rf_*

([SGA 4] XIV 2–4.)

Theorem 7.5.1. *Let X and Y be schemes of finite type over a field k, $f: X\to Y$ an affine k-morphism, $\mathscr{F}$ a torsion sheaf on X, and n an integer. Suppose for any point $a\in X$ with the property $\dim\overline{\{a\}}>n$, we have $\mathscr{F}_{\bar a}=0$. Then for any point $b\in Y$ with the property $\dim\overline{\{b\}}>n-q$, we have $(R^qf_*\mathscr{F})_{\bar b}=0$.*

Taking Y to be the spectrum of a separably closed field, we get the following:

Corollary 7.5.2. *Let X be an affine scheme of finite type over a separably closed field k, $\mathscr{F}$ a torsion sheaf on X, and n an integer. Suppose for any point $a\in X$ with the property $\dim\overline{\{a\}}>n$, we have $\mathscr{F}_{\bar a}=0$. Then $H^q(X,\mathscr{F})=0$ for any $q>n$. In particular, we have $H^q(X,\mathscr{F})=0$ for any $q>\dim X$ and any torsion sheaf $\mathscr{F}$ on X.*

To prove 7.5.1, we need the following lemma.

Lemma 7.5.3. *Let k be a field, $X\to \mathrm{Spec}\, k[t_1,\ldots,t_d]$ a morphism of finite type, K an algebraic extension of $k(t_1,\ldots,t_d)$, and $X_K=X\otimes_{k[t_1,\ldots,k_d]}K$.*

For any point $a' \in X_K$, let a be its image in X. We have

$$\dim \overline{\{a\}} = \dim \overline{\{a'\}} + d,$$

where $\overline{\{a\}}$ (resp. $\overline{\{a'\}}$) denotes the closure of $\{a\}$ (resp. $\{a'\}$) in X (resp. X_K). Moreover, we have $\dim X_K \leq \dim X - d$.

$$\begin{array}{ccc} X_K = X \otimes_{k[t_1,\ldots,t_d]} K \to & & \operatorname{Spec} K \\ \downarrow & & \downarrow \\ X & & \to \operatorname{Spec} k[t_1,\ldots,t_d] \to \operatorname{Spec} k. \end{array}$$

Proof. We have

$$\dim \overline{\{a'\}} = \operatorname{tr.deg}(k(a')/K), \quad \dim \overline{\{a\}} = \operatorname{tr.deg}(k(a)/k),$$

where tr.deg denotes the transcendental degree. Note that $k(a')$ is algebraic over $k(a)$. So we have

$$\begin{aligned} \operatorname{tr.deg}(k(a)/k) &= \operatorname{tr.deg}(k(a')/k) \\ &= \operatorname{tr.deg}(k(a')/K) + \operatorname{tr.deg}(K/k) \\ &= \operatorname{tr.deg}(k(a')/K) + d, \end{aligned}$$

and hence

$$\dim \overline{\{a\}} = \dim \overline{\{a'\}} + d.$$

Let η_i' be all the generic points of X_K, and let η_i be their images in X. We have

$$\dim X_K = \max_i \dim \overline{\{\eta_i'\}} = \max_i \dim \overline{\{\eta_i\}} - d \leq \dim X - d.$$

□

Proof of 7.5.1. Making the base change from k to its algebraic closure, we may assume that k is algebraically closed. Replacing X and Y by X_{red} and Y_{red}, we may assume that X and Y are reduced. The problem is local with respect to Y. We may assume that Y is affine. Then X is also affine. We have $\mathscr{F} = \varinjlim_\lambda \mathscr{F}_\lambda$, where $\mathscr{F}_\lambda$ are constructible subsheaves of $\mathscr{F}$. $\mathscr{F}_\lambda$ satisfy the same condition as $\mathscr{F}$, and

$$R^q f_* \mathscr{F} = \varinjlim_\lambda R^q f_* \mathscr{F}_\lambda.$$

So we may assume that $\mathscr{F}$ is constructible. There is a decomposition $X = \cup_i X_i$ such that X_i are finitely many irreducible locally closed subschemes of X and $\mathscr{F}|_{X_i}$ are locally constant. If $\mathscr{F}_{\bar{x}} \neq 0$ for one $x \in X_i$, then $\mathscr{F}_{\bar{x}} \neq 0$ for all $x \in X_i$. We then have $\dim X_i \leq n$ by our assumption. This shows that $\dim(\operatorname{supp} \mathscr{F}) \leq n$, where $\operatorname{supp} \mathscr{F}$ is the support of $\mathscr{F}$.

We use induction on n. Suppose $n = 0$. Then $\operatorname{supp}\mathscr{F}$ consists of finitely many closed points in X. Put the reduced closed subscheme structure on $\operatorname{supp}\mathscr{F}$, and let $i : \operatorname{supp}\mathscr{F} \to X$ be the closed immersion. We have

$$R^q f_*\mathscr{F} \cong R^q(fi)_* i^*\mathscr{F}.$$

But fi is a finite morphism. So $R^q(fi)_*i^*\mathscr{F} = 0$ for all $q \geq 0$ and $(fi)_*i^*\mathscr{F}$ is supported on finitely many closed points of Y. In particular, we have $(R^q f_*\mathscr{F})_{\bar{b}} = 0$ for any $b \in Y$ satisfying $\dim \overline{\{b\}} > -q$.

Suppose $n = 1$. Replacing X by $\operatorname{supp}\mathscr{F}$, we may assume $\dim X = 1$. We have

$$(R^q f_*\mathscr{F})_{\bar{b}} \cong \varinjlim_{U} H^q(X \times_Y U, \mathscr{F}|_{X\times_Y U}),$$

where $U \to Y$ goes over the family of affine etale neighborhood of $\bar{b}$. Since f is affine, each $X \times_Y U$ is affine. We have $\dim (X \times_Y U) \leq 1$. By 7.2.13, we have

$$H^q(X \times_Y U, \mathscr{F}|_{X\times_Y U}) = 0$$

for any $q \geq 2$. So $(R^q f_*\mathscr{F})_{\bar{b}} = 0$ for any $q > 1 - \dim \overline{\{b\}}$ if $\dim \overline{\{b\}} = 0$. We have

$$\operatorname{supp}(R^q f_*\mathscr{F}) \subset \overline{f(X)}.$$

Note that

$$\dim \overline{f(X)} \leq 1.$$

To prove this, let X_i be an arbitrary irreducible component of X. Then $\overline{f(X_i)}$ is also irreducible. The generic point ξ_i of X_i is mapped to the generic point of $\overline{f(X_i)}$. It follows that

$$\dim \overline{f(X_i)} = \operatorname{tr.deg}(k(f(\xi_i))/k) \leq \operatorname{tr.deg}(k(\xi_i)/k) = \dim X_i \leq 1.$$

So we have $\dim \overline{f(X)} \leq 1$. If $\dim \overline{f(X)} = 0$, we have $(R^q f_*\mathscr{F})_{\bar{b}} = 0$ for all q and those b with $\dim \overline{\{b\}} > 0$. Suppose $\dim \overline{f(X)} = 1$. Then $(R^q f_*\mathscr{F})_{\bar{b}} = 0$ for all q if $\dim \overline{\{b\}} > 1$. It remains to show that for any $b \in \overline{f(X)}$ with $\dim \overline{\{b\}} = 1$, we have $(R^1 f_*\mathscr{F})_{\bar{b}} = 0$. Put the reduced closed subscheme structure on $\overline{f(X)}$. Let $g : X \to \overline{f(X)}$ be the morphism induced by f and let $i : \overline{f(X)} \to Y$ be the closed immersion. We have

$$R^1 f_*\mathscr{F} = i_* R^1 g_*\mathscr{F}.$$

We claim that there exists an open subset U of $\overline{f(X)}$ such that $\overline{f(X)} - U$ consists of finitely many closed point and that $g_U : g^{-1}(U) \to U$ is finite. Then $(R^1 g_*\mathscr{F})|_U = 0$. For any $b \in \overline{f(X)}$ with $\dim \overline{\{b\}} = 1$, we have $b \in U$

and hence $(R^1 f_* \mathscr{F})_{\bar{b}} = 0$. Let $\eta_1, \ldots, \eta_m$ be all the generic points of $\overline{f(X)}$ with $\dim \overline{\{\eta_j\}} = 1$. By 1.10.10 (iv), to prove our claim, it suffices to show that $g^{-1}(\eta_j) \to \eta_j$ is finite for each j. Since g maps closed points of X to closed points of $\overline{f(X)}$ and none of the η_j is a closed point, points in $g^{-1}(\eta_j)$ are generic points of irreducible components of X of dimension 1. The residue fields of X at such kind of generic points and the residue field of Y at η_j have transcendental degree 1 over k. It follows that the residue field of X at any point in $g^{-1}(\eta_j)$ is algebraic and hence finite over the residue field of Y at η_j. So $g^{-1}(\eta_j) \to \eta_j$ is finite. This proves the theorem for the case $n = 1$.

Suppose $n \geq 2$, and suppose that the theorem holds if n is replaced by any integer strictly lesser than n. Since X and Y are affine, $f : X \to Y$ can be factorized as a composite

$$X \xrightarrow{i} \mathbb{A}^m_Y \xrightarrow{\pi} Y$$

such that i is a closed immersion and π is the projection. We have

$$R^q f_* \mathscr{F} \cong R^q \pi_* (i_* \mathscr{F})$$

and $i_* \mathscr{F}$ satisfies the same condition as $\mathscr{F}$. We are thus reduced to the case where f is the projection $\pi : \mathbb{A}^n_Y \to Y$. Note that if $f = gh$ and if the theorem holds for g and h, then it holds for f. Indeed, we have a biregular spectral sequence

$$E_2^{pq} = R^p g_* R^q h_* \mathscr{F} \Rightarrow R^{p+q} f_* \mathscr{F}.$$

To prove $(R^q f_* \mathscr{F})_{\bar{b}} = 0$ for any b with $\dim \overline{\{b\}} > n - q$, it suffices to prove

$$(R^p g_* R^q h_* \mathscr{F})_{\bar{b}} = 0$$

for any b with $\dim \overline{\{b\}} > n - (p+q)$. Since the theorem holds for h, we have $(R^q h_* \mathscr{F})_{\bar{c}} = 0$ for any c with $\dim \overline{\{c\}} > n - q$. Since the theorem holds for g, we have $(R^p g_* R^q h_* \mathscr{F})_{\bar{b}} = 0$ for any b with $\dim \overline{\{b\}} > n - q - p$. This proves our assertion. The projection $\pi : \mathbb{A}^m_Y \to Y$ is the composite of projections

$$\mathbb{A}^m_Y \to \mathbb{A}^{m-1}_Y \to \cdots \to \mathbb{A}^1_Y \to Y.$$

So it suffices to consider the case where f is the projection $\mathbb{A}^1_Y \to Y$.

Let $Y = \operatorname{Spec} B$ for some finitely generated k-algebra B, let $b \in Y$, and let $\mathfrak{q}$ be the prime ideal of B corresponding to b. We need to show that $(R^q f_* \mathscr{F})_{\bar{b}} = 0$ for any $q > n - \dim \overline{\{b\}}$. First consider the case where $\dim \overline{\{b\}} > 0$. We have

$$\dim \overline{\{b\}} = \text{tr.deg}_k (B_{\mathfrak{q}} / \mathfrak{q} B_{\mathfrak{q}}),$$

where tr.deg_k denotes the transcendental degree over k. Replacing Y by an affine open neighborhood of b, we may assume that there exist $t_1, \ldots, t_d \in B$ such that their images in $B_\mathfrak{q}/\mathfrak{q}B_\mathfrak{q}$ form a separating transcendental basis for $B_\mathfrak{q}/\mathfrak{q}B_\mathfrak{q}$ over k, where $d = \dim \overline{\{b\}} > 0$. (We have reduced to the case where k is algebraically closed. Hence a separating transcendental basis exists.) We have

$$\mathfrak{q} \cap k[t_1, \ldots, t_d] = 0.$$

Let K be a separable closure of $B_\mathfrak{q}/\mathfrak{q}B_\mathfrak{q}$ and let $Y' = \text{Spec}\,(B \otimes_{k[t_1,\ldots,t_d]} K)$. There exists a point $b' \in Y'$ lying above b. Let $X' = \mathbb{A}^1_{Y'}$, let $f' : X' \to Y'$ be the projection, and let $\mathscr{F}'$ be the inverse image of $\mathscr{F}$ in X'.

$$\begin{array}{ccccc} X' = \mathbb{A}^1_{Y'} & \xrightarrow{f'} & Y' = \text{Spec}(B \otimes_{k[t_1,\ldots,t_d]} K) \to & \text{Spec}\, K & \\ \downarrow & & \downarrow & \downarrow & \\ X = \mathbb{A}^1_Y & \xrightarrow{f} & Y = \text{Spec}\, B & \to \text{Spec}\, k[t_1, \ldots, t_d] & \to \text{Spec}\, k. \end{array}$$

Since K is separable over $k(t_1, \ldots, t_d)$, the strict localization of Y' at $\bar{b}'$ is isomorphic to the strict localization of Y at $\bar{b}$. So by 5.9.5, we have

$$(R^q f_* \mathscr{F})_{\bar{b}} \cong (R^q f'_* \mathscr{F}')_{\bar{b}'}.$$

For any $a' \in X'$, let a be its image in X. We have

$$\dim \overline{\{a\}} = \dim \overline{\{a'\}} + d$$

by 7.5.3. Since $\mathscr{F}_{\bar{a}} = 0$ for any $a \in X$ with $\dim \overline{\{a\}} > n$, we have $\mathscr{F}'_{\bar{a}'} = 0$ for any $a' \in X'$ with $\dim \overline{\{a'\}} > n - d$. But $n - d \leq n - 1$. So we can apply the induction hypothesis to $\mathscr{F}'$ and to the morphism $f' : X' \to Y'$, and we get $(R^q f'_* \mathscr{F}')_{\bar{b}'} = 0$ if $\dim \overline{\{b'\}} > n - d - q$. So we have $(R^q f_* \mathscr{F})_{\bar{b}} = 0$ if $\dim \overline{\{b\}} > n - q$.

Next we consider the case where $\dim \overline{\{b\}} = 0$. We need to show that $(R^q f_* \mathscr{F})_{\bar{b}} = 0$ for any $q > n$. Let $j : X = \mathbb{A}^1_Y \hookrightarrow \mathbb{P}^1_Y$ be the open immersion and let $\bar{f} : \mathbb{P}^1_Y \to Y$ be the projection. We have a biregular spectral sequence

$$E_2^{pq} = R^p \bar{f}_* R^q j_* \mathscr{F} \Rightarrow R^{p+q} f_* \mathscr{F}.$$

It suffices to show that

$$(R^p \bar{f}_* R^q j_* \mathscr{F})_{\bar{b}} = 0$$

for any $p + q > n$. Recall that $n \geq 2$. If $q = 0$, we have

$$R^p \bar{f}_* R^q j_* \mathscr{F} = R^p \bar{f}_* j_* \mathscr{F} = 0$$

for any $p > 2$ by 7.4.5. For any $q > 0$, we have

$$(R^q j_* \mathscr{F})|_{\mathbb{A}^1_Y} = 0.$$

Since $\bar{f}|_{\mathbb{P}^1_Y - \mathbb{A}^1_Y}$ is an isomorphism, we have

$$R^p \bar{f}_* R^q j_* \mathscr{F} = 0$$

for any $p, q > 0$, and $\bar{f}_* R^q j_* \mathscr{F}$ can be identified with $(R^q j_* \mathscr{F})|_{\mathbb{P}^1_Y - \mathbb{A}^1_Y}$ for any $q > 0$. To prove our assertion, it suffices to prove $(R^q j_* \mathscr{F})_{\bar{a}} = 0$ for any $q > n$ and any point a in $\mathbb{P}^1_Y - \mathbb{A}^1_Y$.

Put the reduced closed subscheme structure on the closure $\overline{\operatorname{supp} \mathscr{F}}$ of $\operatorname{supp} \mathscr{F}$ in $\mathbb{P}^1_Y$, and let $j' : \overline{\operatorname{supp} \mathscr{F}} \cap \mathbb{A}^1_Y \hookrightarrow \overline{\operatorname{supp} \mathscr{F}}$ be the base change of j. If $a \notin \overline{\operatorname{supp} \mathscr{F}}$, we have $(R^q j_* \mathscr{F})_{\bar{a}} = 0$ for all q. If $a \in \overline{\operatorname{supp} \mathscr{F}}$, we have

$$(R^q j_* \mathscr{F})_{\bar{a}} \cong (R^q j'_* (\mathscr{F}|_{\overline{\operatorname{supp} \mathscr{F}} \cap \mathbb{A}^1_Y}))_{\bar{a}}.$$

Let R be the strict henselization of $\mathscr{O}_{\overline{\operatorname{supp} \mathscr{F}}, a}$. We have

$$(R^q j_* \mathscr{F})_{\bar{a}} \cong H^q(\mathbb{A}^1_Y \times_{\mathbb{P}^1_Y} \operatorname{Spec} R, \mathscr{F}) \cong H^q(\mathbb{A}^1_k \times_{\mathbb{P}^1_k} \operatorname{Spec} R, \mathscr{F}).$$

Here we denote also by $\mathscr{F}$ the inverse images of $\mathscr{F}$ on $\mathbb{A}^1_Y \times_{\mathbb{P}^1_Y} \operatorname{Spec} R$ and on $\mathbb{A}^1_k \times_{\mathbb{P}^1_k} \operatorname{Spec} R$. We need to prove that $H^q(\mathbb{A}^1_k \times_{\mathbb{P}^1_k} \operatorname{Spec} R, \mathscr{F}) = 0$ for any $q > n$.

Choose an inverse system $\{U_\alpha\}$ of a sufficiently small affine etale neighborhood of $\bar{a}$ in $\overline{\operatorname{supp} \mathscr{F}}$ such that $R = \varinjlim_\alpha \Gamma(U_\alpha, \mathscr{O}_{U_\alpha})$. We have

$$H^q(\mathbb{A}^1_k \times_{\mathbb{P}^1_k} \operatorname{Spec} R, \mathscr{F}) \cong \varinjlim_{\alpha} H^q(\mathbb{A}^1_k \times_{\mathbb{P}^1_k} U_\alpha, \mathscr{F}).$$

The point a is above the point ∞ of $\mathbb{P}^1_k$. Let $\widetilde{\mathbb{P}}^1_{k\infty}$ be the strict localization of $\mathbb{P}^1_k$ at ∞, and let $\tilde{\eta}_\infty$ be the generic point of $\widetilde{\mathbb{P}}^1_{k\infty}$. We have a commutative diagram

$$\begin{array}{ccc} \operatorname{Spec} R & \rightarrow & \widetilde{\mathbb{P}}^1_{k\infty} \\ \downarrow & & \downarrow \\ U_\alpha & \rightarrow & \mathbb{P}^1_k. \end{array}$$

Making the base change $\mathbb{A}^1_k \to \mathbb{P}^1_k$, we obtain a commutative diagram

$$\begin{array}{ccc} \mathbb{A}^1_k \times_{\mathbb{P}^1_k} \operatorname{Spec} R & \rightarrow & \tilde{\eta}_\infty \\ \downarrow & & \downarrow \\ \mathbb{A}^1_k \times_{\mathbb{P}^1_k} U_\alpha & \rightarrow & \mathbb{A}^1_k. \end{array}$$

The homomorphism

$$H^q(\mathbb{A}^1_k \times_{\mathbb{P}^1_k} U_\alpha, \mathscr{F}) \to H^q(\mathbb{A}^1_k \times_{\mathbb{P}^1_k} \operatorname{Spec} R, \mathscr{F})$$

factors through

$$H^q(\tilde{\eta}_\infty \times_{\mathbb{A}^1_k} (\mathbb{A}^1_k \times_{\mathbb{P}^1_k} U_\alpha), \mathscr{F}) \to H^q(\mathbb{A}^1_k \times_{\mathbb{P}^1_k} \operatorname{Spec} R, \mathscr{F})$$

To prove our assertion, it suffices to show $H^q(\tilde{\eta}_\infty \times_{\mathbb{A}^1_k} (\mathbb{A}^1_k \times_{\mathbb{P}^1_k} U_\alpha), \mathscr{F}) = 0$ for all $q > n$.

Since $\dim \overline{\operatorname{supp} \mathscr{F}} \leq n$, we have $\dim U_\alpha \leq n$, and hence $\dim(\mathbb{A}^1_k \times_{\mathbb{P}^1_k} U_\alpha) \leq n$. By 7.5.3, we have

$$\dim(\tilde{\eta}_\infty \times_{\mathbb{A}^1_k} (\mathbb{A}^1_k \times_{\mathbb{P}^1_k} U_\alpha)) \leq n - 1.$$

Let

$$g : \tilde{\eta}_\infty \times_{\mathbb{A}^1_k} (\mathbb{A}^1_k \times_{\mathbb{P}^1_k} U_\alpha) \to \tilde{\eta}_\infty$$

be the projection. Note that g is an affine morphism. Indeed, $\mathbb{A}^1_k \to \mathbb{P}^1_k$ is an affine morphism. So the projection $\mathbb{A}^1_k \times_{\mathbb{P}^1_k} U_\alpha \to U_\alpha$ is an affine morphism. Since U_α is affine, $\mathbb{A}^1_k \times_{\mathbb{P}^1_k} U_\alpha$ is also affine. So the morphism $\mathbb{A}^1_k \times_{\mathbb{P}^1_k} U_\alpha \to \mathbb{A}^1_k$ is affine, and hence g is affine. By the induction hypothesis, we have

$$R^q g_*(\mathscr{F}|_{\tilde{\eta}_\infty \times_{\mathbb{A}^1_k} (\mathbb{A}^1_k \times_{\mathbb{P}^1_k} U_\alpha)}) = 0$$

for all $q > n - 1$. Since the residue field of $\tilde{\eta}_\infty$ has transcendental degree 1 over the algebraically closed field k, by 4.5.11, we have

$$H^p(\tilde{\eta}_\infty, R^q g_*(\mathscr{F}|_{\tilde{\eta}_\infty \times_{\mathbb{A}^1_k} (\mathbb{A}^1_k \times_{\mathbb{P}^1_k} U_\alpha)})) = 0$$

for all $p > 1$. We have a biregular spectral sequence

$$E_2^{pq} = H^p(\tilde{\eta}_\infty, R^q g_*(\mathscr{F}|_{\tilde{\eta}_\infty \times_{\mathbb{A}^1_k} (\mathbb{A}^1_k \times_{\mathbb{P}^1_k} U_\alpha)})) \Rightarrow H^{p+q}(\tilde{\eta}_\infty \times_{\mathbb{A}^1_k} (\mathbb{A}^1_k \times_{\mathbb{P}^1_k} U_\alpha), \mathscr{F}).$$

It follows that $H^q(\tilde{\eta}_\infty \times_{\mathbb{A}^1_k} (\mathbb{A}^1_k \times_{\mathbb{P}^1_k} U_\alpha), \mathscr{F}) = 0$ for all $q > n$. This proves our assertion. □

Lemma 7.5.4. *Let k be an algebraically closed field, X an integral scheme of finite type over k, $n = \dim X$, η the generic point of X, $\mathscr{F}$ a torsion sheaf on η, and $j : \eta \to X$ the canonical morphism. Then we have $(R^q j_* \mathscr{F})_{\bar{a}} = 0$ for any $a \in X$ with $\dim \overline{\{a\}} > n - q$.*

Proof. Replacing X by an open affine neighborhood of a does not change $(R^q j_* \mathscr{F})_{\bar{a}}$. So we may assume that $X = \operatorname{Spec} A$ for some finitely generated k-algebra A. Let $\mathfrak{p}$ be the prime ideal of A corresponding to a, and let $d = \dim \overline{\{a\}}$. We have $\operatorname{tr.deg}_k(A_\mathfrak{p}/\mathfrak{p}A_\mathfrak{p}) = d$. Shrinking X, we may assume that there exist $t_1, \ldots, t_d \in A$ such that their images in $A_\mathfrak{p}/\mathfrak{p}A_\mathfrak{p}$ form a separating transcendental basis for $A_\mathfrak{p}/\mathfrak{p}A_\mathfrak{q}$ over k. We have

$$\mathfrak{p} \cap k[t_1, \ldots, t_d] = 0.$$

Let K be a separable closure of $A_\mathfrak{p}/\mathfrak{p}A_\mathfrak{p}$ and let $X' = \operatorname{Spec}(A\otimes_{k[t_1,\ldots,t_d]}K)$. There exists a point $a' \in X'$ lying above a. Since K is separable over $k(t_1,\ldots,t_d)$, the strict localization $\widetilde{X}'_{\bar{a}'}$ of X' at $\bar{a}'$ is isomorphic to the strict localization $\widetilde{X}_{\bar{a}}$ of X at $\bar{a}$. By 5.9.5, we have

$$(R^q j_*\mathscr{F})_{\bar{a}} \cong H^q(\eta\times_X \widetilde{X}_{\bar{a}},\mathscr{F}) \cong H^q(\eta\times_X \widetilde{X}'_{\bar{a}'},\mathscr{F}).$$

Let $K_1,\ldots,K_m$ be the residue fields of $\widetilde{X}'_{\bar{a}'}$ at its generic points. We have

$$\eta\times_X \widetilde{X}'_{\bar{a}'} \cong \operatorname{Spec} K_1 \coprod\cdots\coprod \operatorname{Spec} K_m,$$

and

$$\begin{aligned}\text{tr.deg}(K_i/K) &= \text{tr.deg}(K_i/k) - \text{tr.deg}(K/k)\\ &= \text{tr.deg}(k(\eta)/k) - \text{tr.deg}(K/k)\\ &= n-d.\end{aligned}$$

By 4.5.10, we have

$$H^q(\operatorname{Spec} K_i,\mathscr{F}) = 0$$

for all $q > n-d$. So we have $(R^q j_*\mathscr{F})_{\bar{a}} = 0$ for all $q > n-d$.

Theorem 7.5.5. *Let k be a separably closed field, X a scheme of finite type over k, n an integer, $\mathscr{F}$ a torsion sheaf on X such that $\mathscr{F}_{\bar{a}} = 0$ for any $a\in X$ with $\dim\overline{\{a\}} > n$. Then $H^q(X,\mathscr{F}) = 0$ for any $q > 2n$. In particular, we have $H^q(X,\mathscr{F}) = 0$ for any $q > 2\dim X$ and any torsion sheaf $\mathscr{F}$ on X.*

Proof. We may assume that k is algebraically closed and $\mathscr{F}$ is constructible. We use induction on n. If $n = 0$, then $\operatorname{supp}\mathscr{F}$ consists of finitely many closed point of X, and we have $H^q(X,\mathscr{F}) = 0$ for all $q > 0$. Suppose that the theorem holds if n is replaced by any integer strictly less than n. Let η_v be those points in $\operatorname{supp}\mathscr{F}$ with $\dim\overline{\{\eta_v\}} = n$, let $j_v : \operatorname{Spec} k(\eta_v)\to X$ be the canonical morphisms, and let $\mathscr{K}$ (resp. $\mathscr{C}$) be the kernel (resp. cokernel) of the canonical morphism

$$\mathscr{F}\to\prod_v j_{v*}j_v^*\mathscr{F}.$$

We have

$$\mathscr{K}_{\bar{a}} = \mathscr{C}_{\bar{a}} = 0$$

for any a in X with $\dim\overline{\{a\}} > n-1$. Applying the induction hypothesis to $\mathscr{K}$ and $\mathscr{C}$, we see that $H^q(X,\mathscr{K})$ and $H^q(X,\mathscr{C})$ vanish for any $q > 2(n-1)$. To prove $H^q(X,\mathscr{F}) = 0$ for any $q > 2n$, it suffices to show that

$$H^q(X, j_{v*}j_v^*\mathscr{F}) = 0$$

for any $q > 2n$. Replacing X by the reduced closed subscheme $\overline{\{\eta_v\}}$, we are reduced to prove the following statement: Let X be an integral scheme of finite type over an algebraically closed field k of dimension n, let η be the generic point of X, and let $j : \eta \to X$ be the canonical morphism. Then for any torsion sheaf $\mathscr{F}$ on η, we have $H^q(X, j_*\mathscr{F}) = 0$ for any $q > 2n$. We have a biregular spectral sequence

$$E_2^{pq} = H^p(X, R^q j_*\mathscr{F}) \Rightarrow H^{p+q}(\eta, \mathscr{F}).$$

By 7.5.4, we have $(R^q j_*\mathscr{F})_{\bar{a}} = 0$ if $\dim \overline{\{a\}} > n - q$. For each $q > 0$, we can apply the induction hypothesis to $R^q j_*\mathscr{F}$, and we get

$$E_2^{pq} = H^p(X, R^q j_*\mathscr{F}) = 0$$

if $q > 0$ and $p > 2(n - q)$. This implies that

$$E_2^{p0} \cong H^p(\eta, \mathscr{F})$$

for any $p > 2n$. We have $\mathrm{tr.deg}(k(\eta)/k) = n$. By 4.5.10, we have

$$H^p(\eta, \mathscr{F}) = 0$$

for any $p > n$. So we have

$$H^p(X, j_*\mathscr{F}) = E_2^{p0} = 0$$

for any $p > 2n$. □

Corollary 7.5.6. *Let k be a field, X and Y schemes of finite type over k, and $f : X \to Y$ a k-morphism. Then $R^q f_*\mathscr{F} = 0$ for any $q > 2\dim X$ and any torsion sheaf $\mathscr{F}$ on X. In particular, Rf_* has finite cohomological dimension on the category of torsion sheaves.*

Proof. We may assume that k is algebraically closed. $R^q f_*\mathscr{F}$ is the sheaf associated to the presheaf

$$V \mapsto H^q(V \times_Y X, \mathscr{F})$$

for any etale X-scheme V. By 7.5.5, we have

$$H^q(V \times_Y X, \mathscr{F}) = 0$$

for any $q > 2\dim X$. □

Corollary 7.5.7. *Let $f : X \to S$ be a morphism of finite type, $x \in X$, $\widetilde{S}_{f(\bar{x})}$ the strict localization of S at $f(\bar{x})$, $j : t \to \widetilde{S}_{f(\bar{x})}$ a geometric point, and $\tilde{j} : X \times_S t \to X \times_S \widetilde{S}_{\bar{f}(x)}$ the morphism induced by j by base change. Then $R^q \tilde{j}_*\mathscr{F} = 0$ for any $q > \dim(X \times_S t)$ and any torsion sheaf $\mathscr{F}$ on $X \times_S t$. In particular, $R\tilde{j}_*$ has finite cohomological dimension on the category of torsion sheaves.*

Proof. For each q, $R^q\tilde{j}_*\mathscr{F}$ is the sheaf associated to the presheaf

$$V \mapsto H^q(V \times_{\widetilde{S}_{f(\bar{x})}} t, \mathscr{F})$$

for any etale $X \times_S \widetilde{S}_{f(\bar{x})}$-scheme V. We have

$$\dim(V \times_{\widetilde{S}_{f(\bar{x})}} t) \leq \dim(X \times_S t).$$

If V is affine, $V \times_{\widetilde{S}_{f(\bar{x})}} t$ is also affine, and we have

$$H^q(V \times_{\widetilde{S}_{f(\bar{x})}} t, \mathscr{F}) = 0$$

for any $q > \dim(X \times_S t)$ by 7.5.1. So we have $R^q\tilde{j}_*\mathscr{F} = 0$ for any $q > \dim(X \times_S t)$. □

7.6 Local Acyclicity

([SGA $4\frac{1}{2}$] Arcata V 1, Th. finitude 2.12–2.16, Appendice 2.9, 2.10.)

Consider a Cartesian diagram

$$\begin{array}{ccc} X \times_S S' & \xrightarrow{g'} & X \\ f' \downarrow & & \downarrow f \\ S' & \xrightarrow{g} & S. \end{array}$$

Let A be a ring. Suppose that one of the following conditions holds:

(a) Rg_* and Rg'_* have finite cohomological dimension, $K \in \operatorname{ob} D^-(X, A)$ and $L \in \operatorname{ob} D^-(S', A)$.

(b) $K \in \operatorname{ob} D^b_{\mathrm{tf}}(X, A)$ has finite Tor-dimension and $L \in \operatorname{ob} D^+(S', A)$.

(c) Rg_* and Rg'_* have finite cohomological dimension, $K \in \operatorname{ob} D^b_{\mathrm{tf}}(X, A)$ has finite Tor-dimension and $L \in \operatorname{ob} D(S', A)$.

Then we define a canonical morphism

$$K \otimes^L_A f^* Rg_* L \to Rg'_*(g'^* K \otimes^L_A f'^* L)$$

as the composite of canonical morphisms

$$K \otimes^L_A f^* Rg_* L \to K \otimes^L_A Rg'_* f'^* L \to Rg'_*(g'^* K \otimes^L_A f'^* L).$$

By 7.4.8, this canonical morphism is an isomorphism if g is proper. In this section, we give conditions on f under which this canonical morphism is an isomorphism.

Let X be a scheme and let $t \to X$ be a geometric point in X, where t is the spectrum of a separably closed field. We often denote by $k(t)$ the separably closed field such that $t = \operatorname{Spec} k(t)$. Let x be the image of t in

X. We say that t is an *algebraic geometric point* if $k(t)$ is algebraic over the residue field $k(x)$.

Let $f : X \to S$ be a morphism. For any $x \in X$ (resp. $s \in S$), let $\widetilde{X}_{\bar{x}}$ (resp. $\widetilde{S}_{\bar{s}}$) be the strict localization of X (resp. S) at $\bar{x}$ (resp. $\bar{s}$). Let $K \in \operatorname{ob} D^+(X)$. We say that f is *locally acyclic at $x \in X$ relative to K* if for any algebraic geometric point t of $\widetilde{S}_{f(\bar{x})}$, the canonical morphism

$$K_{\bar{x}} = R\Gamma(\widetilde{X}_{\bar{x}}, K) \to R\Gamma(\widetilde{X}_{\bar{x}} \times_{\widetilde{S}_{f(\bar{x})}} t, K)$$

is an isomorphism. By 5.7.2 (ii), this is equivalent to saying that the above canonical morphism is an isomorphism if t is of the form $\bar{y}$ for any $y \in \widetilde{S}_{f(\bar{x})}$. We say f is *locally acyclic relative to K* if it is locally acyclic relative to K at every point of X. Let $s \in S$ and let t be an algebraic geometric point of $\widetilde{S}_{\bar{s}}$. Fix notation by the following diagram:

$$\begin{array}{ccccc} X_t = X \times_S t & \xrightarrow{\tilde{j}} & X \times_S \widetilde{S}_{\bar{s}} & \xleftarrow{\tilde{i}} & X_{\bar{s}} = X \times_S \bar{s} \\ {\scriptstyle f_t} \downarrow & & \tilde{f} \downarrow & & \downarrow {\scriptstyle f_{\bar{s}}} \\ t & \xrightarrow{j} & \widetilde{S}_{\bar{s}} & \xleftarrow{i} & \bar{s}. \end{array}$$

where all vertical arrows are induced by f by base change. Let $\widetilde{K}$ be the inverse image of K on $X \times_S \widetilde{S}_{\bar{s}}$. For any $x \in f^{-1}(s)$, the strict localization of $X \times_S \widetilde{S}_{\bar{s}}$ at $\bar{x}$ is isomorphic to the strict localization of X at $\bar{x}$, and we have

$$\begin{aligned} &(\tilde{i}^* \widetilde{K})_{\bar{x}} \cong K_{\bar{x}}, \\ &(\tilde{i}^* R\tilde{j}_* \tilde{j}^* \widetilde{K})_{\bar{x}} \cong (R\tilde{j}_* \tilde{j}^* \widetilde{K})_{\bar{x}} \cong R\Gamma(\widetilde{X}_{\bar{x}} \times_{\widetilde{S}_{f(\bar{x})}} t, K). \end{aligned}$$

So f is locally acyclic relative to K if and only if for any $s \in S$ and any algebraic geometric point t of $\widetilde{S}_{\bar{s}}$, the canonical morphism

$$\tilde{i}^* \widetilde{K} \to \tilde{i}^* R\tilde{j}_* \tilde{j}^* \widetilde{K}$$

is an isomorphism. We say that f is *universally locally acyclic relative to K* if for any morphism $S' \to S$, the base change $X \times_S S' \to S'$ of f is locally acyclic relative to the inverse image $K|_{X \times_S S'}$ of K. By 5.9.3 or 5.9.6, f is universally locally acyclic relative to K if and only if for any morphism $S' \to S$ locally of finite type, the base change $X \times_S S' \to S'$ of f is locally acyclic relative to the inverse image $K|_{X \times_S S'}$ of K.

Let A be a ring, and let $K \in \operatorname{ob} D^b_{\mathrm{tf}}(X, A)$. We say that f is *strongly locally acyclic at x relative to K* if for any algebraic geometric point t of $\widetilde{S}_{f(\bar{x})}$ and any A-module M, the canonical morphism

$$K_{\bar{x}} \otimes^L_A M = R\Gamma(\widetilde{X}_x, K) \otimes^L_A M \to R\Gamma(\widetilde{X}_x \times_{\widetilde{S}_{f(x)}} t, K \otimes^L_A M)$$

is an isomorphism. We say that f is *strongly locally acyclic relative to* K if it is strongly locally acyclic relative to K at every point of X. For any $s \in S$ and any algebraic geometric point t of $\widetilde{S}_{\bar{s}}$, define a canonical morphism

$$\tilde{i}^*(\widetilde{K} \otimes_A^L \tilde{f}^* Rj_* M) \to \tilde{i}^* R\tilde{j}_*(\tilde{j}^* \widetilde{K} \otimes_A^L f_t^* M)$$

to be the composite of canonical morphisms

$$\tilde{i}^*(\widetilde{K} \otimes_A^L \tilde{f}^* Rj_* M) \to \tilde{i}^*(\widetilde{K} \otimes_A^L R\tilde{j}_* f_t^* M) \to \tilde{i}^* R\tilde{j}_*(\tilde{j}^* \widetilde{K} \otimes_A^L f_t^* M).$$

Lemma 7.6.1. *In the above notation, f is strongly locally acyclic relative to K if and only if for any $s \in S$ and any algebraic geometric point t of $\widetilde{S}_{\bar{s}}$, the canonical morphism*

$$\tilde{i}^*(\widetilde{K} \otimes_A^L \tilde{f}^* Rj_* M) \to \tilde{i}^* R\tilde{j}_*(\tilde{j}^* \widetilde{K} \otimes_A^L f_t^* M)$$

is an isomorphism.

Proof. In the following, we denote also by M the constant sheaf associated to M on any scheme. Note that the canonical morphism

$$i^* M \to i^* Rj_* j^* M$$

is an isomorphism. Indeed, for any q, $R^q j_* j^* M$ is the sheaf associated to the presheaf

$$V \mapsto H^q(V \times_{\widetilde{S}_{\bar{s}}} t, M)$$

for any etale $\widetilde{S}_{\bar{s}}$-scheme V. Since t is the spectrum of a separably closed field, $V \times_{\widetilde{S}_{\bar{s}}} t$ is a disjoint union of copies of t and hence

$$H^q(V \times_{\widetilde{S}_{\bar{s}}} t, M) = 0$$

for any $q \geq 1$. So

$$R^q j_* j^* M = 0$$

for any $q \geq 1$. Moreover, we have

$$(i^* j_* j^* M)_{\bar{s}} \cong \Gamma(\widetilde{S}_{\bar{s}}, j_* j^* M) \cong M.$$

So we have

$$i^* M \cong i^* Rj_* j^* M.$$

This implies that the canonical morphism

$$\tilde{i}^*(\widetilde{K} \otimes_A^L \tilde{f}^* M) \to \tilde{i}^*(\widetilde{K} \otimes_A^L \tilde{f}^* Rj_* j^* M)$$

is an isomorphism.

We have a commutative diagram

$$\begin{array}{ccccc}
 & & \tilde{i}^*(R\tilde{j}_*\tilde{j}^*\widetilde{K}\otimes^L_A \tilde{f}^*M) & & \\
 & \nearrow & & \searrow & \\
\tilde{i}^*(\widetilde{K}\otimes^L_A \tilde{f}^*M) & \to & \tilde{i}^*(\widetilde{K}\otimes^L_A R\tilde{j}_*\tilde{j}^*\tilde{f}^*M) & \to & \tilde{i}^*R\tilde{j}_*(\tilde{j}^*\widetilde{K}\otimes^L_A \tilde{j}^*\tilde{f}^*M) \\
\cong\downarrow & & \cong\downarrow & & \cong\downarrow \\
\tilde{i}^*(\widetilde{K}\otimes^L_A \tilde{f}^*Rj_*j^*M) & \to & \tilde{i}^*(\widetilde{K}\otimes^L_A R\tilde{j}_*f_t^*j^*M) & \to & \tilde{i}^*R\tilde{j}_*(\tilde{j}^*\widetilde{K}\otimes^L_A f_t^*j^*M),
\end{array}$$

where all vertical arrows are isomorphisms. By definition, f is strongly locally acyclic relative to K if and only if the composite

$$\tilde{i}^*(\widetilde{K}\otimes^L_A \tilde{f}^*M)\to \tilde{i}^*(R\tilde{j}_*\tilde{j}^*\widetilde{K}\otimes^L_A \tilde{f}^*M)\to \tilde{i}^*R\tilde{j}_*(\tilde{j}^*\widetilde{K}\otimes^L_A \tilde{j}^*\tilde{f}^*M)$$

on the upper part of the above diagram defines an isomorphism

$$\tilde{i}^*(\widetilde{K}\otimes^L_A \tilde{f}^*M)\to \tilde{i}^*R\tilde{j}_*(\tilde{j}^*\widetilde{K}\otimes^L_A \tilde{j}^*\tilde{f}^*M).$$

By the commutativity of the above diagram, this condition is equivalent to saying that the composite

$$\tilde{i}^*(\widetilde{K}\otimes^L_A \tilde{f}^*Rj_*j^*M)\to \tilde{i}^*(\widetilde{K}\otimes^L_A R\tilde{j}_*f_t^*j^*M)\to \tilde{i}^*R\tilde{j}_*(\tilde{j}^*\widetilde{K}\otimes^L_A f_t^*j^*M)$$

on the lower part of the above diagram defines an isomorphism

$$\tilde{i}^*(\widetilde{K}\otimes^L_A \tilde{f}^*Rj_*j^*M)\to \tilde{i}^*R\tilde{j}_*(\tilde{j}^*\widetilde{K}\otimes^L_A f_t^*j^*M).$$

This proves our assertion. □

If $R\tilde{j}_*$ has finite cohomological dimension for any $s\in S$ and any algebraic geometric point $j:t\to\widetilde{S}_{\bar{s}}$, then in the above discussion, we do not need to assume that K has finite Tor-dimension and can talk about the strong local acyclicity of f relative to K for any $K\in D^-(X,A)$. Many results in this section also hold under this assumption.

We say that f is *universally strongly locally acyclic relative to* K if for any morphism $S'\to S$, the base change $X\times_S S'\to S'$ of f is strongly locally acyclic relative to the inverse image $K|_{X\times_S S'}$ of K. By 5.9.3 or 5.9.6, f is universally strongly locally acyclic relative to K if and only if for any morphism $S'\to S$ locally of finite type, the base change $X\times_S S'\to S'$ of f is strongly locally acyclic relative to the inverse image $K|_{X\times_S S'}$ of K.

Proposition 7.6.2. *Let $f:X\to S$ be a morphism, A a ring and $K\in\mathrm{ob}\,D^-(X,A)$. Suppose for any $x\in X$ and any algebraic geometric point $j:t\to\widetilde{S}_{f(\bar{x})}$, that the functor $R\tilde{j}_*$ has finite cohomological dimension, where $\tilde{j}:X\times_S t\to X\times_S\widetilde{S}_{\bar{f}(x)}$ is the morphism induced by j by base change. If f is locally acyclic relative to K, then f is strongly locally acyclic relative to K.*

Note that if f is of finite type and A is a torsion ring, then $R\tilde{j}_*$ has finite cohomological dimension by 7.5.7.

Proof. Keep the notation in the proof of 7.6.1. If f is locally acyclic relative to K, we have

$$\tilde{i}^*\widetilde{K} \cong \tilde{i}^*R\tilde{j}_*\tilde{j}^*\widetilde{K}.$$

So for any A-module M, we have

$$\tilde{i}^*(\widetilde{K} \otimes_A^L \tilde{f}^*M) \cong \tilde{i}^*(R\tilde{j}_*\tilde{j}^*\widetilde{K} \otimes_A^L \tilde{f}^*M).$$

By 6.5.5, we have

$$R\tilde{j}_*\tilde{j}^*\widetilde{K} \otimes_A^L \tilde{f}^*M \cong R\tilde{j}_*(\tilde{j}^*\widetilde{K} \otimes_A^L \tilde{j}^*\tilde{f}^*M).$$

It follows that

$$\tilde{i}^*(\widetilde{K} \otimes_A^L \tilde{f}^*M) \cong \tilde{i}^*R\tilde{j}_*(\tilde{j}^*\widetilde{K} \otimes_A^L \tilde{j}^*\tilde{f}^*M).$$

So f is strongly locally acyclic relative to K. □

Proposition 7.6.3. *Let A be a ring, let $f : X \to S$ be a morphism, and let $K \in \mathrm{ob}\, D^b_{\mathrm{tf}}(X, A)$. Suppose that f is locally acyclic (resp. strongly locally acyclic) relative to K. Then for any quasi-finite morphism $g : S' \to S$, the base change $f' : X' = X \times_S S' \to S'$ of f is locally acyclic (resp. strongly locally acyclic) relative to the inverse image $K' = K|_{X'}$ of K. Conversely, if f' is locally acyclic (resp. strongly locally acyclic) relative to K' and g is surjective, then f is locally acyclic (resp. strongly locally acyclic) relative to K.*

Proof. We prove the statements for local acyclicity. The problem is local with respect to S'. So we may assume that g is separated and S is quasi-compact and quasi-separated. By the Zariski Main Theorem in [EGA] IV 18.12.13, g is the composite of an open immersion and a finite morphism. Let $x' \in X'$ and let x, s', s be its images in X, S', S, respectively. By 2.8.20, $\widetilde{S}'_{\bar{s}'}$ is a connected component of $S' \times_S \widetilde{S}_{\bar{s}}$ and $\widetilde{S}'_{\bar{s}'} \to \widetilde{S}_{\bar{s}}$ is finite. Similarly $\widetilde{X}'_{\bar{x}'}$ is a connected component of $\widetilde{X}_{\bar{x}} \times_{\widetilde{S}_{\bar{s}}} \widetilde{S}'_{\bar{s}'}$. But $k(\bar{s}')$ is a purely inseparable extension of $k(\bar{s})$, so there exists only one point in $\widetilde{X}_{\bar{x}} \times_{\widetilde{S}_{\bar{s}}} \widetilde{S}'_{\bar{s}'}$ above the closed point of $\widetilde{X}_{\bar{x}}$. Hence we must have

$$\widetilde{X}'_{\bar{x}'} \cong \widetilde{X}_{\bar{x}} \times_{\widetilde{S}_{\bar{s}}} \widetilde{S}'_{\bar{s}'}.$$

For any algebraic geometric point $t \to \widetilde{S}'_{\bar{s}'}$, t is also an algebraic geometric point of $\widetilde{S}_{\bar{s}}$, and we have

$$\widetilde{X}'_{\bar{x}'} \times_{\widetilde{S}'_{\bar{s}'}} t \cong \widetilde{X}_{\bar{x}} \times_{\widetilde{S}_{\bar{s}}} t.$$

If f is locally acyclic relative to K at x, then we have an isomorphism

$$K_{\bar{x}} \cong R\Gamma(\widetilde{X}_{\bar{x}} \times_{\widetilde{S}_{\bar{s}}} t, K).$$

It follows that

$$K'_{\bar{x}'} \cong R\Gamma(\widetilde{X}'_{\bar{x}'} \times_{\widetilde{S}'_{\bar{s}'}} t, K').$$

So f' is locally acyclic relative to K' at x'. The same argument also shows that if f' is locally acyclic relative to K' at x', then f is locally acyclic relative to K at x. Our assertion follows. □

Corollary 7.6.4. *Let $f : X \to S$ be a morphism, A a ring, $K \in \mathrm{ob}\, D^b_{\mathrm{tf}}(X, A)$, $(S_\lambda, u_{\lambda\mu})$ an inverse system of schemes affine and quasi-finite over S, $S' = \varprojlim_\lambda S_\lambda$, $f' : X' = X \times_S S' \to S'$ and $K' = K|_{X'}$ the base changes of f and K, respectively. If f is locally acyclic (resp. strongly locally acyclic) relative to K, then f' is locally acyclic (resp. strongly locally acyclic) relative to K'. Conversely, if f' is locally acyclic (resp. strongly locally acyclic) relative to K' and $S' \to S$ is surjective, then is f locally acyclic (resp. strongly locally acyclic) relative to K.*

Proof. We prove the statements for local acyclicity. Let $x' \in X'$ and let $s', x_\lambda, s_\lambda, x, s$ be its images in $S', X_\lambda = X \times_S S_\lambda, S_\lambda, X, S$, respectively. Then we have

$$\widetilde{S}'_{\bar{s}'} \cong \varprojlim_\lambda \widetilde{S}_{\lambda, \bar{s}_\lambda}, \quad \widetilde{X}_{\lambda, \bar{x}_\lambda} \cong \widetilde{X}_{\bar{x}} \times_{\widetilde{S}_{\bar{s}}} \widetilde{S}_{\lambda, \bar{s}_\lambda}, \quad \widetilde{X}'_{\bar{x}'} \cong \varprojlim_\lambda \widetilde{X}_{\lambda, \bar{x}_\lambda} \cong \widetilde{X}_{\bar{x}} \times_{\widetilde{S}_{\bar{s}}} \widetilde{S}'_{\bar{s}'}.$$

Any algebraic geometric point t of $\widetilde{S}'_{\bar{s}'}$ is also an algebraic geometric point of $\widetilde{S}_{\bar{s}}$. We then use the same argument as in the proof of 7.6.3. □

Lemma 7.6.5. *Let $(A, \mathfrak{m})$ be a normal local integral domain, K its fraction field, and $k = A/\mathfrak{m}$. If K is separably closed, then A is strictly local.*

Proof. Let $\overline{K}$ be an algebraic closure of K, let $\overline{A}$ be the integral closure of A in $\overline{K}$, let $\overline{\mathfrak{m}}$ be a maximal ideal of $\overline{A}$, and let $\bar{k} = \overline{A}/\overline{\mathfrak{m}}$. Then $\overline{\mathfrak{m}}$ is above the maximal ideal $\mathfrak{m}$ of A, and hence $\bar{k}$ is an algebraic extension of k. First let us prove k is separably closed. Let $f(t) \in k[t]$ be a monic polynomial and let $F(t) \in A[t]$ be a monic polynomial lifting $f(t)$. Then we have

$$F(t) = (t - a_1) \cdots (t - a_n)$$

for some $a_1, \ldots, a_n \in \overline{A}$, and hence

$$f(t) = (t - \bar{a}_1) \cdots (t - \bar{a}_n),$$

where for each i, $\bar{a}_i$ is the image of a_i in $\bar{k}$. If $\bar{a}_i$ is a simple root of $f(t)$, then a_i is a simple root of $F(t)$. Since K is separably closed, we must have $a_i \in K$. Since A is normal, we must have $a_i \in A$. So we have $\bar{a}_i \in k$. Thus any simple root of $f(t)$ lies in k, and hence k is separably closed.

Next we show that A is henselian. Let $F(t) \in A[t]$ be a monic polynomial, $f(t) \in k[t]$ the reduction of $F(t)$ mod $\mathfrak{m}$, and

$$f(t) = (t - \bar{a})g(t)$$

a factorization of $f(t)$ in $k[t]$ such that $g(t)$ is monic and relatively prime to $t - \bar{a}$. Again let

$$F(t) = (t - a_1) \cdots (t - a_n)$$

be a factorization of $F(t)$ in $\overline{A}[t]$. It induces a factorization

$$f(t) = (t - \bar{a}_1) \cdots (t - \bar{a}_n)$$

of $f(t)$ in $\bar{k}[t]$. Without loss of generality, assume $\bar{a} = \bar{a}_1$. Since $g(t)$ is relatively prime to $t - \bar{a}$, $\bar{a}_1$ is a simple root of $f(t)$. As before, this implies that $a_1 \in A$. We thus have a factorization

$$F(t) = (t - a_1)G(t)$$

in $A[t]$ lifting the factorization $f(t) = (t - a)g(t)$. By Nakayama's lemma, the ideal of $A[t]$ generated by $t - a_1$ and $G(t)$ is $A[t]$. By 2.8.3 (vi), A is henselian. □

Lemma 7.6.6. *Let $f : X \to S$ be a morphism, A a ring, $K \in \mathrm{ob}\, D^b_{\mathrm{tf}}(X, A)$, and $\gamma : t \to S$ an algebraic geometric point. Fix notation by the following Cartesian diagram:*

$$\begin{array}{ccc} X \times_S t & \xrightarrow{\gamma'} & X \\ {\scriptstyle f_t}\downarrow & & \downarrow{\scriptstyle f} \\ t & \xrightarrow{\gamma} & S. \end{array}$$

(i) *If f is locally acyclic relative to K, then we have a canonical isomorphism*

$$K \otimes^L_A f^* R\gamma_* A \xrightarrow{\cong} R\gamma'_* \gamma'^* K.$$

(ii) *If f is strongly locally acyclic relative to K, then for any A-module M, we have a canonical isomorphism*

$$K \otimes^L_A f^* R\gamma_* M \xrightarrow{\cong} R\gamma'_* (\gamma'^* K \otimes^L_A f_t^* M).$$

Proof. We prove (ii). Let S' be the normalization in t of $\overline{\{\gamma(t)\}}$ with the reduced closed subscheme structure. Fix notation by the following commutative diagram:

$$\begin{array}{ccccc} X_t & \stackrel{\beta'}{\to} & X\times_S S' & \stackrel{\alpha'}{\to} & X \\ {\scriptstyle f_t}\downarrow & & {\scriptstyle f'}\downarrow & & {\scriptstyle f}\downarrow \\ t & \stackrel{\beta}{\to} & S' & \stackrel{\alpha}{\to} & S, \end{array}$$

where $\alpha\beta = \gamma$. Note that $\alpha : S' \to S$ is an integral morphism. Using 5.9.7 and the same argument as the proof of 7.4.7, one can show that

$$K \otimes_A^L f^* R\alpha_* R\beta_* M \cong R\alpha'_*(\alpha'^* K \otimes_A^L f'^* R\beta_* M).$$

We have

$$\begin{aligned} K \otimes_A^L f^* R\gamma_* M &\cong K \otimes_A^L f^* R\alpha_* R\beta_* M \\ &\cong R\alpha'_*(\alpha'^* K \otimes_A^L f'^* R\beta_* M) \\ R\gamma'_*(\gamma'^* K \otimes_A^L f_t^* M) &\cong R\alpha'_* R\beta'_*(\beta'^*\alpha'^* K \otimes_A^L f_t^* M). \end{aligned}$$

To prove our assertion, it suffices to show

$$\alpha'^* K \otimes_A^L f'^* R\beta_* M \cong R\beta'_*(\beta'^*\alpha'^* K \otimes_A^L f_t^* M).$$

For any $s' \in S'$, $\mathscr{O}_{S',s'}$ is a normal local integral domain, and its fraction field is $k(t)$, which is separably closed. By 7.6.5, $\mathscr{O}_{S',s'}$ is strictly local. So the strict localization $\widetilde{S}'_{\bar{s}'}$ of S' at $\bar{s}'$ can be identified with $\operatorname{Spec}\mathscr{O}_{S',s'}$. It follows that the squares in the following commutative diagram are Cartesian:

$$\begin{array}{ccccc} t & \stackrel{j}{\to} & \widetilde{S}'_{\bar{s}'} & \leftarrow & \bar{s}' \\ \| & & \downarrow & & \| \\ t & \stackrel{\beta}{\to} & S' & \leftarrow & \bar{s}'. \end{array}$$

So the squares in the following commutative diagram are Cartesian:

$$\begin{array}{ccccc} X_t & \stackrel{\tilde{j}}{\to} & X\times_S \widetilde{S}'_{\bar{s}'} & \stackrel{\tilde{i}}{\leftarrow} & X_{\bar{s}'} \\ \| & & \downarrow & & \| \\ X_t & \stackrel{\beta'}{\to} & X\times_S S' & \stackrel{i'}{\leftarrow} & X_{\bar{s}'}. \end{array}$$

Let

$$\tilde{f} : X\times_S \widetilde{S}'_{\bar{s}'} \to \widetilde{S}'_{\bar{s}'}$$

be the base change of f. We then have

$$\begin{aligned} i'^*(\alpha'^* K \otimes_A^L f'^* R\beta_* M) &\cong \tilde{i}^*((K|_{X\times_S \widetilde{S}'_{\bar{s}'}}) \otimes_A^L \tilde{f}^* Rj_* M), \\ i'^* R\beta'_*(\beta'^*\alpha'^* K \otimes_A^L f_t^* M) &\cong \tilde{i}^* R\tilde{j}_*(\tilde{j}^*(K|_{X\times_S \widetilde{S}'_{\bar{s}'}}) \otimes_A^L f_t^* M). \end{aligned}$$

By 7.6.4, f' is strongly locally acyclic relative to α'^*K. So we have

$$\tilde{i}^*((K|_{X\times_S\tilde{S}'_{\bar{s}'}})\otimes^L_A \tilde{f}^*Rj_*M)\cong \tilde{i}^*R\tilde{j}_*(\tilde{j}^*(K|_{X\times_S\tilde{S}'_{\bar{s}'}})\otimes^L_A f_t^*M).$$

It follows that

$$i'^*(\alpha'^*K\otimes^L_A f'^*R\beta_*M)\cong i'^*R\beta'_*(\beta'^*\alpha'^*K\otimes^L_A f_t^*M).$$

This is true for any $s'\in S'$. So we have

$$\alpha'^*K\otimes^L_A f'^*R\beta_*M\cong R\beta'_*(\beta'^*\alpha'^*K\otimes^L_A f_t^*M). \qquad \square$$

Lemma 7.6.7. *Consider a Cartesian diagram*

$$\begin{array}{ccc} X\times_S S' & \stackrel{g'}{\to} & X \\ {\scriptstyle f'}\downarrow & & \downarrow{\scriptstyle f} \\ S' & \stackrel{g}{\to} & S. \end{array}$$

Let A be a noetherian ring and let $K\in \mathrm{ob}\, D^b_{\mathrm{tf}}(X,A)$. Suppose that S is a noetherian scheme, g is an open immersion, and one of the following conditions holds:

(a) *f is locally acyclic relative to K, and every finitely generated A-module can be embedded into a free A-module.*

(b) *f is strongly locally acyclic relative to K.*

Then for any $L\in \mathrm{ob}\, D^+(S',A)$, we have a canonical isomorphism

$$K\otimes^L_A f^*Rg_*L\stackrel{\cong}{\to} Rg'_*(g'^*K\otimes^L_A f'^*L).$$

Remark 7.6.8. Note that if $A=R/I$ for a discrete valuation ring R and a nonzero ideal I of R, then every finitely generated A-module can be embedded into a free A-module. Every finitely generated $\mathbb{Z}/n$-module can also be embedded into a free $\mathbb{Z}/n$-module. Such coefficient rings A are enough for application.

Proof. First we prove that under the condition (b) (resp. (a)), we have

$$K\otimes^L_A f^*Rg_*\gamma_*M\cong Rg'_*(g'^*K\otimes^L_A f'^*\gamma_*M)$$

for any algebraic geometric point $\gamma: t\to S'$ and any A-module M (resp. $M=A$). We have

$$\gamma_*M\cong R\gamma_*M.$$

If f is strongly locally acyclic relative to K, then f' is strongly locally acyclic relative to g'^*K since g is an open immersion. Fix notation by the following commutative diagram:

$$\begin{array}{ccccc} X_t & \stackrel{\gamma'}{\to} & X\times_S S' & \stackrel{g'}{\to} & X \\ {\scriptstyle f_t}\downarrow & & {\scriptstyle f'}\downarrow & & {\scriptstyle f}\downarrow \\ t & \stackrel{\gamma}{\to} & S' & \stackrel{g}{\to} & S. \end{array}$$

By 7.6.6, we have isomorphisms

$$\begin{aligned} K \otimes_A^L f^* R(g\gamma)_* M &\cong R(g'\gamma')_*((g'\gamma')^* K \otimes_A^L f_t^* M), \\ g'^* K \otimes_A^L f'^* R\gamma_* M &\cong R\gamma'_*(\gamma'^* g'^* K \otimes_A^L f_t^* M). \end{aligned}$$

It follows that

$$\begin{aligned} K \otimes_A^L f^* Rg_*\gamma_* M &\cong K \otimes_A^L f^* Rg_* R\gamma_* M \\ &\cong Rg'_* R\gamma'_*(\gamma'^* g'^* K \otimes_A^L f_t^* M) \\ &\cong Rg'_*(g'^* K \otimes_A^L f'^* R\gamma_* M) \\ &\cong Rg'_*(g'^* K \otimes_A^L f'^* \gamma_* M). \end{aligned}$$

This proves our assertion. If f is locally acyclic, the above argument works for $M = A$.

By 6.5.2, to prove $K \otimes_A^L f^* Rg_* L \cong Rg'_*(g'^* K \otimes_A^L f'^* L)$ for any $L \in \operatorname{ob} D^+(S', A)$, it suffices to treat the case where $L = \mathscr{F}$ for a sheaf $\mathscr{F}$ of A-modules on S'. For convenience, let

$$S^q(\mathscr{F}) = \mathscr{H}^q(K \otimes_A^L f^* Rg_* \mathscr{F}), \quad T^q(\mathscr{F}) = R^q g'_*(g'^* K \otimes_A^L f'^* \mathscr{F}),$$

and let $\phi^q_{\mathscr{F}} : S^q(\mathscr{F}) \to T^q(\mathscr{F})$ be the homomorphisms induced by the canonical morphism $K \otimes_A^L f^* Rg_* \mathscr{F} \to Rg'_*(g'^* K \otimes_A^L f'^* \mathscr{F})$. Let us prove that $\phi^q_{\mathscr{F}}$ are isomorphisms by induction on q. If K can be represented by a bounded complex of flat sheaves of A-modules such that $K^i = 0$ for any $i \notin [a, b]$, then both $S^q(\mathscr{F})$ and $T^q(\mathscr{F})$ vanish for any $q < a$. So $\phi^q_{\mathscr{F}}$ is an isomorphism for any $q < a$. Suppose we have shown that $\phi^q_{\mathscr{F}}$ is an isomorphism for any $q < n$, and let us prove that $\phi^n_{\mathscr{F}}$ is an isomorphism. By 5.8.8 and 5.9.6, it suffices to prove that $\phi^n_{\mathscr{F}}$ is an isomorphism for any constructible $\mathscr{F}$. In this case, there exists a decomposition $S' = \cup_\lambda U_\lambda$ of S' by finitely many irreducible locally closed subsets such that $\mathscr{F}|_{U_\lambda}$ are locally constant. For each λ, let $\gamma_\lambda : t_\lambda \to S'$ be an algebraic geometric point lying above the generic point of U_λ, and let $M_\lambda = \mathscr{F}_{t_\lambda}$. Note that the canonical morphism $\mathscr{F}|_{U_\lambda} \to (\gamma_{\lambda *} M_\lambda)|_{U_\lambda}$ is injective. So the canonical morphism

$$\mathscr{F} \to \prod_\lambda \gamma_{\lambda *} M_\lambda$$

is injective. Thus we can find a short exact sequence of the form

$$0 \to \mathscr{F} \to \mathscr{G} \to \mathscr{H} \to 0$$

such that $\mathscr{G} = \bigoplus \gamma_* M$ for some algebraic geometric points $\gamma : t \to S'$ and some finitely generated A-modules M, and we can take $M = A$ under the condition (a). We have a commutative diagram

$$\begin{array}{ccccccccc} S^{n-1}(\mathscr{G}) & \to & S^{n-1}(\mathscr{H}) & \to & S^n(\mathscr{F}) & \to & S^n(\mathscr{G}) & \to & S^n(\mathscr{H}) \\ \phi^{n-1}_{\mathscr{G}} \downarrow & & \phi^{n-1}_{\mathscr{H}} \downarrow & & \phi^n_{\mathscr{F}} \downarrow & & \phi^n_{\mathscr{G}} \downarrow & & \phi^n_{\mathscr{H}} \downarrow \\ T^{n-1}(\mathscr{G}) & \to & T^{n-1}(\mathscr{H}) & \to & T^n(\mathscr{F}) & \to & T^n(\mathscr{G}) & \to & T^n(\mathscr{H}), \end{array}$$

where the horizontal lines are exact. By the induction hypothesis, $\phi^{n-1}_{\mathscr{G}}$ and $\phi^{n-1}_{\mathscr{H}}$ are bijective. By what we have shown at the beginning, $\phi^n_{\mathscr{G}}$ is bijective. These facts imply that $\phi^n_{\mathscr{F}}$ is injective for any constructible sheaf $\mathscr{F}$, and hence for any sheaf $\mathscr{F}$. In particular, $\phi^n_{\mathscr{H}}$ is injective. These facts imply that $\phi^n_{\mathscr{F}}$ is bijective for any constructible sheaf $\mathscr{F}$, and hence for any sheaf $\mathscr{F}$. □

Theorem 7.6.9. *Let A be a noetherian ring, S a noetherian scheme, $f : X \to S$ a morphism, and $K \in \mathrm{ob}\, D^b_{\mathrm{tf}}(X, A)$. The following conditions are equivalent:*

(i) *f is universally strongly locally acyclic relative to K.*

(ii) *For any Cartesian diagram*

$$\begin{array}{ccc} X \times_S S' & \stackrel{g'}{\to} & X \\ f' \downarrow & & \downarrow f \\ S' & \stackrel{g}{\to} & S \end{array}$$

such that g is quasi-compact and quasi-separated, and for any $L \in \mathrm{ob}\, D^+(S', A)$, we have a canonical isomorphism

$$K \otimes^L_A f^* Rg_* L \stackrel{\cong}{\to} Rg'_*(g'^* K \otimes^L_A f'^* L).$$

This remains true if we replace f by $f_T : X \times_S T \to T$ for any base change $T \to S$ with T noetherian.

Suppose furthermore that every finitely generated A-module can be embedded into a free A-module. Then the above two conditions are equivalent to the following:

(iii) *f is universally locally acyclic relative to K.*

Proof.

(i)$\Rightarrow$ (ii) The second statement follows from the first one. Let us prove the first statement. The problem is local with respect to S, so we may assume that S is affine. Cover S' by finitely many affine open subsets U_α.

Similar to 5.6.10, we have spectral sequences

$$E_1^{pq} = \prod_{\alpha_0,\ldots,\alpha_p} \mathscr{H}^q\Big(K \otimes_A^L f^* R(g|_{U_{\alpha_0}\times_{S'}\cdots\times_{S'}U_{\alpha_p}})_*(L|_{U_{\alpha_0}\times_{S'}\cdots\times_{S'}U_{\alpha_p}})\Big)$$
$$\Rightarrow \mathscr{H}^{p+q}(K \otimes_A^L f^* Rg_* L)$$
$$E_1^{pq} = \prod_{\alpha_0,\ldots,\alpha_p} R^q(g'|_{U_{\alpha_0}\times_{S'}\cdots\times_{S'}U_{\alpha_p}\times_S X})_*$$
$$((g'^* K \otimes_A^L f'^* L)|_{U_{\alpha_0}\times_{S'}\cdots\times_{S'}U_{\alpha_p}\times_S X})$$
$$\Rightarrow R^{p+q} g'_*(g'^* K \otimes_A^L f'^* L).$$

It suffices to prove

$$K \otimes_A^L f^* R(g|_{U_{\alpha_0}\times_{S'}\cdots\times_{S'}U_{\alpha_p}})_*(L|_{U_{\alpha_0}\times_{S'}\cdots\times_{S'}U_{\alpha_p}})$$
$$\cong R(g'|_{U_{\alpha_0}\times_{S'}\cdots\times_{S'}U_{\alpha_p}\times_S X})_*((g'^* K \otimes_A^L f'^* L)|_{U_{\alpha_0}\times_{S'}\cdots\times_{S'}U_{\alpha_p}\times_S X}).$$

We are thus reduced first to the case where $S' \to S$ is separated, and then to the case where $S = \operatorname{Spec} B$ and $S' = \operatorname{Spec} B'$ are affine. Write $B' = \varinjlim_\lambda B_\lambda$, where B_λ goes over the family of subalgebras of B' finitely generated over B. Let $S_\lambda = \operatorname{Spec} B_\lambda$, and fix notation by the following diagram:

$$\begin{array}{ccccc} X\times_S S' & \stackrel{\pi'_\lambda}{\to} & X\times_S S_\lambda & \stackrel{g'_\lambda}{\to} & X \\ {\scriptstyle f'}\downarrow & & {\scriptstyle f_\lambda}\downarrow & & {\scriptstyle f}\downarrow \\ S' & \stackrel{\pi_\lambda}{\to} & S_\lambda & \stackrel{g_\lambda}{\to} & S. \end{array}$$

By 5.9.6, we have

$$L \cong \varinjlim_\lambda \pi_\lambda^* R\pi_{\lambda *} L.$$

Hence

$$\begin{aligned} g'^* K \otimes_A^L f'^* L &\cong \varinjlim_\lambda (g'^* K \otimes_A^L f'^* \pi_\lambda^* R\pi_{\lambda *} L) \\ &\cong \varinjlim_\lambda (g'^* K \otimes_A^L \pi_\lambda'^* f_\lambda^* R\pi_{\lambda *} L) \\ &\cong \varinjlim_\lambda \pi_\lambda'^* (g_\lambda'^* K \otimes_A^L f_\lambda^* R\pi_{\lambda *} L). \end{aligned}$$

By 5.9.6 again, we have

$$Rg'_*(g'^* K \otimes_A^L f'^* L) \cong \varinjlim_\lambda Rg'_{\lambda *}(g_\lambda'^* K \otimes_A^L f_\lambda^* R\pi_{\lambda *} L).$$

If (ii) holds in the case where g is affine of finite type, then we have

$$K \otimes_A^L f^* Rg_* L \cong K \otimes_A^L f^* Rg_{\lambda *} R\pi_{\lambda *} L \cong Rg'_{\lambda *}(g_\lambda'^* K \otimes_A^L f_\lambda^* R\pi_{\lambda *} L)$$

for all λ, and hence

$$K \otimes_A^L f^* Rg_* L \cong Rg'_*(g'^* K \otimes_A^L f'^* L).$$

We are thus reduced to the case where B' is a finitely generated B-algebra. We then have an epimorphism

$$B[T_1, \ldots, T_n] \to B'.$$

It defines a closed immersion

$$i : S' \to \mathbb{A}_S^n = \operatorname{Spec} B[T_1, \ldots, T_n].$$

Let

$$j : \mathbb{A}_S^n \hookrightarrow \mathbb{P}_S^n = \operatorname{Proj} B[T_0, \ldots, T_n]$$

be the canonical open immersion and let $\pi : \mathbb{P}_S^n \to S$ be the projection. We have $g = \pi j i$. Fix notation by the following commutative diagram of Cartesian squares:

$$\begin{array}{ccccccc} X \times_S S' & \stackrel{i'}{\to} & \mathbb{A}_X^n & \stackrel{j'}{\to} & \mathbb{P}_X^n & \stackrel{\pi'}{\to} & X \\ f' \downarrow & & f_1 \downarrow & & f_2 \downarrow & & f \downarrow \\ S' & \stackrel{i}{\to} & \mathbb{A}_S^n & \stackrel{j}{\to} & \mathbb{P}_S^n & \stackrel{\pi}{\to} & S. \end{array}$$

Since π and i are proper, by 7.4.8, we have

$$K \otimes_A^L f^* R\pi_* Rj_* Ri_* L \cong R\pi'_*(\pi'^* K \otimes_A^L f_2^* Rj_* Ri_* L),$$
$$j'^* \pi'^* K \otimes_A^L f_1^* Ri_* L \cong Ri'_*(i'^* j'^* \pi'^* K \otimes_A^L f'^* L).$$

Since f is universally strongly locally acyclic relative to K, f_2 is strongly locally acyclic relative to $\pi'^* K$. Applying 7.6.7 to the middle Cartesian square in the above diagram, we get

$$\pi'^* K \otimes_A^L f_2^* Rj_* Ri_* L \cong Rj'_*(j'^* \pi'^* K \otimes_A^L f_1^* Ri_* L).$$

So we have

$$\begin{aligned} K \otimes_A^L f^* Rg_* L &\cong K \otimes_A^L f^* R\pi_* Rj_* Ri_* L \\ &\cong R\pi'_*(\pi'^* K \otimes_A^L f_2^* Rj_* Ri_* L) \\ &\cong R\pi'_* Rj'_*(j'^* \pi'^* K \otimes_A^L f_1^* Ri_* L) \\ &\cong R\pi'_* Rj'_* Ri'_*(i'^* j'^* \pi'^* K \otimes_A^L f'^* L) \\ &\cong Rg'_*(g'^* K \otimes_A^L f'^* L). \end{aligned}$$

This proves (ii).

(ii)$\Rightarrow$(i) Let $S' \to S$ be a base change with S' being noetherian and let $s' \in S'$. Then the strict localization $\widetilde{S}'_{\bar{s}'}$ of S' at $\bar{s}'$ is noetherian. For any algebraic geometric point $\gamma : t \to \widetilde{S}'_{\bar{s}'}$, applying (ii) to the Cartesian diagram

$$\begin{array}{ccc} X \times_S t & \xrightarrow{\gamma'} & X \times_S \widetilde{S}'_{\bar{s}'} \\ f_t \downarrow & & \downarrow \tilde{f} \\ t & \xrightarrow{\gamma} & \widetilde{S}'_{\bar{s}'}, \end{array}$$

we get a canonical isomorphism

$$K|_{X \times_S \widetilde{S}'_{\bar{s}'}} \otimes_A^L \tilde{f}^* R\gamma_* M \cong R\gamma'_*(\gamma'^*(K|_{X \times_S \widetilde{S}'_{\bar{s}'}}) \otimes_A^L f_t^* M)$$

for any A-module M. Taking the restriction to the closed fiber $X_{\bar{s}'}$ of $X \times_S \widetilde{S}'_{\bar{s}'}$, we get

$$K|_{X_{\bar{s}'}} \otimes_A^L M \cong (R\gamma'_*(\gamma'^*(K|_{X \times_S \widetilde{S}'_{\bar{s}'}}) \otimes_A^L M))|_{X_{\bar{s}'}}.$$

Hence $X \times_S S' \to S'$ is strongly locally acyclic relative to $K|_{X \times_S S'}$.

Now suppose that every finitely generated A-module can be embedded into a free A-module. (i)$\Rightarrow$(iii) is trivial. (iii)$\Rightarrow$(ii) can be proved by the same argument as (i)$\Rightarrow$(ii). □

Suppose that $f : X \to S$ is locally acyclic relative to $K \in \mathrm{ob}\, D^+(X)$. For any q, any $s \in S$ and any algebraic geometric point $j : t \to \widetilde{S}_{\bar{s}}$, we define the *cospecialization homomorphism* to be the composite

$$\begin{aligned} H^q(X_t, K|_{X_t}) &\cong H^q(X \times_S \widetilde{S}_{\bar{s}}, R\tilde{j}_*\tilde{j}^*\widetilde{K}) \\ &\to H^q(X_{\bar{s}}, \tilde{i}^* R\tilde{j}_*\tilde{j}^*\widetilde{K}) \\ &\cong H^q(X_{\bar{s}}, K|_{X_{\bar{s}}}). \end{aligned}$$

Proposition 7.6.10. *Let A be a noetherian ring, S a noetherian scheme, and $f : X \to S$ a morphism locally acyclic relative to A. Suppose that every finitely generated A-module can be embedded into a free A-module, and for any $s \in S$ and any geometric point $j : t \to \widetilde{S}_{\bar{s}}$, cospecialization homomorphisms*

$$H^q(X_t, A) \to H^q(X_{\bar{s}}, A)$$

are bijective for all q. Then for any sheaf $\mathscr{F}$ of A-modules on S, the canonical morphism

$$(Rf_* f^* \mathscr{F})_{\bar{s}} \to R\Gamma(X_{\bar{s}}, (f^*\mathscr{F})|_{X_{\bar{s}}})$$

is an isomorphism.

Proof. It suffices to prove that the canonical morphism

$$(R\tilde{f}_*\tilde{f}^*\widetilde{\mathscr{F}})_{\bar{s}} \to R\Gamma(X_{\bar{s}}, (\tilde{f}^*\widetilde{\mathscr{F}})|_{X_{\bar{s}}})$$

is an isomorphism for any sheaf of A-modules $\widetilde{\mathscr{F}}$ on $\widetilde{S}_{\bar{s}}$. We have

$$\tilde{f}^*j_*A \cong \tilde{f}^*Rj_*A \cong R\tilde{j}_*A$$

by 7.6.6. Since cospecialization homomorphisms are isomorphisms, the canonical morphism

$$R\Gamma(X \times_S \widetilde{S}_{\bar{s}}, R\tilde{j}_*A) \to R\Gamma(X_{\bar{s}}, \tilde{i}^*R\tilde{j}_*A)$$

is an isomorphism. So the canonical morphism

$$(R\tilde{f}_*\tilde{f}^*\widetilde{\mathscr{F}})_{\bar{s}} \to R\Gamma(X_{\bar{s}}, (\tilde{f}^*\widetilde{\mathscr{F}})|_{X_{\bar{s}}})$$

is an isomorphism for $\widetilde{\mathscr{F}} = j_*A$. Since any sheaf of A-modules on $\widetilde{S}_{\bar{s}}$ is a direct limit of constructible sheaves of A-modules, and any constructible sheaf of A-modules can be embedded into a sheaf of the form $\bigoplus_j j_*A$, we have

$$(R\tilde{f}_*\tilde{f}^*\widetilde{\mathscr{F}})_{\bar{s}} \cong R\Gamma(X_{\bar{s}}, (\tilde{f}^*\widetilde{\mathscr{F}})|_{X_{\bar{s}}})$$

for any sheaf of A-modules $\widetilde{\mathscr{F}}$ on $\widetilde{S}_{\bar{s}}$. (Confer the proof of 7.6.7.) □

7.7 Smooth Base Change Theorem

([SGA 4] XVI, [SGA $4\frac{1}{2}$] Arcata V 2, 3)

The main result of this section is the following.

Theorem 7.7.1. *Let $f : X \to S$ be a smooth morphism, and let n be an integer invertible on S. Then f is universally locally acyclic relative to $\mathbb{Z}/n$.*

Before proving this theorem, we give several applications.

Theorem 7.7.2 (Smooth base change theorem). *Consider a Cartesian diagram*

$$\begin{array}{ccc} X \times_S S' & \xrightarrow{g'} & X \\ {\scriptstyle f'}\downarrow & & \downarrow{\scriptstyle f} \\ S' & \xrightarrow{g} & S. \end{array}$$

If f is smooth and g is quasi-compact and quasi-separated, then for any integer n invertible on S and any $L \in D^+(S', Z/n)$, we have

$$f^*Rg_*L \cong Rg'_*f'^*L.$$

Proof. The problem is local with respect to S and X. We may assume that X and S are affine. As in the beginning of the proof of 7.6.9, we may assume that S' is also affine. We can find an inverse system of noetherian affine schemes $\{S_\lambda\}$ such that $S = \varprojlim_\lambda S_\lambda$ and $f : X \to S$ can be descended down to an inverse system of smooth morphisms $f_\lambda : X_\lambda \to S_\lambda$. Fix notation by the following commutative diagram of Cartesian squares:

$$\begin{array}{ccccc} X \times_S S' & \xrightarrow{g'} & X & \xrightarrow{u'_\lambda} & X_\lambda \\ f' \downarrow & & \downarrow f & & \downarrow f_\lambda \\ S' & \xrightarrow{g} & S & \xrightarrow{u_\lambda} & S_\lambda. \end{array}$$

By 7.7.1, each f_λ is universally locally acyclic relative to $\mathbb{Z}/n$. Note that every finitely generated $\mathbb{Z}/n$-module can be embedded into a free $\mathbb{Z}/n$-module. By 7.6.9, we have

$$f_\lambda^* R(u_\lambda g)_* L \cong R(u'_\lambda g')_* f'^* L.$$

By 5.9.6, we have

$$\begin{aligned} f^* Rg_* L &\cong f^*(\varinjlim_\lambda u_\lambda^* R(u_\lambda g)_* L) \\ &\cong \varinjlim_\lambda f^* u_\lambda^* R(u_\lambda g)_* L \\ &\cong \varinjlim_\lambda u_\lambda'^* f_\lambda^* R(u_\lambda g)_* L, \\ Rg'_* f'^* L &\cong \varinjlim_\lambda u_\lambda'^* R(u'_\lambda g')_* f'^* L. \end{aligned}$$

It follows that $f^* Rg_* L \cong Rg'_* f'^* L$. □

Corollary 7.7.3. *Let K/k be an extension of a separably closed field, X a quasi-compact quasi-separated k-scheme, n an integer relatively prime to the characteristic of k, and $\mathscr{F}$ a sheaf of $\mathbb{Z}/n$-modules on X. Then the canonical homomorphisms*

$$H^q(X, \mathscr{F}) \to H^q(X \otimes_k K, \mathscr{F}|_{X \otimes_k K})$$

are isomorphisms.

Proof. By 5.7.2 (ii) and 5.9.2, we may assume that k and K are algebraically closed. We have $K = \varinjlim_\lambda K_\lambda$, where K_λ are subfields of K finitely generated over k. For each λ, choose a separating transcendental basis $\{x_1, \ldots, x_{n_\lambda}\}$ for K_λ over k. Then K_λ is finite and separable over $k(x_1, \ldots, x_{n_\lambda})$. Fix notation by the following commutative diagram

$$\begin{array}{ccccccccc} X \otimes_k K & \xrightarrow{e'_\lambda} & X \otimes_k K_\lambda & \xrightarrow{g'_\lambda} & X \otimes_k k(x_1, \ldots, x_{n_\lambda}) & \xrightarrow{h'_U} & X \times_k U & \xrightarrow{g'_U} & X \\ f' \downarrow & & f'_\lambda \downarrow & & f_\lambda \downarrow & & f_U \downarrow & & f \downarrow \\ \operatorname{Spec} K & \xrightarrow{e_\lambda} & \operatorname{Spec} K_\lambda & \xrightarrow{g_\lambda} & \operatorname{Spec} k(x_1, \ldots, x_{n_\lambda}) & \xrightarrow{h_U} & U & \xrightarrow{g_U} & \operatorname{Spec} k, \end{array}$$

where U goes over the set of nonempty affine open subsets of $\operatorname{Spec} k[x_1, \ldots, x_{n_\lambda}]$. By the smooth base change theorem 7.7.2, we have

$$g_U^* R^q f_* \mathscr{F} \cong R^q f_{U*} g_U'^* \mathscr{F}.$$

By 5.9.6, we have

$$R^q f_{\lambda *} h_U'^* g_U'^* \mathscr{F} \cong \varinjlim_U h_U^* R^q f_{U*} g_U'^* \mathscr{F}.$$

Since g_λ is etale, we have

$$g_\lambda^* R^q f_{\lambda *} h_U'^* g_U'^* \mathscr{F} \cong R^q f'_{\lambda *} g_\lambda'^* h_U'^* g_U'^* \mathscr{F}.$$

Denote by g the morphism $\operatorname{Spec} K \to \operatorname{Spec} k$ and by $g' : X \otimes_k K \to X$ its base change. By 5.9.6, we have

$$\begin{aligned} R^q f'_* g'^* \mathscr{F} &\cong \varinjlim_\lambda e_\lambda^* R^q f'_{\lambda *} g_\lambda'^* h_U'^* g_U'^* \mathscr{F} \\ &\cong \varinjlim_\lambda e_\lambda^* g_\lambda^* R^q f_{\lambda *} h_U'^* g_U'^* \mathscr{F} \\ &\cong \varinjlim_\lambda \varinjlim_U e_\lambda^* g_\lambda^* h_U^* R^q f_{U*} g_U'^* \mathscr{F} \\ &\cong \varinjlim_\lambda \varinjlim_U e_\lambda^* g_\lambda^* h_U^* g_U^* R^q f_* \mathscr{F} \\ &\cong g^* R^q f_* \mathscr{F}. \end{aligned}$$

Since k and K are algebraically closed, this is equivalent to saying that

$$H^q(X \otimes_k K, \mathscr{F}|_{X \otimes_k K}) \cong H^q(X, \mathscr{F}).$$

□

Corollary 7.7.4. *Let S be a scheme, $\pi : \mathbb{A}_S^k = \mathbf{Spec}\, \mathscr{O}_S[T_1, \ldots, T_k] \to S$ the canonical morphism, n an integer invertible on S, and $K \in \operatorname{ob} D^+(S, \mathbb{Z}/n)$. Then the canonical morphism*

$$K \to R\pi_* \pi^* K$$

is an isomorphism.

Proof. It suffices to treat the case where $k = 1$ by applying this special case to each projection in the sequence

$$\mathbb{A}_S^k \to \mathbb{A}_S^{k-1} \to \cdots \to \mathbb{A}_S^1 \to S.$$

For any $s \in S$, let $\overline{k(s)}$ be a separable closure of the residue field $k(s)$. We have $\operatorname{Pic}(\mathbb{A}^1_{\overline{k(s)}}) = 0$ since $\overline{k(s)}[T]$ is a unique factorization domain. By 7.2.9 (i), we have

$$H^q(\mathbb{A}^1_{\overline{k(s)}}, \mathbb{Z}/n) \cong H^q(\mathbb{A}^1_{\overline{k(s)}}, \boldsymbol{\mu}_n) \cong \begin{cases} \mathbb{Z}/n & \text{if } q = 0, \\ 0 & \text{if } q \geq 1. \end{cases}$$

By 7.7.1, π is universally locally acyclic relative to $\mathbb{Z}/n$. The cospecialization homomorphisms for $\mathbb{Z}/n$ are isomorphisms. By 7.6.10, for any sheaf of $\mathbb{Z}/n$-modules $\mathscr{F}$ on S, we have an isomorphism

$$(R\pi_*\pi^*\mathscr{F})_{\bar{s}} \to R\Gamma(\mathbb{A}^1_{\overline{k(s)}}, \pi^*\mathscr{F}|_{\mathbb{A}^1_{\overline{k(s)}}}).$$

But

$$H^q(\mathbb{A}^1_{\overline{k(s)}}, \pi^*\mathscr{F}|_{\mathbb{A}^1_{\overline{k(s)}}}) \cong \begin{cases} \mathscr{F}_{\bar{s}} & \text{if } q = 0, \\ 0 & \text{if } q \geq 1. \end{cases}$$

It follows that

$$\mathscr{F} \cong R\pi_*\pi^*\mathscr{F}.$$

Hence

$$K \cong R\pi_*\pi^* K$$

for any $K \in \operatorname{ob} D^+(S, \mathbb{Z}/n)$. □

Corollary 7.7.5. *Let $f : X \to S$ be a smooth morphism and let n be an integer invertible on S. Then for any locally constant sheaf of $\mathbb{Z}/n$-modules $\mathscr{F}$ on X, f is universally strongly locally acyclic relative to $\mathscr{F}$.*

Proof. By 7.5.7 and 7.6.2, it suffices to show that f is universally locally acyclic relative to $\mathscr{F}$. Since base changes of smooth morphisms are smooth, it suffices to prove that f is locally acyclic relative to $\mathscr{F}$. For any $s \in S$, and any algebraic geometric point $j : t \to \widetilde{S}_{\bar{s}}$, fix notation by the following diagram:

$$\begin{array}{ccccc} V_t = V \times_{\widetilde{S}_{\bar{s}}} t & \stackrel{j'}{\to} & V & \stackrel{i'}{\leftarrow} & V_{\bar{s}} = V \times_{\widetilde{S}_{\bar{s}}} \bar{s} \\ p_t \downarrow & & p \downarrow & & \downarrow p_{\bar{s}} \\ X_t = X \times_S t & \stackrel{\tilde{j}}{\to} & X \times_S \widetilde{S}_{\bar{s}} & \stackrel{\tilde{i}}{\leftarrow} & X_{\bar{s}} = X \times_S \bar{s} \\ f_t \downarrow & & \tilde{f} \downarrow & & \downarrow f_{\bar{s}} \\ t & \stackrel{j}{\to} & \widetilde{S}_{\bar{s}} & \stackrel{i}{\leftarrow} & \bar{s}. \end{array}$$

where $p : V \to X \times_S \widetilde{S}_{\bar{s}}$ is a morphism that will be used later. Set $\widetilde{\mathscr{F}} = \mathscr{F}|_{X \times_S \widetilde{S}_{\bar{s}}}$. We need to show that the canonical morphism

$$\tilde{i}^*\widetilde{\mathscr{F}} \to \tilde{i}^* R\tilde{j}_*\tilde{j}^*\widetilde{\mathscr{F}}$$

is an isomorphism. For any $x \in X_{\bar{s}}$, we need to check

$$(\tilde{i}^*\widetilde{\mathscr{F}})_{\bar{x}} \to (\tilde{i}^* R\tilde{j}_*\tilde{j}^*\widetilde{\mathscr{F}})_{\bar{x}}$$

is an isomorphism. Since $\mathscr{F}$ is locally constant, we can find an etale morphism $p : V \to X \times_S \widetilde{S}_{\bar{s}}$ whose image contains the point x so that $\widetilde{\mathscr{F}}|_V$ is a constant sheaf, say associated to a $\mathbb{Z}/n$-module M. It suffices to show that

$$p_{\bar{s}}^* \tilde{i}^* \widetilde{\mathscr{F}} \to p_{\bar{s}}^* \tilde{i}^* R\tilde{j}_* \tilde{j}^* \widetilde{\mathscr{F}}$$

is an isomorphism. We have

$$\begin{aligned} p_{\bar{s}}^* \tilde{i}^* \widetilde{\mathscr{F}} &\cong i'^* M, \\ p_{\bar{s}}^* \tilde{i}^* R\tilde{j}_* \tilde{j}^* \widetilde{\mathscr{F}} &\cong i'^* p^* R\tilde{j}_* \tilde{j}^* \widetilde{\mathscr{F}} \\ &\cong i'^* Rj'_* p_t^* \tilde{j}^* \widetilde{\mathscr{F}} \\ &\cong i'^* Rj'_* M. \end{aligned}$$

So it suffices to show that the canonical morphism

$$i'^* M \to i'^* Rj'_* M$$

is an isomorphism. Applying the smooth base change theorem 7.7.2 to the smooth morphism $\tilde{f}p$, we get isomorphisms

$$(\tilde{f}p)^* Rj_* M \cong Rj'_* (f_t p_t)^* M \cong Rj'_* M.$$

So we have

$$(\tilde{f}_{\bar{s}} p_{\bar{s}})^* i^* Rj_* M \cong i'^* (\tilde{f}p)^* Rj_* M \cong i'^* Rj'_* M.$$

Our assertion then follows from the fact that

$$M \cong i^* Rj_* M.$$

□

We now prepare to prove 7.7.1.

Lemma 7.7.6. *Let $X \xrightarrow{f_1} Y \xrightarrow{f_2} S$ be morphisms of noetherian schemes, let A be a noetherian ring, and let $K \in \mathrm{ob}\, D_{\mathrm{tf}}^b(X, A)$. Suppose that f_1 is universally strongly locally acyclic relative to K, and f_2 is universally strongly locally acyclic relative to A. Then $f_2 f_1$ is universally strongly locally acyclic relative to K. If either both f_1 and f_2 are of finite type, or any finitely generated A-module can be embedded into a free A-module, then we have the same conclusion under the assumption that f_1 is universally locally acyclic relative to K, and f_2 is universally locally acyclic relative to A.*

Proof. The second part follows from the first part by 7.6.2 and 7.6.9. Consider a commutative diagram of Cartesian squares

$$\begin{array}{ccc} X \times_S S' & \xrightarrow{g'} & X \\ {\scriptstyle f_1'}\downarrow & & \downarrow{\scriptstyle f_1} \\ Y \times_S S' & \xrightarrow{g''} & Y \\ {\scriptstyle f_2'}\downarrow & & \downarrow{\scriptstyle f_2} \\ S' & \xrightarrow{g} & S \end{array}$$

such that g is quasi-compact and quasi-separated. Since f_1 is universally strongly locally acyclic relative to K and f_2 is universally strongly locally acyclic relative to A, we have

$$K \otimes_A^L f_1^* Rg''_* f_2'^* L \cong Rg'_*(g'^* K \otimes_A^L f_1'^* f_2'^* L),$$
$$f_2^* Rg_* L \cong Rg''_* f_2'^* L$$

for any $L \in \mathrm{ob}\, D^+(S', A)$ by 7.6.9. So we have

$$\begin{aligned} K \otimes_A^L f_1^* f_2^* Rg_* L &\cong K \otimes_A^L f_1^* Rg''_* f_2'^* L \\ &\cong Rg'_*(g'^* K \otimes_A^L f_1'^* f_2'^* L). \end{aligned}$$

This remains true if we replace f_1 and f_2 by their base changes with respect to any $T \to S$ with T noetherian. So $f_2 f_1$ is universally strongly locally acyclic relative to K by 7.6.9. □

Lemma 7.7.7. *Let $(A, \mathfrak{m})$ be a local ring, $k = A/\mathfrak{m}$, and K a finite extension of k. Then there exists a local ring $(B, \mathfrak{n})$ and a local homomorphism $A \to B$ such that B is finite and faithfully flat over A, and $B/\mathfrak{n}$ is k-isomorphic to K.*

Proof. We may reduce to the case where K is a simple extension of k. Then K is k-isomorphic to $k[t]/(f(t))$ for some monic irreducible polynomial $f(t)$. Let $F(t) \in A[t]$ be a monic polynomial whose reduction mod $\mathfrak{m}$ is $f(t)$ and let $B = A[t]/(F(t))$. Then B is finite and faithfully flat over A. Moreover, we have

$$B/\mathfrak{m}B \cong k[t]/(f(t)) \cong K.$$

So $\mathfrak{m}B$ is a maximal ideal of B. Since B is finite over A, any maximal ideal of B is above the unique maximal ideal $\mathfrak{m}$ of A. So any maximal ideal of B contains $\mathfrak{m}B$. It follows that $\mathfrak{m}B$ is the only maximal ideal of B. So B is local and its residue field is k-isomorphic to K. □

Let $f : X \to S$ be a smooth morphism. Let us prove that f is universally locally acyclic relative to $\mathbb{Z}/n$, where n is an integer invertible on S. The problem is local on S and on X. We may assume that $S = \operatorname{Spec} A$ is affine. We have $A = \varinjlim_\lambda A_\lambda$, where A_λ are subrings of A finitely generated over $\mathbb{Z}$. We may assume that f can be descended down to a smooth morphism $f_\lambda : X_\lambda \to S_\lambda$ for some λ. It suffices to prove that f_λ is universally locally acyclic relative to $\mathbb{Z}/n$. We are thus reduced to the case where $S = \operatorname{Spec} A$ for some finitely generated $\mathbb{Z}$-algebra A. We may assume that f can be factorized as

$$X \xrightarrow{j} \mathbb{A}^k_S \to S,$$

where j is an etale morphism. It suffices to prove that $\mathbb{A}^k_S \to S$ is universally locally acyclic relative to $\mathbb{Z}/n$. By induction on k and 7.7.6, we are reduced to the case where $k = 1$. Let $s \in S$ and let $j : t \to \widetilde{S}_{\bar{s}}$ be an algebraic geometric point. Fix notation by the following diagram:

$$\begin{array}{ccccc} \mathbb{A}^1_{k(t)} & \xrightarrow{\tilde{j}} & \mathbb{A}^1_{\widetilde{S}_{\bar{s}}} & \xleftarrow{\tilde{i}} & \mathbb{A}^1_{k(\bar{s})} \\ \downarrow & & \downarrow & & \downarrow \\ t & \xrightarrow{j} & \widetilde{S}_{\bar{s}} & \xleftarrow{i} & \bar{s}. \end{array}$$

Let us prove that the canonical morphism

$$\mathbb{Z}/n \to \tilde{i}^* R\tilde{j}_* \mathbb{Z}/n$$

is an isomorphism, and note that this remains to be true if S is replaced by a base S' of finite type over S. It suffices to prove that for any Zariski closed point x in $\mathbb{A}^1_{k(\bar{s})}$, we have

$$\mathbb{Z}/n \cong (\tilde{i}^* R\tilde{j}_* \mathbb{Z}/n)_{\bar{x}},$$

that is,

$$H^q(\widetilde{\mathbb{A}}^1_{\widetilde{S}_{\bar{s}},\bar{x}} \times_{\widetilde{S}_{\bar{s}}} t, \mathbb{Z}/n) \cong \begin{cases} \mathbb{Z}/n & \text{if } q = 0, \\ \mathbb{Z}/n & \text{if } q \geq 1, \end{cases}$$

where $\widetilde{\mathbb{A}}^1_{\widetilde{S}_{\bar{s}},\bar{x}}$ is the strict localization of $\mathbb{A}^1_{\widetilde{S}_{\bar{s}}}$ at $\bar{x}$. The residue field $k(x)$ of x in $\mathbb{A}^1_{k(\bar{s})}$ is finite over $k(\bar{s})$. By 7.7.7, there exists a local ring $(B, \mathfrak{n})$ finite and faithfully flat over $\mathscr{O}_{\widetilde{S}_{\bar{s}},\bar{s}}$ such that $B/\mathfrak{n}$ is $k(\bar{s})$-isomorphic to $k(x)$. B is strictly henselian and isomorphic to the strict henselization at a point of a scheme of finite type over $\mathbb{Z}$. Replacing $k(t)$ by a purely inseparable algebraic extension, we may assume that the algebraic geometric point $j : t \to \widetilde{S}_{\bar{s}}$ can be lifted to an algebraic geometric point $t \to \operatorname{Spec} B$. The $k(\bar{s})$-isomorphism $k(x) \cong B/\mathfrak{n}$ defines an $\bar{s}$-morphism

$$\operatorname{Spec} B/\mathfrak{n} \to \mathbb{A}^1_{k(\bar{s})}$$

with image x. The composite

$$\mathrm{Spec}\, B/\mathfrak{n} \to \mathbb{A}^1_{k(\bar{s})} \to \mathbb{A}^1_{\widetilde{S}_{\bar{s}}}$$

and the morphism

$$\mathrm{Spec}\, B/\mathfrak{n} \to \mathrm{Spec}\, B$$

induce a morphism

$$\mathrm{Spec}\, B/\mathfrak{n} \to \mathbb{A}^1_{\widetilde{S}_{\bar{s}}} \times_{\widetilde{S}_{\bar{s}}} \mathrm{Spec}\, B = \mathbb{A}^1_B.$$

Let $y \in \mathbb{A}^1_B$ be the image of this morphism. Then y lies above x, and $k(y) \cong B/\mathfrak{n}$. Note that $B/\mathfrak{n}$ is a separably closed field and finite purely inseparable over the separable closed field $k(\bar{s})$. By 2.8.20, we have

$$\widetilde{\mathbb{A}}^1_{\widetilde{S}_{\bar{s}},\bar{x}} \times_{\widetilde{S}_{\bar{s}}} \mathrm{Spec}\, B \cong \widetilde{\mathbb{A}}^1_{B,\bar{y}}.$$

Hence

$$\widetilde{\mathbb{A}}^1_{\widetilde{S}_{\bar{s}},\bar{x}} \times_{\widetilde{S}_{\bar{s}}} t \cong \widetilde{\mathbb{A}}^1_{B,\bar{y}} \times_{\mathrm{Spec}\, B} t.$$

We are thus reduced to proving

$$H^q(\widetilde{\mathbb{A}}^1_{B,\bar{y}} \times_{\mathrm{Spec}\, B} t, \mathbb{Z}/n) \cong \begin{cases} \mathbb{Z}/n & \text{if } q = 0, \\ 0 & \text{if } q \geq 1. \end{cases}$$

Since $y \in \mathbb{A}^1_B$ lies above the closed point of $\mathrm{Spec}\, B$, it lies in the closed subscheme $\mathrm{Spec}\,(B/\mathfrak{n})[T]$ of $\mathbb{A}^1_B = \mathrm{Spec}\, B[T]$. Since $k(y) \cong B/\mathfrak{n}$, y corresponds to a prime ideal of $(B/\mathfrak{n})[T]$ of the form $(T - \bar{b})$ for some $\bar{b} \in B/\mathfrak{n}$. Let $b \in B$ be a lifting of $\bar{b}$. Then y corresponds to the prime ideal of $B[T]$ generated by $\mathfrak{n}$ and $T - b$. Making the change of variable $T \mapsto T - b$, we may assume that y corresponds to the prime ideal of $B[T]$ generated by $\mathfrak{n}$ and T. We are thus reduced to proving the following lemma.

Lemma 7.7.8. *Let $(B, \mathfrak{n})$ be the strict henselization at a point of a scheme of finite type over $\mathbb{Z}$, n an integer invertible in B, $S = \mathrm{Spec}\, B$, $B\{T\}$ the strict henselization of $B[T]$ at the prime ideal generated by $\mathfrak{n}$ and T, $X = \mathrm{Spec}\, B\{T\}$, and $j : t \to S$ an algebraic geometric point. Then*

$$H^q(X_t, \mathbb{Z}/n) \cong \begin{cases} \mathbb{Z}/n & \text{if } q = 0, \\ 0 & \text{if } q \geq 1. \end{cases}$$

Proof. Let S' be the normalization of $\overline{\{j(t)\}}$ with the reduced closed subscheme structure. By [Matsumura (1970)] (31.H) Theorem 72, S' is finite over S, and hence is strictly local. It is also the strict localization at a point of a scheme of finite type over $\mathbb{Z}$. Replacing S by S', we are reduced to the case where S is normal and t lies above the generic point of S. Then X is also normal.

X_t is the inverse limit of an inverse system of smooth affine curves over $k(t)$. By 5.9.3 and 7.2.10, we have

$$H^q(X_t, \mathbb{Z}/n) = 0$$

for any $q \geq 2$. It remains to prove

$$H^0(X_t, \mathbb{Z}/n) \cong \mathbb{Z}/n, \quad H^1(X_t, \mathbb{Z}/n) = 0.$$

As $\mathbb{Z}/n \cong \boldsymbol{\mu}_{n,X_t}$ on X_t, it suffices to prove

$$H^0(X_t, \mathbb{Z}/n) \cong \mathbb{Z}/n, \quad H^1(X_t, \boldsymbol{\mu}_{n,X_t}) = 0.$$

We have $k(t) = \varinjlim_L L$, where L goes over the family of subfields of $k(t)$ finite over the fraction field of B. By 5.9.3, it suffices to prove

$$\varinjlim_L H^0(X_L, \mathbb{Z}/n) \cong \mathbb{Z}/n, \quad \varinjlim_L H^1(X_L, \boldsymbol{\mu}_{n,X_L}) = 0,$$

where $X_L = X \otimes_B L$. By Kummer's theory 7.2.1, we have a short exact sequence

$$\begin{aligned} 0 \to \Gamma(X_L, \mathscr{O}_{X_L}^*)/\Gamma(X_L, \mathscr{O}_{X_L}^*)^n &\to H^1(X_L, \boldsymbol{\mu}_{n,X_L}) \\ &\to \mathrm{Ker}\,(n : \mathrm{Pic}(X_L) \to \mathrm{Pic}(X_L)) \to 0. \end{aligned}$$

For each L, let B_L be the integral closure of B in L. It is also the strict henselization at a point of a scheme of finite type over $\mathbb{Z}$. We have

$$X_L \cong \mathrm{Spec}\,(B_L\{T\} \otimes_{B_L} L).$$

It suffices to prove the following:

(i) X_L is connected and nonempty for each L.

(ii) $\varinjlim_L (B_L\{T\} \otimes_{B_L} L)^*/((B_L\{T\} \otimes_{B_L} L)^*)^n = 0$, where $(B_L\{T\} \otimes_{B_L} L)^*$ is the group of units of $B_L\{T\} \otimes_{B_L} L$.

(iii) $\mathrm{Ker}\,(n : \mathrm{Pic}(X_L) \to \mathrm{Pic}(X_L)) = 0$ for each L.

Note that if A is a strictly henselian local ring, then $A[T]/T^m A[T]$ and $A\{T\}/T^m A\{T\}$ are strictly henselian. We have

$$A\{T\}/T^m A\{T\} \cong (A[T]/T^m A[T]) \otimes_{A[T]} A\{T\}.$$

By 2.8.20, $A\{T\}/T^m A\{T\}$ is isomorphic to the strict henselization of $A[T]/T^m A[T]$. It follows that

$$A[T]/T^m A[T] \cong A\{T\}/T^m A\{T\},$$

and hence

$$\varprojlim_m A\{T\}/T^m A\{T\} \cong A[[T]].$$

Since $A\{T\}$ is noetherian and T lies in the maximal ideal of $A\{T\}$, the canonical homomorphism

$$A\{T\} \to \varprojlim_{m} A\{T\}/T^m A\{T\}$$

is injective. If A is an integral domain, then so are $A[[T]]$ and $A\{T\}$.

Since B_L is an integral domain, $B_L\{T\}$ and hence $B_L\{T\} \otimes_B L$ are integral domains. In particular, $X_L = \mathrm{Spec}\,(B_L\{T\} \otimes_B L)$ is connected. It is clear that $B_L\{T\} \otimes_B L$ is a nontrivial ring. So X_L is nonempty. This proves (i).

To prove (ii), it suffices to show every unit of $B_L\{T\} \otimes_{B_L} L$ is a product of a unit in $B_L\{T\}$ and a nonzero element in L. This is because every unit of the strictly local ring $B_L\{T\}$ has an n-th root by 2.8.3 (v), and every nonzero element α in L has an n-th root in $L[\sqrt[n]{\alpha}] \subset k(t)$.

Let f be a unit of $B_L\{T\} \otimes_{B_L} L$. Then f is a product of an element in $B_L\{T\}$ and a nonzero element in L. To prove that f is a product of a unit in $B_L\{T\}$ and a nonzero element in L, it suffices to consider the case where $f \in B_L\{T\} \cap (B_L\{T\} \otimes_{B_L} L)^*$.

Let V be the subset of $\mathrm{Spec}\, B_L$ of regular points. Since B_L is the strict henselization at a point of a scheme of finite type over $\mathbb{Z}$, V is an open subset of $\mathrm{Spec}\, B_L$ ([Matsumura (1970)] Chapter 13). Since B_L is normal, V contains all the prime ideals of B_L of height 1. Let U be the inverse image of V under the canonical morphism

$$\pi_L : \mathrm{Spec}\, B_L\{T\} \to \mathrm{Spec}\, B_L.$$

Then U is regular. Let $\mathfrak{q}$ be a prime ideal of $B_L\{T\}$ and let $\mathfrak{p} = \mathfrak{q} \cap B_L$. By [Matsumura (1970)] (13.B) Theorem 19 (2), we have

$$\mathrm{ht}\, \mathfrak{q} = \mathrm{ht}\, \mathfrak{p} + \mathrm{ht}(\mathfrak{q}/\mathfrak{p}B_L\{T\}).$$

In particular, we have $\mathrm{ht}\, \mathfrak{q} \geq \mathrm{ht}\, \mathfrak{p}$. If $\mathrm{ht}\, \mathfrak{q} = 1$, then $\mathrm{ht}\, \mathfrak{p} \leq 1$, and hence $\mathfrak{p}$ lies in V. So U contains all prime ideals of $B_L\{T\}$ of height 1. Let (f) be the principle Weil divisor on U defined by $f \in B_L\{T\}$, and let $\mathfrak{q}$ be a prime ideal of $B_L\{T\}$ of height 1 such that $v_{\mathfrak{q}}(f) \neq 0$, where $v_{\mathfrak{q}}$ is the valuation defined by $\mathfrak{q}$. As f is a unit in $B_L\{T\} \otimes_{B_L} L$, $\mathfrak{q}$ does not lie in the generic fiber of $\pi_L : \mathrm{Spec}\, B_L\{T\} \to \mathrm{Spec}\, B_L$. So $\mathfrak{p} \neq 0$, and hence $\mathrm{ht}\, \mathfrak{p} = 1$. Note that $\mathfrak{p}B_L\{T\}$ is a prime ideal of $B_L\{T\}$ since

$$B_L\{T\}/\mathfrak{p}B_L\{T\} \cong (B_L/\mathfrak{p})\{T\}$$

and $(B_L/\mathfrak{p})\{T\}$ is an integral domain. So we must have $\mathfrak{q} = \mathfrak{p}B_L\{T\}$. So the divisor (f) is of the form

$$(f) = \sum_{\mathrm{ht}\, \mathfrak{p}=1} n_{\mathfrak{p}} \overline{\{\mathfrak{p}B_L\{T\}\}}.$$

Let D be the Weil divisor $\sum_{\mathrm{ht}\,\mathfrak{p}=1} n_{\mathfrak{p}}\overline{\{\mathfrak{p}\}}$ on V, and let $\mathscr{L}(D)$ be the invertible $\mathscr{O}_V$-module corresponding to D. Then $\pi_L^*\mathscr{L}(D)$ is isomorphic to the invertible $\mathscr{O}_U$-module corresponding to the divisor (f). So we have $\pi_L^*\mathscr{L}(D) \cong \mathscr{O}_U$. Let

$$i_L : \operatorname{Spec} B_L \to \operatorname{Spec} B_L\{T\}$$

be the morphism corresponding to the B_L-algebra homomorphism

$$B_L\{T\} \to B_L, \quad T \mapsto 0.$$

We have $\pi_L \circ i_L = \mathrm{id}$. It follows that

$$\mathscr{L}(D) \cong i_L^*\pi_L^*\mathscr{L}(D) \cong i_L^*\mathscr{O}_U \cong \mathscr{O}_V.$$

So D is a principle divisor. Let $f_0 \in L^*$ such that $D = (f_0)$. For any prime ideal $\mathfrak{q}$ of $B_L\{T\}$ of height 1, if $\mathfrak{p} = \mathfrak{q} \cap B_L \neq 0$, then we have

$$v_{\mathfrak{q}}(f) = v_{\mathfrak{q}}(f_0) = n_{\mathfrak{p}}.$$

If $\mathfrak{q} \cap B_L = 0$, we have

$$v_{\mathfrak{q}}(f) = v_{\mathfrak{q}}(f_0) = 0.$$

So $v_{\mathfrak{q}}(f) = v_{\mathfrak{q}}(f_0)$ for all prime ideals $\mathfrak{q}$ of $B_L\{T\}$ of height 1. It follows that

$$\frac{f}{f_0}, \frac{f_0}{f} \in \bigcap_{\mathrm{ht}\,\mathfrak{q}=1} (B_L\{T\})_{\mathfrak{q}}.$$

Since $B_L\{T\}$ is normal, we have

$$B_L\{T\} = \bigcap_{\mathrm{ht}\,\mathfrak{q}=1} (B_L\{T\})_{\mathfrak{q}}$$

by [Matsumura (1970)] (17.H) Theorem 38. So $\frac{f}{f_0}, \frac{f_0}{f} \in B_L\{T\}$, and hence $\frac{f}{f_0}$ is a unit in $B_L\{T\}$. We have $f = \frac{f}{f_0} \cdot f_0$ and $f_0 \in L^*$. This proves (ii).

To prove (iii), let $\mathscr{L}$ be an invertible $\mathscr{O}_{X_L}$-module satisfying $\mathscr{L}^n \cong \mathscr{O}_{X_L}$. We need to show $\mathscr{L} \cong \mathscr{O}_{X_L}$. We have

$$X_L \cong U \times_V \operatorname{Spec} L.$$

Since every Weil divisor on X_L can be extended to a Weil divisor on U, the invertible $\mathscr{O}_{X_L}$-module $\mathscr{L}$ can be extended to an invertible $\mathscr{O}_U$-module $\mathscr{L}_U$. Replacing $\mathscr{L}_U$ by $\mathscr{L}_U \otimes \pi_L^* i_L^*(\mathscr{L}_U^{-1})$, we may assume

$$i_L^*\mathscr{L}_U \cong \mathscr{O}_V.$$

Let D_U be a divisor on U defining $\mathscr{L}_U^n$. Since $\mathscr{L}^n \cong \mathscr{O}_{X_L}$, the restriction $D_U|_{X_L}$ of D_U to X_L is a principle divisor. So

$$D_U|_{X_L} = (f)|_{X_L}$$

for some f in the fraction field of $B_L\{T\}$. Replacing D_U by $D_U - (f)$, we may assume that $\mathscr{L}_U^n$ is defined by a divisor D_U satisfying $D_U|_{X_L} = 0$. Using the argument in the proof of (ii), we see that D_U is of the form

$$D_U = \sum_{\mathrm{ht}\,\mathfrak{p}=1} n_{\mathfrak{p}} \overline{\{\mathfrak{p} B_L\{T\}\}}.$$

Let $\mathscr{M}_V$ be the invertible $\mathscr{O}_V$-module defined by the divisor $\sum_{\mathrm{ht}\,\mathfrak{p}=1} n_{\mathfrak{p}} \overline{\{\mathfrak{p}\}}$. Then

$$\pi_L^* \mathscr{M}_V \cong \mathscr{L}_U^n.$$

Since $i_L^* \mathscr{L}_U \cong \mathscr{O}_V$, we have

$$\mathscr{M}_V \cong i_L^* \pi_L^* \mathscr{M}_V \cong i_L^* \mathscr{L}_U^n \cong \mathscr{O}_V.$$

So

$$\mathscr{L}_U^n \cong \mathscr{O}_U.$$

Let us prove that the conditions $i_L^* \mathscr{L}_U \cong \mathscr{O}_V$ and $\mathscr{L}_U^n \cong \mathscr{O}_U$ imply that $\mathscr{L}_U \cong \mathscr{O}_U$.

For each nonnegative integer m, let

$$i_m : \mathrm{Spec}(B_L\{T\}/T^{m+1} B_L\{T\}) \to \mathrm{Spec}\, B_L\{T\}$$

be the canonical morphism and let $U_m = i_m^{-1}(U)$. We have seen that

$$B_L\{T\}/T^{m+1} B_L\{T\} \cong B_L[T]/T^{m+1} B_L[T].$$

In particular, $B_L\{T\}/T^{m+1} B_L\{T\}$ is a free B_L-module of finite rank. Let

$$\pi_m : \mathrm{Spec}(B_L\{T\}/T^{m+1} B_L\{T\}) \to \mathrm{Spec}\, B_L$$

be the canonical morphism. We have

$$\begin{aligned}\Gamma(U_m, \mathscr{O}_{U_m}) &\cong \Gamma(V, \pi_{m*} \mathscr{O}_{\mathrm{Spec}(B_L\{T\}/T^{m+1} B_L\{T\})}) \\ &\cong \Gamma(V, (B_L\{T\}/T^{m+1} B_L\{T\})^{\sim}).\end{aligned}$$

Since B_L is normal, we have

$$B_L = \bigcap_{\mathrm{ht}\,\mathfrak{p}=1} (B_L)_{\mathfrak{p}}$$

by [Matsumura (1970)] (17.H) Theorem 38. Since V contains all prime ideals of B_L of height 1, we have

$$\Gamma(V, \mathscr{O}_{\operatorname{Spec} B_L}) \cong \Gamma(\operatorname{Spec} B_L, \mathscr{O}_{\operatorname{Spec} B_L}).$$

Since $(B_L\{T\}/T^{m+1}B_L\{T\})^{\sim}$ is a free $\mathscr{O}_{\operatorname{Spec} B_L}$-module of finite rank, we also have

$$\begin{aligned}\Gamma(V, (B_L\{T\}/T^{m+1}B_L\{T\})^{\sim}) &\cong \Gamma(\operatorname{Spec} B_L, (B_L\{T\}/T^{m+1}B_L\{T\})^{\sim})\\ &\cong B_L\{T\}/T^{m+1}B_L\{T\}.\end{aligned}$$

We thus have

$$\Gamma(U_m, \mathscr{O}_{U_m}) \cong B_L\{T\}/T^{m+1}B_L\{T\}.$$

Hence

$$H^0(U_m, \mathscr{O}^*_{U_m}) \cong (B_L\{T\}/T^{m+1}B_L\{T\})^*.$$

As $B_L\{T\}/T^{m+1}B_L\{T\}$ is strictly henselian, units in $B_L\{T\}/T^{m+1}B_L\{T\}$ have n-th roots. Using the long exact sequence of cohomology groups associated to the Kummer short exact sequence in 7.2.1, we see that

$$H^1(U_m, \boldsymbol{\mu}_{n,U_m}) \cong \ker(n : \operatorname{Pic}(U_m) \to \operatorname{Pic}(U_m)).$$

On the other hand, U contains all prime ideals of $B_L\{T\}$ of height 1, and $B_L\{T\}$ is normal. So we have

$$\Gamma(U, \mathscr{O}_U) = B_L\{T\}, \quad H^0(U, \mathscr{O}^*_U) = B_L\{T\}^*.$$

As $B_L\{T\}$ is strictly henselian, all units in $B_L\{T\}$ have n-th roots. It follows that

$$H^1(U, \boldsymbol{\mu}_{n,U}) \cong \ker(n : \operatorname{Pic}(U) \to \operatorname{Pic}(U)).$$

Note that $i_0 : \operatorname{Spec}(B_L\{T\}/TB_L\{T\}) \to \operatorname{Spec} B_L\{T\}$ can be identified with $i_L : \operatorname{Spec} B_L \to \operatorname{Spec} B_L\{T\}$. As $\mathscr{L}_U$ is an element in $\ker(n : \operatorname{Pic}(U) \to \operatorname{Pic}(U))$ satisfying $i_0^*\mathscr{L}_U \cong \mathscr{O}_{U_0}$, it corresponds to an element $c \in H^1(U, \boldsymbol{\mu}_{n,U})$ whose image under the canonical homomorphism

$$H^1(U, \boldsymbol{\mu}_{n,U}) \to H^1(U_0, \boldsymbol{\mu}_{n,U_0})$$

is 0. Since

$$H^1(U_m, \boldsymbol{\mu}_{n,U_m}) \cong H^1(U_0, \boldsymbol{\mu}_{n,U_0})$$

by 5.7.2 (i), the image of c under the canonical homomorphism

$$H^1(U, \boldsymbol{\mu}_{n,U}) \to H^1(U_m, \boldsymbol{\mu}_{n,U_m})$$

is also 0. To prove $\mathscr{L}_U \cong \mathscr{O}_U$, it suffices to prove $c = 0$. Taking into account of 5.7.19, this follows from 7.7.9 below. □

Lemma 7.7.9. *Let B be a normal strictly henselian noetherian local ring, V an open subset of $\mathrm{Spec}\, B$ containing all prime ideals of height 1, U the inverse image of V under the canonical morphism $\pi : \mathrm{Spec}\, B\{T\} \to \mathrm{Spec}\, B$, U_m the inverse images of U under the canonical closed immersions $\mathrm{Spec}\, B[T]/T^{m+1}B[T] \to \mathrm{Spec}\, B\{T\}$, and $f : U' \to U$ an etale covering space. If the morphisms $f_m : U'_m \to U_m$ induced by f by base change are trivial etale covering spaces for all m, then $f : U' \to U$ is a trivial etale covering space.*

Proof. Let $j : U \hookrightarrow \mathrm{Spec}\, B\{T\}$ be the open immersion. By [Fu (2006)] 1.4.9 (iii), or [Hartshorne (1977)] II 5.8, or [EGA] I 9.2.1, $j_*f_*\mathscr{O}_{U'}$ is a quasi-coherent $\mathscr{O}_{\mathrm{Spec}B\{T\}}$-module. Let

$$C = \Gamma(\mathrm{Spec}\, B\{T\}, j_*f_*\mathscr{O}_{U'}) = \Gamma(U', \mathscr{O}_{U'}).$$

We then have

$$j_*f_*\mathscr{O}_{U'} \cong C^{\sim}.$$

Since f is affine, we have

$$U' \cong \mathbf{Spec}\, f_*\mathscr{O}_{U'} \cong \mathbf{Spec}\, j^*j_*f_*\mathscr{O}_{U'}.$$

It follows that

$$U' \cong U \times_{\mathrm{Spec}\, B\{T\}} \mathrm{Spec}\, C.$$

To prove that $f : U' \to U$ is a trivial etale covering space, it suffices to prove that $\mathrm{Spec}\, C \to \mathrm{Spec}\, B\{T\}$ is a trivial etale covering space. Since $B\{T\}$ is strictly henselian, it suffices to prove that C is finite and etale over $B\{T\}$. As $B[[T]]$ is faithfully flat over $B\{T\}$, it suffices to prove that $C \otimes_{B\{T\}} B[[T]]$ is finite and etale over $B[[T]]$. Fix notation by the following commutative diagram, where all squares are Cartesian:

$$\begin{array}{ccccc}
U'_m & \stackrel{f_m}{\to} & U_m & \hookrightarrow & \mathrm{Spec}\, B[T]/T^{m+1}B[T] \\
\downarrow & & \downarrow & & \downarrow \\
\widehat{U}' & \stackrel{\hat{f}}{\to} & \widehat{U} & \stackrel{\hat{j}}{\hookrightarrow} & \mathrm{Spec}\, B[[T]] \\
g' \downarrow & & \downarrow & & \downarrow g \\
U' & \stackrel{f}{\to} & U & \stackrel{j}{\hookrightarrow} & \mathrm{Spec}\, B\{T\} \\
 & & \downarrow & & \downarrow \\
 & & V & \hookrightarrow & \mathrm{Spec}\, B.
\end{array}$$

Since $g : \mathrm{Spec}\, B[[T]] \to \mathrm{Spec}\, B\{T\}$ is flat, we have

$$g^*j_*f_*\mathscr{O}_{U'} \cong \hat{j}_*\hat{f}_*g'^*\mathscr{O}_{U'} \cong \hat{j}_*\hat{f}_*\mathscr{O}_{\widehat{U}'}$$

by [Fu (2006)] 2.4.10, or [Hartshorne (1977)] III 9.3, or [EGA] III 1.4.15. Hence

$$g^* C^\sim \cong \hat{j}_* \hat{f}_* \mathscr{O}_{\widehat{U}'}.$$

We thus have

$$\Gamma(\operatorname{Spec} B[[T]], g^* C^\sim) \cong \Gamma(\operatorname{Spec} B[[T]], \hat{j}_* \hat{f}_* \mathscr{O}_{\widehat{U}'}),$$

that is,

$$C \otimes_{B\{T\}} B[[T]] \cong \Gamma(\widehat{U}', \mathscr{O}_{\widehat{U}'}).$$

Similarly, we have

$$\Gamma(\widehat{U}, \mathscr{O}_{\widehat{U}}) \cong B[[T]].$$

Since f_m are trivial etale covering spaces, we have canonical isomorphisms

$$\Gamma(U'_m, \mathscr{O}_{U'_m}) \cong \Gamma(U_m, \mathscr{O}_{U_m})^k,$$

where k is the degree of the finite etale morphism f. The canonical homomorphisms $\Gamma(\widehat{U}', \mathscr{O}_{\widehat{U}'}) \to \Gamma(U'_m, \mathscr{O}_{U'_m})$ thus induce a homomorphism of $B[[T]]$-algebras

$$\Gamma(\widehat{U}', \mathscr{O}_{\widehat{U}'}) \to \varprojlim_k \Gamma(U_m, \mathscr{O}_{U_m})^k.$$

Let $h : \operatorname{Spec} B[T]/T^{m+1}B[T] \to \operatorname{Spec} B$ be the canonical morphism. Note that

$$h_* \mathscr{O}_{\operatorname{Spec} B[T]/T^{m+1}B[T]} \cong (B[T]/T^{m+1}B[T])^\sim$$

is a free $\mathscr{O}_{\operatorname{Spec} B}$-module. Since B is normal and V contains all prime ideals of height 1, we have

$$\Gamma(V, h_* \mathscr{O}_{\operatorname{Spec} B[T]/T^{m+1}B[T]}) \cong \Gamma(\operatorname{Spec} B, h_* \mathscr{O}_{\operatorname{Spec} B[T]/T^{m+1}B[T]})$$

by [Matsumura (1970)] (17.H) Theorem 38. So we have

$$\Gamma(U_m, \mathscr{O}_{U_m}) \cong B[T]/T^{m+1}B[T].$$

We thus have a homomorphism of $B[[T]]$-algebras

$$\Gamma(\widehat{U}', \mathscr{O}_{\widehat{U}'}) \to \varprojlim_k (B[T]/T^{m+1}B[T])^k = B[[T]]^k.$$

The sheaves $\hat{j}_* \hat{f}_* \mathscr{O}_{\widehat{U}'}$ and $\hat{j}_* \mathscr{O}_{\widehat{U}}$ are quasi-coherent $\mathscr{O}_{\operatorname{Spec} B[[T]]}$-modules. We have

$$\hat{j}_* \hat{f}_* \mathscr{O}_{\widehat{U}'} \cong \Gamma(\operatorname{Spec} B[[T]], \hat{j}_* \hat{f}_* \mathscr{O}_{\widehat{U}'})^\sim \cong \Gamma(\widehat{U}', \mathscr{O}_{\widehat{U}'})^\sim,$$

$$\hat{j}_* \mathscr{O}_{\widehat{U}} \cong \Gamma(\operatorname{Spec} B[[T]], \hat{j}_* \mathscr{O}_{\widehat{U}})^\sim \cong \Gamma(\widehat{U}, \mathscr{O}_{\widehat{U}})^\sim.$$

The homomorphism of $B[[T]]$-algebras $\Gamma(\widehat{U}', \mathcal{O}_{\widehat{U}'}) \to B[[T]]^k$ constructed above thus defines a morphism of $\mathcal{O}_{\mathrm{Spec}\,B[[T]]}$-algebras

$$\hat{j}_*\hat{f}_*\mathcal{O}_{\widehat{U}'} \to (\hat{j}_*\mathcal{O}_{\widehat{U}})^k.$$

Taking its restriction to the open subscheme $\widehat{U}$ of $\mathrm{Spec}\,B[[T]]$, we get a morphism of $\mathcal{O}_{\widehat{U}}$-algebras

$$\hat{f}_*\mathcal{O}_{\widehat{U}'} \to \mathcal{O}_{\widehat{U}}^k.$$

Since $\hat{f}$ is affine, we have

$$\widehat{U}' \cong \mathbf{Spec}\,\hat{f}_*\mathcal{O}_{\widehat{U}'}.$$

The above morphism thus defines k sections of the morphism $\hat{f} : \widehat{U}' \to \widehat{U}$, and these sections are different since they induce different sections for $f_m : U'_m \to U_m$. It follows that $\hat{f}$ is a trivial etale covering space and

$$\Gamma(\widehat{U}', \mathcal{O}_{\widehat{U}'}) \cong \Gamma(\widehat{U}, \mathcal{O}_{\widehat{U}})^k \cong B[[T]]^k.$$

So we have

$$C \otimes_{B\{T\}} B[[T]] \cong \Gamma(\widehat{U}', \mathcal{O}_{\widehat{U}'}) \cong B[[T]]^k.$$

In particular, $C \otimes_{B\{T\}} B[[T]]$ is finite and etale over $B[[T]]$. This proves our assertion. □

7.8 Finiteness of $Rf_!$

([SGA 4] XIV 1, XVI 2, [SGA $4\frac{1}{2}$] Arcata IV 6, V 3.)

The main result of this section is the following.

Theorem 7.8.1. *Let A be a torsion noetherian ring, S a scheme, and $f : X \to Y$ an S-compactifiable morphism between noetherian schemes. For any constructible sheaf of A-modules $\mathscr{F}$ on X, $R^q f_!\mathscr{F}$ are constructible for all q.*

Proof. We only treat the case where $nA = 0$ for some integer n invertible in S. This is enough for our application. The problem is local on Y. We may assume that Y is affine. Let

$$\mathscr{S} = \{Z|Z \subset X,\ \text{Z is closed},\ R^q(f|_Z)_!(\mathscr{F}|_Z) \text{ is not constructible for some } q\}.$$

If $\mathscr{S}$ is not empty, then there exists a minimal member Z in $\mathscr{S}$. Put the reduced closed subscheme structure on Z. Let U be an affine open subset of Z. By 7.4.4 (iii), we have an exact sequence

$$\cdots \to R^q(f|_U)_!(\mathscr{F}|_U) \to R^q(f|_Z)_!(\mathscr{F}|_Z) \to R^q(f|_{Z-U})_!(\mathscr{F}|_{Z-U}) \to \cdots.$$

By the minimality of Z, $R^q(f|_{Z-U})_!(\mathscr{F}|_{Z-U})$ are constructible for all q. If we can prove $R^q(f|_U)_!(\mathscr{F}|_U)$ are constructible for all q, then $R^q(f|_Z)_!(\mathscr{F}|_Z)$ are constructible, which contradicts $Z \in \mathscr{S}$. Thus $\mathscr{S}$ is empty and $R^q f_!\mathscr{F}$ are constructible. Therefore it suffices to treat the case where X and Y are affine. Then $f : X \to Y$ can be factorized as the composite of a closed immersion $i : X \to \mathbb{A}^n_Y$ and the projection $\pi : \mathbb{A}^n_Y \to Y$. We have

$$R^q f_!\mathscr{F} \cong R^q \pi_! i_* \mathscr{F}.$$

Note that $i_*\mathscr{F}$ is constructible. So it suffices to consider the case where $f = \pi$. Factorizing π as the composites of projections

$$\mathbb{A}^n_Y \to \mathbb{A}^{n-1}_Y \to \cdots \to \mathbb{A}^1_Y \to Y$$

and applying 7.4.4 (ii), we are reduced to the case where f is the projection $\pi : \mathbb{A}^1_Y \to Y$.

Let $\mathscr{F}$ be a constructible sheaf of A-modules on $\mathbb{A}^1_Y$. By 5.8.5, we can find a resolution

$$\cdots \to \mathscr{F}^{-1} \to \mathscr{F}^0 \to 0$$

of $\mathscr{F}$ such that each $\mathscr{F}^i$ is of the form $p_{i!}A$ for some affine etale morphism $p_i : U_i \to \mathbb{A}^1_Y$. By 7.4.6, we have a biregular spectral sequence

$$E_1^{pq} = R^q \pi_! \mathscr{F}^p \Rightarrow R^{p+q}\pi_!\mathscr{F}.$$

To prove that $R^q\pi_!\mathscr{F}$ are constructible, it suffices to prove that $R^q\pi_!\mathscr{F}^i$ are constructible. We have

$$R^q\pi_!\mathscr{F}^i = R^q(\pi p_i)_! A.$$

On the other hand, we have

$$R(\pi p_i)_! A \cong A \otimes^L_{\mathbb{Z}/n} R(\pi p_i)_! \mathbb{Z}/n.$$

We are thus reduced to prove that $R^q f_! \mathbb{Z}/n$ are constructible for $f = \pi p$, where $p : U \to \mathbb{A}^1_Y$ is an affine etale morphism. By 5.8.3, it suffices to show that for any irreducible closed subset Z of Y, there exists a nonempty open subset V of Z such that $(R^q f_!\mathbb{Z}/n)|_V$ are locally constant with finite stalks. Put the reduced closed subscheme structure on Z. By the proper base change theorem 7.4.4 (i), we have

$$(R^q f_!\mathbb{Z}/n)|_Z \cong R^q f_{Z!}\mathbb{Z}/n, \quad (R^q f_!\mathbb{Z}/n)|_V \cong R^q f_{V!}\mathbb{Z}/n,$$

where $f_Z : f^{-1}(Z) \to Z$ and $f_V : f^{-1}(V) \to V$ are the base changes of f. Replacing Y by Z, we may assume that Y is an integral scheme and we need to prove that there exists a nonempty open subset V of Y such that $(R^q f_!\mathbb{Z}/n)|_V$ are locally constant with finite stalks.

Let K be the function field of Y. Then $U_K = U \times_Y \operatorname{Spec} K$ is a smooth curve over K. We can find a compactification

$$U_K \hookrightarrow \overline{U}_K \to \operatorname{Spec} K$$

of U_K such that $\overline{U}_K$ is a smooth projective curve over K, and $\overline{U}_K - U_K$ is finite over K. By 1.10.9 and 1.10.10, there exists an open subset W of Y and a compactification

$$U_W = f^{-1}(W) \hookrightarrow \overline{U}_W \xrightarrow{g} W$$

of $f_W : f^{-1}(W) \to W$ inducing the above compactification of U_K such that $g|_{\overline{U}_W - U_W}$ is finite and g is smooth projective and pure of relative dimension 1. We have

$$R^q(g|_{\overline{U}_W - U_W})_!\mathbb{Z}/n = \begin{cases} (g|_{\overline{U}_W - U_W})_*\mathbb{Z}/n & \text{if } q = 0, \\ 0 & \text{if } q \neq 0. \end{cases}$$

By 5.8.11 (i), $(g|_{\overline{U}_W - U_W})_*\mathbb{Z}/n$ is constructible. By 7.4.4 (iii), we have an exact sequence

$$\cdots \to R^{q-1}(g|_{\overline{U}_W - U_W})_!\mathbb{Z}/n \to R^q f_{W!}\mathbb{Z}/n \to R^q g_*\mathbb{Z}/n \to \cdots.$$

If we can prove that $R^q g_*\mathbb{Z}/n$ are constructible, then $R^q f_{W!}\mathbb{Z}/n$ are constructible, that is, $(R^q f_!\mathbb{Z}/n)|_W$ are constructible. So there exists an open subset V of W such that $(R^q f_!\mathbb{Z}/n)|_V$ are locally constant with finite stalks.

It remains to show that $R^q g_*\mathbb{Z}/n$ are constructible. First note that the stalks of $R^q g_*\mathbb{Z}/n$ are finite. Indeed, for any $t \in W$, we have

$$(R^q g_*\mathbb{Z}/n)_{\bar{t}} \cong H^q(\overline{U}_W \otimes_{\mathcal{O}_W} \overline{k(t)}, \mathbb{Z}/n)$$

by the proper base change theorem 7.3.1. The groups $H^q(\overline{U}_W \otimes_{\mathcal{O}_W} \overline{k(t)}, \mathbb{Z}/n)$ are finite by 7.2.9 (ii). By 7.8.2 below and 5.8.9, $R^q g_*\mathbb{Z}/n$ are locally constant, and hence constructible. □

Lemma 7.8.2. *Let $f : X \to Y$ be a smooth proper morphism, and let $\mathscr{F}$ be a locally constant sheaf of $\mathbb{Z}/n$-modules on X for some n invertible on Y. Then for all points $s, t \in Y$ such that $s \in \overline{\{t\}}$, the specialization homomorphisms*

$$(R^q f_*\mathscr{F})_{\bar{s}} \to (R^q f_*\mathscr{F})_{\bar{t}}$$

are isomorphisms for all q.

Proof. Let $\widetilde{Y}_{\bar{s}}$ be the strict localization of Y at $\bar{s}$. Fix notation by the following commutative diagram

$$\begin{array}{ccccc} X_{\bar{t}} = X \times_Y \bar{t} & \xrightarrow{\tilde{j}} & X \times_Y \widetilde{Y}_{\bar{s}} & \xleftarrow{\tilde{i}} & X_{\bar{s}} = X \times_Y \bar{s} \\ f_{\bar{t}} \downarrow & & \tilde{f} \downarrow & & \downarrow f_{\bar{s}} \\ \bar{t} & \xrightarrow{j} & \widetilde{Y}_{\bar{s}} & \xleftarrow{i} & \bar{s}, \end{array}$$

where vertical arrows are base changes of f. Let $\widetilde{\mathscr{F}} = \mathscr{F}|_{X \times_Y \widetilde{Y}_{\bar{s}}}$. For each q, we have a commutative diagram of canonical morphisms

$$\begin{array}{ccccc} H^q(X_{\bar{t}}, \tilde{j}^*\widetilde{\mathscr{F}}) & \leftarrow & H^q(X \times_Y \widetilde{Y}_{\bar{s}}, R\tilde{j}_*\tilde{j}^*\widetilde{\mathscr{F}}) & \rightarrow & H^q(X_{\bar{s}}, \tilde{i}^*R\tilde{j}_*\tilde{j}^*\widetilde{\mathscr{F}}) \\ \| & & \uparrow & & \uparrow \\ H^q(X_{\bar{t}}, \tilde{j}^*\widetilde{\mathscr{F}}) & \leftarrow & H^q(X \times_Y \widetilde{Y}_{\bar{s}}, \widetilde{\mathscr{F}}) & \rightarrow & H^q(X_{\bar{s}}, \tilde{i}^*\widetilde{\mathscr{F}}). \end{array}$$

By 7.7.5, we have an isomorphism

$$\tilde{i}^*\widetilde{\mathscr{F}} \xrightarrow{\cong} \tilde{i}^*R\tilde{j}_*\tilde{j}^*\widetilde{\mathscr{F}}.$$

So the rightmost vertical arrow is an isomorphism. By the proper base change theorem 7.3.3, the two horizontal arrows in the square on the right of the diagram are isomorphisms. The canonical morphism

$$H^q(X \times_Y \widetilde{Y}_{\bar{s}}, R\tilde{j}_*\tilde{j}^*\widetilde{\mathscr{F}}) \rightarrow H^q(X_{\bar{t}}, \tilde{j}^*\widetilde{\mathscr{F}})$$

is also an isomorphism. It follows that all arrows in the above diagram are isomorphisms. By 5.9.5, we have

$$(R^q f_*\mathscr{F})_{\bar{s}} \cong H^q(X \times_Y \widetilde{Y}_{\bar{s}}, \widetilde{\mathscr{F}}),$$

and by the proper base change theorem 7.3.1, we have

$$(R^q f_*\mathscr{F})_{\bar{t}} \cong H^q(X_{\bar{t}}, \tilde{j}^*\widetilde{\mathscr{F}}).$$

Through these isomorphisms, the specialization homomorphism

$$(R^q f_*\mathscr{F})_{\bar{s}} \rightarrow (R^q f_*\mathscr{F})_{\bar{t}}$$

is identified with the canonical morphism

$$H^q(X \times_Y \widetilde{Y}_{\bar{s}}, \widetilde{\mathscr{F}}) \rightarrow H^q(X_{\bar{t}}, \tilde{j}^*\widetilde{\mathscr{F}})$$

in the above diagram. So the specialization homomorphism is an isomorphism. □

Corollary 7.8.3. *Let $f : X \rightarrow Y$ be a smooth proper morphism between noetherian schemes, and let $\mathscr{F}$ be a constructible locally constant sheaf of A-modules on X for some noetherian ring A such that $nA = 0$ for some integer n invertible on Y. Then $R^q f_*\mathscr{F}$ are locally constant for all q.*

Proof. Apply 7.8.1, 7.8.2 and 5.8.9. □

Chapter 8

Duality

8.1 Extensions of Henselian Discrete Valuation Rings

Let R be a henselian discrete valuation ring, K its fraction field, $\mathfrak{m}$ its maximal ideal, and $k = R/\mathfrak{m}$ its residue field. Denote by $v : K^* \to \mathbb{Z}$ the valuation on K defined by R. Any element π in R with valuation 1 is called a *uniformizer*. It is a generator of $\mathfrak{m}$. Elements with valuation 0 are units of R. Every element in R can be uniquely written as $u\pi^n$ for some unit u and some nonnegative integer n. So R is a unique factorization domain. In particular, R is normal.

Let K' be a finite separable extension of K. Then the integral closure R' of R in K' is finite over R. It is also a henselian discrete valuation ring. Denote by $\mathfrak{m}', k', v'$ the maximal ideal, the residue field, and the valuation of R'. Define the *ramification index* of the extension K'/K to be the integer e such that $\mathfrak{m}R' = \mathfrak{m}'^e$. We have $e = v'(\pi)$ for any uniformizer π of R. Define the *degree of inertia* of the extension K'/K to be $f = [k' : k]$.

Proposition 8.1.1. *Notation as above. We have*
(i) $ef = [K' : K]$.
(ii) *For any* $a' \in K'^*$, *we have* $v(N_{K'/K}(a')) = fv'(a')$.

Proof.
(i) Using 1.1.3 (ii) and 1.2.9, one can show that R' is free of finite rank as an R-module. We have

$$R' \otimes_R K \cong K'.$$

So the rank of R' as an R-module is $[K' : K]$. We have

$$R' \otimes_R R/\mathfrak{m} \cong R'/\mathfrak{m}'^e.$$

The rank of $R'/\mathfrak{m}'^e$ as an $R/\mathfrak{m}$-module is equal to

$$\sum_{i=1}^{e} \mathrm{rank}(\mathfrak{m}'^{i-1}/\mathfrak{m}'^{i}) = \sum_{i=1}^{e} \mathrm{rank}(R'/\mathfrak{m}') = ef.$$

So we have $[K' : K] = ef$.

(ii) Both sides of the equality define homomorphisms of groups from K'^* to $\mathbb{Z}$. If a' is a unit in R', then $N_{K'/K}(a)$ is a unit in R, and both sides are zero. Let π be a uniformizer of K. We have

$$v(N_{K'/K}(\pi)) = v(\pi^{[K':K]}) = [K' : K] = ef = fv'(\pi).$$

Let π' be a uniformizer K'. We have $\pi = u'\pi'^e$ for some unit u'. The equality holds for $a' = u'$ and for $a' = \pi$. So it holds for $a' = \pi'^e$ and hence for $a' = \pi'$. As K'^* is generated by π' and units in R', the equality holds for any $a' \in K'^*$. □

We say that the finite separable extension K'/K is *unramified* if $e = 1$ and k' is separable over k. This is equivalent to saying that $\mathrm{Spec}\, R' \to \mathrm{Spec}\, R$ is unramified. Note that $\mathrm{Spec}\, R' \to \mathrm{Spec}\, R$ is always flat. So K'/K is unramified if and only if $\mathrm{Spec}\, R' \to \mathrm{Spec}\, R$ is etale. We say K'/K is *totally ramified* if $e = [K' : K]$. This is equivalent to saying that $f = 1$, that is, $k' = k$. Let p be the characteristic of k. We say that K'/K is *tamely ramified* if e is relatively prime to p, and k' is separable over k.

Lemma 8.1.2. *Suppose R is a henselian discrete valuation ring with fraction field K and residue field k.*

(i) *Let K'/K be a finite unramified extension, K''/K a finite separable extension, R' (resp. R'') the integral closure of R in K' (resp. K''), and k' (resp. k'') the residue field of R' (resp. R''). Then the canonical map*

$$\mathrm{Hom}_R(R', R'') \to \mathrm{Hom}_k(k', k'')$$

is bijective.

(ii) *For any finite separable extension k' of k, there exists a finite unramified extension K'/K such that the residue field of the integral closure R' of R in K' is isomorphic to k'.*

Proof.

(i) The set $\mathrm{Hom}_R(R', R'')$ can be identified with the set of R-morphisms $\mathrm{Hom}_{\mathrm{Spec}\, R}(\mathrm{Spec}\, R'', \mathrm{Spec}\, R')$. Identifying an R-morphism with its graph, the set $\mathrm{Hom}_{\mathrm{Spec}\, R}(\mathrm{Spec}\, R'', \mathrm{Spec}\, R')$ can be identified with the set of sections of the projection $\mathrm{Spec}\,(R' \otimes_R R'') \to \mathrm{Spec}\, R''$. Since K'/K is unramified, this projection is etale. Since R'' is henselian, by 2.8.3 (vii) and

2.3.10 (i), the set of sections of the projection $\mathrm{Spec}\,(R'\otimes_R R'')\to\mathrm{Spec}\,R''$ is in one-to-one correspondence with the set of sections of the projection $\mathrm{Spec}\,(k'\otimes_k k'')\to\mathrm{Spec}\,k''$. The set of sections of the projection $\mathrm{Spec}\,(k'\otimes_k k'')\to\mathrm{Spec}\,k''$ can be identified with the set $\mathrm{Hom}_k(k',k'')$. Our assertion follows.

(ii) We have $k'\cong k[t]/(f_0(t))$ for some monic irreducible polynomial $f_0(t)\in k[t]$ such that $f_0(t)$ and $f_0'(t)$ are relatively prime. Let $f(t)\in R[t]$ be a monic polynomial lifting $f_0(t)$. Applying Nakayama's lemma to the R-module $R[t]/(f(t))$, one can show that the ideal generated by $f(t)$ and $f'(t)$ is $R[t]$. By 2.3.3, $R[t]/(f(t))$ is etale over R. Since $f_0(t)$ is irreducible, $f(t)$ is irreducible as a polynomial in $R[t]$. By the Gauss lemma, $f(t)$ is irreducible as a polynomial in $K[t]$. Let $K'=K[t]/(f(t))$, let R' be the integral closure of R in K, and let $k(R')$ be the residue field of R'. Then K' is a field finite and separable over K. The R-algebra homomorphism

$$R[t]/(f(t))\to R'$$

induces a k-homomorphism

$$k[t]/(f_0(t))\to k(R').$$

It follows that

$$[k[t]/(f_0(t)):k]\leq[k(R'):k].$$

But we have

$$[K':K]=\deg(f)=[k[t]/(f_0(t)):k].$$

So we have

$$[K':K]\leq[k(R'):k].$$

Combined with 8.1.1 (i), we see that K'/K must be an unramified extension and $[K':K]=[k(R'):k]$. It has the required property. $\square$

Let R be a henselian discrete valuation ring with fraction field K and residue field k. Fix a separable closure $\overline{K}$ of K. By 8.1.2, for any finite separable extension k'/k, there exists a finite unramified extension K' of K contained in $\overline{K}$ such that the residue field of the integral closure R' of R in K' is k-isomorphic to k'. Let

$$\widetilde{K}=\varinjlim_{K'/K}K',\quad \widetilde{R}=\varinjlim_{K'/K}R',$$

where K' goes over the set of all finite unramified extensions of K contained in $\overline{K}$. Then the valuation $v:K^*\to\mathbb{Z}$ can be uniquely extended to a valuation $v : \widetilde{K}^* \to \mathbb{Z}$, $\widetilde{R}$ is the corresponding discrete valuation ring, and the residue field $\tilde{k}$ of $\widetilde{R}$ is a separable closure of k. Moreover, $\widetilde{K}$ is an (infinite) galois extension of K, and we have a canonical isomorphism

$$\mathrm{Gal}(\widetilde{K}/K) \cong \mathrm{Gal}(\tilde{k}/k).$$

We call $I = \mathrm{Gal}(\overline{K}/\widetilde{K})$ the *inertia subgroup* of $\mathrm{Gal}(\overline{K}/K)$. It is the kernel of the canonical epimorphism

$$\mathrm{Gal}(\overline{K}/K) \to \mathrm{Gal}(\tilde{k}/k).$$

Suppose that $U \to \operatorname{Spec} R$ is a separated etale morphism of finite type such that U is connected. Then by 2.9.2, U is an open subscheme of $\operatorname{Spec} R'$, where R' is the integral closure of R in the function field of U. As $\operatorname{Spec} R'$ consists of only two points, if $U \to \operatorname{Spec} R$ is surjective, then the open immersion $U \hookrightarrow \operatorname{Spec} R'$ is also surjective, and hence $U \cong \operatorname{Spec} R'$. Suppose that A is a local essentially etale R-algebra. The above discussion shows that the fraction field $K(A)$ of A is a finite unramified extension of K and A is the integral closure of R in $K(A)$. So $\widetilde{R} = \varinjlim_{K'/K} R'$ is the strict henselization of R with respect to the separable closure $\tilde{k}$ of k.

Proposition 8.1.3. *Suppose that R is a strictly henselian discrete valuation ring with fraction field K. Let π be a uniformizer of K, and let p be the characteristic of the residue field of R.*

(i) *For any positive integer n relatively prime to p, $K[t]/(t^n - \pi)$ is a finite tamely ramified extension of K. Any finite tamely ramified extension of K is isomorphic to $K[t]/(t^n - \pi)$ for some positive integer n relatively prime to p.*

(ii) *Let K'/K be a finite separable extension. There exists an extension K'' of K contained in K' such that K''/K is tamely ramified and $[K' : K'']$ is a power of p.*

Proof.

(i) By the Eisenstein criterion, $t^n - \pi$ is an irreducible polynomial in $K[t]$. So $K[t]/(t^n - \pi)$ is a field. One can verify that the canonical homomorphism

$$R[t]/(t^n - \pi) \to K[t]/(t^n - \pi)$$

is injective. Hence $R[t]/(t^n - \pi)$ is an integral domain. It is finite over R. If $\mathfrak{m}'$ is a maximal ideal of $R[t]/(t^n - \pi)$, then $\mathfrak{m}'$ lies above the maximal ideal of R. Hence $\pi \in \mathfrak{m}'$. Since $t^n = \pi$ in $R[t]/(t^n - \pi)$, we have $t \in \mathfrak{m}'$. On the other hand, we have

$$R[t]/(t^n - \pi, t) \cong R[t]/(\pi, t) \cong R/\pi R.$$

Hence $R[t]/(t^n - \pi)$ is a local integral domain and its maximal ideal is generated by t. This implies that $R[t]/(t^n - \pi)$ is a discrete valuation ring and hence integrally closed. It is the integral closure of R in $K[t]/(t^n - \pi)$. It has the same residue field as R, and the ramification index is n. So $K[t]/(t^n - \pi)$ is a tamely ramified extension of K.

Suppose that K'/K is a finite tamely ramified extension. Since the residue field of R is separably closed, K'/K is a totally ramified extension. Let $n = [K' : K]$, let R' be the integral closure of R in K', and let π' be a uniformizer of R'. The valuation of π in K' is n. We have $\pi = u\pi'^n$ for some unit u in R'. Since K'/K is tamely ramified, n is relatively prime to p. Let $\mathfrak{m}'$ be the maximal ideal of R' and let $k' = R'/\mathfrak{m}'$. Then k' is isomorphic to the residue field of R and hence is separably closed. So $t^n - u$ mod $\mathfrak{m}'$ is a product of distinct linear polynomials in $k'[t]$. By 2.8.3 (v), $t^n - u$ is a product of linear polynomials in $R'[t]$. So u has an n-th root $\sqrt[n]{u}$ in R'. Replacing π' by $\pi'\sqrt[n]{u}$, we may assume $\pi = \pi'^n$. Since $t^n - \pi$ is an irreducible polynomial in $K[t]$, it is the minimal polynomial of π'. We thus have $[K[\pi'] : K] = n$. Hence

$$K' = K[\pi'] \cong K[t]/(t^n - \pi).$$

(ii) Let R' be the integral closure of R in K', let k and k' be the residue fields of R and R', respectively, let e be the ramification index and let f be the degree of inertia. Since k is separably closed, k'/k is a purely inseparable extension. Hence $f = [k' : k]$ is a power of p. Write $e = e'p^i$ for some nonnegative integer i and some integer e' relatively prime to p. Let π' be a uniformizer of R'. We have $\pi = u\pi'^e$ for some unit u in R'. As k' is separably closed, the same argument as the proof of (i) shows that $\sqrt[e']{u}$ exists in R'. We have $\pi = (\sqrt[e']{u}\pi'^{p^i})^{e'}$. So $\sqrt[e']{\pi}$ exists in R'. Let $K'' = K[\sqrt[e']{\pi}]$. Then K''/K is tamely ramified. We have

$$[K' : K''] = \frac{[K' : K]}{e'} = \frac{ef}{e'} = p^i f.$$

So $[K' : K'']$ is a power of p. □

Let R be a strictly henselian discrete valuation ring, K its fraction field, $\overline{K}$ a separable closure of K, π a uniformizer of R, and p the characteristic of the residue field of R. For each positive integer n relatively prime to p, choose an n-th root $\sqrt[n]{\pi}$ of π in $\overline{K}$. Let

$$K_t = \bigcup_{(n,p)=1} K[\sqrt[n]{\pi}].$$

Then by 8.1.3, $P = \mathrm{Gal}(\overline{K}/K_t)$ is a pro-p-group. We call P the *wild inertia subgroup* of I. Let $I = \mathrm{Gal}(\overline{K}/K)$. The canonical homomorphism

$$I \to \varprojlim_{(n,p)=1} \boldsymbol{\mu}_n(K), \quad \sigma \mapsto \left(\frac{\sigma(\sqrt[n]{\pi})}{\sqrt[n]{\pi}}\right)$$

induces an isomorphism

$$I/P \cong \varprojlim_{(n,p)=1} \boldsymbol{\mu}_n(K),$$

where

$$\boldsymbol{\mu}_n(K) = \{\zeta \in K | \zeta^n = 1\}$$

is the group of n-th roots of unity in K. Note that by 2.8.3 (v), $\boldsymbol{\mu}_n(K)$ is isomorphic to the group $\boldsymbol{\mu}_n(k) = \{\zeta \in k | \zeta^n = 1\}$ of n-th roots of unity in k.

Proposition 8.1.4. *Let R be a strictly henselian discrete valuation ring with fraction field K and residue field k, let $\overline{K}$ be a separable closure of K, and let $I = \mathrm{Gal}(\overline{K}/K)$. For any torsion I-module M with torsion relatively prime to $p = \mathrm{char}\, k$, we have canonical isomorphisms*

$$H^q(I, M) \cong \begin{cases} M^I & \text{if } q = 0, \\ M_I(-1) & \text{if } q = 1, \\ 0 & \text{if } q \neq 0, 1 \end{cases}$$

where for any torsion abelian group A with torsion relatively prime to p, we set

$$A(-1) = \mathrm{Hom}(\varprojlim_{(n,p)=1} \boldsymbol{\mu}_n(k), A).$$

Proof. Let P be the wild inertia subgroup. Taking P-invariant is an exact functor in the category of P-modules with torsion prime to p, and we have

$$M^P \cong M_P$$

for any module M in this category. Indeed, suppose that

$$0 \to M' \to M \to M'' \to 0$$

is an exact sequence in this category. For any $x'' \in M''^P$, let $x \in M$ be a preimage of x'' in M. Then

$$\frac{1}{\#(P/P_x)} \sum_{gP_x \in P/P_x} gx$$

is a preimage of x'' in M^P, where P_x is the stabilizer of x in P. Here multiplication by $\frac{1}{\#(P/P_x)}$ makes sense since M has torsion relatively prime to p so that multiplication by p induces an isomorphism on M. It follows that the sequence

$$0 \to M'^P \to M^P \to M''^P \to 0$$

is exact. The homomorphism

$$M \to M, \quad x \mapsto \frac{1}{\#(P/P_x)} \sum_{gP_x \in P/P_x} gx$$

induces a homomorphism

$$M_P \to M^P.$$

One can show it is the inverse of the homomorphism $M^P \to M_P$ induced by id_M. So $M^P \cong M_P$. Therefore the Hochschild–Serre spectral sequence

$$E_2^{uv} = H^u(I/P, H^v(P, M)) \Rightarrow H^{u+v}(I, M)$$

degenerates, and we have

$$H^q(I, M) \cong H^q(I/P, M^P) \cong H^q(I/P, M_P).$$

By 4.3.9, we have

$$\begin{aligned} &H^0(I/P, M^P) \cong (M^P)^{I/P} \cong M^I, \\ &H^1(I/P, M_P) \cong (M_P)_{I/P}(-1) \cong M_I(-1), \\ &H^q(I/P, M^P) = 0 \text{ for } q \neq 0, 1. \end{aligned}$$

□

At the end of this section, we give a proof of 7.1.6, which relies only on 8.1.1.

Proof of 7.1.6. Since X is smooth over k, the function field $K(X)$ is separably generated over k. So $K(X)$ can be regarded as a finite separable extension of the rational function field $k(t)$. Since X is proper, the extension $k(t) \hookrightarrow K(X)$ defines a finite k-morphism

$$\phi : X \to \mathbb{P}^1_k,$$

and X can be identified with the normalization of $\mathbb{P}^1_k$ in $K(X)$. For any Zariski closed point $s \in |\mathbb{P}^1_k|$ in $\mathbb{P}^1_k$, choose an affine open neighborhood $V = \operatorname{Spec} A$ of s in $\mathbb{P}^1_k$. Then $\phi^{-1}(V) \cong \operatorname{Spec} B$ for some finite A-algebra B. For any Zariski closed point $x \in |X|$ in X, denote by v_x (resp. v_s) the valuation of $K(X)$ (resp. $k(t)$) corresponding to the point x (resp. s), denote by $k(x)$ (resp. $k(s)$) the residue field of $\mathscr{O}_{X,x}$ (resp. $\mathscr{O}_{\mathbb{P}^1_k,s}$), denote by $\widetilde{\mathscr{O}}_{X,x}$ (resp. $\widetilde{\mathscr{O}}_{\mathbb{P}^1_k,s}$) the henselization of $\mathscr{O}_{X,x}$ (resp. $\mathscr{O}_{\mathbb{P}^1_k,s}$), and denote by $K(\widetilde{\mathscr{O}}_{X,x})$ (resp. $K(\widetilde{\mathscr{O}}_{\mathbb{P}^1_k,s})$) the fraction field of $\widetilde{\mathscr{O}}_{X,x}$ (resp. $\widetilde{\mathscr{O}}_{\mathbb{P}^1_k,s}$). By 2.8.12, we have

$$B \otimes_A \widetilde{\mathscr{O}}_{\mathbb{P}^1_k,s} \cong \prod_{\substack{x \in |X| \\ \phi(x) = s}} \widetilde{\mathscr{O}}_{X,x}.$$

So we have

$$K(X) \otimes_{k(t)} K(\widetilde{\mathscr{O}}_{\mathbb{P}^1_k,s}) \cong \prod_{\substack{x \in |X| \\ \phi(x) = s}} K(\widetilde{\mathscr{O}}_{X,x}).$$

For any $g \in K(X)$, we thus have

$$\mathrm{N}_{K(X)/k(t)}(g) = \prod_{\substack{x \in |X| \\ \phi(x) = s}} N_{K(\widetilde{\mathscr{O}}_{X,x})/K(\widetilde{\mathscr{O}}_{\mathbb{P}^1_k,s})}(g).$$

By 8.1.1 (ii), we have

$$v_s(\mathrm{N}_{K(X)/k(t)}(g)) = \sum_{\substack{x \in |X| \\ \phi(x) = s}} [k(x) : k(s)] v_x(g).$$

It follows that

$$\begin{aligned}
\deg(g) &= \deg\Big(\sum_{x \in |X|} v_x(g) x\Big) \\
&= \sum_{x \in |X|} [k(x) : k] v_x(g) \\
&= \sum_{s \in |\mathbb{P}^1_k|} \sum_{\substack{x \in |X| \\ \phi(x) = s}} [k(s) : k][k(x) : k(s)] v_x(g) \\
&= \sum_{s \in |\mathbb{P}^1_k|} [k(s) : k] v_s(\mathrm{N}_{K(X)/k(t)}(g)) \\
&= \deg(\mathrm{N}_{K(X)/k(t)}(g)).
\end{aligned}$$

To prove $\deg(g) = 0$, it suffices to show $\deg(\mathrm{N}_{K(X)/k(t)}(g)) = 0$. We are thus reduced to the case where $X = \mathbb{P}^1_k$. In this case, any nonzero $g \in k(t)$ can be uniquely written as

$$g = \lambda \prod_{i=1}^{n} q_i^{k_i},$$

where $\lambda \in k$, $q_i \in k[t]$ are monic irreducible polynomials, and k_i are nonzero integers. Let d_i be the degrees of the polynomials q_i. A Zariski closed point

s in $\mathbb{P}^1_k$ is either the point ∞, in which case $-v_\infty(g)$ is the degree of the polynomial g, or a Zariski closed point in $\mathbb{A}^1_k$ corresponding to the maximal ideal defined by a monic irreducible polynomial q, in which case we have

$$v_s(g) = \begin{cases} 0 & \text{if } q \neq q_i \text{ for all } i, \\ k_i & \text{if } q = q_i. \end{cases}$$

It follows that

$$\begin{aligned} \deg(g) &= \sum_{s \in |\mathbb{P}^1_k|} [k(s):k] v_s(g) \\ &= \sum_{i=1}^{k} d_i k_i + v_\infty(g) \\ &= 0. \end{aligned}$$

This proves our assertion. □

8.2 Trace Morphisms

([SGA 4] XVIII 1.1, 2.)

In this section, we fix a scheme S and an integer n invertible on S. For any S-scheme X, we denote the sheaf $\boldsymbol{\mu}_{n,X}$ defined in 7.2.1 by $\boldsymbol{\mu}_n$ for simplicity. For any integer d and any sheaf $\mathscr{F}$ of $\mathbb{Z}/n$-modules on X, define

$$\begin{aligned} \mathbb{Z}/n(d) &= \begin{cases} \boldsymbol{\mu}_n^{\otimes d} & \text{if } d \geq 0, \\ \mathscr{H}om(\boldsymbol{\mu}_n^{\otimes(-d)}, \mathbb{Z}/n) & \text{if } d \leq 0, \end{cases} \\ \mathscr{F}(d) &= \mathscr{F} \otimes \mathbb{Z}/n(d). \end{aligned}$$

Let $f : X \to Y$ be a smooth S-compactifiable morphism pure of relative dimension d for some integer d. In this section, we define a canonical morphism

$$R^{2d} f_! f^* \mathscr{F}(d) \to \mathscr{F},$$

called the *trace morphism* denoted by $\mathrm{Tr}_{X/Y}$ or Tr_f.

Suppose that $f : X \to Y$ is etale. Then $f_!$ is left adjoint to f^*. We define $\mathrm{Tr}_f : f_! f^* \mathscr{F} \to \mathscr{F}$ to be the adjunction morphism.

Suppose that X is a smooth irreducible projective curve over an algebraically closed field k. By 7.2.9, we have a canonical isomorphism

$$H^2(X, \boldsymbol{\mu}_n) \cong \mathrm{Pic}(X)/n\mathrm{Pic}(X).$$

The homomorphism $\deg : \mathrm{Pic}(X) \to \mathbb{Z}$ induces an isomorphism

$$\mathrm{Pic}(X)/n\mathrm{Pic}(X) \cong \mathbb{Z}/n.$$

We define $\mathrm{Tr}_{X/k} : H^2(X, \boldsymbol{\mu}_n) \to \mathbb{Z}/n$ to be the composite

$$H^2(X, \boldsymbol{\mu}_n) \cong \mathrm{Pic}(X)/n\mathrm{Pic}(X) \stackrel{\deg}{\to} \mathbb{Z}/n.$$

Suppose that X is a smooth irreducible curve over an algebraically closed field, and let $\overline{X}$ be its smooth compactification. We have an exact sequence

$$\cdots \to H^1_c(\overline{X}-X, \boldsymbol{\mu}_n) \to H^2_c(X, \boldsymbol{\mu}_n) \to H^2(\overline{X}, \boldsymbol{\mu}_n) \to H^2_c(\overline{X}-X, \boldsymbol{\mu}_n) \to \cdots.$$

Since $\overline{X} - X$ is finite, we have

$$H^1_c(\overline{X} - X, \boldsymbol{\mu}_n) = H^2_c(\overline{X} - X, \boldsymbol{\mu}_n) = 0.$$

So we have an isomorphism

$$H^2_c(X, \boldsymbol{\mu}_n) \cong H^2(\overline{X}, \boldsymbol{\mu}_n).$$

We define $\mathrm{Tr}_{X/k} : H^2_c(X, \boldsymbol{\mu}) \to \mathbb{Z}/n$ to be the composite of this isomorphism and $\mathrm{Tr}_{\overline{X}/k}$.

Suppose that X is a smooth curve over an algebraically closed field. Let $X_1, \ldots, X_m$ be its irreducible components. We have

$$H^2_c(X, \boldsymbol{\mu}_n) = \bigoplus_i H^2_c(X_i, \boldsymbol{\mu}_n)$$

We defined $\mathrm{Tr}_{X/k} : H^2_c(X, \boldsymbol{\mu}_n) \to \mathbb{Z}/n$ to be the sum of $\mathrm{Tr}_{X_i/k}$.

Let Y be a smooth curve over an algebraically closed field k and let $f : X \to Y$ be an etale morphism. $\mathrm{Tr}_{X/Y} : f_!\boldsymbol{\mu}_n \to \boldsymbol{\mu}_n$ induces a homomorphism $H^2_c(Y, f_!\boldsymbol{\mu}_n) \to H^2_c(Y, \boldsymbol{\mu}_n)$. Let $S_{X/Y}$ be the composite

$$H^2_c(X, \boldsymbol{\mu}_n) \cong H^2_c(Y, f_!\boldsymbol{\mu}_n) \to H^2_c(Y, \boldsymbol{\mu}_n).$$

Lemma 8.2.1. *Let Y be a smooth curve over an algebraically closed field k and let $f : X \to Y$ be an etale morphism. We have* $\mathrm{Tr}_{Y/k} \circ S_{X/Y} = \mathrm{Tr}_{X/k}$.

Proof. We may reduce to the case where X and Y are irreducible. Let $\overline{X}$ and $\overline{Y}$ be the smooth compactifications of X and Y, respectively, and let $\bar{f} : \overline{X} \to \overline{Y}$ be the morphism induced by f. Then $\bar{f}$ is finite and flat. (The flatness can be proved using 1.1.3 (ii).) So $\bar{f}_*\mathscr{O}_{\overline{X}}$ is a locally free $\mathscr{O}_{\overline{Y}}$-module of finite rank. If V is an open subset of Y such that $(\bar{f}_*\mathscr{O}_{\overline{X}})|_V$ is a free $\mathscr{O}_{\overline{Y}}|_V$-module, then multiplication by a section $s \in (\bar{f}_*\mathscr{O}_{\overline{X}})(V)$ defines an endomorphism on the free $\mathscr{O}_{\overline{Y}}(V)$-module $(\bar{f}_*\mathscr{O}_{\overline{X}})(V)$. Its determinant $\det(s)$ is a section in $\mathscr{O}_{\overline{Y}}(V)$. We thus get a morphism of sheaves

$$\det : \bar{f}_*\mathscr{O}^*_{\overline{X}} \to \mathscr{O}^*_{\overline{Y}}.$$

Applying this construction after etale base changes, we can define a morphism of etale sheaves

$$\det : \bar{f}_* \mathscr{O}^*_{\overline{X}_{\mathrm{et}}} \to \mathscr{O}^*_{\overline{Y}_{\mathrm{et}}}.$$

Consider the diagram

$$\begin{array}{ccccccc} 0 \to & \bar{f}_* \boldsymbol{\mu}_n & \to & \bar{f}_* \mathscr{O}^*_{\overline{X}_{\mathrm{et}}} & \stackrel{n}{\to} & \bar{f}_* \mathscr{O}^*_{\overline{X}_{\mathrm{et}}} & \to 0 \\ & \mathrm{Tr}_{\overline{X}/\overline{Y}} \downarrow & & \downarrow \det & & \downarrow \det & \\ 0 \to & \boldsymbol{\mu}_n & \to & O^*_{\overline{Y}_{\mathrm{et}}} & \stackrel{n}{\to} & \mathscr{O}^*_{\overline{Y}_{\mathrm{et}}} & \to 0, \end{array}$$

where the horizontal lines are exact by Kummer's theory and the fact that $\bar{f}$ is finite, and the morphism

$$\mathrm{Tr}_{\overline{X}/\overline{Y}} : \bar{f}_* \boldsymbol{\mu}_n \to \boldsymbol{\mu}_n$$

is defined to be the morphism that makes the diagram commute. More explicitly, for any $y \in Y$, choose a separable closure $\overline{k(y)}$ of $k(y)$. For any $x \in \overline{X} \otimes_{\mathscr{O}_{\overline{Y}}} \overline{k(y)}$, let $\widetilde{\mathscr{O}}_{\overline{X},\bar{x}}$ (resp. $\widetilde{\mathscr{O}}_{\overline{Y},\bar{y}}$) be the strict henselization of $\overline{X}$ (resp. $\overline{Y}$) at $\bar{x}$ (resp. $\bar{y}$). We have

$$(\bar{f}_* \boldsymbol{\mu}_n)_{\bar{y}} \cong \Gamma(\overline{X} \otimes_{\mathscr{O}_{\overline{Y}}} \widetilde{\mathscr{O}}_{\overline{Y},\bar{y}}, \boldsymbol{\mu}_n) \cong \bigoplus_{x \in \overline{X} \otimes_{\mathscr{O}_{\overline{Y}}} \overline{k(y)}} \Gamma(\mathrm{Spec}\, \widetilde{\mathscr{O}}_{\overline{X},\bar{x}}, \boldsymbol{\mu}_n).$$

For any $(\lambda_x) \in \bigoplus_{x \in \overline{X} \otimes_{\mathscr{O}_{\overline{Y}}} \overline{k(y)}} \Gamma(\mathrm{Spec}\, \widetilde{\mathscr{O}}_{\overline{X},\bar{x}}, \boldsymbol{\mu}_n)$, we have

$$\mathrm{Tr}_{\overline{X}/\overline{Y}}((\lambda_x)) = \prod_{x \in \overline{X} \otimes_{\mathscr{O}_{\overline{Y}}} \overline{k(y)}} \lambda_x^{n_x},$$

where $n_x = \mathrm{rank}_{\widetilde{\mathscr{O}}_{\overline{Y},\bar{y}}} \widetilde{\mathscr{O}}_{\overline{X},\bar{x}}$. Let $i : X \hookrightarrow \overline{X}$ and $j : Y \hookrightarrow \overline{Y}$ be the open immersions. One can show the diagram

$$\begin{array}{ccccc} j_! f_! \boldsymbol{\mu}_n \cong \bar{f}_* i_! \boldsymbol{\mu}_n & & \stackrel{\bar{f}_*(\mathrm{Tr}_i)}{\to} & & \bar{f}_* \boldsymbol{\mu}_n \\ j_!(\mathrm{Tr}_{X/Y}) \searrow & \downarrow & & & \downarrow \mathrm{Tr}_{\overline{X}/\overline{Y}} \\ & j_! \boldsymbol{\mu}_n & \stackrel{\mathrm{Tr}_j}{\to} & & \boldsymbol{\mu}_n \end{array}$$

commutes by showing it commutes on stalks. Applying $H^2(\overline{Y}, -)$ to this diagram, we get a commutative diagram

$$\begin{array}{ccccc} H^2(\overline{Y}, j_! f_! \boldsymbol{\mu}_n) \cong & H^2(\overline{Y}, \bar{f}_* i_! \boldsymbol{\mu}_n) & \to & H^2(\overline{Y}, \bar{f}_* \boldsymbol{\mu}_n) \\ \searrow & \downarrow & & \downarrow \\ & H^2(\overline{Y}, j_! \boldsymbol{\mu}_n) & \to & H^2(\overline{Y}, \boldsymbol{\mu}_n). \end{array}$$

So we have a commutative diagram

$$\begin{array}{ccc} H^2_c(X, \boldsymbol{\mu}_n) & \stackrel{\cong}{\to} & H^2(\overline{X}, \boldsymbol{\mu}_n) \\ S_{X/Y} \downarrow & & \downarrow S_{\overline{X}/\overline{Y}} \\ H^2_c(Y, \boldsymbol{\mu}_n) & \stackrel{\cong}{\to} & H^2(\overline{Y}, \boldsymbol{\mu}_n), \end{array}$$

where $S_{\overline{X}/\overline{Y}}$ is the composite

$$H^2(\overline{X}, \boldsymbol{\mu}_n) \cong H^2(\overline{Y}, \bar{f}_*\boldsymbol{\mu}_n) \stackrel{H^2(\overline{Y}, \mathrm{Tr}_{\overline{X}/\overline{Y}})}{\rightarrow} H^2(\overline{Y}, \boldsymbol{\mu}_n).$$

Consider the diagram

$$\begin{array}{ccccc} H^2_c(X, \boldsymbol{\mu}_n) & & \stackrel{\cong}{\rightarrow} & & H^2(\overline{X}, \boldsymbol{\mu}_n) \\ & \mathrm{Tr}_{X/k} \searrow & & \swarrow \mathrm{Tr}_{\overline{X}/k} & \\ S_{X/Y} \downarrow & & \mathbb{Z}/n & & \downarrow S_{\overline{X}/\overline{Y}} \\ & \mathrm{Tr}_{Y/k} \nearrow & & \nwarrow \mathrm{Tr}_{\overline{Y}/k} & \\ H^2_c(Y, \boldsymbol{\mu}_n) & & \stackrel{\cong}{\rightarrow} & & H^2(\overline{Y}, \boldsymbol{\mu}_n). \end{array}$$

We need to prove that the triangle on the left commutes. We have seen that the outer loop commutes. So it suffices to show that the triangle on the right commutes. By the definition of $S_{\overline{X}/\overline{Y}}$ and Kummer's theory, we have a commutative diagram

$$\begin{array}{ccccc} \cdots \rightarrow \mathrm{Pic}(\overline{X}) & \rightarrow & H^2(\overline{X}, \boldsymbol{\mu}_n) & \rightarrow & 0 \\ \det \downarrow & & \downarrow S_{\overline{X}/\overline{Y}} & & \\ \cdots \rightarrow \mathrm{Pic}(\overline{Y}) & \rightarrow & H^2(\overline{Y}, \boldsymbol{\mu}_n) & \rightarrow & 0, \end{array}$$

where $\det : \mathrm{Pic}(\overline{X}) \rightarrow \mathrm{Pic}(\overline{Y})$ coincides with the homomorphism

$$H^1(\overline{Y}, f_*\mathscr{O}^*_{\overline{X}_{\mathrm{et}}}) \stackrel{H^1(\overline{Y}, \det)}{\rightarrow} H^1(\overline{Y}, \mathscr{O}^*_{\overline{Y}_{\mathrm{et}}})$$

through the identifications

$$\mathrm{Pic}(\overline{X}) \cong H^1(\overline{Y}, f_*\mathscr{O}^*_{\overline{X}_{\mathrm{et}}}), \quad \mathrm{Pic}(\overline{Y}) \cong H^1(\overline{Y}, \mathscr{O}^*_{\overline{Y}_{\mathrm{et}}}).$$

So it suffices to prove that the following diagram commutes:

$$\begin{array}{ccc} \mathrm{Pic}(\overline{X}) & & \\ & \searrow \deg & \\ \det \downarrow & & \mathbb{Z}. \\ & \nearrow \deg & \\ \mathrm{Pic}(\overline{Y}) & & \end{array}$$

Let $\mathscr{L}$ be an invertible $\mathscr{O}_{\overline{X}}$-module. We claim that $\mathscr{L}$ is defined by a Cartier divisor of the form $(s_i, \bar{f}^{-1}(V_i))$, where s_i are nonzero elements in the function field $K(\overline{X})$ of $\overline{X}$, $\{V_i\}$ is an open covering of $\overline{Y}$, and $\frac{s_i}{s_j} \in \mathscr{O}^*_{\overline{X}}(\bar{f}^{-1}(V_i) \cap \bar{f}^{-1}(V_j))$. To see that such a Cartier divisor exists, it suffices to find an open covering $\{V_i\}$ of $\overline{Y}$ such that $\mathscr{L}|_{\bar{f}^{-1}(V_i)}$ is free for each i. Let us prove that for any closed point y in Y, there exists an open neighborhood V of y in Y such that $\mathscr{L}|_{\bar{f}^{-1}(V)}$ is free. Let $W = \mathrm{Spec}\, A$ be an affine open neighborhood of y in Y. Then

$$\bar{f}^{-1}(W) = \mathrm{Spec}\, B$$

for some finite A-algebra B. Let L be a B-module such that

$$\mathscr{L}|_{\bar{f}^{-1}(W)} \cong L^{\sim},$$

and let $\mathfrak{m}$ be the maximal ideal of A corresponding to y. Since $L^{\sim}$ is invertible and $\bar{f}^{-1}(y) \cong \mathrm{Spec}\,(B \otimes_A A/\mathfrak{m})$ is discrete, we can find an isomorphism of $(B \otimes_A A/\mathfrak{m})$-modules

$$B \otimes_A A/\mathfrak{m} \stackrel{\cong}{\to} L \otimes_A A/\mathfrak{m}.$$

Choose $t \in L$ so that $t \otimes 1 \in L \otimes_A A/\mathfrak{m}$ is the image of 1 in $B \otimes_A A/\mathfrak{m}$. Let $\phi : B \to L$ be the B-module homomorphism mapping 1 to t. By Nakayama's lemma, the homomorphism

$$\phi \otimes \mathrm{id}_{A_\mathfrak{m}} : B \otimes_A A_\mathfrak{m} \to L \otimes_A A_\mathfrak{m}$$

is surjective. So there exists $s \in A - \mathfrak{m}$ such that the homomorphism

$$\phi \otimes \mathrm{id}_{A_s} : B \otimes_A A_s \to L \otimes_A A_s$$

is surjective. Let $V = D(s)$. Then ϕ induces an epimorphism

$$\mathscr{O}_{\bar{f}^{-1}(V)} \to \mathscr{L}|_{\bar{f}^{-1}(V)}.$$

It is necessarily an isomorphism since its kernel is torsion free and the rank of the stalk of the kernel at the generic point of $\bar{f}^{-1}(V)$ is 0. This proves our assertion.

Let $(s_i, \bar{f}^{-1}(V_i))$ be a Cartier divisor as above. $\det(\mathscr{L})$ is defined by the Cartier divisor $(\det(s_i), V_i)$. To prove $\deg(\det(\mathscr{L})) = \deg(\mathscr{L})$, it suffices to show

$$v_y(\det(s)) = \sum_{x \in \bar{f}^{-1}(y)} v_x(s)$$

for any $s \in K(X)^*$ and any closed point y of Y, where v_y (resp. v_x) is the valuation of the function field $K(Y)$ (resp. $K(X)$) of Y (resp. X) at y (resp. x). Making the base change $\mathrm{Spec}\,\widehat{\mathscr{O}}_{Y,y} \to Y$, our assertion follows from 8.1.1 (ii).

Given an invertible $\mathscr{O}_X$-module $\mathscr{L}$ on a scheme X, denote by $c_1(\mathscr{L})$ the image of $\mathscr{L}$ under the canonical homomorphism

$$\mathrm{Pic}(X) \to H^2(X, \boldsymbol{\mu}_n)$$

defined via Kummer's theory. For any morphism $f : X \to Y$, denote by $c_{1_{X/Y}}(\mathscr{L})$ the image of $c_1(\mathscr{L})$ under the canonical homomorphism

$$H^2(X, \boldsymbol{\mu}_n) \to \Gamma(Y, R^2 f_* \boldsymbol{\mu}_n).$$

Lemma 8.2.2. *Let $\bar{f} : \mathbb{P}^1_Y \to Y$ be the projection. The morphism*

$$\mathbb{Z}/n \to R^2 \bar{f}_* \boldsymbol{\mu}_n$$

mapping 1 to $c_{1_{\mathbb{P}^1_Y/Y}}(\mathscr{O}_{\mathbb{P}^1_Y/Y}(1))$ is an isomorphism.

Proof. Using the proper base change theorem 7.3.1, we may reduce to the case where $Y = \mathrm{Spec}\, k$ for an algebraically closed field k. We have a canonical isomorphism

$$\mathrm{Pic}(\mathbb{P}^1_k)/n\mathrm{Pic}(\mathbb{P}^1_k) \cong H^2(\mathbb{P}^1_k, \boldsymbol{\mu}_n)$$

and $\deg : \mathrm{Pic}(\mathbb{P}^1_k) \to \mathbb{Z}$ induces an isomorphism

$$\mathrm{Pic}(\mathbb{P}^1_k)/n\mathrm{Pic}(\mathbb{P}^1_k) \cong \mathbb{Z}/n.$$

Our assertion follows from the fact that $\mathscr{O}_{\mathbb{P}^1_k}(1)$ has degree 1. □

We define $\mathrm{Tr}_{\mathbb{P}^1_Y/Y} : R^2\bar{f}_*\boldsymbol{\mu}_n \to \mathbb{Z}/n$ to be the inverse of the isomorphism in 8.2.2. Let $j : \mathbb{A}^1_Y \to \mathbb{P}^1_Y$ be the open immersion and let $f : \mathbb{A}^1_Y \to Y$ be the projection. Define $\mathrm{Tr}_{\mathbb{A}^1_Y/Y} : R^2 f_!\boldsymbol{\mu}_n \to \mathbb{Z}/n$ to be the composite

$$R^2 f_!\boldsymbol{\mu}_n \cong R^2\bar{f}_* j_!\boldsymbol{\mu}_n \stackrel{R^2\bar{f}_*(\mathrm{Tr}_j)}{\to} R^2\bar{f}_*\boldsymbol{\mu}_n \stackrel{\mathrm{Tr}_{\mathbb{P}^1_Y/Y}}{\to} \mathbb{Z}/n.$$

Note that $\mathrm{Tr}_{\mathbb{A}^1_Y/Y}$ commutes with any base change.

Let $g : X \to Y$ and $h : Y \to Z$ be two smooth and S-compactifiable morphisms pure of relative dimensions d and e, respectively. Then $f = hg$ is smooth, S-compactifiable, and pure of relative dimension $d+e$. By 7.4.5, we have $R^q f_!\mathscr{F} = 0$ (resp. $R^q g_!\mathscr{F} = 0$, resp. $R^q h_!\mathscr{F} = 0$) for any $q > 2(d+e)$ (resp. $q > 2d$, resp. $q > 2e$) and any torsion sheaf $\mathscr{F}$ on X (resp. X, resp. Y). Suppose that we have defined

$$\mathrm{Tr}_{X/Y} : R^{2d} g_!\mathbb{Z}/n(d) \to \mathbb{Z}/n, \quad \mathrm{Tr}_{Y/Z} : R^{2e} h_!\mathbb{Z}/n(e) \to \mathbb{Z}/n.$$

They induce morphisms

$$\mathrm{Tr}_{X/Y} : Rg_!\mathbb{Z}/n(d)[2d] \to \mathbb{Z}/n, \quad \mathrm{Tr}_{Y/Z} : Rh_!\mathbb{Z}/n(e)[2e] \to \mathbb{Z}/n$$

in the derived categories $D(Y, \mathbb{Z}/n)$ and $D(Z, \mathbb{Z}/n)$, respectively. We have

$$\mathbb{Z}/n(d+e) \cong \mathbb{Z}/n(d) \otimes^L_{\mathbb{Z}/n} g^*\mathbb{Z}/n(e).$$

By the projection formula 7.4.7, we have

$$Rg_!\big(\mathbb{Z}/n(d) \otimes^L_{\mathbb{Z}/n} g^*\mathbb{Z}/n(e)\big) \cong Rg_!\mathbb{Z}/n(d) \otimes^L_{\mathbb{Z}/n} \mathbb{Z}/n(e).$$

We define

$$\mathrm{Tr}_{X/Z} : Rf_!\mathbb{Z}/n(d+e)[2(d+e)] \to \mathbb{Z}/n$$

to be the composite

$$\begin{array}{rcl} Rf_!\mathbb{Z}/n(d+e)[2(d+e)] & \cong & Rh_!Rg_!\big(\mathbb{Z}/n(d) \otimes^L_{\mathbb{Z}/n} g^*\mathbb{Z}/n(e)\big)[2(d+e)] \\ & \cong & Rh_!\big(Rg_!\mathbb{Z}/n(d)[2d] \otimes^L_{\mathbb{Z}/n} \mathbb{Z}/n(e)[2e]\big) \\ & \stackrel{\mathrm{Tr}_{X/Y}}{\to} & Rh_!\big(\mathbb{Z}/n \otimes^L_{\mathbb{Z}/n} \mathbb{Z}/n(e)[2e]\big) \\ & \cong & Rh_!\mathbb{Z}/n(e)[2e] \\ & \stackrel{\mathrm{Tr}_{Y/Z}}{\to} & \mathbb{Z}/n. \end{array}$$

We define

$$\mathrm{Tr}_{X/Z} : Rf_!^{2(d+e)}\mathbb{Z}/n(d+e) \to \mathbb{Z}/n$$

to be the morphism induced by $\mathrm{Tr}_{X/Z} : Rf_!\mathbb{Z}/n(d+e)[2(d+e)] \to \mathbb{Z}/n$. We denote the above way of defining $\mathrm{Tr}_{X/Z}$ by

$$\mathrm{Tr}_{X/Z} = \mathrm{Tr}_{Y/Z} \diamond \mathrm{Tr}_{X/Y}.$$

Next we define $\mathrm{Tr}_{\mathbb{A}^d_Y/Y}$. This has been done for $d = 0, 1$. We can factorize the projection $\mathbb{A}^d_Y \to Y$ as the composite of projections

$$\mathbb{A}^d_Y \to \mathbb{A}^{d-1}_Y \to \cdots \to \mathbb{A}^1_Y \to Y,$$

where for each i, $\mathbb{A}^i_Y \to \mathbb{A}^{i-1}_Y$ is the projection

$$\mathbf{Spec}\, \mathscr{O}_Y[t_1, \ldots, t_i] \to \mathbf{Spec}\, \mathscr{O}_Y[t_1, \ldots, t_{i-1}]$$

to the first $i-1$ coordinates. We define

$$\mathrm{Tr}_{\mathbb{A}^d_Y/Y} = \mathrm{Tr}_{\mathbb{A}^1_Y/Y} \diamond \cdots \diamond \mathrm{Tr}_{\mathbb{A}^d_Y/\mathbb{A}^{d-1}_Y},$$

where for each i, we define

$$\mathrm{Tr}_{\mathbb{A}^i_Y/\mathbb{A}^{i-1}_Y} = \mathrm{Tr}_{\mathbb{A}^1_{\mathbb{A}^{i-1}_Y}/\mathbb{A}^{i-1}_Y}.$$

If an S-compactifiable morphism $f : X \to Y$ can be factorized as

$$X \xrightarrow{g} \mathbb{A}^d_Y \to Y$$

with g being an etale Y-morphism, then we define

$$\mathrm{Tr}_{X/Y} = \mathrm{Tr}_{\mathbb{A}^d_Y/Y} \diamond \mathrm{Tr}_g.$$

The following lemma shows that $\mathrm{Tr}_{X/Y}$ is independent of the choice of the factorization.

Lemma 8.2.3. *Let $f : X \to Y$ be an S-compactifiable morphism, let $x_1, \ldots, x_d \in \Gamma(X, \mathscr{O}_X)$ be a sequence such that the Y-morphism*

$$g : X \to \mathbb{A}^d_Y = \mathbf{Spec}\, \mathscr{O}_Y[t_1, \ldots, t_d]$$

corresponding to the $\mathscr{O}_Y$-algebra homomorphism

$$g^\natural : \mathscr{O}_Y[t_1, \ldots, t_n] \to f_*\mathscr{O}_X, \quad t_i \mapsto x_i$$

is etale, and let

$$T_{(x_1,\ldots,x_d)} = \mathrm{Tr}_{\mathbb{A}^d_Y/Y} \diamond \mathrm{Tr}_g.$$

Then $T_{(x_1,\ldots,x_d)}$ is independent of the choice of the sequence $x_1, \ldots, x_d$ such that the corresponding morphism $g : X \to \mathbb{A}^n_Y$ is etale.

Proof. By base change, we may assume that $Y = \mathrm{Spec}\, k$ for some algebraically closed field k. If $d = 1$, then X is a smooth curve over k. By 8.2.1, T_{x_1} coincides with $\mathrm{Tr}_{X/k}$ and hence does not depend on the choice of x_1. In general, we may assume that X is irreducible. For any nonempty open subset U of X, we have

$$H_c^q(X - U, \mathbb{Z}/n(d)) = 0$$

for all $q > 2(d-1)$ by 7.4.5. It follows that the canonical homomorphism

$$H_c^{2d}(U, \mathbb{Z}/n(d)) \to H_c^{2d}(X, \mathbb{Z}/n(d))$$

is an isomorphism. To prove our assertion, we may replace X by its nonempty open subset.

First we prove that $T_{(x_1,\ldots,x_d)}$ is independent of the order of $x_1, \ldots, x_d$. In fact, we prove that, more generally, for any matrix $(a_{ij}) \in \mathrm{GL}_d(k)$, we have

$$T_{(x_1,\ldots,x_d)} = T_{(\sum_i a_{i1}x_i,\ldots,\sum_i a_{id}x_i)}.$$

Let $g : X \to \mathbb{A}_k^d$ (resp. $g' : X \to \mathbb{A}_k^d$) be the k-morphism defined by the sections $x_1, \ldots, x_d$ (resp. $\sum_i a_{i1}x_i, \ldots, \sum_i a_{id}x_i$), and let $\phi : \mathbb{A}_k^d \to \mathbb{A}_k^d$ be the k-isomorphism corresponding to the k-algebra isomorphism

$$k[t_1, \ldots, t_n] \to k[t_1, \ldots, t_n], \quad t_j \mapsto \sum_i a_{ij}t_i.$$

By definition, we have

$$\begin{aligned} T_{(x_1,\ldots,x_d)} &= \mathrm{Tr}_{A_k^d/k} \circ R^{2d}\pi_!(\mathrm{Tr}_g), \\ T_{(\sum_i a_{i1}x_i,\ldots,\sum_i a_{id}x_i)} &= \mathrm{Tr}_{A_k^d/k} \circ R^{2d}\pi_!(\mathrm{Tr}_{g'}), \end{aligned}$$

where $\pi : \mathbb{A}_k^d \to \mathrm{Spec}\, k$ is the projection. To prove our assertion, it suffices to show

$$R^{2d}\pi_!(\mathrm{Tr}_g) = R^{2d}\pi_!(\mathrm{Tr}_{g'}).$$

We have $g' = \phi g$. So $\mathrm{Tr}_{g'}$ is the composite

$$g'_!\mathbb{Z}/n \cong \phi_!g_!\mathbb{Z}/n \stackrel{\phi_!(\mathrm{Tr}_g)}{\to} \phi_!\mathbb{Z}/n \stackrel{\mathrm{Tr}_\phi}{\to} \mathbb{Z}/n.$$

We have $\pi\phi = \pi$. To prove $R^{2d}\pi_!(\mathrm{Tr}_g) = R^{2d}\pi_!(\mathrm{Tr}_{g'})$, it suffices to show that the composite

$$R^{2d}\pi_!\mathbb{Z}/n \cong R^{2d}\pi_!\phi_!\mathbb{Z}/n \stackrel{R^{2d}\pi_!(\mathrm{Tr}_\phi)}{\to} R^{2d}\pi_!\mathbb{Z}/n$$

is the identity. Indeed, let Φ be the isomorphism

$$\Phi : \mathrm{GL}_d(k) \times_k \mathbb{A}_k^d \to \mathrm{GL}_d(k) \times_k \mathbb{A}_k^d, \quad ((x_{ij}), (x_j)) \mapsto ((x_{ij}), (\sum_i x_{ij}x_i)).$$

Fix notation by the following commutative diagram:

$$\begin{array}{ccccc} \mathrm{GL}_d(k)\times_k \mathbb{A}_k^d & \stackrel{\Phi}{\to} & \mathrm{GL}_d(k)\times_k \mathbb{A}_k^d & \stackrel{p_2}{\to} & \mathbb{A}_k^d \\ & p_1 \searrow & p_1 \downarrow & & \pi\downarrow \\ & & \mathrm{GL}_d(k) & \to & \operatorname{Spec} k. \end{array}$$

The above composite is the stalk at the geometric point of $\mathrm{GL}_d(k)$ defined by (a_{ij}) of the following composite:

$$R^{2d}p_{1!}\mathbb{Z}/n \cong R^{2d}p_{1!}\Phi_!\mathbb{Z}/n \stackrel{R^{2d}p_{1!}(\mathrm{Tr}_\Phi)}{\to} R^{2d}p_{1!}\mathbb{Z}/n.$$

But $R^{2d}p_{1!}\mathbb{Z}/n$ is isomorphic to the inverse image of $R^{2d}\pi_!\mathbb{Z}/n$ under the morphism $\mathrm{GL}_d(k)\to \operatorname{Spec} k$, and hence is a constant sheaf. As $\mathrm{GL}_d(k)$ is connected, the stalk of the last composite at the geometric point defined by (a_{ij}) can be identified with the stalk at the geometric point defined by the identity matrix. Our assertion follows.

Let x be a closed point of X and let $\mathfrak{m}$ be the maximal ideal of $\mathscr{O}_{X,x}$. For any two sequences $(x_1,\ldots,x_d)$ and $(y_1,\ldots,y_d)$ of $\Gamma(X,\mathscr{O}_X)$ such that the k-morphisms $X\to\mathbb{A}_k^d$ are etale, we claim that after replacing X by an open neighborhood of x, there exists a finite chain of sequences

$$(x_1,\ldots,x_d)=(x_1^0,\ldots,x_d^0),(x_1^1,\ldots,x_d^1),\cdots,(x_1^m,\ldots,x_d^m)=(y_1,\ldots,y_n)$$

such that the k-morphisms $X\to\mathbb{A}_k^d$ defined by these sequences are etale, and for each $v\in\{0,1,\ldots,m-1\}$, the sequence $(x_1^v,\ldots,x_d^v)$ differs from $(x_1^{v+1},\ldots,x_d^{v+1})$ by only one element. We may assume that $x_1,\ldots,x_d$ lie in $\mathfrak{m}$ by replacing x_i by x_i-a_i for some $a_i\in k$ with $x_i\equiv a_i \bmod \mathfrak{m}$. Then $(x_1,\ldots,x_d)$ is a regular system of parameters for the regular local ring $\mathscr{O}_{X,x}$. Indeed, since the morphism $X\to\mathbb{A}_k^d$ is etale, $dx_1,\ldots,dx_d$ form a basis for the free $\mathscr{O}_{X,x}$-module $\Omega_{\mathscr{O}_{X,x}/k}$. By 2.1.5, the canonical homomorphism

$$\mathfrak{m}/\mathfrak{m}^2\to\Omega_{\mathscr{O}_{X,x}/k}\otimes_{\mathscr{O}_{X,x}}\mathscr{O}_{X,x}/\mathfrak{m}$$

is an isomorphism since $\mathscr{O}_{X,x}/\mathfrak{m}\cong k$. It follows that the images of $x_1,\ldots,x_d$ in $\mathfrak{m}/\mathfrak{m}^2$ form a basis for the $(\mathscr{O}_{X,x}/\mathfrak{m})$-vector space $\mathfrak{m}/\mathfrak{m}^2$. Similarly, we may assume that $y_1,\ldots,y_n$ lie in $\mathfrak{m}$ and form a regular system of parameters for $\mathscr{O}_{X,x}$. Any basis of $\mathfrak{m}/\mathfrak{m}^2$ can be changed to another basis by a chain of bases so that each successive pair of bases differ only by one element. So we have a chain of regular system of parameters

$$(x_1,\ldots,x_d)=(x_1^0,\ldots,x_d^0),(x_1^1,\ldots,x_d^1),\cdots,(x_1^m,\ldots,x_d^m)=(y_1,\ldots,y_n)$$

of $\mathscr{O}_{X,x}$ such that for each $v\in\{0,1,\ldots,m-1\}$, the sequence $(x_1^v,\ldots,x_d^v)$ differs from $(x_1^{v+1},\ldots,x_d^{v+1})$ by just one element. Replacing X by an open neighborhood of x, we may assume that elements of these sequences can be extended to sections in $\Gamma(X, \mathscr{O}_X)$, such that the k-morphisms $X \to \mathbb{A}_k^n$ defined by them are etale. This proves our claim.

By the above discussion, to prove that $T_{(x_1,\ldots,x_d)}$ is independent of the choice of $(x_1, \ldots, x_d)$, it suffices to show that $T_{(x_1,\ldots,x_d)}$ remains unchanged if we change x_d but keep $x_1, \ldots, x_{d-1}$ unchanged. Consider the commutative diagram

$$\begin{array}{ccc} & & \mathbb{A}_Y^d \cong \mathbb{A}^1_{\mathbb{A}_Y^{d-1}} \\ & \overset{g}{\nearrow} & \downarrow \\ X & \xrightarrow{h} & \mathbb{A}_Y^{d-1} \\ & \underset{f}{\searrow} & \downarrow \\ & & Y, \end{array}$$

where $g : X \to \mathbb{A}_Y^d$ (resp. $h : X \to \mathbb{A}_Y^{d-1}$) is the Y-morphism defined by the sections $x_1, \ldots, x_d$ (resp. $x_1, \ldots, x_{d-1}$). By our discussion in the $d = 1$ case, $\mathrm{Tr}_{\mathbb{A}_Y^d/\mathbb{A}_Y^{d-1}} \diamond \mathrm{Tr}_g$ is unchanged if we only change x_d. We have

$$\begin{aligned} T_{(x_1,\ldots,x_d)} &= \mathrm{Tr}_{\mathbb{A}_Y^d/Y} \diamond \mathrm{Tr}_g \\ &= \mathrm{Tr}_{\mathbb{A}_Y^{d-1}/Y} \diamond \mathrm{Tr}_{A_Y^d/\mathbb{A}_Y^{d-1}} \diamond \mathrm{Tr}_g. \end{aligned}$$

The last expression is independent of x_d. So $T_{(x_1,\ldots,x_d)}$ is independent of the choice of x_d. □

Finally, let $f : X \to Y$ be a smooth S-compactifiable morphism pure of relative dimension d. There exists an open covering $\{U_\alpha\}$ of X such that $f|_{U_\alpha} : U_\alpha \to Y$ can be factorized as

$$U_\alpha \overset{g_\alpha}{\to} \mathbb{A}_Y^d \to Y,$$

with g_α being etale Y-morphisms. Let $j_\alpha : U_\alpha \hookrightarrow X$ and $j_{\alpha\beta} : U_\alpha \cap U_\beta \to X$ be the open immersions. For any sheaf of $\mathbb{Z}/n$-modules $\mathscr{F}$ on X, we have a canonical exact sequence

$$\bigoplus_{\alpha,\beta} j_{\alpha\beta!}(\mathscr{F}|_{U_\alpha\cap U_\beta}) \to \bigoplus_{\alpha} j_{\alpha!}(\mathscr{F}|_{U_\alpha}) \to \mathscr{F} \to 0.$$

This follows from the fact that for any sheaf of $\mathbb{Z}/n$-modules $\mathscr{G}$ on X, we have an exact sequence

$$0 \to \mathrm{Hom}(\mathscr{F}, \mathscr{G}) \to \mathrm{Hom}\Big(\bigoplus_{\alpha} j_{\alpha!}(\mathscr{F}|_{U_\alpha}), \mathscr{G}\Big) \to \mathrm{Hom}\Big(\bigoplus_{\alpha,\beta} j_{\alpha\beta!}(\mathscr{F}|_{U_\alpha\cap U_\beta}), \mathscr{G}\Big),$$

which can be identified with the canonical exact sequence

$$0 \to \mathrm{Hom}(\mathscr{F}, \mathscr{G}) \to \prod_{\alpha} \mathrm{Hom}(\mathscr{F}|_{U_\alpha}, \mathscr{G}|_{U_\alpha}) \to \prod_{\alpha,\beta} \mathrm{Hom}(\mathscr{F}|_{U_\alpha \cap U_\beta}, \mathscr{G}|_{U_\alpha \cap U_\beta}).$$

By 7.4.5, we have $R^q f_! = 0$ for all $q > 2d$. So $R^{2d} f_!$ is right exact. So we have an exact sequence

$$\bigoplus_{\alpha,\beta} R^{2d}(f|_{U_\alpha \cap U_\beta})_!(\mathscr{F}|_{U_\alpha \cap U_\beta}) \to \bigoplus_{\alpha} R^{2d}(f|_{U_\alpha})_!(\mathscr{F}|_{U_\alpha}) \to R^{2d} f_! \mathscr{F} \to 0.$$

We have defined

$$\mathrm{Tr}_{U_\alpha/Y} : R^{2d}(f|_{U_\alpha})_! \mathbb{Z}/n(d) \to \mathbb{Z}/n.$$

By 8.2.3, $\mathrm{Tr}_{U_\alpha/Y}$ and $\mathrm{Tr}_{U_\beta/Y}$ both induce $\mathrm{Tr}_{U_\alpha \cap U_\beta/Y}$. Using the above exact sequence, we can define

$$\mathrm{Tr}_{X/Y} : R^{2d} f_! \mathbb{Z}/n(d) \to \mathbb{Z}/n$$

so that it induces $\mathrm{Tr}_{U_\alpha/Y}$ for each α. The definition of $\mathrm{Tr}_{X/Y}$ is independent of the choice of the open covering $\{U_\alpha\}$ of X.

Since $R^i f_! = 0$ for $i > 2d$, $\mathrm{Tr}_{X/Y}$ induces a morphism

$$Rf_! \mathbb{Z}/n(d)[2d] \to \mathbb{Z}/n$$

in the derived category $D(Y, \mathbb{Z}/n)$. For any $K \in \mathrm{ob}\, D(Y, \mathbb{Z}/n)$, we have

$$Rf_! \mathbb{Z}/n(d)[2d] \otimes^L_{\mathbb{Z}/n} K \cong Rf_!(f^* K(d)[2d])$$

by the projection formula 7.4.7. So $\mathrm{Tr}_{X/Y}$ induces a morphism

$$Rf_! f^* K(d)[2d] \to K,$$

which we also denote by $\mathrm{Tr}_{X/Y}$ or Tr_f. It is functorial in K.

Proposition 8.2.4. *Let S be a scheme and let n be an integer invertible on S.*

(i) *Consider a Cartesian diagram*

$$\begin{array}{ccc} X \times_Y Y' & \xrightarrow{g'} & X \\ {\scriptstyle f'} \downarrow & & \downarrow {\scriptstyle f} \\ Y' & \xrightarrow{g} & Y. \end{array}$$

Suppose that f is a smooth S-compactifiable morphism pure of relative dimension d, and Y' is quasi-compact and quasi-separated. For any $K \in \mathrm{ob}\, D(Y, \mathbb{Z}/n)$, the following diagram commutes:

$$\begin{array}{ccccc} g^* Rf_! f^* K(d)[2d] & \cong & Rf'_! g'^* f^* K(d)[2d] & \cong & Rf'_! f'^* g^* K(d)[2d] \\ & {\scriptstyle g^*(\mathrm{Tr}_f)} \searrow & & \swarrow {\scriptstyle \mathrm{Tr}_{f'}} & \\ & & g^* K. & & \end{array}$$

(ii) *Let $g : X \to Y$ and $h : Y \to Z$ be two smooth S-compactifiable morphisms pure of relative dimensions d and e, respectively and let $f = hg$. For any $K \in \mathrm{ob}\, D(Z, \mathbb{Z}/n)$, the following diagram commutes:*

$$\begin{array}{ccc} Rh_!Rg_!g^*h^*K(d+e)[2(d+e)] & \stackrel{Rh_!(\mathrm{Tr}_g)}{\to} & Rh_!h^*K(e)[2e] \\ \cong\downarrow & & \downarrow \mathrm{Tr}_h \\ Rf_!f^*K(d+e)[2(d+e)] & \stackrel{\mathrm{Tr}_f}{\to} & K. \end{array}$$

(iii) *Consider a Cartesian diagram*

$$\begin{array}{ccc} X \times_Z Y & \stackrel{p}{\to} & X \\ {\scriptstyle q}\downarrow & & \downarrow {\scriptstyle f} \\ Y & \stackrel{g}{\to} & Z. \end{array}$$

Let $h = fp = gq$. Suppose that f and g are smooth S-compactifiable morphisms pure of relative dimensions d and e, respectively. For any $K, L \in \mathrm{ob}\, D^-(Z, \mathbb{Z}/n)$, the following diagram commutes:

$$\begin{array}{ccc} Rf_!f^*K(d)[2d] \otimes^L_{\mathbb{Z}/n} Rg_!g^*L(e)[2e] & \cong & Rh_!h^*(K \otimes^L_{\mathbb{Z}/n} L)(d+e)[2(d+e)] \\ {\scriptstyle \mathrm{Tr}_f \otimes^L \mathrm{Tr}_g} \searrow & & \swarrow {\scriptstyle \mathrm{Tr}_h} \\ & K \otimes^L_{\mathbb{Z}/n} L, & \end{array}$$

where the top line is given by the Künneth formula.

(iv) *Suppose that $S = \mathrm{Spec}\, k$ for an algebraically closed field k, X and Y are proper smooth schemes over k pure of dimensions d and e, respectively. Then for any $s \in H^{2d}(X, \mathbb{Z}/n(d))$ and $t \in H^{2e}(Y, \mathbb{Z}/n(e))$, we have*

$$\mathrm{Tr}_{X\times_k Y/k}(p^*s \cup q^*t) = \mathrm{Tr}_{X/k}(s)\mathrm{Tr}_{Y/k}(t),$$

where $p : X \times_k Y \to X$ and $q : X \times_k Y \to Y$ are projections.

Proof. (i) follows from the definition of Tr. (iv) follows from (iii) and the fact that the isomorphism defined by the Künneth formula can be defined via the cup product. (Confer the proof of 7.4.11.)

(ii) We can reduce to the case where we have a commutative diagram

$$\begin{array}{ccccc} X & \stackrel{a}{\to} & \mathbb{A}^d_Y & \stackrel{b'}{\to} & \mathbb{A}^{d+e}_Z \\ & {\scriptstyle g}\searrow & {\scriptstyle \pi''}\downarrow & & \downarrow {\scriptstyle \pi'} \\ & & Y & \stackrel{b}{\to} & \mathbb{A}^e_Z \\ & & & {\scriptstyle h}\searrow & \downarrow {\scriptstyle \pi} \\ & & & & Z \end{array}$$

such that the square in the diagram is Cartesian, a, b, b' are etale, and π, π', π'' are the projections. Consider the diagram

$$\begin{array}{ccc}
R\pi_! b_! R\pi''_! a_! \mathbb{Z}/n(d+e)[2(d+e)] & \cong & R\pi_! R\pi'_! b'_! a_! \mathbb{Z}/n(d+e)[2(d+e)] \\
\mathrm{Tr}_a \downarrow & & \downarrow \mathrm{Tr}_a \\
R\pi_! b_! R\pi''_! \mathbb{Z}/n(d+e)[2(d+e)] & \cong & R\pi_! R\pi'_! b'_! \mathbb{Z}/n(d+e)[2(d+e)] \\
& & \downarrow \mathrm{Tr}_{b'} \\
\mathrm{Tr}_{\pi''} \downarrow & (1) & R\pi_! R\pi'_! \mathbb{Z}/n(d+e)[2(d+e)] \\
& & \downarrow \mathrm{Tr}_{\pi'} \\
R\pi_! b_! \mathbb{Z}/n(e)[2e] & \stackrel{\mathrm{Tr}_b}{\to} & R\pi_! \mathbb{Z}/n(e)[2e] \quad \stackrel{\mathrm{Tr}_\pi}{\to} \mathbb{Z}/n.
\end{array}$$

To prove that the square (1) commutes, it suffices to prove that the diagram

$$\begin{array}{ccc}
b_! R\pi''_! \mathbb{Z}/n(d)[2d)] & \cong & R\pi'_! b'_! \mathbb{Z}/n(d)[2d] \\
& & \downarrow \mathrm{Tr}_{b'} \\
\mathrm{Tr}_{\pi''} \downarrow & (2) & R\pi'_! \mathbb{Z}/n(d)[2d] \\
& & \downarrow \mathrm{Tr}_{\pi'} \\
b_! \mathbb{Z}/n & \stackrel{\mathrm{Tr}_b}{\to} & \mathbb{Z}/n
\end{array}$$

commutes if b, b', π', π'' come from the following Cartesian diagram

$$\begin{array}{ccc}
\mathbb{A}^d_Y & \stackrel{b'}{\to} & \mathbb{A}^d_W \\
\pi'' \downarrow & & \downarrow \pi' \\
Y & \stackrel{b}{\to} & W
\end{array}$$

in which b is etale. Let us check that (2) gives rise to a commutative diagram on stalks at each geometric point $s \to W$ of W. Making the base change $s \to W$, we are reduced to the case where W is the spectrum of a separably closed field. We are then reduced to the case where b is the identity morphism. In this case the diagram (2) trivially commutes.

(iii) follows from the commutativity of the following diagram:

$$\begin{array}{ccc}
Rf_! f^* K(d)[2d] \otimes^L_{\mathbb{Z}/n} Rg_! g^* L(e)[2e] & \stackrel{\mathrm{Tr}_f \otimes^L \mathrm{id}}{\to} & K \otimes^L_{\mathbb{Z}/n} Rg_! g^* L(e)[2e] \\
\wr\| & & \wr\| \\
Rg_!(g^* Rf_! f^* K(d)[2d] \otimes^L_{\mathbb{Z}/n} g^* L(e)[2e]) & \stackrel{Rg_!(g^* \mathrm{Tr}_f \otimes^L \mathrm{id})}{\to} & Rg_!(g^* K \otimes^L_{\mathbb{Z}/n} g^* L(e)[2e]) \\
\wr\| & & \\
Rg_!(Rq_! p^* f^* K(d)[2d] \otimes^L_{\mathbb{Z}/n} g^* L(e)[2e]) & & \| \\
\wr\| & & \\
Rg_!(Rq_! q^* g^* K(d)[2d] \otimes^L_{\mathbb{Z}/n} g^* L(e)[2e]) & \stackrel{Rg_!(\mathrm{Tr}_q \otimes^L \mathrm{id})}{\to} & Rg_!(g^* K \otimes^L_{\mathbb{Z}/n} g^* L(e)[2e]) \\
\wr\| & & \wr\| \\
Rg_! Rq_! q^* g^* (K(d)[2d] \otimes^L_{\mathbb{Z}/n} L(e)[2e]) & \stackrel{Rg_!(\mathrm{Tr}_q)}{\to} & Rg_! g^* (K \otimes^L_{\mathbb{Z}/n} L(e)[2e]) \\
\wr\| & & \downarrow \mathrm{Tr}_g \\
Rh_! h^* (K \otimes^L_{\mathbb{Z}/n} L)(d+e)[2(d+e)] & \stackrel{\mathrm{Tr}_h}{\to} & K \otimes^L_{\mathbb{Z}/n} L.
\end{array}$$

□

8.3 Duality for Curves

([SGA $4\frac{1}{2}$] Dualité.)

In this section, we fix an integer n invertible in all schemes that we consider, and we denote $\mathrm{Ext}_{\mathbb{Z}/n}(-,-)$ and $\mathscr{E}xt_{\mathbb{Z}/n}(-,-)$ by $\mathrm{Ext}(-,-)$ and $\mathscr{E}xt(-,-)$, respectively. Let X be a scheme over an algebraically closed field k. For any sheaves of $\mathbb{Z}/n$-modules $\mathscr{F}$ and $\mathscr{G}$ on X, we have canonical homomorphisms

$$\begin{aligned}\mathrm{Ext}^i(\mathscr{F},\mathscr{G}) &\cong \mathrm{Hom}_{D(X,\mathbb{Z}/n)}(\mathscr{F},\mathscr{G}[i])\\ &\to \mathrm{Hom}_{D(\mathbb{Z}/n)}\big(R\Gamma_c(X,\mathscr{F}), R\Gamma_c(X,\mathscr{G}[i])\big)\\ &\to \mathrm{Hom}\big(H_c^j(X,\mathscr{F}), H_c^{i+j}(X,\mathscr{G})\big).\end{aligned}$$

We thus get a pairing

$$H_c^j(X,\mathscr{F})\times \mathrm{Ext}^i(\mathscr{F},\mathscr{G})\to H_c^{i+j}(X,\mathscr{G}).$$

Suppose that X is smooth pure of dimension d. We have a pairing

$$H_c^r(X,\mathscr{F})\times \mathrm{Ext}^{2d-r}(\mathscr{F},\mathbb{Z}/n(d))\to H_c^{2d}(X,\mathbb{Z}/n(d)).$$

Taking its composite with

$$\mathrm{Tr}_{X/k}: H_c^{2d}(X,\mathbb{Z}/n(d))\to \mathbb{Z}/n,$$

we get a pairing

$$H_c^r(X,\mathscr{F})\times \mathrm{Ext}^{2d-r}(\mathscr{F},\mathbb{Z}/n(d))\to \mathbb{Z}/n.$$

In 8.5.3, we prove that this last pairing is perfect. In this section, we prove this result in the case where X is a smooth curve over k and $\mathscr{F}$ is constructible.

Theorem 8.3.1. *Let X be a smooth curve over an algebraically closed field k, and let $\mathscr{F}$ be a constructible sheaf of $\mathbb{Z}/n$-modules on X. The canonical pairing*

$$H_c^r(X,\mathscr{F})\times \mathrm{Ext}^{2-r}(\mathscr{F},\boldsymbol{\mu}_n)\to \mathbb{Z}/n$$

is perfect for each r.

A direct consequence of 8.3.1 is the following:

Corollary 8.3.2. *Let X be a smooth curve over an algebraically closed field k, let $\mathscr{F}$ be a locally constant constructible sheaf of $\mathbb{Z}/n$-modules on X, and let $\mathscr{F}^\vee = \mathscr{H}om(\mathscr{F},\mathbb{Z}/n)$. Then we have a perfect pairing*

$$H_c^r(X,\mathscr{F})\times H^{2-r}(X,\mathscr{F}^\vee(1))\to \mathbb{Z}/n$$

for each r.

Proof. It suffices to show

$$\mathrm{Ext}^{2-r}(\mathscr{F}, \boldsymbol{\mu}_n) \cong H^{2-r}(X, \mathscr{H}om(\mathscr{F}, \boldsymbol{\mu}_n)).$$

We have a biregular spectral sequence

$$E_2^{pq} = H^p(X, \mathscr{E}xt^q(\mathscr{F}, \boldsymbol{\mu}_n)) \Rightarrow \mathrm{Ext}^{p+q}(\mathscr{F}, \boldsymbol{\mu}_n).$$

By 8.3.3 below, we have

$$\mathscr{E}xt^q(\mathscr{F}, \boldsymbol{\mu}_n) = 0$$

for all $q \geq 1$. The spectral sequence degenerates. Our assertion follows. □

Lemma 8.3.3. *Let X be a noetherian scheme, let A be a ring such that A is an injective A-module, and let $\mathscr{F}$ be a locally constant sheaf of A-modules on X. Then $\mathscr{E}xt_A^q(\mathscr{F}, A) = 0$ for all $q \geq 1$.*

Remark 8.3.4. Note that $\mathbb{Z}/n$ is an injective $\mathbb{Z}/n$-module. To prove this, it suffices to show that any homomorphism from an ideal I of $\mathbb{Z}/n$ to $\mathbb{Z}/n$ can be extended to an endomorphism of $\mathbb{Z}/n$. This follows from the fact that I is generated by an element $\bar{i}$ for some integer i dividing n. Similarly, one can show that if $A = R/I$ for a discrete valuation ring R and a nonzero ideal I of R, then A is an injective A-module.

Proof of 8.3.3. The problem is local with respect to the etale topology. We may assume that $\mathscr{F}$ is a constant constructible sheaf associated to some A-module M. Let

$$\cdots \to L_1 \to L_0 \to 0$$

be a resolution of M by free A-modules. Denote the constant sheaf on X associated to L_i also by L_i. Then

$$\mathscr{E}xt_A^q(L_i, \mathscr{G}) = 0$$

for any $q \geq 1$ and any sheaf of A-modules $\mathscr{G}$ on X. It follows that $\mathscr{E}xt_A^q(\mathscr{F}, A)$ is the q-th cohomology sheaf of the complex

$$0 \to \mathscr{H}om_A(L_0, A) \to \mathscr{H}om_A(L_1, A) \to \cdots$$

for each q. The q-th cohomology sheaf of this complex is the constant sheaf associated to the q-th cohomology of the complex of A-modules

$$0 \to \mathrm{Hom}_A(L_0, A) \to \mathrm{Hom}_A(L_1, A) \to \cdots .$$

Since A is an injective A-module, for any $q \geq 1$, the q-th cohomology of this complex is 0. Our assertion follows. □

A scheme is called a *trait* (resp. *strictly local trait*) if it is isomorphic to the spectrum of a henselian discrete valuation ring (resp. strictly henselian discrete valuation ring).

Lemma 8.3.5. *Let S be a strictly local trait, s its closed point, η its generic point, $I = \mathrm{Gal}(\bar{\eta}/\eta)$, $i : s \to S$ and $j : \eta \to S$ the immersions. For any sheaf of $\mathbb{Z}/n$-modules $\mathscr{F}$ on η, we have canonical isomorphisms*

$$(R^q j_* \mathscr{F})_s = \begin{cases} (\mathscr{F}_{\bar{\eta}})^I & \text{if } q = 0, \\ (\mathscr{F}_{\bar{\eta}})_I(-1) & \text{if } q = 1, \\ 0 & \text{if } q \neq 0, 1. \end{cases}$$

Proof. Since S is strictly local, we have

$$(Rj_*\mathscr{F})_s \cong R\Gamma(S, Rj_*\mathscr{F}) \cong R\Gamma(\eta, \mathscr{F}).$$

It follows that

$$(R^q j_* \mathscr{F})_s = H^q(I, \mathscr{F}_{\bar{\eta}}).$$

We then apply 8.1.4.

Lemma 8.3.6. *Let X be a noetherian regular scheme pure of dimension 1, x a closed point of X, and $i : \mathrm{Spec}\, k(x) \to X$ the closed immersion. For any constant sheaf of $\mathbb{Z}/n$-module M on X, we have canonical isomorphisms*

$$R^q i^! M \cong \begin{cases} M(-1) & \text{if } q = 2, \\ 0 & \text{if } q \neq 2. \end{cases}$$

Proof. Let $\widetilde{X}_{\bar{x}}$ be the strict localization of X at $\bar{x}$, and let

$$j : X - \{x\} \to X, \quad \tilde{j} : \widetilde{X}_{\bar{x}} \times_X (X - \{x\}) \to \widetilde{X}_{\bar{x}}$$

be the open immersions. Note that $\widetilde{X}_{\bar{x}}$ is a strictly local trait, and $\widetilde{X}_{\bar{x}} \times_X (X - \{x\})$ is the generic point of $\widetilde{X}_{\bar{x}}$. By 8.3.5, we have

$$(R^q j_* j^* M)_{\bar{x}} \cong (R^q \tilde{j}_* \tilde{j}^* M)_{\bar{x}} = \begin{cases} M & \text{if } q = 0, \\ M(-1) & \text{if } q = 1, \\ 0 & \text{if } q \neq 0, 1. \end{cases}$$

Note that the canonical morphism

$$M \to j_* j^* M$$

is an isomorphism. This is clear when restricted to $X - x$. At x, this follows from the above formula for $(j_* j^* M)_{\bar{x}}$. We have an exact sequence

$$0 \to i_* i^! M \to M \to j_* j^* M \to i_* R^1 i^! M \to 0$$

and isomorphisms

$$i_* R^{q+1} i^! M \cong R^q j_* j^* M$$

for all $q \geq 1$. Our assertion follows. ◻

Lemma 8.3.7. *Let X be a smooth curve over an algebraically closed field k, let $\pi : X' \to X$ be an etale morphism and let $\mathscr{F}'$ be a constructible sheaf of $\mathbb{Z}/n$-modules on X'. For each r, the pairing*

$$H_c^r(X', \mathscr{F}') \times \operatorname{Ext}^{2-r}(\mathscr{F}', \boldsymbol{\mu}_n) \to \mathbb{Z}/n$$

is perfect if and only if the pairing

$$H_c^r(X, \pi_! \mathscr{F}') \times \operatorname{Ext}^{2-r}(\pi_! \mathscr{F}', \boldsymbol{\mu}_n) \to \mathbb{Z}/n$$

is perfect.

Proof. Since π is etale, we have

$$\begin{aligned} H_c^r(X, \pi_! \mathscr{F}') &\cong H_c^r(X', \mathscr{F}'), \\ \operatorname{Ext}^{2-r}(\mathscr{F}', \boldsymbol{\mu}_n) &\cong \operatorname{Hom}_{D(X', \mathbb{Z}/n)}(\mathscr{F}', \pi^* \boldsymbol{\mu}_n[2-r]) \\ &\cong \operatorname{Hom}_{D(X', \mathbb{Z}/n)}(\pi_! \mathscr{F}', \boldsymbol{\mu}_n[2-r]) \\ &\cong \operatorname{Ext}^{2-r}(\pi_! \mathscr{F}', \boldsymbol{\mu}_n). \end{aligned}$$

Our assertion follows from the commutativity of the following diagram:

$$\begin{array}{ccccccc} H_c^r(X', \mathscr{F}') & \times & \operatorname{Ext}^{2-r}(\mathscr{F}', \boldsymbol{\mu}_n) & \to & H_c^2(X', \boldsymbol{\mu}_n) & & \\ \wr\| & & \downarrow & & \wr\| & \operatorname{Tr}_{X'/k} \searrow & \\ H_c^r(X, \pi_! \mathscr{F}') & \times & \operatorname{Ext}^{2-r}(\pi_! \mathscr{F}', \pi_! \boldsymbol{\mu}_n) & \to & H_c^2(X, \pi_! \boldsymbol{\mu}_n) & & \mathbb{Z}/n. \\ \| & & \downarrow \operatorname{Tr}_\pi & & \downarrow \operatorname{Tr}_\pi & \operatorname{Tr}_{X/k} \nearrow & \\ H_c^r(X, \pi_! \mathscr{F}') & \times & \operatorname{Ext}^{2-r}(\pi_! \mathscr{F}', \boldsymbol{\mu}_n) & \to & H_c^2(X, \boldsymbol{\mu}_n) & & \end{array}$$

□

Lemma 8.3.8. *8.3.1 holds if $\mathscr{F}$ is supported on a finite closed subset of X.*

Proof. Working with each irreducible component of X, we may assume that X is irreducible. If $\mathscr{F}$ is supported on a finite closed subset, then $\mathscr{F}$ is a direct sum of sheaves supported on a single point. So it suffices to consider the case where $\mathscr{F} = i_* M$ for some finite $\mathbb{Z}/n$-module M and some closed immersion $i : \{x\} \to X$. By 8.3.7, we may replace X by its smooth compactification. So we may assume that X is projective. We have

$$H^r(X, i_* M) \cong \begin{cases} M & \text{if } r = 0, \\ 0 & \text{if } r \neq 0, \end{cases}$$

and

$$\begin{aligned} \operatorname{Ext}^{2-r}(i_* M, \boldsymbol{\mu}_n) &\cong \operatorname{Hom}_{D(X, \mathbb{Z}/n)}(i_* M, \boldsymbol{\mu}_n[2-r]) \\ &\cong \operatorname{Hom}_{D(\mathbb{Z}/n)}(M, Ri^! \boldsymbol{\mu}_n[2-r]) \\ &\cong \operatorname{Hom}_{D(Z/n)}(M, \mathbb{Z}/n[-r]) \\ &\cong \begin{cases} \operatorname{Hom}(M, \mathbb{Z}/n) & \text{if } r = 0, \\ 0 & \text{if } r \neq 0. \end{cases} \end{aligned}$$

Here we use 8.3.6 and the fact that $\mathbb{Z}/n$ is an injective $\mathbb{Z}/n$-module. Since the canonical pairing

$$M \times \mathrm{Hom}(M, \mathbb{Z}/n) \to \mathbb{Z}/n$$

is perfect, the pairing

$$H^0(x, M) \times \mathrm{Ext}^2(M, Ri^!\boldsymbol{\mu}_n) \to H^2(x, Ri^!\boldsymbol{\mu}_n)$$

is perfect. We claim that each morphism in the composite

$$H^2(x, Ri^!\boldsymbol{\mu}_n) \overset{\cong}{\to} H^2(X, i_*Ri^!\boldsymbol{\mu}_n) \cong H^2_x(X, \boldsymbol{\mu}_n) \to H^2(X, \boldsymbol{\mu}_n) \overset{\mathrm{Tr}_{X/k}}{\to} \mathbb{Z}/n$$

is an isomorphism. Indeed, since X is smooth projective and irreducible, the homomorphism

$$\mathrm{Tr}_{X/k} : H^2(X, \boldsymbol{\mu}_n) \to \mathbb{Z}/n$$

is an isomorphism. We have a long exact sequence

$$\cdots \to H^1(X - \{x\}, \boldsymbol{\mu}_n) \to H^2_x(X, \boldsymbol{\mu}_n) \to H^2(X, \boldsymbol{\mu}_n) \to H^2(X - \{x\}, \boldsymbol{\mu}_n) \to \cdots.$$

By 7.2.13, we have

$$H^2(X - \{x\}, \boldsymbol{\mu}_n) = 0.$$

So the canonical homomorphism

$$H^2_x(X, \boldsymbol{\mu}_n) \to H^2(X, \boldsymbol{\mu}_n)$$

is surjective. Since both $H^2_x(X, \boldsymbol{\mu}_n)$ and $H^2(X, \boldsymbol{\mu}_n)$ are free $\mathbb{Z}/n$-modules of rank 1, this homomorphism is an isomorphism. Our claim follows. 8.3.1 then follows from the commutativity of the following diagram:

$$\begin{array}{ccccccc}
H^0(X, i_*M) & \times & \mathrm{Ext}^2(i_*M, \boldsymbol{\mu}_n) & \to & H^2(X, \boldsymbol{\mu}_n) & & \\
\| & & \uparrow & & \uparrow & \overset{\mathrm{Tr}_{X/k}}{\searrow} & \\
H^0(X, i_*M) & \times & \mathrm{Ext}^2(i_*M, i_*Ri^!\boldsymbol{\mu}_n) & \to & H^2(X, i_*Ri^!\boldsymbol{\mu}_n) & & \mathbb{Z}/n. \\
\wr\| & & \uparrow & & \wr\| & \overset{\cong}{\nearrow} & \\
H^0(x, M) & \times & \mathrm{Ext}^2(M, Ri^!\boldsymbol{\mu}_n) & \to & H^2(x, Ri^!\boldsymbol{\mu}_n) & &
\end{array}$$

□

Lemma 8.3.9. *8.3.1 holds for $\mathscr{F} = \mathbb{Z}/n$.*

Proof. Working with each irreducible component of X, we may assume that X is irreducible. Let $\overline{X}$ be a smooth compactification of X, and let $j : X \hookrightarrow \overline{X}$ and $i : \overline{X} - X \to \overline{X}$ be the immersions. By 8.3.7, it suffices to prove that the pairing

$$H^r(\overline{X}, j_!\mathbb{Z}/n) \times \mathrm{Ext}^{2-r}(j_!\mathbb{Z}/n, \boldsymbol{\mu}_n) \to \mathbb{Z}/n$$

is perfect for each r. We have an exact sequence

$$0 \to j_!\mathbb{Z}/n \to \mathbb{Z}/n \to i_*\mathbb{Z}/n \to 0.$$

For any $\mathbb{Z}/n$-module M, let

$$M^\vee = \mathrm{Hom}(M, \mathbb{Z}/n).$$

Since $\mathbb{Z}/n$ is an injective Z/n-module, the functor $M \mapsto M^\vee$ is exact. Consider the commutative diagram

$$\begin{array}{ccccccc}
0 \to & H^0(\overline{X}, j_!\mathbb{Z}/n) & \to & H^0(\overline{X}, \mathbb{Z}/n) & \to & H^0(X, i_*\mathbb{Z}/n) & \to \\
& \downarrow & & \downarrow & & \downarrow & \\
\cdots \to & (\mathrm{Ext}^2(j_!\mathbb{Z}/n, \boldsymbol{\mu}_n))^\vee & \to & (\mathrm{Ext}^2(\mathbb{Z}/n, \boldsymbol{\mu}_n))^\vee & \to & (\mathrm{Ext}^2(i_*\mathbb{Z}/n, \boldsymbol{\mu}_n))^\vee & \to \\
 & & & & & & \\
\to & H^1(\overline{X}, j_!\mathbb{Z}/n) & \to & H^1(\overline{X}, \mathbb{Z}/n) & \to & H^1(X, i_*\mathbb{Z}/n) & \to \cdots \\
& \downarrow & & \downarrow & & \downarrow & \\
\to & (\mathrm{Ext}^1(j_!\mathbb{Z}/n, \boldsymbol{\mu}_n))^\vee & \to & (\mathrm{Ext}^1(\mathbb{Z}/n, \boldsymbol{\mu}_n))^\vee & \to & (\mathrm{Ext}^1(i_*\mathbb{Z}/n, \boldsymbol{\mu}_n))^\vee & \to \cdots
\end{array}$$

where the vertical arrows are induced by the pairings, and the horizontal lines are exact. By 8.3.8, the homomorphisms

$$H^r(\overline{X}, i_*\mathbb{Z}/n) \to (\mathrm{Ext}^{2-r}(i_*\mathbb{Z}/n, \boldsymbol{\mu}_n))^\vee$$

are bijective. To prove our assertion, it suffices to show that the homomorphisms

$$H^r(\overline{X}, \mathbb{Z}/n) \to (\mathrm{Ext}^{2-r}(\mathbb{Z}/n, \boldsymbol{\mu}_n))^\vee$$

are bijective. We are thus reduced to the case where X is projective over k.

We have seen in the proof of 8.3.2 that we have

$$\mathrm{Ext}^{2-r}(\mathbb{Z}/n, \boldsymbol{\mu}_n) \cong H^{2-r}(X, \mathscr{H}om(\mathbb{Z}/n, \boldsymbol{\mu}_n)) \cong H^{2-r}(X, \boldsymbol{\mu}_n).$$

So we have

$$\mathrm{Ext}^{2-r}(\mathbb{Z}/n, \boldsymbol{\mu}_n) = 0$$

for $r \neq 0, 1, 2$. Moreover, combined with 7.2.9, we see that $\mathrm{Ext}^{2-r}(\mathbb{Z}/n, \boldsymbol{\mu}_n)$ has the same number of elements as $H^r(X, \mathbb{Z}/n)$ for each r.

Consider the commutative diagram

$$\begin{array}{ccc}
\mathrm{Hom}_{D(X,\mathbb{Z}/n)}(\mathbb{Z}/n, \mathbb{Z}/n) \times \mathrm{Hom}_{D(X,\mathbb{Z}/n)}(\mathbb{Z}/n, \boldsymbol{\mu}_n[2]) & \to & \mathrm{Hom}_{D(X,\mathbb{Z}/n)}(\mathbb{Z}/n, \boldsymbol{\mu}_n[2]) \\
\cong\downarrow & & \cong\downarrow \\
H^0(R\Gamma(X, \mathbb{Z}/n)) \times \mathrm{Ext}^2(\mathbb{Z}/n, \boldsymbol{\mu}_n) & \to & H^0(R\Gamma(X, \boldsymbol{\mu}_n[2])) \\
\cong\downarrow & & \cong\downarrow \\
H^0(X, \mathbb{Z}/n) \times \mathrm{Ext}^2(\mathbb{Z}/n, \boldsymbol{\mu}_n) & \to & H^2(X, \boldsymbol{\mu}_n) \\
 & & \cong\downarrow \mathrm{Tr}_{X/k} \\
 & & \mathbb{Z}/n,
\end{array}$$

where the pairing on the top line is

$$(\phi, \psi) \mapsto \psi\phi.$$

For any $\psi \in \mathrm{Hom}_{D(X,\mathbb{Z}/n)}(\mathbb{Z}/n, \boldsymbol{\mu}_n[2])$, if $\psi\phi = 0$ for all $\phi \in \mathrm{Hom}_{D(X,\mathbb{Z}/n)}(\mathbb{Z}/n, \mathbb{Z}/n)$, then we have $\psi = 0$ by taking $\phi = \mathrm{id}$. Since $H^0(X, \mathbb{Z}/n)$ and $\mathrm{Ext}^2(\mathbb{Z}/n, \boldsymbol{\mu}_n)$ have the same number of elements, the pairing

$$H^0(X, \mathbb{Z}/n) \times \mathrm{Ext}^2(\mathbb{Z}/n, \boldsymbol{\mu}_n) \to \mathbb{Z}/n$$

is perfect.

Since $H^1(X, \mathbb{Z}/n)$ and $\mathrm{Ext}^1(\mathbb{Z}/n, \boldsymbol{\mu}_n)$ have the same number of elements, to prove that the pairing

$$H^1(X, \mathbb{Z}/n) \times \mathrm{Ext}^1(\mathbb{Z}/n, \boldsymbol{\mu}_n) \to \mathbb{Z}/n$$

is perfect, it suffices to show that the homomorphism

$$H^1(X, \mathbb{Z}/n) \to (\mathrm{Ext}^1(\mathbb{Z}/n, \boldsymbol{\mu}_n))^\vee$$

induced by the pairing is a monomorphism. Given $\alpha \in H^1(X, \mathbb{Z}/n)$, let $\pi : X' \to X$ be the $\mathbb{Z}/n$-torsor defined by α. The image of α under the canonical homomorphism

$$H^1(X, \mathbb{Z}/n) \to H^1(X', \mathbb{Z}/n)$$

is 0 since the base change of the above torsor by $X' \to X$ has a global section and is hence trivial. It follows that the image of α under the canonical homomorphism

$$H^1(X, \mathbb{Z}/n) \to H^1(X, \pi_*\pi^*\mathbb{Z}/n) \cong H^1(X, \pi_*\mathbb{Z}/n)$$

is 0. Since π is surjective, the canonical morphism $\mathbb{Z}/n \to \pi_*\mathbb{Z}/n$ is injective. Let $\mathscr{F}$ be its cokernel. Consider the commutative diagram

$$\begin{array}{cccc}
H^0(X,\pi_*\mathbb{Z}/n) \to & H^0(X,\mathscr{F}) \to & H^1(X,\mathbb{Z}/n) \to & H^1(X,\pi_*\mathbb{Z}/n) \\
\downarrow & \downarrow & \downarrow & \downarrow \\
(\mathrm{Ext}^2(\pi_*\mathbb{Z}/n,\boldsymbol{\mu}_n))^\vee \to & (\mathrm{Ext}^2(\mathscr{F},\boldsymbol{\mu}_n))^\vee \to & (\mathrm{Ext}^1(\mathbb{Z}/n,\boldsymbol{\mu}_n))^\vee \to & (\mathrm{Ext}^1(\pi_*\mathbb{Z}/n,\boldsymbol{\mu}_n))^\vee
\end{array}$$

We have proved that the pairing

$$H^0(X, \mathbb{Z}/n) \times \mathrm{Ext}^2(\mathbb{Z}/n, \boldsymbol{\mu}_n) \to \mathbb{Z}/n$$

is perfect for any smooth projective curve X. In particular, the pairing

$$H^0(X', \mathbb{Z}/n) \times \mathrm{Ext}^2(\mathbb{Z}/n, \boldsymbol{\mu}_n) \to \mathbb{Z}/n$$

is perfect. By 8.3.7, the pairing

$$H^0(X, \pi_*\mathbb{Z}/n) \times \mathrm{Ext}^2(\pi_*\mathbb{Z}/n, \boldsymbol{\mu}_n) \to \mathbb{Z}/n$$

is perfect. So the first vertical arrow in the above diagram is an isomorphism. Suppose α lies in the kernel of $H^1(X, \mathbb{Z}/n) \to (\mathrm{Ext}^1(\mathbb{Z}/n, \boldsymbol{\mu}_n))^\vee$. If we can show that $H^0(X, \mathscr{F})) \to (\mathrm{Ext}^2(\mathscr{F}, \boldsymbol{\mu}_n))^\vee$ is injective, then by diagram chasing and taking into account the fact that α lies in the kernel of $H^1(X, \mathbb{Z}/n) \to H^1(X, \pi_*\mathbb{Z}/n)$, we can show $\alpha = 0$. Hence to show that

$$H^1(X, \mathbb{Z}/n) \to (\mathrm{Ext}^1(\mathbb{Z}/n, \boldsymbol{\mu}_n))^\vee$$

is injective, it suffices to show that

$$H^0(X, \mathscr{F}) \to (\mathrm{Ext}^2(\mathscr{F}, \boldsymbol{\mu}_n))^\vee$$

is injective.

Since π is finite and etale, $\pi_*\mathbb{Z}/n$ is locally constant by 7.8.3. So $\mathscr{F}$ is locally constant. It is constructible by 5.8.11. Using 5.8.1 (i), one can show that there exists a finite surjective etale morphism $\pi' : X'' \to X$ such that $\pi'^*\mathscr{F}$ is constant. Since every finitely generated $\mathbb{Z}/n$-module can be embedded into a free $\mathbb{Z}/n$-module of finite rank, there exists a free $\mathbb{Z}/n$-module L of finite rank such that we have a monomorphism $\pi'^*\mathscr{F} \to L$. As $\mathscr{F} \to \pi'_*\pi'^*\mathscr{F}$ is injective, we have a monomorphism $\mathscr{F} \to \pi'_*L$. Let $\mathscr{G}$ be its cokernel. Consider the commutative diagram

$$\begin{array}{ccccccc}
0 \to & H^0(X, \mathscr{F}) & \to & H^0(X, \pi'_*L) & \to & H^0(X, \mathscr{G}) & \to \cdots \\
& \downarrow & & \downarrow & & \downarrow & \\
\cdots \to & (\mathrm{Ext}^2(\mathscr{F}, \boldsymbol{\mu}_n))^\vee & \to & (\mathrm{Ext}^2(\pi'_*L, \boldsymbol{\mu}_n))^\vee & \to & (\mathrm{Ext}^2(\mathscr{G}, \boldsymbol{\mu}_n))^\vee & \to \cdots .
\end{array}$$

By our previous discussion, the pairing

$$H^0(X'', \mathbb{Z}/n) \times \mathrm{Ext}^2(\mathbb{Z}/n, \boldsymbol{\mu}_n) \to \mathbb{Z}/n$$

is perfect. By 8.3.7, this implies that the homomorphism

$$H^0(X, \pi'_*L) \to (\mathrm{Ext}^2(\pi'_*L, \boldsymbol{\mu}_n))^\vee$$

is bijective. It follows from the above commutative diagram that

$$H^0(X, \mathscr{F}) \to (\mathrm{Ext}^2(\mathscr{F}, \boldsymbol{\mu}_n))^\vee$$

is injective.

Finally we prove that the homomorphism

$$H^2(X, \mathbb{Z}/n) \to (\mathrm{Hom}(\mathbb{Z}/n, \boldsymbol{\mu}_n))^\vee,$$

induced by the paring

$$H^2(X, \mathbb{Z}/n) \times \mathrm{Hom}(\mathbb{Z}/n, \boldsymbol{\mu}_n) \to \mathbb{Z}/n,$$

is injective. Let $\alpha \in H^2(X, \mathbb{Z}/n)$ be an element in the kernel of the above homomorphism. For any morphism $\phi : \mathbb{Z}/n \to \boldsymbol{\mu}_n$, denote by $[\phi](\alpha) \in H^2(X, \boldsymbol{\mu}_n)$ the image of α under the homomorphism

$$[\phi] : H^2(X, \mathbb{Z}/n) \to H^2(X, \boldsymbol{\mu}_n)$$

induced by ϕ. Then we have $\mathrm{Tr}_{X/k}([\phi](\alpha)) = 0$. As X is smooth projective and irreducible, $\mathrm{Tr}_{X/k}$ is an isomorphism. So we have $[\phi](\alpha) = 0$. If we take ϕ to be an isomorphism from $\mathbb{Z}/n$ to $\boldsymbol{\mu}_n$, we get $\alpha = 0$. □

Proof of 8.3.1. Since $\mathscr{F}$ is constructible, we can find an etale morphism $\pi : X' \to X$ together with an epimorphism $\pi_!\mathbb{Z}/n \to \mathscr{F}$. Let $\mathscr{G}$ be the kernel of this epimorphism. It is also constructible. Consider the commutative diagram

$$\begin{array}{ccccccc} 0 \to \mathrm{Ext}^0(\mathscr{F}, \boldsymbol{\mu}_n) & \to & \mathrm{Ext}^0(\pi_!\mathbb{Z}/n, \boldsymbol{\mu}_n) & \to & \mathrm{Ext}^0(\mathscr{G}, \boldsymbol{\mu}_n) & \to \\ \downarrow (1) & & \downarrow (2) & & \downarrow (3) & \\ 0 \to H_c^2(X, \mathscr{F})^\vee & \to & H_c^2(X, \pi_!\mathbb{Z}/n)^\vee & \to & H_c^2(X, \mathscr{G})^\vee & \to \end{array}$$

$$\begin{array}{ccccccc} \to \mathrm{Ext}^1(\mathscr{F}, \boldsymbol{\mu}_n) & \to & \mathrm{Ext}^1(\pi_!\mathbb{Z}/n, \boldsymbol{\mu}_n) & \to & \mathrm{Ext}^1(\mathscr{G}, \boldsymbol{\mu}_n) & \to \cdots \\ \downarrow (4) & & \downarrow (5) & & \downarrow (6) & \\ \to H_c^1(X, \mathscr{F})^\vee & \to & H_c^1(X, \pi_!\mathbb{Z}/n)^\vee & \to & H_c^1(X, \mathscr{G})^\vee & \to \cdots \end{array}$$

By 8.3.7 and 8.3.9, (2) is bijective. So (1) is injective. This is true for any constructible sheaf $\mathscr{F}$. So (3) is injective. It then follows that (1) is bijective. This is true for any constructible sheaf $\mathscr{F}$. So (3) is bijective. By 8.3.7 and 8.3.9, (5) is bijective. So (4) is injective. This is true for any constructible sheaf $\mathscr{F}$. So (6) is injective. It then follows that (4) is bijective. This is true for any constructible sheaf $\mathscr{F}$. So (6) is bijective. Using this argument repeatedly, we can prove that the homomorphism

$$\mathrm{Ext}^{2-r}(\mathscr{F}, \boldsymbol{\mu}_n) \to H^r(X, \mathscr{F})^\vee$$

is bijective for each r. □

Lemma 8.3.10. *Let S be a regular scheme pure of dimension 1, U a dense open subset of S, $j : U \hookrightarrow S$ the open immersion, A a noetherian ring such that $nA = 0$ and A is an injective A-module, and $\mathscr{F}$ a locally constant constructible sheaf of A-modules on U. Then we have*

$$\mathscr{H}om(j_*\mathscr{F}, A) \cong j_*\mathscr{H}om(\mathscr{F}, A)$$

and

$$\mathscr{E}xt^q(j_*\mathscr{F}, A) = 0$$

for any $q \geq 1$.

Proof. By 8.3.3, we have

$$j^*\mathscr{E}xt^q(j_*\mathscr{F}, A) \cong \mathscr{E}xt^q(\mathscr{F}, A) = 0$$

for all $q \geq 1$. The isomorphism

$$j^*\mathscr{H}om(j_*\mathscr{F}, A)) \cong \mathscr{H}om(\mathscr{F}, A)$$

induces a morphism

$$\mathscr{H}om(j_*\mathscr{F}, A) \to j_*\mathscr{H}om(\mathscr{F}, A)$$

whose restriction to U is an isomorphism. To prove the lemma, we need to show that for any $s \in S - U$, we have $(\mathscr{E}xt^q(j_*\mathscr{F}, A))_{\bar{s}} = 0$ for all $q \geq 1$, and

$$(\mathscr{H}om(j_*\mathscr{F}, A))_{\bar{s}} \to (j_*\mathscr{H}om(\mathscr{F}, A))_{\bar{s}}$$

is an isomorphism. Since S is pure of dimension 1 and U is dense, s is a closed point of X. By 5.9.11, we may replace S by the strict localization of S at s. So we may assume that S is a strictly local trait, s is the closed point of S, and $U = \{\eta\}$ is the generic point of S. Let $i : s \to S$ and $j : \eta \to S$ be the immersions, and let $I = \mathrm{Gal}(\bar{\eta}/\eta)$. We have an exact sequence

$$0 \to j_!\mathscr{F} \to j_*\mathscr{F} \to i_*\mathscr{F}_{\bar{\eta}}^I \to 0.$$

It gives rise to a distinguished triangle

$$\left(R\mathscr{H}om(i_*\mathscr{F}_{\bar{\eta}}^I, A)\right)_{\bar{s}} \to \left(R\mathscr{H}om(j_*\mathscr{F}, A)\right)_{\bar{s}} \to \left(R\mathscr{H}om(j_!\mathscr{F}, A)\right)_{\bar{s}} \to .$$

One can verify

$$R\mathscr{H}om(i_*\mathscr{F}_{\bar{\eta}}^I, A) \cong i_*R\mathscr{H}om(\mathscr{F}_{\bar{\eta}}^I, Ri^!A).$$

By 8.3.6, we have

$$Ri^!A \cong A(-1)[-2].$$

As A is an injective A-module, we have

$$R\mathscr{H}om(\mathscr{F}_{\bar{\eta}}^I, Ri^!A) \cong \mathscr{H}om(\mathscr{F}_{\bar{\eta}}^I, A(-1)[-2]) = (\mathscr{F}_{\bar{\eta}}^I)^\vee(-1)[-2],$$

where for any A-module M, we set

$$M^\vee = \mathrm{Hom}_A(M, A).$$

We thus have

$$R\mathscr{H}om(i_*\mathscr{F}_{\bar{\eta}}^I, A) \cong i_*(\mathscr{F}_{\bar{\eta}}^I)^\vee(-1)[-2].$$

One can verify

$$R\mathscr{H}om(j_!\mathscr{F}, A) \cong Rj_*R\mathscr{H}om(\mathscr{F}, A).$$

By 8.3.3, we have

$$R\mathscr{H}om(\mathscr{F}, A) = \mathscr{H}om(\mathscr{F}, A).$$

So we have

$$R\mathscr{H}om(j_!\mathscr{F}, A) \cong Rj_*\mathscr{H}om(\mathscr{F}, A).$$

We thus have a distinguished triangle

$$(\mathscr{F}_{\bar{\eta}}^I)^\vee(-1)[-2] \to \big(R\mathscr{H}om(j_*\mathscr{F}, A)\big)_{\bar{s}} \to \big(Rj_*\mathscr{H}om(\mathscr{F}, A)\big)_{\bar{s}} \to .$$

Taking the long exact sequence of cohomology associated to this triangle, we get an exact sequence

$$\begin{aligned} 0 &\to (\mathscr{E}xt^1(j_*\mathscr{F}, A))_{\bar{s}} \to (R^1j_*\mathscr{H}om(\mathscr{F}, A))_{\bar{s}} \to (\mathscr{F}_{\bar{\eta}}^I)^\vee(-1) \to \\ &\to (\mathscr{E}xt^2(j_*\mathscr{F}, A))_{\bar{s}} \to (R^2j_*\mathscr{H}om(\mathscr{F}, A))_{\bar{s}} \to \quad 0 \end{aligned}$$

and isomorphisms

$$(\mathscr{E}xt^q(j_*\mathscr{F}, A))_{\bar{s}} \cong (R^qj_*\mathrm{Hom}(\mathscr{F}, A))_{\bar{s}}$$

for all $q \neq 1, 2$. By 8.3.5, we have

$$(R^qj_*\mathscr{H}om(\mathscr{F}, A))_{\bar{s}} \cong \begin{cases} (\mathscr{F}_{\bar{\eta}}^\vee)^I & \text{if } q = 0, \\ (\mathscr{F}_{\bar{\eta}}^\vee)_I(-1) & \text{if } q = 1, \\ 0 & \text{if } q \neq 0, 1. \end{cases}$$

It follows that

$$(\mathscr{E}xt^q(j_*\mathscr{F}, A))_{\bar{s}} = 0$$

for any $q \geq 3$ and

$$(\mathscr{H}om(j_*\mathscr{F}, A))_{\bar{s}} \cong (j_*\mathscr{H}om(\mathscr{F}, A))_{\bar{s}}.$$

In the above discussion, we can replace $\mathscr{F}$ by its constant subsheaf $\mathscr{F}_{\bar{\eta}}^I$. Note that $j_*\mathscr{F}_{\bar{\eta}}^I$ is a constant sheaf on S. So we have

$$\mathscr{E}xt^q(j_*\mathscr{F}_{\bar{\eta}}^I, A) = 0$$

for all $q \geq 1$ by 8.3.3. Moreover, we have a commutative diagram

$$\begin{array}{ccccccccc} & & & & (\mathscr{F}_{\bar{\eta}}^\vee)_I(-1) & & & & \\ & & & & \wr\| & & & & \\ 0 \to & (\mathscr{E}xt^1(j_*\mathscr{F}, A))_{\bar{s}} & \to & & (R^1j_*\mathscr{H}om(\mathscr{F}, A))_{\bar{s}} & \to (\mathscr{F}_{\bar{\eta}}^I)^\vee(-1) \to & (\mathscr{E}xt^2(j_*\mathscr{F}, A))_{\bar{s}} & \to 0 \\ & \downarrow & & & \downarrow & \downarrow & \downarrow & \\ 0 \to & (\mathscr{E}xt^1(j_*\mathscr{F}_{\bar{\eta}}^I, A))_{\bar{s}} & \to & & (R^1j_*\mathscr{H}om(\mathscr{F}_{\bar{\eta}}^I, A))_{\bar{s}} & \to (\mathscr{F}_{\bar{\eta}}^I)^\vee(-1) \to & (\mathscr{E}xt^2(j_*\mathscr{F}_{\bar{\eta}}^I, A))_{\bar{s}} & \to 0 \\ & \| & & & \wr\| & & \| & \\ & 0 & & & (\mathscr{F}_{\bar{\eta}}^\vee)_I(-1) & & 0 & \end{array}$$

It follows that the homomorphism

$$(R^1j_*\mathscr{H}om(\mathscr{F}, A))_{\bar{s}} \to (\mathscr{F}_{\bar{\eta}}^I)^\vee(-1)$$

is an isomorphism. So

$$(\mathscr{E}xt^q(j_*\mathscr{F}, A))_{\bar{s}} = 0$$

for $q = 0, 1$. □

Theorem 8.3.11. *Let X be a smooth curve over an algebraically closed field k, U a dense open subset of X, $j : U \to X$ the open immersion, $\mathscr{F}$ a locally constant constructible sheaf of $\mathbb{Z}/n$-modules on U, and $\mathscr{F}^\vee = \mathscr{H}om(\mathscr{F}, \mathbb{Z}/n)$. Then we have a perfect pairing*

$$H^r_c(X, j_*\mathscr{F}) \times H^{2-r}(X, j_*\mathscr{F}^\vee(1)) \to \mathbb{Z}/n$$

for each r.

Proof. By 8.3.10, the spectral sequence

$$E_2^{pq} = H^p(X, \mathscr{E}xt^q(j_*\mathscr{F}, \mathbb{Z}/n)) \Rightarrow \mathrm{Ext}^{p+q}(j_*\mathscr{F}, \mathbb{Z}/n)$$

degenerates and

$$\mathscr{H}om(j_*\mathscr{F}, \mathbb{Z}/n) \cong j_*\mathscr{F}^\vee.$$

So we have

$$\mathrm{Ext}^p(j_*\mathscr{F}, \mathbb{Z}/n) \cong H^p(X, \mathscr{H}om(j_*\mathscr{F}, \mathbb{Z}/n)) \cong H^p(X, j_*\mathscr{F}^\vee).$$

Our assertion then follows from 8.3.1. □

8.4 The Functor $Rf^!$

([SGA 4] XVIII 3.1.)

Fix a noetherian scheme S. In this section, all schemes are of finite type over S. Let $f : X \to Y$ be an S-compactifiable morphism between S-schemes of finite type, and let A be a noetherian torsion ring. In this section, we construct a functor $Rf^! : D^+(Y, A) \to D^+(X, A)$ that is right adjoint to the functor $Rf_! : D(X, A) \to D(Y, A)$.

For any sheaf of A-modules $\mathscr{F}$ on X, we can write

$$\mathscr{F} = \varinjlim_{i \in I} \mathscr{F}_i,$$

where $(\mathscr{F}_i)_{i \in I}$ is a direct system of constructible sheaves of A-modules on X. Define

$$\mathscr{C}^\cdot_l(\mathscr{F}) = \varinjlim_{i \in I} \mathscr{C}^\cdot(\mathscr{F}_i),$$

where $\mathscr{C}^\cdot(-)$ denotes the functorial Godement resolution constructed at the end of 5.6. Note that $\mathscr{C}^\cdot_l(\mathscr{F})$ is independent of the choice of the direct system $(\mathscr{F}_i)_{i\in I}$. Indeed, suppose $\mathscr{F} = \varinjlim_{j\in J} \mathscr{G}_j$, where $\mathscr{G}_j$ are constructible. Let $\mathscr{F}'_i$ (resp. $\mathscr{G}'_j$) be the images of the morphisms $\mathscr{F}_i \to \mathscr{F}$

(resp. $\mathscr{G}_j \to \mathscr{F}$), and let $\mathscr{K}_i$ (resp. $\mathscr{L}_j$) be the kernels of the morphisms $\mathscr{F}_i \to \mathscr{F}$ (resp. $\mathscr{G}_j \to \mathscr{F}$). These sheaves are constructible. Using 5.9.8, one can show that for each i (resp. j), there exists $i' \geq i$ (resp. $j' \geq j$) such that the morphism $\mathscr{K}_i \to \mathscr{K}_{i'}$ (resp. $\mathscr{L}_j \to \mathscr{L}_{j'}$) is zero. It follows that

$$\varinjlim_i \mathscr{C}^\cdot(\mathscr{K}_i) = 0, \quad \varinjlim_j \mathscr{C}^\cdot(\mathscr{L}_j) = 0.$$

So we have

$$\varinjlim_i \mathscr{C}^\cdot(\mathscr{F}_i) \cong \varinjlim_i \mathscr{C}^\cdot(\mathscr{F}_i'), \quad \varinjlim_j \mathscr{C}^\cdot(\mathscr{G}_j) \cong \varinjlim_j \mathscr{C}^\cdot(\mathscr{G}_j').$$

To prove that $\mathscr{C}_l^\cdot(\mathscr{F})$ is independent of the choice of the direct system $(\mathscr{F}_i)_{i\in I}$, it suffices to show

$$\varinjlim_i \mathscr{C}^\cdot(\mathscr{F}_i') \cong \varinjlim_j \mathscr{C}^\cdot(\mathscr{G}_j').$$

For each i, we have $\mathscr{F}_i' = \cup_j(\mathscr{F}_i' \cap \mathscr{G}_j')$. It follows that $\mathscr{F}_i' = \mathscr{F}_i' \cap \mathscr{G}_j'$ for some j, that is, $\mathscr{F}_i' \subset \mathscr{G}_j'$. Similarly, for each j, we can find i such that $\mathscr{G}_j' \subset \mathscr{F}_i'$. Our assertion follows.

Lemma 8.4.1.

(i) *$\mathscr{C}_l^\cdot(\mathscr{F})$ is a resolution of $\mathscr{F}$.*

(ii) *For each q, $\mathscr{C}_l^q(-)$ is an exact functor and commutes with the direct limit.*

(iii) *If $f : X' \to X$ is etale, we have a canonical isomorphism $f^*\mathscr{C}_l^\cdot(\mathscr{F}) \stackrel{\cong}{\to} \mathscr{C}_l^\cdot(f^*\mathscr{F})$.*

Proof. It follows from the definition that $\mathscr{C}_l^\cdot(\mathscr{F})$ is a resolution of $\mathscr{F}$, and $\mathscr{C}_l^q(-)$ commutes with the direct limit. Let

$$0 \to \mathscr{F} \to \mathscr{G} \to \mathscr{H} \to 0$$

be a short exact sequence of sheaves of A-modules. Write $\mathscr{G} = \varinjlim_i \mathscr{G}_i$, where $\mathscr{G}_i$ are constructible subsheaves of $\mathscr{G}$. We have $\mathscr{F} = \varinjlim_i \mathscr{F} \cap \mathscr{G}_i$, $\mathscr{H} = \varinjlim_i \mathscr{G}_i/\mathscr{F} \cap \mathscr{G}_i$, and $\mathscr{F} \cap \mathscr{G}_i$, $\mathscr{G}_i/\mathscr{F} \cap \mathscr{G}_i$ are constructible. We have a short exact sequence

$$0 \to \mathscr{C}^q(\mathscr{F} \cap \mathscr{G}_i) \to \mathscr{C}^q(\mathscr{G}_i) \to \mathscr{C}^q(\mathscr{G}_i/\mathscr{F} \cap \mathscr{G}_i) \to 0.$$

Taking direct limit, we get an exact sequence

$$0 \to \mathscr{C}_l^q(\mathscr{F}) \to \mathscr{C}_l^q(\mathscr{G}) \to \mathscr{C}_l^q(\mathscr{H}) \to 0.$$

So $\mathscr{C}_l^q(-)$ is an exact functor.

For any S-morphism $f : X' \to X$ between S-schemes of finite type and any sheaf $\mathscr{F}$ on X, we have a canonical morphism

$$f^*\mathscr{C}^\cdot(\mathscr{F}) \to \mathscr{C}^\cdot(f^*\mathscr{F}).$$

If f is etale, this is an isomorphism. Using this fact, one proves (iii). □

Let $f : X \to Y$ be an S-compactifiable morphism. Fix a compactification

$$X \overset{j}{\hookrightarrow} \overline{X} \overset{\bar{f}}{\to} Y,$$

where j is an open immersion and $\bar{f}$ is a proper S-compactifiable morphism. Let d be an integer such that all the fibers of $\bar{f}$ have dimensions $\leq d$. For any sheaf of A-modules $\mathscr{F}$ on X, the complex $\tau_{\leq 2d}\mathscr{C}_l^{\cdot}(j_!\mathscr{F})$ is a resolution of $j_!\mathscr{F}$ by $R\bar{f}_*$-acyclic objects. Indeed, let $\mathscr{F}_q$ be the q-th component of $\tau_{\leq 2d}\mathscr{C}_l^{\cdot}(j_!\mathscr{F})$. When $q \neq 2d$, $\mathscr{F}_q$ is a direct limit of flasque sheaves. Moreover, we have

$$R^p\bar{f}_*\mathscr{F}_{2d} \cong R^{p+2d}\bar{f}_*j_!\mathscr{F} \cong R^{p+2d}f_!\mathscr{F} = 0$$

for any $p \geq 1$ by 7.4.5. For any complex K of sheaves of A-modules on X, denote the complex associated to the bicomplex $\left(\left(\tau_{\leq 2d}\mathscr{C}_l^{\cdot}(j_!K^p)\right)^q\right)_{p,q}$ by $\tau_{\leq 2d}\mathscr{C}_l^{\cdot}(j_!K)$. We then have

$$Rf_!K \cong \bar{f}_*\tau_{\leq 2d}\mathscr{C}_l^{\cdot}(j_!K)$$

in $D(Y, A)$. Define a functor $f_!^{\cdot}$ from the category of complex of sheaves of A-modules on X to the category of complexes of sheaves of A-modules on Y by

$$f_!^{\cdot}(K) = \bar{f}_*\tau_{\leq 2d}\mathscr{C}_l^{\cdot}(j_!K).$$

For any sheaf of A-modules $\mathscr{F}$ on X, let $f_!^q(\mathscr{F})$ be the q-th component of the complex $f_!^{\cdot}(\mathscr{F})$. Then $f_!^q$ is a functor from the category of sheaves of A-modules on X to the category of sheaves of A-modules on Y.

Lemma 8.4.2.

(i) *For each q, the functor $\mathscr{F} \to f_!^q(\mathscr{F})$ is exact and commutes with the direct limit. We have $f_!^q(\mathscr{F}) = 0$ for $q \notin [0, 2d]$.*

(ii) *The functor $f_!^{\cdot}$ induces a functor $Rf_!^{\cdot} : D(X, A) \to D(Y, A)$, and we have a canonical isomorphism $Rf_!^{\cdot} \cong Rf_!$.*

Proof. (i) follows from the definition of $f_!^{\cdot}$. As each $f_!^q$ is exact, the functor $f_!^{\cdot}$ maps acyclic complexes to acyclic complexes, and hence defines a functor on the derived category. □

Lemma 8.4.3. *For each q, the functor $f_!^q$ has a right adjoint functor $f_{-q}^!$. The functor $f_{-q}^!$ is left exact, transforms injective objects to injective objects, and vanishes if $-q \notin [-2d, 0]$.*

Proof. Given a sheaf of A-modules $\mathscr{G}$ on Y, define a presheaf $f^!_{-q}\mathscr{G}$ on X as follows: For any etale X-scheme $\phi : U \to X$, let

$$(f^!_{-q}\mathscr{G})(U) = \mathrm{Hom}(f^q_! A_U, \mathscr{G}),$$

where $A_U = \phi_! A$. For any etale morphism $\psi : V \to U$, define the restriction

$$(f^!_{-q}\mathscr{G})(U) \to (f^!_{-q}\mathscr{G})(V)$$

to be the map

$$\mathrm{Hom}(f^q_! A_U, \mathscr{G}) \to \mathrm{Hom}(f^q_! A_V, \mathscr{G})$$

induced by the canonical morphism

$$A_V = \phi_!\psi_! A \to \phi_! A = A_U.$$

We claim that $f^!_{-q}\mathscr{G}$ is a sheaf. Indeed, if $\{U_\alpha \to U\}_\alpha$ is an etale covering of U, then for any sheaf $\mathscr{F}$ on X, we have the canonical exact sequence

$$0 \to \mathscr{F}(U) \to \prod_\alpha \mathscr{F}(U_\alpha) \to \prod_{\alpha,\beta} \mathscr{F}(U_\alpha \times_U U_\beta).$$

This can be identified with an exact sequence

$$0 \to \mathrm{Hom}(A_U, \mathscr{F}) \to \mathrm{Hom}\Big(\bigoplus_\alpha A_{U_\alpha}, \mathscr{F}\Big) \to \mathrm{Hom}\Big(\bigoplus_{\alpha,\beta} A_{U_\alpha \times_U U_\beta}, \mathscr{F}\Big).$$

We thus have an exact sequence

$$\bigoplus_{\alpha,\beta} A_{U_\alpha \times_U U_\beta} \to \bigoplus_\alpha A_{U_\alpha} \to A_U \to 0.$$

Since $f^q_!$ is exact, we have an exact sequence

$$\bigoplus_{\alpha,\beta} f^q_! A_{U_\alpha \times_U U_\beta} \to \bigoplus_\alpha f^q_! A_{U_\alpha} \to f^q_! A_U \to 0,$$

and hence an exact sequence

$$0 \to \mathrm{Hom}(f^q_! A_U, \mathscr{G}) \to \prod_\alpha \mathrm{Hom}(f^q_! A_{U_\alpha}, \mathscr{G}) \to \prod_{\alpha,\beta} \mathrm{Hom}(f^q_! A_{U_\alpha \times_U U_\beta}, \mathscr{G}).$$

So the sequence

$$0 \to (f^!_{-q}\mathscr{G})(U) \to \prod_\alpha (f^!_{-q}\mathscr{G})(U_\alpha) \to \prod_{\alpha,\beta} (f^!_{-q}\mathscr{G})(U_\alpha \times_U U_\beta)$$

is exact. This proves $f^!_{-q}\mathscr{G}$ is a sheaf.

Next we define a morphism $\mathscr{F} \to f^!_{-q} f^q_! \mathscr{F}$ for any sheaf of A-modules $\mathscr{F}$ on X. It suffices to define a homomorphism

$$\mathscr{F}(U) \to (f^!_{-q} f^q_! \mathscr{F})(U) = \mathrm{Hom}(f^q_! A_U, f^q_! \mathscr{F})$$

functorial in U for any etale X-scheme U. Given $s \in \mathscr{F}(U)$, it defines a morphism $A_U \to \mathscr{F}$ and hence a morphism $f_!^q A_U \to f_!^q \mathscr{F}$. We assign this last morphism to s. The morphism $\mathscr{F} \to f_{-q}^! f_!^q \mathscr{F}$ defines a map

$$\mathrm{Hom}(f_!^q \mathscr{F}, \mathscr{G}) \to \mathrm{Hom}(\mathscr{F}, f_{-q}^! \mathscr{G}).$$

By our construction, we have

$$\mathrm{Hom}(f_!^q A_U, \mathscr{G}) \cong \mathrm{Hom}(A_U, f_{-q}^! \mathscr{G})$$

for any etale X-scheme U. If $\mathscr{F}$ is constructible, then we can find an exact sequence of the form

$$A_V \to A_U \to \mathscr{F} \to 0$$

for some etale X-schemes U and V. Since $f_!^q$ is exact, the sequence

$$f_!^q A_V \to f_!^q A_U \to f_!^q \mathscr{F} \to 0$$

is exact. We have a commutative diagram

$$\begin{array}{ccccccc} 0 \to & \mathrm{Hom}(f_!^q \mathscr{F}, \mathscr{G}) & \to & \mathrm{Hom}(f_!^q A_U, \mathscr{G}) & \to & \mathrm{Hom}(f_!^q A_V, \mathscr{G}) \\ & \downarrow & & \downarrow & & \downarrow \\ 0 \to & \mathrm{Hom}(\mathscr{F}, f_{-q}^! \mathscr{G}) & \to & \mathrm{Hom}(A_U, f_{-q}^! \mathscr{G}) & \to & \mathrm{Hom}(A_V, f_{-q}^! \mathscr{G}). \end{array}$$

By the above discussion, the last two vertical arrows are bijective. It follows from the five lemma that

$$\mathrm{Hom}(f_!^q \mathscr{F}, \mathscr{G}) \cong \mathrm{Hom}(\mathscr{F}, f_{-q}^! \mathscr{G})$$

if $\mathscr{F}$ is constructible. In general, we can write $\mathscr{F} = \varinjlim_i \mathscr{F}_i$, where $\mathscr{F}_i$ are constructible. Since $f_!^q$ commutes with the direct limit, we have

$$\begin{aligned} \mathrm{Hom}(f_!^q \mathscr{F}, \mathscr{G}) &\cong \mathrm{Hom}(f_!^q (\varinjlim_i \mathscr{F}_i), \mathscr{G}) \\ &\cong \mathrm{Hom}(\varinjlim_i (f_!^q \mathscr{F}_i), \mathscr{G}) \\ &\cong \varprojlim_i \mathrm{Hom}(f_!^q \mathscr{F}_i, \mathscr{G}) \\ &\cong \varprojlim_i \mathrm{Hom}(\mathscr{F}_i, f_{-q}^! \mathscr{G}) \\ &\cong \mathrm{Hom}(\varinjlim_i \mathscr{F}_i, f_{-q}^! \mathscr{G}) \\ &\cong \mathrm{Hom}(\mathscr{F}, f_{-q}^! \mathscr{G}). \end{aligned}$$

So

$$\mathrm{Hom}(f_!^q \mathscr{F}, \mathscr{G}) \cong \mathrm{Hom}(\mathscr{F}, f_{-q}^! \mathscr{G})$$

for any sheaf of A-modules $\mathscr{F}$ on X. Therefore $f_{-q}^!$ is right adjoint to $f_!^q$. Since $f_!^q$ is exact, $f_{-q}^!$ transforms injective objects to injective objects. As a right adjoint functor, it is left exact. □

Let $d^{q*} : f^!_{-q-1} \to f^!_{-q}$ be the functor induced by $d^q : f^q_! \to f^{q+1}_!$ by adjunction. We then get a complex of functors $f^!_{\cdot}$ with differentials

$$(-1)^q d^{q*} : f^!_{-q-1} \to f^!_{-q}.$$

For any complex of sheaves of A-modules L on Y, denote the complex associated to the bicomplex $(f^!_q L^p)_{p,q}$ by $f^!_{\cdot} L^{\cdot}$. Then for any complex of sheaves of A-modules K on X, we have a canonical isomorphism of complexes of A-modules

$$\mathrm{Hom}^{\cdot}(f_!^{\cdot} K, L) \cong \mathrm{Hom}^{\cdot}(K, f^!_{\cdot} L).$$

One can show that $f^!_{\cdot} : K(Y, A) \to K(X, A)$ is an exact functor. Let

$$Rf^! : D^+(Y, A) \to D^+(X, A)$$

be its right derived functor. Then $Rf^!$ is right adjoint to $Rf_!$. Indeed, if L is a bounded below complex of injective sheaves of A-modules on Y, then we have

$$\mathrm{Hom}(Rf_! K, L) \cong \mathrm{Hom}(f_!^{\cdot} K, L) \cong H^0(\mathrm{Hom}^{\cdot}(f_!^{\cdot} K, L)) \cong H^0(\mathrm{Hom}^{\cdot}(K, f^!_{\cdot} L)).$$

But $f^!_{\cdot} L$ is a complex of injective sheaves of A-modules on X, so we have

$$H^0(\mathrm{Hom}^{\cdot}(K, f^!_{\cdot} L)) \cong \mathrm{Hom}(K, f^!_{\cdot} L) \cong \mathrm{Hom}(K, Rf^! L).$$

Hence

$$\mathrm{Hom}(Rf_! K, L) \cong \mathrm{Hom}(K, Rf^! L).$$

We thus have the following:

Theorem 8.4.4. *Let S be a noetherian scheme, let $f : X \to Y$ be an S-compactifiable morphism between S-schemes of finite type, and let A be a noetherian torsion ring. Then the functor $Rf^! : D^+(Y, A) \to D^+(X, A)$ is right adjoint to the functor $Rf_! : D(X, A) \to D(Y, A)$, that is, for any $K \in$ $\mathrm{ob}\, D(X, A)$ and $L \in \mathrm{ob}\, D^+(Y, A)$, we have a one-to-one correspondence*

$$\mathrm{Hom}(Rf_! K, L) \cong \mathrm{Hom}(K, Rf^! L)$$

functorial in K and L. Moreover, $Rf^!$ is an exact functor and the following diagram commutes:

$$\begin{array}{ccccc}
\mathrm{Hom}(Rf_!(K[1]), L) & \cong & \mathrm{Hom}(K[1], Rf^! L) & \cong & \mathrm{Hom}(K, (Rf^! L)[-1]) \\
\cong \downarrow & & & & \downarrow \cong \\
\mathrm{Hom}((Rf_! K)[1], L) & \cong & \mathrm{Hom}(Rf_! K, L[-1]) & \cong & \mathrm{Hom}(K, Rf^!(L[-1])).
\end{array}$$

Proposition 8.4.5. *Let S be a noetherian scheme, and let $f : X \to Y$ be a separated quasi-finite morphism between S-schemes of finite type.*

(i) *The functor $f_!$ from the category of sheaves of A-modules on X to the category of sheaves of A-modules on Y has a right adjoint $f^!$, and $Rf^!$ can be identified with the right derived functor of $f^!$.*

(ii) *If f is a closed immersion, then $f^!$ can be identified with the functor defined in 5.4.*

(iii) *If f is etale, then $f^!$ can be identified with f^*.*

Proof. (i) follows from the construction of $f^!$ in the case $d = 0$. (ii) follows from (i) since if f is a closed immersion, then $f_! = f_*$, and the functor $f^!$ constructed in 5.4 is right adjoint to f_*. (iii) follows from (i) since if f is etale, then f^* is right adjoint to $f_!$. □

Let $f : X \to Y$ be an S-compactifiable morphism between S-schemes of finite type, $K \in \operatorname{ob} D^-(X, A)$ and $L \in \operatorname{ob} D^+(X, A)$. We define a canonical morphism

$$Rf_*R\mathscr{H}om(K, L) \to R\mathscr{H}om(Rf_!K, Rf_!L)$$

as follows. Represent L by a bounded below complex of injective sheaves of A-modules. Then

$$R\mathscr{H}om(K, L) \cong \mathscr{H}om^{\cdot}(K, L).$$

By 5.6.8, $\mathscr{H}om^{\cdot}(K, L)$ is a complex of flasque sheaves. So we have

$$Rf_*R\mathscr{H}om(K, L) = f_*\mathscr{H}om^{\cdot}(K, L).$$

Fix a compactification

$$X \overset{j}{\hookrightarrow} \overline{X} \overset{\bar{f}}{\to} Y$$

of f and an upper bound d of dimensions of fibers of $\bar{f}$. For each etale Y-scheme $k : V \to Y$, fix notation by the following commutative diagram

$$\begin{array}{ccccc} X_V & \overset{j_V}{\to} & \overline{X}_V & \overset{\bar{f}_V}{\to} & V \\ k_X \downarrow & & k_{\overline{X}} \downarrow & & k \downarrow \\ X & \overset{j}{\to} & \overline{X} & \overset{\bar{f}}{\to} & Y, \end{array}$$

and let $f_V = \bar{f}_V j_V$ be the base change of f. By 8.4.1 (iii), we have an isomorphism of complexes of functors

$$k^* f_! \overset{\cong}{\to} f_{V!} k_X^*.$$

It follows that

$$\begin{aligned}\mathscr{H}om^{\cdot}(f_!^{\cdot}K, f_!^{\cdot}L)(V) &= \mathrm{Hom}(k^* f_!^{\cdot}K, k^* f_!^{\cdot}L) \\ &\cong \mathrm{Hom}(f_{V!}^{\cdot}k_X^* K, f_{V!}^{\cdot}k_X^* L).\end{aligned}$$

We have a canonical morphism

$$\mathscr{H}om^{\cdot}(K, L)(X_V) = \mathrm{Hom}^{\cdot}(k_X^* K, k_X^* L) \to \mathrm{Hom}^{\cdot}(f_{V!}^{\cdot}k_X^* K, f_{V!}^{\cdot}k_X^* L).$$

We thus have a morphism of complexes of sheaves

$$f_*\mathscr{H}om^{\cdot}(K, L) \to \mathscr{H}om^{\cdot}(f_!^{\cdot}K, f_!^{\cdot}L).$$

It induces a morphism

$$Rf_*R\mathscr{H}om(K, L) \to R\mathscr{H}om(Rf_!K, Rf_!L)$$

in the derived category. Now assume that $K \in \mathrm{ob}\, D^-(X, A)$ and $L \in \mathrm{ob}\, D^+(Y, A)$. Taking the composite of the morphism

$$Rf_*R\mathscr{H}om(K, Rf^!L) \to R\mathscr{H}om(Rf_!K, Rf_!Rf^!L)$$

and the morphism

$$R\mathscr{H}om(Rf_!K, Rf_!Rf^!L) \to R\mathscr{H}om(Rf_!K, L)$$

induced by the canonical morphism $Rf_!Rf^!L \to L$, we get a morphism

$$Rf_*R\mathscr{H}om(K, Rf^!L) \to R\mathscr{H}om(Rf_!K, L).$$

Theorem 8.4.6. *Let S be a noetherian scheme, let $f : X \to Y$ be an S-compactifiable morphism between S-schemes of finite type, let A be a noetherian torsion ring, and let $K \in \mathrm{ob}\, D^-(X, A)$ and $L \in \mathrm{ob}\, D^+(Y, A)$. The morphism*

$$Rf_*R\mathscr{H}om(K, Rf^!L) \to R\mathscr{H}om(Rf_!K, L)$$

defined above is an isomorphism.

Proof. Represent L by a bounded below complex of injective sheaves of A-modules on Y. We have

$$\begin{aligned}Rf_*R\mathscr{H}om(K, Rf^!L) &\cong f_*\mathscr{H}om^{\cdot}(K, f_{\cdot}^!L), \\ R\mathscr{H}om(Rf_!K, L) &\cong \mathscr{H}om^{\cdot}(f_!^{\cdot}K, L).\end{aligned}$$

Let us prove that for any etale Y-scheme $k : V \to Y$, that the morphism

$$\big(\mathscr{H}om^{\cdot}(K, f_{\cdot}^!L)\big)(X_V) \to \big(\mathscr{H}om^{\cdot}(f_!^{\cdot}K, L)\big)(V)$$

is a quasi-isomorphism, that is, the morphism

$$\mathrm{Hom}^{\cdot}(k_X^* K, k_X^* f_{\cdot}^!L) \to \mathrm{Hom}^{\cdot}(k^* f_!^{\cdot}K, k^* L)$$

is a quasi-isomorphism. Without loss of generality, let us prove that

$$H^0(\mathrm{Hom}^{\cdot}(k_X^*K, k_X^*f_{\cdot}^!L)) \to H^0(\mathrm{Hom}^{\cdot}(k^*f_!^{\cdot}K, k^*L))$$

is an isomorphism, that is,

$$\mathrm{Hom}(k_X^*K, k_X^*f_{\cdot}^!L) \cong \mathrm{Hom}(k^*f_!^{\cdot}K, k^*L).$$

Here $\mathrm{Hom}(-,-)$ denotes the spaces of homotopy classes of morphisms of complexes. When $V = Y$, this follows from the adjointness of $f_!^{\cdot}$ and $f_{\cdot}^!$. For general V, we need to introduce a morphism $k_X^*f_{\cdot}^! \to f_{V\cdot}^!k^*$. We have an isomorphism of complexes of functors

$$k^*f_!^{\cdot} \overset{\cong}{\to} f_{V!}^{\cdot}k_X^*.$$

For any q, we get a morphism

$$k_!f_{V!}^q \to f_!^qk_{X!}$$

by taking the composite

$$k_!f_{V!}^q \to k_!f_{V!}^qk_X^*k_{X!} \overset{\cong}{\to} k_!k^*f_!^qk_{X!} \to f_!^qk_{X!}.$$

By the adjointness of the pairs of functors $(k_!, k^*)$, $(k_{X!}, k_X^*)$, $(f_{V!}^q, f_{V,-q}^!)$ and $(f_!^q, f_{-q}^!)$, it induces a morphism

$$k_X^*f_{-q}^! \to f_{V,-q}^!k^*.$$

In this way, we get a morphism of complex of functors

$$k_X^*f_{\cdot}^! \to f_{V\cdot}^!k^*.$$

It induces an isomorphism in the derived category since $k_!f_{V!}^{\cdot} \to f_!^{\cdot}k_{X!}$ induces an isomorphism in the derived category. Since L is a bounded below complex of injective sheaves,

$$k_X^*f_{\cdot}^!L \to f_{V\cdot}^!k^*L$$

is an isomorphism in $K(X_V, A)$ by 6.2.7. Consider the diagram

$$\begin{array}{ccccc}
\mathrm{Hom}(k_X^*K, k_X^*f_{\cdot}^!L) & \overset{\cong}{\to} & \mathrm{Hom}(k_X^*K, f_{V\cdot}^!k^*L) & & \\
\downarrow & (1) & \downarrow & (2)\searrow\cong & \\
\mathrm{Hom}(f_{V!}^{\cdot}k_X^*K, f_{V!}^{\cdot}k_X^*f_{\cdot}^!L) & \overset{\cong}{\to} & \mathrm{Hom}(f_{V!}^{\cdot}k_X^*K, f_{V!}^{\cdot}f_{V\cdot}^!k^*L) & \to & \mathrm{Hom}(f_{V!}^{\cdot}k_X^*K, k^*L) \\
\cong\downarrow & (3) & \downarrow\cong & (4) & \downarrow\cong \\
\mathrm{Hom}(k^*f_!^{\cdot}K, f_{V!}^{\cdot}k_X^*f_{\cdot}^!L) & \overset{\cong}{\to} & \mathrm{Hom}(k^*f_!^{\cdot}K, f_{V!}^{\cdot}f_{V\cdot}^!k^*L) & \to & \mathrm{Hom}(k^*f_!^{\cdot}K, k^*L), \\
 & \nwarrow\cong & (5) & \nearrow & \\
 & & \mathrm{Hom}(k^*f_!^{\cdot}K, k^*f_!^{\cdot}f_{\cdot}^!L) & &
\end{array}$$

where $\mathrm{Hom}(-,-)$ denotes the space of homotopy classes of morphisms of complexes, the horizontal arrows in the squares (1) and (3) are induced by the isomorphism $k_X^* f_\cdot^! L \overset{\cong}{\to} f_{V\cdot}^! k^* L$, the vertical arrows in the squares (3) and (4) and the left slant arrow in the triangle (5) are induced by the isomorphism $k^* f_{!}^{\cdot} K \overset{\cong}{\to} f_{V!}^{\cdot} k_X^* K$, the horizontal arrows in the square (4) and the right slant arrow in the triangle (5) are induced by the canonical morphisms $f_{V!}^{\cdot} f_{V\cdot}^! \to \mathrm{id}$ and $f_!^{\cdot} f_\cdot^! \to \mathrm{id}$, respectively. By the adjointness of $(f_{V!}^{\cdot}, f_{V\cdot}^!)$, the slant arrow in the triangle (2) is an isomorphism. The morphism

$$\mathrm{Hom}(k_X^* K, k_X^* f_\cdot^! L) \to \mathrm{Hom}(k^* f_!^{\cdot} K, k^* L)$$

defined at the beginning is the composite of those morphisms on the left and the bottom part of the boundary of the above diagram. (In the composite, we take the inverse of the left slant arrow in the triangle (5).) On the other hand, the morphisms on the top and the right part of the boundary of the above diagram are isomorphisms. So to prove that $\mathrm{Hom}(k_X^* K, k_X^* f_\cdot^! L) \to \mathrm{Hom}(k^* f_!^{\cdot} K, k^* L)$ is an isomorphism, it suffices to show that the above diagram commutes. It is clear that (1), (2), (3) and (4) commute. To prove that (5) commutes, it suffices to show that the following diagram commutes:

$$\begin{array}{ccc} k^* f_!^{\cdot} f_\cdot^! L & \overset{k^*(\mathrm{adj})}{\to} & k^* L \\ \cong\downarrow & (6) & \uparrow\mathrm{adj} \\ f_{V!}^{\cdot} k_X^* f_\cdot^! L & \overset{\cong}{\to} & f_{V!}^{\cdot} f_{V\cdot}^! k^* L. \end{array}$$

Recall that we define $k_! f_{V!}^{\cdot} \to f_!^{\cdot} k_{X!}$ as the composite

$$k_! f_{V!}^{\cdot} \to k_! f_{V!}^{\cdot} k_X^* k_{X!} \overset{\cong}{\to} k_! k^* f_!^{\cdot} k_{X!} \to f_!^{\cdot} k_{X!}.$$

We then define the morphism $k_X^* f_\cdot^! \to f_{V\cdot}^! k^*$ as the composite

$$k_X^* f_\cdot^! \to f_{V\cdot}^! k^* k_! f_{V!}^{\cdot} k_X^* f_\cdot^! \to f_{V\cdot}^! k^* f_!^{\cdot} k_{X!} k_X^* f_\cdot^! \to f_{V\cdot}^! k^*.$$

It follows that $k_X^* f_\cdot^! \to f_{V\cdot}^! k^*$ is the composite of those morphisms on the left part of the boundary of the following diagram.

$$\begin{array}{ccccc} & & k_X^* f_\cdot^! & & \\ & \swarrow (7) & \downarrow & \searrow & \\ f_{V\cdot}^! k^* k_! f_{V!}^{\cdot} k_X^* k_{X!} k_X^* f_\cdot^! & \to & f_{V\cdot}^! k^* k_! f_{V!}^{\cdot} k_X^* f_\cdot^! & \leftarrow & f_{V\cdot}^! f_{V!}^{\cdot} k_X^* f_\cdot^! \\ \downarrow & & \downarrow & & \downarrow \\ f_{V\cdot}^! k^* k_! k^* f_!^{\cdot} k_{X!} k_X^* f_\cdot^! & \to & f_{V\cdot}^! k^* k_! k^* f_!^{\cdot} f_\cdot^! & \leftarrow & f_{V\cdot}^! k^* f_!^{\cdot} f_\cdot^! \\ & \searrow & \downarrow & (8) \swarrow & \\ & & f_{V\cdot}^! k^* & & \end{array}$$

.

In the above diagram, the triangles (7) and (8) commute because the following two diagrams commute.

$$\begin{array}{ccccccc} k_X^* & \stackrel{\text{adj}}{\to} & (k_X^* k_{X!})k_X^* & \quad & k^* & \stackrel{\text{adj}}{\to} & (k^* k_!)k^* \\ \| & (9) & \| & & \| & (10) & \| \\ k_X^* & \stackrel{k_X^*(\text{adj})}{\leftarrow} & k_X^*(k_{X!}k_X^*), & & k^* & \stackrel{k^*(\text{adj})}{\leftarrow} & k^*(k_! k^*). \end{array}$$

(The commutativity of (9) can be seen as follows: By the adjointness of the pair $(k_{X!}, k_X^*)$, the composite of the morphisms

$$k_X^* \stackrel{\text{adj}}{\to} (k_X^* k_{X!})k_X^* = k_X^*(k_{X!}k_X^*) \stackrel{k_X^*(\text{adj})}{\to} k_X^*$$

corresponds to the composite of the morphisms

$$k_{X!}k_X^* \stackrel{\text{id}}{\to} k_{X!}k_X^* \stackrel{\text{adj}}{\to} \text{id}.$$

The second composite is adj : $k_{X!}k_X^* \to \text{id}$, and it corresponds to id : $k_X^* \to k_X^*$ by the adjointness of $(k_{X!}, k_X^*)$. Similarly (10) commutes.) It is clear that the other parts of the above diagram commute. So the morphism $k_X^* f^! \to f_V^!.k^*$ is also the composite

$$k_X^* f^! \to f_V^!.f_{V!}k_X^* f^! \stackrel{\cong}{\to} f_V^!.k^* f_! f^! \to f_V^!.k^*.$$

The commutativity of (6) then follows from the commutativity of the following diagram:

$$\begin{array}{ccccccc} & & k^* f_! f^! & \to & k^* & & \\ & \swarrow \cong & & & \uparrow & \nwarrow & \\ f_{V!}k_X^* f^! & \stackrel{\text{id}}{\to} & f_{V!}k_X^* f^! & \stackrel{\cong}{\to} & k^* f_! f^! & & f_{V!}f_V^!.k^*. \\ & \searrow (11) & \uparrow & & \uparrow & \nearrow & \\ & & f_{V!}f_V^!.f_{V!}k_X^* f^! & \to & f_{V!}f_V^!.k^* f_! f^! & & \end{array}$$

In this last diagram, the triangle (11) commutes because the following diagram commutes

$$\begin{array}{ccc} f_{V!} & \stackrel{f_{V!}(\text{adj})}{\to} & f_{V!}(f_V^!.f_{V!}) \\ \| & & \| \\ f_{V!} & \stackrel{\text{adj}}{\leftarrow} & (f_{V!}f_V^!.)f_{V!}. \end{array}$$

□

Theorem 8.4.7. *Let S be a noetherian scheme, $f : X \to Y$ an S-compactifiable morphism between S-schemes of finite type, A a noetherian ring, $K \in \text{ob}\, D^-(Y, A)$ and $L \in D^+(Y, A)$. Then we have a canonical isomorphism*

$$R\mathscr{H}om(f^*K, Rf^!L) \stackrel{\cong}{\to} Rf^! R\mathscr{H}om(K, L).$$

Proof. Roughly speaking, the proof goes as follows. For any $M \in \mathrm{D}^-(X, A)$, by the projection formula 7.4.7, we have

$$Rf_!M \otimes_A^L K \xrightarrow{\cong} Rf_!(M \otimes f^*K).$$

Hence

$$\mathrm{Hom}(Rf_!(M \otimes_A^L f^*K), L) \xrightarrow{\cong} \mathrm{Hom}(Rf_!M \otimes_A^L K, L).$$

By the proof of 6.4.7, we have

$$\begin{aligned}\mathrm{Hom}(Rf_!(M \otimes_A^L f^*K), L) &\cong \mathrm{Hom}(M \otimes_A^L f^*K, Rf^!L)\\ &\cong \mathrm{Hom}(M, R\mathscr{H}om(f^*K, Rf^!L)),\\ \mathrm{Hom}(Rf_!M \otimes_A^L K, L) &\cong \mathrm{Hom}(Rf_!M, R\mathscr{H}om(K, L))\\ &\cong \mathrm{Hom}(M, Rf^!R\mathscr{H}om(K, L)).\end{aligned}$$

So we have a one-to-one correspondence

$$\mathrm{Hom}(M, R\mathscr{H}om(f^*K, Rf^!L)) \xrightarrow{\cong} \mathrm{Hom}(M, Rf^!R\mathscr{H}om(K, L)).$$

It induces an isomorphism $R\mathscr{H}om(f^*K, Rf^!L) \xrightarrow{\cong} Rf^!R\mathscr{H}om(K, L)$.

More explicitly, we construct the isomorphism as follows. Represent K by a bounded above complex of flat sheaves. We have

$$Rf_!M \otimes_A^L K = f_!^{\cdot} M \otimes_A K, \quad Rf_!(M \otimes_A^L f^*K) \cong f_!^{\cdot}(M \otimes_A f^*K).$$

We first construct a functorial morphism of complexes

$$f_!^{\cdot} M \otimes_A K \to f_!^{\cdot}(M \otimes_A f^*K).$$

Let

$$X \overset{j}{\hookrightarrow} \overline{X} \overset{\bar{f}}{\to} Y$$

be a compactification of f, where j is an open immersion and $\bar{f}$ is a proper S-compactifiable morphism. For sheaves of A-modules $\mathscr{H}_1$ and $\mathscr{H}_2$ on $\overline{X}$, we construct a morphism

$$\mathscr{C}^{\cdot}(\mathscr{H}_1) \otimes_A \mathscr{H}_2 \to \mathscr{C}^{\cdot}(\mathscr{H}_1 \otimes_A \mathscr{H}_2)$$

as follows. Let $P_{\overline{X}}$ be the set of geometric points in $\overline{X}$ that we use to construct the Godement resolution. For any geometric point $t : \mathrm{Spec}\, k \to \overline{X}$ in $P_{\overline{X}}$, we have a canonical morphism

$$t_*t^*\mathscr{H}_1 \otimes_A \mathscr{H}_2 \to t_*(t^*\mathscr{H}_1 \otimes_A t^*\mathscr{H}_2).$$

Taking direct product over all $t \in P_{\overline{X}}$, we can define a morphism

$$\mathscr{C}^0(\mathscr{H}_1) \otimes_A \mathscr{H}_2 \to \mathscr{C}^0(\mathscr{H}_1 \otimes_A \mathscr{H}_2).$$

Note that the following diagram commutes:

$$\begin{array}{ccc} \mathscr{H}_1 \otimes_A \mathscr{H}_2 & \to & \mathscr{C}^0(\mathscr{H}_1) \otimes_A \mathscr{H}_2 \\ \| & & \downarrow \\ \mathscr{H}_1 \otimes_A \mathscr{H}_2 & \to & \mathscr{C}^0(\mathscr{H}_1 \otimes_A \mathscr{H}_2). \end{array}$$

Suppose we have defined $\mathscr{C}^q(\mathscr{H}_1) \otimes_A \mathscr{H}_2 \to \mathscr{C}^q(\mathscr{H}_1 \otimes_A \mathscr{H}_2)$ for all $q \leq n$ such that the following diagram commute:

$$\begin{array}{ccc} \mathscr{C}^{q-1}(\mathscr{H}_1) \otimes_A \mathscr{H}_2 & \to & \mathscr{C}^q(\mathscr{H}_1) \otimes_A \mathscr{H}_2 \\ \downarrow & & \downarrow \\ \mathscr{C}^{q-1}(\mathscr{H}_1 \otimes_A \mathscr{H}_2) & \to & \mathscr{C}^q(\mathscr{H}_1 \otimes_A \mathscr{H}_2). \end{array}$$

These morphisms induce a morphism

$$\begin{aligned} &\operatorname{coker}(\mathscr{C}^{n-1}(\mathscr{H}_1) \to \mathscr{C}^n(\mathscr{H}_1)) \otimes_A \mathscr{H}_2 \\ &\to \operatorname{coker}(\mathscr{C}^{n-1}(\mathscr{H}_1 \otimes_A \mathscr{H}_2) \to \mathscr{C}^n(\mathscr{H}_1 \otimes_A \mathscr{H}_2)). \end{aligned}$$

We define $\mathscr{C}^{n+1}(\mathscr{H}_1) \otimes_A \mathscr{H}_2 \to \mathscr{C}^{n+1}(\mathscr{H}_1 \otimes_A \mathscr{H}_2)$ as the composite of the following morphisms:

$$\begin{aligned} \mathscr{C}^{n+1}(\mathscr{H}_1) \otimes_A \mathscr{H}_2 &= \mathscr{C}^0(\operatorname{coker}(\mathscr{C}^{n-1}(\mathscr{H}_1) \to \mathscr{C}^n(\mathscr{H}_1))) \otimes_A \mathscr{H}_2 \\ &\to \mathscr{C}^0(\operatorname{coker}(\mathscr{C}^{n-1}(\mathscr{H}_1) \to \mathscr{C}^n(\mathscr{H}_1)) \otimes_A \mathscr{H}_2) \\ &\to \mathscr{C}^0(\operatorname{coker}(\mathscr{C}^{n-1}(\mathscr{H}_1 \otimes_A \mathscr{H}_2) \to \mathscr{C}^n(\mathscr{H}_1 \otimes_A \mathscr{H}_2))) \\ &= \mathscr{C}^{n+1}(\mathscr{H}_1 \otimes_A \mathscr{H}_2). \end{aligned}$$

For any sheaf of A-modules $\mathscr{F}$ on X and any sheaf of A-modules $\mathscr{G}$ on Y, write $\mathscr{F} = \varinjlim_i \mathscr{F}_i$ and $\mathscr{G} = \varinjlim_j \mathscr{G}_j$, where $\mathscr{F}_i$ and $\mathscr{G}_j$ are constructible sheaves of A-modules. We define a morphism of complexes

$$f_! \mathscr{F} \otimes_A \mathscr{G} \to f_!(\mathscr{F} \otimes_A f^* \mathscr{G})$$

as the composite of the following morphisms:

$$\begin{aligned} f_! \mathscr{F} \otimes_A \mathscr{G} &\cong \bar{f}_* \tau_{\leq 2d}(\varinjlim_i \mathscr{C}^\cdot(j_! \mathscr{F}_i)) \otimes_A \varinjlim_j \mathscr{G}_j \\ &\to \bar{f}_* \left(\tau_{\leq 2d}(\varinjlim_i \mathscr{C}^\cdot(j_! \mathscr{F}_i)) \otimes_A \bar{f}^*(\varinjlim_j \mathscr{G}_j) \right) \\ &\to \bar{f}_* \tau_{\leq 2d} \left(\varinjlim_i \mathscr{C}^\cdot(j_! \mathscr{F}_i) \otimes_A \bar{f}^*(\varinjlim_j \mathscr{G}_j) \right) \\ &\cong \bar{f}_* \tau_{\leq 2d} \varinjlim_{i,j} (\mathscr{C}^\cdot(j_! \mathscr{F}_i) \otimes_A \bar{f}^* \mathscr{G}_j) \\ &\to \bar{f}_* \tau_{\leq 2d} \varinjlim_{i,j} \mathscr{C}^\cdot(j_! \mathscr{F}_i \otimes_A \bar{f}^* \mathscr{G}_j) \\ &\to \bar{f}_* \tau_{\leq 2d} \varinjlim_{i,j} \mathscr{C}^\cdot(j_!(\mathscr{F}_i \otimes_A f^* \mathscr{G}_j)) \\ &\cong f_!(\mathscr{F} \otimes_A f^* \mathscr{G}), \end{aligned}$$

where d is an upper bound for the dimensions of fibers of $\bar{f}$. We then use this morphism to define $f_! M \otimes_A K \to f_! (M \otimes_A f^* K)$.

For any q and any sheaf of A-modules $\mathscr{H}$ on S, the morphism

$$f_!^q \mathscr{F} \otimes_A \mathscr{G} \to f_!^q (\mathscr{F} \otimes_A f^* \mathscr{G})$$

induces a homomorphism

$$\mathrm{Hom}(f_!^q(\mathscr{F} \otimes f^*\mathscr{G}), \mathscr{H}) \to \mathrm{Hom}(f_!^q \mathscr{F} \otimes \mathscr{G}, \mathscr{H}).$$

We have

$$\begin{aligned} \mathrm{Hom}(f_!^q(\mathscr{F} \otimes_A f^*\mathscr{G}), \mathscr{H}) &\cong \mathrm{Hom}(\mathscr{F} \otimes_A f^*\mathscr{G}, f^!_{-q}\mathscr{H}) \\ &\cong \mathrm{Hom}(\mathscr{F}, \mathscr{H}om(f^*\mathscr{G}, f^!_{-q}\mathscr{H})), \\ \mathrm{Hom}(f_!^q \mathscr{F} \otimes_A \mathscr{G}, \mathscr{H}) &\cong \mathrm{Hom}(f_!^q \mathscr{F}, \mathscr{H}om(\mathscr{G}, \mathscr{H})) \\ &\cong \mathrm{Hom}(\mathscr{F}, f^!_{-q}\mathscr{H}om(\mathscr{G}, \mathscr{H})). \end{aligned}$$

So we have a homomorphism

$$\mathrm{Hom}(\mathscr{F}, \mathscr{H}om(f^*\mathscr{G}, f^!_{-q}\mathscr{H})) \to \mathrm{Hom}(\mathscr{F}, f^!_{-q}\mathscr{H}om(\mathscr{G}, \mathscr{H})).$$

It gives rise to a morphism

$$\mathscr{H}om(f^*\mathscr{G}, f^!_{-q}\mathscr{H}) \to f^!_{-q}\mathscr{H}om(\mathscr{G}, \mathscr{H}).$$

From this, we can define a morphism of complexes

$$\mathscr{H}om^\cdot(f^*K, f^!_\cdot L) \to f^!_\cdot \mathscr{H}om^\cdot(K, L).$$

Represent K by a bounded above complex of flat sheaves, and L by a bounded below complex of injective sheaves. We have

$$\begin{aligned} Rf_! M \otimes_A^L K &\cong f_! M \otimes_A K, \\ Rf_!(M \otimes_A^L f^*K) &\cong f_! (M \otimes_A f^*K), \\ R\mathscr{H}om(f^*K, Rf^!L) &\cong \mathscr{H}om^\cdot(f^*K, f^!_\cdot L), \\ Rf^! R\mathscr{H}om(K, L) &\cong f^!_\cdot \mathscr{H}om^\cdot(K, L). \end{aligned}$$

We thus get a morphism

$$R\mathscr{H}om(f^*K, Rf^!L) \to Rf^! R\mathscr{H}om(K, L)$$

in the derived category. The following diagram commutes:

$$\begin{array}{ccc} \mathrm{Hom}(Rf_! M \otimes_A^L K, L) & \cong & \mathrm{Hom}(M, Rf^! R\mathscr{H}om(K, L)) \\ \wr\| & & \wr\| \\ H^0(\mathrm{Hom}^\cdot(f_! M \otimes_A K, L)) & \cong & H^0(\mathrm{Hom}^\cdot(M, f^!_\cdot \mathscr{H}om^\cdot(K, L))) \\ \uparrow & & \uparrow \\ H^0(\mathrm{Hom}^\cdot(f_!(M \otimes_A f^*K), L)) & \cong & H^0(\mathrm{Hom}^\cdot(M, \mathscr{H}om^\cdot(f^*K, f^!_\cdot L))) \\ \wr\| & & \wr\| \\ \mathrm{Hom}(Rf_!(M \otimes_A^L f^*K), L) & \cong & \mathrm{Hom}(M, R\mathscr{H}om(f^*K, Rf^!L)). \end{array}$$

Since $Rf_!M \otimes_A^L K \cong Rf_!(M \otimes_A f^*K)$, the left vertical arrow in the above diagram is bijective. So we have an isomorphism

$$\mathrm{Hom}(M, R\mathscr{H}om(f^*K, Rf^!L)) \stackrel{\cong}{\to} \mathrm{Hom}(M, Rf^!R\mathscr{H}om(K, L)).$$

We need to prove that $R\mathscr{H}om(f^*K, Rf^!L) \to Rf^!R\mathscr{H}om(K, L)$ is an isomorphism. Let

$$R\mathscr{H}om(f^*K, Rf^!L) \to Rf^!R\mathscr{H}om(K, L) \to \Delta \to$$

be a distinguished triangle. It suffices to show that Δ is acyclic. By 6.1.1 (ii), we have

$$\mathrm{Hom}(M, \Delta) = 0$$

for any $M \in \mathrm{ob}\, D^-(X, A)$. Representing Δ by a bounded below complex of injective sheaves and take $M = \mathscr{G}[q]$ for any q and any sheaf of A-modules $\mathscr{G}$ on X, we see that the complex $\mathrm{Hom}^{\cdot}(\mathscr{G}, \Delta)$ is acyclic. By 8.4.8 below, Δ is acyclic. □

Lemma 8.4.8. *Let $\mathscr{F}' \stackrel{d'}{\to} \mathscr{F} \stackrel{d}{\to} \mathscr{F}''$ be morphisms of sheaves. If for any sheaf $\mathscr{G}$, the sequence*

$$\mathrm{Hom}(\mathscr{G}, \mathscr{F}') \to \mathrm{Hom}(\mathscr{G}, \mathscr{F}) \to \mathrm{Hom}(\mathscr{G}, \mathscr{F}'')$$

is exact, then $\mathscr{F}' \stackrel{d'}{\to} \mathscr{F} \stackrel{d}{\to} \mathscr{F}''$ is exact.

Proof. The sequence

$$\mathrm{Hom}(\mathscr{F}', \mathscr{F}') \to \mathrm{Hom}(\mathscr{F}', \mathscr{F}) \to \mathrm{Hom}(\mathscr{F}', \mathscr{F}'')$$

is exact. The morphism $\mathrm{id}_{\mathscr{F}'}$ is mapped to 0 in $\mathrm{Hom}(\mathscr{F}', \mathscr{F}'')$ under the composite of the above homomorphisms. So we have $dd' = 0$. The sequence

$$\mathrm{Hom}(\ker d, \mathscr{F}') \to \mathrm{Hom}(\ker d, \mathscr{F}) \to \mathrm{Hom}(\ker d, \mathscr{F}'')$$

is exact. The inclusion $i : \ker d \hookrightarrow \mathscr{F}$ is mapped to 0 in $\mathrm{Hom}(\ker d, \mathscr{F}'')$. So there exists a morphism $\phi : \ker d \to \mathscr{F}'$ such that $d'\phi = i$. This implies that $\ker d \cong \mathrm{im}\, d'$. □

Proposition 8.4.9. *Let S be a noetherian scheme. Consider a Cartesian diagram*

$$\begin{array}{ccc} X \times_Y Y' & \stackrel{g'}{\to} & X \\ f' \downarrow & & \downarrow f \\ Y' & \stackrel{g}{\to} & Y. \end{array}$$

Suppose that X, Y, and Y' are S-schemes of finite type, and f is S-compactifiable. For any $K \in \mathrm{ob}\, D^+(Y', A)$, there exists a canonical isomorphism

$$Rg'_*Rf'^!K \stackrel{\cong}{\to} Rf^!Rg_*K.$$

Proof. For any $L \in \operatorname{ob} D(X, A)$, we have a canonical morphism

$$g^* Rf_! L \stackrel{\cong}{\to} Rf'_! g'^* L.$$

It induces a one-to-one correspondence

$$\operatorname{Hom}(Rf'_! g'^* L, K) \stackrel{\cong}{\to} \operatorname{Hom}(g^* Rf_! L, K).$$

We have

$$\begin{aligned}\operatorname{Hom}(Rf'_! g'^* L, K) &\cong \operatorname{Hom}(g'^* L, Rf'^! K) \cong \operatorname{Hom}(L, Rg'_* Rf'^! K), \\ \operatorname{Hom}(g^* Rf_! L, K) &\cong \operatorname{Hom}(Rf_! L, Rg_* K) \cong \operatorname{Hom}(L, Rf^! Rg_* K).\end{aligned}$$

So we have a one-to-one correspondence

$$\operatorname{Hom}(L, Rg'_* Rf'^! K) \stackrel{\cong}{\to} \operatorname{Hom}(L, Rf^! Rg_* K).$$

It gives rise to an isomorphism

$$Rg'_* Rf'^! K \stackrel{\cong}{\to} Rf^! Rg_* K.$$

□

Corollary 8.4.10. *Let S be a noetherian scheme. Consider a Cartesian diagram*

$$\begin{array}{ccc} X \times_Y Y' & \stackrel{g'}{\to} & X \\ {\scriptstyle f'}\downarrow & & \downarrow{\scriptstyle f} \\ Y' & \stackrel{g}{\to} & Y. \end{array}$$

Suppose that X, Y, and Y' are S-schemes of finite type, and f is S-compactifiable. For any $K \in \operatorname{ob} D^-(Y', A)$ and $L \in \operatorname{ob} D^+(Y', A)$, we have a canonical isomorphism

$$Rg'_* R\mathscr{H}om(f'^* K, Rf'^! L) \stackrel{\cong}{\to} Rf^! Rg_* R\mathscr{H}om(K, L).$$

Proof. Follows from 8.4.7 and 8.4.9. □

Lemma 8.4.11. *Let X be a scheme and let A be a ring. Suppose $\mathscr{F} = \bigoplus_j j_! A$, where j goes over a set of etale morphisms $j : U \to X$. Then for any flasque sheaf of A-modules $\mathscr{G}$ on X, we have $\operatorname{Ext}^q_A(\mathscr{F}, \mathscr{G}) = 0$ and $\mathscr{E}xt^q_A(\mathscr{F}, \mathscr{G}) = 0$ for all $q \geq 1$.*

Proof. Since $\mathscr{E}xt^q_A(\mathscr{F}, \mathscr{G})$ is the sheaf associated to the presheaf $V \mapsto \operatorname{Ext}^q_A(\mathscr{F}|_V, \mathscr{G}|_V)$, it suffices to treat Ext. Let $\mathscr{I}^\cdot$ be a resolution of $\mathscr{G}$ by injective sheaves of A-modules. For any $q \geq 1$, we have

$$\begin{aligned}\mathrm{Ext}^q_A(\mathscr{F},\mathscr{G}) &\cong H^q\Big(\mathrm{Hom}_A(\bigoplus_j j_! A, \mathscr{I}^{\cdot})\Big)\\ &\cong H^q\Big(\prod_j \mathrm{Hom}_A(j_! A, \mathscr{I}^{\cdot})\Big)\\ &\cong H^q\Big(\prod_j \Gamma(U, \mathscr{I}^{\cdot})\Big)\\ &\cong \prod_j H^q(U,\mathscr{G})\\ &= 0.\end{aligned}$$

□

Let X be a scheme, let A be a ring, and let B be an A-algebra. Consider the functor

$$\rho : K(X,B) \to K(X,A)$$

that maps each complex of sheaves of B-modules on X to the same complex considered as a complex of sheaves of A-modules on X. Then ρ transforms quasi-isomorphisms to quasi-isomorphisms. So it induces a functor

$$\rho : D(X,B) \to D(X,A)$$

on the derived categories.

Lemma 8.4.12. *Let X be a scheme, A a ring, B an A-algebra, and $\rho : D(X,B) \to D(X,A)$ the functor defined above. For any $K \in \mathrm{ob}\, D^-(X,A)$ and $L \in \mathrm{ob}\, D^+(X,B)$, we have*

$$\mathrm{Hom}_{D(X,B)}(K \otimes^L_A B, L) \cong \mathrm{Hom}_{D(X,A)}(K, \rho(L)).$$

Proof. We may assume that L is a bounded below complex of injective sheaves of B-modules, and K is a bounded above complex whose components are of the form $\bigoplus_j j_! A$, where j goes over a set of etale morphisms to X. We then have

$$\begin{aligned}\mathrm{Hom}_{D(X,B)}(K \otimes^L_A B, L) &\cong H^0(\mathrm{Hom}^{\cdot}(K \otimes_A B, L))\\ &\cong H^0(\mathrm{Hom}^{\cdot}(K, \rho(L))).\end{aligned}$$

To prove our assertion, it suffices to show

$$H^0(\mathrm{Hom}^{\cdot}(K, \rho(L))) \cong \mathrm{Hom}_{D(X,A)}(K, \rho(L)).$$

Let $I^{\cdot\cdot}$ be a Cartan–Eilenberg resolution of $\rho(L)$. For each q, $I^{\cdot q}$ is a resolution of $\rho(L^q)$ by injective sheaves of A-modules. By 8.4.11, we have

$$\mathrm{Ext}^v(K^{-u}, \rho(L^q)) = 0$$

for all $v \geq 1$ and all u, q. It follows that

$$H^v(\mathrm{Hom}(K^{-u}, I^{\cdot q})) = \begin{cases} \mathrm{Hom}(K^{-u}, \rho(L^q)) & \text{if } v = 0, \\ 0 & \text{if } v \neq 0. \end{cases}$$

So the biregular spectral sequence

$$E_1^{uv} = H^v(\mathrm{Hom}(K^{-u}, I^{\cdot q})) \Rightarrow H^{u+v}(\mathrm{Hom}(K^{\cdot}, I^{\cdot q}))$$

degenerates at the level $E_1^{\cdot\cdot}$, and we have

$$H^u(\mathrm{Hom}(K^{\cdot}, \rho(L^q))) \cong H^u(\mathrm{Hom}(K^{\cdot}, I^{\cdot q})).$$

Consider the bicomplexes $C_1^{\cdot\cdot}$ and $C_2^{\cdot\cdot}$ defined by

$$\begin{aligned} C_1^{pq} &= \mathrm{Hom}(K^{-p}, \rho(L^q)), \\ C_2^{pq} &= \bigoplus_{u+v=p} \mathrm{Hom}(K^{-u}, I^{vq}). \end{aligned}$$

We have a morphism of bicomplexes $C_1^{\cdot\cdot} \to C_2^{\cdot\cdot}$. It induces a morphism between the following two spectral sequences

$$\begin{aligned} E_1^{pq} &= H^q(C_1^{\cdot p}) \Rightarrow H^{p+q}(C_1^{\cdot\cdot}), \\ E_1^{pq} &= H^q(C_2^{\cdot p}) \Rightarrow H^{p+q}(C_2^{\cdot\cdot}). \end{aligned}$$

We have just shown that

$$H^q(C_1^{\cdot p}) \cong H^q(C_2^{\cdot p}).$$

It follows that

$$H^p(C_1^{\cdot\cdot}) \cong H^p(C_2^{\cdot\cdot})$$

for all p. We have

$$\begin{aligned} H^0(C_1^{\cdot\cdot}) &= H^0(\mathrm{Hom}^{\cdot}(K, \rho(L))), \\ H^0(C_2^{\cdot\cdot}) &= \mathrm{Hom}_{D(X,A)}(K, \rho(L)). \end{aligned}$$

We thus have

$$H^0(\mathrm{Hom}^{\cdot}(K, \rho(L))) \cong \mathrm{Hom}_{D(X,A)}(K, \rho(L)).$$

□

Proposition 8.4.13. *Let S be a noetherian ring, $f : X \to Y$ an S-compactifiable morphism between S-schemes of finite type, A a noetherian ring, B a noetherian A-algebra, $L \in \operatorname{ob} D^+(Y, B)$, $\rho : D(X, B) \to D(X, A)$ and $\rho : D(Y, B) \to D(Y, A)$ the functors defined above. Then we have a canonical isomorphism*

$$\rho Rf^! L \stackrel{\cong}{\to} Rf^! \rho L.$$

Proof. By 7.4.7, we have

$$Rf_! K \otimes_A^L B \cong Rf_!(K \otimes_A^L B)$$

for any $K \in \operatorname{ob} D^-(X, A)$. Combined with 8.4.12, we have

$$\begin{aligned} \mathrm{Hom}_{D(X,A)}(K, \rho Rf^! L) &\cong \mathrm{Hom}_{D(X,B)}(K \otimes_A^L B, Rf^! L) \\ &\cong \mathrm{Hom}_{D(Y,B)}(Rf_!(K \otimes_A^L B), L) \\ &\cong \mathrm{Hom}_{D(Y,B)}(Rf_! K \otimes_A^L B, L) \\ &\cong \mathrm{Hom}_{D(Y,A)}(Rf_! K, \rho L) \\ &\cong \mathrm{Hom}_{D(X,A)}(K, Rf^! \rho L). \end{aligned}$$

It follows that we have an isomorphism $\rho Rf^! L \cong Rf^! \rho L$.

More explicitly, we construct the isomorphism as follows. As in the proof of 8.4.7, we can construct a morphism of complexes

$$f_!^{\cdot} \mathscr{F} \otimes_A B \to f_!^{\cdot}(\mathscr{F} \otimes_A B)$$

for any sheaf of A-modules $\mathscr{F}$ on X. We then use this morphism to define

$$f_!^{\cdot} K \otimes B \to f_!^{\cdot}(K \otimes B)$$

for any complex K of sheaves of A-modules on X. Moreover, for any q and any sheaf of B-modules $\mathscr{H}$ on Y, the morphism

$$f_!^q \mathscr{F} \otimes_A B \to f_!^q(\mathscr{F} \otimes_A B)$$

induces a homomorphism

$$\mathrm{Hom}(f_!^q(\mathscr{F} \otimes_A B), \mathscr{H}) \to \mathrm{Hom}(f_!^q \mathscr{F} \otimes_A B, \mathscr{H}).$$

We have

$$\begin{aligned} \mathrm{Hom}(f_!^q(\mathscr{F} \otimes_A B), \mathscr{H}) &\cong \mathrm{Hom}(\mathscr{F} \otimes_A B, f_{-q}^! \mathscr{H}) \\ &\cong \mathrm{Hom}(\mathscr{F}, \rho f_{-q}^! \mathscr{H}), \\ \mathrm{Hom}(f_!^q \mathscr{F} \otimes_A B, \mathscr{H}) &\cong \mathrm{Hom}(f_!^q \mathscr{F}, \rho \mathscr{H}) \\ &\cong \mathrm{Hom}(\mathscr{F}, f_{-q}^! \rho \mathscr{H}). \end{aligned}$$

So we have a homomorphism

$$\mathrm{Hom}(\mathscr{F}, \rho f^{!}_{-q}\mathscr{H}) \to \mathrm{Hom}(\mathscr{F}, f^{!}_{-q}\rho\mathscr{H}).$$

It gives rise to a morphism

$$\rho f^{!}_{-q}\mathscr{H} \to f^{!}_{-q}\rho\mathscr{H}.$$

For any $L \in \mathrm{ob}\, D^{+}(Y, B)$, let $L \to I$ and $\rho I \to J$ be quasi-isomorphisms such that I (resp. J) is a bounded below complex of injective sheaves of B-modules (resp. A-modules) on Y. We define a morphism

$$\rho R f^{!} L \to R f^{!} \rho L$$

in the derived category to be the composite

$$\rho R f^{!} L \cong \rho f^{!}_{\cdot} I \to f^{!}_{\cdot}\rho I \to f^{!}_{\cdot} J \cong R f^{!} \rho L.$$

Represent K by a bounded above complex whose components are of the form $\bigoplus_j j_! A$, where j goes over a set of etale morphisms to X. The following diagram commutes:

$$\begin{array}{ccc}
\mathrm{Hom}(Rf_!K \otimes^{L}_{A} B, L) & \cong & \mathrm{Hom}(K, Rf^{!}\rho L) \\
\uparrow & & \uparrow \\
H^0(\mathrm{Hom}^{\cdot}(f_{!}^{\cdot} K \otimes_A B, I)) & \cong & H^0(\mathrm{Hom}^{\cdot}(K, f^{!}_{\cdot}\rho I)) \\
\uparrow & & \uparrow \\
H^0(\mathrm{Hom}^{\cdot}(f_{!}^{\cdot}(K \otimes_A B), I)) & \cong & H^0(\mathrm{Hom}^{\cdot}(K, \rho f^{!}_{\cdot} I)) \\
\wr\| & & \wr\| \\
\mathrm{Hom}(Rf_!(K \otimes^{L}_{A} B), L) & \cong & \mathrm{Hom}(K, \rho R f^{!} L).
\end{array}$$

Since $Rf_!K \otimes^{L}_{A} B \cong Rf_!(K \otimes_A B)$, the composite of vertical arrows on the left in the above diagram is bijective. So the morphism $\rho R f^{!} L \to R f^{!} \rho L$ induces a bijection

$$\mathrm{Hom}(K, \rho R f^{!} L) \stackrel{\cong}{\to} \mathrm{Hom}(K, R f^{!} \rho L)$$

for any $K \in \mathrm{ob}\, D^{-}(X, A)$. As in the proof of 8.4.7, this implies that $\rho R f^{!} L \cong R f^{!} \rho L$. □

8.5 Poincaré Duality

([SGA 4] XVIII 3.2.)

Let S be a noetherian scheme, n an integer invertible on S, A a noetherian ring with $nA = 0$, and $f : X \to Y$ a smooth S-compactifiable morphism pure of relative dimension d between S-schemes of finite type. Fix a compactification

$$X \hookrightarrow \overline{X} \xrightarrow{\bar{f}} Y$$

of f. Consider a commutative diagram

$$\begin{array}{ccc} U & \xrightarrow{k} & X \\ {\scriptstyle g}\downarrow & & \downarrow{\scriptstyle f} \\ V & \xrightarrow{j} & Y \end{array}$$

such that j and k are etale. We have morphisms of complexes

$$\begin{array}{rcl} f_!(k_!\mathbb{Z}/n) & \to & \mathscr{H}^{2d}(f_!(k_!\mathbb{Z}/n))[-2d] \\ & \cong & R^{2d}f_!k_!\mathbb{Z}/n[-2d] \\ & \cong & j_!R^{2d}g_!\mathbb{Z}/n[-2d] \\ & \stackrel{j_!(\mathrm{Tr}_g)}{\to} & j_!\mathbb{Z}/n(-d)[-2d]. \end{array}$$

For any bounded below complex of sheaves of A-modules K on Y, the composite of the above morphisms induces a homomorphism

$$\mathrm{Hom}(j_!\mathbb{Z}/n(-d)[-2d], K) \to \mathrm{Hom}(f_!k_!\mathbb{Z}/n, K).$$

We thus get a morphism of complexes

$$K(V)(d)[2d] \to (f^!K)(U).$$

It induces a morphism

$$t_f : f^*K(d)[2d] \to Rf^!K$$

in the derived category $D(X, A)$.

Lemma 8.5.1. *Notations as above. The following diagram commutes:*

$$\begin{array}{ccc} Rf_!f^*K(d)[2d] & \stackrel{Rf_!(t_f)}{\to} & Rf_!Rf^!K. \\ {\scriptstyle \mathrm{Tr}_f}\searrow & & \swarrow{\scriptstyle \mathrm{adj}} \\ & K & \end{array}$$

Proof. Represent K by a bounded below complex of injective sheaves of A-modules on S. Consider a Cartesian diagram

$$\begin{array}{ccc} U = V \times_Y X & \xrightarrow{j'} & X \\ {\scriptstyle f'}\downarrow & & \downarrow{\scriptstyle f} \\ V & \xrightarrow{j} & Y, \end{array}$$

where j is etale. By 8.2.4 (i), the following diagram commutes:

$$\begin{array}{ccccccc} R^{2d}f_!f^*j_!\mathbb{Z}/n(d) & \stackrel{\cong}{\to} & R^{2d}f_!\mathbb{Z}/n(d)\otimes_{\mathbb{Z}/n} j_!\mathbb{Z}/n & \stackrel{\cong}{\to} & j_!j^*R^{2d}f_!\mathbb{Z}/n(d) & \stackrel{\cong}{\to} & j_!R^{2d}f'_!\mathbb{Z}/n(d) \\ \mathrm{Tr}_f\downarrow & & \mathrm{Tr}_f\otimes\mathrm{id}\downarrow & & j_!j^*(\mathrm{Tr}_f)\downarrow & & j_!(\mathrm{Tr}_{f'})\downarrow \\ j_!\mathbb{Z}/n & \stackrel{\cong}{\to} & \mathbb{Z}/n\otimes_{\mathbb{Z}/n} j_!\mathbb{Z}/n & \stackrel{\cong}{\to} & j_!j^*\mathbb{Z}/n(d) & \stackrel{\cong}{\to} & j_!\mathbb{Z}/n. \end{array}$$

It follows that the top square in the following diagram commutes:

$$\begin{array}{ccc} \mathrm{Hom}^\cdot(j_!\mathbb{Z}/n, K) & \stackrel{\mathrm{Tr}_f}{\to} & \mathrm{Hom}^\cdot(f_!f^*j_!\mathbb{Z}/n(d)[2d], K) \\ \| & & \| \\ \mathrm{Hom}^\cdot(j_!\mathbb{Z}/n, K) & \stackrel{\mathrm{Tr}_{f'}}{\to} & \mathrm{Hom}^\cdot(f_!f^*j_!\mathbb{Z}/n(d)[2d], K) \\ \downarrow & & \wr\| \\ \mathrm{Hom}^\cdot(f^*j_!\mathbb{Z}/n, f^*K) & \stackrel{t_f}{\to} & \mathrm{Hom}^\cdot(f^*j_!\mathbb{Z}/n(d)[2d], f^!K), \end{array}$$

where the first horizontal arrow is induced by the composite

$$f_!f^*j_!\mathbb{Z}/n(d)[2d] \to R^{2d}f_!f^*j_!\mathbb{Z}/n(d) \stackrel{\mathrm{Tr}_f}{\to} j_!\mathbb{Z}/n,$$

and the second horizontal arrow is induced by the composite

$$f_!f^*j_!\mathbb{Z}/n(d)[2d] \to R^{2d}f_!f^*j_!\mathbb{Z}/n(d) \cong j_!R^{2d}f'_!\mathbb{Z}/n(d) \stackrel{j_!(\mathrm{Tr}_{f'})}{\to} j_!\mathbb{Z}/n.$$

One can verify that the bottom square in the above diagram commutes. So we have a commutative diagram

$$\begin{array}{ccc} \mathrm{Hom}^\cdot(j_!\mathbb{Z}/n, K) & \stackrel{\mathrm{Tr}_f}{\to} & \mathrm{Hom}^\cdot(f_!f^*j_!\mathbb{Z}/n(d)[2d], K) \\ \downarrow & & \downarrow \\ \mathrm{Hom}^\cdot(f^*j_!\mathbb{Z}/n, f^*K) & \stackrel{t_f}{\to} & \mathrm{Hom}^\cdot(f^*j_!\mathbb{Z}/n(d)[2d], f^!K). \end{array}$$

This is true for any etale Y-scheme $j : V \to Y$. One deduces from this first for any sheaf of A-modules L on Y, and then for any complex of sheaves of A-modules L on Y, that the following diagram commutes:

$$\begin{array}{ccc} \mathrm{Hom}^\cdot(L, K) & \stackrel{\mathrm{Tr}_f}{\to} & \mathrm{Hom}^\cdot(f_!f^*L(d)[2d], K) \\ \downarrow & & \downarrow \\ \mathrm{Hom}^\cdot(f^*L, f^*K) & \stackrel{t_f}{\to} & \mathrm{Hom}^\cdot(f^*L(d)[2d], f^!K). \end{array}$$

Taking $L = K$ and calculating the images of id $\in \mathrm{Hom}(L, K)$ under morphisms in the above commutative diagram, we see that Tr_f corresponds to t_f by the adjointness of $(Rf_!, Rf^!)$. □

The main result of this section is the following theorem:

Theorem 8.5.2. *Let S be a noetherian scheme, n an integer invertible on S, A a noetherian ring such that $nA = 0$, and $f : X \to Y$ a smooth S-compactifiable morphism pure of relative dimension d between S-schemes of finite type. Then the morphism*

$$t_f : f^*L(d)[2d] \to Rf^!L$$

is an isomorphism for any $L \in \mathrm{ob}\, D^+(Y, A)$. For any $K \in \mathrm{ob}\, D(X, A)$, the morphism

$$\mathrm{Tr}_f : Rf_!f^*L(d)[2d] \to L$$

induces a one-to-one correspondence

$$\mathrm{Hom}(K, f^*L(d)[2d]) \stackrel{\cong}{\to} \mathrm{Hom}(Rf_!K, L), \quad \phi \mapsto \mathrm{Tr}_f \circ Rf_!(\phi).$$

Before proving the theorem, we deduce some of its consequences.

Corollary 8.5.3 (Poincaré Duality). *Let X be a smooth compactifiable scheme pure of dimension d over a separably closed field, let n be an integer invertible on X, and let A be a noetherian ring such that $nA = 0$ and A is an injective A-module. For any sheaf of A-modules $\mathscr{F}$ on X, we have an isomorphism*

$$\mathrm{Ext}_A^{2d-q}(\mathscr{F}, A(d)) \cong \mathrm{Hom}_A(H_c^q(X, \mathscr{F}), A)$$

for any q. If $\mathscr{F}$ is locally constant, then we have an isomorphism

$$H^{2d-q}(X, \mathscr{H}om_A(\mathscr{F}, A(d))) \cong \mathrm{Hom}_A(H_c^q(X, \mathscr{F}), A).$$

Proof. We have

$$\begin{aligned}&\mathrm{Hom}(\mathscr{F}, f^*A(d)[2d-i]) \cong \mathrm{Ext}_A^{2d-i}(\mathscr{F}, A(d)),\\ &\mathrm{Hom}(Rf_!\mathscr{F}, A[-i]) = \mathrm{Ext}_A^{-i}(R\Gamma_c(X, \mathscr{F}), A) \cong \mathrm{Hom}_A(H_c^i(X, \mathscr{F}), A).\end{aligned}$$

Here for the second equation, we use the assumption that A is an injective A-module. By 8.5.2, we have

$$\mathrm{Hom}(\mathscr{F}, f^*A(d)[2d-i]) \cong \mathrm{Hom}(Rf_!\mathscr{F}, A[-i]).$$

Our assertion follows. If $\mathscr{F}$ is locally constant, then by 8.3.3, we have

$$\mathscr{E}xt_A^q(\mathscr{F}, A) = 0$$

for any $q \geq 1$. It follows that

$$\mathrm{Ext}_A^q(\mathscr{F}, A) \cong H^q(X, \mathscr{H}om_A(\mathscr{F}, A)).$$

□

Lemma 8.5.4. *Let A be a ring. For any A-module M, let $M^\vee = \mathrm{Hom}_A(M,A)$. If A is an injective A-module and M has finite presentation, then the canonical homomorphism $M \to M^{\vee\vee}$ is an isomorphism.*

Proof. Since A is an injective A-module, the functor $M \to M^\vee$ is an exact functor. Suppose M has finite presentation. Then we can find an exact sequence

$$A^m \to A^n \to M \to 0.$$

The following diagram commutes:

$$\begin{array}{ccccccc} A^m & \to & A^n & \to & M & \to 0 \\ \downarrow & & \downarrow & & \downarrow & \\ (A^m)^{\vee\vee} & \to & (A^n)^{\vee\vee} & \to & M^{\vee\vee} & \to 0. \end{array}$$

The first two vertical arrows are isomorphisms. By the five lemma, $M \to M^{\vee\vee}$ is an isomorphism. $\square$

Proposition 8.5.5 (Weak Lefschetz Theorem). *Let k be a separably closed field, n an integer relatively prime to the characteristic of k, A a noetherian ring such that $nA = 0$ and A is an injective A-module, X a closed subscheme of $\mathbb{P}^N_k$, and H a hyperplane of $\mathbb{P}^N_k$. Suppose that $X - X\cap H$ is smooth. For any sheaf of A-modules $\mathscr{F}$ on X such that $\mathscr{F}|_{X-X\cap H}$ is locally constant and constructible, the canonical homomorphism*

$$H^q(X,\mathscr{F}) \to H^q(X\cap H, \mathscr{F}|_{X\cap H})$$

is bijective for $i < \dim X - 1$ and injective for $i = \dim X - 1$.

Proof. By 7.5.2, we have

$$H^q(X - X\cap H, \mathscr{H}om_A(\mathscr{F}, A(d))) = 0$$

for any $q > \dim X$. By 8.5.3 and 8.5.4, we have $H^q_c(X - X\cap H, \mathscr{F}) = 0$ for any $q < \dim X$. Our assertion follows from the long exact sequence

$$\cdots \to H^i_c(X - X\cap H, \mathscr{F}) \to H^i(X,\mathscr{F}) \to H^i(X\cap H, \mathscr{F}|_{X\cap H}) \to \cdots. \square$$

Corollary 8.5.6 (Relative Purity Theorem). *Consider a commutative diagram*

$$\begin{array}{ccccc} U & \stackrel{j}{\hookrightarrow} & X & \stackrel{i}{\hookleftarrow} & Z, \\ & {\scriptstyle f|_U}\searrow & \downarrow {\scriptstyle f} & \swarrow {\scriptstyle f|_Z} & \\ & & Y & & \end{array}$$

where i is a closed immersion and $U = X - Z$. Let $c = \mathrm{codim}(Z, X)$, let A be a noetherian ring such that $nA = 0$ for some integer n invertible on S, and let $\mathscr{F}$ be a sheaf of A-modules on X. Suppose that S is noetherian, f and $f|Z$ are smooth S-compactifiable morphisms between S-schemes of finite type, $c \geq 1$, and locally with respect to the etale topology, and $\mathscr{F}$ is isomorphic to $f^*\mathscr{G}$ for some sheaf of A-modules $\mathscr{G}$ on Y. Then we have

$$R^q i^! \mathscr{F} \cong \begin{cases} i^*\mathscr{F}(-c) & \text{if } q = 2c, \\ 0 & \text{if } q \neq 2c, \end{cases}$$

$$R^q j_* j^* \mathscr{F} \cong \begin{cases} \mathscr{F} & \text{if } q = 0, \\ i_* i^* \mathscr{F}(-c) & \text{if } q = 2c-1, \\ 0 & \text{if } q \neq 0, 2c-1, \end{cases}$$

$$R^q f_* \mathscr{F} \cong R^q (f|_U)_* (j^* \mathscr{F}) \quad \text{if } 0 \leq q \leq 2c-2$$

and we have a long exact sequence

$$\begin{aligned} 0 \to R^{2c-1} f_* \mathscr{F} \to R^{2c-1}(f|_U)_* j^* \mathscr{F} \to \quad (f|_Z)_* i^* \mathscr{F}(-c) \quad &\to \cdots \\ \to R^{q-1} f_* \mathscr{F} \to R^{q-1}(f|_U)_* j^* \mathscr{F} \to R^{q-2c}(f|_Z)_* i^* \mathscr{F}(-c) &\to \cdots . \end{aligned}$$

Proof. The problem is local with respect to the etale topology on X. We may assume $\mathscr{F} = f^*\mathscr{G}$. We have

$$Rf_! Ri_! \cong R(f|_Z)_!.$$

It follows that

$$R(f|_Z)^! \cong Ri^! Rf^!.$$

Let d_1 and d_2 be the relative dimensions of f and $f|_Z$, respectively. We have $c = d_1 - d_2$, and by 8.5.2, we have

$$Rf^! \mathscr{G} \cong f^* \mathscr{G}(d_1)[2d_1], \quad R(f|_Z)^! \mathscr{G} \cong (f|_Z)^* \mathscr{G}(d_2)[2d_2].$$

So we have

$$Ri^! f^* \mathscr{G}(d_1)[2d_1] \cong (f|_Z)^* \mathscr{G}(d_2)[2d_2].$$

Hence we have an isomorphism

$$Ri^! \mathscr{F} \cong i^* \mathscr{F}(-c)[-2c].$$

We have an exact sequence

$$0 \to i_* i^! \mathscr{F} \to \mathscr{F} \to j_* j^* \mathscr{F} \to i_* R^1 i^! \mathscr{F} \to 0$$

and $R^q j_* j^* \mathscr{F} = i_* R^{q+1} i^! \mathscr{F}$ for any $q \geq 1$. This implies the equations in the corollary. The long exact sequence comes from the spectral sequence

$$E_2^{pq} = R^p f_* R^q j_* (j^* \mathscr{F}) \Rightarrow R^{p+q} (f|_U)_* j^* \mathscr{F}.$$

□

To prove 8.5.2, we need the following lemma.

Lemma 8.5.7. *Let S be a noetherian scheme, let $f : X \to Y$ be a smooth S-compactifiable morphism pure of relative dimension d, and let n be an integer invertible on S. Then for any $x \in X$, there exists an etale neighborhood V of $f(\bar{x})$ in Y and an etale neighborhood U of $\bar{x}$ in $X \times_Y V$ such that the canonical morphism*

$$R^q f'_{V!}\mathbb{Z}/n \to R^q f_{V!}\mathbb{Z}/n$$

vanishes for any $q < 2d$ and the restriction of

$$\mathrm{Tr}_{f_V} : R^{2d} f_{V!}\mathbb{Z}/n(d) \to \mathbb{Z}/n$$

to the image of the canonical morphism

$$R^{2d} f'_{V!}\mathbb{Z}/n \to R^{2d} f_{V!}\mathbb{Z}/n$$

is an isomorphism, where the notation is given by the following commutative diagram:

$$\begin{array}{ccccc} U & \xrightarrow{j} & X \times_Y V & \to & X \\ & {\scriptstyle f'_V}\searrow & {\scriptstyle f_V}\downarrow & & \downarrow{\scriptstyle f} \\ & & V & \to & Y. \end{array}$$

We give the proof of 8.5.7 after the proof of 8.5.2.

Proof of 8.5.2. It suffices to show that t_f is an isomorphism. By 8.4.13, we may assume $A = \mathbb{Z}/n$. For any $x \in X$, let $\mathscr{S}$ be the category defined as follows: Objects in $\mathscr{S}$ are triples (U, V, g), where $k : U \to X$ is an etale neighborhood of $\bar{x}$, $j : V \to Y$ is an etale neighborhood of $f(\bar{x})$, and $g : U \to V$ is a morphism satisfying $fk = jg$; morphisms in $\mathscr{S}$ from an object (U, V, g) to an object (U', V', g') is a pair (ϕ, ψ) such that $\phi : U \to U'$ is an X-morphism, $\psi : V \to V'$ is a Y-morphism, and $g'\phi = \psi g$. For any $K \in \mathrm{ob}\, D^+(Y, \mathbb{Z}/n)$, we have

$$\mathscr{H}^q(f^*K(d)[2d])_{\bar{x}} \cong \varinjlim_{(U,V,g)\in \mathrm{ob}\,\mathscr{S}^\circ} \mathrm{Ext}^q(j_!\mathbb{Z}/n(-d)[-2d], K),$$

$$\mathscr{H}^q(Rf^!K)_{\bar{x}} \cong \varinjlim_{(U,V,g)\in \mathrm{ob}\,\mathscr{S}^\circ} \mathrm{Ext}^q(f_! k_!\mathbb{Z}/n, K).$$

So we have biregular spectral sequences

$$E_2^{pq} = \varinjlim_{(U,V,g)\in \mathrm{ob}\,\mathscr{S}^\circ} \mathrm{Ext}^p(\mathscr{H}^{-q}(j_!\mathbb{Z}/n(-d)[-2d]), K) \Rightarrow \mathscr{H}^{p+q}(f^*K(d)[2d])_{\bar{x}},$$

$$E_2^{pq} = \varinjlim_{(U,V,g)\in \mathrm{ob}\,\mathscr{S}^\circ} \mathrm{Ext}^p(R^{-q} f_! k_!\mathbb{Z}/n, K) \Rightarrow \mathscr{H}^{p+q}(Rf^!K)_{\bar{x}},$$

and the morphism of complexes

$$f_! k_! \mathbb{Z}/n \to j_! \mathbb{Z}/n(-d)[-2d]$$

introduced at the beginning of this section defines a morphism from the first spectral sequence to the second one. To prove the theorem, it suffices to show that the induced homomorphism

$$\begin{aligned}&\varinjlim_{(U,V,g)\in \mathrm{ob}\,\mathscr{S}^\circ} \mathrm{Ext}^p(\mathscr{H}^{-q}(j_!\mathbb{Z}/n(-d)[-2d]), K)\\ \to \; &\varinjlim_{(U,V,g)\in \mathrm{ob}\,\mathscr{S}^\circ} \mathrm{Ext}^p(R^{-q} f_! k_! \mathbb{Z}/n, K)\end{aligned}$$

is an isomorphism for any pair p, q. Given $(U, V, g) \in \mathrm{ob}\,\mathscr{S}$, by 8.5.7, we can find a commutative diagram

$$\begin{array}{ccccccc} U' & \stackrel{k''}{\to} & U \times_V V' & \stackrel{k'}{\to} & U & \stackrel{k}{\to} & X \\ & {\scriptstyle g''}\searrow & \downarrow {\scriptstyle g'} & & {\scriptstyle g}\downarrow & & {\scriptstyle f}\downarrow \\ & & V' & \stackrel{j'}{\to} & V & \stackrel{j}{\to} & Y \end{array}$$

such that U' and V' are etale neighborhoods of $\bar{x}$ and $f(\bar{x})$, respectively, the canonical morphisms $R^q g''_! \mathbb{Z}/n \to R^q g'_! \mathbb{Z}/n$ vanishes for any $q < 2d$, and the restriction of $\mathrm{Tr}_{g'} : R^{2d} g'_! \mathbb{Z}/n(d) \to \mathbb{Z}/n$ to the image of $R^{2d} g''_! \mathbb{Z}/n(d) \to R^{2d} g'_! \mathbb{Z}/n(d)$ is an isomorphism. So we can find a morphism

$$\alpha : \mathbb{Z}/n \to R^{2d} g'_! \mathbb{Z}/n(d)$$

which is a section of $\mathrm{Tr}_{g'}$ such that its composite with $\mathrm{Tr}_{g''}$ coincides with $R^{2d} g''_! \mathbb{Z}/n(d) \to R^{2d} g'_! \mathbb{Z}/n(d)$.

$$\begin{array}{ccc} R^{2d} g''_! \mathbb{Z}/n(d) & \stackrel{\mathrm{Tr}_{g''}}{\to} & \mathbb{Z}/n \\ \downarrow & & \| \\ R^{2d} g'_! \mathbb{Z}/n(d) & \underset{\mathrm{Tr}_{g'}}{\overset{\alpha}{\leftrightarrows}} & \mathbb{Z}/n. \end{array}$$

We have a commutative diagram

$$\begin{array}{ccc} R^q f_! (kk'k'')_! \mathbb{Z}/n & \cong & (jj')_! R^q g''_! \mathbb{Z}/n \\ & & \downarrow 0 \\ \downarrow & & (jj')_! R^q g'_! \mathbb{Z}/n \\ & & \downarrow \\ R^q f_! k_! \mathbb{Z}/n & \cong & j_! R^q g_! \mathbb{Z}/n \end{array}$$

for any $q < 2d$, and a commutative diagram

$$\begin{array}{ccccc} R^{2d}f_!(kk'k'')_!\mathbb{Z}/n \cong (jj')_!R^{2d}g''_!\mathbb{Z}/n & \stackrel{(jj')_!(\mathrm{Tr}_{g''})}{\to} & (jj')_!\mathbb{Z}/n(-d) \\ \downarrow & & \| \\ \downarrow \qquad (jj')_!R^{2d}g'_!\mathbb{Z}/n & \underset{(jj')_!(\mathrm{Tr}_{g'})}{\stackrel{(jj')_!(\alpha)}{\leftrightarrows}} & (jj')_!\mathbb{Z}/n(-d) \\ \downarrow & & \downarrow \\ R^{2d}f_!k_!\mathbb{Z}/n \cong j_!R^{2d}g_!\mathbb{Z}/n & \stackrel{j_!(\mathrm{Tr}_g)}{\to} & j_!\mathbb{Z}/n(-d). \end{array}$$

These facts imply that the homomorphism

$$\begin{aligned} &\varinjlim_{(U,V,g)\in \mathrm{ob}\,\mathscr{S}^\circ} \mathrm{Ext}^p(\mathscr{H}^{-q}(j_!\mathbb{Z}/n(-d)[-2d]), K) \\ \to &\varinjlim_{(U,V,g)\in \mathrm{ob}\,\mathscr{S}^\circ} \mathrm{Ext}^p(R^{-q}f_!k_!\mathbb{Z}/n, K) \end{aligned}$$

is an isomorphism for any p, q. Indeed, when $q \neq -2d$, both sides are 0. When $q = -2d$, we have a commutative diagram

$$\begin{array}{ccc} \mathrm{Ext}^p((jj')_!R^{2d}g''_!\mathbb{Z}/n, K) & \stackrel{(jj')_!(\mathrm{Tr}_{g''})}{\leftarrow} & \mathrm{Ext}^p((jj')_!\mathbb{Z}/n(-d), K) \\ \uparrow & & \| \\ \mathrm{Ext}^p((jj')_!R^{2d}g'_!\mathbb{Z}/n, K) & \underset{(jj')_!(\mathrm{Tr}_{g'})}{\stackrel{(jj')_!(\alpha)}{\rightleftarrows}} & \mathrm{Ext}^p((jj')_!\mathbb{Z}/n(-d), K) \\ \uparrow & & \uparrow \\ \mathrm{Ext}^p(j_!R^{2d}g_!\mathbb{Z}/n, K) & \stackrel{j_!(\mathrm{Tr}_g)}{\leftarrow} & \mathrm{Ext}^p(j_!\mathbb{Z}/n(-d), K). \end{array}$$

If $e \in \mathrm{Ext}^p(j_!\mathbb{Z}/n(-d), K)$ is mapped to 0 in

$$\mathrm{Ext}^p(j_!R^{2d}g_!\mathbb{Z}/n, K) \cong \mathrm{Ext}^p(R^{2d}f_!k_!\mathbb{Z}/n, K),$$

the above commutative diagram shows that it is mapped to 0 in $\mathrm{Ext}^p((jj')_!\mathbb{Z}/n(-d), K)$. This proves the homomorphism

$$\varinjlim_{(U,V,g)\in \mathrm{ob}\,\mathscr{S}^\circ} \mathrm{Ext}^p(j_!\mathbb{Z}/n(-d), K) \to \varinjlim_{(U,V,g)\in \mathrm{ob}\,\mathscr{S}^\circ} \mathrm{Ext}^p(R^{2d}f_!k_!\mathbb{Z}/n, K)$$

is injective. For any

$$e' \in \mathrm{Ext}^p(j_!R^{2d}g_!\mathbb{Z}/n, K) \cong \mathrm{Ext}^p(R^{2d}f_!k_!\mathbb{Z}/n, K),$$

let $e \in \mathrm{Ext}^p((jj')_!\mathbb{Z}/n(-d), K)$ be the image of e' under the composite

$$\begin{aligned} \mathrm{Ext}^p(j_!R^{2d}g_!\mathbb{Z}/n, K) &\to \mathrm{Ext}^p((jj')_!R^{2d}g'_!\mathbb{Z}/n, K) \\ &\stackrel{(jj')_!(\alpha)}{\to} \mathrm{Ext}^p((jj')_!\mathbb{Z}/n(-d), K). \end{aligned}$$

Then e and e' have the same image in

$$\mathrm{Ext}^p((jj')_!R^{2d}g''_!\mathbb{Z}/n, K) \cong \mathrm{Ext}^p(R^{2d}f_!(kk'k'')_!\mathbb{Z}/n, K).$$

This proves that the homomorphism

$$\varinjlim_{(U,V,g)\in \mathrm{ob}\,\mathscr{S}^\circ} \mathrm{Ext}^p(j_!\mathbb{Z}/n(-d), K) \to \varinjlim_{(U,V,g)\in \mathrm{ob}\,\mathscr{S}^\circ} \mathrm{Ext}^p(R^{2d}f_!k_!\mathbb{Z}/n, K)$$

is surjective. □

Proof of 8.5.7. We use induction on d. Let $y = f(x)$. If $d = 0$, then f is etale and separated. Let $\widetilde{Y}_{\bar{y}}$ be the strict localization of Y at $\bar{y}$. By 2.8.14, there exists a section $\tilde{j} : \widetilde{Y}_{\bar{y}} \to X \times_Y \widetilde{Y}_{\bar{y}}$ of the base change $f_{\widetilde{Y}_{\bar{y}}} : X \times_Y \widetilde{Y}_{\bar{y}} \to \widetilde{Y}_{\bar{y}}$ of f that maps $\bar{y}$ to $\bar{x}$. By 1.10.9, we can find an etale neighborhood V of $\bar{y}$ in Y and a section $j : V \to X \times_Y V$ of the base change $f_V : X \times_Y V \to V$ that maps $\bar{y}$ to $\bar{x}$. By 2.3.9, j is an open and closed immersion. Taking $U = V$, this proves the lemma for the case $d = 0$.

Suppose $d = 1$. By 7.7.1, f is locally acyclic relative to $\mathbb{Z}/n$. So for any algebraic geometric point $t \to \widetilde{Y}_{\bar{y}}$, we have

$$H^q(\widetilde{X}_{\bar{x}} \times_{\widetilde{Y}_{\bar{y}}} t, \mathbb{Z}/n) \cong \begin{cases} \mathbb{Z}/n & \text{if } q = 0, \\ 0 & \text{if } q \neq 0. \end{cases}$$

We thus have

$$\varinjlim_{\widetilde{U}} H^q(\widetilde{U} \times_{\widetilde{Y}_{\bar{y}}} t, \mathbb{Z}/n) \cong \begin{cases} \mathbb{Z}/n & \text{if } q = 0, \\ 0 & \text{if } q \neq 0, \end{cases}$$

where $\widetilde{U}$ goes over the family of etale neighborhoods of $\bar{x}$ in $X \times_Y \widetilde{Y}_{\bar{y}}$. By 7.2.10, $H^q(X_t, \mathbb{Z}/n)$ is finite for any q, where $X_t = X \times_Y t$. So there exists an etale neighborhood $\widetilde{U}$ of $\bar{x}$ in $X \times_Y \widetilde{Y}_{\bar{y}}$ such that the canonical homomorphisms

$$H^q(X_t, \mathbb{Z}/n) \to H^q(\widetilde{U} \times_{\widetilde{Y}_{\bar{y}}} t, \mathbb{Z}/n)$$

are 0 for $q = 1, 2$. Let $\widetilde{W}' \to \widetilde{W}$ be a $\widetilde{U}$-morphism between etale neighborhoods of $\bar{x}$ in $\widetilde{U}$, and let $K_{\widetilde{W}}$ (resp. $K_{\widetilde{W}'}$) be the kernel of the canonical homomorphism

$$H^0(X_t, \mathbb{Z}/n) \to H^0(\widetilde{W} \times_{\widetilde{Y}_{\bar{y}}} t, \mathbb{Z}/n) \text{ (resp. } H^0(X_t, \mathbb{Z}/n) \to H^0(\widetilde{W}' \times_{\widetilde{Y}_{\bar{y}}} t, \mathbb{Z}/n)).$$

We have $K_{\widetilde{W}} \subset K_{\widetilde{W}'}$. Since $H^0(X_t, \mathbb{Z}/n)$ is finite, there exists an etale neighborhood $\widetilde{W}$ of $\bar{x}$ in $\widetilde{U}$ such that for any etale neighborhood $\widetilde{W}'$ of $\bar{x}$ in $\widetilde{W}$, we have $K_{\widetilde{W}} = K_{\widetilde{W}'}$. As $\varinjlim_{\widetilde{W}'} H^0(\widetilde{W}' \times_{\widetilde{Y}_{\bar{y}}} t, \mathbb{Z}/n) \cong \mathbb{Z}/n$, this implies that the composite

$$\mathbb{Z}/n \to H^0(X_t, \mathbb{Z}/n) \to H^0(X_t, \mathbb{Z}/n)/K_{\widetilde{W}}$$

is an isomorphism. Since $\widetilde{W}$ is an etale neighborhood of $\bar{x}$ in $\widetilde{U}$, the canonical homomorphisms

$$H^q(X_t, \mathbb{Z}/n) \to H^q(\widetilde{W} \times_{\widetilde{Y}_{\bar{y}}} t, \mathbb{Z}/n)$$

are 0 for $q = 1, 2$. By 8.3.2, the canonical homomorphisms

$$H^q_c(\widetilde{W} \times_{\widetilde{Y}_{\bar{y}}} t, \boldsymbol{\mu}_n) \to H^q_c(X_t, \boldsymbol{\mu}_n)$$

are 0 for $q = 0, 1$, and the restriction of

$$\mathrm{Tr}_{X_t/t} : H^2_c(X_t, \boldsymbol{\mu}_n) \to \mathbb{Z}/n$$

to the image of

$$H^2_c(\widetilde{W} \times_{\widetilde{S}_{\bar{s}}} t, \boldsymbol{\mu}_n) \to H^2_c(X_t, \boldsymbol{\mu}_n)$$

is an isomorphism. Here we use the fact that for any etale k-morphism $\pi : U' \to U$ between smooth algebraic curves over a separably closed field k, the transposes of the canonical homomorphisms

$$H^q(U, \mathbb{Z}/n) \to H^q(U', \mathbb{Z}/n)$$

through the pairing in 8.3.2 are the canonical homomorphisms

$$H^{2-q}_c(U', \boldsymbol{\mu}_n) \to H^{2-q}_c(U, \boldsymbol{\mu}_n).$$

This follows from the commutativity of the following diagram

$$\begin{array}{ccccccc}
H^{2-q}_c(U', \boldsymbol{\mu}_n) & \times & \mathrm{Ext}^q_{U'}(\boldsymbol{\mu}_n, \boldsymbol{\mu}_n) & \to & H^2_c(U', \boldsymbol{\mu}_n) & & \\
\wr\| & & \downarrow & & \wr\| & {\scriptstyle \mathrm{Tr}_{U'/k}} \searrow & \\
H^{2-q}_c(U, \pi_!\boldsymbol{\mu}_n) & \times & \mathrm{Ext}^q_U(\pi_!\boldsymbol{\mu}_n, \pi_!\boldsymbol{\mu}_n) & \to & H^2_c(U, \pi_!\boldsymbol{\mu}_n) & & \mathbb{Z}/n, \\
\| & & \downarrow {\scriptstyle \mathrm{Tr}_\pi} & & \downarrow {\scriptstyle \mathrm{Tr}_\pi} & {\scriptstyle \mathrm{Tr}_{U/k}} \nearrow & \\
H^{2-q}_c(U, \pi_!\boldsymbol{\mu}_n) & \times & \mathrm{Ext}^q_U(\pi_!\boldsymbol{\mu}_n, \boldsymbol{\mu}_n) & \to & H^2_c(U, \boldsymbol{\mu}_n) & & \\
{\scriptstyle \mathrm{Tr}_\pi} \downarrow & & \uparrow {\scriptstyle \mathrm{Tr}^*_\pi} & & \| & & \\
H^{2-q}_c(U, \boldsymbol{\mu}_n) & \times & \mathrm{Ext}^q_U(\boldsymbol{\mu}_n, \boldsymbol{\mu}_n) & \to & H^2_c(U, \boldsymbol{\mu}_n) & &
\end{array}$$

the fact that the composite $\phi : \mathrm{Ext}^q_{U'}(\boldsymbol{\mu}_n, \boldsymbol{\mu}_n) \to \mathrm{Ext}^q_U(\pi_!\boldsymbol{\mu}_n, \boldsymbol{\mu}_n)$ of

$$\mathrm{Ext}^q_{U'}(\boldsymbol{\mu}_n, \boldsymbol{\mu}_n) \to \mathrm{Ext}^q_U(\pi_!\boldsymbol{\mu}_n, \pi_!\boldsymbol{\mu}_n) \stackrel{\mathrm{Tr}_\pi}{\to} \mathrm{Ext}^q_U(\pi_!\boldsymbol{\mu}_n, \boldsymbol{\mu}_n)$$

is an isomorphism, and the fact that the composite

$$\mathrm{Ext}^q_U(\boldsymbol{\mu}_n, \boldsymbol{\mu}_n) \stackrel{\mathrm{Tr}^*_\pi}{\to} \mathrm{Ext}^q_U(\pi_!\boldsymbol{\mu}_n, \boldsymbol{\mu}_n) \stackrel{\phi^{-1}}{\to} \mathrm{Ext}^q_{U'}(\boldsymbol{\mu}_n, \boldsymbol{\mu}_n)$$

can be identified with the canonical homomorphism $H^q(U, \mathbb{Z}/n) \to H^q(U', \mathbb{Z}/n)$.

For any etale neighborhood $\widetilde{U}$ of $\bar{x}$ in $X\times_Y\widetilde{Y}_{\bar{y}}$, denote by $\mathscr{F}^q_{\widetilde{U}}$ the images of the canonical morphisms

$$R^q(\tilde{f}|_{\widetilde{U}})_!\boldsymbol{\mu}_n \to R^q\tilde{f}_!\boldsymbol{\mu}_n,$$

where $\tilde{f}: X\times_Y\widetilde{Y}_{\bar{y}} \to \widetilde{Y}_{\bar{y}}$ is the base change of f. We use noetherian induction to show that there exists an etale neighborhood $\widetilde{U}$ such that $\mathscr{F}^q_{\widetilde{U}}=0$ for $q=0,1$ and the restriction of

$$\mathrm{Tr}_{\tilde{f}}: R^2\tilde{f}_!\boldsymbol{\mu}_n \to \mathbb{Z}/n$$

to $\mathscr{F}^2_{\widetilde{U}}$ is an isomorphism. Let $\mathscr{S}$ be the set of those closed subsets A of $\widetilde{Y}_{\bar{y}}$ such that for any etale neighborhood $\widetilde{U}$ of $\bar{x}$ in $X\times_Y\widetilde{Y}_{\bar{y}}$, either $\mathscr{F}^0_{\widetilde{U}}|_A\neq 0$, or $\mathscr{F}^1_{\widetilde{U}}|_A\neq 0$, or the restriction of $\mathrm{Tr}_{\tilde{f}}|_A$ to $\mathscr{F}^2_{\widetilde{U}}|_A$ is not an isomorphism. If $\mathscr{S}$ is not empty, then there exists a minimal element A in $\mathscr{S}$, and A must be irreducible. Let t be an algebraic geometric point lying above the generic point of A. By our previous discussion, there exists an etale neighborhood $\widetilde{W}$ of $\bar{x}$ in $X\times_Y\widetilde{Y}_{\bar{y}}$ such that $H^q_c(\widetilde{W}\times_{\widetilde{Y}_{\bar{y}}}t,\boldsymbol{\mu}_n)\to H^q_c(X_t,\boldsymbol{\mu}_n)$ are 0 for $q=0,1$, and the restriction of $\mathrm{Tr}_{X_t/t}: H^2_c(X_t,\boldsymbol{\mu}_n)\to\mathbb{Z}/n$ to the image of the canonical homomorphism $H^2_c(\widetilde{W}\times_{\widetilde{Y}_{\bar{y}}}t,\boldsymbol{\mu}_n)\to H^2_c(X_t,\boldsymbol{\mu}_n)$ is an isomorphism. So we have $(\mathscr{F}^q_{\widetilde{W}})_t=0$ for $q=0,1$, and the restriction of $(\mathrm{Tr}_{\tilde{f}})_t: (R^2\tilde{f}_!\boldsymbol{\mu}_n)_t\to\mathbb{Z}/n$ to $(\mathscr{F}^2_{\widetilde{W}})_t$ is an isomorphism. Since $\mathscr{F}^q_{\widetilde{W}}$ $(q=0,1)$, $\ker(\mathrm{Tr}_{\tilde{f}})\cap\mathscr{F}^2_{\widetilde{W}}$ and $\mathrm{coker}(\mathrm{Tr}_{\tilde{f}})\cap\mathscr{F}^2_{\widetilde{W}}$ are constructible by 7.8.1, there exists a nonempty open subset O of A such that $\mathscr{F}^q_{\widetilde{W}}|_O$ $(q=0,1)$, $(\ker(\mathrm{Tr}_{\tilde{f}})\cap\mathscr{F}^2_{\widetilde{W}})|_O$ and $(\mathrm{coker}(\mathrm{Tr}_{\tilde{f}})\cap\mathscr{F}^2_{\widetilde{W}})|_O$ are locally constant. As t lies in O, all these locally constant sheaves are 0. By the minimality of A, there exists an etale neighborhood $\widetilde{W}'$ of $\bar{x}$ in $X\times_Y\widetilde{Y}_{\bar{y}}$ such that $\mathscr{F}^q_{\widetilde{W}'}|_{A-O}=0$ for $q=0,1$ and the restriction of $\mathrm{Tr}_{\tilde{f}}|_{A-O}$ to $\mathscr{F}^2_{\widetilde{W}'}|_{A-O}$ is an isomorphism. Let $\widetilde{W}''=\widetilde{W}\times_{(X\times_Y\widetilde{Y}_{\bar{y}})}\widetilde{W}'$. We have $\mathscr{F}^q_{\widetilde{W}''}\subset\mathscr{F}^q_{\widetilde{W}}\cap\mathscr{F}^q_{\widetilde{W}'}$ for all q. So $\mathscr{F}^q_{\widetilde{W}''}|_A$ $(q=0,1)$, $(\ker(\mathrm{Tr}_{\tilde{f}})\cap\mathscr{F}^2_{\widetilde{W}''})|_A$ and $(\mathrm{coker}(\mathrm{Tr}_{\tilde{f}})\cap\mathscr{F}^2_{\widetilde{W}''})|_A$ are all 0. This contradicts $A\in\mathscr{S}$. So $\mathscr{S}$ is empty. Therefore there exists an etale neighborhood $\widetilde{U}$ of $\bar{x}$ in $X\times_Y\widetilde{Y}_{\bar{y}}$ such that $\mathscr{F}^q_{\widetilde{U}}=0$ for $q=0,1$ and the restriction of $\mathrm{Tr}_{\tilde{f}}: R^2\tilde{f}_!\boldsymbol{\mu}_n\to\mathbb{Z}/n$ to $\mathscr{F}^2_{\widetilde{U}}$ is an isomorphism. By 5.9.8, there exists an etale neighborhood V of $\bar{y}$ in Y and an etale neighborhood U of $\bar{x}$ in $X\times_Y V$ such that $\widetilde{U}\cong U\times_V\widetilde{Y}_{\bar{y}}$ and such that the assertions in 8.5.7 hold. This proves 8.5.7 in the case where $d=1$.

Suppose that 8.5.7 holds for smooth morphisms of relative dimensions $<d$, and let us prove that it holds for any smooth morphism $f: X\to Y$ pure of relative dimension d. Note that if we can find an etale neighborhood

X' of $\bar{x}$ in X such that 8.5.7 holds for the morphism $f|_{X'}$, then it holds for f. So to prove our assertion, we may assume there exists an etale Y-morphism

$$X \to \mathbb{A}^n_Y = \mathbf{Spec}\, \mathscr{O}_Y[t_1, \ldots, t_n].$$

Let $Y' = \mathbb{A}^1_Y = \mathbf{Spec}\, \mathscr{O}_Y[t_1]$. We can factorize f as a composite

$$X \stackrel{h}{\to} Y' \stackrel{g}{\to} Y$$

such that h and g are smooth pure of relative dimensions $d' = d - 1$ and 1, respectively. By the induction hypothesis, we can find a commutative diagram

$$\begin{array}{ccccc} U & \to & X \times_{Y'} V & \to & X \\ & h_0' \searrow & \downarrow h_0 & & \downarrow h \\ & & V & \to & Y' \end{array}$$

such that V is an etale neighborhood of $h(\bar{x})$ in Y', U is an etale neighborhood of $\bar{x}$ in $X \times_{Y'} V$, the canonical morphisms

$$R^q h'_{0!}\mathbb{Z}/n \to R^q h_{0!}\mathbb{Z}/n$$

are 0 for $q \neq 2d'$, and the restriction of

$$\mathrm{Tr}_{h_0} : R^{2d'} h_{0!}\mathbb{Z}/n(d') \to \mathbb{Z}/n$$

to the image of the canonical morphism

$$R^{2d'} h'_{0!}\mathbb{Z}/n(d') \to R^{2d'} h_{0!}\mathbb{Z}/n(d')$$

is an isomorphism. Let $g_1 : V \to Y$ be the composite of $V \to Y'$ and $g : Y' \to Y$. Applying the $d = 1$ case to g_1, we get a commutative diagram

$$\begin{array}{ccccc} Z & \stackrel{j}{\to} & V \times_Y W & \to & V \\ & g_1'' \searrow & \downarrow g_1' & & \downarrow g_1 \\ & & W & \to & Y \end{array}$$

such that W is an etale neighborhood of $\bar{y}$ in Y, Z is an etale neighborhood of $h_0(\bar{x})$ in $V \times_Y W$, the canonical morphisms

$$R^q g''_{1!}\mathbb{Z}/n \to R^q g'_{1!}\mathbb{Z}/n$$

are 0 for $q \neq 2$, and the restriction of

$$\mathrm{Tr}_{g_1'} : R^2 g'_{1!}\mathbb{Z}/n(1) \to \mathbb{Z}/n$$

to the image of the canonical morphism

$$R^2 g''_{1!}\mathbb{Z}/n(1) \to R^2 g'_{1!}\mathbb{Z}/n(1)$$

is an isomorphism. Consider the following commutative diagram:

$$\begin{array}{ccccccc}
 & & & & U & & \to X\times_{Y'} V \\
 & & & \nearrow & & & \nearrow\downarrow h_0 \\
 & U\times_Y W & & \to X\times_{Y'} V\times_Y W & & & V \\
 & \nearrow & & \nearrow & \downarrow h_1 & & \nearrow\downarrow g_1 \\
U\times_V Z & \to X\times_{Y'} V\times_V Z & & & V\times_Y W & & Y \\
 h_2'\searrow & \downarrow h_2 & & \overset{j}{\nearrow} & \downarrow g_1' & & \nearrow \\
 & Z & & \underset{g_1''}{\to} & W & & ,
\end{array}$$

where h_1 and h_2 are the base changes of h_0, and h_2' is the base change of h_0'. When $q \neq 2d'$, the canonical morphisms

$$R^p g''_{1!} R^q h'_{2!}\mathbb{Z}/n(d') \to R^p g'_{1!} R^q h_{1!}\mathbb{Z}/n(d')$$

are 0 since they are the composites of the canonical morphisms

$$R^p g''_{1!} R^q h'_{2!}\mathbb{Z}/n(d') \to R^p g''_{1!} R^q h_{2!}\mathbb{Z}/n(d') \to R^p g'_{1!} R^q h_{1!}\mathbb{Z}/n(d')$$

and the canonical morphisms

$$R^q h'_{2!}\mathbb{Z}/n(d') \to R^q h_{2!}\mathbb{Z}/n(d')$$

are 0. When $p \neq 2$, the canonical morphisms

$$R^p g''_{1!} R^{2d'} h'_{2!}\mathbb{Z}/n(d') \to R^p g'_{1!} R^{2d'} h_{1!}\mathbb{Z}/n(d')$$

are 0. To prove this, let $h_1' : U\times_Y W \to V\times_Y W$ be the base change of h_0'. The restrictions of Tr_{h_i} $(i = 1, 2)$ to the images of the canonical morphisms

$$R^{2d'} h'_{i!}\mathbb{Z}/n(d') \to R^{2d'} h_{i!}\mathbb{Z}/n(d')$$

are isomorphisms. Denote the inverses of these isomorphisms by $\mathrm{Tr}_{h_i^{-1}}$. Then these canonical morphisms can be factorized as

$$R^{2d'} h'_{i!}\mathbb{Z}/n(d') \overset{\mathrm{Tr}_{h_i'}}{\to} \mathbb{Z}/n \overset{\mathrm{Tr}_{h_i}^{-1}}{\to} R^{2d'} h_{i!}\mathbb{Z}/n(d').$$

The following diagram commutes:

$$\begin{array}{ccccc}
R^p g''_{1!} R^{2d'} h'_{2!}\mathbb{Z}/n(d') & \overset{\mathrm{Tr}_{h_2'}}{\to} & R^p g''_{1!}\mathbb{Z}/n & \overset{\mathrm{Tr}_{h_2}^{-1}}{\to} & R^p g''_{1!} R^{2d'} h_{2!}\mathbb{Z}/n(d') \\
\wr\| & & \wr\| & & \wr\| \\
R^p g'_{1!} j_! j^* R^{2d'} h'_{1!}\mathbb{Z}/n(d') & \overset{\mathrm{Tr}_{h_1'}}{\to} & R^p g'_{1!} j_! j^*\mathbb{Z}/n & \overset{\mathrm{Tr}_{h_1}^{-1}}{\to} & R^p g'_{1!} j_! j^* R^{2d'} h_{1!}\mathbb{Z}/n(d') \\
\mathrm{Tr}_j\downarrow & & \mathrm{Tr}_j\downarrow & & \mathrm{Tr}_j\downarrow \\
R^p g'_{1!} R^{2d'} h'_{1!}\mathbb{Z}/n(d') & \overset{\mathrm{Tr}_{h_1'}}{\to} & R^p g'_{1!}\mathbb{Z}/n & \overset{\mathrm{Tr}_{h_1}^{-1}}{\to} & R^p g'_{1!} R^{2d'} h_{1!}\mathbb{Z}/n(d').
\end{array}$$

When $p \neq 2$, the canonical morphisms $R^p g''_{1!}\mathbb{Z}/n \to R^p g'_{1!}\mathbb{Z}/n$ are 0. So the canonical morphisms $R^p g''_{1!} R^{2d'} h'_{2!}\mathbb{Z}/n(d') \to R^p g'_{1!} R^{2d'} h_{1!}\mathbb{Z}/n(d')$ are 0. Hence when $p+q \neq 2d$, the canonical morphisms

$$R^p g''_{1!} R^q h'_{2!}\mathbb{Z}/n(d') \to R^p g'_{1!} R^q h_{1!}\mathbb{Z}/n(d')$$

are 0. We have biregular spectral sequences

$$E_2^{pq} = R^p g''_{1!} R^q h'_{2!}\mathbb{Z}/n(d') \Rightarrow R^{p+q}(g''_1 h'_2)_!\mathbb{Z}/n(d'),$$
$$E_2^{pq} = R^p g'_{1!} R^q h_{1!}\mathbb{Z}/n(d') \Rightarrow R^{p+q}(g'_1 h_1)_!\mathbb{Z}/n(d').$$

It follows that the canonical morphisms

$$R^q(g''_1 h'_2)_!\mathbb{Z}/n(d') \to R^q(g'_1 h_1)_!\mathbb{Z}/n(d')$$

are 0 when $q \neq 2d$.

Let $\mathscr{F} \overset{\phi}{\to} \mathscr{G} \overset{\psi}{\to} \mathscr{H}$ be morphisms of sheaves. One can verify that the following three statements are equivalent:

(a) ψ induces an isomorphism from $\operatorname{im}\phi$ to $\mathscr{H}$.

(b) $\psi\phi$ is surjective and $\ker(\psi\phi) \subset \ker\phi$.

(c) $\psi\phi$ is surjective and there exists a morphism $\theta : \mathscr{H} \to \mathscr{G}$ such that $\psi\theta = \mathrm{id}_{\mathscr{H}}$ and $\phi = \theta\psi\phi$.

If these conditions holds, then for any right exact additive functor F, $F(\psi)$ induces an isomorphism from $\operatorname{im}(F(\phi))$ to $F(\mathscr{H})$. Since the restriction of

$$\mathrm{Tr}_{h_2} : R^{2d'} h_{2!}\mathbb{Z}/n(d') \to \mathbb{Z}/n$$

to the image of the canonical morphism

$$R^{2d'} h'_{2!}\mathbb{Z}/n(d') \to R^{2d'} h_{2!}\mathbb{Z}/n(d')$$

is an isomorphism and the functor $R^2 g''_{1!}$ is right exact, the restriction of the morphism

$$R^2 g''_{1!} R^{2d'} h_{2!}\mathbb{Z}/n(d') \to R^2 g''_{1!}\mathbb{Z}/n$$

induced by Tr_{h_2} to the image of the canonical morphism

$$R^2 g''_{1!} R^{2d'} h'_{2!}\mathbb{Z}/n(d') \to R^2 g''_{1!} R^{2d'} h_{2!}\mathbb{Z}/n(d')$$

is an isomorphism. Moreover, the restriction of

$$\mathrm{Tr}_{g'_1} : R^2 g'_{1!}\mathbb{Z}/n(1) \to \mathbb{Z}/n$$

to the image of the canonical morphism

$$R^2 g''_{1!}\mathbb{Z}/n(1) \to R^2 g'_{1!}\mathbb{Z}/n(1)$$

is an isomorphism. Using these facts and the following commutative diagram,

$$\begin{array}{ccccccc} R^2g''_{1!}R^{2d'}h'_{2!}\mathbb{Z}/n(d) & \to & R^2g''_{1!}R^{2d'}h_{2!}\mathbb{Z}/n(d) & \stackrel{\mathrm{Tr}_{h_2}}{\to} & R^2g''_{1!}\mathbb{Z}/n(1) & & \\ \wr\| & & \wr\| & & \wr\| & \mathrm{Tr}_{g''_1}\searrow & \\ R^2g'_{1!}j_!j^*R^{2d'}h'_{1!}\mathbb{Z}/n(d) & \to & R^2g'_{1!}j_!j^*R^{2d'}h_{1!}\mathbb{Z}/n(d) & \stackrel{\mathrm{Tr}_{h_1}}{\to} & R^2g'_{1!}j_!j^*\mathbb{Z}/n(1) & & \mathbb{Z}/n, \\ \mathrm{Tr}_j\downarrow & & \mathrm{Tr}_j\downarrow & & \mathrm{Tr}_j\downarrow & \mathrm{Tr}_{g'_1}\nearrow & \\ R^2g'_{1!}R^{2d'}h'_{1!}\mathbb{Z}/n(d) & \to & R^2g'_{1!}R^{2d'}h_{1!}\mathbb{Z}/n(d) & \stackrel{\mathrm{Tr}_{h_1}}{\to} & R^2g'_{1!}\mathbb{Z}/n(1) & & \end{array}$$

one can check that the morphism

$$R^2g''_{1!}R^{2d'}h'_{2!}\mathbb{Z}/n(d) \to \mathbb{Z}/n$$

is surjective, and its kernel is contained in the kernel of the morphism

$$R^2g''_{1!}R^{2d'}h'_{2!}\mathbb{Z}/n(d) \to R^2g'_{1!}R^{2d'}h_{1!}\mathbb{Z}/n(d).$$

So the restriction of

$$R^2g'_{1!}R^{2d'}h_{1!}\mathbb{Z}/n(d) \to \mathbb{Z}/n$$

to the image of $R^2g''_{1!}R^{2d'}h'_{2!}\mathbb{Z}/n(d) \to R^2g'_{1!}R^{2d'}h_{1!}\mathbb{Z}/n(d)$ is an isomorphism. We have

$$R^{2d}(g''_1h'_2)_!\mathbb{Z}/n(d) \cong R^2g''_{1!}R^{2d'}h'_{2!}\mathbb{Z}/n(d),$$
$$R^{2d}(g'_1h_1)_!\mathbb{Z}/n(d) \cong R^2g'_{1!}R^{2d'}h_{1!}\mathbb{Z}/n(d).$$

It follows that the restriction of

$$\mathrm{Tr}_{g'_1h_1} : R^{2d}(g'_1h_1)_!\mathbb{Z}/n(d) \to \mathbb{Z}/n$$

to the image of the canonical morphism

$$R^{2d}(g''_1h'_2)_!\mathbb{Z}/n(d) \to R^{2d}(g'_1h_1)_!\mathbb{Z}/n(d)$$

is an isomorphism. Recall that we have shown the canonical morphisms

$$R^q(g''_1h'_2)_!\mathbb{Z}/n(d) \to R^q(g'_1h_1)_!\mathbb{Z}/n(d)$$

are 0 for $q \neq 2d$. Let $g_W : Y' \times_Y W \to W$ and $h_W : X \times_Y W \to Y \times_Y W$ be the base changes of g and h, respectively. The above facts imply that the canonical morphisms

$$R^q(g''_1h'_2)_!\mathbb{Z}/n(d) \to R^q(g_Wh_W)_!\mathbb{Z}/n(d)$$

are 0 for any $q \neq 2d$, and the restriction of

$$\mathrm{Tr}_{g_Wh_W} : R^{2d}(g_Wh_W)_!\mathbb{Z}/n(d) \to \mathbb{Z}/n$$

to the image of the canonical morphism

$$R^{2d}(g''_1h'_2)_!\mathbb{Z}/n(d) \to R^{2d}(g_Wh_W)_!\mathbb{Z}/n(d)$$

is an isomorphism. This finishes the proof of the lemma.

$$\begin{array}{ccccc} U \times_V Z & \to & X \times_{Y'} V \times_Y W & \to & X \times_Y W \\ h'_2\downarrow & & h_1\downarrow & & \downarrow h_W \\ Z & \to & V \times_Y W & \to & Y' \times_Y W \\ & \stackrel{g''_1}{\searrow} & g'_1\downarrow & \stackrel{g_W}{\swarrow} & \\ & & W & & \end{array}$$

□

8.6 Cohomology Classes of Algebraic Cycles

([SGA $4\frac{1}{2}$] Cycle.)

Throughout this section, we fix a noetherian ring A such that A is an injective A-module, and $nA = 0$ for some integer n invertible on schemes we consider. Let S be a noetherian scheme, and let $f : Y \to Z$ be a smooth S-compactifiable morphism pure of relative dimension d between S-schemes of finite type. We have the trace morphism

$$\mathrm{Tr}_f : R^{2d} f_! A(d) \to A.$$

On the other hand, we have $R^q f_! A(d) = 0$ for all $q > 2d$ and $R^q f^! A(-d) = 0$ for all $q < -2d$. So we have

$$\begin{aligned}\mathrm{Hom}_A(R^{2d} f_! A(d), A) &\cong \mathrm{Hom}_{D(Z,A)}(Rf_! A(d), A[-2d]) \\ &\cong \mathrm{Hom}_{D(Y,A)}(A(d), Rf^! A[-2d]) \\ &\cong H^0(Y, Rf^! A(-d)[-2d]) \\ &\cong H^0(Y, R^{-2d} f^! A(-d)).\end{aligned}$$

Hence Tr_f:$R^{2d} f_! A(d) \to A$ corresponds to an element in $H^0(Y, Rf^! A(-d)$ $[-2d])$. Suppose that we have a smooth S-compactifiable morphism $g : X \to Z$ pure of relative dimension N and an immersion $i : Y \to X$ such that $gi = f$. Let $c = N - d$ be the codimension of Y in X. We have

$$Rf^! A \cong Ri^! Rg^! A \cong Ri^! A(N)[2N].$$

So we have

$$Rf^! A(-d)[-2d] \cong Ri^! A(c)[2c].$$

It follows that $R^q i^! A = 0$ for any $q < 2c$ and

$$H^0(Y, Rf^! A(-d)[-2d]) \cong H^0(Y, Ri^! A(c)[2c]) \cong H^{2c}_Y(X, A(c)).$$

The element in $H^{2c}_Y(X, A(c))$ corresponding to Tr_f is called the *cohomology class* associated to Y and is denoted by $\mathrm{cl}'(Y)$. The image of $\mathrm{cl}'(Y)$ under the canonical homomorphism

$$H^{2c}_Y(X, A(c)) \to H^{2c}(X, A(c))$$

is also called the cohomology class of Y and is denoted by $\mathrm{cl}(Y)$.

The functors $H^q_Y(X, -)$ that we used above are defined as follows. Choose an open subset U of X containing $i(Y)$ as a closed subset. We define

$$\Gamma_Y(X, \mathscr{F}) = \Gamma_Y(U, \mathscr{F}|_U)$$

for any sheaf $\mathscr{F}$ on X. The functor $\Gamma_Y(X,-)$ is independent of the choice of U and is left exact. Let $R\Gamma_Y(X,-)$ be the right derived functor of $\Gamma_Y(X,-)$. We define

$$H^q_Y(X,\mathscr{F}) = H^q(R\Gamma_Y(X,\mathscr{F})) = H^q_Y(U,\mathscr{F}).$$

Let $k : Y \to U$ and $j : U \to X$ be the immersions such that $i = jk$. We have

$$\Gamma_Y(X,\mathscr{F}) = \Gamma(Y, k^! j^* \mathscr{F}).$$

So we have

$$R\Gamma_Y(X,\mathscr{F}) = R\Gamma(Y, R(k^! j^*)\mathscr{F}) \cong R\Gamma(Y, Ri^! \mathscr{F}).$$

The next proposition shows that $\mathrm{cl}'(Y) \in H^{2c}_Y(X, A(c))$ depends only on $i : Y \to X$ and not on the base Z.

Proposition 8.6.1. *Let S be a noetherian scheme. Consider a commutative diagram*

$$\begin{array}{ccc} & & X \\ & \overset{i}{\nearrow} & \downarrow g' \\ Y & \xrightarrow{f'} & Z' \\ & \underset{f}{\searrow} & \downarrow h \\ & & Z, \end{array}$$

where i is an immersion, f, f', $g = hg'$, g' and h are smooth S-compactifiable morphisms between S-schemes of finite type pure of relative dimensions d, d', N, N' and $N-N'$, respectively. Let

$$c = N - d = N' - d'.$$

The element in $H^{2c}_Y(X, A(c))$ corresponding to $\mathrm{Tr}_f \in \mathrm{Hom}(R^{2d} f_! A(d), A)$ coincides with the element corresponding to $\mathrm{Tr}_{f'} \in \mathrm{Hom}(R^{2d'} f'_! A(d'), A)$.

Proof. Use the commutativity of the following diagram:

$$\begin{array}{ccccc} \mathrm{Hom}(Rf'_! A(d'), A[-2d']) & \to & \mathrm{Hom}(Rh_! Rf'_! A(d'), Rh_! A[-2d']) & \overset{\mathrm{Tr}_h}{\to} & \mathrm{Hom}(Rf_! A(d), A[-2d]) \\ \cong\downarrow & (1) & \cong\downarrow & & \cong\downarrow \\ \mathrm{Hom}(A(d'), Rf'^! A[-2d']) & \to & \mathrm{Hom}(A(d'), Rf'^! Rh^! Rh_! A[-2d']) & \overset{\mathrm{Tr}_h}{\to} & \mathrm{Hom}(A(d), Rf^! A[-2d]) \\ \cong\downarrow & & \cong\downarrow & & \cong\downarrow \\ \mathrm{Hom}(A, Ri^! Rg'^! A(-d')[-2d']) & \to & \mathrm{Hom}(A, Ri^! Rg'^! Rh^! Rh_! A(-d')[-2d']) & \overset{\mathrm{Tr}_h}{\to} & \mathrm{Hom}(A, Ri^! Rg^! A(-d)[-2d]) \\ & \cong\searrow & (2) & \swarrow\cong & \\ & & \mathrm{Hom}(A, Ri^! A(c)[2c]), & & \end{array}$$

where (1) commutes because for any $\phi \in \mathrm{Hom}(Rf'_!A(d'), A[-2d'])$, the following diagram commutes:

$$\begin{array}{cccc} A(d') \to & Rf'^!Rf'_!A(d') & \stackrel{Rf'^!(\phi)}{\to} & Rf'^!A[-2d'] \\ \searrow & \downarrow & & \downarrow \\ & Rf'^!Rh^!Rh_!Rf'_!A(d') & \stackrel{Rf'^!Rh^!Rh_!(\phi)}{\to} & Rf'^!Rh^!Rh_!A[-2d'], \end{array}$$

and (2) commutes because

$$A(-d')[-2d'] \cong Rh^!A(-d)[-2d]$$

and the following diagram commutes:

$$\begin{array}{ccc} Rh^!A(-d)[-2d] & = & Rh^!A(-d)[-2d] \\ \mathrm{adj}\downarrow & & \uparrow \mathrm{adj} \\ (Rh^!Rh_!)Rh^!A(-d)[-2d] & = & Rh^!(Rh_!Rh^!)A(-d)[-2d]. \end{array}$$

Proposition 8.6.2. *Let S be a noetherian scheme, $g : X \to Z$ and $f : Y \to Z$ smooth S-compactifiable morphisms between S-schemes of finite type pure of relative dimensions N and d, respectively, $i : Y \to X$ a closed immersion such that $f = gi$, and $c = N - d$. Then $\mathrm{Tr}_f : R^{2d}f_!A(d) \to A$ is mapped to $\mathrm{cl}(Y) \in H^{2c}(X, A(c))$ under the composite of the following homomorphisms:*

$$\begin{aligned} \mathrm{Hom}(R^{2d}f_!A(d), A) &\cong \mathrm{Hom}(Rf_!A(d), A[-2d]) \\ &\cong \mathrm{Hom}(Rg_!i_*i^*A(d), A[-2d]) \\ &\to \mathrm{Hom}(Rg_!A(d), A[-2d]) \\ &\cong \mathrm{Hom}(A, Rg^!A(-d)[-2d]) \\ &\cong \mathrm{Hom}(A, A(c)[2c]) \\ &\cong H^{2c}(X, A(c)). \end{aligned}$$

Equivalently, through the isomorphisms

$$\mathrm{Hom}(Rg_!A(d), A[-2d]) \cong \mathrm{Hom}(A, Rg^!A(-d)[-2d]) \cong H^{2c}(X, A(c)),$$

$\mathrm{cl}(Y) \in H^{2c}(X, A(c))$ *corresponds to the composite*

$$Rg_!A(d) \to Rf_!A(d) \stackrel{\mathrm{Tr}_f}{\to} A[-2d].$$

Proof. Use the commutativity of the following diagram:

$$\begin{array}{ccccc}
\mathrm{Hom}(Rf_!A(d),A[-2d]) & & & & \\
\cong\downarrow & & & & \\
\mathrm{Hom}(Rg_!i_*i^*A(d),A[-2d]) & = & \mathrm{Hom}(Rg_!i_*i^*A(d),A[-2d]) & \to & \mathrm{Hom}(Rg_!A(d),A[-2d]) \\
\cong\downarrow & & \cong\downarrow & (1) & \cong\downarrow \\
\mathrm{Hom}(i^*A(d),Ri^!Rg^!A[-2d]) & \cong & \mathrm{Hom}(A(d),i_*Ri^!Rg^!A[-2d]) & \to & \mathrm{Hom}(A(d),Rg^!A[-2d]) \\
\cong\downarrow & & \cong\downarrow & & \cong\downarrow \\
\mathrm{Hom}(i^*A,Ri^!A(c)[2c]) & \cong & \mathrm{Hom}(A,i_*Ri^!A(c)[2c]) & \to & \mathrm{Hom}(A,A(c)[2c]) \\
\cong\downarrow & & & & \cong\downarrow \\
H^{2c}_Y(X,A(c)) & & \to & & H^{2c}(X,A(c)).
\end{array}$$

The commutativity of (1) follows from the adjointness of $(Rg_!, Rg^!)$ and the following fact: For any sheaves $\mathscr{F}$ and $\mathscr{G}$ on X and any morphism

$$\psi : i_*i^*\mathscr{F} \to \mathscr{G},$$

if

$$\psi' : \mathscr{F} \to i_*i^!\mathscr{G}$$

is the morphism induced by ψ by adjunction, then the composite

$$\mathscr{F} \stackrel{\text{adj}}{\to} i_*i^*\mathscr{F} \stackrel{\psi}{\to} \mathscr{G}$$

coincides with the composite

$$\mathscr{F} \stackrel{\psi'}{\to} i_*i^!\mathscr{G} \stackrel{\text{adj}}{\to} \mathscr{G}.$$

To prove this fact, note that ψ induces a morphism

$$\phi : i^*\mathscr{F} \to i^!\mathscr{G}$$

by adjunction such that ψ coincides with the composite

$$i_*i^*\mathscr{F} \stackrel{i_*\phi}{\to} i_*i^!\mathscr{G} \stackrel{\text{adj}}{\to} \mathscr{G},$$

and ψ' coincides with the composite

$$\mathscr{F} \stackrel{\text{adj}}{\to} i_*i^*\mathscr{F} \stackrel{i_*\phi}{\to} i_*i^!\mathscr{G}.$$

Our assertion follows. □

Let X be a smooth compactifiable scheme pure of dimension N over an algebraically closed field k, let Y be an integral closed subscheme of X pure of dimension d, and let $c = N - d$. There exists an open subset U of Y smooth over k. We can define $\mathrm{cl}'(U) \in H^{2c}_U(X, A(c))$ as above. We claim that

$$H^{2c}_U(X, A(c)) \cong H^{2c}_Y(X, A(c)).$$

We call the element in $H^{2c}_Y(X, A(c))$ corresponding to $\mathrm{cl}'(U)$ the *cohomology class* associated to Y and denote it by $\mathrm{cl}'(Y)$.

Let us prove our claim. Let $Z = Y - U$, and fix notation by the following commutative diagram

$$\begin{array}{ccccc} & & Z & & \\ & & \downarrow l & \searrow h & \\ U & \xrightarrow{j} & Y & \xrightarrow{i} & X, \end{array}$$

where h, i, j, l are all immersions. We have a long exact sequence

$$\cdots \to H^q_Z(Y, Ri^! A(c)) \to H^q(Y, Ri^! A(c)) \to H^q(U, j^* Ri^! A(c)) \to \cdots .$$

Moreover, we have

$$\begin{aligned} &H^q_Z(Y, Ri^! A(c)) \cong H^q(Z, Rl^! Ri^! A(c)) \cong H^q(Z, Rh^! A(c)),\\ &H^q(Y, Ri^! A(c)) \cong H^q_Y(X, A(c)),\\ &H^q(U, j^* Ri^! A(c)) \cong H^q_U(X, A(c)). \end{aligned}$$

To prove our claim, it suffices to show

$$R^q h^! A(c) = 0$$

for any $q < 2(c+1)$. Let $g : X \to \operatorname{Spec} k$ be the structure morphism. We have

$$Rh^! A(c) \cong Rh^! Rg^! A(c-N)[-2N] \cong R(gh)^! A(c-N)[-2N].$$

Since $\dim Z \leq d-1$, we have

$$R^q (gh)^! A(c-N) = 0$$

for any $q < -2(d-1)$. Hence

$$R^q h^! A(c) \cong R^{q-2N} (gh)^! A(c-N) = 0$$

for any $q - 2N < -2(d-1)$, or equivalently, $q < 2(c+1)$.

Next suppose that Y is a closed subscheme of X pure of dimension d, and let $Y_1, \ldots, Y_m$ be all the irreducible components of Y with the reduced closed subscheme structure. We can define $\mathrm{cl}'(Y_i) \in H^{2c}_{Y_i}(X, A(c))$ as above. Denote their images in $H^{2c}_Y(X, A(c))$ also by $\mathrm{cl}'(Y_i)$. Let η_i be the generic point of Y_i. We define the *cohomology class* associated to Y to be

$$\mathrm{cl}'(Y) = \sum_{i=1}^m \mathrm{length}(\mathscr{O}_{Y,\eta_i})\, \mathrm{cl}'(Y_i) \in H^{2c}_Y(X, A(c)).$$

The image of $\mathrm{cl}'(Y)$ in $H^{2c}(X, A(c))$ is also called the cohomology class associated to Y and is denoted by $\mathrm{cl}(Y)$.

Finally, let $\gamma = \sum_i n_i Y_i$ be an algebraic cycle of codimension c. We define the *cohomology class* associated to γ to be

$$\mathrm{cl}(\gamma) = \sum_i n_i \mathrm{cl}(Y_i) \in H^{2c}(X, A(c)).$$

Lemma 8.6.3. *Consider a commutative diagram*

$$\begin{array}{ccc} Y' & \stackrel{i'}{\to} & X' \\ {\scriptstyle j'}\downarrow & & \downarrow{\scriptstyle j} \\ Y & \stackrel{i}{\to} & X \\ & {\scriptstyle f}\searrow & \downarrow{\scriptstyle g} \\ & & \mathrm{Spec}\, k, \end{array}$$

where k is an algebraically closed field, X and Y are smooth compactifiable k-schemes pure of dimensions N and d, respectively, i is a closed immersion, j is etale, and the square in the diagram is Cartesian. Let $c = N - d$. Then the image of $\mathrm{cl}'(Y)$ under the canonical homomorphism

$$H^{2c}_Y(X, A(c)) \to H^{2c}_{Y'}(X', A(c))$$

is $\mathrm{cl}'(Y')$.

Proof. Let $f' = fj'$ and $g' = gj$. We use the commutativity of the following diagram:

$$\begin{array}{ccccc} \mathrm{Hom}(Rf_!A(d), A[-2d]) & \to & \mathrm{Hom}(Rf_!j'_!j'^*A(d), A[-2d]) & \cong & \mathrm{Hom}(Rf'_!A(d), A[-2d]) \\ \cong\downarrow & (1) & \cong\downarrow & & \cong\downarrow \\ \mathrm{Hom}(A, Rf^!A(-d)[-2d]) & \to & \mathrm{Hom}(j'^*A(d), j'^*Rf^!A[-2d]) & \cong & \mathrm{Hom}(A, Rf'^!A(-d)[-2d]) \\ \cong\downarrow & & \cong\downarrow & & \cong\downarrow \\ \mathrm{Hom}(A, Ri^!A(c)[2c]) & \to & \mathrm{Hom}(j'^*A, j'^*Ri^!A(c)[2c]) & \cong & \mathrm{Hom}(A, Ri'^!A(c)[2c]). \end{array}$$

The commutativity of (1) follows from the fact that for any $\phi \in \mathrm{Hom}(Rf_!A, A(-d)[-2d])$, the following diagram commutes:

$$\begin{array}{ccccc} & & j'^*A & \stackrel{\mathrm{adj}}{\to} & (j'^*Rf^!Rf_!j'_!)j'^*A \\ & & {\scriptstyle j'^*(\mathrm{adj})}\downarrow & (2) & \| \\ j'^*Rf^!A(-d)[-2d] & \stackrel{j'^*Rf^!(\phi)}{\leftarrow} & j'^*Rf^!Rf_!A & \stackrel{j'^*Rf^!Rf_!(\mathrm{adj})}{\leftarrow} & j'^*Rf^!Rf_!(j'_!j'^*A). \end{array}$$

To prove that (2) commutes, we use the adjointness of $(j'_!, j'^*)$ to reduce it to the commutativity of the following diagram:

$$\begin{array}{ccc} j'_!j'^*A & \stackrel{\mathrm{adj}}{\to} & Rf^!Rf_!j'_!j'^*A \\ {\scriptstyle \mathrm{adj}}\downarrow & & \downarrow{\scriptstyle Rf^!Rf_!(\mathrm{adj})} \\ A & \stackrel{\mathrm{adj}}{\to} & Rf^!Rf_!A. \end{array}$$

$\square$

Lemma 8.6.4. *Let S be a noetherian scheme. Consider a commutative diagram of Cartesian squares*

$$\begin{array}{ccccc} Y' & \stackrel{i'}{\to} & X' & \stackrel{g'}{\to} & Z' \\ h''\downarrow & & h'\downarrow & & h\downarrow \\ Y & \stackrel{i}{\to} & X & \stackrel{g}{\to} & Z. \end{array}$$

Suppose all the schemes in the diagram are S-schemes of finite type, $f = gi$ and g are smooth S-compactifiable morphisms pure of relative dimensions d and N, respectively, and i is a closed immersion. Let $c = N - d$. Then the image of $\mathrm{cl}'(Y)$ under the canonical homomorphism

$$H^{2c}_Y(X, A(c)) \to H^{2c}_{Y'}(X', A(c))$$

is $\mathrm{cl}'(Y')$.

Proof. Let $f' = g'i'$. The following diagram commutes:

$$\begin{array}{ccccc} \mathrm{Hom}(Rf_!A(d), A[-2d]) & \cong & \mathrm{Hom}(A, Rf^!A(-d)[-2d]) & \cong & \mathrm{Hom}(A, Ri^!Rg^!A(-d)[-2d]) \\ \downarrow & & \downarrow & & \downarrow \\ \mathrm{Hom}(h^*Rf_!A(d), h^*A[-2d]) & \cong & \mathrm{Hom}(A, Rf^!Rh_*h^*A(-d)[-2d]) & \cong & \mathrm{Hom}(A, Ri^!Rg^!Rh_*h^*A(-d)[-2d]) \\ \cong\downarrow & & \cong\downarrow & & \cong\downarrow \\ \mathrm{Hom}(Rf'_!h''^*A(d), h^*A[-2d]) & \cong & \mathrm{Hom}(A, Rh''_*Rf'^!h^*A(-d)[-2d]) & \cong & \mathrm{Hom}(A, Rh''_*Ri'^!Rg'^!h^*A(-d)[-2d]) \\ \cong\downarrow & & \cong\downarrow & & \cong\downarrow \\ \mathrm{Hom}(Rf'_!A(d), A[-2d]) & \cong & \mathrm{Hom}(A, Rf'^!A(-d)[-2d]) & \cong & \mathrm{Hom}(A, Ri'^!Rg'^!A(-d)[-2d]). \end{array}$$

To prove our assertion, we need to show that the composite of the rightmost vertical arrows can be identified with the canonical homomorphism

$$H^{2c}_Y(X, A(c)) \to H^{2c}_{Y'}(X', A(c)).$$

It suffices to show that the following diagram commutes:

$$\begin{array}{ccc} Ri^!Rg^!A(-d)[-2d] & & \\ \downarrow & \searrow & \\ Ri^!Rg^!Rh_*h^*A(-d)[2d] & (1) & Ri^!Rh'_*h'^*Rg^!A(-d)[-2d] \\ \wr\| & \nearrow & \downarrow \\ Ri^!Rh'_*Rg'^!h^*A(-d)[-2d] & & Rh''_*Ri'^!h'^*Rg^!A(-d)[-2d], \\ \downarrow & \nearrow & \\ Rh''_*Ri'^!Rg'^!h^*A(-d)[-2d] & & \end{array}$$

where the slant arrows in the parallelogram in the lower part of the diagram is induced by the composite of the following isomorphisms

$$Rg'^!h^* \stackrel{t^{-1}_{g'}}{\to} g'^*h^*(N)[2N] \cong h'^*g^*(N)[2N] \stackrel{t_g}{\to} h'^*Rg^!.$$

Here

$$t_g : g^*(N)[2N] \overset{\cong}{\to} Rg^!, \quad t_{g'} : g'^*(N)[2N] \overset{\cong}{\to} Rg'^!$$

are the isomorphisms in 8.5.2. The difficulty is to prove that the diagram (1) commutes. We will show in a moment that for any $K \in \mathrm{ob}\, D^+(Z', A)$, the following diagram commutes:

$$\begin{array}{ccc} g^*Rh_*K(N)[2N] & \overset{t_g}{\underset{\cong}{\to}} & Rg^!Rh_*K \\ \cong\downarrow & & \uparrow\cong \\ Rh'_*g'^*K(N)[2N] & \overset{Rh'_*(t_{g'})}{\underset{\cong}{\to}} & Rh'_*Rg'^!K, \end{array} \tag{2}$$

where the vertical arrows are the canonical isomorphisms in 7.7.2 and 8.4.9, respectively. Thus through the isomorphisms t_g and $t_{g'}$, the canonical morphism

$$g^*Rh_*(N)[2N] \to Rh'_*g'^*(N)[2N]$$

can be identified with the inverse of the canonical morphism

$$Rh'_*Rg'^! \overset{\cong}{\to} Rg^!Rh_*.$$

The commutativity of the diagram (1) then follows from the commutativity of the following diagram:

$$\begin{array}{ccc} g^* & \to & Rh'_*h'^*g^* \\ \downarrow & & \wr\| \\ g^*Rh_*h^* & \to & Rh'_*g'^*h^*. \end{array}$$

Let us prove that the diagram (2) commutes. By the adjointness of the functors (g^*, Rg_*), it suffices to show that the following diagram commutes:

$$\begin{array}{ccccc} Rh_*K(N)[2N] & & \to & & Rg_*Rg^!Rh_*K \\ \downarrow & \searrow & & & \\ Rh_*Rg'_*g'^*K(N)[2N] & \overset{Rh_*Rg'_*(t_{g'})}{\to} & Rh_*Rg'_*Rg'^!K & & \| \\ \wr\| & & \wr\| & & \\ Rg_*Rh'_*g'^*K(N)[2N] & \overset{Rg_*Rh'_*(t_{g'})}{\to} & Rg_*Rh'_*Rg'^!K & \overset{\cong}{\to} & Rg_*Rg^!Rh_*K, \end{array}$$

where the horizontal arrow on the top of the diagram is obtained from

$$t_g : g^*Rh_*K(N)[2N] \to Rg^!Rh_*K$$

through the adjoint functors (g^*, Rg_*), and the slant arrow is induced by the morphism

$$K(N)[2N] \to Rg'_*Rg'^!K$$

obtained from $t_{g'}$ through the adjoint functors (g'^*, Rg'_*). It is clear that the triangle and the square in this diagram commute. We are thus reduced to show the right part of the above diagram, that is, the following diagram, commutes:

$$\begin{array}{ccc} Rh_*K(N)[2N] & \to & Rg_*Rg^!Rh_*K \\ \downarrow & & \uparrow \cong \\ Rh_*Rg'_*Rg'^!K & \cong & Rg_*Rh'_*Rg'^!K. \end{array} \tag{3}$$

Let

$$\begin{array}{ccc} U & \xrightarrow{k} & X \\ {\scriptstyle g_1}\downarrow & & \downarrow{\scriptstyle g} \\ V & \xrightarrow{j} & Z \end{array}$$

be a Cartesian diagram such that j is etale. Taking its base change with respect to $h : Z' \to Z$, and denote the resulting Cartesian diagram by

$$\begin{array}{ccc} U' & \xrightarrow{k'} & X' \\ {\scriptstyle g'_1}\downarrow & & \downarrow{\scriptstyle g'} \\ V' & \xrightarrow{j'} & Z'. \end{array}$$

Represent K by a bounded below complex of injective sheaves. By the definition of the isomorphisms t_g and $t_{g'}$ at the beginning of 8.5, to prove that the diagram (3) commutes, it suffices to show that the following diagram commutes:

$$\begin{array}{ccc} \mathrm{Hom}(j_!\mathbb{Z}/n(-N)[-2N], h_*K) & \to & \mathrm{Hom}(g_!k_!\mathbb{Z}/n, h_*K) \\ \wr\| & & \uparrow \\ \mathrm{Hom}(j'_!\mathbb{Z}/n(-N)[-2N], K) & \to & \mathrm{Hom}(g'_!k'_!\mathbb{Z}/n, K), \end{array} \tag{4}$$

where the first horizontal arrow is induced by the composite

$$\begin{array}{rcl} g_!k_!\mathbb{Z}/n & \to & \mathscr{H}^{2N}(g_!k_!\mathbb{Z}/n)[-2N] \\ & \cong & R^{2N}g_!k_!\mathbb{Z}/n[-2N] \\ & \cong & j_!R^{2N}g_{1!}\mathbb{Z}/n[-2N] \\ & \stackrel{j_!(\mathrm{Tr}_{g_1})}{\to} & j_!\mathbb{Z}/n(-N)[-2N], \end{array}$$

the second horizontal arrow is induced by the composite defined in the same way with g, k, j, g_1 replaced by g', k', j', g'_1, respectively, and the right vertical arrow is induced by the canonical morphism

$$h^*g_!k_!\mathbb{Z}/n \to g'_!k'_!\mathbb{Z}/n.$$

The commutativity of the diagram (4) follows from 8.2.4 (i) applied to the Cartesian diagram

$$\begin{array}{ccc} U' & \to & U \\ {\scriptstyle g'_1}\downarrow & & \downarrow{\scriptstyle g_1} \\ V' & \to & V. \end{array}$$

□

Suppose that X is a smooth compactifiable scheme pure of dimension N over an algebraically closed field k. Any Weil divisor D on X defines an element in $H^1(X, \mathscr{O}^*_{X_{\mathrm{et}}})$ represented by the invertible $\mathscr{O}_X$-module $\mathscr{L}(D)$ by 7.1.4 and 7.1.5. Kummer's theory defines a homomorphism

$$H^1(X, \mathscr{O}^*_{X_{\mathrm{et}}}) \to H^2(X, \boldsymbol{\mu}_n).$$

Let $\mathrm{cl}(\mathscr{L}(D)) \in H^2(X, \boldsymbol{\mu}_n)$ be the image of $\mathscr{L}(D)$ under this homomorphism.

Proposition 8.6.5. *Under the above assumption, we have* $\mathrm{cl}(D) = \mathrm{cl}(\mathscr{L}(D))$ *in* $H^2(X, \boldsymbol{\mu}_n)$.

Proof. It suffices to treat the case where $D = Y$ is an integral closed subscheme of X of codimension 1. Let $\mathscr{K}^*$ be the etale sheaf on X so that $\mathscr{K}^*(U)$ is the multiplicative group of the function field of U for any etale X-scheme U, and let $\mathscr{O}^*$ be the etale sheaf $\mathscr{O}^*_{X_{\mathrm{et}}}$. The Weil divisor D defines a Cartier divisor, that is, a section $f_D \in \Gamma(X, \mathscr{K}^*/\mathscr{O}^*)$. It is clear that f_D lies in $\Gamma_Y(X, \mathscr{K}^*/\mathscr{O}^*)$. Let

$$\delta : \Gamma_Y(X, \mathscr{K}^*/\mathscr{O}^*) \to H^1_Y(X, \mathscr{O}^*)$$

be the homomorphism on cohomology groups arising from the short exact sequence

$$0 \to \mathscr{O}^* \to \mathscr{K}^* \to \mathscr{K}^*/\mathscr{O}^* \to 0.$$

Then the image of $\delta(\frac{1}{f_D}) \in H^1_Y(X, \mathscr{O}^*)$ under the canonical homomorphism

$$H^1_Y(X, \mathscr{O}^*) \to H^1(X, \mathscr{O}^*)$$

is $\mathscr{L}(D)$. Kummer's theory defines a homomorphism

$$H^1_Y(X, \mathscr{O}^*) \to H^2_Y(X, \boldsymbol{\mu}_n).$$

Let $\mathrm{cl}'(\mathscr{L}(D)) \in H^2_Y(X, \boldsymbol{\mu}_n)$ be the image of $\delta(\frac{1}{f_D})$ under this homomorphism. Then the image of $\mathrm{cl}'(\mathscr{L}(D))$ under the canonical homomorphism

$$H^2_Y(X, \boldsymbol{\mu}_n) \to H^2(X, \boldsymbol{\mu}_n)$$

is $\mathrm{cl}(\mathscr{L}(D))$. To prove the proposition, it suffices to show $\mathrm{cl}'(D) = \mathrm{cl}'(\mathscr{L}(D))$ in $H^2_Y(X, \boldsymbol{\mu}_n)$.

Let U be an open subset of X so that $U \cap Y$ is nonempty and smooth. Taking U sufficiently small, we may assume that there exists an etale morphism

$$U \to \mathbb{A}^N_k = \operatorname{Spec} k[t_1, \ldots, t_N]$$

such that

$$U \cap Y \cong U \otimes_{k[t_1,\ldots,t_N]} k[t_1,\ldots,t_{N-1}].$$

On the other hand, we have

$$H^2_Y(X,\boldsymbol{\mu}_n) \cong H^2_{Y\cap U}(X,\boldsymbol{\mu}_n) \cong H^2_{Y\cap U}(U,\boldsymbol{\mu}_n).$$

So we may replace X by U. We then have a Cartesian diagram

$$\begin{array}{ccc} Y & \to & X \\ \downarrow & & \downarrow \\ \mathbb{A}_k^{N-1} & \stackrel{i}{\to} & \mathbb{A}_k^N, \end{array}$$

where $i : \mathbb{A}_k^{N-1} \to \mathbb{A}_k^N$ is the closed immersion corresponding to the coordinate plane $t_N = 0$, and $X \to \mathbb{A}_k^N$ is etale. Regard $\mathbb{A}_k^{N-1}$ as a Weil divisor of $\mathbb{A}_k^N$ through this closed immersion. One can verify that $\mathrm{cl}'(\mathscr{L}(D))$ is the image of $\mathrm{cl}'(\mathscr{L}(\mathbb{A}_k^{N-1}))$ under the canonical homomorphism

$$H^2_{\mathbb{A}_k^{N-1}}(\mathbb{A}_k^N,\boldsymbol{\mu}_n) \to H^2_Y(X,\boldsymbol{\mu}_n).$$

By 8.6.3, $\mathrm{cl}'(D)$ is also the image of $\mathrm{cl}'(\mathbb{A}_k^{N-1})$ under this homomorphism. So we are reduced to the case where $Y \to X$ is $\mathbb{A}_k^{N-1} \to \mathbb{A}_k^N$. We have a commutative diagram

$$\begin{array}{ccc} \mathbb{A}_k^{N-1} & \stackrel{i}{\to} & \mathbb{A}_k^N \\ & \mathrm{id}\searrow & \downarrow \pi \\ & & \mathbb{A}_k^{N-1}, \end{array}$$

where $\pi : \mathbb{A}_k^N \to \mathbb{A}_k^{N-1}$ is the projection to the first $N-1$ coordinates. By 8.6.1, when we define $\mathrm{cl}'(Y)$, we may use $\mathbb{A}_k^{N-1}$ as the base scheme instead of $\mathrm{Spec}\, k$. The above commutative diagram can be obtained by base change from the diagram

$$\begin{array}{ccc} \{0\} & \to & \mathbb{A}_k^1 \\ & \cong\searrow & \downarrow \\ & & \mathrm{Spec}\, k. \end{array}$$

Using 8.6.4, we are reduced to the case where $Y \to X$ is the closed immersion $\{0\} \to \mathbb{A}_k^1$, and then to the case where $Y \to X$ is the closed immersion $\{0\} \to \mathbb{P}_k^1$. We have a long exact sequence

$$\cdots \to H^1(\mathbb{A}_k^1,\boldsymbol{\mu}_n) \to H^2_{\{0\}}(\mathbb{P}_k^1,\boldsymbol{\mu}_n) \to H^2(\mathbb{P}_k^1,\boldsymbol{\mu}_n) \to \cdots.$$

One can show $H^1(\mathbb{A}_k^1,\boldsymbol{\mu}_n) = 0$. (Use 7.7.4, or Kummer's theory and 7.1.2–5.) So

$$H^2_{\{0\}}(\mathbb{P}_k^1,\boldsymbol{\mu}_n) \to H^2(\mathbb{P}_k^1,\boldsymbol{\mu}_n)$$

is injective. To prove the proposition, it suffices to show $\mathrm{cl}(D) = \mathrm{cl}(\mathscr{L}(D))$. Since

$$\mathrm{Tr}_{\mathbb{P}^1_k/k} : H^2(\mathbb{P}^1_k, \boldsymbol{\mu}_n) \to \mathbb{Z}/n$$

is an isomorphism, it suffices to show

$$\mathrm{Tr}_{\mathbb{P}^1_k/k}(\mathrm{cl}(D)) = \mathrm{Tr}_{\mathbb{P}^1_k/k}(\mathrm{cl}(\mathscr{L}(D))).$$

We have

$$\mathrm{Tr}_{\mathbb{P}^1_k/k}(\mathrm{cl}(\mathscr{L}(D))) = \deg(\mathscr{L}(D)) = 1.$$

By 8.6.2, through the Poincaré duality

$$\mathrm{Hom}(H^0(\mathbb{P}^1_k, \mathbb{Z}/n), \mathbb{Z}/n) \cong H^2(\mathbb{P}^1_k, \boldsymbol{\mu}_n),$$

$\mathrm{cl}(D)$ corresponds to the composite

$$H^0(\mathbb{P}^1_k, \mathbb{Z}/n) \to H^0(\{0\}, \mathbb{Z}/n) \stackrel{\mathrm{Tr}_{\{0\}/k}}{\to} \mathbb{Z}/n.$$

On the other hand, the homomorphism $\phi : H^0(\mathbb{P}^1_k, \mathbb{Z}/n) \to \mathbb{Z}/n$ corresponding to $\mathrm{cl}(D) \in H^2(\mathbb{P}^1_k, \boldsymbol{\mu}_n)$ has the property $\phi(1) = \mathrm{Tr}_{\mathbb{P}^1_k/k}(\mathrm{cl}(D))$. Using these facts, one proves $\mathrm{Tr}_{\mathbb{P}^1_k/k}(\mathrm{cl}(D)) = 1$. □

Lemma 8.6.6. *Let X be a smooth proper scheme of dimension d over an algebraically closed field k. The perfect pairing*

$$H^q(X, \mathbb{Z}/n) \times H^{2d-q}(X, \mathbb{Z}/n(d)) \to \mathbb{Z}/n$$

defined by 8.5.3 coincides with the pairing

$$H^q(X, \mathbb{Z}/n) \times H^{2d-q}(X, \mathbb{Z}/n(d)) \to \mathbb{Z}/n, \quad (s,t) \mapsto \mathrm{Tr}_{X/k}(t \cup s).$$

Proof. Let $\mathscr{C}^{\cdot}(\mathbb{Z}/n)$ be the Godement resolution of $\mathbb{Z}/n$ on X, and let $\mathscr{I}^{\cdot}$ be a resolution of $\mathbb{Z}/n(d)$ by injective sheaves of $\mathbb{Z}/n$-modules on X. The canonical quasi-isomorphism

$$\mathbb{Z}/n \to \mathscr{C}^{\cdot}(\mathbb{Z}/n)$$

induces a quasi-isomorphism

$$\mathscr{H}om^{\cdot}(\mathscr{C}^{\cdot}(\mathbb{Z}/n), \mathscr{I}^{\cdot}) \to \mathscr{H}om^{\cdot}(\mathbb{Z}/n, \mathscr{I}^{\cdot}).$$

By 6.4.9, the stalk of $\mathscr{C}^{\cdot}(\mathbb{Z}/n)$ at each geometric point is homotopically equivalent to $\mathbb{Z}/n$. So the above quasi-isomorphism induces a quasi-isomorphism

$$\mathscr{H}om^{\cdot}(\mathscr{C}^{\cdot}(\mathbb{Z}/n), \mathscr{I}^{\cdot}) \otimes_{\mathbb{Z}/n} \mathscr{C}^{\cdot}(\mathbb{Z}/n) \to \mathscr{H}om^{\cdot}(\mathbb{Z}/n, \mathscr{I}^{\cdot}) \otimes_{\mathbb{Z}/n} \mathscr{C}^{\cdot}(\mathbb{Z}/n),$$

and the stalkwise homotopic equivalence $\mathbb{Z}/n \to \mathscr{C}^\cdot(\mathbb{Z}/n)$ induces a quasi-isomorphism

$$\mathscr{K}^\cdot \otimes_{\mathbb{Z}/n} \mathbb{Z}/n \to \mathscr{K}^\cdot \otimes_{\mathbb{Z}/n} \mathscr{C}^\cdot(\mathbb{Z}/n)$$

for any complex of sheaves $\mathscr{K}^\cdot$. One can verify that the following diagram commutes:

$$\begin{array}{ccc}
\mathscr{H}om^\cdot(\mathscr{C}^\cdot(\mathbb{Z}/n), \mathscr{I}^\cdot) \otimes_{\mathbb{Z}/n} \mathbb{Z}/n \to \mathscr{H}om^\cdot(\mathscr{C}^\cdot(\mathbb{Z}/n), \mathscr{I}^\cdot) \otimes_{\mathbb{Z}/n} \mathscr{C}^\cdot(\mathbb{Z}/n) & \stackrel{\mathrm{Ev}}{\to} & \mathscr{I}^\cdot \\
\downarrow & & \| \\
\mathscr{H}om^\cdot(\mathbb{Z}/n, \mathscr{I}^\cdot) \otimes_{\mathbb{Z}/n} \mathbb{Z}/n & \stackrel{\mathrm{Ev}}{\to} & \mathscr{I}^\cdot,
\end{array}$$

where Ev are the evaluation morphisms. The evaluation morphism on the second line is an isomorphism. It follows that the evaluation morphism on the first line is a quasi-isomorphism since all other morphisms in the above diagram are quasi-isomorphisms. We can find a commutative diagram (up to homotopy)

$$\begin{array}{ccccc}
 & & \mathscr{H}om^\cdot(\mathscr{C}^\cdot(\mathbb{Z}/n), \mathscr{I}^\cdot) \otimes_{\mathbb{Z}/n} \mathscr{C}^\cdot(\mathbb{Z}/n) & \stackrel{\mathrm{Ev}}{\to} & \mathscr{I}^\cdot \\
 & & \downarrow & & \\
 & & \mathscr{H}om^\cdot(\mathbb{Z}/n, \mathscr{I}^\cdot) \otimes_{\mathbb{Z}/n} \mathscr{C}^\cdot(\mathbb{Z}/n) & & \downarrow \\
 & & \wr\| & & \\
\mathscr{I}^\cdot \cong \mathscr{I}^\cdot \otimes_{\mathbb{Z}/n} \mathbb{Z}/n & \to & \mathscr{I}^\cdot \otimes_{\mathbb{Z}/n} \mathscr{C}^\cdot(\mathbb{Z}/n) & \to & \mathscr{J}^\cdot
\end{array}$$

such that $\mathscr{J}^\cdot$ is a complex of injective sheaves of $\mathbb{Z}/n$-modules, and all arrows in the diagram are quasi-isomorphisms. We have

$$\mathbb{Z}/n(d) \cong \mathbb{Z}/n \otimes^L_{\mathbb{Z}/n} \mathbb{Z}/n(d) \cong \mathscr{C}^\cdot(\mathbb{Z}/n) \otimes_{\mathbb{Z}/n} \mathscr{I}^\cdot$$

in $D(X, \mathbb{Z}/n)$. The cup product

$$R\Gamma(X, \mathbb{Z}/n(d)) \otimes^L_{\mathbb{Z}/n} R\Gamma(X, \mathbb{Z}/n) \to R\Gamma(X, \mathbb{Z}/n(d))$$

is the composite of the following morphisms

$$\begin{aligned}
& R\Gamma(X, \mathbb{Z}/n(d)) \otimes^L_{\mathbb{Z}/n} R\Gamma(X, \mathbb{Z}/n) \\
\cong\ & \Gamma(X, \mathscr{I}^\cdot) \otimes^L_{\mathbb{Z}/n} \Gamma(X, \mathscr{C}^\cdot(\mathbb{Z}/n)) \\
\to\ & \Gamma(X, \mathscr{I}^\cdot) \otimes_{\mathbb{Z}/n} \Gamma(X, \mathscr{C}^\cdot(\mathbb{Z}/n)) \\
\to\ & \Gamma(X, \mathscr{I}^\cdot \otimes_{\mathbb{Z}/n} \mathscr{C}^\cdot(\mathbb{Z}/n)) \\
\to\ & \Gamma(X, \mathscr{J}^\cdot) \\
\cong\ & R\Gamma(X, \mathbb{Z}/n(d)).
\end{aligned}$$

The pairing in 8.5.3 is the composite

$$\begin{aligned}
&H^q(X, \mathbb{Z}/n) \times H^{2d-q}(X, \mathbb{Z}/n(d)) \\
\cong\ & H^q\big(\Gamma(X, \mathscr{C}^\cdot(\mathbb{Z}/n))\big) \times H^{2d-q}\Big(\Gamma\big(X, \mathscr{H}om^\cdot(\mathscr{C}^\cdot(\mathbb{Z}/n), \mathscr{I}^\cdot)\big)\Big) \\
\to\ & H^{2d}\Big(\Gamma\big(X, \mathscr{H}om^\cdot(\mathscr{C}^\cdot(\mathbb{Z}/n), \mathscr{I}^\cdot) \otimes_{\mathbb{Z}/n} \mathscr{C}^\cdot(\mathbb{Z}/n)\big)\Big) \\
\stackrel{\mathrm{Ev}}{\to}\ & H^{2d}(\Gamma(X, \mathscr{I}^\cdot)) \\
\cong\ & H^{2d}(X, \mathbb{Z}/n(d)) \\
\stackrel{\mathrm{Tr}_{X/k}}{\to}\ & \mathbb{Z}/n.
\end{aligned}$$

Our assertion follows from the commutativity of the following diagram:

$$\begin{array}{ccccc}
\begin{array}{c} H^{2d-q}(\Gamma(X, \mathscr{H}om^\cdot(\mathscr{C}^\cdot(\mathbb{Z}/n), \mathscr{I}^\cdot))) \\ \times H^q(\Gamma(X, \mathscr{C}^\cdot(\mathbb{Z}/n))) \end{array} & \to & \begin{array}{c} H^{2d}(\Gamma(X, \mathscr{H}om^\cdot(\mathscr{C}^\cdot(\mathbb{Z}/n), \mathscr{I}^\cdot) \\ \otimes_{\mathbb{Z}/n}\mathscr{C}^\cdot(\mathbb{Z}/n))) \end{array} & \stackrel{\mathrm{Ev}}{\to} & H^{2d}(\Gamma(X, \mathscr{I}^\cdot)) \\
\cong\downarrow & & \downarrow\cong & & \\
\begin{array}{c} H^{2d-q}(\Gamma(X, \mathscr{H}om^\cdot(\mathbb{Z}/n, \mathscr{I}^\cdot))) \\ \times H^q(\Gamma(X, \mathscr{C}^\cdot(\mathbb{Z}/n))) \end{array} & \to & \begin{array}{c} H^{2d}(\Gamma(X, \mathscr{H}om^\cdot(\mathbb{Z}/n, \mathscr{I}^\cdot) \\ \otimes_{\mathbb{Z}/n}\mathscr{C}^\cdot(\mathbb{Z}/n))) \end{array} & & \downarrow \\
\cong\downarrow & & \downarrow\cong & & \\
\begin{array}{c} H^{2d-q}(\Gamma(X, \mathscr{I}^\cdot)) \\ \times H^q(\Gamma(X, \mathscr{C}^\cdot(\mathbb{Z}/n))) \end{array} & \to & H^{2d}(\Gamma(X, \mathscr{I}^\cdot \otimes_{\mathbb{Z}/n} \mathscr{C}^\cdot(\mathbb{Z}/n))) & \to & H^{2d}(\Gamma(X, \mathscr{I}^\cdot)).
\end{array}$$

□

Proposition 8.6.7 (Lefschetz Fixed Point Formula). *Let X be a smooth projective curve over an algebraically closed field k, $h : X \to X$ a k-morphism, $\Gamma_h : X \to X \times_k X$ the graph of h, Δ the divisor of $X \times_k X$ defined by the diagonal morphism $\Delta : X \to X \times_k X$, and $\Gamma_h^*(\Delta)$ the pulling back of the divisor Δ. Then we have*

$$\sum_{q=0}^{2} (-1)^q \mathrm{Tr}(h^*, H^q(X, \mathbb{Z}/n)) \equiv \deg(\Gamma_h^*(\Delta)) \bmod n.$$

Note that by 7.2.9 (ii), $H^q(X, \mathbb{Z}/n)$ are free $\mathbb{Z}/n$-modules for all q so that we can talk about $\mathrm{Tr}(h^*, H^q(X, \mathbb{Z}/n))$.

Proof. Fix a primitive n-th root of unity and use it to define isomorphisms $\boldsymbol{\mu}_n \cong \mathbb{Z}/n$ and $\mathbb{Z}/n(d) \cong \mathbb{Z}/n$ for all d. The homomorphism

$$\mathrm{Tr}_{X/k} : H^2(X, \mathbb{Z}/n(1)) \to \mathbb{Z}/n$$

can be identified with a homomorphism

$$H^2(X, \mathbb{Z}/n) \to \mathbb{Z}/n$$

which we still denote by $\mathrm{Tr}_{X/k}$. Similarly we have a homomorphism

$$\mathrm{Tr}_{X \times_k X/k} : H^4(X \times_k X, \mathbb{Z}/n) \to \mathbb{Z}/n.$$

For any $s, t \in H^2(X, \mathbb{Z}/n)$, we have

$$\mathrm{Tr}_{X\times_k X/k}(\pi_1^* s \cup \pi_2^* t) = \mathrm{Tr}_{X/k}(s)\mathrm{Tr}_{X/k}(t)$$

by 8.2.4 (iv), where $\pi_1, \pi_2 : X \times_k X \to X$ are the projections. By 8.5.3 and 8.6.6, we have perfect pairings

$$H^q(X, \mathbb{Z}/n) \times H^{2-q}(X, \mathbb{Z}/n) \to \mathbb{Z}/n, \ \ (s,t) \mapsto \mathrm{Tr}_{X/k}(s \cup t),$$

$$H^q(X \times_k X, \mathbb{Z}/n) \times H^{4-q}(X \times_k X, \mathbb{Z}/n) \to \mathbb{Z}/n, \ \ (s,t) \mapsto \mathrm{Tr}_{X\times_k X/k}(s \cup t).$$

By 7.4.11, we have isomorphisms

$$\bigoplus_{u+v=q} H^u(X, \mathbb{Z}/n) \otimes H^v(X, \mathbb{Z}/n) \to H^q(X \times_k X, \mathbb{Z}/n),$$

$$s \otimes t \mapsto \pi_1^* s \cup \pi_2^* t.$$

For each u, choose a basis $\{e_{u1}, \ldots, e_{ub_u}\}$ for the free $\mathbb{Z}/n$-module $H^u(X, \mathbb{Z}/n)$. Let $\{f_{u1}, \ldots, f_{ub_u}\}$ be the basis of $H^{2-u}(X, \mathbb{Z}/n)$ so that

$$\mathrm{Tr}_{X/k}(e_{u\alpha} \cup f_{u\beta}) = \delta_{\alpha\beta}.$$

Then $\{\pi_1^* e_{u\alpha} \cup \pi_2^* f_{u\beta}\}_{u,\alpha,\beta}$ is a basis of $H^2(X \times_k X, \mathbb{Z}/n)$. We claim that

$$\mathrm{cl}(\Delta) = \sum_{v,\gamma} \pi_1^* f_{v\gamma} \cup \pi_2^* e_{v\gamma}.$$

Indeed, by 8.6.2, through the isomorphism

$$\mathrm{Hom}(H^2(X \times_k X, \mathbb{Z}/n), \mathbb{Z}/n) \cong H^2(X \times_k X, \mathbb{Z}/n)$$

induced by the perfect pairing

$$H^2(X \times_k X, \mathbb{Z}/n) \times H^2(X \times_k X, \mathbb{Z}/n) \to \mathbb{Z}/n, \quad (s,t) \mapsto \mathrm{Tr}_{X\times_k X/k}(s \cup t),$$

$\mathrm{cl}(\Delta) \in H^2(X \times_k X, \mathbb{Z}/n)$ corresponds to the composite

$$H^2(X \times_k X, \mathbb{Z}/n) \xrightarrow{\Delta^*} H^2(X, \mathbb{Z}/n) \xrightarrow{\mathrm{Tr}_{X/k}} \mathbb{Z}/n.$$

So for any $s \in H^2(X \times_k X, \mathbb{Z}/n)$, we have

$$\mathrm{Tr}_{X\times_k X/k}(\mathrm{cl}(\Delta) \cup s) = \mathrm{Tr}_{X/k}(\Delta^*(s)).$$

In particular, we have

$$\begin{aligned}\mathrm{Tr}_{X\times_k X/k}\Big(\mathrm{cl}(\Delta) \cup (\pi_1^* e_{u\alpha} \cup \pi_2^* f_{u\beta})\Big) &= \mathrm{Tr}_{X/k}\Big(\Delta^*(\pi_1^* e_{u\alpha} \cup \pi_2^* f_{u\beta})\Big) \\ &= \mathrm{Tr}_{X/k}(\Delta^*\pi_1^* e_{u\alpha} \cup \Delta^*\pi_2^* f_{u\beta}) \\ &= \mathrm{Tr}_{X/k}(e_{u\alpha} \cup f_{u\beta}) \\ &= \delta_{\alpha\beta}.\end{aligned}$$

On the other hand, we have

$$
\begin{aligned}
&\mathrm{Tr}_{X\times_k X/k}\Big(\Big(\sum_{v,\gamma}\pi_1^*f_{v\gamma}\cup\pi_2^*e_{v\gamma}\Big)\cup(\pi_1^*e_{u\alpha}\cup\pi_2^*f_{u\beta})\Big)\\
&=\mathrm{Tr}_{X\times_k X/k}\Big(\sum_{v,\gamma}\pi_1^*f_{v\gamma}\cup\pi_2^*e_{v\gamma}\cup\pi_1^*e_{u\alpha}\cup\pi_2^*f_{u\beta}\Big)\\
&=\mathrm{Tr}_{X\times_k X/k}\Big(\sum_{v,\gamma}(-1)^{u(2-v+v)}\pi_1^*e_{u\alpha}\cup\pi_1^*f_{v\gamma}\cup\pi_2^*e_{v\gamma}\cup\pi_2^*f_{u\beta}\Big)\\
&=\mathrm{Tr}_{X\times_k X/k}\Big(\sum_{v,\gamma}\pi_1^*(e_{u\alpha}\cup f_{v\gamma})\cup\pi_2^*(e_{v\gamma}\cup f_{u\beta})\Big).
\end{aligned}
$$

When $u\neq v$, we have either $u+2-v\geq 3$ or $v+2-u\geq 3$. So we have either $e_{u\alpha}\cup f_{v\gamma}=0$ or $e_{v\gamma}\cup f_{u\beta}=0$. Hence

$$
\begin{aligned}
&\mathrm{Tr}_{X\times_k X/k}\Big(\Big(\sum_{v,\gamma}\pi_1^*f_{v\gamma}\cup\pi_2^*e_{v\gamma}\Big)\cup(\pi_1^*e_{u\alpha}\cup\pi_2^*f_{u\beta})\Big)\\
&=\mathrm{Tr}_{X\times_k X/k}\Big(\sum_{\gamma}\pi_1^*(e_{u\alpha}\cup f_{u\gamma})\cup\pi_2^*(e_{u\gamma}\cup f_{u\beta})\Big)\\
&=\sum_{\gamma}\mathrm{Tr}_{X/k}(e_{u\alpha}\cup f_{u\gamma})\mathrm{Tr}_{X/k}(e_{u\gamma}\cup f_{u\beta})\\
&=\sum_{\gamma}\delta_{\alpha\gamma}\delta_{\beta\gamma}\\
&=\delta_{\alpha\beta}.
\end{aligned}
$$

Therefore

$$
\begin{aligned}
&\mathrm{Tr}_{X\times_k X/k}(\mathrm{cl}(\Delta)\cup(\pi_1^*e_{u\alpha}\cup\pi_2^*f_{u\beta}))\\
&=\mathrm{Tr}_{X\times_k X/k}\Big(\Big(\sum_{v,\gamma}\pi_1^*f_{v\gamma}\cup\pi_2^*e_{v\gamma}\Big)\cup(\pi_1^*e_{u\alpha}\cup\pi_2^*f_{u\beta})\Big).
\end{aligned}
$$

So we have $\mathrm{cl}(\Delta)=\sum_{v,\gamma}\pi_1^*f_{v\gamma}\cup\pi_2^*e_{v\gamma}$. This proves our claim. Write

$$h^*(e_{v\gamma})=\sum_{\delta}a_{v\gamma\delta}e_{v\delta},$$

where $a_{v\gamma\delta}\in\mathbb{Z}/n$. Then we have

$$
\begin{aligned}
\Gamma_h^*(\mathrm{cl}(\Delta))&=\Gamma_h^*\Big(\sum_{v,\gamma}\pi_1^*f_{v\gamma}\cup\pi_2^*e_{v\gamma}\Big)\\
&=\sum_{v,\gamma}\Gamma_h^*\pi_1^*f_{v\gamma}\cup\Gamma_h^*\pi_2^*e_{v\gamma}\\
&=\sum_{v,\gamma}f_{v\gamma}\cup h^*e_{v\gamma}\\
&=\sum_{v,\gamma,\delta}a_{v\gamma\delta}f_{v\gamma}\cup e_{v\delta}.
\end{aligned}
$$

So we have

$$\begin{aligned}\mathrm{Tr}_{X/k}\Big(\Gamma_h^*(\mathrm{cl}(\Delta))\Big) &= \sum_{v,\gamma,\delta} a_{v\gamma\delta}\mathrm{Tr}_{X/k}(f_{v\gamma}\cup e_{v\delta})\\ &= \sum_{v,\gamma,\delta}(-1)^{v(2-v)}a_{v\gamma\delta}\mathrm{Tr}_{X/k}(e_{v\delta}\cup f_{v\gamma})\\ &= \sum_{v,\gamma,\delta}(-1)^{v}a_{v\gamma\delta}\delta_{\gamma\delta}\\ &= \sum_{v,\gamma}(-1)^{v}a_{v\gamma\gamma}\\ &= \sum_{v}(-1)^{v}\mathrm{Tr}(h^*, H^v(X,\mathbb{Z}/n)).\end{aligned}$$

On the other hand, we have

$$\Gamma_h^*(\mathrm{cl}(\Delta)) = \Gamma_h^*\big(\mathrm{cl}(\mathscr{L}(\Delta))\big) = \mathrm{cl}\big(\mathscr{L}(\Gamma_h^*(\Delta))\big).$$

So

$$\mathrm{Tr}_{X/k}\Big(\Gamma_h^*(\mathrm{cl}(\Delta))\Big) \equiv \deg(\Gamma_h^*(\Delta)) \bmod n.$$

Therefore

$$\sum_{v}(-1)^{v}\mathrm{Tr}(h^*, H^v(X,\mathbb{Z}/n)) \equiv \deg(\Gamma_h^*(\Delta)) \bmod n.$$

□

Corollary 8.6.8. *Let X be a smooth projective curve over an algebraically closed field k and let $h : X \to X$ be a k-morphism with isolated fixed points $x_1, \ldots, x_m$. For each x_i, let v_{x_i} be the valuation on the function field of X defined by x_i, and let $\pi_{x_i} \in \mathscr{O}_{X,x_i}$ be a uniformizer for v_{x_i}. Then we have*

$$\sum_{j=0}^{2}(-1)^{j}\mathrm{Tr}(h^*, H^j(X,\mathbb{Z}/n)) \equiv \sum_{i=1}^{m} v_{x_i}(\pi_{x_i} - h^*(\pi_{x_i})) \bmod n.$$

Remark 8.6.9. Note that the number $v_{x_i}(\pi_{x_i} - h^*(\pi_{x_i}))$ is independent of the choice of the uniformizer π_{x_i} of v_{x_i}. In fact, it is the largest integer j so that h is the identity on $\mathscr{O}_{X,x_i}/\mathfrak{m}_{x_i}^j$. We call $v_{x_i}(\pi_{x_i} - h^*(\pi_{x_i}))$ the *multiplicity of the fixed point x_i*.

Proof. Let x be a closed point of X. Let us calculate the multiplicity of the divisor $\Gamma_h^*(\Delta)$ at x. If x is not a fixed point of h, then $(x, h(x))$ does not lie in Δ, and hence the multiplicity of $\Gamma_h^*(\Delta)$ at x is 0. Suppose that x is one of the isolated fixed point x_i. Since X is smooth, we can find an open neighborhood U of x in X admitting an etale morphism $f : U \to \mathbb{A}_k^1$

that maps x to the origin of $\mathbb{A}_k^1$. Let Δ_U and $\Delta_{\mathbb{A}_k^1}$ be the diagonal divisors of $U \times_k U$ and $\mathbb{A}_k^1 \times_k \mathbb{A}_k^1$, respectively. Consider the commutative diagram

$$\begin{array}{ccccc} U & \stackrel{\Delta_f}{\to} & U \times_{\mathbb{A}_k^1} U & \to & U \times_k U \\ & f \searrow & \downarrow & & \downarrow f \times f \\ & & \mathbb{A}_k^1 & \stackrel{\Delta_{\mathbb{A}_k^1}}{\to} & \mathbb{A}_k^1 \times_k \mathbb{A}_k^1. \end{array}$$

The square in the diagram is Cartesian, and the diagonal morphism Δ_f is an open immersion by 2.3.9. Δ_U coincides with the composite

$$U \stackrel{\Delta_f}{\to} U \times_{\mathbb{A}_k^1} U \to U \times_k U.$$

It follows that the divisors Δ_U and $(f \times f)^*(\Delta_{\mathbb{A}_k^1})$ coincide in a neighborhood of (x, x) in $U \times_k U$. Hence the divisors $\Gamma_h^*(\Delta)$ and $(f, fh)^{-1}(\Delta_{\mathbb{A}_k^1})$ coincide in a neighborhood of x in X. The divisor $\Delta_{\mathbb{A}_k^1}$ on $\mathbb{A}_k^1 \times_k \mathbb{A}_k^1 \cong \mathrm{Spec}\,(k[t] \otimes_k k[t])$ is the principle divisor defined by $t \otimes 1 - 1 \otimes t$. So in a neighborhood of x in X, $\Gamma_h^*(\Delta)$ is the principle divisor defined by $f^*(t) - h^* f^*(t)$. Since f is etale, $f^*(t)$ is a uniformizer for v_x. For any uniformizer π_x of v_x, we have

$$v_x(f^*(t) - h^* f^*(t)) = v_x(\pi_x - h^*(\pi_x)).$$

So we have

$$\deg(\Gamma_h^*(\Delta)) = \sum_{i=1}^m v_{x_i}(\pi_{x_i} - h^*(\pi_{x_i})).$$

We then apply 8.6.7. □

Chapter 9

Finiteness Theorems

9.1 Sheaves with Group Actions

Let G be a group acting on a scheme X on the right. For any etale X-scheme $p : V \to X$, and any $\sigma \in G$, let V_σ be the etale X-scheme $\sigma^{-1}p : V \to X$, and let $V \times_{X,\sigma} X$ be the etale X-scheme obtained from $p : V \to X$ by the base change $\sigma : X \to X$. Then the projection $V \times_{X,\sigma} X \to V_\sigma$ is an isomorphism of etale X-schemes.

Let $\mathscr{F}$ be a sheaf on X. An action of G on $\mathscr{F}$ compatible with the action of G on X is a family of isomorphisms

$$\phi_\sigma : \mathscr{F} \to \sigma_* \mathscr{F} \quad (\sigma \in G)$$

such that $\phi_e = \mathrm{id}$ and for any $\sigma, \tau \in G$, the following diagram commutes:

$$\begin{array}{ccc} \mathscr{F} & \stackrel{\phi_\sigma}{\to} & \sigma_* \mathscr{F} \\ \phi_{\tau\sigma} \downarrow & & \downarrow \sigma_*(\phi_\tau) \\ (\tau\sigma)_* \mathscr{F} & \cong & \sigma_*(\tau_* \mathscr{F}). \end{array}$$

Equivalently, an action of G on $\mathscr{F}$ is a family of isomorphisms

$$\phi_{\sigma,V} : \mathscr{F}(V) \to \mathscr{F}(V_\sigma)$$

for any $\sigma \in G$ and any etale X-scheme V such that for any X-morphism $V' \to V$ of etale X-schemes and any $\sigma, \tau \in G$, the following diagrams commute:

$$\begin{array}{ccc} \mathscr{F}(V) & \stackrel{\phi_{\sigma,V}}{\to} & \mathscr{F}(V_\sigma) \\ \downarrow & & \downarrow \\ \mathscr{F}(V') & \stackrel{\phi_{\sigma,V'}}{\to} & \mathscr{F}(V'_\sigma), \end{array} \qquad \begin{array}{ccc} \mathscr{F}(V) & \stackrel{\phi_{\sigma,V}}{\to} & \mathscr{F}(V_\sigma) \\ \phi_{\tau\sigma,V} \downarrow & & \downarrow \phi_{\tau,V_\sigma} \\ \mathscr{F}(V_{\tau\sigma}) & = & \mathscr{F}((V_\sigma)_\tau). \end{array}$$

We can also define an action of G on $\mathscr{F}$ to be a family of isomorphisms

$$\psi_\sigma : \sigma^* \mathscr{F} \to \mathscr{F} \quad (\sigma \in G)$$

such that $\psi_e = \mathrm{id}$ and for any $\sigma, \tau \in G$, the following diagram commutes:

$$\begin{array}{ccc} \tau^*(\sigma^*\mathscr{F}) & \stackrel{\tau^*(\psi_\sigma)}{\to} & \tau^*\mathscr{F} \\ \wr\| & & \downarrow \psi_\tau \\ (\tau\sigma)^*\mathscr{F} & \stackrel{\psi_{\tau\sigma}}{\to} & \mathscr{F}. \end{array}$$

A sheaf $\mathscr{F}$ on X with an action of G is called a G-sheaf. We define morphisms of G-sheaves to be morphisms of sheaves commuting with the group actions. For all G-sheaves $\mathscr{F}$ and $\mathscr{G}$ on X, denote by $\mathrm{Hom}_G(\mathscr{F}, \mathscr{G})$ the set of G-morphisms from $\mathscr{F}$ to $\mathscr{G}$.

Let $\mathscr{F}$ be a sheaf on X. Define a G-sheaf $\mathscr{F}[G]$ by

$$\mathscr{F}[G] = \bigoplus_{\sigma \in G} \sigma_* \mathscr{F},$$

where for any $\tau \in G$,

$$\phi_\tau : \mathscr{F}[G] \to \tau_*(\mathscr{F}[G])$$

is the composite

$$\mathscr{F}[G] \cong \bigoplus_{\sigma \in G} (\sigma\tau)_* \mathscr{F} \cong \bigoplus_{\sigma \in G} \tau_*(\sigma_* \mathscr{F}) \cong \tau_*(\mathscr{F}[G]).$$

For any G-sheaf $\mathscr{G}$ on X, we have a one-to-one correspondence

$$\mathrm{Hom}(\mathscr{F}, \mathscr{G}) \cong \mathrm{Hom}_G(\mathscr{F}[G], \mathscr{G})$$

which is functorial in $\mathscr{F}$ and $\mathscr{G}$. So the functor $\mathscr{F} \to \mathscr{F}[G]$ from the category of sheaves to the category of G-sheaves is left adjoint to the forgetful functor from the category of G-sheaves to the category of sheaves. As the functor $\mathscr{F} \to \mathscr{F}[G]$ is exact, if $\mathscr{G}$ is injective in the category of G-sheaves, then it is injective in the category of sheaves.

For any etale X-scheme $p : V \to X$, let $\mathbb{Z}_V = p_!\mathbb{Z}$. We have

$$\mathrm{Hom}_G(\mathbb{Z}_V[G], \mathscr{G}) \cong \mathrm{Hom}(\mathbb{Z}_V, \mathscr{G}) \cong \mathscr{G}(V).$$

It follows that $\mathbb{Z}_V[G]$ form a family of generators for the category of G-sheaves. Using this fact, one proves that the category of G-sheaves has enough injective objects. (Confer [Grothendieck (1957)] 1.10.1, or [Fu (2006)] 2.1.6.)

Let Y be another scheme on which G acts on the right, and let $f : X \to Y$ be a G-equivariant morphism. For any G-sheaf $\mathscr{F}$ on X and G-sheaf $\mathscr{G}$ on Y, $f_*\mathscr{F}$ and $f^*\mathscr{G}$ are G-sheaves on Y and on X, respectively, and the canonical morphisms $\mathscr{G} \to f_*f^*\mathscr{G}$ and $f^*f_*\mathscr{F} \to \mathscr{F}$ are G-morphisms. This follows from the commutativity of the following diagrams:

$$\begin{array}{ccccc} \mathscr{G} \to f_*f^*\mathscr{G} & \to & & f_*\sigma_*\sigma^*f^*\mathscr{G} \cong f_*\sigma_*f^*\sigma^*\mathscr{G} \\ \downarrow & & & \downarrow \\ \sigma_*\mathscr{G} & \to & \sigma_*f_*f^*\mathscr{G} \quad \cong & f_*\sigma_*f^*\mathscr{G}, \end{array}$$

$$\begin{array}{ccccc} f^*\sigma^*f_*\mathscr{F} & \cong & \sigma^*f^*f_*\mathscr{F} & \to & \sigma^*\mathscr{F} \\ \downarrow & & & & \downarrow \\ f^*\sigma^*f_*\sigma_*\mathscr{F} \cong f^*\sigma^*\sigma_*f_*\mathscr{F} & & \to & & f^*f_*\mathscr{F} \to \mathscr{F}. \end{array}$$

To prove the commutativity of these diagrams, note that they are the outer loops of the following diagrams:

$$\begin{array}{ccccccc} \mathscr{G} & \to & f_*f^*\mathscr{G} & \to & f_*\sigma_*\sigma^*f^*\mathscr{G} & \cong & f_*\sigma_*f^*\sigma^*\mathscr{G} \\ \downarrow & & \downarrow & & \downarrow & & \downarrow \\ \sigma_*\mathscr{G} & \to & f_*f^*\sigma_*\mathscr{G} & \to & f_*\sigma_*\sigma^*f^*\sigma_*\mathscr{G} & \cong & f_*\sigma_*f^*\sigma^*\sigma_*\mathscr{G} \\ \| & & & (1) & & & \downarrow \\ \sigma_*\mathscr{G} & & \to & \sigma_*f_*f^*\mathscr{G} & \cong & & f_*\sigma_*f^*\mathscr{G}, \end{array}$$

$$\begin{array}{ccccccc} f^*\sigma^*f_*\mathscr{F} & & \cong & \sigma^*f^*f_*\mathscr{F} & \to & & \sigma^*\mathscr{F} \\ \downarrow & & & (2) & & & \| \\ f^*\sigma^*f_*\sigma_*\sigma^*\mathscr{F} & \cong & f^*\sigma^*\sigma_*f_*\sigma^*\mathscr{F} & \to & f^*f_*\sigma^*\mathscr{F} & \to & \sigma^*\mathscr{F} \\ \downarrow & & \downarrow & & \downarrow & & \downarrow \\ f^*\sigma^*f_*\sigma_*\mathscr{F} & \cong & f^*\sigma^*\sigma_*f_*\mathscr{F} & \to & f^*f_*\mathscr{F} & \to & \mathscr{F}. \end{array}$$

The diagrams (1) and (2) commute by 7.3.16. Since f^* is an exact functor, f_* maps injective G-sheaves on X to injective G-sheaves on Y.

If $f : X \to Y$ is an etale G-equivariant morphism, and $\mathscr{F}$ a G-sheaf on X, then $f_!\mathscr{F}$ is a G-sheaf on Y, and the canonical morphism $\mathscr{F} \to f^*f_!\mathscr{F}$ is a morphism of G-sheaves. This follows from the commutativity of the following diagram:

$$\begin{array}{ccccccc} \sigma^*\mathscr{F} & & = & & \sigma^*\mathscr{F} & \to & \mathscr{F} \\ \downarrow & & & & \downarrow & & \downarrow \\ \sigma^*f^*f_!\mathscr{F} & \to & f^*\sigma^*f_!\mathscr{F} & \overset{\cong}{\to} & f^*f_!\sigma^*\mathscr{F} & \to & f^*f_!\mathscr{F}. \end{array}$$

For any G-sheaf $\mathscr{G}$ on Y, the canonical morphism $f_!f^*\mathscr{G} \to \mathscr{G}$ is a morphism of G-sheaves. This follows from the commutativity of the following diagram:

$$\begin{array}{ccccccc} \sigma^*f_!f^*\mathscr{G} & \overset{\cong}{\to} & f_!\sigma^*f^*\mathscr{G} & \to & f_!f^*\sigma^*\mathscr{G} & \to & f_!f^*\mathscr{G} \\ \downarrow & & & & \downarrow & & \downarrow \\ \sigma^*\mathscr{G} & & = & & \sigma^*\mathscr{G} & \to & \mathscr{G}. \end{array}$$

$f_!$ is left adjoint to f^*. Since $f_!$ is an exact functor, f^* maps injective G-sheaves on Y to injective G-sheaves on X.

Suppose G acts trivially on a scheme S. For any G-sheaf $\mathscr{F}$ on S, let $\mathscr{F}^G$ be the subsheaf formed by G-invariant sections. Let $\mathscr{G}$ be a sheaf on S. Put the trivial G-action on $\mathscr{G}$. Then we have

$$\mathrm{Hom}_G(\mathscr{G}, \mathscr{F}) \cong \mathrm{Hom}(\mathscr{G}, \mathscr{F}^G).$$

So the functor that maps $\mathscr{G}$ to the G-sheaf $\mathscr{G}$ with the trivial G-action is left adjoint to the functor $\mathscr{F} \mapsto \mathscr{F}^G$, and it is exact. If $\mathscr{F}$ is injective in the category of G-sheaves, then $\mathscr{F}^G$ is injective in the category of sheaves. We denote the q-th right derived functor of $\mathscr{F} \mapsto \mathscr{F}^G$ by $\mathscr{H}^q(G, -)$ or $R^q\underline{\Gamma}^G$.

Let X be a scheme on which G acts on the right, let Y be a scheme with the trivial G-action, and let $f : X \to Y$ be a G-equivariant morphism. For any G-sheaf $\mathscr{F}$ on X, let

$$f_*^G \mathscr{F} = (f_*\mathscr{F})^G.$$

The functor f_*^G is left exact. Since f_* maps injective G-sheaves on X to injective G-sheaves on Y, we have a biregular spectral sequence

$$E_2^{pq} = \mathscr{H}^p(G, R^q f_*\mathscr{F}) \Rightarrow R^{p+q} f_*^G \mathscr{F}.$$

Let S be a scheme with the trivial G-action, let $g : Y \to S$ be a morphism, and let $h = gf$. We have

$$h_*^G \mathscr{F} = g_* f_*^G \mathscr{F}.$$

Since f_*^G maps injective G-sheaves on X to injective sheaves on Y, we have a biregular spectral sequence

$$E_2^{pq} = R^p g_* R^q f_*^G \mathscr{F} \Rightarrow R^{p+q} h_*^G \mathscr{F}.$$

Consider the case where $f : X \to Y$ is a galois etale covering with galois group $G = \mathrm{Aut}(X/Y)^\circ$. G acts on the right of X, and acts trivially on Y. For any sheaf $\mathscr{G}$ on Y, the canonical morphism $\mathscr{G} \to f_* f^* \mathscr{G}$ induces an isomorphism

$$\mathscr{G} \cong f_*^G f^* \mathscr{G}.$$

To prove this, we choose a surjective etale morphism $Y' \to Y$ such that $X \times_Y Y' \to Y'$ is a trivial etale covering space. We then use the fact that

$$\mathscr{G}|_{Y'} \cong (f_* f^* \mathscr{G})^G|_{Y'}.$$

Let

$$0 \to \mathscr{I}^0 \to \mathscr{I}^1 \to \cdots$$

be an injective resolution of $\mathscr{G}$. Then

$$0 \to f^*\mathscr{I}^0 \to f^*\mathscr{I}^1 \to \cdots$$

is an injective resolution of $f^*\mathscr{G}$ in the category of G-sheaves. So $R^q f_*^G(f^*\mathscr{G})$ is the q-th cohomology sheaf of the complex

$$0 \to f_*^G f^*\mathscr{I}^0 \to f_*^G f^*\mathscr{I}^1 \to \cdots.$$

But we have $\mathscr{I}^{\cdot} \cong f_*^G f^*\mathscr{I}$. So we have

$$R^q f_*^G(f^*\mathscr{G}) = \begin{cases} \mathscr{G} & \text{if } q = 0, \\ 0 & \text{if } q \geq 1. \end{cases}$$

Let $g : Y \to S$ be a morphism of schemes with trivial G-actions, and let $h = gf$. The biregular spectral sequence

$$E_2^{pq} = R^p g_* R^q f_*^G(f^*\mathscr{G}) \Rightarrow R^{p+q} h_*^G(f^*\mathscr{G})$$

degenerates, and we have

$$R^p g_* \mathscr{G} \cong R^p h_*^G(f^*\mathscr{G}).$$

The biregular spectral sequence

$$E_2^{pq} = \mathscr{H}^p(G, R^q h_*(f^*\mathscr{G})) \Rightarrow R^{p+q} h_*^G(f^*\mathscr{G})$$

can be identified with a biregular spectral sequence

$$E_2^{pq} = \mathscr{H}^p(G, R^q h_*(f^*\mathscr{G})) \Rightarrow R^{p+q} g_*\mathscr{G},$$

which is called the *Hochschild–Serre spectral sequence.*

Let X be a scheme on which G acts on the right, and let A be a ring. We define $\mathrm{Ext}^{\cdot}_{A[G]}(-,-)$ and $\mathscr{E}xt^{\cdot}_{A[G]}(-,-)$ to be the derived functors of $\mathrm{Hom}_{A[G]}(-,-)$ and $\mathscr{H}om_{A[G]}(-,-)$ on the category of G-sheaves of A-modules on X, respectively. Suppose that G acts trivially on X and $\mathscr{F}$ is a G-sheaf of A-modules on X. We have

$$\mathscr{F}^G = \mathscr{H}om_{A[G]}(A, \mathscr{F}),$$

where A is the constant sheaf on which G acts trivially. So we have

$$\mathscr{H}^p(G, \mathscr{F}) = \mathscr{E}xt^p_{A[G]}(A, \mathscr{F}).$$

Let

$$\cdots \to L^{-1} \to L^0 \to 0$$

be a resolution of A by free $A[G]$-modules of finite rank. Then we have

$$\mathscr{H}^p(G, \mathscr{F}) = \mathscr{H}^p(\mathscr{H}om_G(L^{\cdot}, \mathscr{F})).$$

Using this fact, one can show that if $\mathscr{F}$ is locally constant, then $\mathscr{H}^p(G, \mathscr{F})$ are locally constant, and that if $f : X' \to X$ is a morphism of schemes and G acts trivially on X', then

$$f^*\mathscr{H}^p(G, \mathscr{F}) \cong \mathscr{H}^p(G, f^*\mathscr{F}).$$

In particular, we have

$$\mathscr{H}^p(G, \mathscr{F})_{\bar{x}} \cong H^p(G, \mathscr{F}_{\bar{x}})$$

for any $x \in X$.

9.2 Nearby Cycle and Vanishing Cycle

([SGA 7] I 2, XIII.)

Let k be a field, $\bar{k}$ a separable closure of k, Y a scheme over k, and $\overline{Y} = Y \otimes_k \bar{k}$. Then $\mathrm{Gal}(\bar{k}/k)$ acts on $\overline{Y}$. Let $\overline{\mathscr{F}}$ be a $\mathrm{Gal}(\bar{k}/k)$-sheaf on $\overline{Y}$. For any finite galois extension k' of k contained in $\bar{k}$, let $Y' = Y \otimes_k k'$. Then for any etale morphism $U' \to Y'$, $\mathrm{Gal}(\bar{k}/k')$ acts on $\overline{\mathscr{F}}(U' \times_{Y'} \overline{Y})$. If this action is continuous with respect to the discrete topology on $\overline{\mathscr{F}}(U' \times_{Y'} \overline{Y})$ for any finite extension k' of k contained in $\bar{k}$ and any quasi-compact scheme U' etale over Y', we say that $\mathrm{Gal}(\bar{k}/k)$ acts continuously on $\overline{\mathscr{F}}$. To check $\mathrm{Gal}(\bar{k}/k)$ acts continuously on $\overline{\mathscr{F}}$, it suffices to check that $\mathrm{Gal}(\bar{k}/k)$ acts continuously on $\overline{\mathscr{F}}(U' \times_{Y'} \overline{Y})$ for any etale Y'-scheme U' such that U' is affine and its image in Y' is contained in the inverse image in Y' of an affine open subset of Y.

Proposition 9.2.1. *Let k be a field, $\bar{k}$ a separable closure of k, Y a scheme over k, $\mathscr{F}$ a sheaf on Y, $\overline{Y} = Y \otimes_k \bar{k}$, and $\overline{\mathscr{F}}$ the inverse image of $\mathscr{F}$ on $\overline{Y}$. Then $\mathrm{Gal}(\bar{k}/k)$ acts on $\overline{\mathscr{F}}$.*

(i) *The action of $\mathrm{Gal}(\bar{k}/k)$ on $\overline{\mathscr{F}}$ is continuous.*

(ii) *The functor $\mathscr{F} \mapsto \overline{\mathscr{F}}$ is an equivalence from the category of sheaves on Y to the category of sheaves on $\overline{Y}$ with continuous $\mathrm{Gal}(\bar{k}/k)$-action.*

Proof.

(i) Let k' be a finite extension of k, let $Y' = Y \otimes_k k'$, and let U' be an affine etale Y'-scheme such that its image in Y' is contained in an affine open subset of Y'. We have $\bar{k} = \varinjlim k_\lambda$, where k_λ are finite extensions of k containing k' and contained in $\bar{k}$. For each λ, let $\mathscr{F}_\lambda$ be the inverse image of $\mathscr{F}$ on $Y \otimes_k k_\lambda$. By 5.9.3, we have

$$\overline{\mathscr{F}}(U' \times_{Y'} \overline{Y}) = \varinjlim_{\lambda} \mathscr{F}_\lambda(U' \otimes_{k'} k_\lambda).$$

For each λ, $\mathrm{Gal}(\bar{k}/k_\lambda)$ acts trivially on $\mathscr{F}_\lambda(U' \otimes_{k'} k_\lambda)$. It follows that $\mathrm{Gal}(\bar{k}/k')$ acts continuously on $\overline{\mathscr{F}}(U' \times_{Y'} \overline{Y})$.

(ii) We have $\bar{k} = \varinjlim k_\lambda$, where k_λ are finite galois extensions of k contained in $\bar{k}$. Let $Y_\lambda = Y \otimes_k k_\lambda$, and let $\pi : \overline{Y} \to Y$, $\pi_\lambda : Y_\lambda \to Y$ and $\bar{\pi}_\lambda : \overline{Y} \to Y_\lambda$ and be the projections. Denote the functor $\mathscr{F} \mapsto \overline{\mathscr{F}}$ by S, and define a functor T from the category of sheaves on $\overline{Y}$ with continuous $\mathrm{Gal}(\bar{k}/k)$-action to the category of sheaves on Y by

$$T(\overline{\mathscr{F}}) = (\pi_* \overline{\mathscr{F}})^{\mathrm{Gal}(\bar{k}/k)}.$$

To prove that S is an equivalence of categories, it suffices to show that the canonical morphisms

$$\mathscr{F} \to (\pi_*\pi^*\mathscr{F})^{\mathrm{Gal}(\bar{k}/k)}, \quad \pi^*\left((\pi_*\overline{\mathscr{F}})^{\mathrm{Gal}(\bar{k}/k)}\right) \to \overline{\mathscr{F}}$$

are isomorphisms for any sheaf $\mathscr{F}$ on Y, and any sheaf $\overline{\mathscr{F}}$ on $\overline{Y}$ with continuous $\mathrm{Gal}(\bar{k}/k)$-action. By 5.9.6, we have

$$(\pi_*\pi^*\mathscr{F})^{\mathrm{Gal}(\bar{k}/k)} = \varinjlim(\pi_{\lambda *}\pi_\lambda^*\mathscr{F})^{\mathrm{Gal}(k_\lambda/k)}.$$

Note that $\mathscr{F} \to (\pi_{\lambda *}\pi_\lambda^*\mathscr{F})^{\mathrm{Gal}(k_\lambda/k)}$ is an isomorphism for each λ. This can be seen by taking the base change from k to k_λ. It follows that $\mathscr{F} \to (\pi_*\pi^*\mathscr{F})^{\mathrm{Gal}(\bar{k}/k)}$ is an isomorphism.

To prove that the morphism $\pi^*\left((\pi_*\overline{\mathscr{F}})^{\mathrm{Gal}(\bar{k}/k)}\right) \to \overline{\mathscr{F}}$ is an isomorphism for any sheaf $\overline{\mathscr{F}}$ on $\overline{Y}$ with continuous $\mathrm{Gal}(\bar{k}/k)$-action, it suffices to prove that the above morphism is surjective for any $\overline{\mathscr{F}}$. Indeed, suppose that this can be done and let $\overline{\mathscr{K}}$ be the kernel of the above morphism. Since T is left exact, we have an exact sequence

$$0 \to T(\overline{\mathscr{K}}) \to T(ST(\overline{\mathscr{F}})) \to T(\overline{\mathscr{F}}).$$

We have shown that the canonical morphism $T(\overline{\mathscr{F}}) \to (TS)T(\overline{\mathscr{F}})$ is an isomorphism. One can show that it is a right inverse of the canonical morphism $T(ST(\overline{\mathscr{F}})) \to T(\overline{\mathscr{F}})$. It follows that $T(\overline{\mathscr{K}}) = 0$. But the canonical morphism $ST(\overline{\mathscr{K}}) \to \overline{\mathscr{K}}$ is surjective. So we have $\overline{\mathscr{K}} = 0$ and hence the canonical morphism $ST(\overline{\mathscr{F}}) \to \overline{\mathscr{F}}$ is an isomorphism.

First let us show that the canonical morphism

$$\varinjlim \bar{\pi}_\lambda^*\left((\bar{\pi}_{\lambda *}\overline{\mathscr{F}})^{\mathrm{Gal}(\bar{k}/k_\lambda)}\right) \to \overline{\mathscr{F}}$$

is surjective. Let $\overline{U}$ be an affine etale $\overline{Y}$-scheme so that its image in $\overline{Y}$ is contained in the inverse image in $\overline{Y}$ of an affine open subset of Y. We can find λ_0 and a quasi-compact etale Y_{λ_0}-scheme U_{λ_0} such that $\overline{U} \cong \overline{Y} \times_{Y_{\lambda_0}} U_{\lambda_0}$. For any $\lambda \geq \lambda_0$, let $U_\lambda = Y_\lambda \times_{Y_{\lambda_0}} U_{\lambda_0}$. Given $s \in \overline{\mathscr{F}}(\overline{U})$, there exists $\lambda_1 \geq \lambda_0$ such that s is fixed by $\mathrm{Gal}(\bar{k}/k_{\lambda_1})$. Then s lies in the image of

$$\bar{\pi}_{\lambda_1}^*\left((\bar{\pi}_{\lambda_1 *}\overline{\mathscr{F}})^{\mathrm{Gal}(\bar{k}/k_{\lambda_1})}\right)(\overline{U}) \to \overline{\mathscr{F}}(\overline{U}).$$

This proves our assertion. We have

$$\begin{aligned}\pi^*\left((\pi_*\overline{\mathscr{F}})^{\mathrm{Gal}(\bar{k}/k)}\right) &\cong \bar{\pi}_\lambda^*\pi_\lambda^*\left((\pi_{\lambda *}\bar{\pi}_{\lambda_*}\overline{\mathscr{F}})^{\mathrm{Gal}(\bar{k}/k)}\right)\\ &\cong \bar{\pi}_\lambda^*\pi_\lambda^*\left(\left(\pi_{\lambda *}\left((\bar{\pi}_{\lambda_*}\overline{\mathscr{F}})^{\mathrm{Gal}(\bar{k}/k_\lambda)}\right)\right)^{\mathrm{Gal}(k_\lambda/k)}\right).\end{aligned}$$

For any sheaf $\mathscr{F}_\lambda$ on Y_λ provided with a $\mathrm{Gal}(k_\lambda/k)$-action, the canonical morphism

$$\pi_\lambda^*\Big((\pi_{\lambda *}\mathscr{F}_\lambda)^{\mathrm{Gal}(k_\lambda/k)}\Big) \to \mathscr{F}_\lambda$$

is an isomorphism. This can be seen by taking the base change from k to k_λ. So we have

$$\pi^*\Big((\pi_*\overline{\mathscr{F}})^{\mathrm{Gal}(\bar{k}/k)}\Big) \cong \bar{\pi}_\lambda^*\Big((\bar{\pi}_{\lambda_*}\overline{\mathscr{F}})^{\mathrm{Gal}(\bar{k}/k_\lambda)}\Big).$$

Since $\varinjlim \bar{\pi}_\lambda^*\Big((\bar{\pi}_{\lambda *}\overline{\mathscr{F}})^{\mathrm{Gal}(\bar{k}/k_\lambda)}\Big) \to \overline{\mathscr{F}}$ is surjective, $\pi^*\Big((\pi_*\overline{\mathscr{F}})^{\mathrm{Gal}(\bar{k}/k)}\Big) \to \overline{\mathscr{F}}$ is also surjective. □

Let S be a trait, s its closed point, η its generic point, $\widetilde{S}$ the strict localization of S at $\bar{s}$, and $\tilde{\eta}$ the generic point of $\widetilde{S}$. We have an exact sequence

$$0 \to I \to \mathrm{Gal}(\bar{\eta}/\eta) \to \mathrm{Gal}(\bar{s}/s) \to 0,$$

where $I = \mathrm{Gal}(\bar{\eta}/\tilde{\eta})$ is the inertia subgroup. We have a commutative diagram

$$\begin{array}{ccccc} \bar{\eta} & \to & \tilde{S} & \leftarrow & \bar{s} \\ \downarrow & & \downarrow & & \downarrow \\ \eta & \to & S & \leftarrow & s. \end{array}$$

For any S-scheme X, taking the base change $X \to S$ to the above diagram, we get

$$\begin{array}{ccccc} X_{\bar{\eta}} & \xrightarrow{\tilde{j}} & \widetilde{X} = X \times_S \tilde{S} & \xleftarrow{\tilde{i}} & X_{\bar{s}} \\ \downarrow & & \downarrow & & \downarrow \\ X_\eta & \to & X & \leftarrow & X_s. \end{array}$$

For any sheaf $\mathscr{F}_\eta$ on X_η, denote its inverse image in $X_{\bar{\eta}}$ by $\mathscr{F}_{\bar{\eta}}$. By 9.2.1, $\mathscr{F}_\eta$ is completely determined by the sheaf $\mathscr{F}_{\bar{\eta}}$ with $\mathrm{Gal}(\bar{\eta}/\eta)$-action. The group $\mathrm{Gal}(\bar{\eta}/\eta)$ acts on the scheme $X_{\bar{s}}$ though its quotient group $\mathrm{Gal}(\bar{s}/s)$. The sheaf $\tilde{i}^*\tilde{j}_*\mathscr{F}_{\bar{\eta}}$ is a $\mathrm{Gal}(\bar{\eta}/\eta)$-sheaf on $X_{\bar{s}}$. Indeed, given $\sigma \in \mathrm{Gal}(\bar{\eta}/\eta)$, it gives rise to an element in $\mathrm{Aut}(\tilde{S}/S)$ and an element in $\mathrm{Gal}(\bar{s}/s)$. So it gives rise to elements in $\mathrm{Aut}(X_{\bar{\eta}}/X_\eta)$, $\mathrm{Aut}(\tilde{X}/X)$, and $\mathrm{Aut}(X_{\bar{s}}/X_s)$, which we still denote by σ. The action of σ on $\tilde{i}^*\tilde{j}_*\mathscr{F}_{\bar{\eta}}$ is given by the isomorphism

$$\sigma^*\tilde{i}^*\tilde{j}_*\mathscr{F}_{\bar{\eta}} \to \tilde{i}^*\tilde{j}_*\mathscr{F}_{\bar{\eta}}$$

defined by the composite

$$\sigma^*\tilde{i}^*\tilde{j}_*\mathscr{F}_{\bar{\eta}} \cong \tilde{i}^*\sigma^*\tilde{j}_*\mathscr{F}_{\bar{\eta}} \to \tilde{i}^*\tilde{j}_*\sigma^*\mathscr{F}_{\bar{\eta}} \to \tilde{i}^*\tilde{j}_*\mathscr{F}_{\bar{\eta}}.$$

Define

$$\Psi_\eta(\mathscr{F}_\eta) = \tilde{i}^* \tilde{j}_* \mathscr{F}_{\bar{\eta}}.$$

Ψ_η is a functor from the category of sheaves on X_η to the category of sheaves on $X_{\bar{s}}$ with $\mathrm{Gal}(\bar{\eta}/\eta)$-action. Its derived functor $R\Psi_\eta$ is called the *nearby cycle functor*. We have

$$R\Psi_\eta(K_\eta) = \tilde{i}^* R\tilde{j}_* K_{\bar{\eta}}$$

for any $K_\eta \in \mathrm{ob}\, D^+(X_\eta)$.

Proposition 9.2.2. *Keep the above notation.*

(i) *For any $x \in X_{\bar{s}}$, let $\widetilde{X}_{\bar{x}}$ be the strict localization of X at $\bar{x}$. It is a scheme over $\widetilde{S}$. We have $(R^q\Psi_\eta \mathscr{F}_\eta)_{\bar{x}} \cong H^q(\widetilde{X}_{\bar{x}} \times_{\widetilde{S}} \bar{\eta}, \mathscr{F}_{\bar{\eta}})$.*

(ii) *If $f : X \to S$ is of finite type, then $R\Psi_\eta$ has finite cohomological dimension on $D^+_{\mathrm{tor}}(X_\eta)$. More precisely, we have $R^q\Psi_\eta(\mathscr{F}_\eta) = 0$ for any $q > \dim X_\eta$ and any torsion sheaf $\mathscr{F}_\eta$ on X_η.*

(iii) *If $X = S$, then we have $R\Psi_\eta(K_\eta) = K_{\bar{\eta}}$, where on the right-hand side, $K_{\bar{\eta}}$ is considered as a complex of sheaves on $\bar{s}$ with* $\mathrm{Gal}(\bar{\eta}/\eta)$ *action.*

Proof.

(i) We have

$$(R^q\Psi_\eta \mathscr{F}_\eta)_{\bar{x}} = (\tilde{i}^* R^q \tilde{j}_* \mathscr{F}_{\bar{\eta}})_{\bar{x}} \cong (R^q \tilde{j}_* \mathscr{F}_{\bar{\eta}})_{\bar{x}} \cong H^q(\widetilde{X}_{\bar{x}} \times_{\widetilde{S}} \bar{\eta}, \mathscr{F}_{\bar{\eta}}).$$

(ii) Use 7.5.7.

(iii) This has been noted in the proof of 7.6.1. Let $i : \bar{s} \to \tilde{S}$ and $j : \bar{\eta} \to \tilde{S}$ be the canonical morphisms. We have $Rj_* K_{\bar{\eta}} = j_* K_{\bar{\eta}}$, and $i^* j_* K_{\bar{\eta}} = K_{\bar{\eta}}$. □

Let $\mathscr{F}$ be a sheaf on X. Denote its inverse images on X_s, $X_{\bar{s}}$, X_η, $X_{\bar{\eta}}$, and $\widetilde{X}$ by $\mathscr{F}_s$, $\mathscr{F}_{\bar{s}}$, $\mathscr{F}_\eta$, $\mathscr{F}_{\bar{\eta}}$, and $\widetilde{\mathscr{F}}$, respectively. By 9.2.1, $\mathscr{F}_s$ is completely determined by $\mathscr{F}_{\bar{s}}$ with $\mathrm{Gal}(\bar{s}/s)$-action. Let $\mathrm{Gal}(\bar{\eta}/\eta)$ act on $\mathscr{F}_{\bar{s}}$ though its quotient $\mathrm{Gal}(\bar{s}/s)$. Then the canonical morphism

$$\mathscr{F}_{\bar{s}} = \tilde{i}^* \widetilde{\mathscr{F}} \to \tilde{i}^* \tilde{j}_* \tilde{j}^* \widetilde{\mathscr{F}} = \tilde{i}^* \tilde{j}_* \mathscr{F}_{\bar{\eta}}$$

is a morphism of $\mathrm{Gal}(\bar{\eta}/\eta)$-sheaves on $X_{\bar{s}}$. If $\mathscr{F}$ is an injective sheaf, then $R^q\Psi_\eta(\mathscr{F}_\eta) = 0$ for all $q > 0$. This follows from 5.9.9 and the fact that $X_{\bar{\eta}}$ is the inverse limit of an inverse system of etale X-schemes.

For any complex of sheaves K on X, let K_s, $K_{\bar{s}}$, K_η, $K_{\bar{\eta}}$, and $\widetilde{K}$ be its inverse images on X_s, $X_{\bar{s}}$, X_η, $X_{\bar{\eta}}$, and $\widetilde{X}$ respectively. Suppose that K is bounded below and let $K \to J$ be a quasi-isomorphism such that J is a bounded below complex of injective sheaves. We have

$$R\Psi_\eta(K_\eta) \cong \tilde{i}^* \tilde{j}_* \tilde{j}^* \widetilde{J}.$$

We have a canonical morphism of complexes of $\mathrm{Gal}(\bar{\eta}/\eta)$-sheaves on $X_{\bar{s}}$:

$$J_{\bar{s}} \to \tilde{i}^* \tilde{j}_* \tilde{j}^* \widetilde{J}.$$

Let $R\Phi(K)$ be its mapping cone. Then $R\Phi$ defines a functor from $D^+(X)$ to the derived category of $\mathrm{Gal}(\bar{\eta}/\eta)$-sheaves on $X_{\bar{s}}$. It is called the *vanishing cycle functor*. We have a distinguished triangle

$$K_{\bar{s}} \to R\Psi_\eta(K_\eta) \to R\Phi(K) \to$$

for any $K \in \mathrm{ob}\, D^+(X)$. Suppose that $\sigma \in \mathrm{Gal}(\bar{\eta}/\eta)$ lies in the inertia subgroup I. Then σ acts trivially on $K_{\bar{s}}$. There exists a morphism

$$\mathrm{var}(\sigma) : R\Phi(K) \to R\Psi_\eta(K_\eta),$$

which we call the *variation*, such that the following diagram commute:

$$\begin{array}{ccc} R\Psi_\eta(K_\eta) & \stackrel{\sigma-\mathrm{id}}{\to} & R\Psi_\eta(K_\eta) \\ \downarrow & \nearrow \mathrm{var}(\sigma) & \downarrow \\ R\Phi(K) & \stackrel{\sigma-\mathrm{id}}{\to} & R\Phi(K). \end{array}$$

For each q, the composite of the canonical homomorphisms

$$H^q(X_{\bar{\eta}}, K_{\bar{\eta}}) \cong H^q(\widetilde{X}, R\tilde{j}_* K_{\bar{\eta}}) \to H^q(X_{\bar{s}}, \tilde{i}^* R\tilde{j}_* K_{\bar{\eta}}) = H^q(X_{\bar{s}}, R\Psi_\eta(K_\eta))$$

defines a homomorphism

$$H^q(X_{\bar{\eta}}, K_{\bar{\eta}}) \to H^q(X_{\bar{s}}, R\Psi_\eta(K_\eta)),$$

which is an isomorphism if X is proper over S and $K \in \mathrm{ob}\, D^+_{\mathrm{tor}}(X)$ by the proper base change theorem 7.3.3. We have a long exact sequence

$$\cdots \to H^q(X_{\bar{s}}, K_{\bar{s}}) \to H^q(X_{\bar{s}}, R\Psi_\eta(K_\eta)) \to H^q(X_{\bar{s}}, R\Phi(K)) \to \cdots.$$

If X is proper over S and $K \in \mathrm{ob}\, D^+_{\mathrm{tor}}(X)$, we then have a long exact sequence

$$\cdots \to H^q(X_{\bar{s}}, K_{\bar{s}}) \to H^q(X_{\bar{\eta}}, K_{\bar{\eta}}) \to H^q(X_{\bar{s}}, R\Phi(K)) \to \cdots.$$

So $H^\cdot(X_{\bar{s}}, R\Phi(K))$ measures the difference between $H^\cdot(X_{\bar{s}}, K_{\bar{s}})$ and $H^\cdot(X_{\bar{\eta}}, K_{\bar{\eta}})$ if X is proper over S.

Let $f : X \to X'$ be an S-morphism of S-schemes. Fix notation by the following commutative diagram:

$$\begin{array}{ccccc} X_{\bar{\eta}} & \stackrel{\tilde{j}}{\to} & \widetilde{X} = X \times_S \tilde{S} & \stackrel{\tilde{i}}{\leftarrow} & X_{\bar{s}} \\ f_{\bar{\eta}} \downarrow & & \downarrow \tilde{f} & & \downarrow f_{\bar{s}} \\ X'_{\bar{\eta}} & \stackrel{\tilde{j}'}{\to} & \widetilde{X}' = X' \times_S \tilde{S} & \stackrel{\tilde{i}'}{\leftarrow} & X'_{\bar{s}}. \end{array}$$

For any $K \in \operatorname{ob} D^+(X, A)$, we define a morphism

$$R\Psi_\eta((Rf_*K)_\eta) \to Rf_{\bar{s}*}R\Psi_\eta(K_\eta)$$

to be the composite of the canonical morphisms

$$\begin{aligned} R\Psi_\eta((Rf_*K)_\eta) &\cong \tilde{i}'^*R\tilde{j}'_*\tilde{j}'^*R\tilde{f}_*\widetilde{K} \\ &\to \tilde{i}'^*R\tilde{j}'_*Rf_{\bar{\eta}*}\tilde{j}^*\widetilde{K} \\ &\cong \tilde{i}'^*R\tilde{f}_*R\tilde{j}_*\tilde{j}^*\widetilde{K} \\ &\to Rf_{\bar{s}*}\tilde{i}^*R\tilde{j}_*\tilde{j}^*\widetilde{K} \\ &= Rf_{\bar{s}*}R\Psi_\eta(K_\eta). \end{aligned}$$

When f is proper and $K \in \operatorname{ob} D^+_{\mathrm{tor}}(X)$, this is an isomorphism by the proper base change theorem 7.3.1.

Suppose that $f : X \to X'$ is an open immersion. Define a morphism

$$f_{\bar{s}!}R\Psi_\eta(K_\eta) \to R\Psi_\eta((f_!K)_\eta)$$

to be the composite of the canonical morphisms

$$\begin{aligned} f_{\bar{s}!}R\Psi_\eta(K_\eta) &\cong f_{\bar{s}!}\tilde{i}^*R\tilde{j}_*\tilde{j}^*\widetilde{K} \\ &\cong \tilde{i}'^*\tilde{f}_!R\tilde{j}_*\tilde{j}^*\widetilde{K} \\ &\to \tilde{i}'^*R\tilde{j}'_*f_{\bar{\eta}!}\tilde{j}^*\widetilde{K} \\ &\cong \tilde{i}'^*R\tilde{j}'_*\tilde{j}'^*\tilde{f}_!\widetilde{K} \\ &\cong R\Psi_\eta((f_!K)_\eta). \end{aligned}$$

Suppose that $K \in \operatorname{ob} D^+_{\mathrm{tor}}(X)$, $f : X \to X'$ is an S-compactifiable morphism, and

$$X \stackrel{k}{\hookrightarrow} \overline{X} \stackrel{\bar{f}}{\to} X'$$

is a compactification for f, where k is an open immersion, and $\bar{f}$ is a proper S-compactifiable morphism. Define a morphism

$$Rf_{\bar{s}!}R\Psi_\eta(K_\eta) \to R\Psi_\eta((Rf_!K)_\eta)$$

to be the composite of

$$\begin{aligned} Rf_{\bar{s}!}R\Psi_\eta(K_\eta) &\cong R\bar{f}_{\bar{s}*}k_{\bar{s}!}R\Psi_\eta(K_\eta) \\ &\to R\bar{f}_{\bar{s}*}R\Psi_\eta((k_!K)_\eta) \\ &\cong R\Psi_\eta((R\bar{f}_*k_!K)_\eta) \\ &\cong R\Psi_\eta((Rf_!K)_\eta). \end{aligned}$$

If f is proper, this is the inverse of the morphism

$$R\Psi_\eta((Rf_*K)_\eta) \to Rf_{\bar{s}*}R\Psi_\eta(K_\eta)$$

defined above.

Suppose that $X \to S$ is compactifiable. Applying the above construction to the structure morphism $X \to S$, we get a homomorphism

$$H_c^q(X_{\bar{s}}, R\Psi_\eta(K_\eta)) \to H_c^q(X_{\bar{\eta}}, K_{\bar{\eta}})$$

for each q. If X is proper over S, it is the inverse of the canonical isomorphism

$$H^q(X_{\bar{\eta}}, K_{\bar{\eta}}) \to H^q(X_{\bar{s}}, R\Psi_\eta(K_\eta)).$$

For any $\sigma \in I$ in the inertia group, we have commutative diagrams

$$\begin{array}{ccc} H^q(X_{\bar{\eta}}, K_{\bar{\eta}}) & \stackrel{\sigma-\mathrm{id}}{\to} & H^i(X_{\bar{\eta}}, K_{\bar{\eta}}) \\ \downarrow & & \downarrow \\ H^q(X_{\bar{s}}, R\Psi_\eta(K_\eta)) \to H^q(X_{\bar{s}}, R\Phi(K)) & \stackrel{\mathrm{var}(\sigma)}{\to} & H^q(X_{\bar{s}}, R\Psi_\eta(K_\eta)), \end{array}$$

$$\begin{array}{ccc} H_c^q(X_{\bar{\eta}}, K_{\bar{\eta}}) & \stackrel{\sigma-\mathrm{id}}{\to} & H_c^i(X_{\bar{\eta}}, K_{\bar{\eta}}) \\ \uparrow & & \uparrow \\ H_c^q(X_{\bar{s}}, R\Psi_\eta(K_\eta)) \to H_c^q(X_{\bar{s}}, R\Phi(K)) & \stackrel{\mathrm{var}(\sigma)}{\to} & H_c^q(X_{\bar{s}}, R\Psi_\eta(K_\eta)). \end{array}$$

For any $L \in D^+(X')$, define a morphism

$$f_{\bar{s}}^* R\Psi_\eta(L_\eta) \to R\Psi_\eta((f^*L)_\eta)$$

to be the composite of the canonical morphisms

$$\begin{aligned} f_{\bar{s}}^* R\Psi_\eta(L_\eta) &= f_{\bar{s}}^* \tilde{i}'^* R\tilde{j}'_* \tilde{j}'^* \widetilde{L} \\ &\cong \tilde{i}^* \tilde{f}^* R\tilde{j}'_* \tilde{j}'^* \widetilde{L} \\ &\to \tilde{i}^* R\tilde{j}_* f_{\bar{\eta}}^* \tilde{j}'^* \widetilde{L} \\ &\cong \tilde{i}^* R\tilde{j}_* \tilde{j}^* \tilde{f}^* \widetilde{L} \\ &\cong R\Psi_\eta((f^*L)_\eta). \end{aligned}$$

If f is smooth and $L \in D^+(X', \mathbb{Z}/n)$ for some n invertible on S, this is an isomorphism by the smooth base change theorem 7.7.2.

Proposition 9.2.3. *Suppose that X is smooth over S, and $\mathscr{H}^q(K)$ are locally constant sheaves of $\mathbb{Z}/n$-modules for some n invertible on S. Then $R\Phi(K) = 0$.*

Proof. We need to show that $K_{\bar{s}} \to R\Psi_\eta(K_\eta)$ is an isomorphism. The problem is local with respect to the etale topology on X. So we may assume that K is a constant sheaf associated to a $\mathbb{Z}/n$-module M. Fix notation by the following diagram

$$\begin{array}{ccccc} X_{\bar{\eta}} & \stackrel{\tilde{j}}{\to} & \tilde{X} = X \times_S \tilde{S} & \stackrel{\tilde{i}}{\leftarrow} & X_{\bar{s}} \\ f_{\bar{\eta}} \downarrow & & \tilde{f} \downarrow & & \downarrow f_{\bar{s}} \\ \bar{\eta} & \stackrel{j}{\to} & \tilde{S} & \stackrel{i}{\leftarrow} & \bar{s}. \end{array}$$

Denote by M the constant sheaf on S. We have a commutative diagram

$$\begin{array}{ccc} f_{\bar{s}}^* i^* M & \to & f_{\bar{s}}^* R\Psi_\eta(M) \\ \cong \downarrow & & \downarrow \\ \tilde{i}^* \tilde{f}^* M & \to & R\Psi_\eta(\tilde{f}^* M). \end{array}$$

The top horizontal arrow is an isomorphism by 9.2.2 (iii). The right vertical arrow is an isomorphism by the smooth base change theorem 7.7.2. So

$$\tilde{i}^* \tilde{f}^* M \to R\Psi_\eta(\tilde{f}^* M)$$

is an isomorphism. □

9.3 Generic Base Change Theorem and Generic Local Acyclicity

([SGA $4\frac{1}{2}$] Th. finitude 2.)

Throughout this section, A is a noetherian ring such that $nA = 0$ for some integer n invertible on a base scheme S. The main results of this section are the following theorems.

Theorem 9.3.1. *Let S be a noetherian scheme, X and Y two S-schemes of finite type, $f : X \to Y$ an S-morphism, and $\mathscr{F}$ a constructible sheaf of A-modules on X. Then there exists a dense open subset U of S such that the following conditions hold:*

(i) *$(R^q f_* \mathscr{F})|_{Y_U}$ are constructible and are nonzero only for finitely many q, where Y_U is the inverse image of U in Y.*

(ii) *The formation of $R^q f_* \mathscr{F}$ commutes with any base change $S' \to U \subset S$, or equivalently, the formation of $R^q f_{U*}(\mathscr{F}|_{X_U})$ commutes with any base change $S' \to S$, where X_U is the inverse image of U in X, and $f_U : X_U \to Y_U$ is the morphism induced by f.*

Theorem 9.3.2. *Let S be a noetherian scheme, $f : X \to S$ a morphism of finite type, and $K \in \mathrm{ob}\, D^b_{\mathrm{ctf}}(X, A)$. Then there exists a dense open subset U of S such that $f|_{f^{-1}(U)} : f^{-1}(U) \to U$ is universally locally acyclic relative to $K|_{f^{-1}(U)}$.*

When S is the spectrum of a field, we must have $U = S$. So we have the following:

Corollary 9.3.3. *Let X and Y be schemes of finite type over a field k, let $f : X \to Y$ be a k-morphism, and let $\mathscr{F}$ be a constructible sheaf of A-modules. Then $R^q f_* \mathscr{F}$ are constructible for all q. If k is separably closed, then $H^q(X, \mathscr{F})$ are finitely generated A-modules.*

Corollary 9.3.4. *Let k be a field, and let $f : X \to \operatorname{Spec} k$ be a morphism of finite type. Then f is universally strongly locally acyclic relative to any $K \in \operatorname{ob} D^b_{\mathrm{ctf}}(X, A)$.*

Corollary 9.3.5 (Künneth formula). *Let k be a field, let X_i and Y_i $(i = 1, 2)$ be k-schemes of finite type, let $f_i : X_i \to Y_i$ be k-morphisms, and let $p_i : X_1 \times_k X_2 \to X_i$ and $q_i : Y_1 \times_k Y_2 \to Y_i$ be projections. For any $K_i \in \operatorname{ob} D^b_{\mathrm{ctf}}(X_i, A)$, we have*

$$q_1^* Rf_{1*} K_1 \otimes^L_A q_2^* Rf_{2*} K_2 \cong R(f_1 \times f_2)_* (p_1^* K_1 \otimes^L_A p_2^* K_2).$$

Proof. Fix notation by the following commutative diagram of Cartesian squares:

$$\begin{array}{ccccc} X_1 \times_k X_2 & \stackrel{f_1''}{\to} & Y_1 \times_k X_2 & \stackrel{b_1''}{\to} & X_2 \\ {\scriptstyle f_2''} \downarrow & & {\scriptstyle f_2'} \downarrow & & \downarrow {\scriptstyle f_2} \\ X_1 \times_k Y_2 & \stackrel{f_1'}{\to} & Y_1 \times_k Y_2 & \stackrel{q_2}{\to} & Y_2 \\ {\scriptstyle b_2''} \downarrow & & {\scriptstyle q_1} \downarrow & & \downarrow {\scriptstyle b_2} \\ X_1 & \stackrel{f_1}{\to} & Y_1 & \stackrel{b_1}{\to} & \operatorname{Spec} k. \end{array}$$

By 6.5.6, 7.5.6 and 9.3.3, we have $Rf_{1*} K_1 \in \operatorname{ob} D^b_{\mathrm{ctf}}(Y_1, A)$. By 7.5.7 and 7.6.2 and 9.3.4, b_1 is universally strongly locally acyclic relative to $Rf_{1*} K$. By 7.6.9, we have an isomorphism

$$q_1^* Rf_{1*} K_1 \otimes^L_A q_2^* Rf_{2*} K_2 \cong Rf'_{2*} (f_2'^* q_1^* Rf_{1*} K_1 \otimes^L_A b_1''^* K_2).$$

Similarly, $b_2 f_2$ is universally strongly locally acyclic relative to K_2 and we have an isomorphism

$$f_2'^* q_1^* Rf_{1*} K_1 \otimes^L_A b_1''^* K_2 \cong Rf''_{1*} (f_2''^* b_2''^* K_1 \otimes^L_A f_1''^* b_1''^* K_2),$$

and hence

$$Rf'_{2*} (f_2'^* q_1^* Rf_{1*} K_1 \otimes^L_A b_1''^* K_2) \cong Rf'_{2*} Rf''_{1*} (f_2''^* b_2''^* K_1 \otimes^L_A f_1''^* b_1''^* K_2).$$

It follows that

$$q_1^* Rf_{1*} K_1 \otimes^L_A q_2^* Rf_{2*} K_2 \cong Rf'_{2*} Rf''_{1*} (f_2''^* b_2''^* K_1 \otimes^L_A f_1''^* b_1''^* K_2). \qquad \square$$

Proof of 9.3.1. Working with each irreducible component of S, we may assume that S is an integral scheme. Let η be its generic point. First consider the case where X is smooth over S pure of relative dimension d, $A = \mathbb{Z}/m$ for some integer m invertible on S, $\mathscr{F}$ and $R^q f_! \mathscr{F}^\vee$ are locally constant for all q, and $Y = S$, where $\mathscr{F}^\vee = \mathscr{H}om(\mathscr{F}, \mathbb{Z}/m)$. By the Poincaré duality (8.4.6 and 8.5.2), we have

$$Rf_* R\mathscr{H}om(\mathscr{F}^\vee, \mathbb{Z}/m(d)[2d]) \cong R\mathscr{H}om(Rf_! \mathscr{F}^\vee, \mathbb{Z}/m).$$

Since $\mathscr{F}$ and $R^q f_! \mathscr{F}^\vee$ are locally constant, and $\mathbb{Z}/m$ is an injective $\mathbb{Z}/m$-module, we have

$$R\mathscr{H}om(\mathscr{F}^\vee, \mathbb{Z}/m(d)[2d]) \cong \mathscr{H}om^\cdot(\mathscr{F}^\vee, \mathbb{Z}/m(d)[2d]) \cong \mathscr{F}(d)[2d],$$
$$R^{-q}\mathscr{H}om(Rf_! \mathscr{F}^\vee, \mathbb{Z}/m) \cong \mathscr{H}om(R^q f_! \mathscr{F}^\vee, \mathbb{Z}/m).$$

It follows that

$$R^{2d-q} f_* \mathscr{F} \cong \mathscr{H}om(R^q f_! \mathscr{F}^\vee, \mathbb{Z}/m)(-d).$$

For each q, the right-hand side is locally constant and constructible (7.8.1), commutes with any base change (7.4.4), and is nonzero only for finitely many q (7.4.5). So 9.3.1 holds for $U = S$.

Next we consider the case where X is smooth over S, $Y = S$, $\mathscr{F}$ is locally constant, and there exists a galois etale covering space $X_1 \to X$ such that the inverse image $\mathscr{F}_1$ of $\mathscr{F}$ in X_1 is a constant sheaf, say associated to an A-module F. Working with each connected component of X, we may assume that X is pure of relative dimension d over S. We have

$$\mathscr{F} = \bigoplus_\ell \mathscr{F}_\ell,$$

where ℓ goes over the set of prime numbers invertible on S, and $\mathscr{F}_\ell$ is the ℓ-torsion part of $\mathscr{F}$. Working with each direct factor $\mathscr{F}_\ell$, we may assume that $\ell^k \mathscr{F} = 0$ for some integer k and some prime number ℓ invertible on S. Filtrating $\mathscr{F}$ by $l^i \mathscr{F}$ $(i \in \mathbb{N} \cup \{0\})$ and working with each successive quotient $\ell^i \mathscr{F} / \ell^{i+1} \mathscr{F}$, we may assume $\ell \mathscr{F} = 0$. Replacing A by $A/\ell A$, we may assume $\ell A = 0$. Let $f_1 : X_1 \to S$ be the composite of $X_1 \to X$ and f. By 6.5.5, we have

$$R^q f_{1*} \mathscr{F}_1 \cong R^q f_{1*}(\mathbb{Z}/\ell \otimes^L_{\mathbb{Z}/\ell} f_1^* F) \cong R^q f_{1*} \mathbb{Z}/\ell \otimes^L_{\mathbb{Z}/\ell} F \cong R^q f_{1*} \mathbb{Z}/\ell \otimes_{\mathbb{Z}/\ell} F.$$

Since $R^q f_{1!} \mathbb{Z}/\ell$ are constructible, and are nonzero only for finitely many q, by shrinking S, we may assume they are locally constant. By our previous discussion, $R^q f_{1*} \mathbb{Z}/\ell$ are locally constant, constructible, nonzero for only finitely many q, and commute with any base change. $R^q f_{1*} \mathscr{F}_1$ have the same property. Let $G = \mathrm{Aut}(X_1/X)$. We have the Hochschild–Serre spectral sequence

$$E_2^{pq} = \mathscr{H}^p(G, R^q f_{1*}\mathscr{F}_1) \Rightarrow R^{p+q} f_* \mathscr{F}.$$

So $R^q f_* \mathscr{F}$ are locally constant, constructible, and commute with any base change. In particular, we have

$$(R^q f_* \mathscr{F})_{\bar{\eta}} \cong H^q(X_{\bar{\eta}}, \mathscr{F}).$$

Since $H^q(X_{\bar{\eta}}, \mathscr{F})$ are nonzero only for finitely many q (7.5.5), $R^q f_* \mathscr{F}$ have the same property.

To prove 9.3.1 in the general case, the problem is local on Y and we may assume that Y is affine. Covering X by finitely many affine open subsets and applying 5.6.10, we are reduced to the case where X is also affine. Then we can write $f = \bar{f} j$ with j being an open immersion and $\bar{f}$ being proper. 9.3.1 holds for the proper morphism $\bar{f}$. So it suffices to prove 9.3.1 for any open immersion f. As 9.3.1 holds for any closed immersion, it suffices to consider the case where f is an open immersion with dense image. We prove this by induction on $\dim X_\eta$.

Suppose that $\dim X_\eta \leq 0$ and f is an open immersion with dense image. We then have $X_\eta = Y_\eta$. Indeed, let $Y_1, \ldots, Y_m$ be the irreducible components of Y, and let $\eta_1, \ldots, \eta_m$ be the corresponding generic points. Since X is dense in Y, $\eta_1, \ldots, \eta_m$ all lie in X. If $Y_{i\eta} \neq \emptyset$, then η_i must lie in $Y_{i\eta}$. Note that η_i is the generic point of $Y_{i\eta}$. As $\eta_i \in X$, the morphism $(X \cap Y_i)_\eta \to Y_{i\eta}$ is an open immersion with dense image. But $\dim (X \cap Y_i)_\eta = 0$. So we must have $(X \cap Y_i)_{i\eta} = Y_{i\eta}$. It follows that $X_\eta = Y_\eta$. Shrinking S, we may assume $X = Y$. 9.3.1 holds trivially in this case.

Suppose that 9.3.1 holds for any immersion so that the fiber over η of its domain has dimension $\leq d-1$. Let us prove that it holds for any immersion $f : X \to Y$ with $\dim X_\eta \leq d$. We first show that after shrinking S, there exists an open subset $V \subset Y$ such that $Y - V \to S$ is finite, and that $R^q f_{V*} \mathscr{F}$ are constructible, nonzero only for finitely many q, and commute with any base change $S' \to S$, where $f_V : f^{-1}(V) \to V$ is the morphism induced by f. Cover Y by finitely many affine open subsets Y_i. Choose immersions $Y_i \hookrightarrow \mathbb{A}_S^k$. Let $\pi_j : \mathbb{A}_S^k \to \mathbb{A}_S^1$ $(j = 1, \ldots, k)$ be the projections. By 7.5.3, the generic fiber of $(\pi_j|_{Y_i}) \circ (f_{Y_i})$ has dimension $\leq d-1$ for each pair (i, j), and we can apply the induction hypothesis to the morphism $f_{Y_i} : f^{-1}(Y_i) \to Y_i$ over the base $\mathbb{A}_S^1$, where Y_i is considered as an $\mathbb{A}_S^1$-scheme through the morphism $\pi_j|_{Y_i}$. We can find dense open subsets U_{ij} of $\mathbb{A}^1_S$ so that $(R^q f_* \mathscr{F})|_{Y_i \cap \pi_j^{-1}(U_{ij})}$ are constructible, nonzero only for finitely many q, and commute with any base change $S' \to \mathbb{A}^1_S$. Let $V = \bigcup_{i,j}(Y_i \cap \pi_j^{-1}(U_{ij}))$. Then $R^q f_{V*}\mathscr{F}$ are constructible, nonzero only for finitely many q, and commute with any base change $S' \to S$. We have

$$\begin{aligned} Y - V &\subset \bigcup_i \Big(Y_i \cap (\mathbb{A}^k_S - \cup_j \pi_j^{-1}(U_{ij}))\Big) \\ &= \bigcup_i \Big(Y_i \cap ((\mathbb{A}^1_S - U_{i1}) \times_S \cdots \times_S (\mathbb{A}^1_S - U_{ik}))\Big). \end{aligned}$$

As the fiber of $(\mathbb{A}^1_S - U_{i1}) \times_S \cdots \times_S (\mathbb{A}^1_S - U_{ik}) \to S$ over η is finite, we may assume that $Y - V \to S$ is finite by shrinking S.

Let us prove that 9.3.1 holds for the immersion $f : X \to Y$ under the extra condition that X is smooth over S, $\mathscr{F}$ is locally constant, and there exists a galois etale covering space $X_1 \to X$ such that the inverse image of $\mathscr{F}$ in X_1 is a constant sheaf. The problem is local with respect to Y. We may assume that Y is affine. Then Y is an open subscheme of a scheme projective over S. We are thus reduced to the case where Y is projective over S. Shrinking S, we may find an open subset V of Y such that $Y - V \to S$ is finite, and $R^q f_{V*}\mathscr{F}$ are constructible, nonzero only for finitely many q, and commute with any base change $S' \to S$. Fix notation by the following commutative diagram:

$$\begin{array}{ccccc} & & V & & \\ & & j \downarrow & & \\ X & \xrightarrow{f} & Y & \xleftarrow{i} & Y - V, \\ & a \searrow & b \downarrow & \swarrow c & \\ & & S & & \end{array}$$

where i and j are immersions, and a, b, c are the structure morphisms. By the discussion at the beginning of the proof, under our extra condition, after shrinking S, $R^q a_* \mathscr{F}$ are constructible, nonzero only for finitely many q, and commute with any base change. By the choice of V, the proper base change theorem 7.3.1, 7.4.5, and 7.8.1, after shrinking S, $R^q b_* j_! j^* R f_* \mathscr{F}$ are constructible, nonzero only for finitely many q, and commute with any base change. Applying Rb_* to the distinguished triangle

$$j_! j^* R f_* \mathscr{F} \to R f_* \mathscr{F} \to i_* i^* R f_* \mathscr{F} \to,$$

we get a distinguished triangle

$$R b_* j_! j^* R f_* \mathscr{F} \to R a_* \mathscr{F} \to c_* i^* R f_* \mathscr{F} \to .$$

It follows that after shrinking S, $c_*i^*R^qf_*\mathscr{F}$ are constructible, nonzero only for finitely many q, and commute with any base change. As c is finite, $i^*R^qf_*\mathscr{F}$ have the same properties. By our choice of V, $j_!j^*R^qf_*\mathscr{F}$ also have these properties. So $R^qf_*\mathscr{F}$ have these properties. This proves that the theorem holds for f under our extra condition.

Finally we prove that 9.3.1 holds for the immersion $f : X \to Y$ unconditionally. We may assume that there exists an open dense subset W of X smooth over S. Indeed, if S is the spectrum of a perfect field, it suffices to replace X by X_{red}. In general, we use the passage to limit argument, and we need to shrink S, make a finite radiciel surjective base change, and replace X by a closed subscheme defined by a nilpotent ideal. These changes have no effect on etale cohomology. Shrinking W, we may assume that $\mathscr{F}|_W$ is locally constant, and there exists a galois etale covering space $W_1 \to W$ such that the inverse image of $\mathscr{F}$ in W_1 is a constant sheaf (5.8.1 (ii)). Let $j : W \hookrightarrow X$ be the open immersion. Choose a distinguished triangle

$$\mathscr{F} \to Rj_*j^*\mathscr{F} \to \Delta \to .$$

The cohomology sheaves of Δ are supported in $X-W$. We have shown that the theorem holds for the open immersion $j : W \to X$. So after shrinking S, we may assume that $R^qj_*j^*\mathscr{F}$ are constructible, nonzero only for finitely many q, and commutes with any base change. It follows that $\mathscr{H}^q(\Delta)$ have the same property. Since $\dim (X-W)_\eta \leq d-1$, by the induction hypothesis applied to $X - W \to Y$ and $\Delta|_{X-W}$, we see that $R^qf_*\Delta$ are constructible, nonzero only for finitely many q, and commute with any base change. We have a distinguished triangle

$$Rf_*\mathscr{F} \to R(fj)_*j^*\mathscr{F} \to Rf_*\Delta \to .$$

We have shown that the theorem holds for the immersion $fj : W \to Y$. So after shrinking S, $R^q(fj)_*j^*\mathscr{F}$ are constructible, nonzero only for finitely many q, and commute with any base change. It follows that $R^qf_*\mathscr{F}$ have the same property. So 9.3.1 holds for f. □

Lemma 9.3.6. *Let $f : X \to S$ be a proper morphism of noetherian schemes, let K be an object in $D^b_{\mathrm{ctf}}(X, A)$ such that R^qf_*K are locally constant for all q, and let W be an open subset of X such that $f|_W : W \to S$ is locally acyclic relative to K and such that $X - W$ is finite over S. Then f is locally acyclic relative to K.*

Proof. Let $s \in S$, let $\widetilde{S}_{\bar{s}}$ be the strict localization of S at $\bar{s}$, and let $j : t \to \widetilde{S}_{\bar{s}}$ be an algebraic geometric point. Fix notation by the following diagram

$$\begin{array}{ccccc} X_t = X \times_S t & \xrightarrow{\tilde{j}} & X \times_S \widetilde{S}_{\bar{s}} & \xleftarrow{\tilde{i}} & X_{\bar{s}} = X \times_S \bar{s} \\ {\scriptstyle f_t}\downarrow & & \tilde{f}\downarrow & & \downarrow{\scriptstyle f_{\bar{s}}} \\ t & \xrightarrow{j} & \widetilde{S}_{\bar{s}} & \xleftarrow{i} & \bar{s}, \end{array}$$

where vertical arrows are base changes of f. Denote the inverse image of K on $X \times_S \widetilde{S}_{\bar{s}}$ by $\widetilde{K}$. We need to show that $\tilde{i}^*\widetilde{K} \to \tilde{i}^* R\tilde{j}_*\tilde{j}^*\widetilde{K}$ is an isomorphism. Let

$$\tilde{i}^*\widetilde{K} \to \tilde{i}^* R\tilde{j}_*\tilde{j}^*\widetilde{K} \to \Delta \to$$

be a distinguished triangle. We need to show that Δ is acyclic. Since f is locally acyclic relative to K on W, the cohomology sheaves of Δ are supported on $(X - W)_{\bar{s}}$, which is a finite set. To prove that Δ is acyclic, it suffices to show that $R\Gamma(X_{\bar{s}}, \Delta)$ is acyclic. We have a distinguished triangle

$$R\Gamma(X_{\bar{s}}, \tilde{i}^*\widetilde{K}) \to R\Gamma(X_{\bar{s}}, \tilde{i}^* R\tilde{j}_*\tilde{j}^*\widetilde{K}) \to R\Gamma(X_{\bar{s}}, \Delta) \to .$$

It suffices to show that the canonical morphism

$$R\Gamma(X_{\bar{s}}, \tilde{i}^*\widetilde{K}) \to R\Gamma(X_{\bar{s}}, \tilde{i}^* R\tilde{j}_*\tilde{j}^*\widetilde{K})$$

is an isomorphism. For each q, since $R^q f_* K$ is locally constant, the specialization homomorphism

$$(R^q f_* K)_{\bar{s}} \to (R^q f_* K)_{\bar{t}}$$

is an isomorphism. By 5.9.5, we have

$$(R^q f_* K)_{\bar{s}} \cong H^q(X \times_S \widetilde{S}_{\bar{s}}, \widetilde{K}),$$

and by the proper base change theorem 7.3.1, we have

$$(R^q f_* K)_{\bar{t}} \cong H^q(X_{\bar{t}}, \tilde{j}^*\widetilde{K}).$$

Through these isomorphisms, the specialization homomorphism is identified with the canonical homomorphism

$$H^q(X \times_S \widetilde{S}_{\bar{s}}, \widetilde{K}) \to H^q(X_{\bar{t}}, \tilde{j}^*\widetilde{K}).$$

Since the specialization homomorphism is an isomorphism, we have

$$H^q(X \times_S \widetilde{S}_{\bar{s}}, \widetilde{K}) \cong H^q(X_{\bar{t}}, \tilde{j}^*\widetilde{K}).$$

We have a commutative diagram of canonical morphisms

$$\begin{array}{ccccc} H^q(X_{\bar{t}}, \tilde{j}^*\widetilde{K}) & \leftarrow & H^q(X \times_S \widetilde{S}_{\bar{s}}, R\tilde{j}_*\tilde{j}^*\widetilde{K}) & \to & H^q(X_{\bar{s}}, \tilde{i}^* R\tilde{j}_*\tilde{j}^*\widetilde{K}) \\ \| & & \uparrow & & \uparrow \\ H^q(X_{\bar{t}}, \tilde{j}^*\widetilde{K}) & \leftarrow & H^q(X \times_S \widetilde{S}_{\bar{s}}, \widetilde{K}) & \to & H^q(X_{\bar{s}}, \tilde{i}^*\widetilde{K}). \end{array}$$

We have just shown that the lower horizontal arrow in the square on the left of the diagram is an isomorphism. By the proper base change theorem 7.3.3, the two horizontal arrows in the square on the right of the diagram are isomorphisms. The canonical morphism

$$H^q(X\times_S \widetilde{S}_{\bar{s}}, R\tilde{j}_*\tilde{j}^*\widetilde{K})\to H^q(X_{\bar{t}}, \tilde{j}^*\widetilde{K})$$

is also an isomorphism. It follows that all arrows in the above diagram are isomorphisms. So

$$H^q(X_{\bar{s}}, \tilde{i}^*\widetilde{K})\to H^q(X_{\bar{s}}, \tilde{i}^*R\tilde{j}_*\tilde{j}^*\widetilde{K})$$

is an isomorphism. Hence

$$R\Gamma(X_{\bar{s}}, \tilde{i}^*\widetilde{K})\to R\Gamma(X_{\bar{s}}, \tilde{i}^*R\tilde{j}_*\tilde{j}^*\widetilde{K})$$

is an isomorphism.

Proof of Theorem 9.3.2. We may assume that S is an integral scheme with generic point η, and we use induction on $\dim X_\eta$. First we use the induction hypothesis to prove the following statement:

(*) After shrinking S, we can find an open subset W of X such that $f|_W : W\to S$ is universally locally acyclic relative to $K|_W$ and $X-W\to S$ is finite.

Cover X by finitely many affine open subsets X_i $(i=1,\ldots,m)$ so that there exist immersions $X_i\hookrightarrow \mathbb{A}^k_S$. Let $\pi_j:\mathbb{A}^k_S\to A^1_S$ $(j=1,\ldots,k)$ be the projections. For each pair (i,j), the generic fiber of $\pi_j|_{X_i}$ has dimension $\leq \dim X_\eta - 1$ by 7.5.3. By the induction hypothesis, we can find a dense open subset V_{ij} of $\mathbb{A}^1_S$ such that $\pi_j : X_i\cap\pi_j^{-1}(V_{ij})\to V_{ij}$ is universally locally acyclic relative to K. The projection $V_{ij}\to S$ is smooth and hence universally strongly locally acyclic relative to A by 7.7.5. So by 7.7.6, $f : X_i\cap\pi_j^{-1}(V_{ij})\to S$ is universally locally acyclic relative to K. Let $W=\bigcup_{i,j}(X_i\cap\pi_j^{-1}(V_{ij}))$. We have

$$\begin{aligned} X-W &\subset \bigcup_i \Big(X_i\cap(\mathbb{A}^k_S-\cup_j\pi_j^{-1}(V_{ij}))\Big)\\ &=\bigcup_i\Big(X_i\cap((\mathbb{A}^1_S-V_{i1})\times_S\cdots\times_S(\mathbb{A}^1_S-V_{ik}))\Big). \end{aligned}$$

As the generic fiber of $(\mathbb{A}^1_S-V_{i1})\times_S\cdots\times_S(\mathbb{A}^1_S-V_{ik})\to S$ is finite, we may assume that $X-W\to S$ is finite by shrinking S. This proves (*) under the induction hypothesis.

Now let us prove the theorem. The problem is local. We may assume that S is affine. Covering X by finitely many affine open subsets, and working with each of them, we may assume that X is affine. Then X is an open subscheme of a proper S-scheme. So it suffices prove the theorem for any proper S-scheme X. By 7.4.5 and 7.8.1, after shrinking S, we may assume that $R^q f_* K$ are locally constant for all q. By (*), we may assume that there exists an open subset W of X such that $f|_W : W \to S$ is universally locally acyclic relative to K, and $X - W \to S$ is finite. We then apply 9.3.6. □

9.4 Finiteness of $R\Psi_\eta$

([SGA $4\frac{1}{2}$] Th. finitude 3.)

The main result of this section is the following:

Theorem 9.4.1. *Let S be a strictly local trait, s its closed point, η its generic point, and A a noetherian ring such that $nA = 0$ for some integer n invertible on S. For any S-scheme X of finite type and any constructible sheaf of A-modules $\mathscr{F}$ on X_η, $R^q\Psi_\eta\mathscr{F}$ are constructible for all q.*

Before proving the theorem, we make some preparations. Let $S = \mathrm{Spec}\, R$, where R is a strictly henselian discrete valuation ring with maximal ideal $\mathfrak{m}$. Let s' be the generic point of $\mathbb{A}^1_s = \mathrm{Spec}\,(R/\mathfrak{m})[t]$, and denote the image of s' in $\mathbb{A}^1_S = \mathrm{Spec}\, R[t]$ also by s'. It corresponds to the prime ideal $\mathfrak{m}[t]$ of $R[t]$. Let S' be the strict localization of $\mathbb{A}^1_S$ at $\bar{s}'$. Then $S' = \mathrm{Spec}\, R'$ for a strictly henselian discrete valuation ring R', and any uniformizer of R is also a uniformizer of R'. Let η' be the generic point of S'. We have

$$\mathrm{Spec}\, k(\eta') \cong \mathrm{Spec}\, k(\eta) \times_S S'.$$

Let $K' = k(\bar{\eta}) \otimes_{k(\eta)} k(\eta')$. Then K' is a field. To show this, it suffices to show that for any finite galois extension K_1 of $k(\eta)$ contained in $k(\bar{\eta})$, $K_1 \otimes_{k(\eta)} k(\eta')$ is a field. Let R_1 be the integral closure of R in K_1 and let $S_1 = \mathrm{Spec}\, R_1$. Then R_1 is a strictly henselian discrete valuation ring with fraction field K_1. Since the residue field of R_1 is purely inseparable over the residue field of R, there exists one and only one point in $S_1 \times_S S'$ lying above the closed point of S'. On the other hand, $S_1 \times_S S'$ is finite over S'. It follows that $S_1 \times_S S'$ is strictly local, and it is the strict localization of $\mathbb{A}^1_{S_1}$ at the generic point of the special fiber of $\mathbb{A}^1_{S_1} \to S_1$. We have

$$\begin{aligned}\mathrm{Spec}\,(K_1 \otimes_{k(\eta)} k(\eta')) &\cong \mathrm{Spec}\, K_1 \times_{\mathrm{Spec}\, k(\eta)} (\mathrm{Spec}\, k(\eta) \times_S S')\\ &\cong \mathrm{Spec}\, K_1 \times_{S_1} (S_1 \times_S S').\end{aligned}$$

So $K_1 \otimes_{k(\eta)} k(\eta')$ is the function field of $S_1 \times_S S'$. We have

$$\mathrm{Gal}(K_1 \otimes_{k(\eta)} k(\eta')/k(\eta')) \cong \mathrm{Gal}(K_1/k(\eta)).$$

It follows that $K' = k(\bar{\eta}) \otimes_{k(\eta)} k(\eta')$ is a field and

$$\mathrm{Gal}(K'/k(\eta')) \cong \mathrm{Gal}(k(\bar{\eta})/k(\eta)).$$

Let $\overline{K}'$ be a separable closure of K' and let $\bar{\eta}' = \mathrm{Spec}\,\overline{K}'$. Note that $\mathrm{Gal}(\overline{K}'/K')$ is a pro-p-group, where p is the characteristic of the residue field of R. To see this, we need to show that any finite tamely ramified extension L of $k(\eta')$ is contained in K'. Let π be a uniformizer of R. It is also a uniformizer of R'. By 8.1.3, we have $L = k(\eta')[\sqrt[n]{\pi}]$, where $n = [L : k(\eta')]$. Since $\sqrt[n]{\pi} \in k(\bar{\eta})$, we have $L \subset K'$.

Lemma 9.4.2. *Notation as above. Let X' be an S'-scheme, let $\mathscr{F}$ be a sheaf of $\mathbb{Z}/n$-modules on $X'_{\eta'} \cong X'_\eta$ for some integer n invertible on S, and let $R\Psi_{\eta'}(\mathscr{F})$ and $R\Psi_\eta(\mathscr{F})$ be the nearby cycles of $\mathscr{F}$ with respect to X'/S' and X'/S, respectively. We have*

$$R^q\Psi_\eta(\mathscr{F}) \cong R^q\Psi_{\eta'}(\mathscr{F})^{\mathrm{Gal}(\overline{K}'/K')}.$$

Proof. Fix notation by the following commutative diagram

$$\begin{array}{ccccccccc}
X'_{\bar{\eta}'} & \stackrel{j_1}{\to} & X'_{\bar{\eta}} & \stackrel{j_2}{\to} & X'_\eta \cong X'_{\eta'} & \stackrel{j_3}{\to} & X' & \stackrel{i}{\leftarrow} & X'_s \cong X'_{s'} \\
\downarrow & & \downarrow & & \downarrow & & \downarrow & & \downarrow \\
\bar{\eta}' & \to & \mathrm{Spec}\,K' & \to & \eta' & \to & S' & \leftarrow & S' \times_S s \cong s' \\
 & & \downarrow & & \downarrow & & \downarrow & & \downarrow \\
 & & \bar{\eta} & \to & \eta & \to & S & \leftarrow & s.
\end{array}$$

We have

$$R^u\Psi_\eta(\mathscr{F}) = i^* R^u(j_3j_2)_* j_2^* \mathscr{F}, \quad R^u\Psi_{\eta'}(\mathscr{F}) = i^* R^u(j_3j_2j_1)_* j_1^* j_2^* \mathscr{F}.$$

We have the Hochschild–Serre spectral sequence

$$E_2^{uv} = \mathscr{H}^u(\mathrm{Gal}(\overline{K}'/K'), R^v\Psi_{\eta'}(\mathscr{F})) \Rightarrow R^{u+v}\Psi_\eta(\mathscr{F}).$$

Since $\mathrm{Gal}(\overline{K}'/K')$ is a pro-p-group and $R^v\Psi_{\eta'}(\mathscr{F})$ are n-torsion sheaves and n is relatively prime to p, we have

$$\mathscr{H}^u(\mathrm{Gal}(\overline{K'}/K'), R^v\Psi_{\eta'}(\mathscr{F})) = 0$$

for all $u \geq 1$. (The functor $\mathscr{G} \to \mathscr{G}^{\mathrm{Gal}(\overline{K}'/K')}$ is exact in the category of torsion sheaves with continuous $\mathrm{Gal}(\overline{K}'/K')$-action and with torsion prime to p.) So the spectral sequence degenerates, and we have

$$R^q\Psi_\eta(\mathscr{F}) \cong R^q\Psi_{\eta'}(\mathscr{F})^{\mathrm{Gal}(\overline{K}'/K')}.$$

Lemma 9.4.3. *Let k be a field, $\pi_i : \mathbb{A}_k^m \to \mathbb{A}_k^1$ $(i = 1, \ldots, m)$ the projections, $\bar{\eta}$ an algebraic geometric point over the generic point of $\mathbb{A}_k^1$, X a subscheme of $\mathbb{A}_k^m$, and $\mathscr{F}$ a sheaf of A-modules on X. For each i, let $X_{i\bar{\eta}}$ be the fiber of $\pi_i|_X$ over $\bar{\eta}$, and let $\mathscr{F}_{i\bar{\eta}}$ be the inverse image of $\mathscr{F}$ on $X_{i\bar{\eta}}$. Suppose $\mathscr{F}_{i\bar{\eta}}$ are constructible. Then there exists a constructible subsheaf $\mathscr{F}'$ of $\mathscr{F}$ such that every section of the sheaf $\mathscr{F}/\mathscr{F}'$ has finite support.*

Proof. By 5.9.8, for each i, we can find an etale neighborhood U of $\bar{\eta}$ in $\mathbb{A}_k^1$, a constructible sheaf $\mathscr{G}_{iU}$ on $X_{iU} = X \times_{\pi_i|_X, \mathbb{A}_k^1} U$ such that the inverse image $\mathscr{G}_{i\bar{\eta}}$ of $\mathscr{G}_{iU}$ on $X_{i\bar{\eta}}$ is isomorphic to $\mathscr{F}_{i\bar{\eta}}$. Moreover, we may assume that the isomorphism is induced by a morphism

$$\mathscr{G}_{iU} \to \mathscr{F}_{iU},$$

where $\mathscr{F}_{iU}$ is the inverse image of $\mathscr{F}$ on X_{iU}. Let $\phi : X_{iU} \to X$ be the projection. Then the above morphism induces a morphism

$$\phi_! \mathscr{G}_{iU} \to \mathscr{F}.$$

Let $\mathscr{F}_i'$ be the image of this morphism. It is constructible. The image of the morphism

$$\phi^* \phi_! \mathscr{G}_{iU} \to \mathscr{F}_{iU}$$

is $\phi^* \mathscr{F}_i'$. Since the morphism $\mathscr{G}_{iU} \to \mathscr{F}_{iU}$ can be factorized as the composite

$$\mathscr{G}_{iU} \to \phi^* \phi_! \mathscr{G}_{iU} \to \mathscr{F}_{iU},$$

$\phi^* \mathscr{F}_i'$ contains the image of $\mathscr{G}_{iU} \to \mathscr{F}_{iU}$. As $\mathscr{G}_{i\bar{\eta}} \to \mathscr{F}_{i\bar{\eta}}$ is an isomorphism, the inverse image of $\mathscr{F}/\mathscr{F}_i'$ on $X_{i\eta}$ is 0. Let $\mathscr{F}'$ be the image of the canonical morphism

$$\bigoplus_{i=1}^n \mathscr{F}_i' \to \mathscr{F}.$$

Then $\mathscr{F}'$ is constructible, and the inverse image of $\mathscr{F}/\mathscr{F}'$ on $X_{i\eta}$ is 0 for each i. Any section of the sheaf $\mathscr{F}/\mathscr{F}'$ has finite support. Indeed, let $f : V \to X$ be an etale X-scheme of finite type and let $s \in (\mathscr{F}/\mathscr{F}')(V)$. By 5.9.3, for each i, we can find a nonempty open subset U_i of $\mathbb{A}_k^1$ such that the restriction of s to $(\pi_i|_X \circ f)^{-1}(U_i)$ is 0. We have

$$V - \bigcup_{i=1}^m (\pi_i|_X \circ f)^{-1}(U_i) = f^{-1}\Big(X \cap \big((\mathbb{A}_k^1 - U_1) \times_k \cdots \times_k (\mathbb{A}_k^1 - U_n)\big)\Big).$$

As $(\mathbb{A}_k^1 - U_1) \times_k \cdots \times_k (\mathbb{A}_k^1 - U_n)$ is finite, the support of s is finite. □

Proof of 9.4.1. We use induction on $\dim X_\eta$. When $\dim X_\eta < 0$, we have $X_\eta = \emptyset$. Then $R^q\Psi_\eta(\mathscr{F}) = 0$ for all q. Suppose that the theorem holds for those S-schemes whose generic fibers have dimensions $< d$. Let X be an S-scheme with $\dim X_\eta = d$. We first prove that for each q, there exists a constructible subsheaf $\mathscr{G}_q$ of $R^q\Psi_\eta(\mathscr{F})$ such that the support of any section of $R^q\Psi_\eta(\mathscr{F})/\mathscr{G}$ is finite. Cover X by finitely many affine open subsets X_i, and let $j_i : X_{is} \to X_s$ be the open immersion. If we can find constructible subsheaves $\mathscr{G}_{qi}$ of $R^q\Psi_\eta(\mathscr{F})|_{X_{is}}$ such that the sections of $\big(R^q\Psi_\eta(\mathscr{F})|_{X_{is}}\big)/\mathscr{G}_{qi}$ have finite support, then we can take $\mathscr{G}_q$ to be the image of the canonical morphism

$$\bigoplus_i j_{i!}\mathscr{G}_{qi} \to R^i\Psi_\eta(\mathscr{F}).$$

So we may assume that X is affine. Then we have an immersion $X \hookrightarrow \mathbb{A}^m_S$. Let π be one of the projections $\pi_i|X : X \hookrightarrow \mathbb{A}^m_S \to \mathbb{A}^1_S$, let S' be the strict localization of $\mathbb{A}^1_S$ at the generic point of the special fiber of $\mathbb{A}^1_S \to S$, let $X' = X \times_{\pi,\mathbb{A}^1_S} S'$, let $\lambda : X' \to X$ be the projection, and let $\mathscr{F}'$ be the inverse image of $\mathscr{F}$ on X'.

$$\begin{array}{ccccc} X' & \to & S' & & \\ \lambda\downarrow & & \downarrow & \searrow & \\ X & \xrightarrow{\pi} & \mathbb{A}^1_S & \to & S. \end{array}$$

Since X' is the inverse limit of an inverse system of etale X-schemes, we have

$$\lambda_s^* R^q\Psi_\eta(\mathscr{F}) = R^q\Psi_\eta(\mathscr{F}').$$

By 9.4.2, we have

$$R^q\Psi_\eta(\mathscr{F}') = R^q\Psi_{\eta'}(\mathscr{F}')^{\mathrm{Gal}(\overline{K}'/K')}.$$

By the induction hypothesis, $R^q\Psi_{\eta'}(\mathscr{F}')$ are constructible. So $\lambda_s^* R^q\Psi_\eta(\mathscr{F})$ are constructible. By 9.4.3, there exists a constructible subsheaf $\mathscr{G}_q$ of $R^q\Psi_\eta(\mathscr{F})$ such that the support of any section of $R^q\Psi_\eta(\mathscr{F})/\mathscr{G}_q$ is finite.

Let us prove that $R^q\Psi_\eta(\mathscr{F})$ are constructible. The problem is local on X. We may assume that X is affine. Then X is an open subscheme of a proper S-scheme. So it suffices to prove the theorem for any proper S-scheme X. We then have a biregular spectral sequence

$$E_2^{pq} = H^p(X_s, R^q\Psi_\eta(\mathscr{F})) \Rightarrow H^{p+q}(X_{\bar\eta}, \mathscr{F}_{\bar\eta}).$$

Let $\mathscr{H}_q = R^q\Psi_\eta(\mathscr{F})/\mathscr{G}_q$. Since sections of $\mathscr{H}_q$ have finite support, we have $H^p(X_s, \mathscr{H}_q) = 0$ for all $p \geq 1$ by 7.2.4. To prove that $R^q\Psi_\eta(\mathscr{F})$ are constructible, it suffices to show that $H^0(X_s, \mathscr{H}_q)$ are finitely generated A-modules. For any two A-modules M and N, we write $M \sim N$ if there exists a finite family of A-modules $M_0, \ldots, M_k$ such that $M_0 = M$, $M_k = N$, and for each $i \in \{0, \ldots, k-1\}$, there exists a homomorphism $M_i \to M_{i+1}$ or a homomorphism $M_{i+1} \to M_i$ with finite kernel and finite cokernel. Since $\mathscr{G}_q$ are constructible, $H^p(X_s, \mathscr{G}_q)$ are finitely generated A-modules by 7.8.1. From the long exact sequences of cohomology groups associated to the short exact sequence

$$0 \to \mathscr{G}_q \to R^q\Psi_\eta(\mathscr{F}) \to \mathscr{H}_q \to 0,$$

we get $E_2^{pq} \sim H^p(X, \mathscr{H}_q)$. In particular, $E_2^{pq} \sim 0$ for all $p \neq 0$, that is, the spectral sequence degenerates modulo the equivalence relation $\sim$. So we have $E_2^{0q} \sim H^q(X_{\bar{\eta}}, \mathscr{F}_{\bar{\eta}})$. By 7.8.1, we have $H^q(X_{\bar{\eta}}, \mathscr{F}_{\bar{\eta}}) \sim 0$. So we have

$$H^0(X_s, \mathscr{H}_q) = E_2^{0q} \sim 0.$$

This proves our assertion. □

9.5 Finiteness Theorems

([SGA $4\frac{1}{2}$] Th. finitude 1, 3.)

Throughout this section, A is a noetherian ring such that $nA = 0$ for some integer n invertible on a base scheme S.

Theorem 9.5.1. *Suppose that S is a regular noetherian scheme of dimension ≤ 1. Let X and Y be S-schemes of finite type, $f : X \to Y$ an S-morphism, and $\mathscr{F}$ a constructible sheaf of A-modules on X. Then $R^q f_* \mathscr{F}$ are constructible for all q, and are nonzero only for finitely many q.*

Proof. First suppose that S is a strictly local trait. Let η and s be the generic point and the closed point of S, respectively. We first prove that for any S-scheme X of finite type and any constructible sheaf of A-modules $\mathscr{F}_\eta$ on X_η, the sheaves $i^* R^q j_* \mathscr{F}_\eta$ are constructible for all q and are nonzero only for finitely many q, where $i : X_s \to X$ and $j : X_\eta \to X$ are immersions. We have the Hochschild–Serre spectral sequence

$$E_2^{uv} = \mathscr{H}^u(\mathrm{Gal}(\bar{\eta}/\eta), R^v\Psi_\eta(\mathscr{F}_\eta)) \Rightarrow i^* R^{u+v} j_* \mathscr{F}.$$

It suffices to show that $\mathscr{H}^u(\mathrm{Gal}(\bar{\eta}/\eta), R^v\Psi_\eta(\mathscr{F}_\eta))$ are constructible for all u, v and are nonzero only for finitely many pairs (u, v). Let P be the wild inertia subgroup of $\mathrm{Gal}(\bar{\eta}/\eta)$. It is a pro-$p$-subgroup of $\mathrm{Gal}(\bar{\eta}/\eta)$ and

$$\mathrm{Gal}(\bar{\eta}/\eta)/P \cong \varprojlim_{(n,p)=1} \boldsymbol{\mu}_n(k(\eta)),$$

where p is the characteristic of the residue field of S. Let $\sigma \in \mathrm{Gal}(\bar{\eta}/\eta)/P$ be a topological generator so that its components in each $\boldsymbol{\mu}_n(k(\eta))$ is a primitive n-th root of unity. Since $R^q\Psi_\eta(\mathscr{F}_\eta)$ are n-torsion sheaves and n is relatively prime to p, we have

$$\mathscr{H}^v(P, R^q\Psi_\eta(\mathscr{F}_\eta)) = 0$$

for any $v \neq 0$. (The functor $\mathscr{G} \to \mathscr{G}^P$ is exact in the category of torsion sheaves with continuous P-action and with torsion prime to p.) So the Hochschild–Serre spectral sequence

$$E_2^{uv} = \mathscr{H}^u(\mathrm{Gal}(\bar{\eta}/\eta)/P, \mathscr{H}^v(P, R^q\Psi_\eta(\mathscr{F}_\eta))) \Rightarrow \mathscr{H}^{u+v}(\mathrm{Gal}(\bar{\eta}/\eta), R^q\Psi_\eta(\mathscr{F}_\eta))$$

degenerates, and we have

$$\mathscr{H}^u(\mathrm{Gal}(\bar{\eta}/\eta), R^q\Psi_\eta(\mathscr{F}_\eta)) \cong \mathscr{H}^u(\mathrm{Gal}(\bar{\eta}/\eta)/P, (R^q\Psi_\eta(\mathscr{F}_\eta))^P).$$

By 4.3.9, we have

$$\mathscr{H}^u(\mathrm{Gal}(\bar{\eta}/\eta)/P, (R^q\Psi_\eta(\mathscr{F}_\eta))^P) \cong \begin{cases} \ker(\sigma - 1, (R^q\Psi_\eta(\mathscr{F}_\eta))^P) & \text{if } u = 0, \\ \mathrm{coker}(\sigma - 1, (R^q\Psi_\eta(\mathscr{F}_\eta))^P) & \text{if } u = 1, \\ 0 & \text{if } u \neq 0, 1. \end{cases}$$

By 9.4.1 and 9.2.2 (ii), $R^q\Psi_\eta(\mathscr{F}_\eta)$ are constructible for all q, and are nonzero only for finite many q. Hence $\ker(\sigma - 1, (R^q\Psi_\eta(\mathscr{F}_\eta))^P)$ and $\mathrm{coker}(\sigma - 1, (R^q\Psi_\eta(\mathscr{F}_\eta))^P)$ are constructible for all q, and are nonzero only for finitely many q. This proves our assertion.

To prove the theorem, we may assume that S is connected and hence integral. If $\dim S = 0$, the theorem follows from 9.3.3. Suppose $\dim S = 1$. Then by 9.3.1, there exists a finite closed subset T of S such that over $S - T$, $R^q f_*\mathscr{F}$ are constructible for all q and are nonzero only for finitely many q. To prove the theorem, it suffices to show that $R^q f_*\mathscr{F}|_{Y_s}$ are constructible for all q and are nonzero only for finitely many q for all $s \in T$. The problem is local on S. Replacing S by the strict localization of S at s, we may assume that S is a strictly local trait. Fix notation by the following commutative diagram.

$$\begin{array}{ccccc} X_\eta & \stackrel{j_1}{\to} & X & \stackrel{i_1}{\leftarrow} & X_s \\ {\scriptstyle f_\eta}\downarrow & & {\scriptstyle f}\downarrow & & \downarrow{\scriptstyle f_s} \\ Y_\eta & \stackrel{j_2}{\to} & Y & \stackrel{i_2}{\leftarrow} & Y_s \\ \downarrow & & \downarrow & & \downarrow \\ \eta & \to & S & \leftarrow & s. \end{array}$$

We need to show that $i_2^* R^q f_*\mathscr{F}$ are constructible for all q and are nonzero only for finitely many q. Choose a distinguished triangle

$$\mathscr{F} \to Rj_{1*}j_1^*\mathscr{F} \to \Delta \to .$$

We then have a distinguished triangle

$$i_2^*Rf_*\mathscr{F} \to i_2^*Rf_*Rj_{1*}j_1^*\mathscr{F} \to i_2^*Rf_*\Delta \to .$$

To prove our assertion, it suffices to prove that $i_2^*R^qf_*\Delta$ and $i_2^*R^qf_*Rj_{1*}j_1^*\mathscr{F}$ are constructible for all q and are nonzero only for finitely many q. We have shown that $i_1^*R^qj_{1*}j_1^*\mathscr{F}$ have these properties. So the cohomology sheaves of $i_1^*\Delta$ are all constructible and only finite many of them are nonzero. By 9.3.3, $R^qf_{s*}i_1^*\Delta$ are constructible for all q and are nonzero only for finite many q. The cohomology sheaves of Δ are supported on X_s. So we have

$$i_2^*Rf_*\Delta \cong i_2^*Rf_*i_{1*}i_1^*\Delta \cong i_2^*i_{2*}Rf_{s*}i_1^*\Delta \cong Rf_{s*}i_1^*\Delta.$$

Hence $i_2^*R^qf_*\Delta$ are constructible for all q and are nonzero only for finitely many q. We have

$$i_2^*Rf_*Rj_{1*}j_1^*\mathscr{F} \cong i_2^*Rj_{2*}Rf_{\eta*}j_1^*\mathscr{F}.$$

By 9.3.3, the cohomology sheaves of $Rf_{\eta*}j_1^*\mathscr{F}$ are all constructible and only finitely many of them are nonzero. By our previous discussion, $i_2^*R^qj_{2*}Rf_{\eta*}j_1^*\mathscr{F}$ are constructible for all q and are nonzero only for finitely many q. This proves our assertion.

Theorem 9.5.2. *Suppose that S is a regular noetherian scheme of dimension ≤ 1. Let X and Y be S-schemes of finite type and let $f : X \to Y$ be an S-morphism. Then $Rf_*, Rf_!, f^*, Rf^!$ have finite cohomological dimensions, Rf_* and $Rf_!$ map objects in $D_c(X,A)$ (resp. $D^b_{\mathrm{tf}}(X,A)$, resp. $D^b_{\mathrm{ctf}}(X,A)$) to objects in $D_c(Y,A)$ (resp. $D^b_{\mathrm{tf}}(Y,A)$, resp. $D^b_{\mathrm{ctf}}(Y,A)$), and f^* and $Rf^!$ map objects in $D_c(Y,A)$ (resp. $D^b_{\mathrm{tf}}(Y,A)$, resp. $D^b_{\mathrm{ctf}}(Y,A)$) to objects in $D_c(X,A)$ (resp. $D^b_{\mathrm{tf}}(X,A)$, resp. $D^b_{\mathrm{ctf}}(X,A)$).*

Proof. Using 7.5.6, 9.2.2 (ii) and the same argument as in the proof of 9.5.1, one can show that Rf_* has finite cohomological dimension. Combined with 9.5.1 and 6.5.3, we see that Rf_* maps objects in $D_c(X,A)$ to objects in $D_c(Y,A)$. By 6.5.6, it maps objects in $D^b_{\mathrm{tf}}(X,A)$ to objects in $D^b_{\mathrm{tf}}(Y,A)$. It follows that it maps objects in $D^b_{\mathrm{ctf}}(X,A)$ to objects in $D^b_{\mathrm{ctf}}(Y,A)$. The assertions for $Rf_!$ follows from 7.4.5, 7.4.7 (ii) and 7.8.1. The assertions for f^* are clear. To prove the assertions for $Rf^!$, the problem is local. We may assume that f can be factorized as the composite $X \xrightarrow{i} \mathbb{A}^d_Y \xrightarrow{\pi} Y$, where i is a closed immersion and π is the projection. We have

$$Rf^!K \cong Ri^!R\pi^!K = Ri^!\pi^*K(d)[2d].$$

So it suffices to prove the assertions for $Ri^!$. Let $j : \mathbb{A}^d_Y - X \hookrightarrow \mathbb{A}^d_Y$ be the open immersion. We have a distinguished triangle

$$i_*Ri^!K \to K \to Rj_*j^*K \to .$$

The assertions for $Ri^!$ follows from those for Rj_*. □

Theorem 9.5.3.

(i) *For any scheme X, the functor $-\otimes_A^L-$ maps objects in $D_c^-(X,A)\times D_c^-(X,A)$ (resp. $D_{\mathrm{tf}}^b(X,A)\times D_{\mathrm{tf}}^b(X,A)$, resp. $D_{\mathrm{ctf}}^b(X,A)\times D_{\mathrm{ctf}}^b(X,A)$) to objects in $D_c^-(X,A)$ (resp. $D_{\mathrm{tf}}^b(X,A)$, resp. $D_{\mathrm{ctf}}^b(X,A)$).*

(ii) *Suppose that S is a regular noetherian scheme of dimension ≤ 1. For any scheme X of finite type over S, the functor $R\mathscr{H}om_A(-,-)$ maps objects in $D_c^-(X,A)\times D_c^+(X,A)$ (resp. $D_{\mathrm{ctf}}^b(X,A)\times D_{\mathrm{tf}}^b(X,A)$, resp. $D_{\mathrm{ctf}}^b(X,A)\times D_{\mathrm{ctf}}^b(X,A)$) to objects in $D_c^+(X,A)$ (resp. $D_{\mathrm{tf}}^b(X,A)$, resp. $D_{\mathrm{ctf}}^b(X,A)$).*

Proof.

(i) Use 6.4.5, 6.4.6 and 6.5.4.

(ii) By 6.5.3, to show that $R\mathscr{H}om$ maps objects in $D_c^-(X,A)\times D_c^+(X,A)$ to objects in $D_c^+(X,A)$, it suffices to show that for any constructible sheaves of A-modules $\mathscr{F}$ and $\mathscr{G}$ on X, $\mathscr{E}xt^q(\mathscr{F},\mathscr{G})$ are constructible for all q. First we prove that $\mathscr{F}$ has a filtration

$$0=\mathscr{F}_0\subset\mathscr{F}_1\subset\cdots\subset\mathscr{F}_k=\mathscr{F}$$

such that $\mathscr{F}_i/\mathscr{F}_{i-1}$ are of the form $j_!\mathscr{F}'$ for some immersion $j:Y\to X$ and some locally constant constructible sheaf of A-modules $\mathscr{F}'$ on Y. We use induction on $\dim X$. This is clear if $\dim X\leq 0$. Suppose that $\dim X=d$ and the assertion holds for schemes of dimension $<d$. Let $\eta_1,\ldots,\eta_m$ be all the generic points of X. For each i, there exists an open neighborhood U_i of η_i such that $\mathscr{F}|_{U_i}$ is locally constant. Let $U=\bigcup_{i=1}^m U_i$ and let $j:U\hookrightarrow X$ be the open immersion. $\mathscr{F}|_U$ is locally constant, and $j_!(\mathscr{F}|_U)$ is a subsheaf $\mathscr{F}$. The quotient $\mathscr{F}/j_!(\mathscr{F}|_U)$ is supported on $X-U$ which has dimension $<n$. We then apply the induction hypothesis to $\mathscr{F}/j_!(\mathscr{F}|_U)$. To prove our assertion, it suffices to show that $\mathscr{E}xt^q(j_!\mathscr{F}',\mathscr{G})$ are constructible for any immersion $j:Y\to X$, any locally constant constructible sheaf of A-modules $\mathscr{F}'$ on Y, and any constructible sheaf of A-modules $\mathscr{G}$ on X. We have

$$R\mathscr{H}om(j_!\mathscr{F}',\mathscr{G})\cong Rj_*R\mathscr{H}om(\mathscr{F}',Rj^!\mathscr{G}).$$

By 9.5.2 applied to Rj_* and $Rj^!$, we are reduced to proving that $\mathscr{E}xt^q(\mathscr{F}',\mathscr{G}')$ are constructible for any locally constant constructible sheaf of A-modules $\mathscr{F}'$, and any constructible sheaf of A-modules $\mathscr{G}'$ on Y. The problem is local with respect to the etale topology on Y. So we may assume that $\mathscr{F}'$ is a constant constructible sheaf of A-modules, say associated to some finitely generated A-module M. Let

$$\cdots\to L_1\to L_0\to 0$$

be a resolution of M by free A-modules of finite rank. Then $\mathscr{E}xt^q(\mathscr{F}', \mathscr{G}')$ is the q-th cohomology sheaf of the complex

$$0 \to \mathscr{H}om_A(L_0, \mathscr{G}') \to \mathscr{H}om_A(L_1, \mathscr{G}') \to \cdots.$$

It is clear that it is constructible.

Next we show that $R\mathscr{H}om(-,-)$ maps objects in $D^b_{\text{ctf}}(X, A) \times D^b_{\text{tf}}(X, A)$ to objects in $D^b_{\text{tf}}(X, A)$. Let K be a bounded complex of flat constructible sheaves of A-modules. There exists a filtration

$$0 = K_0 \subset K_1 \subset \cdots \subset K_k = K$$

such that K_i/K_{i-1} are of the form $j_!K'$ for some immersion $j : Y \to X$ and some complex of locally constant flat constructible sheaves of A-modules K'. We have

$$R\mathscr{H}om(j_!K', L) \cong Rj_*R\mathscr{H}om(K', Rj^!L).$$

By 9.5.2, we are reduced to proving that $R\mathscr{H}om(K, L)$ has finite Tor-dimension for any bounded complex K of locally constant flat constructible sheaves of A-modules and any $L \in \text{ob}\, D^b_{\text{tf}}(X, A)$. The problem is local, we may assume that the components of K are constant sheaves. By 1.6.7, the components of K are constant sheaves of projective A-modules. We then have $R\mathscr{H}om(K, L) = \mathscr{H}om^{\cdot}(K, L)$. As L has finite Tor-dimension, we may assume that L is a bounded complex of flat sheaves of A-modules. Then $\mathscr{H}om^{\cdot}(K, L)$ is a bounded complex of flat sheaves of A-modules. Hence $R\mathscr{H}om(K, L)$ has finite Tor-dimension. □

Using 9.3.1 and the same argument as above, one can prove the following propositions.

Proposition 9.5.4. *Let S be a noetherian scheme, X and Y two S-schemes of finite type, $f : X \to Y$ an S-morphism, and $L \in \text{ob}\, D^b_c(Y, A)$. Then there exists a dense open subset U of S such that the following conditions hold:*

(i) *$(Rf^!L)|_{X_U}$ is an object in $D^b_c(X_U, A)$, where X_U is the inverse image of U in X.*

(ii) *The formation of $Rf^!L$ commutes with any base change $S' \to U \subset S$, or equivalently, the formation of $Rf^!_U(L|_{Y_U})$ commutes with any base change $S' \to S$, where Y_U is the inverse image of U in Y, and $f_U : X_U \to Y_U$ is the restriction of f.*

Proposition 9.5.5. *Let S be a noetherian scheme, X an S-scheme of finite type, $K \in \text{ob}\, D^b_{\text{ctf}}(X, A)$ and $L \in \text{ob}\, D^b_c(Y, A)$. Then there exists a dense open subset U of S such that the following conditions hold:*

(i) $(R\mathscr{H}om_A(K,L))|_{X_U}$ *is an object in* $D^b_c(X_U, A)$, *where* X_U *is the inverse image of* U *in* X.

(ii) *The formation of* $R\mathscr{H}om_A(K,L)$ *commutes with any base change* $S' \to U \subset S$, *or equivalently, the formation of* $(R\mathscr{H}om_A(K,L))|_{X_U}$ *commutes with any base change* $S' \to S$.

9.6 Biduality

([SGA $4\frac{1}{2}$] Th. finitude 4.)

The main result of this section is the following:

Theorem 9.6.1. *Let* S *be a regular noetherian scheme of dimension* ≤ 1 *and let* A *be a noetherian ring such that* A *is an injective* A*-module, and* $nA = 0$ *for some integer* n *invertible on* S. *For any* S*-scheme* $f : X \to S$ *of finite type, define* $K_X = Rf^!A$, *and for any* $K \in \operatorname{ob} D^b_{\mathrm{ctf}}(X, A)$, *define* $DK = R\mathscr{H}om_A(K, K_X)$. *Then* $DK \in \operatorname{ob} D^b_{\mathrm{ctf}}(X, A)$ *and the morphism*

$$K \to DDK = R\mathscr{H}om_A(DK, K_X)$$

corresponding to the canonical morphism

$$K \otimes^L_A DK \cong DK \otimes^L_A K = R\mathscr{H}om_A(K, K_X) \otimes^L_A K \stackrel{\mathrm{Ev}}{\to} K_X$$

is an isomorphism.

Lemma 9.6.2. *9.6.1 holds if* $X = S$.

Proof. By definition, we have $K_S = A$. To prove $K \cong DDK$, the problem is local with respect to the etale topology. If $\dim S = 0$, any constructible sheaf of A-modules $\mathscr{F}$ on S is locally constant. So to prove the lemma in this case, we may assume that K is a bounded complex of finitely generated flat A-modules. By 1.6.7, all components of K are finitely generated projective A-modules. So we have

$$DK = \mathscr{H}om^{\cdot}(K, A), \quad DDK = \mathscr{H}om^{\cdot}(\mathscr{H}om^{\cdot}(K, A)).$$

It is then clear that $DK \in \operatorname{ob} D^b_{\mathrm{ctf}}(S, A)$ and $K \cong DDK$.

Suppose that S is a connected regular scheme of dimension 1. Given a constructible sheaf of A-modules $\mathscr{F}$ on S, let U be an open dense subset of S such that $\mathscr{F}|_U$ is locally constant, and let $j : U \hookrightarrow S$ and $i : S - U \to S$

be the immersions. By 8.3.10, we have

$$\begin{aligned} D(j_*j^*\mathscr{F}) &= R\mathscr{H}om(j_*j^*\mathscr{F}, A) \\ &\cong j_*\mathscr{H}om(j^*\mathscr{F}, A), \\ DD(j_*j^*\mathscr{F}) &\cong R\mathscr{H}om(j_*\mathscr{H}om(j^*\mathscr{F}, A), A) \\ &\cong j_*\mathscr{H}om(\mathscr{H}om(j^*\mathscr{F}, A), A). \end{aligned}$$

It follows that $D(j_*j^*\mathscr{F}) \in \mathrm{ob}\, D^b_{\mathrm{ctf}}(S, A)$ and $j_*j^*\mathscr{F} \cong DD(j_*j^*\mathscr{F})$. Any constructible sheaf of A-modules on S supported on $S - U$ is of the form $i_*\mathscr{G}$ for a constructible sheaf of A-modules $\mathscr{G}$ on $S - U$. By 8.3.6 and the $\dim S = 0$ case treated above, we have

$$\begin{aligned} D(i_*\mathscr{G}) &= R\mathscr{H}om(i_*\mathscr{G}, A) \\ &\cong i_*R\mathscr{H}om(\mathscr{G}, Ri^!A) \\ &\cong i_*R\mathscr{H}om(\mathscr{G}, A(-1)[-2]) \\ &\cong i_*\mathscr{H}om^{\cdot}(\mathscr{G}, A(-1)[-2]). \end{aligned}$$

Using this formula, one can verify that $D(i_*\mathscr{G}) \in \mathrm{ob}\, D^b_{\mathrm{ctf}}(S, A)$ and $i_*\mathscr{G} \cong DD(i_*\mathscr{G})$. Since the kernel and cokernel of the canonical morphism $\mathscr{F} \to j_*j^*F$ are supported on $S - U$, it follows that $D(\mathscr{F}) \in \mathrm{ob}\, D^b_{\mathrm{ctf}}(S, A)$ and $\mathscr{F} \cong DD\mathscr{F}$. We then conclude that for any $K \in \mathrm{ob}\, D^b_{\mathrm{ctf}}(S, A)$, we have $DK \in \mathrm{ob}\, D^b_{\mathrm{ctf}}(S, A)$ and $K \cong DDK$. □

Lemma 9.6.3. *Under the assumption of 9.6.1, if f is proper, then we have $Rf_*K \cong Rf_*DDK$.*

Proof. Since f is proper, we have $Rf_! = Rf_*$, and hence

$$Rf_*DK = Rf_*R\mathscr{H}om(K, Rf^!A) \cong R\mathscr{H}om(Rf_*K, A) = DRf_*K.$$

We thus have a canonical isomorphism

$$Rf_*DK \stackrel{\cong}{\to} DRf_*K.$$

It follows that we have canonical isomorphisms

$$Rf_*DDK \stackrel{\cong}{\to} DRf_*DK, \quad DDRf_*K \stackrel{\cong}{\to} DRf_*DK.$$

By 9.6.2, we have

$$Rf_*K \cong DDRf_*K.$$

To prove $Rf_*K \cong Rf_*DDK$, it suffices to show that the following diagram commutes:

$$\begin{array}{ccc} Rf_*K & \to & Rf_*DDK \\ \cong\downarrow & & \downarrow\cong \\ DDRf_*K & \stackrel{\cong}{\to} & DRf_*DK. \end{array}$$

Note that via 6.4.7, the composite

$$Rf_*K \to DDRf_*K \to DRf_*DK$$

corresponds to the composite

$$Rf_*K \otimes^L_A Rf_*DK \to (Rf_*K) \otimes^L_A D(Rf_*K) \cong D(Rf_*K) \otimes^L_A (Rf_*K) \to A,$$

and the composite

$$Rf_*K \to Rf_*DDK \to DRf_*DK$$

corresponds to the composite

$$Rf_*K\otimes^L_A Rf_*DK \to Rf_*DDK\otimes^L_A Rf_*DK \to D(Rf_*DK)\otimes^L_A(Rf_*DK) \stackrel{\mathrm{Ev}}{\to} A.$$

So it suffices to prove the outer loop of the following diagram commutes:

$$\begin{array}{ccccccc}
(Rf_*K)\otimes^L_A D(Rf_*K) & \leftarrow & Rf_*K\otimes^L_A Rf_*DK & \to & Rf_*DDK\otimes^L_A Rf_*DK & \to & D(Rf_*DK)\otimes^L_A(Rf_*DK) \\
 & & \downarrow & (1) & \downarrow & & \\
\wr\| & & Rf_*(K\otimes^L_A DK) & \to & Rf_*(DDK\otimes^L_A DK) & & \\
 & & \downarrow & & & & \\
D(Rf_*K)\otimes^L_A Rf_*K & & Rf_*(DK\otimes^L_A K) & (2) & \downarrow & & \downarrow \mathrm{Ev} \\
 & & \downarrow & & & & \\
\mathrm{Ev}\downarrow & & Rf_*Rf^!A & = & Rf_*Rf^!A & & \\
 & & \downarrow & (3) & \downarrow & & \\
A & = & A & = & A & & = A.
\end{array}$$

The commutativity of (1) and (3) is clear. The commutativity of (2) follows from the commutativity of the diagram

$$\begin{array}{ccc}
K \otimes^L_A DK & \to & DDK \otimes^L_A DK \\
\downarrow & & \downarrow \mathrm{Ev} \\
DK \otimes^L_A K & \stackrel{\mathrm{Ev}}{\to} & Rf^!A.
\end{array}$$

The leftmost square is of the same type as the rightmost square. Their commutativity follows from the commutativity of the outer loop of the following diagram:

$$\begin{array}{ccc}
Rf_*R\mathscr{H}om(K, Rf^!A) \otimes^L_A Rf_*K & \to & Rf_*(R\mathscr{H}om(K, Rf^!A) \otimes^L_A K) \\
\downarrow & (4) & \downarrow \\
R\mathscr{H}om(Rf_*K, Rf_*Rf^!A) \otimes^L_A Rf_*K & \to & Rf_*Rf^!A \\
\downarrow & (5) & \downarrow \\
R\mathscr{H}om(Rf_*K, A) \otimes^L_A Rf_*K & \to & A.
\end{array}$$

In this diagram, the commutativity of (5) is clear. To prove the commutativity of (4), let $I^\cdot$ be a complex of injective sheaves of A-modules quasi-isomorphic to $Rf^!A$, represent K by a complex of bounded complex of flat sheaves of A-modules, and let $\mathscr{C}^\cdot(K)$ be the Godement resolution. One can check directly that the following diagram commutes:

$$\begin{array}{ccc} f_*\mathscr{H}om^\cdot(\mathscr{C}^\cdot(K), I^\cdot)\otimes_A f_*\mathscr{C}^\cdot(K) & \rightarrow & f_*(\mathscr{H}om^\cdot(\mathscr{C}^\cdot(K), I^\cdot)\otimes_A \mathscr{C}^\cdot(K)) \\ \downarrow & & \downarrow \\ \mathscr{H}om^\cdot(f_*\mathscr{C}^\cdot(K), f_*I^\cdot)\otimes_A f_*\mathscr{C}^\cdot(K) \rightarrow & & f_*I^\cdot. \end{array}$$

□

Proof of 9.6.1. By 9.5.2 and 9.5.3, we have $DK \in \mathrm{ob}\, D^b_{\mathrm{ctf}}(X, A)$. To prove $K \cong DDK$, the problem is local. By 5.9.10, we may assume that S is strictly local.

Suppose that S is the spectrum of a separably closed field. We prove $K \cong DDK$ by induction on $\dim X$. The problem is local. We may assume that X is affine. Then X is a subscheme of a scheme projective over S. We are thus reduced to the case where X is proper over S. Let

$$K \rightarrow DDK \rightarrow \Delta \rightarrow$$

be a distinguished triangle. We need to show that Δ is acyclic. Cover X by finitely many affine open subsets X_i so that we have immersions $X_i \rightarrow \mathbb{A}^m_S$. Let η be the generic point of $\mathbb{A}^1_S$ and let $\pi_j : \mathbb{A}^m_S \rightarrow \mathbb{A}^1_S$ $(j = 1, \ldots, m)$ be the projections. Applying the induction hypothesis to the generic fiber of $\pi_j|_{X_i}$, we see that $\Delta|_{X_i\cap\pi_j^{-1}(\eta)}$ is acyclic for each pair (i, j). Since $\Delta \in \mathrm{ob}\, D^b_c(X, A)$, there exists a nonempty open subset $V_j \subset \mathbb{A}^1_S$ such that $\Delta|_{X_i\cap\pi_j^{-1}(V_j)}$ is acyclic for each i by 5.9.8. We have

$$X_i - \bigcup_{j=1}^{n}(X_i \cap \pi_j^{-1}(V_j)) = X_i \bigcap \big((\mathbb{A}^1_S - V_1)\times_S \cdots \times_S (\mathbb{A}^1_S - V_n)\big),$$

which is finite over S. It follows that the cohomology sheaves of Δ have finite support. On the other hand, we have $Rf_*\Delta = 0$ by 9.6.3. So we have $\Delta = 0$.

Suppose that S is a trait. Let s be the closed point of S. Let us prove that $K \cong DDK$ by induction on $\dim X_s$. Again we may assume that X is proper over S. Let

$$K \rightarrow DDK \rightarrow \Delta \rightarrow$$

be a distinguished triangle. By the above discussion applied to the generic fiber X_η, the cohomology sheaves of Δ are supported on X_s. Cover X by finitely many affine open subsets $X_i \subset \mathbb{A}^m_S$. Let S' be the strict localization of $\mathbb{A}^1_S$ at the generic point s' of the special fiber of $\mathbb{A}^1_S \rightarrow S$. Then

S' is a strictly local trait. Applying the induction hypothesis to the morphisms $X_i \times_{\pi_j|_{X_i}, \mathbb{A}^1_S} S' \to S'$, we see that $\Delta|_{X_i \times_{\pi_j|_{X_i}, \mathbb{A}^1_S} S'}$ are acyclic. Then there exist open neighborhoods V_i of s' in $\mathbb{A}^1_S$ such that $\Delta|_{X_i \times_{\pi_j|_{X_i}, \mathbb{A}^1_S} V_i}$ are acyclic. Together with the fact that the cohomology sheaves of Δ are supported on X_s, this implies that the cohomology sheaves are supported on finitely many points in X_s. Since $Rf_*\Delta = 0$ by 9.6.3, we have $\Delta = 0$. $\square$

Chapter 10

ℓ-adic Cohomology

10.1 Adic Formalism

([SGA 5] V, VI, [Deligne (1980)] 1.1.)

Let ℓ be a prime number, $\mathbb{Z}_\ell = \varprojlim_n \mathbb{Z}/\ell^{n+1}$ the ring of ℓ-adic integers, $\mathbb{Q}_\ell$ its fraction field, E a finite extension of $\mathbb{Q}_\ell$, and R the integral closure of $\mathbb{Z}_\ell$ in E. Then R is a complete discrete valuation ring. Fix a uniformizer λ of R. Then $R/(\lambda^{n+1})$ are finite and $R/(\lambda)$ is a field of characteristic ℓ. Throughout this chapter, we assume that ℓ is invertible in all schemes we consider.

Let X be a noetherian scheme. A sheaf $\mathscr{F}$ of R-modules on X is called a *λ-torsion sheaf* if

$$\mathscr{F} = \bigcup_{v\geq 0} \ker\,(\lambda^v : \mathscr{F} \to \mathscr{F}).$$

Consider the category of inverse systems of λ-torsion sheaves the form

$$\mathscr{F} = (\mathscr{F}_n, u_n)_{n\in\mathbb{Z}}, \quad u_n : \mathscr{F}_n \to \mathscr{F}_{n-1}$$

so that $\mathscr{F}_n = 0$ for all $n \leq n_0$ for some integer n_0. This is an abelian category. Given an integer r, we define $\mathscr{F}[r]$ to be the inverse system defined by

$$(\mathscr{F}[r])_n = \mathscr{F}_{n+r}$$

with the transition morphisms given by

$$u_{n+r} : (\mathscr{F}[r])_n \to (\mathscr{F}[r])_{n-1}.$$

When $r \geq 0$, we have a canonical morphism $\mathscr{F}[r] \to \mathscr{F}$ whose n-th component is

$$u_{n+1} \cdots u_{n+r} : \mathscr{F}_{n+r} \to \mathscr{F}_n.$$

Similarly, we have a canonical morphism $\mathscr{F} \to \mathscr{F}[r]$ for any $r \leq 0$. We say that a system $\mathscr{F} = (\mathscr{F}_n)$ satisfies the *Mittag–Leffler condition (ML)* if for any integer n, there exists an integer $r \geq n$ such that

$$\mathrm{im}(\mathscr{F}_r \to \mathscr{F}_n) = \mathrm{im}(\mathscr{F}_t \to \mathscr{F}_n)$$

for all $t \geq r$. We say that it satisfies the *Artin–Rees–Mittag–Leffler condition (ARML)* if there exists an integer r such that

$$\mathrm{im}(\mathscr{F}[r] \to \mathscr{F}) = \mathrm{im}(\mathscr{F}[t] \to \mathscr{F})$$

for all $t \geq r$. We say that it is a *null system* if there exists an integer $r \geq 0$ such that $\mathscr{F}[r] \to \mathscr{F}$ is 0.

In the category of inverse systems of λ-torsion sheaves, the family of morphisms of the form $\mathscr{F}[r] \to \mathscr{F}$ is multiplicative. By 6.2.1, there exists a category, which we call the *A-R category* of inverse systems of λ-torsion sheaves, so that its objects are inverse systems of λ-torsion sheaves, and for any objects $\mathscr{F}$ and $\mathscr{G}$, we have

$$\mathrm{Hom}_{\mathrm{AR}}(\mathscr{F}, \mathscr{G}) = \varinjlim_{r \geq 0} \mathrm{Hom}(\mathscr{F}[r], \mathscr{G}).$$

A morphism $\mathscr{F} \to \mathscr{G}$ in the A-R category is represented by a morphism $\mathscr{F}[r] \to \mathscr{G}$ of inverse systems for some $r \geq 0$, and the kernel and cokernel of $\mathscr{F}[r] \to \mathscr{G}$ are the kernel and cokernel of $\mathscr{F} \to \mathscr{G}$ in the A-R category, respectively. The A-R category of inverse systems of λ-torsion sheaves is an abelian category. A system $\mathscr{F}$ is a zero object in the A-R category if and only if it is a null system.

An inverse system $\mathscr{F} = (\mathscr{F}_n, u_n)_{n \in \mathbb{Z}}$ of λ-torsion sheaves is called a *λ-adic sheaf* if $\mathscr{F}_n$ are constructible for all n, $\mathscr{F}_n = 0$ for all $n < 0$, $\lambda^{n+1}\mathscr{F}_n = 0$ for all $n \geq 0$, and the morphisms $u_{n+1} : \mathscr{F}_{n+1} \to \mathscr{F}_n$ induce isomorphisms

$$\mathscr{F}_{n+1}/\lambda^{n+1}\mathscr{F}_{n+1} \cong \mathscr{F}_n$$

for all $n \geq 0$. A λ-adic sheaf $\mathscr{F}$ is called *lisse* if $\mathscr{F}_n$ are locally constant for all n. An object $\mathscr{F}$ in the A-R category is called an *A-R λ-adic sheaf* if it is A-R isomorphic to a λ-adic sheaf, that is, we can find a λ-adic sheaf $\mathscr{G}$ and a morphism $\mathscr{G}[r] \to \mathscr{F}$ for some $r \geq 0$ whose kernel and cokernel are null systems.

Let $\mathscr{G}$ be a λ-adic sheaf and let $\mathscr{F}$ be an inverse system of λ-torsion sheaves such that $\lambda^{n+1}\mathscr{F}_n = 0$ for all n. Then we have

$$\mathrm{Hom}(\mathscr{G}, \mathscr{F}) \cong \mathrm{Hom}_{\mathrm{AR}}(\mathscr{G}, \mathscr{F}).$$

So the category of λ-adic sheaves is equivalent to the category of A-R λ-adic sheaves.

Examples.

1. Let M be a finitely generated R-module and let $M/\lambda^{n+1}M$ be the constant sheaf on X associated $M/\lambda^{n+1}M$. Then $M = (M/\lambda^{n+1}M)$ is a λ-adic sheaf.

2. Let $u_n : \boldsymbol{\mu}_{\ell^{n+1},X} \to \boldsymbol{\mu}_{\ell^n,X}$ be the morphism $s \mapsto s^l$. Then $(\boldsymbol{\mu}_{\ell^{n+1},X})$ is an ℓ-adic sheaf. We denote it by $\mathbb{Z}_\ell(1)$. Similarly, we have the ℓ-adic sheaf $\mathbb{Z}_\ell(m) = (\mathbb{Z}/\ell^{n+1}(m))$ for any integer m.

3. Let $\mathscr{F}$ be a constructible λ-torsion sheaf on X. Then $(\mathscr{F}/\lambda^{n+1}\mathscr{F})$ is a λ-adic sheaf. We have $\lambda^{m+1}\mathscr{F} = 0$ for a sufficiently large m and $\mathscr{F}/\lambda^{n+1}\mathscr{F} = \mathscr{F}$ for all $n \geq m$. The functor $\mathscr{F} \to (\mathscr{F}/\lambda^{n+1}\mathscr{F})$ identifies the category of λ-torsion sheaves with a full subcategory of the category of λ-adic sheaves.

4. If $\mathscr{F} = (\mathscr{F}_n)$ and $\mathscr{G} = (\mathscr{G}_n)$ are λ-adic sheaves, then

$$\mathscr{F} \otimes_R \mathscr{G} = (\mathscr{F}_n \otimes_{R/(\lambda^{n+1})} \mathscr{G}_n)$$

is a λ-adic sheaf.

5. If $\mathscr{F} = (\mathscr{F}_n)$ and $\mathscr{G} = (\mathscr{G}_n)$ are λ-adic sheaves, and each $\mathscr{F}_n$ is a locally free sheaf of $R/(\lambda^{n+1})$-modules of finite rank, then

$$\mathscr{H}om_R(\mathscr{F}, \mathscr{G}) = (\mathscr{H}om_{R/\lambda^{n+1}}(\mathscr{F}_n, \mathscr{G}_n))$$

is a λ-adic sheaf. This follows from the fact that for each nonnegative integer n, we have an exact sequence

$$\begin{aligned} &\mathscr{H}om_{R/(\lambda^{n+2})}(\mathscr{F}_{n+1}, \mathscr{G}_{n+1}) \stackrel{\lambda^{n+1}}{\to} \mathscr{H}om_{R/(\lambda^{n+2})}(\mathscr{F}_{n+1}, \mathscr{G}_{n+1}) \\ &\quad \to \mathscr{H}om_{R/(\lambda^{n+2})}(\mathscr{F}_{n+1}, \mathscr{G}_n) \to 0 \end{aligned}$$

and an isomorphism

$$\mathscr{H}om_{R/(\lambda^{n+2})}(\mathscr{F}_{n+1}, \mathscr{G}_n) \cong \mathscr{H}om_{R/(\lambda^{n+1})}(\mathscr{F}_n, \mathscr{G}_n).$$

Proposition 10.1.1. *Let $\mathscr{F} = (\mathscr{F}_n)$ be an inverse system of λ-torsion sheaves on a noetherian scheme X. Suppose that $\mathscr{F}_n$ are constructible for all n, $\lambda^{n+1}\mathscr{F}_n = 0$ for all $n \geq 0$ and $\mathscr{F}_n = 0$ for all $n < 0$. Then $\mathscr{F}$ is an A-R λ-adic sheaf if and only if the following two conditions hold:*

(a) *$\mathscr{F}$ satisfies ARML.*

(b) *Let $r \geq 0$ be an integer such that*

$$\mathrm{im}\,(\mathscr{F}[t] \to \mathscr{F}) = \mathrm{im}\,(\mathscr{F}[r] \to \mathscr{F})$$

for all $t \geq r$. Such an integer exists by (a). *Let*

$$\overline{\mathscr{F}}_n = \mathrm{im}\,(\mathscr{F}_{n+t} \to \mathscr{F}_n)$$

for any n and any $t \geq r$. Then there exists an integer $s \geq 0$ such that for any $t \geq s$, the canonical morphisms $\mathscr{F}_{n+t} \to \mathscr{F}_{n+s}$ induce isomorphisms

$$\overline{\mathscr{F}}_{n+t}/\lambda^{n+1}\overline{\mathscr{F}}_{n+t} \cong \overline{\mathscr{F}}_{n+s}/\lambda^{n+1}\overline{\mathscr{F}}_{n+s}$$

for all $n \geq 0$ and all $t \geq s$.

If these conditions hold, then $(\overline{\mathscr{F}}_{n+s}/\lambda^{n+1}\overline{\mathscr{F}}_{n+s})$ is a λ-adic sheaf A-R isomorphic to $\mathscr{F}$.

Proof.

($\Rightarrow$) Let $\mathscr{G} = (\mathscr{G}_n)$ be a λ-adic sheaf A-R isomorphic to $\mathscr{F} = (\mathscr{F}_n)$. Since we have

$$\mathrm{Hom}(\mathscr{G}, \mathscr{F}) = \mathrm{Hom}_{\mathrm{AR}}(\mathscr{G}, \mathscr{F}),$$

there exists a morphism $\mathscr{G} \to \mathscr{F}$ whose kernel and cokernel are null systems. So there exists an integer $r \geq 0$ such that

$$\ker(\mathscr{G}_{n+r} \to \mathscr{F}_{n+r}) \subset \ker(\mathscr{G}_{n+r} \to \mathscr{G}_n), \ \mathrm{im}(\mathscr{F}_{n+r} \to \mathscr{F}_n) \subset \mathrm{im}(\mathscr{G}_n \to \mathscr{F}_n)$$

for all $n \geq 0$. Since $\mathscr{G}$ is a λ-adic sheaf, we have $\mathscr{G}_n = \mathrm{im}(\mathscr{G}_m \to \mathscr{G}_n)$ for all $m \geq n$. It follows that

$$\mathrm{im}(\mathscr{F}_{n+r} \to \mathscr{F}_n) \subset \mathrm{im}(\mathscr{G}_{n+t} \to \mathscr{G}_n \to \mathscr{F}_n)$$

for all $t \geq r$ and hence

$$\mathrm{im}(\mathscr{F}_{n+r} \to \mathscr{F}_n) \subset \mathrm{im}(\mathscr{F}_{n+t} \to \mathscr{F}_n)$$

for all $t \geq r$. This proves (a).

Note that the morphisms $\mathscr{F}_{n+t} \to \mathscr{F}_{n+r}$ $(t \geq r)$ induce epimorphisms $\overline{\mathscr{F}}_{n+t} \to \overline{\mathscr{F}}_{n+r}$. Let x be a section of $\overline{\mathscr{F}}_{n+t}$ such that its image in $\overline{\mathscr{F}}_{n+r}$ is 0. Since

$$\mathrm{im}(\mathscr{F}_{n+t+r} \to \mathscr{F}_{n+t}) \subset \mathrm{im}(\mathscr{G}_{n+t} \to \mathscr{F}_{n+t}),$$

locally with respect to the etale topology, we can lift x to a section y of $\mathscr{G}_{n+t}$. As

$$\ker(\mathscr{G}_{n+r} \to \mathscr{F}_{n+r}) \subset \ker(\mathscr{G}_{n+r} \to \mathscr{G}_n),$$

the image of y in $\mathscr{G}_n$ is 0, and hence y is a section of $\lambda^{n+1}\mathscr{G}_{n+t}$. Thus x is a section of $\lambda^{n+1}\overline{\mathscr{F}}_{n+t}$. This implies (b) by taking $s = r$.

($\Leftarrow$) Suppose conditions (a) and (b) hold. Let $\overline{\mathscr{F}} = (\overline{\mathscr{F}}_n)$. The canonical morphism $\mathscr{F}[r] \to \overline{\mathscr{F}}$ defines the A-R inverse of the inclusion $\overline{\mathscr{F}} \hookrightarrow \mathscr{F}$. Hence $\overline{\mathscr{F}}$ is A-R isomorphic to $\mathscr{F}$. The canonical morphism $\overline{\mathscr{F}}[s] \to (\overline{\mathscr{F}}_{n+s}/\lambda^{n+1}\overline{\mathscr{F}}_{n+s})$ defines the A-R inverse of the canonical morphism $(\overline{\mathscr{F}}_{n+s}/\lambda^{n+1}\overline{\mathscr{F}}_{n+s}) \to \overline{\mathscr{F}}$. So $\overline{\mathscr{F}}$ is A-R isomorphic to

$(\overline{\mathscr{F}}_{n+s}/\lambda^{n+1}\overline{\mathscr{F}}_{n+s})$. We have

$$\begin{aligned}&(\overline{\mathscr{F}}_{n+1+s}/\lambda^{n+2}\overline{\mathscr{F}}_{n+1+s})\Big/\lambda^{n+1}(\overline{\mathscr{F}}_{n+1+s}/\lambda^{n+2}\overline{\mathscr{F}}_{n+1+s})\\ &\cong \overline{\mathscr{F}}_{n+1+s}/\lambda^{n+1}\overline{\mathscr{F}}_{n+1+s}\\ &\cong \overline{\mathscr{F}}_{n+s}/\lambda^{n+1}\overline{\mathscr{F}}_{n+s}.\end{aligned}$$

So $(\overline{\mathscr{F}}_{n+s}/\lambda^{n+1}\overline{\mathscr{F}}_{n+s})$ is a λ-adic sheaf A-R isomorphic to $\mathscr{F}$.

Corollary 10.1.2. *Let $\mathscr{F} = (\mathscr{F}_n)$ be an inverse system of λ-torsion sheaves on a noetherian scheme X. Suppose that $\mathscr{F}_n$ are constructible for all n, $\lambda^{n+1}\mathscr{F} = 0$ for all $n \geq 0$ and $\mathscr{F}_n = 0$ for all $n < 0$.*

(i) *If $\mathscr{F}$ is A-R λ-adic, then so is $f^*\mathscr{F}$ for any morphism $f : X' \to X$ of schemes.*

(ii) *Let $\{U_i \to X\}_i$ be a finite etale covering of X. If each $\mathscr{F}|_{U_i}$ is A-R λ-adic, then so is $\mathscr{F}$.*

(iii) *Let (X_i) be a finite covering of X by locally closed subsets. If each $\mathscr{F}|_{X_i}$ is A-R λ-adic, then so is $\mathscr{F}$.*

Similarly, one can define the A-R category of inverse systems of λ-torsion R-modules, the λ-adic system of R-modules, and the A-R λ-adic system of R-modules. For example, a λ-adic system of R-modules is an inverse system $F = (F_n)$ such that F_n are finitely generated R-modules for all n, $F_n = 0$ for all $n < 0$, $\lambda^{n+1}F_n = 0$ for all $n \geq 0$, and the homomorphisms $F_{n+1} \to F_n$ induce isomorphisms $F_{n+1}/\lambda^{n+1}F_{n+1} \cong F_n$. The functor

$$F = (F_n) \mapsto \varprojlim_n F_n$$

is well-defined in the A-R category of inverse system of λ-torsion R-modules since the canonical morphism $F[r] \to F$ induces an isomorphism $\varprojlim_n F_{n+r} \cong \varprojlim_n F_n$ for any $r \geq 0$.

Lemma 10.1.3. *Let*

$$0 \to (F_n) \to (G_n) \to (H_n) \to 0$$

be a short exact sequence of inverse system of R-modules. Suppose F_n is finite for all n. Then the sequence

$$0 \to \varprojlim_n F_n \to \varprojlim_n G_n \to \varprojlim_n H_n \to 0$$

is exact.

Proof. It is clear that

$$0 \to \varprojlim_n F_n \to \varprojlim_n G_n \to \varprojlim_n H_n$$

is exact. Let us prove that $\varprojlim_n G_n \to \varprojlim_n H_n$ is onto. Given an element $(h_n) \in \varprojlim_n H_n$, let E_n be the inverse images of h_n in G_n. Then E_n are finite. This implies that $\varprojlim_n E_n \neq \emptyset$. Let $(e_n) \in \varprojlim_n E_n$. Then (e_n) is an element in $\varprojlim_n G_n$ that is mapped to (h_n). □

Proposition 10.1.4.

(i) *If $F = (F_n)$ is a λ-adic system of R-modules, then $M = \varprojlim_n F_n$ is a finitely generated R-module, and the projections $M \to F_n$ induce isomorphisms $M/\lambda^{n+1}M \cong F_n$.*

(ii) *The functor $(F_n) \mapsto \varprojlim_n F_n$ defines an equivalence between the category of A-R λ-adic systems of R-modules and the category of finitely generated R-modules.*

(iii) *Let $f : (F_n) \to (G_n)$ be an A-R morphism of A-R λ-adic systems of R-modules. Then* $\ker f$ *and* $\operatorname{coker} f$ *are A-R λ-adic.*

Proof.

(i) Let $\{x_1^{(0)}, \ldots, x_m^{(0)}\}$ be a finite family of generators of F_0 as an R-module. Choose $x_i^{(n)} \in F_n$ $(i = 1, \ldots, m)$ by induction on n so that they are mapped to $x_i^{(n-1)}$ under the canonical homomorphisms $F_n \to F_{n-1}$. By Nakayama's lemma, $\{x_1^{(n)}, \ldots, x_m^{(n)}\}$ is a family of generators of F_n as an R-module. We have a family of epimorphisms

$$(R/(\lambda^{n+1}))^m \to F_n, \quad (a_1, \ldots, a_m) \mapsto \sum_{i=1}^m a_i x_i^{(n)}.$$

By 10.1.3, we have an epimorphism

$$R^m \cong \varprojlim_n (R/(\lambda^{n+1}))^m \to \varprojlim_n F_n.$$

It follows that $M = \varprojlim_n F_n$ is a finitely generated R-module. For all nonnegative integers m and n, we have an exact sequence

$$F_{n+m} \stackrel{\lambda^{n+1}}{\to} F_{n+m} \to F_n \to 0.$$

Using 10.1.3, one verifies that the sequence

$$\varprojlim_m F_{n+m} \stackrel{\lambda^{n+1}}{\to} \varprojlim_m F_{n+m} \to F_n \to 0$$

is exact. So we have $M/\lambda^{n+1}M \cong F_n$.

(ii) Let us show that the functor is fully faithful. Let $F = (F_n)$ and $G = (G_n)$ be λ-adic systems of R-modules, let $M = \varprojlim_n F_n$ and $N = \varprojlim_n G_n$. We have

$$\begin{aligned}\mathrm{Hom}_{\mathrm{AR}}(F, G) &\cong \mathrm{Hom}(F, G)\\ &= \varprojlim_n \mathrm{Hom}_{R/(\lambda^{n+1})}(F_n, G_n)\\ &\cong \varprojlim_n \mathrm{Hom}_{R/(\lambda^{n+1})}(M/\lambda^{n+1}M, N/\lambda^{n+1}N).\end{aligned}$$

To prove that the functor is fully faithful, it suffices to show

$$\mathrm{Hom}_R(M, N) \cong \varprojlim_n \mathrm{Hom}_{R/(\lambda^{n+1})}(M/\lambda^{n+1}M, N/\lambda^{n+1}N).$$

We can find an exact sequence

$$R^v \to R^u \to M \to 0.$$

It gives rise to a commutative diagram

$$\begin{array}{ccccccc} 0 \to & \mathrm{Hom}_R(M,N) & \to & \mathrm{Hom}_R(R^u, N) & \to & \mathrm{Hom}_R(R^v, N) \\ & \downarrow & & \downarrow & & \downarrow \\ 0 \to & \begin{array}{c}\varprojlim_n \mathrm{Hom}(M/\lambda^{n+1}M, \\ N/\lambda^{n+1}N)\end{array} & \to & \begin{array}{c}\varprojlim_n \mathrm{Hom}((R/(\lambda^{n+1}))^u, \\ N/\lambda^{n+1}N)\end{array} & \to & \begin{array}{c}\varprojlim_n \mathrm{Hom}((R/(\lambda^{n+1}))^v, \\ N/\lambda^{n+1}N).\end{array} \end{array}$$

By the five lemma, to prove that the functor is fully faithful, it suffices to show that the last two vertical arrows are bijective. We are thus reduced to proving

$$\mathrm{Hom}_R(R, N) \cong \varprojlim_n \mathrm{Hom}_{R/(\lambda^{n+1})}(R/(\lambda^{n+1}), N/\lambda^{n+1}N),$$

or equivalently, $N \cong \varprojlim_n N/\lambda^{n+1}N$. This follows from (i).

For any finitely generated R-module M, $(M/\lambda^{n+1}M)$ is a λ-adic system, and $M \cong \varprojlim_n M/\lambda^{n+1}M$. So the functor is essentially surjective.

(iii) We may assume (F_n) and (G_n) are λ-adic systems of R-modules. Then f is given by an element (f_n) in $\varprojlim_n \mathrm{Hom}(F_n, G_n)$. Let $M = \varprojlim_n F_n$ and $N = \varprojlim_n G_n$. Then M and N are finitely generated R-modules, $M/\lambda^{n+1}M \cong F_n$, and $N/\lambda^{n+1}N \cong G_n$. Denote also by f the homomorphism $M \to N$ induced by $f_n : M_n \to N_n$ by passing to limit. We have an exact sequence

$$M \xrightarrow{f} N \to \operatorname{coker} f \to 0.$$

It induces exact sequences

$$M/\lambda^{n+1}M \to N/\lambda^{n+1}N \to \operatorname{coker} f/\lambda^{n+1}\operatorname{coker} f \to 0.$$

These sequences can be identified with

$$F_n \xrightarrow{f_n} G_n \to \operatorname{coker} f/\lambda^{n+1}\operatorname{coker} f \to 0.$$

So we have $\operatorname{coker} f_n \cong \operatorname{coker} f/\lambda^{n+1}\operatorname{coker} f$. Hence $\operatorname{coker} f = (\operatorname{coker} f_n)$ is λ-adic.

We have canonical homomorphisms

$$\ker f/\lambda^{n+1}\ker f \to \ker f_n.$$

Let us show that their kernels and cokernels form null systems. This implies that $\ker f = (\ker f_n)$ is A-R isomorphic to $(\ker f/\lambda^{n+1}\ker f)$ and hence is A-R λ-adic. By the Artin–Rees theorem ([Matsumura (1970)] (11.C) Theorem 15), there exists an integer $r \geq 0$ such that

$$\operatorname{im} f \cap \lambda^n N = \lambda^{n-r}(\operatorname{im} f \cap \lambda^r N)$$

for all $n \geq r$. Suppose

$$x + \lambda^{n+1}M \in \ker(M/\lambda^{n+1}M \to N/\lambda^{n+1}N),$$

where $x \in M$. We have

$$f(x) \in \operatorname{im} f \cap \lambda^{n+1}N = \lambda^{n+1-r}(\operatorname{im} f \cap \lambda^r N).$$

So $f(x) = \lambda^{n+1-r}f(x')$ for some $x' \in M$. Then $x - \lambda^{n+1-r}x' \in \ker f$, and the image of $x + \lambda^{n+1}M$ in $\ker(M/\lambda^{n+1-r}M \to N/\lambda^{n+1-r}N)$ is equal to the image of

$$(x - \lambda^{n+1-r}x') + \lambda^{n+1-r}\ker f \in \ker f/\lambda^{n+1-r}\ker f$$

in $\ker(M/\lambda^{n+1-r}M \to N/\lambda^{n+1-r}N)$. It follows that the cokernels of $\ker f/\lambda^{n+1}\ker f \to \ker f_n$ form a null system.

By the Artin–Rees theorem, there exists an integer $r \geq 0$ such that

$$\ker f \cap \lambda^n M = \lambda^{n-r}(\ker f \cap \lambda^r M)$$

for all $n \geq r$. Let $x + \lambda^{n+1}\ker f$ be an element in the kernel of $\ker f/\lambda^{n+1}\ker f \to \ker f_n$, where $x \in \ker f$. Then

$$x \in \ker f \cap \lambda^{n+1}M = \lambda^{n+1-r}(\ker f \cap \lambda^r M).$$

So $x = \lambda^{n+1-r}x'$ for some $x' \in \ker f$. The image of $x + \lambda^{n+1}\ker f$ in $\ker f/\lambda^{n+1-r}\ker f$ is 0. So the kernels of $\ker f/\lambda^{n+1}\ker f \to \ker f_n$ form a null system. □

Let X be a scheme and let $s \to X$ be a geometric point of X. For any A-R λ-adic sheaf $\mathscr{F} = (\mathscr{F}_n)$ on X, the stalks $(\mathscr{F}_{n,s})$ form an A-R λ-adic system of R-modules. We define the stalk of $\mathscr{F}$ at s to be

$$\mathscr{F}_s = \varprojlim_n \mathscr{F}_{n,s}.$$

Proposition 10.1.5. *Let $\mathscr{F} = (\mathscr{F}_n)$ be a λ-adic sheaf on a noetherian scheme X. There exists a dense open subset U of X such that $\mathscr{F}_n|_U$ are locally constant for all n.*

Proof. For each n, the epimorphism $\lambda^{n+1} : \mathscr{F}_{n+1} \to \lambda^{n+1}\mathscr{F}_{n+1}$ vanishes on $\lambda\mathscr{F}_{n+1}$. Since $\mathscr{F}_{n+1}/\lambda\mathscr{F}_{n+1} \cong \mathscr{F}_0$, it induces an epimorphism $\mathscr{F}_0 \to \lambda^{n+1}\mathscr{F}_{n+1}$. Let $\mathscr{K}_n$ be the kernel of this epimorphism. Since $\mathscr{F}_0$ is constructible, the ascending chain

$$\cdots \subset \mathscr{K}_n \subset \mathscr{K}_{n+1} \subset \cdots$$

in $\mathscr{F}_0$ is stationary by 5.8.5 (i). So we can find a dense open subset U of X such that $\mathscr{F}_0|U$ and $\mathscr{K}_n|_U$ are locally constant for all n. Then $\mathscr{F}_0|_U$ and $\lambda^{n+1}\mathscr{F}_{n+1}|_U$ are locally constant for all n. Using induction on m and the fact that $\mathscr{F}_{m+1}/\lambda^{m+1}\mathscr{F}_{m+1} \cong \mathscr{F}_m$, one can show $\mathscr{F}_{m+1}|_U$ that are locally constant for all m. □

Remark 10.1.6. The same argument as in the proof of 10.1.5 shows that in the definition of an λ-adic sheaf $\mathscr{F} = (\mathscr{F}_n)$, we may replace the condition that $\mathscr{F}_n$ are constructible for all n by the condition that $\mathscr{F}_0$ is constructible.

Proposition 10.1.7.

(i) *Let $f : \mathscr{F} \to \mathscr{G}$ be an A-R morphism between A-R λ-adic sheaves on a noetherian scheme X. Then* $\ker f$ *and* $\operatorname{coker} f$ *are A-R λ-adic. The category of A-R λ-adic sheaves form an abelian subcategory of the abelian A-R category of inverse system of λ-torsion sheaves. The category of λ-adic sheaves is an abelian category.*

(ii) *Let*

$$\mathscr{F} \xrightarrow{f} \mathscr{G} \xrightarrow{g} \mathscr{H}$$

be a sequence of A-R λ-adic sheaves on a noetherian scheme X. It is exact if and only if for any $x \in X$, the sequence

$$\mathscr{F}_{\bar{x}} \xrightarrow{f_{\bar{x}}} \mathscr{G}_{\bar{x}} \xrightarrow{g_{\bar{x}}} \mathscr{H}_{\bar{x}}$$

is exact.

(iii) *Let*

$$0 \to \mathscr{F} \xrightarrow{f} \mathscr{G} \xrightarrow{g} \mathscr{H} \to 0$$

be a short exact sequence in the A-R category of inverse systems of λ-torsion sheaves on a noetherian scheme X. Suppose that $\mathscr{G}_n$ are constructible and there exists an integer t such that $\lambda^{n+t}\mathscr{G}_n = 0$ for all n. If $\mathscr{F}$ and $\mathscr{H}$ are A-R λ-adic, so is $\mathscr{G}$.

Proof.

(i) We may assume that $\mathscr{F} = (\mathscr{F}_n)$ and $\mathscr{G} = (\mathscr{G}_n)$ are λ-adic sheaves. Then $f : \mathscr{F} \to \mathscr{G}$ is given by an element $(f_n) \in \varprojlim_n \operatorname{Hom}(\mathscr{F}_n, \mathscr{G}_n)$. To prove that $\ker f$ and $\operatorname{coker} f$ are A-R λ-adic, we may use noetherian induction and 10.1.2 (iii) to reduce to the case where X is irreducible and prove that there exists a nonempty open subset U of X such that $(\ker f_n|_U)$ and $(\operatorname{coker} f_n|_U)$ are A-R λ-adic. By 10.1.5, there exists a nonempty open subset U of X such that $\mathscr{F}_n|_U$ and $\mathscr{G}_n|_U$ are locally constant. By 10.1.2 (ii) and 10.1.4 (iii), $(\ker f_n|_U)$ and $(\operatorname{coker} f_n|_U)$ are A-R λ-adic.

(ii) We may assume that $\mathscr{F} = (\mathscr{F}_n)$, $\mathscr{G} = (\mathscr{G}_n)$ and $\mathscr{H} = (\mathscr{H}_n)$ are λ-adic sheaves. Then f and g are given by the elements $(f_n) \in \varprojlim_n \operatorname{Hom}(\mathscr{F}_n, \mathscr{G}_n)$ and $(g_n) \in \varprojlim_n \operatorname{Hom}(\mathscr{G}_n, \mathscr{H}_n)$, respectively.

Suppose $\mathscr{F} \xrightarrow{f} \mathscr{G} \xrightarrow{g} \mathscr{H}$ is exact. Then $gf = 0$ in the A-R category. As

$$\operatorname{Hom}_{\mathrm{AR}}(\mathscr{F}, \mathscr{H}) \cong \operatorname{Hom}(\mathscr{F}, \mathscr{H}),$$

we have $(g_n f_n) = 0$. Hence $g_{\bar{x}} f_{\bar{x}} = 0$. The inclusion $(\operatorname{im} f_n) \to (\ker g_n)$ induces an isomorphism in the A-R category. Its cokernel is a null system. Hence the cokernel of $(\operatorname{im} f_{n,\bar{x}}) \to (\ker g_{n,\bar{x}})$ is a null system. By 10.1.3, we have

$$\begin{aligned} \varprojlim_n \operatorname{im} f_{n,\bar{x}} &\cong \varprojlim_n \ker g_{n,\bar{x}}, \\ \varprojlim_n \operatorname{im} f_{n,\bar{x}} &\cong \operatorname{im} f_{\bar{x}}, \\ \varprojlim_n \ker g_{n,\bar{x}} &\cong \ker g_{\bar{x}}, \end{aligned}$$

So we have $\operatorname{im} f_{\bar{x}} \cong \ker g_{\bar{x}}$ and the sequence $\mathscr{F}_{\bar{x}} \xrightarrow{f_{\bar{x}}} \mathscr{G}_{\bar{x}} \xrightarrow{g_{\bar{x}}} \mathscr{H}_{\bar{x}}$ is exact.

Conversely suppose that $\mathscr{F}_{\bar{x}} \xrightarrow{f_{\bar{x}}} \mathscr{G}_{\bar{x}} \xrightarrow{g_{\bar{x}}} \mathscr{H}_{\bar{x}}$ is exact for all $x \in X$. To prove that $\mathscr{F} \xrightarrow{f} \mathscr{G} \xrightarrow{g} \mathscr{H}$ is exact, we may use noetherian induction to reduce to the case where X is irreducible and prove that there exists a nonempty open subset U of X such that $\mathscr{F}|_U \to \mathscr{G}|_U \to \mathscr{H}|_U$ is exact. By 10.1.5, there exists a nonempty open subset U of X such that $\mathscr{F}_n|_U$, $\mathscr{G}_n|_U$ and $\mathscr{H}_n|_U$ are locally constant. Let x be a point in U. As $\mathscr{F}_{\bar{x}} \xrightarrow{f_{\bar{x}}} \mathscr{G}_{\bar{x}} \xrightarrow{g_{\bar{x}}} \mathscr{H}_{\bar{x}}$ is exact, $(\mathscr{F}_{n,\bar{x}}) \to (\mathscr{G}_{n,\bar{x}}) \to (\mathscr{H}_{n,\bar{x}})$ is A-R exact by 10.1.4 (ii). This implies that $\mathscr{F}|_U \to \mathscr{G}|_U \to \mathscr{H}|_U$ is A-R exact.

(iii) Replacing $\mathscr{G}$ by $\mathscr{G}[-t+1]$, we may assume that $\lambda^{n+1}\mathscr{G}_n = 0$ for all n. We may assume that $\mathscr{F} = (\mathscr{F}_n)$ is λ-adic. Then $f : \mathscr{F} \to \mathscr{G}$ is given by an element $(f_n) \in \varprojlim_n \operatorname{Hom}(\mathscr{F}_n, \mathscr{G}_n)$. Let $\mathscr{H}_n = \operatorname{coker} f_n$. By our assumption, $\mathscr{H} = (\mathscr{H}_n)$ is an A-R λ-adic sheaf. So there exist integers $r, s \geq 0$ such that the conditions (a) and (b) in 10.1.1 hold for $\mathscr{H}$. Chasing the following commutative diagram

$$\begin{array}{ccccccc}\mathscr{F}_{n+t} & \to & \mathscr{G}_{n+t} & \to & \mathscr{H}_{n+t} & \to & 0 \\ \downarrow & & \downarrow & & \downarrow & & \\ \mathscr{F}_{n+r} & \to & \mathscr{G}_{n+r} & \to & \mathscr{H}_{n+r} & \to & 0 \\ \downarrow & & \downarrow & & \downarrow & & \\ \mathscr{F}_{n} & \to & \mathscr{G}_{n} & \to & \mathscr{H}_{n} & \to & 0,\end{array}$$

one gets

$$\mathrm{im}\,(\mathscr{G}[t] \to \mathscr{G}) = \mathrm{im}\,(\mathscr{G}[r] \to \mathscr{G})$$

for all $t \geq r$. So condition (a) in 10.1.1 holds for $\mathscr{G}$. Replacing $\mathscr{G}$ by $\overline{\mathscr{G}} = (\overline{\mathscr{G}}_n)$, where $\overline{\mathscr{G}}_n = \mathrm{im}\,(\mathscr{G}_{n+t} \to \mathscr{G}_n)$ for all $t \geq r$ and all n, we may assume that $\mathscr{G}_{n+1} \to \mathscr{G}_n$ are surjective. By our assumption, $\ker\,(\mathscr{F}_n \to \mathscr{G}_n)$ forms a null system. We may choose the integer s such that condition (b) in 10.1.1 holds for $\mathscr{H}$ and

$$\ker\,(\mathscr{F}_n \to \mathscr{G}_n) \subset \ker\,(\mathscr{F}_n \to \mathscr{F}_{n-s})$$

for all $n \geq s$. Let us prove

$$\mathscr{G}_{n+t}/\lambda^{n+1}\mathscr{G}_{n+t} \cong \mathscr{G}_{n+2s}/\lambda^{n+1}\mathscr{G}_{n+2s}$$

for all n and all $t \geq 2s$. Thus the condition (b) in 10.1.1 holds for $\mathscr{G}$, and $\mathscr{G}$ is A-R λ-adic. Let y be a section in $\ker\,(\mathscr{G}_{n+t} \to \mathscr{G}_{n+2s})$. Then the image of y in $\mathscr{H}_{n+t}$ lies in $\ker\,(\mathscr{H}_{n+t} \to \mathscr{H}_{n+2s})$. Since $\mathscr{H}$ satisfies the conditions in 10.1.1, we have

$$\ker\,(\mathscr{H}_{n+t} \to \mathscr{H}_{n+2s}) \subset \lambda^{n+s+1}\mathscr{H}_{n+t}.$$

So locally for the etale topology, there exists a section y' of $\mathscr{G}_{n+t}$ such that

$$y - \lambda^{n+s+1}y' = f_{n+t}(x)$$

for some section x of $\mathscr{F}_{n+t}$. Since y is a section of $\ker\,(\mathscr{G}_{n+t} \to \mathscr{G}_{n+2s})$ and $\lambda^{n+s+1}y'$ is a section of $\ker\,(\mathscr{G}_{n+t} \to \mathscr{G}_{n+s})$, the image of x in $\mathscr{F}_{n+s}$ is a section of $\ker\,(\mathscr{F}_{n+s} \to \mathscr{G}_{n+s})$. By our choice of s, we have

$$\ker\,(\mathscr{F}_{n+s} \to \mathscr{G}_{n+s}) \subset \ker\,(\mathscr{F}_{n+s} \to \mathscr{F}_n).$$

So x is a section of $\ker\,(\mathscr{F}_{n+t} \to \mathscr{F}_n) = \lambda^{n+1}\mathscr{F}_{n+t}$. It follows that $y = \lambda^{n+s+1}y' + f_{n+t}(x)$ is a section of $\lambda^{n+1}\mathscr{G}_{n+t}$. This implies that $\mathscr{G}_{n+t}/\lambda^{n+1}\mathscr{G}_{n+t} \cong \mathscr{G}_{n+2s}/\lambda^{n+1}\mathscr{G}_{n+2s}$. □

Proposition 10.1.8. *Let X be a noetherian scheme, $\mathscr{F}$ an A-R λ-adic sheaf, and*

$$\mathscr{F}^{(0)} \subset \mathscr{F}^{(1)} \subset \cdots$$

an ascending chain of A-R λ-adic subsheaves in $\mathscr{F}$. We have

$$\mathscr{F}^{(m)} = \mathscr{F}^{(m+1)} = \cdots$$

for sufficiently large m.

Proof. Using noetherian induction, to show the chain is stationary, we may assume that X is irreducible and prove that there exists a nonempty open subset U of X such that the chain

$$\mathscr{F}^{(0)}|_U \subset \mathscr{F}^{(1)}|_U \subset \cdots$$

is stationary. Let η be the generic point of X. We have a chain of finitely generated R-modules

$$\mathscr{F}^{(0)}_{\bar{\eta}} \subset \mathscr{F}^{(1)}_{\bar{\eta}} \subset \cdots .$$

This chain is stationary. Let $m \geq 0$ be an integer such that

$$\mathscr{F}^{(m)}_{\bar{\eta}} = \mathscr{F}^{(m+1)}_{\bar{\eta}} = \cdots .$$

By 10.1.5, there exists a nonempty open subset U of X such that $(\mathscr{F}/\mathscr{F}^{(m)})|_U$ is A-R isomorphic to a lisse λ-adic sheaf on U. For any $x \in U$, the specialization homomorphism $\mathscr{F}_{\bar{x}}/\mathscr{F}^{(m)}_{\bar{x}} \to \mathscr{F}_{\bar{\eta}}/\mathscr{F}^{(m)}_{\bar{\eta}}$ is an isomorphism. So the specialization homomorphism $\mathscr{F}^{(i)}_{\bar{x}}/\mathscr{F}^{(m)}_{\bar{x}} \to \mathscr{F}^{(i)}_{\bar{\eta}}/\mathscr{F}^{(m)}_{\bar{\eta}}$ is injective for any $i \geq m$. We have $\mathscr{F}^{(i)}_{\bar{\eta}}/\mathscr{F}^{(m)}_{\bar{\eta}} = 0$. So $\mathscr{F}^{(i)}_{\bar{x}}/\mathscr{F}^{(m)}_{\bar{x}} = 0$ for any $i \geq m$ and any $x \in U$. It follows that $(\mathscr{F}^{(i)}/\mathscr{F}^{(m)})|_U = 0$ for any $i \geq m$, and hence the chain is stationary when restricted to U. □

Corollary 10.1.9. *For any λ-adic sheaf $\mathscr{F}$ on a noetherian scheme X, there exist λ-adic sheaves $\mathscr{F}'$ and $\mathscr{F}''$ and a short exact sequence*

$$0 \to \mathscr{F}' \to \mathscr{F} \to \mathscr{F}'' \to 0$$

in the A-R category such that the following conditions hold:

(a) *$\mathscr{F}' = (\mathscr{F}'_n)$ is a torsion sheaf, that is, there exists an integer $m \geq 0$ such that $\mathscr{F}'_n = \mathscr{F}'_m$ for all $n \geq m$.*

(b) *$\mathscr{F}'' = (\mathscr{F}''_n)$ is torsion free, that is, the morphism $\lambda : \mathscr{F}'' \to \mathscr{F}''$ is A-R injective. $\mathscr{F}''_n$ are flat sheaves of $R/(\lambda^{n+1})$-modules for all n.*

Proof. Let $\mathscr{K}^{(n)}$ be the kernel of $\lambda^n : \mathscr{F} \to \mathscr{F}$. Then $\mathscr{K}^{(n)}$ form an ascending chain of A-R λ-adic subsheaves of $\mathscr{F}$. By 10.1.8, there exists an integer $m \geq 0$ such that

$$\mathscr{K}^{(m)} = \mathscr{K}^{(m+1)} = \cdots .$$

Let $\mathscr{F}'$ and $\mathscr{F}''$ be the λ-adic sheaves A-R isomorphic to $\mathscr{K}^{(m)}$ and $\mathscr{F}/\mathscr{K}^{(m)}$, respectively. Then $\lambda^m : \mathscr{F}' \to \mathscr{F}'$ is 0 in the A-R category. So there exists an integer $r \geq 0$ such that the composite $\mathscr{F}'[r] \to \mathscr{F}' \stackrel{\lambda^m}{\to} \mathscr{F}'$ is 0. As $\mathscr{F}'_{n+r} \to \mathscr{F}'_n$ is onto for all n, $\lambda^m : \mathscr{F}' \to \mathscr{F}'$ is 0, that is, $\lambda^m \mathscr{F}'_n = 0$ for all n. It follows that for any $n \geq m$, we have

$$\mathscr{F}'_n = \mathscr{F}'_n/\lambda^{m+1}\mathscr{F}'_n \cong \mathscr{F}'_m.$$

So the condition (a) holds. The kernel of the composite

$$\mathscr{F} \stackrel{\lambda}{\to} \mathscr{F} \to \mathscr{F}/\mathscr{K}^{(m)}$$

is $\mathscr{K}^{(m+1)}$. So we have an A-R monomorphism

$$\lambda : \mathscr{F}/\mathscr{K}^{(m+1)} \to \mathscr{F}/\mathscr{K}^{(m)}.$$

But $\mathscr{K}^{(m)} = \mathscr{K}^{(m+1)}$. So $\lambda : \mathscr{F}/\mathscr{K}^{(m)} \to \mathscr{F}/\mathscr{K}^{(m)}$ is A-R injective. Hence $\lambda : \mathscr{F}'' \to \mathscr{F}''$ is A-R injective. For any $x \in X$, the homomorphism $\lambda : \mathscr{F}''_{\bar{x}} \to \mathscr{F}''_{\bar{x}}$ is injective. So $\mathscr{F}''_{\bar{x}}$ is a finitely generated torsion free R-module. As R is a discrete valuation ring, this implies that $\mathscr{F}''_{\bar{x}}$ is free of finite rank. Then $\mathscr{F}''_{n,\bar{x}} = \mathscr{F}''_{\bar{x}}/\lambda^{n+1}\mathscr{F}''_{\bar{x}}$ are free of finite rank for all n. Hence $\mathscr{F}''_n$ are flat sheaves of $R/(\lambda^{n+1})$-modules. □

In the following (10.1.10–12), we fix a local noetherian ring A and a proper ideal I of A. Let $A_0 = A/I$. For any A-module M, let $M_0 = M \otimes_A A_0$. For all free A-modules M and N of finite ranks and any homomorphism $\phi_0 : M_0 \to N_0$, there exists a homomorphism $\phi : M \to N$ such that $\phi_0 = \phi \otimes \mathrm{id}_{A_0}$. Moreover, ϕ identifies M with a direct factor of N if and only if ϕ_0 identifies M_0 with a direct factor of N_0. The "only if" part is clear. To prove the "if" part, let $\psi_0 : N_0 \to M_0$ be a homomorphism such that $\psi_0\phi_0 = \mathrm{id}$, and let $\psi : N \to M$ be any lift of ψ_0. Then

$$\det(\psi\phi) \equiv \det(\psi_0\phi_0) \equiv 1 \mod I.$$

It follows that $\det(\psi\phi)$ is a unit, and hence $\psi\phi$ is invertible. Let $\delta : M \to M$ be the inverse of $\psi\phi$. Then $(\delta\psi)\phi = \mathrm{id}$. So ϕ identifies M with a direct factor of N.

Lemma 10.1.10. *Let $M_0^{\cdot}$ be a bounded acyclic complex of free A_0-modules of finite ranks. There exists a bounded acyclic complex $M^{\cdot}$ of free A-modules of finite ranks such that $M^{\cdot} \otimes_A A_0 \cong M_0^{\cdot}$.*

Proof. Without loss of generality, suppose that $M_0^\cdot$ is of the form

$$0 \to M_0^0 \stackrel{d_0^0}{\to} M_0^1 \to \cdots \stackrel{d_0^{n-1}}{\to} M_0^n \to 0.$$

Let $Z_0^i = \ker d_0^i$. We have short exact sequences

$$\begin{array}{l} 0 \to Z_0^{n-1} \to M_0^{n-1} \to M_0^n \to 0, \\ 0 \to Z_0^{n-2} \to M_0^{n-2} \to Z_0^{n-1} \to 0, \\ \qquad\vdots \\ 0 \to M_0^0 \to M_0^1 \to Z_0^2 \to 0. \end{array}$$

Choose free A-modules of finite rank M^i such that $M^i \otimes_A A_0 \cong M_0^i$ for all i. Lift the homomorphism $M_0^{n-1} \to M_0^n$ to a homomorphism $M^{n-1} \to M^n$. By Nakayama's lemma, the lifting is surjective. Let Z^{n-1} be its kernel. The short exact sequence

$$0 \to Z^{n-1} \to M^{n-1} \to M^n \to 0$$

splits. So Z^{n-1} is free of finite rank and $Z^{n-1} \otimes_A A_0 \cong Z_0^{n-1}$. Lift the homomorphism $M_0^{n-2} \to Z_0^{n-1}$ to $M^{n-2} \to Z^{n-1}$ and let Z^{n-2} be its kernel. Then we have a split short exact sequence

$$0 \to Z^{n-2} \to M^{n-2} \to Z^{n-1} \to 0,$$

Z^{n-2} is free of finite rank, and $Z^{n-2} \otimes_A A_0 \cong Z_0^{n-2}$. In this way, we get split short exact sequences

$$\begin{array}{l} 0 \to Z^{n-1} \to M^{n-1} \to M^n \to 0, \\ 0 \to Z^{n-2} \to M^{n-2} \to Z^{n-1} \to 0, \\ \qquad\vdots \\ 0 \to Z^1 \to M^1 \to Z^2 \to 0. \end{array}$$

Finally we take $M^0 = Z^1$. Then the complex

$$0 \to M^0 \to M^1 \to \cdots \to M^n \to 0$$

has the required property. □

Lemma 10.1.11. *Let $\phi : M^\cdot \to N^\cdot$ be a morphism of bounded complexes of free A-modules of finite ranks, and let $\phi_0' : M_0^\cdot \to N_0^\cdot$ be a morphism of complexes homotopic to $\phi_0 : M_0^\cdot \to N_0^\cdot$. Then there exists a morphism of complexes $\phi' : M^\cdot \to N^\cdot$ such that $\phi_0' = \phi' \otimes \mathrm{id}_{A_0}$ and ϕ' is homotopic to ϕ.*

Proof. Let $k_0 : M_0^{\cdot} \to N_0^{\cdot}[-1]$ be a homotopy between ϕ_0 and ϕ_0', that is,

$$dk_0 + k_0 d = \phi_0 - \phi_0'.$$

Let $k : M^{\cdot} \to N^{\cdot}[-1]$ be a lift of k_0 and define $\phi' = \phi - dk - kd$. Then $d\phi' = \phi' d$. So $\phi' : M^{\cdot} \to N^{\cdot}$ is a morphism of complex. It has the required property. □

Lemma 10.1.12. *Let $M^{\cdot}$ (resp. $N_0^{\cdot}$) be a bounded complex of free A-modules (resp. free A_0-modules) of finite ranks and let $\phi_0 : M_0^{\cdot} \to N_0^{\cdot}$ be a quasi-isomorphism. Then there exists a bounded complex $N^{\cdot}$ of free A-modules of finite ranks and a quasi-isomorphism $\phi : M^{\cdot} \to N^{\cdot}$ such that $N^{\cdot} \otimes_A A_0 \cong N_0^{\cdot}$ and $\phi \otimes \mathrm{id}_{A_0} = \phi_0$.*

Proof. Let $C_0^{\cdot}$ be the mapping cone of $\mathrm{id} : N_0^{\cdot} \to N_0^{\cdot}$. It is a bounded acyclic complex of free A_0-modules of finite rank. By 10.1.10, we can lift it to a bounded acyclic complex $C^{\cdot}$ of free A-modules of finite rank. Replacing $M^{\cdot}$ by $M^{\cdot} \oplus C^{\cdot}[-1]$ and replacing ϕ_0 by the morphism $M_0^{\cdot} \oplus C_0^{\cdot}[-1] \to N_0^{\cdot}$ defined by the homomorphisms

$$M_0^p \oplus N_0^p \oplus N_0^{p-1} \to N_0^p, \quad (x, y, z) \mapsto \phi_0(x) + y,$$

we may assume ϕ_0 is surjective. Let $K_0^{\cdot} = \ker \phi_0$. Then $K_0^{\cdot}$ is a bounded acyclic complex of free A_0-modules. By 10.1.10, it can be lifted to a bounded acyclic complex $K^{\cdot}$ of free A-modules. The complex K_0 is split. By 6.4.11, K_0 is homotopic to the zero complex. So the inclusion $i_0 : K_0^{\cdot} \hookrightarrow M_0^{\cdot}$ is homotopic to 0. By 10.1.11, there exists a morphism of complexes $i : K^{\cdot} \to M^{\cdot}$ homotopic to 0 such that $i \otimes \mathrm{id}_{A_0} = i_0$. Since i_0 identifies each K_0^n with a direct factor M_0^n, i identifies each K^n with a direct factor M^n. Set $N^{\cdot} = M^{\cdot}/K^{\cdot}$. Then $N^{\cdot}$ has the required property. □

Lemma 10.1.13. *Let A be a ring, $L^{\cdot}$ and $N^{\cdot}$ complexes of A-modules, and $\pi : N^{\cdot} \to L^{\cdot}$ a quasi-isomorphism. Then for any bounded above complex of projective A-modules $M^{\cdot}$, and any morphism of complexes $\phi : M^{\cdot} \to L^{\cdot}$, there exists a morphism of complexes $\psi : M^{\cdot} \to N^{\cdot}$ such that $\pi\psi$ is homotopic to ϕ.*

Proof. Let $C^{\cdot}$ be the mapping cone of $\mathrm{id} : L^{\cdot} \to L^{\cdot}$. Then $C^{\cdot}$ is homotopic to 0. Hence $N^{\cdot}$ is homotopic to $N^{\cdot} \oplus C^{\cdot}[-1]$. Let

$$\pi' : N^{\cdot} \oplus C^{\cdot}[-1] \to L^{\cdot}, \quad \pi'' : N^{\cdot} \oplus C^{\cdot}[-1] \to N^{\cdot}, \quad i : N^{\cdot} \to N^{\cdot} \oplus C^{\cdot}[-1]$$

be the morphisms of complexes defined by the homomorphisms

$$\begin{aligned} N^p \oplus L^p \oplus L^{p-1} &\to L^p, & (x, y, z) &\mapsto \pi(x) + y, \\ N^p \oplus L^p \oplus L^{p-1} &\to N^p, & (x, y, z) &\mapsto x, \\ N^p &\to N^p \oplus L^p \oplus L^{p-1}, & x &\mapsto (x, 0, 0), \end{aligned}$$

respectively. Then we have $\pi' i = \pi$, and $i\pi''$ is homotopic to $\mathrm{id}_{N^\cdot \oplus C^\cdot[-1]}$. So $\pi\pi'' = \pi' i \pi''$ is homotopic to π'. Suppose we can find a morphism of complexes $\psi' : M^\cdot \to N^\cdot \oplus C^\cdot[-1]$ such that $\pi'\psi'$ is homotopic to ϕ. Taking $\psi = \pi''\psi'$, then $\pi\psi$ is homotopic to ϕ. Replacing $N^\cdot$ by $N^\cdot \oplus C^\cdot[-1]$ and replacing π by π', we are reduced to the case where π is surjective.

Let us prove that under the assumption that π is surjective, we can find a morphism of complexes $\psi : M^\cdot \to N^\cdot$ such that $\pi\psi = \phi$. Let $K^\cdot = \ker \pi$. It is acyclic. Define $\psi^q : M^q \to N^q$ to be 0 for those q such that $M^i = 0$ for all $i \geq q$. Suppose we have defined ψ^q for all $q > r$. Since M^r is projective and $\pi^r : N^r \to L^r$ is onto, there exists a homomorphism $\psi_1^r : M^r \to N^r$ such that $\pi^r \psi_1^r = \phi^r$. We have

$$\pi^{r+1}(d^r_{N^\cdot}\psi_1^r - \psi^{r+1}d^r_{M^\cdot}) = d^r_{L^\cdot}\pi^r\psi_1^r - \phi^{r+1}d^r_{M^\cdot} = d^r_{L^\cdot}\phi^r - \phi^{r+1}d^r_{M^\cdot} = 0.$$

So $d^r_{N^\cdot}\psi_1^r - \psi^{r+1}d^r_{M^\cdot}$ maps M^r to K^{r+1}. We have

$$d^{r+1}_{N^\cdot}(d^r_{N^\cdot}\psi_1^r - \psi^{r+1}d^r_{M^\cdot}) = d^{r+1}_{N^\cdot}d^r_{N^\cdot}\psi_1^r - \psi^{r+2}d^{r+1}_{M^\cdot}d^r_{M^\cdot} = 0.$$

So $d^r_{N^\cdot}\psi_1^r - \psi^{r+1}d^r_{M^\cdot}$ maps M^r to $\ker(d : K^{r+1} \to K^{r+2})$. Since $K^\cdot$ is acyclic and M^r is projective, there exists a homomorphism $\psi_2^r : M^r \to K^r$ such that

$$d^r_{N^\cdot}\psi_2^r = d^r_{N^\cdot}\psi_1^r - \psi^{r+1}d^r_{M^\cdot}.$$

We then define $\psi^r = \psi_1^r - \psi_2^r$. □

Lemma 10.1.14. *Let A be a noetherian local ring.*

(i) *Let $M^\cdot$ be a bounded above complex of A-modules such that $H^i(M^\cdot)$ are finitely generated A-modules. Then there exists a quasi-isomorphism $s : M'^\cdot \to M^\cdot$ such that $M'^\cdot$ is a bounded above complex of free A-modules of finite ranks.*

(ii) *Let $M^\cdot$ be a bounded complex of flat A-modules such that $H^i(M^\cdot)$ are finitely generated A-modules and that $M^i = 0$ for $i \notin [a, b]$ for some integers a and b. Then there exists a quasi-isomorphism $s : M'^\cdot \to M^\cdot$ such that $M^\cdot$ is a bounded complex of free A-modules of finite ranks, and $M'^i = 0$ for $i \notin [a, b]$.*

Proof. Since A is a noetherian local ring, a finitely generated A-module is free if and only if it is flat. Using the same method as the proof of 6.4.5, one can prove (i), and under the assumption of (ii), one can construct a quasi-isomorphism $F^{\cdot} \to M^{\cdot}$ such that F^i are finitely generated flat A-modules, and $F^i = 0$ for all $i > b$. Since $M^{\cdot}$ is a bounded above complex of flat A-modules, by 6.4.4, we have

$$H^i(F^{\cdot} \otimes_A N) \cong H^i(M^{\cdot} \otimes_A N)$$

for any A-module N. Since $M^i = 0$ for all $i < a$, we have $H^i(F^{\cdot} \otimes_A N) = 0$ for all $i < a$. Note that

$$\cdots \to F^{a-1} \to F^a \to 0$$

is a flat resolution of $\mathrm{coker}\,(F^{a-1} \to F^a)$. It follows that

$$\mathrm{Tor}_i^A(\mathrm{coker}\,(F^{a-1} \to F^a), N) \cong H^{a-i}(F^{\cdot} \otimes_A N) = 0$$

for all $i > 0$. So $\mathrm{coker}\,(F^{a-1} \to F^a)$ is flat. Let $M'^{\cdot}$ be the complex

$$0 \to \mathrm{coker}\,(F^{a-1} \to F^a) \to F^{a+1} \to \cdots \to F^b \to 0.$$

Then the quasi-isomorphism $F^{\cdot} \to M^{\cdot}$ induces a quasi-isomorphism $M'^{\cdot} \to M^{\cdot}$, and $M'^{\cdot}$ is a complex of free A-modules of finite ranks. □

Proposition 10.1.15. *Let $(K_n^{\cdot})_{n\in\mathbb{Z}}$ be an inverse system of complexes of R-modules and let $a \leq b$ be integers. Suppose the following conditions hold:*

(a) $K_n^{\cdot} = 0$ *for* $n < 0$, *and* $K_n^i = 0$ *for* $i \notin [a, b]$.

(b) $\lambda^{n+1} K_n^i = 0$ *and* K_n^i *are flat* $R/(\lambda^{n+1})$*-modules for all* $n \geq 0$ *and* i.

(c) $H^i(K_n^{\cdot})$ *are finite.*

(d) *The morphisms*

$$K_{n+1}^{\cdot} \otimes_{R/(\lambda^{n+2})} R/(\lambda^{n+1}) \to K_n^{\cdot}$$

induced by $K_{n+1}^{\cdot} \to K_n^{\cdot}$ are quasi-isomorphism for all $n \geq 0$.

Then we can find a bounded complex $K^{\cdot}$ of free R-modules of finite ranks and quasi-isomorphisms $K^{\cdot}/\lambda^{n+1}K^{\cdot} \to K_n^{\cdot}$ such that the diagrams

$$\begin{array}{ccc} K^{\cdot}/\lambda^{n+2}K^{\cdot} & \to & K_{n+1}^{\cdot} \\ \downarrow & & \downarrow \\ K^{\cdot}/\lambda^{n+1}K^{\cdot} & \to & K_n^{\cdot} \end{array}$$

commute up to homotopy. The inverse systems $(H^i(K_n^{\cdot}))_{n\in\mathbb{Z}}$ are A-R λ-adic, and $H^i(K^{\cdot}) \cong \varprojlim_n H^i(K_n^{\cdot})$.

Proof. By 10.1.14, we can find quasi-isomorphisms $K'^{\cdot}_n \to K^{\cdot}_n$ $(n \geq 0)$ such that $K'^{\cdot}_n$ are complexes of free $R/(\lambda^{n+1})$-modules of finite ranks and $K'^{i}_n = 0$ for $i \notin [a, b]$. Let $K''^{\cdot}_0 = K'^{\cdot}_0$. By 10.1.13, we can find a morphism

$$K'^{\cdot}_1 \otimes_{R/(\lambda^2)} R/(\lambda) \to K''^{\cdot}_0$$

such that the following diagram commutes up to homotopy:

$$\begin{array}{ccc} K'^{\cdot}_1 \otimes_{R/(\lambda^2)} R/(\lambda) & \to & K^{\cdot}_1 \otimes_{R/(\lambda^2)} R/(\lambda) \\ \downarrow & & \downarrow \\ K''^{\cdot}_0 & \to & K^{\cdot}_0. \end{array}$$

Note that $K'^{\cdot}_1 \otimes_{R/(\lambda^2)} R/(\lambda) \to K''^{\cdot}_0$ is necessarily a quasi-isomorphism. By 10.1.12, we can find a bounded complex $K''^{\cdot}_1$ of free $R/(\lambda^2)$-modules of finite rank and a quasi-isomorphism $K'^{\cdot}_1 \to K''^{\cdot}_1$ such that

$$K''^{\cdot}_1 \otimes_{R/(\lambda^2)} R/(\lambda) \cong K''^{\cdot}_0$$

and that the morphism $K'^{\cdot}_1 \otimes_{R/(\lambda^2)} R/(\lambda) \to K''^{\cdot}_0$ is induced by the morphism $K'^{\cdot}_1 \to K''^{\cdot}_1$. By the dual version of 6.2.7 for complexes of projective objects, the quasi-isomorphism $K'^{\cdot}_1 \to K''^{\cdot}_1$ is homotopically invertible. Let $K''^{\cdot}_1 \to K^{\cdot}_1$ be the composite of a homotopic inverse of $K'^{\cdot}_1 \to K''^{\cdot}_1$ with the quasi-isomorphism $K'^{\cdot}_1 \to K^{\cdot}_1$. Then $K''^{\cdot}_1 \to K^{\cdot}_1$ is a quasi-isomorphism, and the following diagram commutes up to homotopy:

$$\begin{array}{ccc} K''^{\cdot}_1 & \to & K^{\cdot}_1 \\ \downarrow & & \downarrow \\ K''^{\cdot}_0 & \to & K^{\cdot}_0. \end{array}$$

Similarly, we can find a quasi-morphism

$$K'^{\cdot}_2 \otimes_{R/(\lambda^3)} R/(\lambda^2) \to K''^{\cdot}_1$$

such that the diagram

$$\begin{array}{ccc} K'^{\cdot}_2 \otimes_{R/(\lambda^3)} R/(\lambda^2) & \to & K^{\cdot}_2 \otimes_{R/(\lambda^3)} R/(\lambda^2) \\ \downarrow & & \downarrow \\ K''^{\cdot}_1 & \to & K^{\cdot}_1 \end{array}$$

commutes up to homotopy, and we can find a bounded complex $K''^{\cdot}_2$ of free $R/(\lambda^3)$-modules of finite ranks and a quasi-isomorphism $K'^{\cdot}_2 \to K''^{\cdot}_2$ such that

$$K''^{\cdot}_2 \otimes_{R/(\lambda^3)} R/(\lambda^2) \cong K''^{\cdot}_1$$

and that the morphism $K'^{\cdot}_2 \otimes_{R/(\lambda^3)} R/(\lambda^2) \to K''^{\cdot}_1$ is induced by the morphism $K'^{\cdot}_2 \to K''^{\cdot}_2$. Let $K''^{\cdot}_2 \to K^{\cdot}_2$ be the composite of a homotopic inverse of $K_2'^{\cdot} \to K_2''^{\cdot}$ with the quasi-isomorphism $K_2'^{\cdot} \to K_2^{\cdot}$. Then $K_2''^{\cdot} \to K_2^{\cdot}$ is a quasi-isomorphism, and the following diagram commutes up to homotopy:

$$\begin{array}{ccc} K_2''^{\cdot} & \to & K_2^{\cdot} \\ \downarrow & & \downarrow \\ K_1''^{\cdot} & \to & K_1^{\cdot}. \end{array}$$

In this way, we can construct a family of bounded complexes of free $R/(\lambda^{n+1})$-modules of finite ranks $K_n''^{\cdot}$ and quasi-isomorphisms $K_n''^{\cdot} \to K_n$ such that

$$K_{n+1}''^{\cdot} \otimes_{R/(\lambda^{n+2})} R/(\lambda^{n+1}) \cong K_n''^{\cdot}$$

and the diagrams

$$\begin{array}{ccc} K_{n+1}''^{\cdot} & \to & K_{n+1}^{\cdot} \\ \downarrow & & \downarrow \\ K_n''^{\cdot} & \to & K_n^{\cdot} \end{array}$$

commute up to homotopy. Since $K_0''^i = 0$ for $i \notin [a,b]$, we have $K_n''^i = 0$ for $i \notin [a,b]$ by Nakayama's lemma. Set $K^{\cdot} = \varprojlim_n K_n''^{\cdot}$. Then $K^{\cdot}$ has the required property. By 10.1.4 (iii), $(H^i(K_n''^{\cdot}))_{n\in\mathbb{Z}}$ are A-R λ-adic, and by 10.1.3, we have $H^i(K^{\cdot}) \cong \varprojlim_n H^i(K_n''^{\cdot})$. It follows that $(H^i(K_n^{\cdot}))_{n\in\mathbb{Z}}$ are A-R λ-adic, and $H^i(K^{\cdot}) \cong \varprojlim_n H^i(K_n^{\cdot})$. □

Proposition 10.1.16. *Let X be a noetherian scheme, and let $K = (K_n, u_n)_{n\geq 0}$ be a family consisting of bounded above complexes of sheaves of $R/(\lambda^{n+1})$-modules K_n on X and isomorphisms*

$$u_n : K_{n+1} \otimes^L_{R/(\lambda^{n+2})} R/(\lambda^{n+1}) \cong K_n$$

in $D^-(X, R/(\lambda^{n+1}))$. If $\mathscr{H}^i(K_0)$ are constructible sheaves of $R/(\lambda)$-modules and are nonzero for only finitely many i, then $K_n \in \operatorname{ob} D^b_{\mathrm{ctf}}(X, R/(\lambda^{n+1}))$ for all n, and the inverse systems $(\mathscr{H}^i(K_n))_{n\in\mathbb{Z}}$ are A-R λ-adic.

Proof. Represent each K_n by a bounded above complex of sheaves of flat $R/(\lambda^{n+1})$-modules. We have

$$K_n \otimes_{R/(\lambda^{n+1})} R/(\lambda) \cong K_n \otimes^L_{R/(\lambda^{n+1})} R/(\lambda^n) \otimes^L_{R/(\lambda^n)} \cdots \otimes^L_{R/(\lambda^2)} R/(\lambda) \cong K_0.$$

On the other hand, we have

$$\lambda^i K_n/\lambda^{i+1} K_n \cong K_n \otimes_{R/(\lambda^{n+1})} (\lambda^i)/(\lambda^{i+1}) \cong K_n \otimes_{R/(\lambda^{n+1})} R/(\lambda) \cong K_0.$$

So each K_n has a finite filtration

$$0 = \lambda^{n+1} K_n \subset \lambda^n K_n \subset \cdots \subset \lambda K_n \subset K_n$$

such that the successive quotients $\lambda^i K_n/\lambda^{i+1}K_n$ are isomorphic to K_0. As $\mathscr{H}^i(K_0)$ are constructible, so are $\mathscr{H}^i(K_n)$. Suppose that $\mathscr{H}^i(K_0) = 0$ for $i \notin [a,b]$. Then we have $\mathscr{H}^i(K_n) = 0$ for $i \notin [a,b]$. The complex

$$\cdots \to K_n^{a-1} \to K_n^a \to 0$$

is a resolution of $\mathrm{coker}\,(K_n^{a-1} \to K_n^a)$ by flat sheaves of $R/(\lambda^{n+1})$-modules. So

$$\begin{aligned}\mathrm{Tor}_i^{R/(\lambda^{n+1})}(\mathrm{coker}\,(K_n^{a-1} \to K_n^a), R/(\lambda)) &\cong \mathscr{H}^{a-i}(K_n \otimes_{R/(\lambda^{n+1})} R/(\lambda)) \\ &\cong \mathscr{H}^{a-i}(K_0).\end{aligned}$$

We have $\mathscr{H}^{a-i}(K_0) = 0$ for all $i > 0$. So

$$\mathrm{Tor}_i^{R/(\lambda^{n+1})}(\mathrm{coker}\,(K_n^{a-1} \to K_n^a), R/(\lambda)) = 0$$

for all $i > 0$. By 1.2.8, $\mathrm{coker}\,(K_n^{a-1} \to K_n^a)$ is a flat sheaf of $R/(\lambda^{n+1})$-modules. The complex K_n is quasi-isomorphic to

$$0 \to \mathrm{coker}\,(K_n^{a-1} \to K_n^a) \to K_n^{a+1} \to \cdots,$$

and the latter is a bounded complex of flat sheaves of $R/(\lambda^{n+1})$-modules. So K_n has finite Tor-dimension. It follows that $K_n \in \mathrm{ob}\, D^b_{\mathrm{ctf}}(X, R/(\lambda^{n+1}))$.

To prove that the inverse systems $(\mathscr{H}^i(K_n))_{n\in\mathbb{Z}}$ are A-R λ-adic, we use noetherian induction, and we may assume that X is irreducible and prove that there exists a nonempty open subset U of X such that $(\mathscr{H}^i(K_n)|_U)_{n\in\mathbb{Z}}$ are A-R λ-adic. There exists a nonempty open subset U such that $\mathscr{H}^i(K_0)|_U$ are locally constant for all i. This implies that $\mathscr{H}^i(K_n)|_U$ are locally constant for all i and n. Let x be an arbitrary point in U. To prove our assertion, it suffices to show that $(\mathscr{H}^i(K_{n,\bar{x}}))_{n\in\mathbb{Z}}$ are A-R λ-adic.

By 10.1.14, each $K_{n,\bar{x}}$ is isomorphic in the derived category to a complex $L_n^{\cdot}$ of free $R/(\lambda^{n+1})$-modules of finite ranks with $L_n^v = 0$ for $v \notin [a,b]$. The isomorphism

$$K_{n+1} \otimes^L_{R/(\lambda^{n+2})} R/(\lambda^{n+1}) \cong K_n$$

induces an isomorphism

$$L_{n+1}^{\cdot} \otimes_{R/(\lambda^{n+2})} R/(\lambda^{n+1}) \cong L_n^{\cdot}$$

in the derived category. By 10.1.13, the latter isomorphism is induced by a quasi-isomorphism $L_{n+1}^{\cdot} \otimes_{R/(\lambda^{n+2})} R/(\lambda^{n+1}) \cong L_n^{\cdot}$ of complexes. Applying 10.1.15 to the inverse system of complexes $(L_n^{\cdot})_{n\in\mathbb{Z}}$, we see that $(\mathscr{H}^i(L_n^{\cdot}))_{n\in\mathbb{Z}}$ are A-R λ-adic. So $(\mathscr{H}^i(K_{n,\bar{x}}))_{n\in\mathbb{Z}}$ are A-R λ-adic. □

Let X be a noetherian scheme. Define a category $D_c^b(X, R)$ as follows: Objects in $D_c^b(X, R)$ are families $K = (K_n, u_n)_{n\geq 0}$ consisting of objects K_n in $D^-(X, R/(\lambda^{n+1}))$ and isomorphisms

$$u_n : K_{n+1} \otimes^L_{R/(\lambda^{n+2})} R/(\lambda^{n+1}) \cong K_n$$

in $D^-(X, R/(\lambda^{n+1}))$ such that $K_0 \in \operatorname{ob} D_c^b(X, R/(\lambda))$. Let $K = (K_n, u_n)_{n\geq 0}$ and $K' = (K'_n, u'_n)_{n\geq 0}$ be two objects in $D_c^b(X, R)$. A morphism $f : K \to K'$ in $D_c^b(X, R)$ is a family (f_n) of morphisms $f_n : K_n \to K_n$ in $D(X, R/(\lambda^{n+1}))$ such that

$$f_n u_n = u'_n(f_{n+1} \otimes^L_{R/(\lambda^{n+2})} \mathrm{id}_{R/(\lambda^{n+1})}).$$

If $\mathscr{F} = (\mathscr{F}_n)$ is a torsion free λ-adic sheaf, then $\mathscr{F}$ defines an object in $D_c^b(X, R)$.

Proposition 10.1.17.

(i) *Let $f : X \to Y$ be a morphism between noetherian schemes. For any $K = (K_n) \in \operatorname{ob} D_c^b(Y, R)$, we have $f^*K = (f^*K_n) \in \operatorname{ob} D_c^b(X, R)$.*

(ii) *Let $f : X \to Y$ be an S-compactifiable morphism between noetherian schemes. For any $K = (K_n) \in \operatorname{ob} D_c^b(X, R)$, we have $Rf_!K = (Rf_!K_n) \in \operatorname{ob} D_c^b(Y, R)$.*

(iii) *Let S be a noetherian regular scheme of dimension ≤ 1, X and Y two S-schemes of finite type, $f : X \to Y$ an S-compactifiable morphism, $K = (K_n)$ and $L = (L_n)$ objects in $D_c^b(X, R)$, and $M = (M_n)$ an object in $D_c^b(Y, R)$. Define*

$$\begin{aligned} Rf_*K &= (Rf_*K_n), \\ f^*M &= (f^*M_n), \\ Rf_!K &= (Rf_!K_n), \\ Rf^!M &= (Rf^!M_n), \\ K \otimes^L_R L &= (K_n \otimes^L_{R/(\lambda^{n+1})} L_n), \\ R\mathscr{H}om(K, L) &= (R\mathscr{H}om(K_n, L_n)). \end{aligned}$$

Then Rf_*K and $Rf_!K$ are objects in $D_c^b(Y, R)$, while f^*M, $Rf^!M$, $K \otimes^L_R L$ and $R\mathscr{H}om(K, L)$ are objects in $D_c^b(X, R)$.

(iv) *Let S be a trait, X an S-scheme of finite type, η the generic point of S, s the closed point of S, $K = (K_n) \in \operatorname{ob} D_c^b(X_\eta, R)$, and $L = (L_n) \in \operatorname{ob} D_c^b(X, R)$. Define*

$$R\Psi_\eta(K) = (R\Psi_\eta(K_n)), \quad R\Phi(L) = (R\Phi(L_n)).$$

Then $R\Psi_\eta(K)$ and $R\Phi(L)$ are objects in $D_c^b(X_{\bar{s}}, R)$.

(v) *Let S be a noetherian regular scheme of dimension ≤ 1 and let $f : X \to S$ be an S-scheme of finite type. For any $K = (K_n) \in \mathrm{ob}\, D_c^b(X, R)$, define*

$$DK = \Big(R\mathscr{H}om(K_n, Rf^!(R/(\lambda^{n+1})))\Big).$$

Then the canonical morphism $K \to DDK$ is an isomorphism.

Proof. We prove (iii) and leave the rest to the reader. By 10.1.16, we have $K_n \in \mathrm{ob}\, D^b_{\mathrm{ctf}}(X, R/(\lambda^{n+1}))$. By 7.4.7 (ii) and 7.8.1, we have $Rf_!K_n \in \mathrm{ob}\, D^b_{\mathrm{ctf}}(Y, R/(\lambda^{n+1}))$. By 7.4.7 (iii), we have

$$Rf_!K_{n+1} \otimes^L_{R/(\lambda^{n+2})} R/(\lambda^{n+1}) \cong Rf_!(K_{n+1} \otimes^L_{R/(\lambda^{n+2})} R/(\lambda^{n+1})) \cong Rf_!K_n.$$

So $Rf_!K \in \mathrm{ob}\, D^b_c(Y, R)$. Similarly, using 6.5.7 and 9.5.2, one can show $Rf_*K \in \mathrm{ob}\, D^b_c(Y, R)$. One can easily show that f^*M and $K \otimes^L_R L$ are objects in $D^b_c(X, R)$. By 9.5.2, we have $Rf^!M_n \in \mathrm{ob}\, D^b_{\mathrm{ctf}}(X, R/(\lambda^{n+1}))$. To prove $Rf^!M \in \mathrm{ob}\, D^b_c(X, R)$, it suffices to show

$$Rf^!M_{n+1} \otimes^L_{R/(\lambda^{n+2})} R/(\lambda^{n+1}) \cong Rf^!(M_{n+1} \otimes^L_{R/(\lambda^{n+2})} R/(\lambda^{n+1})).$$

Let A be a noetherian ring and let B be a noetherian A-algebra. Let us prove more generally that for any $M \in \mathrm{ob}\, D^b_{\mathrm{ctf}}(Y, A)$, we have an isomorphism

$$Rf^!M \otimes^L_A B \cong Rf^!(M \otimes^L_A B)$$

in $D(X, B)$. First we construct a morphism

$$Rf^!M \otimes^L_A B \to Rf^!(M \otimes^L_A B)$$

in $D(X, B)$. Let $\rho : D(X, B) \to D(X, A)$ and $\rho : D(Y, B) \to D(Y, A)$ be the functors in 8.4.12. It suffices to construct a morphism

$$Rf^!M \to \rho Rf^!(M \otimes^L_A B)$$

in $D(X, A)$. By 8.4.13, we have a canonical isomorphism

$$\rho Rf^!(M \otimes^L_A B) \cong Rf^!\rho(M \otimes^L_A B).$$

So we need to construct a morphism

$$Rf^!M \to Rf^!\rho(M \otimes^L_A B).$$

We define it to be the morphism induced by the canonical morphism

$$M \to \rho(M \otimes^L_A B).$$

To prove that the morphism $Rf^!M \otimes^L_A B \to Rf^!(M \otimes^L_A B)$ is an isomorphism is a local problem. So we may assume that f is a composite

$$X \xrightarrow{i} \mathbb{A}^d_Y \xrightarrow{\pi} Y,$$

so that i is a closed immersion and π is the projection. We have

$$\begin{aligned}&Rf^!M\otimes_A^L B\cong Ri^!\pi^*M(d)[2d]\otimes_A^L B,\\ &Rf^!(M\otimes_A^L B)\cong Ri^!\pi^*(M\otimes_A^L B)(d)[2d]\cong Ri^!(\pi^*M(d)[2d]\otimes_A^L B).\end{aligned}$$

To prove our assertion, it suffices to show that

$$Ri^!M\otimes_A^L B\cong Ri^!(M\otimes_A^L B)$$

for any $M\in\operatorname{ob} D^b_{\mathrm{ctf}}(\mathbb{A}^d_Y, A)$. Let $j:\mathbb{A}^n_Y - i(X)\to\mathbb{A}^n_Y$ be the open immersion. We have a morphism of distinguished triangles

$$\begin{array}{ccccccc} i_*Ri^!M\otimes_A^L B & \to & M\otimes_A^L B & \to & Rj_*j^*M\otimes_A^L B & \to \\ \downarrow & & \downarrow & & \downarrow & \\ i_*Ri^!(M\otimes_A^L B) & \to & M\otimes_A^L B & \to & Rj_*j^*(M\otimes_A^L B) & \to . \end{array}$$

The second and the third vertical arrows are isomorphisms. So the first arrow is also an isomorphism. Our assertion follows.

By 9.5.3 (ii), we have $R\mathscr{H}om(K_n, L_n)\in\operatorname{ob} D^b_{\mathrm{ctf}}(X, R/(\lambda^{n+1}))$. To prove $R\mathscr{H}om(K,L)\in\operatorname{ob} D^b_c(X,R)$, we need to show

$$\begin{aligned}&R\mathscr{H}om(K_{n+1}, L_{n+1})\otimes^L_{R/(\lambda^{n+2})} R/(\lambda^{n+1})\\ &\cong R\mathscr{H}om(K_{n+1}\otimes^L_{R/(\lambda^{n+2})} R/(\lambda^{n+1}), L_{n+1}\otimes^L_{R/(\lambda^{n+2})} R/(\lambda^{n+1})).\end{aligned}$$

Let A be a noetherian ring and let B be a noetherian A-algebra. Let us prove more generally that for any $K,L\in\operatorname{ob} D^b_{\mathrm{ctf}}(X,A)$, we have an isomorphism

$$R\mathscr{H}om(K,L)\otimes_A^L B\cong R\mathscr{H}om(K\otimes_A^L B, L\otimes_A^L B).$$

We have a canonical morphism

$$R\mathscr{H}om(K,L)\otimes_A^L K\stackrel{\mathrm{Ev}}{\to} L.$$

It induces a morphism

$$(R\mathscr{H}om(K,L)\otimes_A^L B)\otimes_B^L(K\otimes_A^L B)\stackrel{\mathrm{Ev}}{\to} L\otimes_A^L B.$$

By 6.4.7, this gives rise to a morphism

$$R\mathscr{H}om(K,L)\otimes_A^L B\to R\mathscr{H}om(K\otimes_A^L B, L\otimes_A^L B).$$

To prove that it is an isomorphism, we may assume $K=j_!\mathscr{F}$ such that $j:Y\to X$ is an immersion, and $\mathscr{F}$ is a locally constant constructible sheaf of A-modules on Y. This is because K can be represented by a bounded complex of constructible sheaves of A-modules, and any constructible sheaf of A-modules has a finite filtration such that the successive quotients are of the form $j_!\mathscr{F}$. We have

$$\begin{aligned} R\mathscr{H}om(j_!\mathscr{F}, L) \otimes_A^L B &\cong Rj_*R\mathscr{H}om(\mathscr{F}, Rj^!L) \otimes_A^L B \\ &\cong Rj_*(R\mathscr{H}om(\mathscr{F}, Rj^!L) \otimes_A^L B). \\ R\mathscr{H}om(j_!\mathscr{F} \otimes_A^L B, L \otimes_A^L B) &\cong R\mathscr{H}om(j_!(\mathscr{F} \otimes_A^L B), L \otimes_A^L B) \\ &\cong Rj_*R\mathscr{H}om(\mathscr{F} \otimes_A^L B, Rj^!L \otimes_A^L B). \end{aligned}$$

To prove our assertion, it suffices to show that the canonical morphism

$$R\mathscr{H}om(\mathscr{F}, M) \otimes_A^L B \to R\mathscr{H}om(\mathscr{F} \otimes_A^L B, M \otimes_A^L B)$$

is an isomorphism for any $M \in \operatorname{ob} D^b_{\mathrm{tf}}(Y, A)$. The problem is local with respect to the etale topology. We may assume that $\mathscr{F}$ is a constant sheaf. When $\mathscr{F}$ is the constant sheaf associated to A, the above morphism is clearly an isomorphism. It follows that it is an isomorphism if $\mathscr{F}$ is a constant sheaf associated to a free A-module of finite rank. In general, $\mathscr{F}$ has a resolution by constant sheaves associated to free A-modules of finite rank. So the above morphism is an isomorphism. □

Proposition 10.1.18.

(i) *Let $f : X \to Y$ be a morphism between noetherian schemes. For any λ-adic sheaf $\mathscr{F} = (\mathscr{F}_n)$ on Y, $f^*\mathscr{F} = (f^*\mathscr{F}_n)$ is a λ-adic sheaf on X.*

(ii) *Let $f : X \to Y$ be an S-compactifiable morphism between noetherian schemes. For any λ-adic sheaf $\mathscr{F} = (\mathscr{F}_n)$ on X, $R^if_!\mathscr{F} = (R^if_!\mathscr{F}_n)$ are A-R λ-adic sheaves on Y for all i.*

(iii) *Let S be a noetherian regular scheme of dimension ≤ 1, X and Y two S-schemes of finite type, $f : X \to Y$ an S-compactifiable morphism, $\mathscr{F} = (\mathscr{F}_n)$ and $\mathscr{G} = (\mathscr{G}_n)$ λ-adic sheaves on X, and $\mathscr{H} = (\mathscr{H}_n)$ a λ-adic sheaf on Y. Define*

$$\begin{aligned} R^if_*\mathscr{F} &= (R^if_*\mathscr{F}_n), \\ R^if_!\mathscr{F} &= (R^if_!\mathscr{F}_n), \\ f^*\mathscr{H} &= (f^*\mathscr{H}_n), \\ R^if^!\mathscr{H} &= (R^if^!\mathscr{H}_n), \\ \mathrm{Tor}_i^R(\mathscr{F}, \mathscr{G}) &= (\mathrm{Tor}_i^{R/(\lambda^{n+1})}(\mathscr{F}_n, \mathscr{G}_n)), \\ \mathscr{E}xt^i_R(\mathscr{F}, \mathscr{G}) &= (\mathscr{E}xt^i_{R/(\lambda^{n+1})}(\mathscr{F}_n, \mathscr{G}_n)). \end{aligned}$$

Then $R^if_*\mathscr{F}$ and $R^if_!\mathscr{F}$ are A-R λ-adic sheaves on Y, and $f^*\mathscr{H}$, $R^if^!\mathscr{H}$, $\mathrm{Tor}_i^R(\mathscr{F}, \mathscr{G})$ and $\mathscr{E}xt^i_R(\mathscr{F}, \mathscr{G})$ are A-R λ-adic sheaves on X.

(iv) *Let S be a trait, X an S-scheme of finite type, η the generic point of S, s the closed point of S, $\mathscr{F} = (\mathscr{F}_n)$ a λ-adic sheaf on X_η, and $\mathscr{G} = (\mathscr{G}_n)$ a λ-adic sheaf on X. Define*

$$R^i\Psi_\eta(\mathscr{F}) = (R^i\Psi_\eta(\mathscr{F}_n)), \quad R^i\Phi(\mathscr{G}) = (R^i\Phi(\mathscr{G}_n)).$$

Then $R^i\Psi_\eta(\mathscr{F})$ and $R^i\Phi(\mathscr{G})$ are A-R λ-adic sheaves on $X_{\bar{s}}$.

Proof. We prove (ii) and leave the rest to the reader. By 10.1.9, there exists an A-R exact sequence

$$0 \to \mathscr{F}' \stackrel{\phi}{\to} \mathscr{F} \stackrel{\psi}{\to} \mathscr{F}'' \to 0$$

such that $\mathscr{F}'$ is a torsion λ-adic sheaf, and $\mathscr{F}''$ is a torsion free λ-adic sheaf. We have short exact sequences

$$\begin{array}{ccccccc} 0 \to \ker\phi_n & \to & \mathscr{F}'_n & \to & \operatorname{im}\phi_n & \to 0, \\ 0 \to \operatorname{im}\phi_n & \to & \ker\psi_n & \to & \ker\psi_n/\operatorname{im}\phi_n & \to 0, \\ 0 \to \ker\psi_n & \to & \mathscr{F}_n & \to & \operatorname{im}\psi_n & \to 0, \\ 0 \to \operatorname{im}\psi_n & \to & \mathscr{F}''_n & \to & \mathscr{F}''_n/\operatorname{im}\psi_n & \to 0. \end{array}$$

Moreover, $(\ker\phi_n)$, $(\ker\psi_n/\operatorname{im}\phi_n)$, and $(\mathscr{F}''_n/\operatorname{im}\psi_n)$ are null systems. Hence $(R^if_!\ker\phi_n)$, $(R^if_!(\ker\psi_n/\operatorname{im}\phi_n))$, and $(R^if_!(\mathscr{F}''_n/\operatorname{im}\psi_n))$ are null systems. In particular, they are A-R λ-adic. Since $\mathscr{F}'$ is a torsion λ-adic sheaf, $(R^if_!\mathscr{F}'_n)$ are also A-R λ-adic. By 10.1.7 (iii) and the long exact sequences for $R^{\cdot}f_!$, we see that to prove that $(R^if_!\mathscr{F}_n)$ are A-R λ-adic, it suffices to prove that $(R^if_!\mathscr{F}''_n)$ are A-R λ-adic. But $(\mathscr{F}''_n) \in \operatorname{ob} D_c^b(X, R)$. By 10.1.17, we have $(Rf_!\mathscr{F}''_n) \in \operatorname{ob} D_c^b(Y, R)$. By 10.1.16, $(R^if_!\mathscr{F}''_n)$ are A-R λ-adic. □

Let $(\mathscr{D}_n, T_{n+1})_{n\geq 0}$ be a family such that $\mathscr{D}_n$ are triangulated categories, and $T_{n+1} : \mathscr{D}_{n+1} \to \mathscr{D}_n$ are exact functors. Define the inverse limit $\mathscr{D}$ of $(\mathscr{D}_n, T_n)_{n\geq 0}$ as follows: Objects in $\mathscr{D}$ are families $(K_n, u_n)_{n\geq 0}$, where $K_n \in \operatorname{ob}\mathscr{D}_n$, and u_n are isomorphisms $T_n(K_n) \stackrel{\cong}{\to} K_{n-1}$ in $\mathscr{D}_{n-1}$. Given objects $K = (K_n, u_n)$ and $L = (L_n, v_n)$ in $\mathscr{D}$, a morphism $f = (f_n) : K \to L$ consists of morphisms $f_n : K_n \to L_n$ in $\mathscr{D}_n$ such that the diagrams

$$\begin{array}{ccc} T_n(K_n) & \stackrel{T_n(f_n)}{\to} & T(L_n) \\ u_n \downarrow & & \downarrow v_n \\ K_{n-1} & \stackrel{f_{n-1}}{\to} & L_{n-1} \end{array}$$

commute for all n. A distinguished triangle in $\mathscr{D}$ is a sextuple

$$((K_n, u_n), (L_n, v_n), (M_n, w_n), (\phi_n), (\psi_n), (\delta_n))$$

such that

$$(K_n, u_n), \ (L_n, v_n), \ (M_n, w_n)$$

are objects in $\mathscr{D}$,

$$(\phi_n) : (K_n) \to (L_n), \quad (\psi_n) : (L_n) \to (M_n), \quad (\delta_n) : (M_n) \to (K_n[1])$$

are morphisms in $\mathscr{D}$, and

$$(K_n, L_n, M_n, \phi_n, \psi_n, \delta_n)$$

are distinguished triangles in $\mathscr{D}_n$ for all n.

Lemma 10.1.19. *Notation as above. If for any $n \geq 0$ and any $K, L \in \mathscr{D}_n$,* $\mathrm{Hom}(K, L)$ *is finite, then $\mathscr{D}$ is a triangulated category.*

Proof. Let $K = (K_n)$ and $L = (L_n)$ be objects in $\mathscr{D}$, and let $\phi = (\phi_n)$: $K \to L$ be a morphism in $\mathscr{D}$. For each n, by (TR1) for $\mathscr{D}_n$, we have a distinguished triangle

$$K_n \stackrel{\phi_n}{\to} L_n \to M_n \to$$

in $\mathscr{D}_n$, and by (TR3) for $\mathscr{D}_{n-1}$, we have an isomorphism of distinguished triangles

$$\begin{array}{ccccccc} T_n(K_n) & \stackrel{T(\phi_n)}{\to} & T_n(L_n) & \to & T_n(M_n) & \to \\ \cong\downarrow & & \cong\downarrow & & \cong\downarrow & \\ K_{n-1} & \stackrel{\phi_{n-1}}{\to} & L_{n-1} & \to & M_{n-1} & \to \end{array}$$

in $\mathscr{D}_{n-1}$. Then

$$(K_n) \stackrel{\phi}{\to} (L_n) \to (M_n) \to$$

is a distinguished triangle in $\mathscr{D}$. This proves (TR1) for $\mathscr{D}$. One can easily verify (TR2) for $\mathscr{D}$. Given a commutative diagram

$$\begin{array}{ccccccc} (K_n) & \to & (L_n) & \to & (M_n) & \to \\ {\scriptstyle (f_n)}\downarrow & & {\scriptstyle (g_n)}\downarrow & & & \\ (K'_n) & \to & (L'_n) & \to & (M'_n) & \to \end{array}$$

in $\mathscr{D}$, where the horizontal lines are distinguished triangles in $\mathscr{D}$, let S_n $(n \geq 0)$ be the sets of morphisms $h_n : M_n \to M'_n$ such that (f_n, g_n, h_n) are morphisms of triangles in $\mathscr{D}_n$:

$$\begin{array}{ccccccc} K_n & \to & L_n & \to & M_n & \to \\ {\scriptstyle f_n}\downarrow & & {\scriptstyle g_n}\downarrow & & {\scriptstyle h_n}\downarrow & \\ K'_n & \to & L'_n & \to & M'_n & \to . \end{array}$$

For each $h_n \in S_n$, let $\pi_n(h_n)$ be the element in S_{n-1} so that the following diagram commutes:

$$\begin{array}{ccc} T_n(M_n) & \stackrel{T_n(h_n)}{\to} & T(M'_n) \\ \cong\downarrow & & \downarrow\cong \\ M_{n-1} & \stackrel{\pi_n(h_n)}{\to} & M'_{n-1}. \end{array}$$

By our assumption and (TR3) for $\mathscr{D}_n$, S_n are nonempty finite sets. So $\varprojlim_n (S_n, \pi_n)$ is nonempty. Let (h_n) be an element in $\varprojlim_n (S_n, \pi_n)$. Then $((f_n), (g_n), (h_n))$ is a morphism of triangles. This proves (TR3) for $\mathscr{D}$. Similarly, one can verify (TR4) for $\mathscr{D}$. □

Lemma 10.1.20. *Let X be a scheme of finite type over a field k. Suppose that k is either separably closed or finite. Then for any $K, L \in \operatorname{ob} D^b_{\mathrm{ctf}}(X, R/(\lambda^{n+1}))$, $\operatorname{Hom}(K, L)$ is finite.*

Proof. Let $f : X \to \operatorname{Spec} k$ be the structure morphism. We have

$$\operatorname{Hom}(K, L) \cong R^0\Gamma(\operatorname{Spec} k, Rf_*R\mathscr{H}om(K, L)).$$

By 9.5.2 and 9.5.3, we have $Rf_*R\mathscr{H}om(K, L) \in \operatorname{ob} D^b_{\mathrm{ctf}}(\operatorname{Spec} k, R/(\lambda^{n+1}))$. When k is separably closed, we have

$$R^0\Gamma(\operatorname{Spec} k, Rf_*R\mathscr{H}om(K, L)) \cong \Gamma(\operatorname{Spec} k, R^0f_*R\mathscr{H}om(K, L))$$

which is finite. When k is a finite field, we have $\operatorname{Gal}(\bar{k}/k) \cong \widehat{\mathbb{Z}}$. By 4.3.7,

$$R^0\Gamma(\operatorname{Spec} k, Rf_*R\mathscr{H}om(K, L)) \cong H^0(\operatorname{Gal}(\bar{k}/k), Rf_*R\mathscr{H}om(K, L))$$

is finite. □

Corollary 10.1.21. *Let X be a scheme of finite type over a field k which is either separably closed or finite. Then $D^b_c(X, R)$ is a triangulated category.*

Let k be a separably closed field or a finite field, X and Y two k-schemes of finite type, and $f : X \to Y$ a k-compactifiable morphism. Then

$$Rf_*, Rf_! : D^b_c(X, R) \to D^b_c(Y, R),$$
$$f^*, Rf^! : D^b_c(Y, R) \to D^b_c(X, R),$$
$$- \otimes^L_R -, \quad R\mathscr{H}om(-,-) : D^b_c(X, R) \times D^b_c(X, R) \to D^b_c(X, R)$$

are exact functors. Given a distinguished triangle

$$(K_n) \to (L_n) \to (M_n) \to$$

in $D^b_c(Y, R)$, we have a long A-R exact sequence

$$\cdots \to (\mathscr{H}^i(K_n)) \to (\mathscr{H}^i(L_n)) \to (\mathscr{H}^i(M_n)) \to \cdots$$

of A-R λ-adic sheaves. Using these facts, one easily gets the long exact sequences for $R^{\cdot}f_*$ and $R^{\cdot}f_!$ associated to distinguished triangles in $D^b_c(X, R)$.

Let X be a noetherian scheme. Recall that E is the fraction field of R. In the category A-R λ-adic sheaves on X, morphisms defined by multiplication by λ^m $(m \geq 0)$ form a multiplicative system. By 6.2.1, we can construct a category so that its objects are A-R λ-adic sheaves, and for any two objects $\mathscr{F}$ and $\mathscr{G}$, morphisms from $\mathscr{F}$ to $\mathscr{G}$ are diagrams of the form

$$\begin{array}{ccc} \mathscr{F} & & \\ \lambda^m \downarrow & f \searrow & \\ \mathscr{F} & & \mathscr{G} \end{array}$$

such that $m \geq 0$ and $f : \mathscr{F} \to \mathscr{G}$ are morphisms of A-R λ-adic sheaves. We call this category the *category of E-sheaves.* Objects in this category are called E-sheaves. Denote the space of morphisms from $\mathscr{F}$ to $\mathscr{G}$ in this category by $\mathrm{Hom}_E(\mathscr{F}, \mathscr{G})$. Then we have

$$\mathrm{Hom}_{\mathrm{AR}}(\mathscr{F}, \mathscr{G}) \otimes_R E \cong \mathrm{Hom}_E(\mathscr{F}, \mathscr{G}).$$

Indeed, assign

$$f \otimes \frac{1}{\lambda^m} \in \mathrm{Hom}_{\mathrm{AR}}(\mathscr{F}, \mathscr{G}) \otimes_R E$$

to the morphism in $\mathrm{Hom}_E(\mathscr{F}, \mathscr{G})$ defined by the above diagram. This assignment is well-defined and surjective. If $f_1 \otimes \frac{1}{\lambda^{m_1}}$ and $f_2 \otimes \frac{1}{\lambda^{m_2}}$ are mapped to the same morphism in $\mathrm{Hom}_E(\mathscr{F}, \mathscr{G})$, where $f_1, f_2 \in \mathrm{Hom}_{\mathrm{AR}}(\mathscr{F}, \mathscr{G})$ and $m_1, m_2 \geq 0$, then there exist $f_3 \in \mathrm{Hom}_{\mathrm{AR}}(\mathscr{F}, \mathscr{G})$, $m_3 \geq 0$ and A-R morphisms $\phi_1, \phi_2 : \mathscr{F} \to \mathscr{F}$ such that

$$\begin{aligned} \lambda^{m_1}\phi_1 &= \lambda^{m_2}\phi_2 = \lambda^{m_3}, \\ f_1\phi_1 &= f_2\phi_2 = f_3. \end{aligned}$$

We have

$$\phi_1\lambda^{m_1} = \phi_2\lambda^{m_2} = \lambda^{m_3}.$$

So we have

$$\begin{aligned} f_1 \otimes \frac{1}{\lambda^{m_1}} &= f_1\lambda^{m_3} \otimes \frac{1}{\lambda^{m_1+m_3}} \\ &= f_1\phi_1\lambda^{m_1} \otimes \frac{1}{\lambda^{m_1+m_3}} \\ &= f_3\lambda^{m_1} \otimes \frac{1}{\lambda^{m_1+m_3}} \\ &= f_3 \otimes \frac{1}{\lambda^{m_3}}. \end{aligned}$$

Similarly, we have

$$f_2 \otimes \frac{1}{\lambda^{m_2}} = f_3 \otimes \frac{1}{\lambda^{m_3}}.$$

So we have

$$f_1 \otimes \frac{1}{\lambda^{m_1}} = f_2 \otimes \frac{1}{\lambda^{m_2}}.$$

Hence the above assignment is injective.

The category of E-sheaves is abelian. In this category, multiplications by λ^m $(m \geq 0)$ are isomorphisms. Since on a torsion sheaf, multiplication by some λ^m is 0, any torsion sheaf becomes a zero object in this category. Conversely, if an A-R λ-adic sheaf $\mathscr{F}$ is a zero object in the category of E-sheaves, then $\mathrm{id}_{\mathscr{F}} = 0$ in this category. This means that multiplication on $\mathscr{F}$ by some λ^m is 0. So $\mathscr{F}$ is a torsion sheaf. By 10.1.9, any E-sheaf is isomorphic to a torsion free λ-adic sheaf in the category of E-sheaves. An E-sheaf is called *lisse* if it is isomorphic to a lisse λ-adic sheaf in the category of E-sheaves.

If k is a separably closed field, then the A-R category of λ-adic sheaves on $\mathrm{Spec}\, k$ is equivalent to the category of finitely generated R-modules. For any finitely generated R-modules M and N, we have

$$\mathrm{Hom}_R(M, N) \otimes_R E \cong \mathrm{Hom}_E(M \otimes_R E, N \otimes_R E).$$

It follows that the functor $M \mapsto M \otimes_R E$ defines an equivalence between the category of E-sheaves on $\mathrm{Spec}\, k$ and the category of finite dimensional E-vector spaces.

Let X be a noetherian scheme, $\mathscr{F}$ an E-sheaf on X, and $s \to X$ a geometric point on X. Choose a λ-adic sheaf $\mathscr{F}'$ representing $\mathscr{F}$. We define the stalk of $\mathscr{F}$ at s to be $\mathscr{F}_s = \mathscr{F}'_s \otimes_R E$. A sequence

$$\mathscr{F} \to \mathscr{G} \to \mathscr{H}$$

in the category of E-sheaves is exact if and only if

$$\mathscr{F}_{\bar{x}} \to \mathscr{G}_{\bar{x}} \to \mathscr{H}_{\bar{x}}$$

is exact for any $x \in X$.

Let $\mathscr{F} = (\mathscr{F}_n)$ be a λ-adic sheaf on X. Consider the functor from the category of λ-adic sheaves to the category of E-sheaves that maps each λ-adic sheaf $\mathscr{G} = (\mathscr{G}_n)$ to $\mathscr{F} \otimes \mathscr{G} = (\mathscr{F}_n \otimes_{R/(\lambda^{n+1})} \mathscr{G}_n)$ regarded as an E-sheaf. This functor transforms multiplications by λ^m on $\mathscr{G}$ to isomorphisms in the category of E-sheaves. So it induces a functor on the category of E-sheaves. Applying the same argument to the first variable, we get the tensor product functor on the category of E-sheaves which maps E-sheaves represented by λ-adic sheaves $\mathscr{F} = (\mathscr{F}_n)$ and $\mathscr{G} = (\mathscr{G}_n)$ to the E-sheaf represented by the λ-adic sheaf $\mathscr{F} \otimes \mathscr{G} = (\mathscr{F}_n \otimes_{R/(\lambda^{n+1})} \mathscr{G}_n)$.

Let $\mathscr{F} = (\mathscr{F}_n)$ be a λ-adic sheaf such that $\mathscr{F}_n$ are locally free sheaves of $R/(\lambda^{n+1})$-modules of finite rank. Consider the functor from the category of λ-adic sheaves to the category of E-sheaves that maps each λ-adic sheaf $\mathscr{G} = (\mathscr{G}_n)$ to $\mathscr{H}om(\mathscr{F}, \mathscr{G}) = (\mathscr{H}om(\mathscr{F}_n, \mathscr{G}_n))$ regarded as an E-sheaf. This functor transforms multiplications by λ^m on $\mathscr{G}$ to isomorphisms in the category of E-sheaves. So it induces a functor on the category of E-sheaves. If $\mathscr{F}' = (\mathscr{F}'_n)$ is another λ-adic sheaf such that $\mathscr{F}'_n$ are locally free sheaves of $R/(\lambda^{n+1})$-modules of finite ranks, and if $\mathscr{F}' \to \mathscr{F}$ is an isomorphism in the category of E-sheaves, then we have an isomorphism

$$\mathscr{H}om(\mathscr{F}, \mathscr{G}) \to \mathscr{H}om(\mathscr{F}', \mathscr{G})$$

in the category of E-sheaves.

On the category $D_c^b(X, R)$, morphisms defined by multiplications by λ^m $(m \geq 0)$ form a multiplicative system. By 6.2.1, we can construct a category $D_c^b(X, E)$ so that its objects are objects in $D_c^b(X, R)$, and for any two objects K and L, morphisms from K to L are diagrams of the form

$$\begin{array}{ccc} K & & \\ \lambda^m \downarrow & f \searrow & \\ K & & \mathscr{L} \end{array}$$

such that $m \geq 0$ and $f : K \to L$ are morphisms in $D_c^b(X, R)$. For any $K, L \in \mathrm{ob}\, D_c^b(X, E)$, we have

$$\mathrm{Hom}_{D_c^b(X,R)}(K, L) \otimes_R E \cong \mathrm{Hom}_{D_c^b(X,E)}(K, L).$$

Let S be a noetherian regular scheme of dimension ≤ 1, X and Y two S-schemes of finite type, and $f : X \to Y$ an S-compactifiable morphism. Then we have functors

$$\begin{aligned} Rf_*, Rf_! &: D_c^b(X, E) \to D_c^b(Y, E), \\ f^*, Rf^! &: D_c^b(Y, E) \to D_c^b(X, E), \\ - \otimes_E^L -, \quad R\mathscr{H}om(-,-) &: D_c^b(X, E) \times D_c^b(X, E) \to D_c^b(X, E). \end{aligned}$$

If $S = \mathrm{Spec}\, k$ for a separably closed field or a finite field k, then $D_c^b(X, E)$ and $D_c^b(Y, E)$ are triangulated categories, and the above functors are exact. If X is a scheme of finite type over a trait S with generic point η and closed point s, then we have functors

$$R\Psi_\eta : D_c^b(X_\eta, E) \to D_c^b(X_{\bar{s}}, E), \quad R\Phi : D_c^b(X, E) \to D_c^b(X_{\bar{s}}, E).$$

Let E' be a finite extension of E, let R' be the integral closure of R in E', and let λ' be a uniformizer of R'. For any λ-adic sheaf $\mathscr{F} = (\mathscr{F}_n)$ on a noetherian scheme X, define an inverse system $\mathscr{F} \otimes_R R' = (\mathscr{F}'_n)$ as follows: Let e be the ramification index of R' over R. For any pair $n, i \geq 0$ such that

$$ie < n + 1 \leq (i+1)e,$$

we have

$$\lambda'^{n+1} R' \cap R = \lambda^{i+1} R.$$

We define

$$\mathscr{F}'_n = \mathscr{F}_i \otimes_{R/(\lambda^{i+1})} R'/(\lambda'^{n+1}).$$

Then $\mathscr{F} \otimes_R R'$ is a λ'-adic sheaf. We thus get a functor

$$\mathscr{F} \mapsto \mathscr{F} \otimes_R R'$$

from the category of λ-adic sheaves to the category of λ'-adic sheaves. This functor transforms multiplications by λ^m $(m \geq 0)$ to composites of multiplications by λ'^{me} with isomorphisms in the category of λ'-adic sheaves. So it induces a functor

$$\mathscr{F} \mapsto \mathscr{F} \otimes_E E'$$

from the category of E-sheaves to the category of E'-sheaves.

Let $K = (K_n)$ be an object in $D_c^b(X, R)$. Define $K \otimes_R^L R' = (K'_n)$ by

$$K'_n = K_i \otimes^L_{R/(\lambda^{i+1})} R'/(\lambda'^{n+1})$$

for any pair $n, i \geq 0$ such that

$$ie < n + 1 \leq (i+1)e.$$

Then $K \otimes_R^L R'$ is an object in $D_c^b(X, R')$. We thus get a functor

$$D_c^b(X, R) \to D_c^b(X, R'), \quad K \mapsto K \otimes_R^L R'.$$

It induces a functor

$$D_c^b(X, E) \to D_c^b(X, E'), \quad K \mapsto K \otimes_E^L E'.$$

Fix an algebraic closure of $\overline{\mathbb{Q}}_\ell$ of $\mathbb{Q}_\ell$. The direct limit of the categories of E-sheaves on a noetherian scheme X as E goes over the family of finite extensions of $\mathbb{Q}_\ell$ in $\overline{\mathbb{Q}}_\ell$ is called the *category of* $\overline{\mathbb{Q}}_\ell$*-sheaves*. Objects in this category are called $\overline{\mathbb{Q}}_\ell$-sheaves. Each $\overline{\mathbb{Q}}_\ell$-sheaf is represented by an E-sheaf for some finite extension E of $\mathbb{Q}_\ell$ in $\overline{\mathbb{Q}}_\ell$. Given two $\overline{\mathbb{Q}}_\ell$-sheaves represented by an E_1-sheaf $\mathscr{F}_1$ and an E_2-sheaf $\mathscr{F}_2$, the space of morphisms from $\mathscr{F}_1$ to $\mathscr{F}_2$ in the category of $\overline{\mathbb{Q}}_\ell$-sheaves is

$$\mathrm{Hom}_{\overline{\mathbb{Q}}_\ell}(\mathscr{F}_1, \mathscr{F}_2) = \varinjlim_{E} \mathrm{Hom}(\mathscr{F}_1 \otimes_{E_1} E, \mathscr{F}_2 \otimes_{E_2} E),$$

where the direct limit is taken over those finite extensions E of $\mathbb{Q}_\ell$ in $\overline{\mathbb{Q}}_\ell$ containing both E_1 and E_2. A $\overline{\mathbb{Q}}_\ell$-sheaf is called *lisse* if it is represented by a lisse E-sheaf for some finite extension E of $\mathbb{Q}_\ell$ in $\overline{\mathbb{Q}}_\ell$. Let $\mathscr{F}$ be a $\overline{\mathbb{Q}}_\ell$-sheaf on X and let s be a geometric point. Choose an E-sheaf $\mathscr{F}'$ representing $\mathscr{F}$. We define the stalk of $\mathscr{F}$ at s to be $\mathscr{F}_s = \mathscr{F}'_s \otimes_E \overline{\mathbb{Q}}_\ell$.

The direct limit of the categories $D_c^b(X, E)$ as E goes over the family of finite extensions of $\mathbb{Q}_\ell$ in $\overline{\mathbb{Q}}_\ell$ is denoted by $D_c^b(X, \overline{\mathbb{Q}}_\ell)$. Each object in this category is represented by an object in $D_c^b(X, E)$ for some finite extension E of $\mathbb{Q}_\ell$ in $\overline{\mathbb{Q}}_\ell$. Given two objects in this category, represented by $K_1 \in \mathrm{ob}\, D_c^b(X, E_1)$ and $K_2 \in \mathrm{ob}\, D_c^b(X, E_2)$, we have

$$\mathrm{Hom}_{D_c^b(X, \overline{\mathbb{Q}}_\ell)}(K_1, K_2) = \varinjlim_{E} \mathrm{Hom}(K_1 \otimes_{E_1}^L E, K_2 \otimes_{E_2}^L E),$$

where the direct limit is taken over those finite extensions E of $\mathbb{Q}_\ell$ in $\overline{\mathbb{Q}}_\ell$ containing both E_1 and E_2. Let S be a noetherian regular scheme of dimension ≤ 1, X and Y two S-schemes of finite type, and $f : X \to Y$ an S-compactifiable morphism. Then we have functors

$$\begin{aligned} Rf_*, Rf_! &: D_c^b(X, \overline{\mathbb{Q}}_\ell) \to D_c^b(Y, \overline{\mathbb{Q}}_\ell), \\ f^*, Rf^! &: D_c^b(Y, \overline{\mathbb{Q}}_\ell) \to D_c^b(X, \overline{\mathbb{Q}}_\ell), \\ - \otimes_{\overline{\mathbb{Q}}_\ell}^L -, \quad R\mathscr{H}om(-,-) &: D_c^b(X, \overline{\mathbb{Q}}_\ell) \times D_c^b(X, \overline{\mathbb{Q}}_\ell) \to D_c^b(X, \overline{\mathbb{Q}}_\ell). \end{aligned}$$

If $S = \mathrm{Spec}\, k$ for a separably closed field or a finite field k, then $D_c^b(X, \overline{\mathbb{Q}}_\ell)$ and $D_c^b(Y, \overline{\mathbb{Q}}_\ell)$ are triangulated categories, and the above functors are exact. If X is a scheme of finite type over a trait S with generic point η and closed point s, then we have functors

$$R\Psi_\eta : D_c^b(X_\eta, \overline{\mathbb{Q}}_\ell) \to D_c^b(X_{\bar{s}}, \overline{\mathbb{Q}}_\ell), \quad R\Phi : D_c^b(X, \overline{\mathbb{Q}}_\ell) \to D_c^b(X_{\bar{s}}, \overline{\mathbb{Q}}_\ell).$$

Let M be a finitely generated R-module provided with the λ-adic topology. Any R-endomorphism $\phi : M \to M$ is continuous. We have $M \cong \varprojlim_n M/\lambda^{n+1}M$. So we have

$$\mathrm{Aut}_R(M) \cong \varprojlim_n \mathrm{Aut}_{R/(\lambda^{n+1})}(M/\lambda^{n+1}M).$$

Put the discrete topology on $\mathrm{Aut}_{R/(\lambda^{n+1})}(M/\lambda^{n+1}M)$, and put the inverse limit topology on $\mathrm{Aut}_R(M)$. In the case where $M = R^k$, there is another way to define a topology on $\mathrm{Aut}_R(R^k)$. Representing each element in

$\mathrm{Aut}_R(R^k)$ by the matrix with respect to the standard basis of R^k, we can embed $\mathrm{Aut}_R(R^k)$ into R^{k^2}. Put on $\mathrm{Aut}_R(R^k)$ the topology induced from the product topology on R^{k^2}. These two topologies on $\mathrm{Aut}_R(R^k)$ are the same since the family

$$\{A|A \in \mathrm{Aut}_R(R^k),\ A \equiv I \bmod \lambda^{n+1}\} \quad (n \in \mathbb{N})$$

are base of neighborhoods at the identity for both topologies.

Lemma 10.1.22.

(i) *Let M be a finitely generated R-module and let G be a topological group acting linearly on M. Then the homomorphism $G \to \mathrm{Aut}_R(M)$ is continuous if and only if the action $G \times M \to M$ is continuous.*

(ii) *Let G be a topological group acting linearly on E^k. Then the homomorphism $G \to \mathrm{GL}(E^k)$ is continuous if and only if the action $G \times E^k \to E^k$ is continuous.*

Proof.

(i) The canonical maps

$$\mathrm{Aut}_{R/(\lambda^{n+1})}(M/\lambda^{n+1}M) \times M/\lambda^{n+1}M \to M/\lambda^{n+1}M$$

are continuous with respect to the discrete topology for all n. It follows that

$$\varprojlim_n \mathrm{Aut}_{R/(\lambda^{n+1})}(M/\lambda^{n+1}M) \times \varprojlim_n M/\lambda^{n+1}M \to \varprojlim_n M/\lambda^{n+1}M$$

is continuous, that is, $\mathrm{Aut}_R(M) \times M \to M$ is continuous. If $G \to \mathrm{Aut}_R(M)$ is continuous, then the action $G \times M \to M$ is continuous.

We have commutative diagrams

$$\begin{array}{ccc} G \times M & \to & M \\ \downarrow & & \downarrow \\ G \times M/\lambda^{n+1}M & \to & M/\lambda^{n+1}M. \end{array}$$

The vertical arrows are surjective open maps. If $G \times M \to M$ is continuous, then $G \times M/\lambda^{n+1}M \to M/\lambda^{n+1}M$ are continuous. Since $M/\lambda^{n+1}M$ are finite and discrete, the homomorphisms $G \to \mathrm{Aut}_{R/(\lambda^{n+1})}(M/\lambda^{n+1}M)$ are continuous. So $G \to \mathrm{Aut}_R(M)$ is continuous.

(ii) The canonical map $\mathrm{GL}(E^k) \times E^k \to E^k$ is continuous. If $G \to \mathrm{GL}(E^k)$ is continuous, then $G \times E^k \to E^k$ is continuous.

Let $\{e_1, \ldots, e_n\}$ be the standard basis of E^k, let $g \in G$, let $(a_{ij}) \in \mathrm{GL}(E^k)$ be the matrix of g with respect to the standard basis, and let U_{ij}

be a neighborhood of a_{ij} in E for each pair (i,j). If $G \times E^k \to E^k$ is continuous. then the set

$$G_i = \{h \in G | he_i \in U_{1i} \times \cdots \times U_{ki}\}$$

is open in G and contains g. The map $G \to \mathrm{GL}(E^k)$ maps the open neighborhood $G_1 \cap \cdots \cap G_k$ of g in G to the set $\mathrm{GL}(E^k) \cap \prod_{(i,j)} U_{ij}$. So $G \to \mathrm{GL}(E^k)$ is continuous. □

Proposition 10.1.23. *Let X be a connected noetherian scheme and let x be a point in X. The functor $\mathscr{F} \mapsto \mathscr{F}_{\bar{x}}$ defines an equivalence between the category of lisse λ-adic sheaves (resp. lisse E-sheaves) on X and the category of finitely generated R-modules (resp. finite dimensional E-vector spaces) with continuous $\pi_1(X, \bar{x})$-actions.*

Proof. Let $\mathscr{F} = (\mathscr{F}_n)$ be a lisse λ-adic sheaf. Each $\mathscr{F}_n$ gives rise to a continuous homomorphism

$$\pi_1(X, \bar{x}) \to \mathrm{Aut}_{R/(\lambda^{n+1})}(\mathscr{F}_{n,\bar{x}}).$$

We have $\mathscr{F}_{n,\bar{x}} \cong \mathscr{F}_{\bar{x}}/\lambda^{n+1}\mathscr{F}_{\bar{x}}$. So we have a continuous homomorphism

$$\pi_1(X, \bar{x}) \to \varprojlim_n \mathrm{Aut}_{R/(\lambda^{n+1})}(\mathscr{F}_{\bar{x}}/\lambda^{n+1}\mathscr{F}_{\bar{x}}) \cong \mathrm{Aut}_R(\mathscr{F}_{\bar{x}}).$$

Conversely, if M is a finitely generated R-module and

$$\pi_1(X, \bar{x}) \to \mathrm{Aut}_R(M) \cong \varprojlim_n \mathrm{Aut}_{R/(\lambda^{n+1})}(M/\lambda^{n+1}M)$$

is a continuous homomorphism, then the continuous representations

$$\pi_1(X, \bar{x}) \to \mathrm{Aut}_{R/(\lambda^{n+1})}(M/\lambda^{n+1}M)$$

define locally constant constructible sheaves of $R/(\lambda^{n+1})$-modules $\mathscr{F}_n$ on X, and $\mathscr{F} = (\mathscr{F}_n)$ is a λ-adic sheaf with $\mathscr{F}_{\bar{x}} \cong M$. This proves that the category of lisse λ-adic sheaves is equivalent to the category of finitely generated R-modules with continuous $\pi_1(X, \bar{x})$-actions.

Let $\mathbf{Rep}_R$ (resp. $\mathbf{Rep}_E$) be the category of finitely generated R-modules (resp. finite dimensional E-vector spaces) with continuous $\pi_1(X, \bar{x})$-actions, and let S be the multiplicative system in $\mathbf{Rep}_R$ consisting of multiplications by λ^m $(m \geq 0)$ on objects in $\mathbf{Rep}_R$. Then the category $S^{-1}\mathbf{Rep}_R$ is equivalent to the category of lisse E-sheaves on X. Note that if M is a finitely generated R-module with a continuous $\pi_1(X, \bar{x})$-action, then $M \otimes_R E$ is a finite dimensional E-vector space with a continuous $\pi_1(X, \bar{x})$-action. Indeed, let

$$M_t = \{z \in M | \lambda^n z = 0 \text{ for some } n \geq 0\}$$

be the torsion submodule of M. Then M_t is stable under the action of $\pi_1(X, \bar{x})$. On the other hand, the canonical homomorphism

$$\mathrm{Aut}_R(M) \to \mathrm{Aut}_R(M/M_t)$$

is continuous. It follows that M/M_t is a free R-module of finite rank with a continuous $\pi_1(X, \bar{x})$-action. We have

$$M \otimes_R E \cong M/M_t \otimes_R E.$$

Choose a basis for M/M_t. It induces a basis for $M \otimes_R E$. Let k be the rank of M/M_t. The composite of the canonical homomorphism $\mathrm{GL}(R^k) \to \mathrm{GL}(E^k)$ and the homomorphism $\pi_1(X, \bar{x}) \to \mathrm{GL}(R^k)$ which is defined by the action of $\pi_1(X, \bar{x})$ on M/M_t using the chosen basis of M/M_t, coincides with the homomorphism $\pi_1(X, \bar{x}) \to \mathrm{GL}(E^k)$ defined by the action of $\pi_1(X, \bar{x})$ on $M \otimes_R E$ using the corresponding basis on $M \otimes_R E$. It follows that $\pi_1(X, \bar{x}) \to \mathrm{GL}(E^k)$ is continuous and hence the action of $\pi_1(X, \bar{x})$ on $M \otimes_R E$ is continuous. We thus get a functor

$$\mathbf{Rep}_R \to \mathbf{Rep}_E, \quad M \mapsto M \otimes_R E.$$

It transforms multiplications by λ^m $(m \geq 0)$ to isomorphisms in $\mathbf{Rep}_E$. So it induces a functor $S^{-1}\mathbf{Rep}_R \to \mathbf{Rep}_E$. To prove that the category of lisse E-sheaves is equivalent to the category of finite dimensional E-vector spaces with continuous $\pi_1(X, \bar{x})$-actions, it suffices to show that the functor $S^{-1}\mathbf{Rep}_R \to \mathbf{Rep}_E$ is an equivalence of categories.

Note that any finitely generated R-module M with a continuous $\pi_1(X, \bar{x})$-action is isomorphic to M/M_t in $S^{-1}\mathbf{Rep}_R$. Indeed, there exists an integer $m \geq 0$ such that $\lambda^m M_t = 0$. The canonical projection $M \to M/M_t$ and the homomorphism $M/M_t \to M$ induced by $\lambda^m : M \to M$ define an isomorphism $M \cong M/M_t$ in $S^{-1}\mathbf{Rep}_R$. So to prove that the functor $S^{-1}\mathbf{Rep}_R \to \mathbf{Rep}_E$ is fully faithful, it suffices to show that for any free R-modules M and N of finite ranks with continuous $\pi_1(X, \bar{x})$-actions, we have

$$\mathrm{Hom}_{\mathbf{Rep}_R}(M, N) \otimes_R E \cong \mathrm{Hom}_{\mathbf{Rep}_E}(M \otimes_R E, N \otimes_R E).$$

Note that the canonical homomorphisms $M \to M \otimes_R E$ and $N \to N \otimes_R E$ are injective. Regard M and N as subspaces of $M \otimes_R E$ and $N \otimes_R E$ respectively through these monomorphisms. For any morphism

$$\phi : M \otimes_R E \to N \otimes_R E$$

in $\mathbf{Rep}_E$, there exists an integer $m \geq 0$ such that $\lambda^m \phi(M) \subset N$. Then ϕ is the image of $\lambda^m \phi \otimes \frac{1}{\lambda^m} \in \mathrm{Hom}_{\mathbf{Rep}_R}(M, N) \otimes_R E$. If a morphism

$\psi : M \to N$ in $\mathrm{Hom}_{\mathbf{Rep}_R}(M, N)$ has the property $\psi \otimes \mathrm{id}_E = 0$, then $\psi = 0$. This proves that the functor is fully faithful.

Let V be a finite dimensional E-vector space with a continuous $\pi_1(X, \bar{x})$-action. Fix a basis $\{e_1, \ldots, e_k\}$ of V and let $L = Re_1 + \cdots + Re_k$. If we identify $\mathrm{GL}(V)$ with $\mathrm{GL}(E^k)$ through this basis, then the subgroup $\{g \in \mathrm{GL}(V) | gL = L\}$ of $\mathrm{GL}(V)$ is identified with $\mathrm{GL}(R^k)$ and hence is an open subgroup. So the subgroup $H = \{g \in \pi_1(X, \bar{x}) | gL = L\}$ is open in $\pi_1(X, \bar{x})$, and hence the set $\pi_1(X, \bar{x})/H$ is finite. Let $g_1, \ldots, g_m \in \pi_1(X, \bar{x})$ so that

$$\pi_1(X, \bar{x})/H = \{g_1H, \ldots, g_mH\},$$

and let $M = g_1L + \cdots + g_mL$. For any $g \in \pi_1(X, \bar{x})$ and any $i \in \{1, \ldots, m\}$, there exist $j \in \{1, \ldots, m\}$ and $h \in H$ such that $gg_i = g_jh$. We then have $gg_iL = g_jL$. Hence M is stable under the action of $\pi_1(X, \bar{x})$. It is a free R-module of finite rank with a continuous $\pi_1(X, \bar{x})$-action and $M \otimes_R E \cong V$. So the functor $S^{-1}\mathbf{Rep}_R \to \mathbf{Rep}_E$ is essentially surjective. □

A $\overline{\mathbb{Q}}_\ell$-representation for $\pi_1(X, \bar{x})$ is a homomorphism $\pi_1(X, \bar{x}) \to \mathrm{GL}(V)$ for some finite dimensional $\overline{\mathbb{Q}}_\ell$-vector space V such that we can find a finite extension E of $\mathbb{Q}_\ell$ in $\overline{\mathbb{Q}}_\ell$ and a finite dimensional E-vector space V_E with a continuous $\pi_1(X, \bar{x})$-action with the property that $V \cong V \otimes_E \overline{\mathbb{Q}}_\ell$ and that the homomorphism $\pi_1(X, \bar{x}) \to \mathrm{GL}(V)$ is the composite

$$\pi_1(X, \bar{x}) \to \mathrm{GL}(V_E) \to \mathrm{GL}(V_E \otimes_E \overline{\mathbb{Q}}_\ell) \cong \mathrm{GL}(V).$$

Corollary 10.1.24. *Let X be a noetherian connected scheme and let x be a point in X. The functor $\mathscr{F} \to \mathscr{F}_{\bar{x}}$ defines an equivalence between the category of lisse $\overline{\mathbb{Q}}_\ell$-sheaves and the category of $\overline{\mathbb{Q}}_\ell$-representations of $\pi_1(X, \bar{x})$.*

10.2 Grothendieck–Ogg–Shafarevich Formula

([SGA 5] VIII, X, [SGA $4\frac{1}{2}$] Rapport 4.1–4.4.)

Suppose that Λ is a ring with the identity element 1 but not necessarily commutative. Let $\Lambda^\natural$ be the quotient of the additive group $(\Lambda, +)$ by the subgroup generated by elements of the form $ab - ba$ $(a, b \in \Lambda)$. For any endomorphism $f : \Lambda^n \to \Lambda^n$ on a free (left) Λ-module Λ^n, if f is represented by the matrix (a_{ij}) with respect to the standard basis of Λ^n, we define the *trace* $\mathrm{Tr}(f) \in \Lambda^\natural$ of f to be the image of $\sum\limits_{i=1}^{n} a_{ii}$ in $\Lambda^\natural$. If $f : \Lambda^n \to \Lambda^m$ and

$g : \Lambda^m \to \Lambda^n$ are two homomorphisms between free Λ-modules, we have $\mathrm{Tr}(fg) = \mathrm{Tr}(gf)$.

Let $f : P \to P$ be an endomorphism on a finitely generated projective Λ-module P. We can find an integer $n \geq 0$ and homomorphisms $a : P \to \Lambda^n$ and $b : \Lambda^n \to P$ such that $ba = \mathrm{id}$. We define $\mathrm{Tr}(f)$ to be $\mathrm{Tr}(afb)$. This definition does not depend on the choices of a and b. Indeed, if $c : P \to \Lambda^m$ and $d : \Lambda^m \to P$ are homomorphisms such that $dc = \mathrm{id}$. Then we have

$$\mathrm{Tr}(afb) = \mathrm{Tr}(adcfb) = \mathrm{Tr}(cfbad) = \mathrm{Tr}(cfd).$$

Let P and Q be two finitely generated projective Λ-modules and let $f : P \to Q$ and $g : Q \to P$ be two homomorphisms. Then we have $\mathrm{Tr}(fg) = \mathrm{Tr}(gf)$.

Let $P^\cdot$ be a bounded complex of finitely generated projective Λ-modules, and let $f = (f^i) : P^\cdot \to P^\cdot$ be a morphism of complexes. We define

$$\mathrm{Tr}(f) = \sum_i (-1)^i \mathrm{Tr}(f^i).$$

If f is homotopic to 0, then $\mathrm{Tr}(f) = 0$. Indeed, if $h^i : P^i \to P^{i-1}$ are homomorphisms such that $h^{i+1}d^i + d^{i-1}h^i = f^i$, then we have

$$\begin{aligned}
\mathrm{Tr}(f) &= \sum_i (-1)^i \mathrm{Tr}(f^i) \\
&= \sum_i (-1)^i \mathrm{Tr}(h^{i+1}d^i + d^{i-1}h^i) \\
&= \sum_i (-1)^i \mathrm{Tr}(h^{i+1}d^i) + \sum_i (-1)^i \mathrm{Tr}(d^{i-1}h^i) \\
&= \sum_i (-1)^i \mathrm{Tr}(d^i h^{i+1}) + \sum_i (-1)^i \mathrm{Tr}(d^{i-1}h^i) \\
&= 0.
\end{aligned}$$

Any bounded complex of finitely generated projective Λ-modules is called *perfect*. Let $K_{\mathrm{perf}}(\Lambda)$ be the category whose objects are perfect complexes and whose morphisms are homotopy classes of morphisms between complexes. By the dual version of 6.2.8, the functor $K_{\mathrm{perf}}(\Lambda) \to D^b(\Lambda)$ is fully faithful, where $D^b(\Lambda)$ is the derived category of bounded complexes of Λ-modules. Let $D^b_{\mathrm{perf}}(\Lambda)$ be the essential image of this functor. By the above discussion, for any endomorphism $f : K \to K$ in $D^b_{\mathrm{perf}}(\Lambda)$, we can define $\mathrm{Tr}(f)$.

Proposition 10.2.1. *Let Λ be a left noetherian ring with the identity element and let K be an object in $D^-(\Lambda)$. Then $K \in \mathrm{ob}\, D^b_{\mathrm{perf}}(\Lambda)$ if and only if $H^i(K)$ are finitely generated Λ-modules and K has finite* Tor*-dimension.*

Proof. Use the same argument as 6.4.6 and 10.2.2 below. □

Lemma 10.2.2. *Let Λ be a ring with the identity element, and let P be a flat Λ-module with finite presentation. Then P is projective.*

Proof. Let $M \to N$ be an epimorphism of Λ-modules. We need to show the map

$$\mathrm{Hom}_\Lambda(P, M) \to \mathrm{Hom}_\Lambda(P, N)$$

is surjective. Suppose this is not true. Embed the cokernel of this homomorphism into an injective $\mathbb{Z}$-module I. Then the homomorphism

$$\mathrm{Hom}_\mathbb{Z}(\mathrm{Hom}_\Lambda(P, N), I) \to \mathrm{Hom}_\mathbb{Z}(\mathrm{Hom}_\Lambda(P, M), I)$$

is not injective. We have a commutative diagram

$$\begin{array}{ccc} \mathrm{Hom}_\mathbb{Z}(N, I) \otimes_\Lambda P & \to & \mathrm{Hom}_\mathbb{Z}(M, I) \otimes_\Lambda P \\ \downarrow & & \downarrow \\ \mathrm{Hom}_\mathbb{Z}(\mathrm{Hom}_\Lambda(P, N), I) & \to & \mathrm{Hom}_\mathbb{Z}(\mathrm{Hom}_\Lambda(P, M), I). \end{array}$$

The lower horizontal arrow is not injective. Since $M \to N$ is surjective and P is flat, the top horizontal arrow is injective. We will show that the vertical arrows are isomorphisms. We thus get a contradiction. Choose an exact sequence

$$\Lambda^m \to \Lambda^n \to P \to 0.$$

Consider the commutative diagram

$$\begin{array}{ccccccc} \mathrm{Hom}_\mathbb{Z}(M,I)\otimes_\Lambda \Lambda^m & \to & \mathrm{Hom}_\mathbb{Z}(M,I)\otimes_\Lambda \Lambda^n & \to & \mathrm{Hom}_\mathbb{Z}(M,I)\otimes_\Lambda P & \to 0 \\ \downarrow & & \downarrow & & \downarrow & \\ \mathrm{Hom}_\mathbb{Z}(\mathrm{Hom}_\Lambda(\Lambda^m,M),I) & \to & \mathrm{Hom}_\mathbb{Z}(\mathrm{Hom}_\Lambda(\Lambda^n,M),I) & \to & \mathrm{Hom}_\mathbb{Z}(\mathrm{Hom}_\Lambda(P,M),I) & \to 0. \end{array}$$

The two horizontal lines are exact. The first two vertical arrows are isomorphisms. So

$$\mathrm{Hom}_\mathbb{Z}(M, I) \otimes_\Lambda P \to \mathrm{Hom}_\mathbb{Z}(\mathrm{Hom}_\Lambda(P, M), I)$$

is an isomorphism. □

Let $\mathscr{A}$ be an abelian category. Define the category $F(\mathscr{A})$ of finitely filtered objects as follows: Objects in $F(\mathscr{A})$ are objects A in $\mathscr{A}$ provided with decreasing filtrations

$$\cdots \supset F^i A \supset F^{i+1} A \supset \cdots$$

such that $F^i A = A$ for $i \ll 0$ and $F^i A = 0$ for $i \gg 0$. Given two objects $(A, F^\cdot A)$ and $(B, F^\cdot B)$ in $F(\mathscr{A})$, a morphism $f : (A, F^\cdot A) \to (B, F^\cdot B)$ in

$F(\mathscr{A})$ is a morphism $f : A \to B$ in $\mathscr{A}$ such that $f(F^iA) \subset F^iB$ for all i. Then $F(\mathscr{A})$ is an additive category. Taking the i-th successive quotient defines a functor

$$\mathrm{Gr}^i : F(\mathscr{A}) \to \mathscr{A}, \quad (A, F^\cdot) \mapsto \mathrm{Gr}^i(A) = F^iA/F^{i+1}A.$$

Define the category $KF(\mathscr{A})$ of finitely filtered complexes as follows: Objects in $KF(\mathscr{A})$ are complexes (K, F) of objects in $F(\mathscr{A})$ so that there exist i_0 and i_1 with the property that $F^iK^j = K^j$ for all $i \leq i_0$ and all j, and $F^iK^j = 0$ for all $i \geq i_1$ and all j. Morphisms in $KF(\mathscr{A})$ are morphisms of complexes preserving filtrations, and two morphisms of complexes are considered to be the same morphism in $KF(\mathscr{A})$ if there exists a homotopy that preserves filtrations between these two morphisms. Then $KF(\mathscr{A})$ is a triangulated category. Let $f : K \to L$ be a morphism in $KF(\mathscr{A})$. We say that f is a quasi-isomorphism if f induces quasi-isomorphisms

$$\mathrm{Gr}^i(f) : \mathrm{Gr}^i(K) \to \mathrm{Gr}^i(L)$$

for all i. Let S be the family of all quasi-isomorphisms in $KF(\mathscr{A})$. We define the derived category of finitely filtered complexes $DF(\mathscr{A})$ to be $S^{-1}(KF(\mathscr{A}))$.

Take $\mathscr{A}$ to be the category of Λ-modules. Let $KF_{\mathrm{perf}}(\Lambda)$ be the full subcategory of $KF(\mathscr{A})$ consisting of those finitely filtered bounded complexes (K, F) such that $\mathrm{Gr}^i(K^j)$ are finitely generated projective A-modules for all i and all j. Denote $DF(\mathscr{A})$ by $DF(\Lambda)$. Then the canonical functor

$$KF_{\mathrm{perf}}(\Lambda) \to DF(\Lambda)$$

is fully faithful. Denote its essential image by $DF_{\mathrm{perf}}(\Lambda)$. An object (K, F) in $DF(\Lambda)$ lies in $DF_{\mathrm{perf}}(\Lambda)$ if and only if $\mathrm{Gr}^i(K)$ are objects in $D^b_{\mathrm{perf}}(\Lambda)$ for all i. If (K, F) is an object in $DF_{\mathrm{perf}}(\Lambda)$, then K is an object in $D^b_{\mathrm{perf}}(\Lambda)$. Let $f : (K, F) \to (K, F)$ be an endomorphism of an object (K, F) in $DF_{\mathrm{perf}}(\Lambda)$, then we have

$$\mathrm{Tr}(f, K) = \sum_i \mathrm{Tr}(f, \mathrm{Gr}^i(K)).$$

Let X be a scheme and let A be a commutative noetherian ring such that $nA = 0$ for some integer n. Take $\mathscr{A}$ to be the category of sheaves of A-modules on X, and denote $DF(\mathscr{A})$ by $DF(X, A)$. For any bounded below finitely filtered complex K of sheaves of A-modules on X, the complex $\mathscr{C}^\cdot(K)$ defined by the Godement resolution is a bounded below finitely filtered complex of flasque sheaves of A-modules and each $\mathrm{Gr}^i(\mathscr{C}^\cdot(K))$ is isomorphic to the Godement resolution of $\mathrm{Gr}^i(K)$. We can define the right derived functor

$$R\Gamma(X,-): D^+F(X,A)\to DF(A)$$

of the functor $\Gamma(X,-)$ and we have

$$\begin{aligned} R\Gamma(X,K) &\cong \Gamma(X,\mathscr{C}^{\cdot}(K)),\\ \mathrm{Gr}^i R\Gamma(X,K) &\cong R\Gamma(X,\mathrm{Gr}^i(K)). \end{aligned}$$

If X is a compactifiable scheme over an algebraically closed field, and let $j: X\hookrightarrow \overline{X}$ be a compactification of X, then we have a functor

$$R\Gamma_c(X,-): D^+F(X,A)\to DF(A)$$

defined by

$$R\Gamma_c(X,-) = R\Gamma(\overline{X}, j_!-).$$

Suppose that X is a compactifiable scheme over an algebraically closed field k. Let $j: U\to X$ an open immersion, $l: X-U\to X$ a closed immersion, and $K\in \mathrm{ob}\, D^b_{\mathrm{ctf}}(X,A)$. Provide K with the following filtration

$$F^iK = \begin{cases} K & \text{if } i\leq 0,\\ j_!j^*K & \text{if } i=1,\\ 0 & \text{if } i\geq 2. \end{cases}$$

Then K can be considered as an object in $D^+F(X,A)$, and we have

$$\mathrm{Gr}^i(K)\cong \begin{cases} l_*l^*K & \text{if } i=0,\\ j_!j^*K & \text{if } i=1,\\ 0 & \text{if } i\neq 0,1. \end{cases}$$

So we have

$$\mathrm{Gr}^i R\Gamma_c(X,K)\cong R\Gamma_c(X,\mathrm{Gr}^i(K))\cong \begin{cases} R\Gamma_c(X-U, K|_{X-U}) & \text{if } i=0,\\ R\Gamma_c(U,K|_U) & \text{if } i=1,\\ 0 & \text{if } i\neq 0,1. \end{cases}$$

By 10.2.1, 7.4.7 (ii) and 7.8.1, $R\Gamma_c(X-U,K|_{X-U})$ and $R\Gamma_c(U,K|_U)$ are objects in $D^b_{\mathrm{perf}}(A)$. So $R\Gamma_c(X,K)$ is an object in $DF_{\mathrm{perf}}(A)$. Let $f: X\to X$ be a k-morphism such that $f^{-1}(U)=U$, and let $f^*: f^*K\to K$ be a morphism of complexes. Then f induces a morphism $f^*K\to K$ in $DF(X,A)$ and hence an endomorphism

$$f^*: R\Gamma_c(X,K)\to R\Gamma_c(X,K)$$

in $DF_{\mathrm{perf}}(A)$. We have

$$\begin{aligned}\mathrm{Tr}(f^*, R\Gamma_c(X,K)) &= \sum_i \mathrm{Tr}(f^*, \mathrm{Gr}^i R\Gamma_c(X,K)) \\ &= \mathrm{Tr}(f^*, R\Gamma_c(U, K|_U)) + \mathrm{Tr}(f^*, R\Gamma_c(X-U, K|_{X-U})).\end{aligned}$$

Suppose that $\mathscr{D}$ is a triangulated category so that isomorphic classes of objects in $\mathscr{D}$ form a set. Let L be the free abelian group generated by isomorphic classes of objects in $\mathscr{D}$. For any object X in $\mathscr{D}$, let (X) be the element in L corresponding to the isomorphic class of X. Let R be the subgroup of L generated by elements of the form $(X)-(Y)+(Z)$ whenever there exists a distinguished triangle

$$X \to Y \to Z \to .$$

Define the *K-group* of $\mathscr{D}$ to be $K(\mathscr{D}) = L/R$. For any $X \in \mathrm{ob}\,\mathscr{D}$, denote by $[X]$ the image of (X) in $K(\mathscr{D})$. For any abelian group G and any map

$$\phi : \{\text{isomorphic classes of objects in } \mathscr{D}\} \to G$$

with the property

$$\phi((X)) - \phi((Y)) + \phi((Z)) = 0$$

for any distinguished triangle

$$X \to Y \to Z \to,$$

there exists a unique homomorphism $\psi : K(\mathscr{D}) \to G$ such that $\psi([X]) = \phi((X))$. If

$$\mathscr{D} \to \mathscr{D}', \quad \mathscr{D} \times \mathscr{D}' \to \mathscr{D}''$$

are exact functors for some triangulated categories $\mathscr{D}$, $\mathscr{D}'$ and $\mathscr{D}''$ so that isomorphic classes of objects in these categories form sets, then these functors induce homomorphisms

$$K(\mathscr{D}) \to K(\mathscr{D}'), \quad K(\mathscr{D}) \times K(\mathscr{D}') \to K(\mathscr{D}'').$$

Let Λ be a ring with the identity element and let L' be the free abelian group generated by isomorphic classes of finitely generated projective Λ-modules. For any finitely generated projective Λ-module M, let (M) be the element in L' corresponding to the isomorphic class of M. Let R' be the subgroup of L' generated by elements of the form $(M') - (M) + (M'')$ whenever there exists a short exact sequence of finitely generated projective Λ-modules

$$0 \to M' \to M \to M'' \to 0.$$

Define the *K-group* of Λ to be $K(\Lambda) = L'/R'$. For any finitely generated projective Λ-module M, denote by $[M]$ the image of (M) in $K(\Lambda)$.

Proposition 10.2.3. *Let Λ be a ring with the identity element. We have an isomorphism $K(D^b_{\mathrm{perf}}(\Lambda)) \cong K(\Lambda)$.*

Proof. We know $D^b_{\mathrm{perf}}(\Lambda)$ is equivalent to the category $K_{\mathrm{perf}}(\Lambda)$ of perfect complexes. It suffices to show that $K(K_{\mathrm{perf}}(\Lambda)) \cong K(\Lambda)$. For any perfect complex

$$P^{\cdot} = (\cdots \to P^i \to P^{i+1} \to \cdots),$$

let $\phi(P^{\cdot}) = \sum\limits_i (-1)^i [P^i] \in K(\Lambda)$. We claim that $\phi(P^{\cdot})$ only depends on the isomorphic class of $P^{\cdot}$ in $K_{\mathrm{perf}}(\Lambda)$. Let $f : P^{\cdot} \to Q^{\cdot}$ be an isomorphism in $K_{\mathrm{perf}}(\Lambda)$ and let $C^{\cdot}$ be the mapping cone of f. We have $C^i = P^i \oplus Q^{i+1}$ for all i. It follows that $\phi(P^{\cdot}) - \phi(Q^{\cdot}) = \phi(C^{\cdot})$. To prove our assertion, it suffices to show $\phi(C^{\cdot}) = 0$. Note that $C^{\cdot}$ is a bounded acyclic complex of projective Λ-modules. Suppose that it is of the form

$$0 \to C^a \to C^{a+1} \to \cdots \to C^{a+m} \to 0.$$

Let $Z^i = \mathrm{Ker}\,(C^i \to C^{i+1})$. We have short exact sequences

$$\begin{array}{ccccccc}
0 \to & Z^{a+m-1} & \to & C^{a+m-1} & \to & C^{a+m} & \to 0, \\
0 \to & Z^{a+m-2} & \to & C^{a+m-2} & \to & Z^{a+m-1} & \to 0, \\
 & & & \vdots & & & \\
0 \to & Z^{a+2} & \to & C^{a+2} & \to & Z^{a+3} & \to 0, \\
0 \to & C^{a} & \to & C^{a+1} & \to & Z^{a+2} & \to 0.
\end{array}$$

One can show that Z^i are finitely generated projective Λ-modules. We have

$$\begin{aligned}
\phi(C^{\cdot}) &= \sum_i (-1)^i [C^i] \\
&= \sum_i (-1)^i ([Z^i] + [Z^{i+1}]) \\
&= \sum_i (-1)^i [Z^i] + \sum_i (-1)^i [Z^{i+1}] \\
&= 0.
\end{aligned}$$

This proves our assertion. So ϕ defines a map on the set of isomorphic classes of objects in $K_{\mathrm{perf}}(\Lambda)$. Any distinguished triangle in $K_{\mathrm{perf}}(\Lambda)$ is isomorphic to a distinguished triangle

$$P_1^{\cdot} \to P_2^{\cdot} \to P_3^{\cdot} \to$$

such that $P_3^{\cdot}$ is the mapping cone of the morphism $P_1^{\cdot} \to P_2^{\cdot}$. We have $P_3^i = P_1^i \oplus P_2^{i+1}$ for all i. It follows that $\phi(P_1^{\cdot}) - \phi(P_2^{\cdot}) + \phi(P_3^{\cdot}) = 0$. So ϕ induces a homomorphism

$$\Phi : K(K_{\mathrm{perf}}(\Lambda)) \to K(\Lambda).$$

For any finitely generated projective Λ-module P, let $\psi(P)$ be the element in $K(K_{\mathrm{perf}}(\Lambda))$ corresponding to the isomorphic class of the complex

$$\cdots \to 0 \to P \to 0 \to \cdots .$$

$\psi(P)$ depends only on the isomorphic class of P and ψ induces a homomorphism

$$\Psi : K(\Lambda) \to K(K_{\mathrm{perf}}(\Lambda)).$$

One can easily verify $\Phi\Psi = \mathrm{id}$. Let us prove $\Psi\Phi = \mathrm{id}$, that is, for any perfect complex $P^{\cdot}$, we have the equality

$$[P^{\cdot}] = \sum_i (-1)^i \psi(P^i)$$

in $K(K_{\mathrm{perf}}(\Lambda))$. We say that $P^{\cdot}$ has length $\leq m$ if the exists an integer a such that $P^i = 0$ for $i \notin [a, a+m]$, that is, $P^{\cdot}$ is of the form

$$0 \to P^a \to P^{a+1} \to \cdots \to P^{a+m} \to 0.$$

We prove the above equality using induction on m. First note that we have a distinguished triangle of the form

$$P^{\cdot} \to 0 \to P^{\cdot}[1] \to .$$

It follows that

$$[P^{\cdot}[1]] = -[P^{\cdot}].$$

Suppose $m = 0$. Then $P^{\cdot}$ coincides with the complex

$$(\cdots \to 0 \to P^a \to 0 \to \cdots)[a],$$

where P^a sits in the 0-th components. It follows that $[P^{\cdot}] = (-1)^a \psi(P^a)$. Suppose that we have shown the above equality for perfect complexes of length $\leq m-1$. Let $\sigma_{\geq a+1}(P^{\cdot})$ be the truncated complex

$$0 \to P^{a+1} \to \cdots \to P^{a+m} \to 0.$$

We have a morphism of complex

$$P^a[-(a+1)] \to \sigma_{\geq a+1}(P^{\cdot})$$

whose $(a+1)$-th component is the morphism $d_{P^{\cdot}}^a : P^a \to P^{a+1}$. The mapping cone of this morphism is isomorphic to $P^{\cdot}$, and $\sigma_{\geq a+1}(P^{\cdot})$ has length $\leq m-1$. So we have

$$[P^{\cdot}] = -[P^a(-(a+1))] + [\sigma_{\geq a+1}(P^{\cdot})] = \sum_i (-1)^i \psi(P^i).$$

□

Let X be a smooth irreducible projective curve over an algebraically closed field k, let K be its function field, and let K' be a finite galois extension of K with galois group G. Denote by X' the normalization of X in K, and by $p : X' \to X$ the canonical morphism. Then G acts on X' on the right. For any closed point x' in X', let $G_{x'}$ be the stabilizer of x' in G, let π be a uniformizer of $\mathscr{O}_{X',x'}$, and let $v_{x'}$ be the valuation on K' corresponding to the point x'. The *Swan character* on $G_{x'}$ is defined to be

$$b_{x'}(g) = \begin{cases} 1 - v_{x'}(g\pi - \pi) & \text{if } g \neq e, \\ -\sum\limits_{g \in G_{x'} - \{e\}} (1 - v_{x'}(g\pi - \pi)) & \text{if } g = e. \end{cases}$$

By [Serre (1979)] VI 2 and [Serre (1977)] 19.2 Theorem 44, there exists a projective $\mathbb{Z}_l[G_{x'}]$-module $\mathrm{Sw}_{x'}$ determined up to isomorphism such that b is its character. The *Artin character* of $G_{x'}$ is defined to be

$$a_{x'} = b_{x'} + r_{G_{x'}} - 1 = b_{x'} + u_{G_{x'}},$$

where $r_{G_{x'}}$ is the character of the regular representation $\mathbb{Q}_\ell[G_{x'}]$ of G_x, and $u_{G_{x'}}$ is the character of the kernel of the homomorphism

$$\mathbb{Q}_\ell[G_{x'}] \to \mathbb{Q}_\ell, \quad \sum_{g \in G_{x'}} a_g g \mapsto \sum_{g \in G_{x'}} a_g.$$

We have

$$a_{x'}(g) = \begin{cases} -v_{x'}(g\pi - \pi) & \text{if } g \neq e, \\ \sum\limits_{g \in G_{x'} - \{e\}} v_{x'}(g\pi - \pi) & \text{if } g = e. \end{cases}$$

The Artin character is the character of a $\mathbb{Q}_\ell[G_{x'}]$-module. Let $x = p(x')$. Define

$$\mathrm{Sw}_x = \mathbb{Z}_\ell[G] \otimes_{\mathbb{Z}_\ell[G_{x'}]} \mathrm{Sw}_{x'}.$$

Choose $g_1, \ldots, g_m \in G$ such that $G/G_{x'} = \{g_1 G_{x'}, \ldots, g_m G_{x'}\}$. The character of Sw_x is

$$b_x(g) = \sum_{g_i^{-1} g g_i \in G_{x'}} b_{x'}(g_i^{-1} g g_i).$$

We have

$$p^{-1}(x) = \{x' g_1^{-1}, \ldots, x' g_m^{-1}\}.$$

Set $x_i' = x' g_i^{-1}$. We have $G_{x_i'} = g_i G_{x'} g_i^{-1}$, and $v_{x_i'}(\alpha) = v_{x'}(g_i^{-1}\alpha)$ for any $\alpha \in K$. It follows that $b_{x_i'}(g) = b_{x'}(g_i^{-1} g g_i)$ for any $g \in G_{x_i'}$. Therefore

$$b_x(g) = \sum_{g \in G_{x_i'}} b_{x_i'}(g) = \sum_{p(x') = x,\ gx' = x'} b_{x'}(g).$$

This shows that $\mathrm{Sw}_x = \mathbb{Z}_\ell[G] \otimes_{\mathbb{Z}_\ell[G_{x'}]} \mathrm{Sw}_{x'}$ depends only on x, and not on the choice of $x' \in p^{-1}(x)$.

Lemma 10.2.4. *Let X be a smooth irreducible projective curve of genus g over an algebraically closed field k, K its function field, K' a finite galois extension of K with galois group G, X' the normalization of X in K', $p: X' \to X$ the canonical morphism, U a dense open subset of X such that $U' = p^{-1}(U)$ is etale over U, and $j' : U' \to X'$ the open immersion. Then $R\Gamma(X', j'_!\mathbb{Z}/\ell^n) \in \mathrm{ob}\, D^b_{\mathrm{perf}}(\mathbb{Z}/\ell^n[G])$ and*

$$\begin{aligned}&\big[R\Gamma(X', j'_!\mathbb{Z}/\ell^n)\big]\\ &= (2-2g)\big[\mathbb{Z}/\ell^n[G]\big] - \sum_{x\in X-U}\Big(\big[\mathbb{Z}/\ell^n[G] \otimes_{\mathbb{Z}_\ell[G]} \mathrm{Sw}_x\big] + \big[\mathbb{Z}/\ell^n[G]\big]\Big)\end{aligned}$$

in $K(D^b_{\mathrm{perf}}(\mathbb{Z}/\ell^n[G]))$.

Proof. We have $R\Gamma(X', j'_!\mathbb{Z}/\ell^n) = R\Gamma(X, p_*j'_!\mathbb{Z}/\ell^n)$. By 10.2.1, 7.4.7 (ii) and 7.8.1 applied to $R\Gamma(X, p_*j'_!\mathbb{Z}/\ell^n)$, to prove $R\Gamma(X', j'_!\mathbb{Z}/\ell^n) \in \mathrm{ob}\, D^b_{\mathrm{perf}}(\mathbb{Z}/\ell^n[G])$, it suffices to prove $p_*j'_!\mathbb{Z}/\ell^n \in \mathrm{ob}\, D^b_{\mathrm{ctf}}(X, \mathbb{Z}/\ell^n[G])$. There exists an etale covering $\{V_i \to U\}_i$ such that each $U' \times_U V_i \to V_i$ is a trivial etale covering space. We have

$$(p_*j'_!\mathbb{Z}/\ell^n)|_{V_i} \cong \mathbb{Z}/\ell^n[G].$$

In particular, $(p_*j'_!\mathbb{Z}/\ell^n)|_U$ is locally constant. Moreover, we have $(p_*j'_!\mathbb{Z}/\ell^n)|_{X-U} = 0$. It follows that $p_*j'_!\mathbb{Z}/\ell^n \in \mathrm{ob}\, D^b_{\mathrm{ctf}}(X, \mathbb{Z}/\ell^n[G])$.

By [Serre (1977)] 14.4, we have a canonical isomorphism

$$K(D^b_{\mathrm{perf}}(\mathbb{Z}_\ell[G])) \stackrel{\cong}{\to} K(D^b_{\mathrm{perf}}(\mathbb{Z}/\ell^n[G])).$$

Denote by S the inverse of this isomorphism. By [Serre (1977)] 14.1 and 16.1 Corollary 2 of Theorem 34, two objects in $D^b_{\mathrm{perf}}(\mathbb{Z}_\ell[G])$ have the same image in $K(D^b_{\mathrm{perf}}(\mathbb{Z}_\ell[G]))$ if and only if they have the same character. Denote the left-hand and the right-hand sides of the equality in the lemma by L_n and R_n, respectively. To prove the lemma, it suffices to show that $S(L_n)$ and $S(R_n)$ have the same character. Obviously $S(R_n)$ has character

$$(2-2g)r_G - \sum_{x\in X-U}(b_x + r_G),$$

where r_G is the character of the regular representation $\mathbb{Q}_\ell[G]$ of G. Since

$$R\Gamma(X', j_!\mathbb{Z}/\ell^m) \otimes^L_{\mathbb{Z}/\ell^m} \mathbb{Z}/\ell^n \cong R\Gamma(X', j_!\mathbb{Z}/\ell^n)$$

for all $m \geq n$, the character of $S(L_n)$ is the map

$$G \to Z_\ell = \varprojlim_n \mathbb{Z}/\ell^n, \quad g \mapsto \Big(\mathrm{Tr}(g, R\Gamma(X', j_! Z/\ell^n))\Big).$$

So to prove the lemma, it suffices to show

$$\mathrm{Tr}(g, R\Gamma(X', j'_! Z/\ell^n)) \equiv \Big((2-2g)r_G - \sum_{x\in X-U}(b_x + r_G)\Big)(g) \mod \ell^n$$

for all $n \geq 0$ and all $g \in G$. By 10.2.1, and 7.4.7 (ii) and 7.8.1, the complexes $R\Gamma(X', j'_! \mathbb{Z}/\ell^n)$, $R\Gamma(X', \mathbb{Z}/\ell^n)$ and $R\Gamma(X'-U', \mathbb{Z}/\ell^n)$ are objects in $D^b_{\mathrm{perf}}(\mathbb{Z}/\ell^n)$. We have

$$\mathrm{Tr}(g, R\Gamma(X', j'_! Z/\ell^n)) = \mathrm{Tr}(g, R\Gamma(X', Z/\ell^n)) - \mathrm{Tr}(g, R\Gamma(X'-U', Z/\ell^n))$$

for any $g \in G$. It is clear that

$$\mathrm{Tr}(g, R\Gamma(X'-U', Z/\ell^n)) = \sum_{x'\in X'-U',\ gx'=x'} 1.$$

Since U' is etale over U, any $g \neq e$ has no fixed point on U'. So by 8.6.8, if $g \neq e$, we have

$$\begin{aligned}\mathrm{Tr}(g, R\Gamma(X', Z/\ell^n)) &\equiv \sum_{gx'=x'} (1 - b_{x'}(g)) \mod \ell^n \\ &\equiv \sum_{x'\in X'-U',\ gx'=x'} (1 - b_{x'}(g)) \mod \ell^n.\end{aligned}$$

So for any $g \neq e$, we have

$$\begin{aligned}&\mathrm{Tr}(g, R\Gamma(X', j'_! Z/\ell^n)) \\ \equiv\ & \sum_{x'\in X'-U',\ gx'=x'} (1 - b_{x'}(g)) - \sum_{x'\in X'-U',\ gx'=x'} 1 \mod \ell^n \\ \equiv\ & -\sum_{x'\in X'-U',\ gx'=x'} b_{x'}(g) \mod \ell^n \\ \equiv\ & \Big((2-2g)r_G - \sum_{x\in X-U}(b_x + r_G)\Big)(g) \mod \ell^n.\end{aligned}$$

Let g' be the genus of X'. By 7.2.9 (ii), we have

$$\begin{aligned}&\mathrm{Tr}(e, R\Gamma(X', \mathbb{Z}/\ell^n)) \\ =\ & \mathrm{rank}\, H^0(X', \mathbb{Z}/\ell^n) - \mathrm{rank}\, H^1(X', \mathbb{Z}/\ell^n) + \mathrm{rank}\, H^2(X', \mathbb{Z}/\ell^n) \\ \equiv\ & 2 - 2g' \mod \ell^n.\end{aligned}$$

It is clear that

$$\mathrm{Tr}(e, R\Gamma(X'-U', \mathbb{Z}/\ell^n)) = \sum_{x'\in X'-U'} 1.$$

So we have

$$\mathrm{Tr}(e, R\Gamma(X', j_!\mathbb{Z}/\ell^n)) \equiv 2 - 2g' - \sum_{x' \in X' - U'} 1 \mod \ell^n.$$

On the other hand, we have

$$\begin{aligned}
&\Big((2-2g)r_G - \sum_{x \in X-U} (b_x + r_G)\Big)(e) \\
&= (2-2g)[K' : K] - \sum_{x \in X-U} (b_x(e) + [K' : K]) \\
&= (2-2g)[K' : K] - \sum_{x \in X-U} \sum_{p(x')=x} (b_{x'}(e) + \#G_{x'}) \\
&= (2-2g)[K' : K] - \sum_{x \in X-U} \sum_{p(x')=x} (a_{x'}(e) + 1) \\
&= (2-2g)[K' : K] - \sum_{x' \in X'-U'} a_{x'}(e) - \sum_{x' \in X'-U'} 1.
\end{aligned}$$

By [Serre (1979)] IV §1 Proposition 4, we have

$$\sum_{x' \in X'-U'} a_{x'}(e) = \deg \mathfrak{D}_{K'/K},$$

where $\mathfrak{D}_{K'/K}$ is the different of K'/K. We thus have

$$\Big((2-2g)r_G - \sum_{x \in X-U} (b_x + r_G)\Big)(e) = (2-2g)[K' : K] - \deg \mathfrak{D}_{K'/K} - \sum_{x' \in X'-U'} 1.$$

To prove the lemma, it remains to show

$$2 - 2g' = (2-2g)[K' : K] - \deg \mathfrak{D}_{K'/K}.$$

This is the Hurwitz formula, which can be proved as follows: We have an exact sequence

$$p^*\Omega_{X/k} \to \Omega_{X'/k} \to \Omega_{X'/X} \to 0.$$

Note that $p^*\Omega_{X/k}$ and $\Omega_{X'/k}$ are invertible $\mathscr{O}_{X'}$-modules. Since U' is etale over U, $p^*\Omega_{X/k} \to \Omega_{X'/k}$ induces an isomorphism on stalks at the generic point of X'. Since the kernel of this morphism is torsion free as a subsheaf of $p^*\Omega_{X/k}$, this morphism is injective. So we have an exact sequence

$$0 \to p^*\Omega_{X/k} \to \Omega_{X'/k} \to \Omega_{X'/X} \to 0.$$

The different $\mathfrak{D}_{K'/K}$ is an effective divisor of X'. Let $\mathscr{I}_{\mathfrak{D}_{K'/K}}$ be its ideal sheaf. By the proof of [Serre (1979)] III §7 Proposition 14, the kernel of the canonical morphism $\Omega_{X'/k} \to \Omega_{X'/X}$ is isomorphic to $\mathscr{I}_{\mathfrak{D}_{K'/K}} \otimes_{\mathscr{O}_{X'}} \Omega_{X'/k}$. It follows that

$$p^*\Omega_{X/k} \cong \mathscr{I}_{\mathfrak{D}_{K'/K}} \otimes_{\mathscr{O}_{X'}} \Omega_{X'/k}.$$

Taking degrees on both sides, we get

$$[K' : K]\deg(\Omega_{X/k}) = \deg(\mathscr{I}_{\mathfrak{D}_{K'/K}}) + \deg(\Omega_{X'/k}).$$

We have $\deg(\mathscr{I}_{\mathfrak{D}_{K'/K}}) = -\deg \mathfrak{D}_{K'/K}$. By Riemann–Roch's formula, we have $\deg(\Omega_{X/k}) = 2g-2$ and $\deg(\Omega_{X'/k}) = 2g'-2$. We thus have

$$2 - 2g' = (2-2g)[K' : K] - \deg \mathfrak{D}_{K'/K}.$$

□

Let X be a smooth irreducible curve over an algebraically closed field k, let A be a noetherian $\mathbb{Z}/\ell^n$-algebra, and let $K \in \operatorname{ob} D^b_{\mathrm{ctf}}(X, A)$. We can represent K by a bounded complex of flat constructible sheaves of A-modules on X. Let $K(X)$ be the function field of X, let $\overline{K(X)}$ be a separable closure of $K(X)$, and let η be the generic point of X. There exists a finite galois extension K' of $K(X)$ contained in $\overline{K(X)}$ such that the action of $\mathrm{Gal}(\overline{K(X)}/K(X))$ on $K_{\bar\eta}$ factors through $G = \mathrm{Gal}(K'/K(X))$. Let X' be the normalization of X in K', let $p : X' \to X$ be the canonical morphism, and let x be a closed point of X. One can check that

$$\mathrm{Hom}^{\cdot}_{\mathbb{Z}/\ell^n[G]}(\mathbb{Z}/\ell^n[G] \otimes_{\mathbb{Z}_\ell[G]} \mathrm{Sw}_x, K_{\bar\eta})$$

is a perfect complex of A-modules. This complex is independent of the choice of the galois extension $K'/K(X)$. Indeed, let $K''/K(X)$ be a finite galois extension with galois group G' containing K', and let X'' be the normalization of X in K''. Then we have

$$\mathrm{Sw}_{X'/X,x} \cong \mathbb{Z}_\ell[G] \otimes_{\mathbb{Z}_\ell[G']} \mathrm{Sw}_{X''/X,x}.$$

One can show this by checking that both sides have the same character by using [Serre (1979)] IV §1 Proposition 3. So we have

$$\begin{aligned}&\mathrm{Hom}^{\cdot}_{\mathbb{Z}/\ell^n[G]}(\mathbb{Z}/\ell^n[G] \otimes_{\mathbb{Z}_\ell[G]} \mathrm{Sw}_{X'/X,x}, K_{\bar\eta})\\ &\cong \mathrm{Hom}^{\cdot}_{\mathbb{Z}/\ell^n[G']}(\mathbb{Z}/\ell^n[G'] \otimes_{\mathbb{Z}_\ell[G']} \mathrm{Sw}_{X''/X,x}, K_{\bar\eta}).\end{aligned}$$

Consider the following elements in $K(D^b_{\mathrm{perf}}(A))$:

$$\begin{aligned}\alpha_x(K) &= \big[\mathrm{Hom}^{\cdot}_{\mathbb{Z}/\ell^n[G]}(\mathbb{Z}/\ell^n[G] \otimes_{\mathbb{Z}_\ell[G]} \mathrm{Sw}_x, K_{\bar\eta})\big],\\ \epsilon_x(K) &= \alpha_x(K) + [K_{\bar\eta}] - [K_{\bar x}].\end{aligned}$$

By 5.8.1 (ii), we may find a dense open subset U of X and a finite surjective etale morphism $U' \to U$ such that U' is connected and the components of

$K|_{U'}$ are constant sheaves. Moreover, we can choose U' so that it is galois over U. In the above discussion, taking K' to be the function field of U', we can identify U' with $p^{-1}(U)$. If x is a closed point of U, then $\alpha_x(K) = 0$ and $K_{\bar{x}} \cong K_{\bar{\eta}}$, and hence $\epsilon_x(K) = 0$. In particular, $\epsilon_x(K)$ is nonzero only for finitely many closed point x in X.

Theorem 10.2.5. *Let X be a smooth irreducible projective curve of genus g over an algebraically closed field, η its generic point, A a noetherian $\mathbb{Z}/\ell^n$-algebra, and $K \in \operatorname{ob} D^b_{\mathrm{ctf}}(X, A)$. Then $R\Gamma(X, K) \in \operatorname{ob} D^b_{\mathrm{perf}}(A)$ and*

$$\big[R\Gamma(X, K)\big] = (2-2g)[K_{\bar{\eta}}] - \sum_{x \in |X|} \epsilon_x(K)$$

in $K(D^b_{\mathrm{perf}}(A))$, where $|X|$ is the set of closed points in X.

Proof. By 10.2.1, 7.4.7 (ii) and 7.8.1, we have $R\Gamma(X, K) \in \operatorname{ob} D^b_{\mathrm{perf}}(A)$. Represent K by a bounded complex of flat constructible sheaves of A-modules. Choose a finite galois extension K' of the function field of X and a dense open subset U of X so that if X' is the normalization of X in K' and $p : X' \to X$ is the canonical morphism, then p is etale on $U' = p^{-1}(U)$ and the components of $K|_{U'}$ are constant sheaves. Let $j : U \hookrightarrow X$ be the open immersion. We claim that

$$j_!(K|_U) = R\underline{\Gamma}^G(p_*p^*j_!(K|_U)),$$

where $\underline{\Gamma}^G$ is the functor $\mathscr{F} \mapsto \mathscr{F}^G$ on the category of G-sheaves. Indeed, by the discussion at the end of §9.1, for any $x \in X$, we have

$$\big(R\underline{\Gamma}^G(p_*p^*j_!(K|_U))\big)_{\bar{x}} \cong R\Gamma^G\big((p_*p^*j_!(K|_U))_{\bar{x}}\big),$$

where Γ^G is the functor $M \to M^G$ on the category of G-modules. When $x \in X - U$, we have $\big(p_*p^*j_!(K|_U)\big)_{\bar{x}} = 0$. Hence

$$R\underline{\Gamma}^G(p_*p^*j_!(K|_U))\big|_{X-U} = 0.$$

When $x \in U$, since $p : U' \to U$ is a galois etale covering, components of $\big(p_*p^*j_!(K|_U)\big)_{\bar{x}}$ are induced G-modules and hence

$$\big(R\underline{\Gamma}^G(p_*p^*j_!(K|_U))\big)_{\bar{x}} \cong \big(p_*p^*j_!(K|_U)\big)^G_{\bar{x}} \cong K_{\bar{x}}.$$

Our claim follows. By the discussion in §9.1 and 6.3.3, we have

$$R\Gamma(X, R\underline{\Gamma}^G(-)) \cong R\big(\Gamma(X, \underline{\Gamma}^G(-))\big) = R\big(\Gamma^G\Gamma(X, -)\big) = R\Gamma^G R\Gamma(X, -).$$

We thus have

$$\begin{aligned} R\Gamma(X, j_!(K|_U)) &\cong R\Gamma(X, R\underline{\Gamma}^G(p_*p^*j_!(K|_U))) \\ &\cong R\Gamma^G R\Gamma(X, p_*p^*j_!(K|_U)). \end{aligned}$$

Let $j' : U' \hookrightarrow X'$ be the open immersion. By 7.4.7, we have

$$\begin{aligned} R\Gamma(X, p_*p^*j_!(K|_U)) &\cong R\Gamma(X, p_*j'_!(K|_{U'})) \\ &\cong R\Gamma_c(U', K|_{U'}) \\ &\cong K_{\bar\eta} \otimes^L_{\mathbb{Z}/\ell^n} R\Gamma_c(U', \mathbb{Z}/\ell^n) \\ &\cong K_{\bar\eta} \otimes^L_{\mathbb{Z}/\ell^n} R\Gamma(X', j'_!\mathbb{Z}/\ell^n). \end{aligned}$$

Here the action of G on $K_{\bar\eta} \otimes^L_{\mathbb{Z}/\ell^n} R\Gamma(X', j'_!\mathbb{Z}/\ell^n)$ is the diagonal action. So we have

$$R\Gamma(X, j_!(K|_U)) \cong R\Gamma^G\Big(K_{\bar\eta} \otimes^L_{\mathbb{Z}/\ell^n} R\Gamma(X', j'_!\mathbb{Z}/\ell^n)\Big). \tag{10.1}$$

By 10.2.4, we have $R\Gamma(X', j'_!\mathbb{Z}/\ell^n) \in \mathrm{ob}\, D^b_{\mathrm{perf}}(\mathbb{Z}/\ell^n[G])$. Represent $R\Gamma(X', j'_!\mathbb{Z}/\ell^n)$ by a bounded complex P of finitely generated projective $(\mathbb{Z}/\ell^n[G])$-modules. We claim that the components of the complex $K_{\bar\eta} \otimes_{\mathbb{Z}/\ell^n} P$ are weakly injective G-modules. Indeed, since the components of P are direct factors of free $(\mathbb{Z}/\ell^n[G])$-modules, it suffices to show that $M \otimes_{\mathbb{Z}/\ell^n} \mathbb{Z}/\ell^n[G]$ is an induced G-module for any G-module M, where the G-module structure on $M \otimes_{\mathbb{Z}/\ell^n} \mathbb{Z}/\ell^n[G]$ is defined by

$$g(x \otimes y) = gx \otimes gy$$

for any $x \in M$ and $y \in \mathbb{Z}/\ell^n[G]$. One easily checks

$$M \otimes_{\mathbb{Z}/\ell^n} \mathbb{Z}/\ell^n[G] \to M[G], \quad x \otimes g \to (g^{-1}(x))g$$

is an isomorphism of G-modules. On the other hand, we have an isomorphism of G-modules

$$M[G] \cong \mathrm{Hom}(\mathbb{Z}[G], M), \quad \sum_{g \in G} x_g g \mapsto (g \mapsto x_{g^{-1}}).$$

This proves our claim. So we have

$$R\Gamma^G\Big(K_{\bar\eta} \otimes^L_{\mathbb{Z}/\ell^n} R\Gamma(X', j'_!\mathbb{Z}/\ell^n)\Big) \cong (K_{\bar\eta} \otimes_{\mathbb{Z}/\ell^n} P)^G, \tag{10.2}$$

Let $P^\vee = \mathrm{Hom}^\cdot_{\mathbb{Z}/\ell^n}(P, \mathbb{Z}/\ell^n)$, where the action of G on the i-th component $\mathrm{Hom}_{\mathbb{Z}/\ell^n}(P^{-i}, \mathbb{Z}/\ell^n)$ of $P^\vee$ is given by

$$(g\phi)(x) = \phi(g^{-1}x)$$

for any $g \in G$, $\phi \in \mathrm{Hom}_{\mathbb{Z}/\ell^n}(P^{-i}, \mathbb{Z}/\ell^n)$ and $x \in P^{-i}$. We have an isomorphism of complexes

$$K_{\bar\eta} \otimes_{\mathbb{Z}/\ell^n} P \cong \mathrm{Hom}^\cdot_{\mathbb{Z}/\ell^n}(P^\vee, K_{\bar\eta}). \tag{10.3}$$

¿From the isomorphisms (10.1)–(10.3), we get

$$R\Gamma(X, j_!(K|_U)) \cong \mathrm{Hom}^\cdot_{\mathbb{Z}/\ell^n[G]}(P^\vee, K_{\bar\eta}).$$

By 10.2.4, in $K(D^b_{\mathrm{perf}}(\mathbb{Z}/\ell^n[G]))$, we have

$$\begin{aligned}[P] &= \big[R\Gamma(X', j'_!\mathbb{Z}/\ell^n[G])\big] \\ &= (2-2g)\big[\mathbb{Z}/\ell^n[G]\big] - \sum_{x\in X-U}\Big(\big[Z/\ell^n[G]\otimes_{\mathbb{Z}_\ell[G]}\mathrm{Sw}_x\big] + \big[Z/\ell^n[G]\big]\Big).\end{aligned}$$

This implies that

$$[P^\vee] = (2-2g)\big[\mathbb{Z}/\ell^n[G]\big] - \sum_{x\in X-U}\Big(\big[Z/\ell^n[G]\otimes_{\mathbb{Z}_\ell[G]}\mathrm{Sw}_x\big] + \big[Z/\ell^n[G]\big]\Big).$$

This can be seen by calculating the characters on both sides, and by noting that Sw_x and $\mathrm{Sw}_x^\vee$ have the same character. So we have

$$\begin{aligned}&\big[R\Gamma(X, j_!(K|_U))\big] \\ &= \big[\mathrm{Hom}^\cdot_{\mathbb{Z}/\ell^n[G]}(P^\vee, K_{\bar\eta})\big] \\ &= (2-2g)\big[\mathrm{Hom}^\cdot_{\mathbb{Z}/\ell^n[G]}(\mathbb{Z}/\ell^n[G], K_{\bar\eta})\big] \\ &\quad - \sum_{x\in X-U}\Big(\big[\mathrm{Hom}^\cdot_{\mathbb{Z}/\ell^n[G]}(Z/\ell^n[G]\otimes_{\mathbb{Z}_\ell[G]}\mathrm{Sw}_x, K_{\bar\eta})\big] \\ &\quad + \big[\mathrm{Hom}^\cdot_{\mathbb{Z}/\ell^n[G]}(Z/\ell^n[G], K_{\bar\eta})\big]\Big) \\ &= (2-2g)[K_{\bar\eta}] - \sum_{x\in X-U}\Big(\big[\mathrm{Hom}^\cdot_{\mathbb{Z}/\ell^n[G]}(Z/\ell^n[G]\otimes_{\mathbb{Z}_\ell[G]}\mathrm{Sw}_x, K_{\bar\eta})\big] + [K_{\bar\eta}]\Big).\end{aligned}$$

We have a distinguished triangle

$$R\Gamma(X, j_!(K|_U)) \to R\Gamma(X, K) \to R\Gamma(X-U, K|_{X-U}) \to .$$

Again by 10.2.1, and 7.4.7 (ii) and 7.8.1, all the complexes in this triangle are objects in $D^b_{\mathrm{perf}}(A)$. It is clear that

$$\big[R\Gamma(X-U, K|_{X-U})\big] = \sum_{x\in X-U}[K_{\bar x}].$$

So we have

$$\begin{aligned}&\big[R\Gamma(X, K)\big] \\ &= \big[R\Gamma(X, j_!(K|_U))\big] + \big[R\Gamma(X-U, K|_{X-U})\big] \\ &= (2-2g)[K_{\bar\eta}] - \sum_{x\in X-U}\Big(\big[\mathrm{Hom}^\cdot_{\mathbb{Z}/\ell^n[G]}(Z/\ell^n[G]\otimes_{\mathbb{Z}_\ell[G]}\mathrm{Sw}_x, K_{\bar\eta})\big] + [K_{\bar\eta}]\Big) \\ &\quad + \sum_{x\in X-U}[K_{\bar x}]\end{aligned}$$

$$= (2-2g)[K_{\bar\eta}] - \sum_{x\in X-U}\Big([\mathrm{Hom}^{\cdot}_{\mathbb{Z}/\ell^n[G]}(Z/\ell^n[G]\otimes_{\mathbb{Z}_\ell[G]}\mathrm{Sw}_x, K_{\bar\eta})]$$

$$+[K_{\bar\eta}]-[K_{\bar x}]\Big)$$

$$= (2-2g)[K_{\bar\eta}] - \sum_{x\in X-U}\epsilon_x(K)$$

$$= (2-2g)[K_{\bar\eta}] - \sum_{x\in |X|}\epsilon_x(K).$$

□

Theorem 10.2.6. *Let X be a smooth irreducible projective curve of genus g over an algebraically closed field, η its generic point, A a noetherian $\mathbb{Z}/\ell^n$-algebra, U a dense open subset of X, and $K\in \mathrm{ob}\, D^b_{\mathrm{ctf}}(U,A)$. Then $R\Gamma_c(U,K)$ and $R\Gamma(U,K)$ are objects in $D^b_{\mathrm{perf}}(A)$, and*

$$\big[R\Gamma_c(U,K)\big] = \big[R\Gamma(U,K)\big]$$
$$= (2-2g-\#(X-U))[K_{\bar\eta}] - \sum_{x\in|U|}\epsilon_x(K) - \sum_{x\in X-U}\alpha_x(K).$$

in $K(D^b_{\mathrm{perf}}(A))$, where $|U|$ is the set of closed points in U.

Proof. Let $j: U\hookrightarrow X$ be the open immersion. By 10.2.5, we have $R\Gamma_c(U,K) = R\Gamma(X, j_!K)\in \mathrm{ob}\, D^b_{\mathrm{perf}}(A)$ and

$$\big[R\Gamma_c(U,K)\big] = \big[R\Gamma(X,j_!K)\big]$$
$$= (2-2g)[K_{\bar\eta}] - \sum_{x\in|U|}\epsilon_x(j_!K) - \sum_{x\in X-U}\epsilon_x(j_!K)$$
$$= (2-2g)[K_{\bar\eta}] - \sum_{x\in|U|}\epsilon_x(j_!K) - \sum_{x\in X-U}(\alpha_x(j_!K)+[K_{\bar\eta}])$$
$$= (2-2g-\#(X-U))[K_{\bar\eta}] - \sum_{x\in|U|}\epsilon_x(K) - \sum_{x\in X-U}\alpha_x(K).$$

It remains to prove $R\Gamma(U,K)\in \mathrm{ob}\, D^b_{\mathrm{perf}}(A)$ and $\big[R\Gamma_c(X,K)\big] = \big[R\Gamma(X,K)\big]$. Choose a distinguished triangle

$$j_!K\to Rj_*K\to C\to,$$

where $j_!K\to Rj_*K\cong Rj_*j^*(j_!K)$ is the canonical morphism. We have a distinguished triangle

$$R\Gamma(X,j_!K)\to R\Gamma(X,Rj_*K)\to R\Gamma(X,C)\to,$$

which can be identified with a distinguished triangle

$$R\Gamma_c(U,K)\to R\Gamma(U,K)\to R\Gamma(X,C)\to.$$

To prove our assertion, it suffices to show that $R\Gamma(X, C) \in \operatorname{ob} D^b_{\mathrm{perf}}(A)$ and $[R\Gamma(X, C)] = 0$. The cohomology sheaves of C are supported on $X - U$, so we have

$$[R\Gamma(X, C)] = \sum_{x \in X-U} [C_{\bar{x}}].$$

We are reduced to show that $C_{\bar{x}} \in \operatorname{ob} D^b_{\mathrm{perf}}(A)$ and $[C_{\bar{x}}] = 0$ for any $x \in X - U$. Let $\tilde{\eta}$ be the generic point of the strict localization $\widetilde{X}_{\bar{x}}$ of X at $\bar{x}$. By 5.9.5, we have

$$C_{\bar{x}} = R\Gamma(U \times_X \widetilde{X}_{\bar{x}}, K) = R\Gamma(\tilde{\eta}, K) = R\Gamma^{\mathrm{Gal}(\bar{\eta}/\tilde{\eta})}(K_{\bar{\eta}}).$$

Let P be the wild inertia subgroup of $\mathrm{Gal}(\bar{\eta}/\tilde{\eta})$. Then P is a pro-p-group for $p = \operatorname{char} k$, and we have

$$\mathrm{Gal}(\bar{\eta}/\tilde{\eta})/P \cong \varprojlim_{(m,p)=1} \mathbb{Z}/m,$$

$$R\Gamma^{\mathrm{Gal}(\bar{\eta}/\tilde{\eta})}(K_{\bar{\eta}}) \cong R\Gamma^{\mathrm{Gal}(\bar{\eta}/\tilde{\eta})/P}(K^P_{\bar{\eta}}).$$

We have $K^P_{\bar{\eta}} \in \operatorname{ob} D^b_{\mathrm{perf}}(A)$ since $K^P_{\bar{\eta}}$ is a direct factor of $K_{\bar{\eta}}$. Indeed, for any A-module M with continuous P-action, the inclusion $M^P \hookrightarrow M$ has a left inverse $M \to M^P$ defined by

$$x \mapsto \frac{1}{\#(P/P_x)} \sum_{gP_x \in P/P_x} gx,$$

where P_x is the stabilizer of x in P. Here multiplication by $\frac{1}{\#(P/P_x)}$ makes sense since p is invertible in A. To prove our assertion, it suffices to show that for $G = \varprojlim_{(m,p)=1} \mathbb{Z}/m$ and for any complex $L^\cdot$ of $A[G]$-modules which lies in $D^b_{\mathrm{perf}}(A)$ considered as a complex of A-modules, we have $R\Gamma^G(L^\cdot) \in \operatorname{ob} D^b_{\mathrm{perf}}(A)$ and $[R\Gamma^G(L^\cdot)] = 0$ in $K(D^b_{\mathrm{perf}}(A))$. Represent $L^\cdot$ by a bounded below complex of $R\Gamma^G$-acyclic A-modules with continuous G-action, and let σ be the element in G so that its image in each $\mathbb{Z}/m$ is 1. We have

$$R\Gamma^G(L^\cdot) = \ker(\sigma - 1, L^\cdot).$$

On the other hand, for each component L^i of $L^\cdot$, we have

$$\operatorname{coker}(\sigma - 1, L^i) = H^1(G, L^i) = 0.$$

by 4.3.8. So $\sigma : L^i \to L^i$ is onto. We thus have a distinguished triangle

$$R\Gamma^G(L^\cdot) \to L^\cdot \stackrel{\sigma-1}{\to} L^\cdot \to .$$

It follows that $R\Gamma^G(L^\cdot) \in \operatorname{ob} D^b_{\mathrm{perf}}(A)$ and $[R\Gamma^G(L^\cdot)] = 0$. □

Assume A is a local ring. Then every projective A-module of finite type is free and we may talk about its rank. It defines a homomorphism

$$\mathrm{rank}: K(A) \to \mathbb{Z}.$$

By 10.2.3, we can define a homomorphism

$$\mathrm{rank}: K(D^b_{\mathrm{perf}}(A)) \to \mathbb{Z}.$$

Representing an object in $D^b_{\mathrm{perf}}(A)$ by a bounded complex of projective A-modules K, we have

$$\mathrm{rank}\, K = \sum_i (-1)^i \mathrm{rank}\, K^i.$$

Corollary 10.2.7 (Grothendieck–Ogg–Shafarevich Formula). *Let X be a smooth irreducible projective curve of genus g over an algebraically closed field, η its generic point, U an open dense subset, $K \in \mathrm{ob}\, D^b_c(U, \overline{\mathbb{Q}}_\ell)$, and x any closed point in X. Define*

$$\begin{aligned}
\chi_c(U,K) &= \sum_i (-1)^i \dim H^i_c(U,K),\\
\chi(U,K) &= \sum_i (-1)^i \dim H^i(U,K),\\
\dim K_{\bar\eta} &= \sum_i (-1)^i \dim (\mathscr{H}^i(K))_{\bar\eta},\\
\dim K_{\bar x} &= \sum_i (-1)^i \dim (\mathscr{H}^i(K))_{\bar x},\\
\alpha_x(K) &= \sum_i (-1)^i \dim \mathrm{Hom}_G(\overline{\mathbb{Q}}_\ell[G] \otimes_{\mathbb{Z}_\ell[G]} \mathrm{Sw}_x, \mathscr{H}^i(K)_{\bar\eta}),\\
\epsilon_x(K) &= \alpha_x(K) + \dim K_{\bar\eta} - \dim K_{\bar x}.
\end{aligned}$$

Then we have

$$\chi_c(U,K) = \chi(U,K) = (2-2g-\#(X-U))\dim K_{\bar\eta} - \sum_{x\in|U|} \epsilon_x(K) - \sum_{x\in X-U} \alpha_x(K).$$

Proof. Let E be a finite extension of $\mathbb{Q}_\ell$ such that K is represented by an object in $D^b_c(U,E)$, let R be the integral closure of $\mathbb{Z}_\ell$ in E, and let λ be a uniformizer of R. Represent K by an object (K_n) in $D^b_c(X,R)$, where $K_n \in \mathrm{ob}\, D^b_{\mathrm{ctf}}(U, R/(\lambda^{n+1}))$ and

$$K_{n+1} \otimes^L_{R/(\lambda^{n+2})} R/(\lambda^{n+1}) \cong K_n.$$

We have $R\Gamma_c(U,K_n) \in \mathrm{ob}\, D^b_{\mathrm{perf}}(R/(\lambda^{n+1}))$ and

$$R\Gamma_c(U,K_{n+1}) \otimes^L_{R/(\lambda^{n+2})} R/(\lambda^{n+1}) \cong R\Gamma_c(U,K_n).$$

Representing each $R\Gamma_c(U, K_n)$ by a bounded complex L_n of free $R/(\lambda^{n+1})$-modules of finite rank, we have quasi-isomorphisms

$$L_{n+1} \otimes_{R/(\lambda^{n+2})} R/(\lambda^{n+1}) \cong L_n.$$

By 10.1.15, we may find a bounded complex L of free R-modules of finite ranks and quasi-isomorphisms $L/\lambda^{n+1}L \to L_n$ such that the diagrams

$$\begin{array}{ccc} L/\lambda^{n+2}L & \to & L_{n+1} \\ \downarrow & & \downarrow \\ L/\lambda^{n+1}L & \to & L_n \end{array}$$

commute up to homotopy. Then we have

$$\begin{aligned} \chi_c(U, K) &= \sum_i (-1)^i \dim (L^i \otimes_R E) \\ &= \sum_i (-1)^i \mathrm{rank}\, (L^i/\lambda^{n+1}L^i) \\ &= \mathrm{rank}\, R\Gamma_c(U, K_n) \end{aligned}$$

for all n. Similarly, we have

$$\begin{aligned} \chi(U, K) &= \mathrm{rank}\, R\Gamma(U, K_n), \\ \dim K_{\bar\eta} &= \mathrm{rank}\, K_{n,\bar\eta}, \\ \alpha_x(K) &= \mathrm{rank}\, \alpha_x(K_n), \\ \epsilon_x(K) &= \mathrm{rank}\, \epsilon_x(K_n). \end{aligned}$$

We then apply 10.2.6 to K_n. □

10.3 Frobenius Correspondences

([SGA 5] XIV=XV.)

Let p be a prime number. A scheme is called to be of characteristic p if $p \cdot 1 = 0$ in $\Gamma(X, \mathscr{O}_X)$, or equivalently, the canonical morphism $X \to \mathrm{Spec}\, \mathbb{Z}$ factors through a morphism $X \to \mathrm{Spec}\, \mathbb{Z}/p$. Fix a power q of p. Then the morphism $\phi : \mathscr{O}_X \to \mathscr{O}_X$ that maps each section s of $\mathscr{O}_X$ to s^q is a morphism of sheaves of rings. The pair $(\mathrm{id}, \phi) : (X, \mathscr{O}_X) \to (X, \mathscr{O}_X)$ is a morphism of schemes. We call it the *Frobenius morphism* on X, and denote it by fr_X.

Let X be a scheme of characteristic p. Then every X-scheme is also of characteristic p. For any X-scheme Y, let $Y^{(q/X)} = Y \times_{X, \mathrm{fr}_X} X$. We have a Cartesian diagram

$$\begin{array}{ccc} Y^{(q/X)} & \stackrel{\pi_1}{\to} & Y \\ \pi_2 \downarrow & & \downarrow \\ X & \stackrel{\mathrm{fr}_X}{\to} & X, \end{array}$$

where π_1 and π_2 are the projections. We have a commutative diagram

$$\begin{array}{ccc} Y & \stackrel{\mathrm{fr}_Y}{\to} & Y \\ \downarrow & & \downarrow \\ X & \stackrel{\mathrm{fr}_X}{\to} & X, \end{array}$$

We define the *Frobenius morphism* $\mathrm{Fr}_{Y/X}$ *of* Y *relative to* X to be the X-morphism $\mathrm{Fr}_{Y/X} : Y \to Y^{(q/X)}$ such that $\pi_1 \mathrm{Fr}_{Y/X} = \mathrm{fr}_Y$.

Proposition 10.3.1. *Let* X *be a scheme of characteristic* p *and let* Y *be an* X*-scheme. The relative Frobenius morphism* $\mathrm{Fr}_{Y/X} : Y \to Y^{(q/X)}$ *is integral, radiciel, and surjective. If* Y *is etale over* X*, then* $\mathrm{Fr}_{Y/X}$ *is an isomorphism.*

Proof. It is clear that fr_X and fr_Y and π_1 are integral, radiciel and surjective. It follows that $\mathrm{Fr}_{Y/X}$ has the same property. If Y is etale over X, then $Y^{(q/X)}$ is also etale over X and hence $\mathrm{Fr}_{Y/X}$ is etale. By 2.3.9, $\mathrm{Fr}_{Y/X}$ is an isomorphism. □

Let X be a scheme of characteristic p and let $\mathscr{F}$ be a sheaf on X. For any etale X-scheme U, since $\mathrm{Fr}_{U/X}$ is an isomorphism, the restriction map $\mathscr{F}(U^{(q/X)}) \to \mathscr{F}(U)$ is an isomorphism. We thus have an isomorphism $\mathrm{fr}_{X*}\mathscr{F} \to \mathscr{F}$. Its inverse $\mathscr{F} \to \mathrm{fr}_{X*}\mathscr{F}$ defines a morphism

$$\mathrm{Fr}^*_{\mathscr{F}} : \mathrm{fr}^*_X \mathscr{F} \to \mathscr{F}.$$

The pair $(\mathrm{fr}_X, \mathrm{Fr}^*_{\mathscr{F}})$ is called the *Frobenius correspondence* on $(X, \mathscr{F})$. We can also define a morphism $\mathrm{Fr}^*_K : \mathrm{fr}^*_X K \to K$ for any object K in the derived category $D(X) = D(X, \mathbb{Z})$ of sheaves of abelian groups on X.

Proposition 10.3.2. *Let* X *and* Y *be schemes of characteristic* p*, and let* $f : Y \to X$ *be a morphism.*

(i) *For any sheaf* $\mathscr{G}$ *on* Y*, the composite*

$$f_*\mathscr{G} \to f_*\mathrm{fr}_{Y*}\mathrm{fr}^*_Y\mathscr{G} \stackrel{f_*\mathrm{fr}_{Y*}(\mathrm{Fr}^*_{\mathscr{G}})}{\to} f_*\mathrm{fr}_{Y*}\mathscr{G} \cong \mathrm{fr}_{X*}f_*\mathscr{G}$$

induces the morphism $\mathrm{Fr}^*_{f_*\mathscr{G}} : \mathrm{fr}^*_X f_*\mathscr{G} \to f_*\mathscr{G}$ *by adjunction.*

(ii) *For any sheaf* $\mathscr{F}$ *on* X*, the composite*

$$\mathrm{fr}^*_Y f^*\mathscr{F} \cong f^*\mathrm{fr}^*_X\mathscr{F} \stackrel{f^*(\mathrm{Fr}^*_{\mathscr{F}})}{\to} f^*\mathscr{F}$$

coincides with $\mathrm{Fr}^*_{f^*\mathscr{F}} : \mathrm{fr}^*_Y f^*\mathscr{F} \to f^*\mathscr{F}$.

Proof. We leave it for the reader to prove (i). Let us prove (ii). Consider the diagram

$$\begin{array}{ccccc}
 & & \mathrm{fr}_{X*}\mathrm{fr}_X^*\mathscr{F} & \stackrel{\text{adj}}{\to} & \mathrm{fr}_{X*}f_*f^*\mathrm{fr}_X^*\mathscr{F} \\
 & \text{adj} \nearrow & \mathrm{fr}_{X*}(\mathrm{Fr}_{\mathscr{F}}^*)\downarrow & & \downarrow \mathrm{fr}_{X*}f_*f^*(\mathrm{Fr}_{\mathscr{F}}^*) \\
\mathscr{F} & \to & \mathrm{fr}_{X*}\mathscr{F} & \stackrel{\text{adj}}{\to} & \mathrm{fr}_{X*}f_*f^*\mathscr{F} \\
 & \text{adj} \searrow & & & (1) \wr\| \\
 & & f_*f^*\mathscr{F} & \to & f_*\mathrm{fr}_{Y*}f^*\mathscr{F} \\
 & & & \text{adj} \searrow & \uparrow f_*\mathrm{fr}_{Y*}(\mathrm{Fr}_{f^*\mathscr{F}}^*) \\
 & & & & f_*\mathrm{fr}_{Y*}\mathrm{fr}_Y^*f^*\mathscr{F}.
\end{array}$$

Using the definition of the Frobenius correspondence, one can check that (1) commutes. It is clear that the other parts of the diagram commute. It follows that the composite in the proposition induces the same morphism $\mathscr{F} \to f_*\mathrm{fr}_{Y*}f^*\mathscr{F}$ as $\mathrm{Fr}_{f^*\mathscr{F}}^*$, and hence they coincide. □

Proposition 10.3.3. *Let X be a scheme of characteristic p, and let $K \in \mathrm{ob}\, D^+(X)$. The composite*

$$H^i(X, K) \to H^i(X, \mathrm{fr}_X^*K) \stackrel{\mathrm{Fr}_K^*}{\to} H^i(X, K)$$

is the identity for each i, where the first homomorphism is the composite

$$H^i(X, K) \stackrel{\text{adj}}{\to} H^i(X, R\mathrm{fr}_{X*}\mathrm{fr}_X^*K) \cong H^i(X, \mathrm{fr}_X^*K).$$

Proof. Let I and J be bounded below complexes of injective sheaves on X so that we have quasi-isomorphisms $K \to I$ and $\mathrm{fr}_X^*I \to J$. We can find a diagram

$$\begin{array}{ccc}
\mathrm{fr}_X^*I & \stackrel{\mathrm{Fr}_I^*}{\to} & I \\
\downarrow & & \downarrow \\
J & \to & J'
\end{array}$$

such that J' is a bounded below complex of injective sheaves, $I \to J'$ is a quasi-isomorphism, and the diagram commutes up to homotopy. So we have a diagram

$$\begin{array}{ccccc}
\Gamma(X, I) & \to & \Gamma(X, \mathrm{fr}_X^*I) & \stackrel{\mathrm{Fr}_I^*}{\to} & \Gamma(X, I) \\
 & & \downarrow & & \downarrow \\
 & & \Gamma(X, J) & \to & \Gamma(X, J')
\end{array}$$

which commutes up to homotopy. The composite of the morphisms

$$R\Gamma(X, K) \to R\Gamma(X, \mathrm{fr}_X^*K) \stackrel{\mathrm{Fr}_K^*}{\to} R\Gamma(X, K)$$

can be identified with the composite

$$\Gamma(X, I) \to \Gamma(X, \mathrm{fr}_X^* I) \to \Gamma(X, J) \to \Gamma(X, J').$$

It is homotopic to the composite

$$\Gamma(X, I) \to \Gamma(X, \mathrm{fr}_X^* I) \stackrel{\mathrm{Fr}_I^*}{\to} \Gamma(X, I) \to \Gamma(X, J').$$

We claim that the composite

$$\Gamma(X, I) \to \Gamma(X, \mathrm{fr}_X^* I) \stackrel{\mathrm{Fr}_I^*}{\to} \Gamma(X, I)$$

is the identity. Our assertion then follows. To prove the claim, note that the composite

$$I \to \mathrm{fr}_{X*}\mathrm{fr}_X^* I \stackrel{\mathrm{fr}_*(\mathrm{Fr}_I^*)}{\to} \mathrm{fr}_{X*} I$$

is the inverse of the morphism $\mathrm{fr}_{X*}(I) \to I$ defined by the restriction

$$(\mathrm{fr}_{X*}(I))(U) = I(U^{(q/X)}) \to I(U)$$

for any etale X-scheme U. If $U = X$, the restriction $I(U^{(q/X)}) \to I(U)$ is the identity. So the composite

$$\Gamma(X, I) \to \Gamma(X, \mathrm{fr}_{X*}\mathrm{fr}_X^* I) \stackrel{\mathrm{Fr}_I^*}{\to} \Gamma(X, \mathrm{fr}_{X*} I)$$

is the identity. This proves the claim. □

Let $\mathbb{F}_q$ be a finite field of characteristic p with q elements, and let $\mathbb{F}$ be an algebraic closure of $\mathbb{F}_q$. Denote the Frobenius morphism on $\mathrm{Spec}\,\mathbb{F}$ by $\mathrm{fr}_{\mathbb{F}}$. Let X_0 be a scheme over $\mathrm{Spec}\,\mathbb{F}_q$ and let $X = X_0 \otimes_{\mathbb{F}_q} \mathbb{F}$. Denote the $\mathbb{F}$-morphism

$$\mathrm{fr}_{X_0} \otimes \mathrm{id} : X_0 \otimes_{\mathbb{F}_q} \mathbb{F} \to X_0 \otimes_{\mathbb{F}_q} \mathbb{F}$$

by $F_X : X \to X$ or simply F. If $X_0 = \mathbb{A}^n_{\mathbb{F}_q}$, we have $X = \mathbb{A}^n_{\mathbb{F}}$, and $F : \mathbb{A}^n_{\mathbb{F}} \to \mathbb{A}^n_{\mathbb{F}}$ corresponds to the $\mathbb{F}$-algebra homomorphism

$$\mathbb{F}[t_1, \ldots, t_n] \to \mathbb{F}[t_1, \ldots, t_n], \quad t_i \mapsto t_i^q.$$

So F maps a point in $\mathbb{A}^n_{\mathbb{F}}$ with coordinate $(a_1, \ldots, a_n)$ to the point with coordinate $(a_1^q, \ldots, a_n^q)$.

For any scheme Y over a field k, and any extension K of k, a K-point of Y is a k-morphism $\mathrm{Spec}\,K \to Y$. The set of K-points in Y is denoted by $Y(K)$. We have a canonical bijection $X(\mathbb{F}) \cong X_0(\mathbb{F})$. For any $\mathbb{F}$-point $t : \mathrm{Spec}\,\mathbb{F} \to X_0$, since $\mathrm{fr}_{X_0} \circ t = t \circ \mathrm{fr}_{\mathbb{F}}$, the morphism fr_{X_0} maps t to the $\mathbb{F}$-point $t \circ \mathrm{fr}_{\mathbb{F}}$. So t is fixed by fr_{X_0} if and only if there exists an $\mathbb{F}_q$-point $t_0 : \mathrm{Spec}\,\mathbb{F}_q \to X_0$ such that t is the composite

$$\mathrm{Spec}\,\mathbb{F} \to \mathrm{Spec}\,\mathbb{F}_q \stackrel{t_0}{\to} X_0.$$

It follows that the set of fixed points of F on $X(\mathbb{F})$ can be identified with $X_0(\mathbb{F}_q)$.

Let $\mathscr{F}_0$ be a sheaf on X_0, let $\pi : X \to X_0$ be the projection, and let $\mathscr{F} = \pi^*\mathscr{F}_0$ be the inverse image of $\mathscr{F}_0$. Define

$$F^*_{\mathscr{F}_0} : F^*\mathscr{F} \to \mathscr{F}$$

to be the morphism induced from $\mathrm{Fr}^*_{\mathscr{F}_0} : \mathrm{fr}^*_{X_0}\mathscr{F}_0 \to \mathscr{F}_0$ by base change, that is, the composite

$$F^*\mathscr{F} = F^*\pi^*\mathscr{F}_0 \cong \pi^*\mathrm{fr}^*_{X_0}\mathscr{F}_0 \overset{\pi^*(\mathrm{Fr}^*_{\mathscr{F}_0})}{\to} \pi^*\mathscr{F}_0 = \mathscr{F}.$$

We call the pair $(F, F^*_{\mathscr{F}_0})$ the *geometric Frobenius correspondence*. Similarly, for any $K_0 \in \mathrm{ob}\, D(X_0)$, we can define a morphism $F^*_{K_0} : F^*K \to K$ in $D(X)$, where $K = \pi^*K_0$. On the other hand, we have an isomorphism $(\mathrm{id}_{X_0} \times \mathrm{fr}_{\mathbb{F}})^*\mathscr{F} \cong \mathscr{F}$ defined as the composite

$$(\mathrm{id}_{X_0} \times \mathrm{fr}_{\mathbb{F}})^*\mathscr{F} = (\mathrm{id}_{X_0} \times \mathrm{fr}_{\mathbb{F}})^*\pi^*\mathscr{F}_0 \cong (\pi \circ (\mathrm{id}_{X_0} \times \mathrm{fr}_{\mathbb{F}}))^*\mathscr{F}_0 = \pi^*\mathscr{F}_0 = \mathscr{F}.$$

Proposition 10.3.4. *Notation as above. The composite*

$$\mathrm{fr}^*_X\mathscr{F} \cong F^*(\mathrm{id}_{X_0} \times \mathrm{fr}_{\mathbb{F}})^*\mathscr{F} \cong F^*\mathscr{F} \overset{F^*_{\mathscr{F}_0}}{\to} \mathscr{F}$$

coincides with $\mathrm{Fr}^*_{\mathscr{F}} : \mathrm{fr}^*_X\mathscr{F} \to \mathscr{F}$.

Proof. We have a commutative diagram

$$\begin{array}{ccccccc}
\mathrm{fr}^*_X\mathscr{F} & \cong & F^*(\mathrm{id} \times \mathrm{fr}_{\mathbb{F}})^*\mathscr{F} & \cong & F^*\mathscr{F} & \overset{F^*_{\mathscr{F}_0}}{\to} & \pi^*\mathscr{F}_0 = \mathscr{F} \\
\| & & \| & & \| & & \uparrow \pi^*(\mathrm{Fr}^*_{\mathscr{F}_0}) \\
\mathrm{fr}^*_X\pi^*\mathscr{F}_0 & \cong & F^*(\mathrm{id} \times \mathrm{fr}_{\mathbb{F}})^*\pi^*\mathscr{F}_0 & \cong & F^*\pi^*\mathscr{F}_0 & \cong & \pi^*\mathrm{fr}^*_{X_0}\mathscr{F}_0 \\
\wr\| & & \wr\| & & \wr\| & & \wr\| \\
(\pi \circ \mathrm{fr}_X)^*\mathscr{F}_0 & = & (\pi \circ (\mathrm{id} \times \mathrm{fr}_{\mathbb{F}}) \circ F)^*\mathscr{F}_0 & = & (\pi F)^*\mathscr{F}_0 & = & (\mathrm{fr}_{X_0} \circ \pi)^*\mathscr{F}_0.
\end{array}$$

So the composite in the proposition coincides with the composite

$$\mathrm{fr}^*_X\mathscr{F} = \mathrm{fr}^*_X\pi^*\mathscr{F}_0 \cong \pi^*\mathrm{fr}^*_{X_0}\mathscr{F}_0 \overset{\pi^*(\mathrm{Fr}^*_{\mathscr{F}_0})}{\to} \pi^*\mathscr{F}_0 = \mathscr{F}.$$

By 10.3.2 (ii), this last composite coincides with $\mathrm{Fr}^*_{\mathscr{F}} : \mathrm{fr}^*_X\mathscr{F} \to \mathscr{F}$. □

Corollary 10.3.5. *Notation as above. For any $K_0 \in \mathrm{ob}\, D^+(X_0)$, the composite*

$$H^i(X, K) \to H^i(X, F^*K) \overset{F^*_{K_0}}{\to} H^i(X, K)$$

and the composite

$$H^i(X, K) \to H^i(X, (\mathrm{id}_{X_0} \times \mathrm{fr}_{\mathbb{F}})^*K) \cong H^i(X, K)$$

are inverse to each other.

Proof. Denote the homomorphisms defined by the composites in the corollary by ϕ and ψ, respectively. Consider the diagram

$$\begin{array}{ccccc} H^i(X,K) & \to & H^i(X,(\mathrm{id}_{X_0}\times \mathrm{fr}_{\mathbb{F}})^*K) & \cong & H^i(X,K) \\ \downarrow & & \downarrow & & \downarrow \\ H^i(X,\mathrm{fr}_X^* K) & \cong & H^i(X,F^*(\mathrm{id}_{X_0}\times \mathrm{fr}_{\mathbb{F}})^*K) & \cong & H^i(X,F^*K) \\ {\scriptstyle \mathrm{Fr}_K^*}\downarrow & & & & \downarrow {\scriptstyle F_{K_0}^*} \\ H^i(X,K) & & = & & H^i(X,K). \end{array}$$

By 10.3.4, the lower rectangle in the diagram commutes. It is clear that the other parts of the diagram commute. By 10.3.3, the composite of the vertical arrows on the left is the identity. It follows that $\phi\psi = \mathrm{id}$. It is clear that ψ is an isomorphism. So ϕ and ψ are inverse to each other. □

Proposition 10.3.6. *Let n be a positive integer, let $\mathbb{F}_{q^n}$ be the extension of $\mathbb{F}_q$ of degree n contained in $\mathbb{F}$, let $X_1 = X_0\otimes_{\mathbb{F}_q}\mathbb{F}_{q^n}$, and let $\mathscr{F}_1$ be the inverse image on X_1 of a sheaf $\mathscr{F}_0$ on X_0. Define $\mathrm{fr}_{X_1} : X_1 \to X_1$ so that it is the identity on the underlying topological spaces, and that it maps each section of $\mathscr{O}_{X_1}$ to its q^n-th power. Define $\mathrm{Fr}^*_{\mathscr{F}_1} : \mathrm{fr}^*_{X_1}\mathscr{F}_1 \to \mathscr{F}_1$ as before using q^n in place of q. If we iterate n times the correspondence $(\mathrm{fr}_{X_0}, \mathrm{Fr}^*_{\mathscr{F}_0})$ and take the base change of the iteration under $\mathbb{F}_{q^n} \to \mathbb{F}$, we get the correspondence $(\mathrm{fr}_{X_1}, \mathrm{Fr}^*_{\mathscr{F}_1})$. Let $F_1 = \mathrm{fr}_{X_1}\otimes \mathrm{id}_{\mathbb{F}}$ and let $F^*_{\mathscr{F}_1} : F_1^*\mathscr{F} \to \mathscr{F}$ be the morphism deduced from $\mathrm{Fr}^*_{\mathscr{F}_1}$ by the base change $\mathbb{F}_{q^n} \to \mathbb{F}$. Then the correspondence $(F_1, F^*_{\mathscr{F}_1})$ is the n-th iteration of the correspondence $(F, F^*_{\mathscr{F}_0})$.*

Proof. Let $\pi_1 : X_1 \to X_0$ be the projection. Denote by $(\mathrm{fr}^n_{X_0}, \mathrm{Fr}^{n*}_{\mathscr{F}_0})$ the n-th iteration of $(\mathrm{fr}_{X_0}, \mathrm{Fr}^*_{\mathscr{F}_0})$. By 10.3.2 (ii), the composite

$$\mathrm{fr}^*_{X_1}\pi_1^*\mathscr{F}_0 \cong \pi_1^*(\mathrm{fr}^n_{X_0})^*\mathscr{F}_0 \overset{\pi_1^*(\mathrm{Fr}^{n*}_{\mathscr{F}_0})}{\to} \pi_1^*\mathscr{F}_0$$

coincides with $\mathrm{Fr}^*_{\mathscr{F}_1}$. Our assertion follows. □

Let x be a closed point of X_0 with $[k(x) : \mathbb{F}_q] = n$ and let $\bar{x} \in X(\mathbb{F})$ be an $\mathbb{F}$-point of X whose image in X_0 is x. Then $\bar{x}$ is a fixed point of $F^n = F_1$. The n-th iteration of the correspondence $(F, F^*_{\mathscr{F}_0})$ induces a homomorphism $F^{n*}_{\bar{x}} : \mathscr{F}_{\bar{x}} \to \mathscr{F}_{\bar{x}}$. Denote the closed immersion $\mathrm{Spec}\, k(x) \to X_0$ by i. Then $i^*\mathscr{F}_0$ is completely determined by the galois action of $\mathrm{Gal}(\overline{k(x)}/k(x))$ on $\mathscr{F}_{\bar{x}}$. Let $f_x : \mathscr{F}_{\bar{x}} \to \mathscr{F}_{\bar{x}}$ be the action on $\mathscr{F}_{\bar{x}}$ of the Frobenius substitution $\alpha \mapsto \alpha^{q^n}$ in $\mathrm{Gal}(\overline{k(x)}/k(x))$.

Proposition 10.3.7. *With the above notation, the homomorphisms $F^{n*}_{\bar{x}} : \mathscr{F}_{\bar{x}} \to \mathscr{F}_{\bar{x}}$ and $f_x : \mathscr{F}_{\bar{x}} \to \mathscr{F}_{\bar{x}}$ are inverse to each other.*

Proof. Without loss of generality, assume $n = 1$ and hence $k(x) = \mathbb{F}_q$. Denote the image of $\bar{x}$ in $X_0(\mathbb{F})$ also by $\bar{x}$. Then $\bar{x}$ is a fixed point of fr_{X_0}, and $F_{\bar{x}}^*$ is the homomorphism induced by $\mathrm{Fr}_{\mathscr{F}_0}^* : \mathrm{fr}_{X_0}^* \mathscr{F}_0 \to \mathscr{F}_0$ on stalks at $\bar{x}$. By 10.3.2 (ii), we have a commutative diagram

$$\begin{array}{ccc} \mathrm{fr}_{\mathbb{F}_q}^* i^* \mathscr{F}_0 & \cong & i^* \mathrm{fr}_{X_0}^* \mathscr{F}_0 \\ {\scriptstyle \mathrm{Fr}_{i^*\mathscr{F}_0}} \downarrow & & \downarrow {\scriptstyle i^*(\mathrm{Fr}_{\mathscr{F}_0}^*)} \\ i^* \mathscr{F}_0 & = & i^* \mathscr{F}_0. \end{array}$$

We are thus reduced to the case where $X_0 = \mathrm{Spec}\, \mathbb{F}_q$. One checks $F_x^* = f_{\bar{x}}^{-1}$ directly in this case, or use 10.3.4. (Even though $\mathrm{fr}_{\mathbb{F}_q}$ is the identity, $\mathrm{Fr}_{i^*\mathscr{F}_0}$ may be nontrivial.) □

Remark 10.3.8. Note that 10.3.5 also follows from 10.3.7 applied to the sheaf $\mathscr{F}_0 = R^i f_{0*} K_0$ on $\mathrm{Spec}\, \mathbb{F}_q$, where $f_0 : X_0 \to \mathrm{Spec}\, \mathbb{F}_q$ is the structure morphism. Let $\bar{x}$ be the geometric point $\mathrm{Spec}\, \mathbb{F} \to \mathrm{Spec}\, \mathbb{F}_q$, and let $f = f_0 \otimes \mathrm{id}_{\mathbb{F}}$. Using 10.3.2 (i) and taking base change, we find that the composite

$$Rf_* K \to Rf_* RF_* F^* K \stackrel{F_{K_0}^*}{\to} Rf_* RF_* K \cong RF_* Rf_* K$$

induces the morphism

$$F_{Rf_{0*}K_0}^* : F^* Rf_* K \to Rf_* K$$

by adjunction. But $\mathrm{fr}_{\mathbb{F}_q}$ is the identity morphism. So the above composite can be written as

$$Rf_* K \to Rf_* RF_* F^* K \stackrel{F_{K_0}^*}{\to} Rf_* RF_* K \cong Rf_* K.$$

It follows that the composite

$$H^i(X, K) \to H^i(X, F^* K) \stackrel{F_{K_0}^*}{\to} H^i(X, K)$$

can be identified with the homomorphism $F_{\bar{x}} : (R^i f_* K)_{\bar{x}} \to (R^i f_* K)_{\bar{x}}$. The composite

$$H^i(X, K) \to H^i(X, (\mathrm{id}_{X_0} \times \mathrm{fr}_{\mathbb{F}})^* K) \cong H^i(X, K)$$

can be identified with the action of the Frobenius substitution on $(R^i f_{0*} \mathscr{F}_0)_{\bar{x}}$. So 10.3.7 implies 10.3.5.

10.4 Lefschetz Trace Formula

([SGA $4\frac{1}{2}$] Rapport 4–6.)

Let A be a commutative noetherian ring, let G be a finite group, and let $A[G]^\natural$ be the quotient of the additive group $(A[G],+)$ by the subgroup generated by elements of the form $xy-yx$ $(x,y\in A[G])$. Then $A[G]^\natural$ has an A-module structure, and it is the quotient of the A-module $A[G]$ by the submodule generated by elements of the form $h^{-1}gh-g$ $(g,h\in G)$. For any $g\in G$, let $[g]$ be the subset of G consisting of those elements conjugate to g. Elements in $[g]$ have the same image $\bar{g}$ in $A[G]^\natural$. Choose $g_1,\ldots,g_k\in G$ so that G is the disjoint union of the conjugacy classes $[g_j]$ $(j=1,\ldots,k)$. Then $A[G]^\natural$ is the free A-module with basis $\bar{g}_j$ $(j=1,\ldots,k)$. For any $g\in G$, the map

$$A[G]\to A,\quad \sum_{h\in G}a_h h\mapsto \sum_{h\in[g]}a_h$$

is an A-module homomorphism, and it vanishes on the submodule generated by $h'^{-1}hh'-h$ $(h,h'\in G)$. So it induces a homomorphism

$$\theta_g : A[G]^\natural\to A.$$

Proposition 10.4.1. *Let P be a finitely generated projective $A[G]$-module and let $v:P\to P$ be an $A[G]$-module endomorphism. For any $g\in G$, let $\mathrm{Tr}_A(g^{-1}v,P)$ (resp, $\mathrm{Tr}_{A[G]}(v,P)$) be the trace of $g^{-1}v$ (resp. v) considered as an A-module endomorphism (resp. $A[G]$-module endomorphism). Choose $g_j\in G$ $(j=1,\ldots,k)$ so that G is the disjoint union of the conjugacy classes $[g_j]$. Denote the images of g_j in $A[G]^\natural$ by $\bar{g}_j$. Then we have*

$$\mathrm{Tr}_{A[G]}(v,P)=\sum_{j=1}^{k}\theta_{g_j}(\mathrm{Tr}_{A[G]}(v,P))\bar{g}_j,$$
$$\mathrm{Tr}_A(g^{-1}v,P)=\#C(g)\theta_g(\mathrm{Tr}_{A[G]}(v,P)),$$

where $C(g)=\{h\in G|gh=hg\}$ is the center of g.

Proof. Let $\phi:P\to A[G]^m$ and $\psi:A[G]^m\to P$ be $A[G]$-module homomorphisms such that $\psi\phi=\mathrm{id}$. We have

$$\mathrm{Tr}_A(g^{-1}v,P)=\mathrm{Tr}_A(\phi g^{-1}v\psi,A[G]^m)=\mathrm{Tr}_A(g^{-1}\phi v\psi,A[G]^m),$$
$$\mathrm{Tr}_{A[G]}(v,P)=\mathrm{Tr}_{A[G]}(\phi v\psi,A[G]^m).$$

To prove the proposition, we may assume $P = A[G]^m$. For each $i \in \{1, \ldots, m\}$, Let e_i be the element in $A[G]^m$ whose i-th component is 1 and whose other components are 0. Write

$$v(e_i) = \sum_{j=1}^{m} \Big(\sum_{h' \in G} a_{ij}^{h'} h' \Big) e_j$$

with $a_{ij}^{h'} \in A$. Then $\mathrm{Tr}_{A[G]}(v, P)$ is the image of $\sum_{i=1}^{m} \sum_{h \in G} a_{ii}^{h} h$ in $A[G]^\natural$. Elements in $[g_j]$ have the same image in $A[G]^\natural$. So we have

$$\mathrm{Tr}_{A[G]}(v, P) = \sum_{i=1}^{m} \sum_{j=1}^{k} \sum_{h \in [g_j]} a_{ii}^{h} \bar{g}_j = \sum_{j=1}^{k} \Big(\sum_{i=1}^{m} \sum_{h \in [g_j]} a_{ii}^{h} \Big) \bar{g}_j,$$

$$\theta_g(\mathrm{Tr}_{A[G]}(v, P)) = \sum_{i=1}^{m} \sum_{h \in [g]} a_{ii}^{h}.$$

Hence

$$\mathrm{Tr}_{A[G]}(v, P) = \sum_{j=1}^{k} \theta_{g_j}(\mathrm{Tr}_{A[G]}(v, P)) \bar{g}_j.$$

(Actually we have $x = \sum_{j=1}^{k} \theta_{g_j}(x) \bar{g}_j$ for any $x \in A[G]^\natural$.) On the other hand, we have

$$(g^{-1}v)(he_i) = g^{-1}hv(e_i) = \sum_{j=1}^{m} \sum_{h' \in G} a_{ij}^{h'} g^{-1}hh' e_j = \sum_{j=1}^{m} \sum_{h' \in G} a_{ij}^{h^{-1}gh'} h' e_j.$$

So we have

$$\begin{aligned}
\mathrm{Tr}_A(g^{-1}v, P) &= \sum_{i=1}^{m} \sum_{h \in G} a_{ii}^{h^{-1}gh} \\
&= \#C(g) \sum_{i=1}^{m} \sum_{h \in [g]} a_{ii}^{h} \\
&= \#C(g) \theta_g(\mathrm{Tr}_{A[G]}(v, P)).
\end{aligned}$$

□

Proposition 10.4.2. *Let Λ be a commutative ring, A a commutative Λ-algebra, P a finitely generated projective $\Lambda[G]$-module, and M a finitely generated $A[G]$-module which is projective as an A-module. Then the $A[G]$-module $M \otimes_\Lambda P$ is a finitely generated projective $A[G]$-module, where the G-action on $M \otimes_\Lambda P$ is given by $g(x \otimes y) = gx \otimes gy$ for any $x \in M$ and $y \in P$. Let $v : P \to P$ be a $\Lambda[G]$-module endomorphism. If $\mathrm{Tr}_{\Lambda[G]}(v, P) = 0$, then $\mathrm{Tr}_{A[G]}(\mathrm{id}_M \otimes v, M \otimes_\Lambda P) = 0$.*

Proof. We may reduce to the case where $P = \Lambda[G]^m$. Let $(M\otimes_\Lambda P)'$ be the $A[G]$-module $M\otimes_\Lambda P$ so that the G-action is given by $g(x\otimes y) = x\otimes gy$ for any $x\in M$ and $y\in P$. It is clear that $(M\otimes_\Lambda P)'$ is a finitely generated projective $A[G]$-module. One can check that

$$\phi : M\otimes_\Lambda \Lambda[G]^m \to (M\otimes_\Lambda \Lambda[G]^m)', \quad x\otimes ge_i \mapsto g^{-1}x\otimes ge_i$$

is an $A[G]$-module isomorphism, where e_i $(i = 1,\ldots,m)$ is the element in $\Lambda[G]^m$ whose i-th component is 1 and whose other components are 0. So $M\otimes_\Lambda P$ is a finitely generated projective $A[G]$-module.

Write

$$v(e_i) = \sum_{j=1}^{m}\sum_{g\in G} a_{ij}^g ge_j$$

with $a_{ij}^g\in\Lambda$. We have

$$\begin{aligned}(\phi\circ(\mathrm{id}_M\otimes v)\circ\phi^{-1})(x\otimes e_i) &= \phi((\mathrm{id}_M\otimes v)(x\otimes e_i))\\ &= \phi\Big(x\otimes\sum_{j=1}^{m}\sum_{g\in G} a_{ij}^g ge_j\Big)\\ &= \sum_{j=1}^{m}\sum_{g\in G} a_{ij}^g g^{-1}x\otimes ge_j.\end{aligned}$$

Hence $\mathrm{Tr}_{A[G]}(\mathrm{id}_M\otimes v, M\otimes_\Lambda P)$ is the image of $\sum_{i=1}^{m}\sum_{g\in G} a_{ii}^g \mathrm{Tr}_A(g^{-1},M)g$ in $A[G]^\natural$. Let $g_1,\ldots,g_k\in G$ such that G is the disjoint union of the conjugacy classes $[g_1],\ldots,[g_n]$, and let $\bar{g}_j$ be the image of g_j in $A[G]^\natural$. Then we have

$$\begin{aligned}\mathrm{Tr}_{A[G]}(\mathrm{id}_M\otimes v, M\otimes_\Lambda P) &= \sum_{j=1}^{k}\sum_{i=1}^{m}\sum_{g\in[g_j]} a_{ii}^g\mathrm{Tr}_A(g^{-1},M)\bar{g}_j\\ &= \sum_{j=1}^{k}\Big(\sum_{i=1}^{m}\sum_{g\in[g_j]} a_{ii}^g\Big)\mathrm{Tr}_A(g_j^{-1},M)\bar{g}_j.\end{aligned}$$

On the other hand, we have

$$0 = \mathrm{Tr}_{\Lambda[G]}(v,P) = \sum_{i=1}^{m}\sum_{g\in G} a_{ii}^g\bar{g} = \sum_{j=1}^{k}\Big(\sum_{i=1}^{m}\sum_{g\in[g_j]} a_{ii}^g\Big)\bar{g}_j.$$

Since $\{\bar{g}_1,\ldots,\bar{g}_k\}$ is a basis of $\Lambda[G]^\natural$, this implies that

$$\sum_{i=1}^{m}\sum_{g\in[g_j]} a_{ii}^g = 0.$$

So we have $\mathrm{Tr}_{A[G]}(\mathrm{id}_M\otimes v, M\otimes_\Lambda P) = 0$. □

The kernel of the homomorphism

$$\epsilon : A[G] \to A, \quad \sum_{g\in G} a_g g \mapsto \sum_{g\in G} a_g$$

is generated by $g-e$ ($g \in G$) as an A-module. If we put the trivial G-action on A, then ϵ is a homomorphism of $A[G]$-modules. For any $A[G]$-module P, we have

$$A \otimes_{A[G]} P \cong P_G,$$

where P_G is the space of G-coinvariants, that is, the quotient of G by the A-submodule generated by $gx - x$ ($g \in G, x \in P$). If P is a finitely generated projective $A[G]$-module, then P_G is a finitely generated projective A-module. The homomorphism ϵ induces a homomorphism

$$\delta : A[G]^\natural \to A.$$

Proposition 10.4.3. *Let P be a finitely generated projective $A[G]$-module, let $u : P \to P$ be an $A[G]$-module endomorphism, and let $u' : P_G \to P_G$ be the endomorphism induced by u. Then $\mathrm{Tr}_A(u', P_G) = \delta(\mathrm{Tr}_{A[G]}(u, P))$.*

Proof. We may reduce to the case where $P = A[G]^m$. For each $i \in \{1, \ldots, m\}$, let e_i (resp. e_i') be the element in $A[G]^m$ (resp. A^m) whose i-th component is 1 and whose other components are 0. Write

$$u(e_i) = \sum_{j=1}^m \Big(\sum_{g\in G} a_{ij}^g g\Big) e_j$$

with $a_{ij}^g \in A$. We have

$$\delta(\mathrm{Tr}_{A[G]}(u, P)) = \sum_i \sum_{g\in G} a_{ii}^g.$$

Since $P_G \cong A \otimes_{A[G]} P \cong A^m$, we have

$$u'(e_i') = \sum_j \Big(\sum_{g\in G} a_{ij}^g\Big) e_j'.$$

So

$$\mathrm{Tr}_A(u', P_G) = \sum_i \sum_{g\in G} a_{ii}^g = \delta(\mathrm{Tr}_{A[G]}(u, P)).$$

□

Let X_0 be a compactifiable scheme over a finite field $\mathbb{F}_q$ of characteristic p with q elements, A a noetherian $\mathbb{Z}/\ell^n$-algebra with $(\ell, p) = 1$, $K_0 \in \operatorname{ob} D^b_{\mathrm{ctf}}(X_0, A)$, $\mathbb{F}$ an algebraic closure of $\mathbb{F}_q$, $X = X_0 \otimes_{\mathbb{F}_q} \mathbb{F}$, K the inverse image of K_0 on X, and $F : X \to X$ and $F^*_{K_0} : F^*K \to K$ the morphisms induced by $\mathrm{fr}_{X_0} : X_0 \to X_0$ and $\mathrm{Fr}^*_{K_0} : \mathrm{fr}^*_{X_0} K_0 \to K_0$ respectively by base change. Choose a compactification $\overline{X}_0$ of X_0, let $\overline{X} = \overline{X}_0 \otimes_{\mathbb{F}_q} \mathbb{F}$, and let $j : X \hookrightarrow \overline{X}$ be the open immersion. Denote by $F^* : R\Gamma_c(X, K) \to R\Gamma_c(X, K)$ the composite

$$\begin{aligned} R\Gamma_c(X, K) &\cong R\Gamma(\overline{X}, j_! K) \\ &\to R\Gamma(\overline{X}, F^* j_! K) \\ &\cong R\Gamma(\overline{X}, j_! F^* K) \\ &\overset{F^*_{K_0}}{\to} R\Gamma(\overline{X}, j_! K) \\ &\cong R\Gamma_c(X, K). \end{aligned}$$

This morphism is independent of the choice of the compactification. By 10.2.1, 7.4.7 and 7.8.1, we have $R\Gamma_c(X, K) \in \operatorname{ob} D^b_{\mathrm{perf}}(A)$, and for any geometric point $\bar{x}$ of X_0, we have $K_{\bar{x}} \in \operatorname{ob} D^b_{\mathrm{perf}}(A)$. If $\bar{x}$ is an $\mathbb{F}$-point of X fixed by F, then $F^*_{K_0} : F^*K \to K$ induces a morphism $F^*_{\bar{x}} : K_{\bar{x}} \to K_{\bar{x}}$. The main result of this section is the following:

Theorem 10.4.4 (Lefschetz trace formula). *Notation as above. Let $X^F \cong X_0(\mathbb{F}_q)$ be the set of fixed points of F on $X(\mathbb{F}) \cong X_0(\mathbb{F})$. We have*

$$\sum_{\bar{x} \in X^F} \mathrm{Tr}(F^*_{\bar{x}}, K_{\bar{x}}) = \mathrm{Tr}(F^*, R\Gamma_c(X, K)).$$

We first prove the following corollary of this theorem.

Theorem 10.4.5. *Let X_0 be a compactifiable scheme over* $\operatorname{Spec} \mathbb{F}_q$ *and let $K \in \operatorname{ob} D^b_c(X_0, \overline{\mathbb{Q}}_\ell)$. We have*

$$\sum_{\bar{x} \in X^F} \mathrm{Tr}(F^*_{\bar{x}}, K_{\bar{x}}) = \mathrm{Tr}(F^*, R\Gamma_c(X, K)).$$

Proof. Let E be a finite extension of $\mathbb{Q}_\ell$ such that K is represented by an object in $D^b_c(X, E)$, let R be the integral closure of $\mathbb{Z}_\ell$ in E, and let λ be a uniformizer of R. Represent K by an object (K_n) in $D^b_c(X, R)$, where $K_n \in \operatorname{ob} D^b_{\mathrm{ctf}}(X, R/(\lambda^{n+1}))$ and

$$K_{n+1} \otimes^L_{R/(\lambda^{n+2})} R/(\lambda^{n+1}) \cong K_n.$$

We have $R\Gamma_c(X, K_n) \in \mathrm{ob}\, D^b_{\mathrm{perf}}(R/(\lambda^{n+1}))$ and

$$R\Gamma_c(X, K_{n+1}) \otimes^L_{R/(\lambda^{n+2})} R/(\lambda^{n+1}) \cong R\Gamma_c(X, K_n).$$

Representing each $R\Gamma_c(X, K_n)$ by a bounded complex L_n of free $R/(\lambda^{n+1})$-modules of finite rank. Then we have quasi-isomorphisms

$$L_{n+1} \otimes_{R/(\lambda^{n+2})} R/(\lambda^{n+1}) \cong L_n.$$

By 10.1.15, we can find a bounded complex L of free R-modules of finite rank and quasi-isomorphisms $L/\lambda^{n+1}L \to L_n$ such that the diagrams

$$\begin{array}{ccc} L/\lambda^{n+2}L & \to & L_{n+1} \\ \downarrow & & \downarrow \\ L/\lambda^{n+1}L & \to & L_n \end{array}$$

commute up to homotopy. The morphism F^* on $R\Gamma_c(X, K_n)$ can be represented by a morphism of complexes F_n^* on $L/\lambda^{n+1}L$. By 10.1.11, we may assume

$$F_n^* = F_{n+1}^* \otimes \mathrm{id}_{R/(\lambda^{n+1})}.$$

So the family

$$\Big(\mathrm{Tr}(F^*, R\Gamma_c(X, K_n))\Big) = \Big(\mathrm{Tr}(F_n^*, L/\lambda^{n+1}L)\Big)$$

defines an element in $R = \varprojlim_n R/(\lambda^{n+1})$, and this element is $\mathrm{Tr}(F^*, R\Gamma_c(X, K))$. Similarly, for any $\bar{x} \in X^F$, the family $(\mathrm{Tr}(F_{\bar{x}}^*, K_{n\bar{x}}))$ defines an element in $R = \varprojlim_n R/(\lambda^{n+1})$, and this element is $\mathrm{Tr}(F_{\bar{x}}^*, K_{\bar{x}})$. By 10.4.4, we have

$$\sum_{\bar{x} \in X^F} \mathrm{Tr}(F_{\bar{x}}^*, K_{n\bar{x}}) = \mathrm{Tr}(F^*, R\Gamma_c(X, K_n))$$

for all n. So we have

$$\sum_{\bar{x} \in X^F} \mathrm{Tr}(F_{\bar{x}}^*, K_{\bar{x}}) = \mathrm{Tr}(F^*, R\Gamma_c(X, K_n)).$$

□

To prove 10.4.4, we first consider the special case where $\dim X \leq 1$, and then reduce the general case to the special case.

Lemma 10.4.6. *Let X be a scheme of finite type over an algebraically closed field k, $f : X \to X$ a k-morphism, $K \in \mathrm{ob}\, D^b_{\mathrm{ctf}}(X, A)$, $f^* : f^*K \to K$ a morphism, and X^f the set of fixed points of f in $X(k)$. If $\dim X = 0$, then we have*

$$\sum_{\bar{x} \in X^f} \mathrm{Tr}(f^*, K_{\bar{x}}) = \mathrm{Tr}(f^*, R\Gamma_c(X, K)).$$

In particular, 10.4.4 holds if $\dim X_0 = 0$.

Proof. Use the fact that

$$R\Gamma_c(X,K)=\bigoplus_{\bar{x}\in X(k)} K_{\bar{x}}.$$

□

Lemma 10.4.7. *Let X be an irreducible smooth projective curve over an algebraically closed field k, $f : X \to X$ a k-morphism with isolated fixed points, and U a dense open subset of X such that $f^{-1}(U) = U$. For any fixed point x of f on $X(k)$, the multiplicity of f at x is defined to be $v_x(f^*(\pi_x)-\pi_x)$, where v_x is the valuation of the function field of X defined by x and π_x is a uniformizer for this valuation. Suppose that fixed points of f in $X-U$ have multiplicity 1. Let $v_U(f)$ be the sum of multiplicities of fixed points of f in U. Then we have*

$$\sum_{i=0}^{2}(-1)^i\mathrm{Tr}(f^*,H_c^i(U,\mathbb{Q}_\ell))=v_U(f).$$

Proof. We have

$$\begin{aligned}&\sum_{i=0}^{2}(-1)^i\mathrm{Tr}(f^*,H_c^i(U,\mathbb{Q}_\ell))\\&=\sum_{i=0}^{2}(-1)^i\mathrm{Tr}(f^*,H^i(X,\mathbb{Q}_\ell))-\sum_{i=0}^{2}(-1)^i\mathrm{Tr}(f^*,H^i(X-U,\mathbb{Q}_\ell)).\end{aligned}$$

By 8.6.8, $\sum_{i=0}^{2}(-1)^i\mathrm{Tr}(f^*,H^i(X,\mathbb{Q}_\ell))$ is the sum of multiplicities of fixed points of f in X. By 10.4.6, $\sum_{i=0}^{2}(-1)^i\mathrm{Tr}(f^*,H^i(X-U,\mathbb{Q}_\ell))$ is the number of fixed points of f in $X-U$. Since fixed points of f in $X-U$ have multiplicity 1, $\sum_{i=0}^{2}(-1)^i\mathrm{Tr}(f^*,H_c^i(U,\mathbb{Q}_\ell))$ is the sum of multiplicities of fixed points of f in U. □

Lemma 10.4.8. *Let X_0 be a smooth irreducible curve on $\mathbb{F}_q$, and let $\mathscr{F}_0$ be a locally constant sheaf of A-modules on X_0 such that there exists a galois etale covering space $f : X_0' \to X_0$ with the property that $f^*\mathscr{F}_0$ is a constant sheaf associated to a finitely generated projective A-module. Then we have*

$$\sum_{\bar{x}\in X^F}\mathrm{Tr}(F_{\bar{x}}^*,\mathscr{F}_{\bar{x}})=\mathrm{Tr}(F^*,R\Gamma_c(X,\mathscr{F})),$$

where $\mathscr{F}$ is the inverse image of $\mathscr{F}_0$ in X.

Proof. Let $M=\Gamma(X_0',f^*\mathscr{F}_0)$ and let G be the galois group of the etale covering space $f : X_0' \to X_0$. Then M is a finitely generated projective A-module with a G-action. The sheaf $f_*\mathbb{Z}/\ell^n$ on X_0 is a locally free sheaf of $\mathbb{Z}/\ell^n[G]$-modules. By 10.2.1, 7.4.7 and 7.8.1, we have $R\Gamma_c(X, f_*\mathbb{Z}/\ell^n) \in \operatorname{ob} D^b_{\mathrm{perf}}(\mathbb{Z}/\ell^n[G])$. The canonical morphism

$$\mathrm{adj} : f_*f^*\mathscr{F}_0 = f_!f^*\mathscr{F}_0 \to \mathscr{F}_0$$

is a morphism of G-sheaves, where $\mathscr{F}_0$ is provided with the trivial G-action. It induces an isomorphism

$$(f_*f^*\mathscr{F}_0)_G \stackrel{\cong}{\to} \mathscr{F}_0.$$

Consider the morphism of sheaves

$$M \otimes_{\mathbb{Z}/\ell^n} f_*\mathbb{Z}/\ell^n \to f_*f^*\mathscr{F}_0$$

induced by the morphism of presheaves

$$M \otimes_{\mathbb{Z}/\ell^n} (f_*\mathbb{Z}/\ell^n)(U) \to (f_*f^*\mathscr{F}_0)(U) = (f^*\mathscr{F})(U \times_{X_0} X_0'),$$
$$s \otimes a \mapsto a(s|_{U\times_{X_0}X_0'})$$

for any etale X_0-scheme U, any $a \in (f_*\mathbb{Z}/\ell^n)(U) = \mathbb{Z}/\ell^n(U \times_{X_0} X_0')$, and any $s \in M = \Gamma(X_0', f^*\mathscr{F}_0)$. Since $f^*\mathscr{F}_0 \cong M$, this is an isomorphism of sheaves by 7.4.7. Note that this is an isomorphism of G-sheaves, where the action of G on $M \otimes_{\mathbb{Z}/\ell^n} (f_*\mathbb{Z}/\ell^n)$ is given by

$$g(s \otimes a) = gs \otimes ga.$$

We thus have

$$\mathscr{F}_0 \cong (M \otimes_{\mathbb{Z}/\ell^n} f_*\mathbb{Z}/\ell^n)_G \cong A \otimes_{A[G]} (M \otimes_{\mathbb{Z}/\ell^n} f_*\mathbb{Z}/\ell^n).$$

So we have

$$R\Gamma_c(X, \mathscr{F}) \cong A \otimes^L_{A[G]} (M \otimes^L_{\mathbb{Z}/\ell^n} R\Gamma_c(X, f_*\mathbb{Z}/\ell^n)).$$

Representing $R\Gamma_c(X, f_*\mathbb{Z}/\ell^n)$ by a perfect complex of $\mathbb{Z}/\ell^n[G]$-modules P. We then have

$$R\Gamma_c(X, \mathscr{F}) \cong (M \otimes_{\mathbb{Z}/\ell^n} P)_G.$$

Through this isomorphism, the action of F^* on $R\Gamma_c(X, \mathscr{F})$ is induced by the action of F^* on $R\Gamma_c(X, f_*\mathbb{Z}/\ell^n) \cong R\Gamma_c(X', \mathbb{Z}/\ell^n)$, where $X' = X_0' \otimes_{\mathbb{F}_q} \mathbb{F}$.

Suppose that F has no fixed point in $X(\mathbb{F})$ and let us prove

$$\mathrm{Tr}(F^*, R\Gamma_c(X, \mathscr{F})) = 0,$$

that is,

$$\mathrm{Tr}_A(F^*, (M \otimes_{\mathbb{Z}/\ell^n} P)_G) = 0.$$

By 10.4.3, it suffices to prove

$$\mathrm{Tr}_{A[G]}(F^*, M \otimes_{\mathbb{Z}/\ell^n} P) = 0.$$

By 10.4.2, it suffices to prove

$$\mathrm{Tr}_{\mathbb{Z}/\ell^n[G]}(F^*, P) = 0.$$

By 10.4.1, it suffices to prove

$$\theta_g(\mathrm{Tr}_{\mathbb{Z}/\ell^n[G]}(F^*, P)) = 0$$

for any $g \in G$, that is,

$$\theta_g(\mathrm{Tr}_{\mathbb{Z}/\ell^n[G]}(F^*, R\Gamma_c(X', \mathbb{Z}/\ell^n))) = 0.$$

By 10.4.1, we have

$$\#C(g)\theta_g(\mathrm{Tr}_{\mathbb{Z}/\ell^n[G]}(F^*, R\Gamma_c(X', \mathbb{Z}/\ell^n))) = \mathrm{Tr}_{\mathbb{Z}/\ell^n}((Fg^{-1})^*, R\Gamma_c(X', \mathbb{Z}/\ell^n)).$$

This formula is true for any n. It follows that

$$\begin{aligned}&\theta_g(\mathrm{Tr}_{\mathbb{Z}/\ell^n[G]}(F^*, R\Gamma_c(X', \mathbb{Z}/\ell^n))\\ &\equiv \frac{1}{\#C(g)}\sum_{i=0}^{2}(-1)^i\mathrm{Tr}((Fg^{-1})^*, H_c^i(X', \mathbb{Q}_\ell)) \mod \ell^n.\end{aligned}$$

To prove our assertion, it suffices to show

$$\sum_{i=0}^{2}(-1)^i\mathrm{Tr}((Fg^{-1})^*, H_c^i(X', \mathbb{Q}_\ell)) = 0$$

if F has no fixed points in $X(\mathbb{F})$. Let $\overline{X}_0'$ be the smooth compactification of X_0', let $\overline{X}' = \overline{X}_0' \otimes_{\mathbb{F}_q} \mathbb{F}$, let $\bar{x}$ be a fixed point of Fg^{-1} in $\overline{X}'(\mathbb{F})$, and let $x_0 \in \overline{X}_0'$ be the image of $\bar{x}$ in X_0'. Then we have $g(x_0) = x_0$. Choose a uniformizer π_{x_0} for $\mathscr{O}_{\overline{X}_0', x_0}$. Then π_{x_0} is also a uniformizer for $\mathscr{O}_{\overline{X}', \bar{x}}$. We have

$$v_{\bar{x}}((Fg^{-1})^*(\pi_{x_0}) - \pi_{x_0}) = v_{\bar{x}}(g^{-1}(\pi_{x_0})^q - \pi_{x_0}) = 1.$$

So the multiplicity of Fg^{-1} at $\bar{x}$ is 1. Moreover, if $\bar{x}$ is a fixed point of Fg^{-1} in $X'(\mathbb{F})$, then its image in X is a fixed point of F in $X(\mathbb{F})$. But F has no fixed point in $X(\mathbb{F})$. So Fg^{-1} has no fixed point in $X'(\mathbb{F})$. We thus have

$$\sum_{i=0}^{2}(-1)^i\mathrm{Tr}((Fg^{-1})^*, H_c^i(X', \mathbb{Q}_\ell)) = 0$$

by 10.4.7. This proves $\mathrm{Tr}(F^*, R\Gamma_c(X, \mathscr{F})) = 0$ under the assumption that F has no fixed point in X.

In general, we have

$$\mathrm{Tr}(F^*, R\Gamma_c(X - X^F, \mathscr{F}|_{X - X^F})) = 0$$

by the above discussion. By 10.4.6, we have

$$\mathrm{Tr}(F^*, R\Gamma_c(X^F, \mathscr{F}|_{X^F})) = \sum_{\bar{x}\in X^F} \mathrm{Tr}(F^*_{\bar{x}}, \mathscr{F}_{\bar{x}}).$$

So we have

$$\begin{aligned}
&\mathrm{Tr}(F^*, R\Gamma_c(X, \mathscr{F})) \\
&\quad = \mathrm{Tr}(F^*, R\Gamma_c(X - X^F, \mathscr{F}|_{X-X^F})) + \mathrm{Tr}(F^*, R\Gamma_c(X^F, \mathscr{F}|_{X^F})) \\
&\quad = \sum_{\bar{x}\in X^F} \mathrm{Tr}(F^*_{\bar{x}}, \mathscr{F}_{\bar{x}}).
\end{aligned}$$

□

Lemma 10.4.9. *Let X_0 be a smooth irreducible curve over $\mathbb{F}_q$, and let $K \in \mathrm{ob}\, D^b_{\mathrm{ctf}}(X_0, A)$. We have*

$$\sum_{\bar{x}\in X^F} \mathrm{Tr}(F^*_{\bar{x}}, K_{\bar{x}}) = \mathrm{Tr}(F^*, R\Gamma_c(X, K)).$$

Proof. Represent K by a bounded complex of flat constructible sheaves of A-modules. By 5.8.1 (ii), we can find an open dense subset U_0 of X_0, and an etale covering space $U_0' \to U_0$ such that $K|_{U_0'}$ is a complex of constant sheaves. By 10.4.8, we have

$$\sum_{\bar{x}\in U^F} \mathrm{Tr}(F^*_{\bar{x}}, K_{\bar{x}}) = \mathrm{Tr}(F^*, R\Gamma_c(U, K|_U)),$$

where $U = U_0 \otimes_{\mathbb{F}_q} \mathbb{F}$. By 10.4.6, we have

$$\sum_{\bar{x}\in (X-U)^F} \mathrm{Tr}(F^*_{\bar{x}}, K_{\bar{x}}) = \mathrm{Tr}(F^*, R\Gamma_c(X-U, K|_{X-U})).$$

So we have

$$\begin{aligned}
&\mathrm{Tr}(F^*, R\Gamma_c(X, K)) \\
&\quad = \mathrm{Tr}(F^*, R\Gamma_c(U, K|_U)) + \mathrm{Tr}(F^*, R\Gamma_c(X-U, K|_{X-U})) \\
&\quad = \sum_{\bar{x}\in U^F} \mathrm{Tr}(F^*_{\bar{x}}, K_{\bar{x}}) + \sum_{\bar{x}\in (X-U)^F} \mathrm{Tr}(F^*_{\bar{x}}, K_{\bar{x}}) \\
&\quad = \sum_{\bar{x}\in X^F} \mathrm{Tr}(F^*_{\bar{x}}, K_{\bar{x}}).
\end{aligned}$$

□

Proof of 10.4.4. For every locally closed subset Y of X, let

$$\begin{aligned}
T_1(Y, K|_Y) &= \sum_{\bar{x}\in Y^F} \mathrm{Tr}(F^*_{\bar{x}}, K_{\bar{x}}), \\
T_2(Y, K|_Y) &= \mathrm{Tr}(F^*, R\Gamma_c(Y, K|_Y)),
\end{aligned}$$

and let

$$\mathscr{S} = \{Y|Y \text{ is closed in } X \text{ such that } T_1(Y, K|_Y) \neq T_2(Y, K|_Y)\}.$$

To prove 10.4.4, it suffices to show that $\mathscr{S}$ is empty. If $\mathscr{S}$ is not empty, then we can choose a minimal element Y in $\mathscr{S}$. Let U be an affine open subset of Y. By the minimality of Y, we have

$$T_1(Y-U, K|_{Y-U}) = T_2(Y-U, K|_{Y-U}).$$

We have

$$T_k(Y, K|_Y) = T_k(Y-U, K|_{Y-U}) + T_k(U, K|_U) \quad (k=1,2).$$

Since $T_1(Y, K|_Y) \neq T_2(Y, K|_Y)$, we have

$$T_1(U, K|_U) \neq T_2(U, K|_U).$$

Since U is affine, we can find a closed immersion $i : U \to \mathbb{A}^m_{\mathbb{F}_q}$ for some m. We have

$$T_k(U, K|_U) = T_k(\mathbb{A}^m_{\mathbb{F}_q}, i_*(K|_U)) \quad (k=1,2).$$

Denote $i_*(K|_U)$ also by K. We thus find a complex $K \in \operatorname{ob} D^b_{\mathrm{ctf}}(\mathbb{A}^m_{\mathbb{F}_q}, A)$ such that $T_1(\mathbb{A}^m_{\mathbb{F}_q}, K) \neq T_2(\mathbb{A}^m_{\mathbb{F}_q}, K)$.

Let $\pi : \mathbb{A}^m_{\mathbb{F}_q} \to \mathbb{A}^{m-1}_{\mathbb{F}_q}$ be the projection. We have

$$\begin{aligned} T_1(\mathbb{A}^{m-1}_{\mathbb{F}_q}, R\pi_! K) &= \sum_{x \in \mathbb{A}^{m-1}_{\mathbb{F}_q}(\mathbb{F}_q)} \mathrm{Tr}(F_{\bar{x}}, (R\pi_! K)_{\bar{x}}) \\ &= \sum_{x \in \mathbb{A}^{m-1}_{\mathbb{F}_q}(\mathbb{F}_q)} \mathrm{Tr}\Big(F_{\bar{x}}, R\Gamma_c\big(\pi^{-1}(x) \otimes_{\mathbb{F}_q} \mathbb{F}, K|_{\pi^{-1}(x)\otimes_{\mathbb{F}_q}\mathbb{F}}\big)\Big). \end{aligned}$$

By 10.4.9, we have

$$\mathrm{Tr}\Big(F_{\bar{x}}, R\Gamma_c\big(\pi^{-1}(x) \otimes_{\mathbb{F}_q} \mathbb{F}, K|_{\pi^{-1}(x)\otimes_{\mathbb{F}_q}\mathbb{F}}\big)\Big) = \sum_{y \in \pi^{-1}(x)(\mathbb{F}_q)} \mathrm{Tr}(F_{\bar{y}}, K_{\bar{y}}).$$

So we have

$$\begin{aligned} T_1(\mathbb{A}^{m-1}_{\mathbb{F}_q}, R\pi_! K) &= \sum_{x \in \mathbb{A}^{m-1}_{\mathbb{F}_q}(\mathbb{F}_q)} \sum_{y \in \pi^{-1}(x)(\mathbb{F}_q)} \mathrm{Tr}(F_{\bar{y}}, K_{\bar{y}}) \\ &= T_1(\mathbb{A}^m_{\mathbb{F}_q}, K). \end{aligned}$$

On the other hand, we have

$$\begin{aligned} T_2(\mathbb{A}^m_{\mathbb{F}_q}, K) &= \mathrm{Tr}(F^*, R\Gamma_c(\mathbb{A}^m_{\mathbb{F}}, K)) \\ &= \mathrm{Tr}(F^*, R\Gamma_c(\mathbb{A}^{m-1}_{\mathbb{F}}, R\pi_! K)) \\ &= T_2(\mathbb{A}^{m-1}_{\mathbb{F}_q}, R\pi_! K). \end{aligned}$$

So we have

$$T_k(\mathbb{A}^m_{\mathbb{F}_q}, K) = T_k(\mathbb{A}^{m-1}_{\mathbb{F}_q}, R\pi_! K) \quad (k = 1, 2).$$

We thus have $T_1(\mathbb{A}^{m-1}_{\mathbb{F}_q}, R\pi_! K) \neq T_2(\mathbb{A}^{m-1}_{\mathbb{F}_q}, R\pi_! K)$. Denote $R\pi_! K$ also by K. We thus find $K \in \mathrm{ob}\, D^b_{\mathrm{ctf}}(\mathbb{A}^{m-1}_{\mathbb{F}_q}, A)$ such that $T_1(\mathbb{A}^{m-1}_{\mathbb{F}_q}, K) \neq T_2(\mathbb{A}^{m-1}_{\mathbb{F}_q}, K)$. Using this argument repeatedly, we can find $K \in \mathrm{ob}\, D^b_{\mathrm{ctf}}(\mathbb{A}^1_{\mathbb{F}_q}, A)$ such that $T_1(\mathbb{A}^1_{\mathbb{F}_q}, K) \neq T_2(\mathbb{A}^1_{\mathbb{F}_q}, K)$. This contradicts 10.4.9. So $\mathscr{S}$ is empty. □

10.5 Grothendieck's Formula of L-functions

([SGA $4\frac{1}{2}$] Rapport 3.)

Let X_0 be a compactifiable scheme over $\mathbb{F}_q$, let $X = X_0 \otimes_{\mathbb{F}_q} \mathbb{F}$, let $K_0 \in \mathrm{ob}\, D^b_c(X_0, \overline{\mathbb{Q}}_\ell)$, and let K be the inverse image of K_0 in X. Denote by $|X_0|$ the set of closed point in X_0. For any $x \in |X_0|$, let $N(x)$ be the number of elements of the residue field $k(x)$, and let $\deg(x) = [k(x) : \mathbb{F}_q]$. We have $N(x) = q^{\deg(x)}$. Let $i : \mathrm{Spec}\, k(x) \to X_0$ be the closed immersion. The complex $i^* K_0$ is completely determined by the galois action of $\mathrm{Gal}(\overline{k(x)}/k(x))$ on $K_{\bar{x}}$. Let $f_x : K_{\bar{x}} \to K_{\bar{x}}$ be the action of the Frobenius substitution $\alpha \mapsto \alpha^{q^{\deg(x)}}$ in $\mathrm{Gal}(\overline{k(x)}/k(x))$.Define the *L-function* of K_0 to be

$$L(X_0, K_0, s) = \prod_{x \in |X_0|} \frac{1}{\det\Big(1 - \frac{1}{N(x)^s} f_x^{-1}, K_{\bar{x}}\Big)},$$

where

$$\det\Big(1 - \frac{1}{N(x)^s} f_x^{-1}, K_{\bar{x}}\Big) = \prod_i \det\Big(1 - \frac{1}{N(x)^s} f_x^{-1}, \mathscr{H}^i(K_{\bar{x}})\Big)^{(-1)^i}.$$

Making the change of variable $t = q^{-s}$, we can also define the L-function as

$$L(X_0, K_0, t) = \prod_{x \in |X_0|} \frac{1}{\det(1 - t^{\deg(x)} f_x^{-1}, K_{\bar{x}})}.$$

Let $F : X \to X$ be the base change of $\mathrm{fr}_{X_0} : X_0 \to X_0$. For any $x \in |X_0|$ with $\deg(x) = n$, any $\mathbb{F}$-point of X with image x is a fixed point of F^n. Iterating n times the Frobenius correspondence $F^*_{K_0} : F^* K \to K$ induces a morphism $F^{n*}_{\bar{x}} : K_{\bar{x}} \to K_{\bar{x}}$. By 10.3.5, we have $F^{n*}_{\bar{x}} = f_x^{-1}$. So we have

$$L(X_0, K_0, t) = \prod_{x \in |X_0|} \frac{1}{\deg(1 - t^{\deg(x)} F^{\deg(x)*}_{\bar{x}}, K_{\bar{x}})}.$$

Theorem 10.5.1 (Grothendieck). *Notation as above, we have*

$$L(X_0, K_0, t) = \prod_i \det(1 - F^*t, H_c^i(X, K))^{(-1)^{i+1}}$$

as a formal power series in $\overline{\mathbb{Q}}_\ell[[t]]$.

Remark 10.5.2. Since $H_c^i(X, K)$ is nonzero only for finitely many i, this formula shows that $L(X_0, K_0, t)$ is a rational function.

Proof of 10.5.1. Since both sides of the equation are formal power series with constant term 1, it suffices to show

$$t\frac{d}{dt} \ln L(X_0, K_0, t) = t\frac{d}{dt} \ln \prod_i \det(1 - F^*t, H_c^i(X, K))^{(-1)^{i+1}}.$$

Note that for any endomorphism $\phi : V \to V$ on a finite dimensional $\overline{\mathbb{Q}}_\ell$-vector space V, we have

$$t\frac{d}{dt} \ln \det(1 - \phi t^k)^{-1} = \sum_{n=1}^{\infty} k\mathrm{Tr}(\phi^n)t^{kn}.$$

Indeed, let $\lambda_1, \ldots, \lambda_d$ be the eigenvalues of ϕ, where $d = \dim V$. We have

$$\begin{aligned} t\frac{d}{dt} \ln \det(1 - \phi t^k)^{-1} &= t\frac{d}{dt} \ln \prod_{i=1}^{d} (1 - \lambda_i t^k)^{-1} \\ &= -\sum_{i=1}^{d} t\frac{d}{dt} \ln(1 - \lambda_i t^k) \\ &= \sum_{i=1}^{d} \frac{k\lambda_i t^k}{1 - \lambda_i t^k} \\ &= \sum_{i=1}^{d} \sum_{n=1}^{\infty} k\lambda_i^n t^{kn} \\ &= \sum_{n=1}^{\infty} k\Big(\sum_{i=1}^{d} \lambda_i^n\Big)t^{kn} \\ &= \sum_{n=1}^{\infty} k\mathrm{Tr}(\phi^n)t^{kn}. \end{aligned}$$

So we have

$$\begin{aligned}
& t\frac{d}{dt}\ln L(X_0, K_0, t) \\
&= t\frac{d}{dt}\ln \prod_{x\in|X_0|} \det(1 - t^{\deg(x)}F_{\bar{x}}^{\deg(x)*}, K_{\bar{x}})^{-1} \\
&= \sum_{x\in|X_0|} t\frac{d}{dt}\ln\det(1 - t^{\deg(x)}F_{\bar{x}}^{\deg(x)*}, K_{\bar{x}})^{-1} \\
&= \sum_{x\in|X_0|}\sum_{n=1}^{\infty} \deg(x)\mathrm{Tr}(F_{\bar{x}}^{n\deg(x)*}, K_{\bar{x}})t^{n\deg(x)} \\
&= \sum_{m=1}^{\infty} \sum_{\substack{x\in|X_0| \\ \deg(x)|m}} \deg(x)\mathrm{Tr}(F_{\bar{x}}^{m*}, K_{\bar{x}})t^m \\
&= \sum_{m=1}^{\infty}\sum_{\bar{x}\in X^{F^m}} \mathrm{Tr}(F_{\bar{x}}^{m*}, K_{\bar{x}})t^m.
\end{aligned}$$

Here for the last equality, we use the fact that there are exactly $\deg(x)$ $\mathbb{F}$-points fixed by F^m with image x for any $x \in |X_0|$ satisfying $\deg(x)|m$. Applying 10.4.5 to the scheme $X_0 \otimes_{\mathbb{F}_q} \mathbb{F}_{q^m}$, we get

$$\sum_{\bar{x}\in X^{F^m}} \mathrm{Tr}(F_{\bar{x}}^{m*}, K_{\bar{x}}) = \sum_i (-1)^i \mathrm{Tr}(F^{*m}, H_c^i(X, K)).$$

So we have

$$t\frac{d}{dt}\ln L(X_0, K_0, t) = \sum_{m=1}^{\infty}\sum_i (-1)^i\mathrm{Tr}(F^{*m}, H_c^i(X, K))t^m.$$

On the other hand, we have

$$\begin{aligned}
& t\frac{d}{dt}\ln\prod_i \det(1 - F^*t, H_c^i(X, K))^{(-1)^{i+1}} \\
&= \sum_i (-1)^i t\frac{d}{dt}\ln\det(1 - F^*t, H_c^i(X, K))^{-1} \\
&= \sum_i\sum_{m=1}^{\infty} (-1)^i\mathrm{Tr}(F^{*m}, H_c^i(X, K))t^m.
\end{aligned}$$

So we have

$$t\frac{d}{dt}\ln L(X_0, K_0, t) = t\frac{d}{dt}\ln\prod_i \det(1 - F^*t, H_c^i(X, K))^{(-1)^{i+1}}.$$

□

Corollary 10.5.3. *Let X_0 and Y_0 be $\mathbb{F}_q$-schemes of finite type, and let $f : X_0 \to Y_0$ be a compactifiable $\mathbb{F}_q$-morphism. For any $K_0 \in \mathrm{ob}\, D_c^b(X_0, \overline{\mathbb{Q}}_\ell)$, we have*

$$L(X_0, K_0, t) = L(Y_0, Rf_! K_0, t).$$

Proof. We have

$$\begin{aligned}
&L(Y_0, Rf_! K_0, t)\\
&= \prod_{y \in |Y_0|} \det(1 - t^{\deg(y)} f_y^{-1}, (Rf_! K)_{\bar{y}})^{-1}\\
&= \prod_{y \in |Y_0|} \prod_i \det(1 - t^{\deg(y)} f_y^{-1}, H_c^i(f^{-1}(\bar{y}), K|_{f^{-1}(\bar{y})}))^{(-1)^{i+1}}\\
&= \prod_{y \in |Y_0|} \prod_{x \in |f^{-1}(y)|} \det(1 - t^{\deg(y)[k(x):k(y)]} f_x^{-1}, K_{\bar{x}})\\
&= \prod_{x \in |X_0|} \det(1 - t^{\deg(x)} f_x^{-1}, K_{\bar{x}})\\
&= L(X_0, K_0, t),
\end{aligned}$$

where the third equality follows from 10.3.5 and 10.5.1 applied to the $k(y)$-scheme $f^{-1}(y)$. This proves our assertion.

Bibliography

A. Grothendieck and J. Dieudonné,

Éléments de Géométrie Algébrique.

EGA I Le langage des schémas, *Publ. Math. IHES* 4 (1960).

EGA II Étude globale élémentaire de quelques classes de morphismes, *Ibid.* 8 (1961).

EGA III Étude cohomologique des faisceaux cohérents, *Ibid.* 11 (1961), 17 (1963).

EGA IV Étude locale des schémas et des morphismes de schémas, *Ibid.* 20 (1964), 24 (1965), 28 (1966), 32 (1967).

A. Grothendieck *et al.*

Séminaire de Géometrie Algébrique.

SGA 1 *Revêtements Étales et Groupe Fondamental*, Lecture Notes in Math. 224, Springer-Verlag (1971).

SGA 4 (with M. Artin and J. L. Verdier) *Théorie des Topos et Cohomologie Étale des Schémas*, Lecture Notes in Math. 269, 270, 305, Springer-Verlag (1972–1973).

SGA 4½ (by P. Deligne *et al.*) *Cohomologie Étale*, Lecture Notes in Math. 569, Springer-Verlag (1977).

SGA 5 (with I. Bucur, C. Houzel, L. Illusie, J.-P. Jouanonou and J.-P. Serre) *Cohomologie l-adique et Fonctions L*, Lecture Notes in Math. 589, Springer-Verlag (1977).

SGA 7 (with M. Raynaud, D. S. Rim, P. Deligne, and N. Katz) *Groupes de Monodromie en Géométrie Algébrique*, Lecture Notes in Math. 288, 340, Springer-Verlag (1972–1973).

M. F. Artin, Algebraic approximation of structures over complete local rings, *Publ. Math. IHES* 36 (1969), 23–58.

M. Atiyah and I. G. Macdonald. *Introduction to Commutative Algebra*, Addison-Wesley, Reading, Mass. (1969).

A. Beilinson, J. Bernstein and P. Deligne. Faisceaux Pervers, in *Analyse et Topologie sur les Espace Singuliers (I)*, Astérique 100 (1982).

S. Bosch, W. Lütkebohmert, and M. Raynaud. *Néron Models*, Springer-Verlag (1990).

P. Deligne. La conjecture de Weil I, *Publ. Math. IHES* 43 (1974), 273–307.

P. Deligne. La conjecture de Weil II, *Publ. Math. IHES* 52 (1980), 137–252.

L. Fu. *Algebraic Geometry*, Tsinghua University Press, 2006.

E. Freitag and R. Kiehl. *Etale Cohomology and the Weil Conjecture*, Springer-Verlag (1988).

A. Grothendieck. Sur quelques points d'algèbre homologique, *Tôhoku Math. J.* 9 (1957), 119–221.

R. Hartshorne.*Algebraic Geometry*, Springer-Verlag (1977).

R. Hartshorne. *Residue and Duality*, Lecture Notes in Math. 20, Springer-Verlag (1966).

H. Matsumura. *Commutative algebra*, W. A. Benjamin Co., New York (1970).

J. Milne. *Etale Cohomology*, Princeton University Press 1980.

G. Laumon. Transformation de Fourier, constantes d'équations fontionnelles, et conjecture de Weil, *Publ. Math. IHES* 65 (1987), 131–210.

J.-P. Serre. *Local Fields*, Sringer-Verlag (1979). (English translation of *Corps Locaux*, Hermann, Paris (1962)).

J.-P. Serre. *Cohomologie Galoisienne*, Lecture Notes in Math. 5, Springer-Verlag (1964).

J.-P. Serre. *Linear Representations of Finite Groups*, Springer-Verlag (1977). (English translation of *Représentations Lináires des Groupes Finis*, Hermann, Paris (1971)).

Index

www.ingramcontent.com/pod-product-compliance
Lightning Source LLC
LaVergne TN
LVHW010636300926
849161LV00005B/957
* 9 7 8 9 8 1 4 3 0 7 7 2 7 *